Selected Baryons

Symbol	Spin	Quarks	mc^2 (GeV)	Lifetime (s)
p	$\frac{1}{2}$	uud	0.93827	∞
n	$\frac{1}{2}$	ddu	0.93957	900
Λ	$\frac{1}{2}$	uds	1.11568	2.63×10^{-10}
Σ^+	$\frac{1}{2}$	uus	1.18937	0.80×10^{-10}
Σ^0	$\frac{1}{2}$	uds	1.19264	6×10^{-20}
Σ^-	$\frac{1}{2}$	dds	1.19974	1.48×10^{-10}
Δ^{++}	$\frac{3}{2}$	uuu	1.232	6.6×10^{-24}
Δ^+	$\frac{3}{2}$	uud	1.232	6.6×10^{-24}
Δ^0	$\frac{3}{2}$	udd	1.232	6.6×10^{-24}
Δ^0	$\frac{3}{2}$	ddd	1.232	6.6×10^{-24}
Ξ^0	$\frac{1}{2}$	uss	1.31486	2.90×10^{-10}
Ξ^-	$\frac{1}{2}$	dss	1.32171	1.64×10^{-10}
Ω^-	$\frac{3}{2}$	sss	1.6724	8.2×10^{-11}

RELATIVITY

Lorentz transformation (assuming the primed frame moves with $v_z = \beta$ relative to the unprimed frame):

$$
\Lambda^\mu_{\ \nu} = \begin{pmatrix} \gamma & 0 & 0 & -\gamma\beta \\ 0 & 1 & 0 & 0 \\ 0 & 0 & 1 & 0 \\ -\gamma\beta & 0 & 0 & \gamma \end{pmatrix} = \begin{pmatrix} \cosh\theta & 0 & 0 & -\sinh\theta \\ 0 & 1 & 0 & 0 \\ 0 & 0 & 1 & 0 \\ -\sinh\theta & 0 & 0 & \cosh\theta \end{pmatrix}, \quad \text{where} \quad \begin{aligned} \gamma &\equiv (1-\beta^2)^{-1/2} \\ \theta &= \tanh^{-1}\beta \\ \cosh\theta &= \tfrac{1}{2}(e^\theta + e^{-\theta}) \\ \sinh\theta &= \tfrac{1}{2}(e^\theta - e^{-\theta}). \end{aligned}
$$

(To get the inverse transformation (Λ^{-1}), change $\beta \to -\beta$ or $\theta \to -\theta$.)

$$
\text{Metric: } g_{\mu\nu} = g^{\mu\nu} = \begin{pmatrix} 1 & 0 & 0 & 0 \\ 0 & -1 & 0 & 0 \\ 0 & 0 & -1 & 0 \\ 0 & 0 & 0 & -1 \end{pmatrix}, \qquad \text{Kronecker delta: } \delta^\mu_\nu = g^{\mu\alpha} g_{\alpha\nu} = \begin{pmatrix} 1 & 0 & 0 & 0 \\ 0 & 1 & 0 & 0 \\ 0 & 0 & 1 & 0 \\ 0 & 0 & 0 & 1 \end{pmatrix}
$$

Greek index vs. latin index:	$p^\mu = p^t, p^x, p^y,$ or p^z;	$p^j = p^x, p^y,$ or p^z (spatial only)
Summation convention:	$p_\mu x^\mu \equiv \sum_\mu p_\mu x^\mu$	(one upper with one lower index)
Transformation of upper index(s):	$p'^\mu = \Lambda^\mu_{\ \nu} p^\mu,$	$T'^{\mu\nu} = \Lambda^\mu_{\ \alpha} \Lambda^\nu_{\ \beta} T^{\alpha\beta}$
Transformation of lower index(s):	$\partial'_\mu \phi = (\Lambda^{-1})^\nu_{\ \mu} \partial_\nu \phi,$	$T'_{\mu\nu} = (\Lambda^{-1})^\alpha_{\ \mu} (\Lambda^{-1})^\beta_{\ \nu} T_{\alpha\beta}$
Raising and lowering indexes:	$\partial^\mu \phi = g^{\mu\nu} \partial_\nu \phi,$	$p_\mu = g_{\mu\nu} p^\nu$
Dot product of four-vectors:	$a \cdot b = g_{\mu\nu} a^\mu b^\nu,$ same as	$a \cdot b = a^t b^t - a^x b^x - a^y b^y - a^z b^z$
Effect of Kroneker delta:	$\delta^\mu_\nu p^\nu = p^\mu$	$\delta^\mu_\nu p_\mu = p_\nu$
Four-momentum magnitude:	$p^2 \equiv p \cdot p = m^2,$	$p^2 = m^2 = E^2 - |\vec{p}|^2$
Derivatives of plane waves:	$\partial_\mu e^{ip \cdot x} = ip_\mu e^{ip \cdot x}$	$\partial^\mu e^{ip \cdot x} = ip^\mu e^{ip \cdot x}$

INDEX RENAMING RULES

- A **bound index** is an index in a pair that is summed over. Example: $m^2 = p_\mu p^\mu$.
- A **free index** is an index that is common to all terms in an expression. Example: $p^\mu_{\text{tot}} = p^\mu_1 + p^\mu_2$.
- One may rename any *pair* of bound indices or *all* members in any set of free indices.
- **But one must avoid collisions** where more than one pair of bound indices or a pair of bound indices and a set of free indices have the same name. Example: $\partial_\mu \partial^\mu A^\alpha \to \partial_\mu \partial^\mu A^\mu$ is bad.
- This problem can arise when multiplying terms: $p^\mu_1 p_{1\mu} = m^2_1$ and $p^\mu_2 p_{2\mu} = m^2_1$, but $p^\mu_1 p_{1\mu} p^\alpha_2 p_{2\alpha} = m^2_1 m^2_2$ (we rename the second pair of bound indices to avoid collisions).

A STANDARD MODEL WORKBOOK

A STANDARD MODEL WORKBOOK

Thomas A. Moore

Pomona College

University Science Books

New York

UNIVERSITY SCIENCE BOOKS
An Imprint of AIP Publishing
uscibooks.aip.org

Publisher: *Jane Ellis*
Production and Marketing Manager: *Barbara Dickson*
Project Manager: *Paul C. Anagnostopoulos, Windfall Software*
Acquisitions Editor: *Katerina Heidhausen*
Publishing Associate: *Felicity Henson*
Manuscript Editor: *Kot Copyediting & Proofreading*
Illustrator: *Laurel Muller*
Compositor: *Paul C. Anagnostopoulos*
Cover Design: *Laurel and Hinrich Muller*
Printer & Binder: *Sheridan Books*

About the cover
This artist's rendering shows the fundamental particles of The Standard Model, with bosons on the left representing the possible interactions between the fermions on the right. Each platform represents a type of fundamental interaction: the bosons on the left side of a given platform are carriers of that interaction, and the fermions on the right respond to that interaction as well as all interactions below.

This book is set in Times Roman and Futura. It is printed on acid-free paper. ∞

Library of Congress Cataloging-in-Publication Data

Names: Moore, Thomas A. (Thomas Andrew), author.
Title: A standard model workbook / Thomas A. Moore.
Identifiers: LCCN 2023013750 | ISBN 9781940380179 (paperback) | ISBN 9781940380186 (ebook)
Subjects: LCSH: Standard model (Nuclear physics)—Textbooks. | Particles (Nuclear physics)—Textbooks.
Classification: LCC QC794.6.S75 M66 2023 | DDC 539.7/2—dc23/eng/20231002
LC record available at https://lccn.loc.gov/2023013750

Printed in the United States of America

10 9 8 7 6 5 4 3 2 1

For Ben, for whom the Universe is a source of boundless wonder.

CONTENTS IN BRIEF

CONTENTS

32. BEYOND THE STANDARD MODEL · 543

Appendix A: Potentials from Interactions · 559

Appendix B: Short Answers to Selected Problems · 563

PREFACE

The Standard Model of particle physics represents perhaps the most significant triumph of late twentieth-century physics. Along with the theory of general relativity, the Standard Model provides the foundation of our current understanding of physics. It is an extraordinarily powerful theory, illuminating not only the newsworthy results of experiments in particle-accelerator laboratories but also the interiors of atoms and nuclei, and a broad range of astrophysical phenomena stretching back to the Big Bang itself. Its structure and focus on symmetry principles also provides the inspiration and place for many current efforts to move beyond the Standard Model.

The discovery of the Higgs particle in 2012, and the 2013 Nobel Prizes awarded to those central in the development of the Higgs mechanism, have only intensified students' perennially lively interest in particle physics and the Standard Model. Even so, few physics programs in the USA offer undergraduates any but the most cursory introduction (usually in the context of a "modern physics" class) to this foundation of contemporary physics. This is partly because the mathematics involved is abstract and difficult, and few available textbooks on the Standard Model are accessible to undergraduates.

Audience

The goal of this book is to provide a one-semester introduction to quantum field theory and the Standard Model of particle physics at a level suitable for junior and/or senior undergraduates at college or university in the USA. It assumes that students have taken multivariable calculus and linear algebra, and have at least seen a solid introduction to quantum physics in a modern physics course (at roughly the level of Townsend's *Quantum Physics: A Fundamental Approach to Modern Physics*, University Science Books, 2010). Students who have taken a course in classical mechanics that involves Lagrangian mechanics, electrodynamics, and/or upper-level quantum mechanics will be able to move through the book more quickly and easily, but such courses are not required.

This book has grown out of various iterations of a course on the Standard Model that I taught at Pomona College from 2014 through 2022. These iterations have convinced me that students *can* develop a solid and functional understanding of the Standard Model and also that their intrinsic interest in the topic makes them very willing to work hard. This has been one of the most popular courses I have taught.

Pedagogical Principles

The Standard Model involves some subtle and tricky mathematics. In my experience, junior and senior undergraduates in a well-designed course *can* learn the math. This is truly worth the effort, as it provides the foundation needed for confidence and flexibility in dealing with applications.

The key to developing the necessary mastery is for one to *personally own* the math by working through most of the arguments and derivations *oneself*. I have therefore designed this textbook as a *workbook*. Each chapter opens with a concise presentation that helps one see the big picture without too many mathematical distractions. This presentation is keyed to subsequent "boxes" that guide one in working through the supporting derivations and/or exploring other details or applications whose direct presentation would obscure the main argument. I have found this combination of overview and guided personal effort to be uniquely effective in building a deep understanding of the topic, particularly when it involves challenging mathematics.

Using this book successfully therefore involves working carefully through *all* of the boxes (as well as some homework problems). Doing this gives one a range of experience and depth of understanding that is difficult to obtain any other way.

I have deliberately and selectively included pictures of certain physicists who have made important contributions to the Standard Model (or its possible replacements) to show that not all physicists look like the movie stereotype.

Website

At `https://pages.pomona.edu/~tmoore/smw/` you will find supporting material, errata, and forms to send me error notices, suggestions, and questions.

Information for Instructors

So far in this preface, I have addressed issues relevant to all readers. In this section, I want to specifically speak to instructors designing courses around this book. (If you are teaching *yourself* from this book, please skip ahead to the **Information for Solitary Readers** section below.)

Course pacing.   I have designed this text so that (in my experience) *each chapter can be discussed in a single 70-minute class session,* or two chapters per week, particularly if you use the format for class sessions I describe below. When I teach this material in a schedule that involves three 50-minute sessions per week, I assign one chapter to each of the first two sessions and use the third session to further discuss and develop the ideas in those two chapters. Your mileage may vary (for example, you may need to spend more time on Chapters 2 and 3 if your students' backgrounds are weak on relativity, and perhaps less time on Chapters 7–9 if your students have taken an upper-level quantum mechanics course), but the two-chapters-per-week rule should help you to appropriately pace the course.

Chapter dependencies.   At this pacing, the book contains four to six more chapters than one can hope to cover in a semester. I have done this deliberately to allow you some flexibility in shaping the course to better fit your interests and your students' needs. Table P.1 shows a list of the chapters in the book. Generally, one should go through the chapters in order, but one can skip any or all of the asterisked chapters.

In particular, note that if you are more interested in the Standard Model than in quantum field theory, you can assign Chapter 10 instead of Chapters 11–14. Chapter 10 describes how Feynman diagrams work, and how one uses them to calculate decay rates and cross sections but with only a hand-waving justification of a number of the formulas. Chapters 11–14 go over the same material but in greater depth, so that students better understand how quantum field theory really works and how to justify many of the formulas treated more lightly in Chapter 10. If one assigns Chapter 10 instead of Chapters 11–14, *one should also skip Chapter 18,* which is otherwise strongly recommended.

The final chapter does not have any boxes and could simply be recommended for reading after the course is finished, or to provide the basis for an optional session (maybe with treats!) to celebrate the end of the course. It also provides a nice way to end the course on a more relaxed note. (I have included some simple problems that one might use for class activities.)

The remaining optional chapters mostly explore applications that are valuable if one wants to emphasize the experimental side of particle physics.

	1. Introduction
Relativity	2. Special Relativity
	3. Index Notation
Classical Fields	4. A Classical Scalar Field
	5. The Klein-Gordon Equation
	6. Noether's Theorem
Quantum Mechanics	7. Basic Quantum Mechanics
	8. Operators and Time Evolution
	9. Angular Momentum
(Shortcut to 15)	10. Feynman Diagram Quick Start*
Quantum Field Theory	11. Creation and Annihilation Operators
	12. The Quantized Scalar Field
	13. Interactions and Feynman Diagrams
Connection to Experiment	14. Fermi's Golden Rules
	15. Toy Universe Predictions
Spin-$\frac{1}{2}$ Particles	16. The Dirac Equation
	17. Transforming Dirac Spinors
	18. Quantizing the Dirac Field*
The Electroweak Theory	19. Electromagnetism from U(1)
	20. Quantum Electrodynamics
	21. Renormalization
	22. Applications of QED
	23. The Weak Interaction from SU(2)
	24. The Weak Interaction's Leftist Bias
	25. The Unified Electroweak Theory
Particle Masses	26. The Higgs Mechanism
	27. Fermion Masses and Mixing
	28. Neutrino Masses and Mixing*
	29. Applications of Electroweak Theory*
Chromodynamics	30. The Color Interaction
	31. Applications of Chromodynamics*
And Beyond!	32. Beyond the Standard Model*

TABLE P.1   This table displays the organization of chapters in this text. Chapters marked with an asterisk may be omitted without affecting continuity.

In a pinch, one could also skip the chromodynamics section, but that would leave students without having seen an important part of the Standard Model.

As I mentioned earlier, depending on your students' background in relativity, one may need to spend more time on Chapters 2 and 3. In particular, it is *crucial* that students become adept at using index notation correctly. I have found that student difficulties with later sections are often rooted here.

Conversely, if all your students have a solid background in upper-level quantum mechanics (particularly at the level of Townsend's *A Modern Approach to Quantum Mechanics*, 2nd edition, University Science Books, 2012, or the equivalent), one may be able to spend less time on Chapters 7–9.

Rewarding work on boxes.   The workbook format will push students to gain mastery *only* if your course design somehow rewards students for filling out the boxes. There are several different ways to do this. In the past, I randomly selected several students to submit their workbooks each class session, which I graded for thoughtful *effort*, with special emphasis on the chapter discussed in class that day. Each student's average grade for these random samples counted between 10 and 15% of their course grade. I also arranged things so that each student was selected five or six times during the semester.

However, almost every student now has a smartphone, so I currently use the following scheme. By noon the day before a chapter will be discussed in class, students submit a scan (or other electronic copy) of their work for that chapter's boxes. I quickly grade each submission on a two-point scale (0 = nothing done, 1 = partial effort, 2 = adequate effort, which can include explaining why they are stuck). Meanwhile, I post solutions to the boxes in a secure way online. Students use these to correct their box work using a pen with a different color. I drop the two lowest scores per semester as a way of recognizing that sometimes life will intrude on one's ability to get things done on time.

This approach has several advantages. This method is better at prodding students to stay on top of the material, which is very important in this highly sequential course. For class sizes smaller than 20 students, it takes less than 20 minutes to do the grading, which is not much more time than it takes to look through multiple chapters in several randomly submitted workbooks. But the most important advantage is that I am able to see what students are having difficulty with *before class*, so that I can target class activities to addressing those difficulties. (This is a technique known as Just-in-Time Teaching, or JiTT, which you can read more about online.)

I also recommend that you (the professor) work though all the boxes *yourself* before class the first few times you teach the course. This will help refresh your memory, isolate issues that you may need to resolve for yourself before class, and (most importantly) help you anticipate and appreciate the difficulties that students may have with the boxes.

What to do in class.   Over the years, I have found it most effective to spend class time doing the following:

- Addressing observed or reported difficulties with the box exercises.
- Answering general questions about the material.
- Providing a large-scale perspective on the role that the current chapter plays in the course structure.
- Working examples (sometimes).
- (Most important!) Designing supervised activities that strengthen students in the areas where you see they are having problems.

I have found the last item to be *especially* important, as students are not always willing or even able to articulate what they find challenging. When I see that students are struggling with certain issues in the box exercises, I create a short (10- to 15-minute) in-class activity addressing that issue and charge them to work on it in groups of two to four (ideally three). I then circulate around the room to monitor the progress that groups are making. Such a task often exposes quite clearly the stumbling blocks that individual students are encountering, and allows me to help them directly without having them embarrass themselves in front of the whole class. This is especially important in preventing weaker or more marginalized students from falling through the cracks.

I have specifically designed some of the problems in the homework section of each chapter to be suitable as class activities. You may also be able to take an example that you are thinking about presenting and repurpose it as a class activity. I often tell my students that if you want to learn how to play a piece on the piano, it is *sometimes* helpful to hear an expert play it, but it is better (and ultimately essential) that you play it yourself *for* an expert and get their feedback. That is where functional learning really takes place.

Homework.   I typically assign three homework problems per chapter: this is enough to keep students quite busy. The homework problems for this class can be challenging, and even the best students may not get them right the first time. Homework-grading schemes that focus *only* on the final results can therefore make students anxious. However, one can devise grading schemes that (1) allow students to engage difficult problems with less anxiety, (2) provide them with opportunities for feedback and further learning, and (3) make grading easier for you or your TAs. Consult the "Course Design" section of the book's website at `https://pages.pomona.edu/~tmoore/smw/` for up-to-date suggestions on how to do this.

Resources for instructors.   Instructors adopting the text for classes should contact us using the form provided on the book's website (see above) to request access to problem solutions and other resources for instructors only. Please do *not* use this form unless you are someone who is at least thinking about adopting this book for a course.

Information for Solitary Readers

This book is designed primarily for use in the context of a college or university class, but its workbook format also makes it useful for teaching oneself. I can completely understand why solitary readers would like to have access to box and problem solutions. However, neither the publisher nor I can tell who is a solitary reader and who is a student who wants such access to cheat or to post the solutions online. (The latter is not an idle concern, as I caught one of my own students in a different course submitting homework copied from solutions illegally posted online by someone else.) Therefore, please do not ask us for access to the full solutions unless you are someone we can independently verify the instructor of a course.

That said, we do want the book to be useful to solitary readers. Please note that many of the box exercises (and some homework problems) provide answers and/or fairly detailed instructions about how to arrive at answers. So, in many cases, you will know when you have done things correctly.

The website above also offers carefully designed resources for readers in general that provide extra guidance, some problem answers, and other aids that don't enable students to cheat. Consult the website if you need help.

Appreciation

I am grateful to many people who have helped bring this book to fruition. First, I would like to thank authors whose excellent books have helped me understand the subject and how to teach it, especially books by Stephen Blundell, Sidney Coleman, Dave Goldberg, David J. Griffiths, Robert Klauber, Tom Lancaster, Michael Peskin, Daniel V. Schroeder, and Paul Teller.

I also want to thank my students in Pomona College's Physics 170 class (especially Adam Dvorak, Jacob Hauser, Derek Li, and Adele Myers) for documenting errors in early versions and offering useful feedback. I am grateful to Jonas Mureika, James Overduin, Leo Rodriguez, Brian Shuve, and especially Daniel Schroeder, who read a working draft and offered detailed error corrections and valuable suggestions. Dan has been a wonderful mentor and thoughtful critic on this project.

Thanks also to William Loinaz, who read the final manuscript and provided helpful feedback.

I am grateful to the book's production team, including Laurel Muller of Coho-Graphics, Kot Copyediting and Proofreading, Inc., and especially Project Manager Paul Anagnostopoulos of Windfall Software and Production Manager Barbara Dickson, for their excellent work, care, and patience in dealing with a difficult book (and sometimes difficult author); and to Jane Ellis, my publisher at University Science Books, for her support, enthusiasm, and hard work in bringing this book to print.

Finally, I am grateful to my wife, Joyce, whose unfailing support for my writing habit is loving and gracious beyond the call of duty.

Heartfelt thanks to all!

Thomas A. Moore
Claremont, CA
October 2023

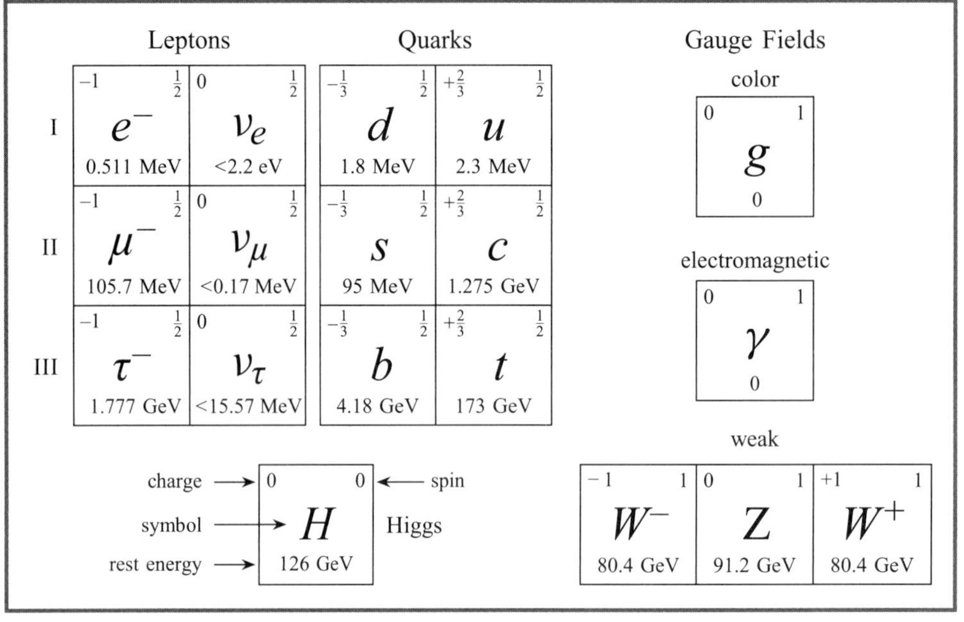

Leptons
Quarks
Gauge Fields
color
electromagnetic
weak
I
II
III
−1   1/2   0   1/2
e⁻   νₑ
0.511 MeV   <2.2 eV
−1   1/2   0   1/2
μ⁻   νμ
105.7 MeV   <0.17 MeV
−1   1/2   0   1/2
τ⁻   ντ
1.777 GeV   <15.57 MeV
−1/3   1/2   +2/3   1/2
d   u
1.8 MeV   2.3 MeV
−1/3   1/2   +2/3   1/2
s   c
95 MeV   1.275 GeV
−1/3   1/2   +2/3   1/2
b   t
4.18 GeV   173 GeV
0   1
g
0
0   1
γ
0
charge
symbol
rest energy
spin
0   0
H   Higgs
126 GeV
−1   1   0   1   +1   1
W⁻   Z   W⁺
80.4 GeV   91.2 GeV   80.4 GeV

1. INTRODUCTION

The so-called **Standard Model of Particle Physics** provides a unified theoretical structure for answering many of our most basic questions about the universe. This theory describes the fundamental particles (electrons, photons, and quarks) out of which all known matter is constructed, as well as the plethora of short-lived exotic particles seen in other contexts. It clarifies the structure of atoms and nuclei, explains the results we see in high-energy accelerator experiments, helps us analyze what cosmic-ray particles tell us about the universe, guides us in creating models of high-energy astrophysical events, and illuminates the physics involved in the early stages of the Big Bang itself. Indeed, the Standard Model is the foundation on which our current understanding of all physics (other than that of gravitation) rests. Its development represents perhaps the most significant triumph of late-twentieth-century physics.

This chapter provides an overview of the entire Standard Model, establishing both the larger context and the frames into which future chapters insert their more detailed pictures. This overview will also help us see what kinds of questions the Standard Model answers and how it arrives at those answers.

1.1 Relativistic Quantum Field Theory

The Standard Model is itself founded on an approach to quantum mechanics that physicists call **relativistic quantum field theory**. Non-relativistic quantum mechanics (the version of quantum mechanics most people learn first) assumes the permanent existence of the particles it describes. But in the world of particle physics, this assumption is routinely violated: particles decay into other particles, particles and antiparticles annihilate into photons, and colliding particles may transform themselves into completely different particles or create new particle-antiparticle pairs from their kinetic energy.

Relativistic quantum field theory handles these issues by approaching quantum physics from a different direction that takes its inspiration from electrodynamics. We can consider a light wave to be a traveling disturbance in an electromagnetic field, but we also know that light consists of photons. Indeed, we can think of a photon as a quantized "excitation" of the electromagnetic field, a concept that we will better define in Box 1.1.

Quantum field theory posits that *all* the particles we observe in nature are similar "excitations" of some kind of field: electrons are the quantized excitations of an electron field, quarks are excitations of a quark field, and so on. This implies that if we want to understand how such particles behave, we first must understand the fields out of which they arise. We must determine how to mathematically describe the field corresponding to each particle, and discover the field equations (analogous to Maxwell's equations for the electromagnetic field) that the field must obey. To understand how particles interact, we must also understand how fields connect to and influence each other.

The beauty of this approach is that it focuses our attention on what is truly fundamental (the fields and their characteristics), rather than on the particles that are transitory excitations of those fields. This enables quantum field theory to handle the creation and destruction of particles with equanimity.

In this text, we will begin our study of quantum fields in Chapter 4 by considering the simplest possible field: a spinless **scalar field** $\phi(t, x, y, z)$ that attaches a single number to every point in space and time. We will learn in Chapter 5 how to describe its field equations in a way consistent with relativity, and how to calculate quantities such as currents and energy densities from those equations in Chapter 6. After a review of some crucial concepts from non-relativistic quantum mechanics in Chapters 7–9, we will see how to mathematically link the field to its quantized particles in Chapter 11.

This basic structure will provide a foundation for understanding more complicated fields later in the book, such as the electromagnetic field (where we must assign a four-component *vector* to every point in space and time) and fields for describing spin-$\frac{1}{2}$ particles (which attach a different kind of mathematical object called a **spinor** to every point in space and time). We will also see that fields have two *types* of excitations— a "particle" excitation and an **"antiparticle"** excitation—that have the same mass but are distinct in other ways. This background will help us in Chapter 18 to understand the **spin-statistics theorem**—the reason that particles with spin 0, 1, 2, and so on *can* occupy the same quantum state and behave as **bosons**, but particles with spin $\frac{1}{2}$, $\frac{3}{2}$, and so on *cannot* and behave as **fermions** (something that the well-known **Pauli exclusion principle** states but does not explain).

1.2 Interactions

In this scheme, material objects and the forces between them are not as distinct as they seem in Newtonian physics: the universe is really constructed entirely of fields, some of which are interconnected in certain ways. The most basic form of the quantum field theory known as **quantum electrodynamics (QED)** describes the interaction of two fields—an electron field and an electromagnetic field (whose excitations are the electron and the photon, respectively)—by including a term in the expression for the fields' energy density that is the product of two powers of the electron field, one power of the electromagnetic field, and a constant that tells us how strongly the two kinds of fields interact (a quantity that is proportional to the electron's *charge*). This term tells us that at any point where the electron and photon fields are both nonzero, the system has an extra energy density beyond what each field itself would bring to that point. It turns out that how an electron field (and thus an electron) behaves in an electromagnetic field is completely determined by this interaction term.

How do we calculate the effects of such an interaction? In quantum physics in general, we calculate the *probability* that a quantum system in a given initial state ends up in a given final state by calculating the **quantum amplitude** for each possible subprocess that takes us from the initial state to the final state, adding the amplitudes, and taking the absolute square of the total (see Chapter 7). In quantum field theory, the amplitudes we sum essentially represent terms in a power-series expansion of the total amplitude. We can pictorially represent each amplitude term in the sum using a schematic diagram called a **Feynman diagram**. Figure 1.1 shows the diagrams depicting three different amplitudes

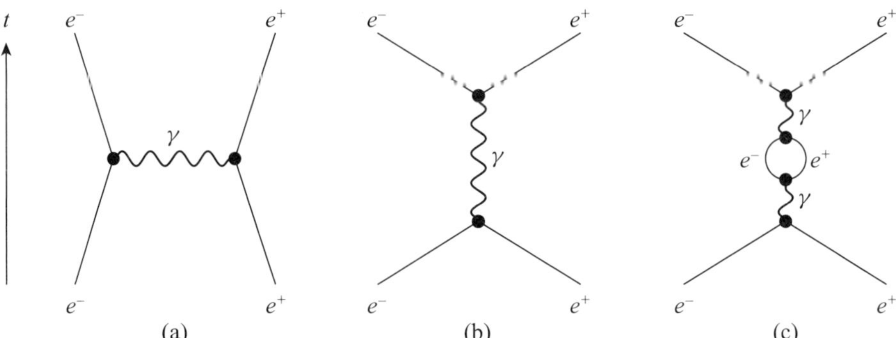

FIGURE 1.1   These Feynman diagrams depict three different electron-positron scattering amplitude terms. (Time progresses vertically here.)

for the situation where our initial state is an electron and an antielectron approaching each other and our final state is an electron and antielectron departing with different momenta. In the most basic form of QED, all diagrams involve electron/antielectron lines (solid lines) and photon lines (squiggly lines) that meet at vertices (solid dots) where two electron/antielectron lines meet one photon line, reflecting how the electron-photon interaction term in QED has two factors of the electron field strength but just one factor of the photon field strength.

In this text (though this is not a universal convention), we draw these diagrams to look like spacetime diagrams (that is, with a *vertical* time axis) so that the system's initial state appears at the bottom and the system's final state at the top. Read as a spacetime diagram, each Feynman diagram *appears* to tell a story. For example, the middle diagram seems to say that the electron and positron approach each other, collide, and annihilate to a photon at rest. The photon subsequently decays back into an electron and positron that move away from each other.

In spite of this, one should *not*, repeat *not* take such a diagram as literally describing a physical process we might observe. The middle diagram in fact describes something impossible: though an electron and positron colliding with equal and opposite momenta must produce a photon at rest to conserve momentum, a photon can never be at rest. The left-hand diagram shows the photon traveling at infinite speed, which is also absurd. Internal lines in a Feynman diagram (lines that are not part of the system's initial or final state) do not represent real particles; we say that the lines depict **virtual particles**.

You should instead think of a Feynman diagram as a map for calculating a numerical quantum amplitude: each vertex, each internal line, and each external line corresponds to a distinct factor in that amplitude. Calculating the amplitude also involves integrating over all vertex positions, so not only do the slopes of internal lines mean nothing physically, but it is meaningless to ask (for example, in the left diagram) which particle emits and which absorbs the photon. The time axis' *sole* purpose is to distinguish the initial and final states.

Another reason for not interpreting a diagram literally is that each depicted term is only part of an infinite series that *collectively* describes what physically happens. The right diagram shows one of eleven possible 4-vertex diagrams, and we can draw many more 6-vertex diagrams, 8-vertex diagrams, and so on. Fortunately, a term's magnitude usually decreases as the number of diagram vertices increases, so adding just the terms corresponding to diagrams with the fewest vertices often yields a good approximation for the total amplitude.

Since almost all everyday interactions between objects involve electromagnetic interactions at some level, quantum electrodynamics describes virtually all aspects of our common experience (Atomic structure! Emission and absorption of light! Why objects that are mostly empty space resist merging!), while also delivering accurate predictions at all experimentally accessible energies. QED provides the inspiration for the other field theories in the Standard Model.

1.3 Fundamental Fields

According to the Standard Model, the basic building blocks of matter are twelve fundamental fields whose excitations are spin-$\frac{1}{2}$ particles known as **leptons** and **quarks**. The leptons consist of six particles arranged in three families of particle pairs: the **electron** (symbol $= e^-$) and **electron neutrino** (ν_e), the **muon** (μ^-) and **muon neutrino** (ν_μ), and the **tau** (τ^-) and **tau neutrino** (ν_τ). The six quarks also come in three families of paired particles: the **down** (d) and **up** (u), the **strange** (s) and **charm** (c), and the **bottom** (b) and **top** (t).

Figure 1.2 shows the leptons and quarks arranged in the form of a periodic table somewhat analogous to the periodic table of the chemical elements. Each box shows

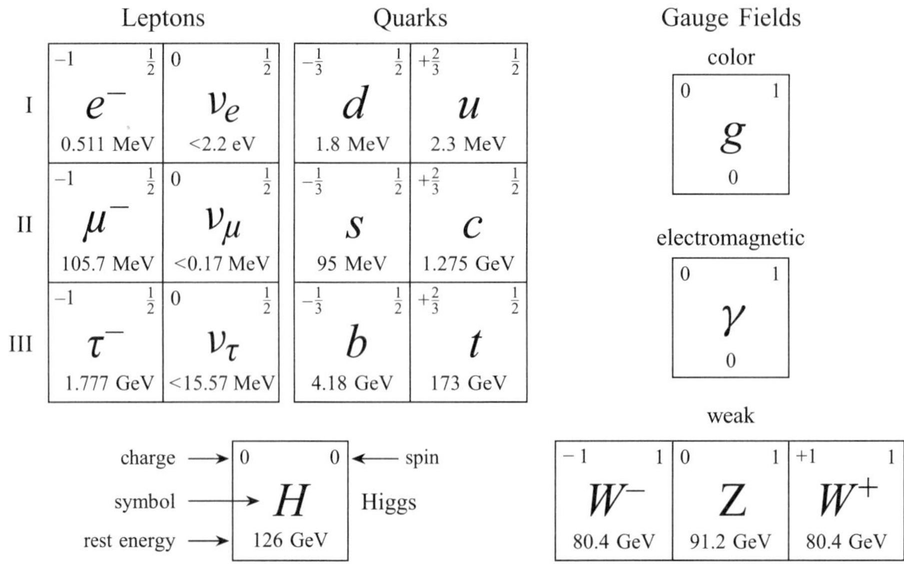

FIGURE 1.2   This table displays the fundamental particles in the Standard Model, where e^- represents the electron, ν_e the electron neutrino, μ^- the muon, ν_μ the muon neutrino, τ the tau, ν_τ the tau neutrino, and u, d, s, c, b, and t the up, down, strange, charm, bottom, and top quarks, respectively. The box for the Higgs particle (H) also provides the key for interpreting the numbers.

the particle's standard symbol in the center, its electrical charge (in units of the proton charge) in the upper left corner, its spin quantum number in the upper right corner, and its **rest energy** (defined as mc^2, where m is the particle's mass) at the bottom. As in the table of the elements, mass increases as we go downward, and the matter particles in each vertical column interact in similar ways with other particles.

In addition to the twelve matter particles, the Standard Model includes twelve so-called **gauge fields** that emerge because the field theories for the quarks and leptons obey certain symmetry principles. These fields are a group of eight **color** fields, the **electromagnetic field**, and a group of three **weak** fields. Excitations of the color fields are called **gluons** (g); excitations of the three weak fields are **weak vector bosons** called the W^-, the W^+, and the Z, respectively; and excitations of the electromagnetic field are, of course, photons (γ). These gauge field particles all have spin 1, and all but the weak vector bosons are massless. Figure 1.2 does not display all eight gluons because they are identical with respect to properties listed on the table. (We will discuss the property that distinguishes the gluons in Chapter 30.)

Electromagnetic fields interact with every field whose excitation is a charged particle (including the W^+ and W^- bosons). The fields whose excitations are the W^+, W^- and Z particles also interact with all fields whose excitations are fermions. Gluon fields interact only with quark fields and each other.

The remaining field in the table is **Higgs field**, whose excitation is the spin-0 **Higgs boson** (H). The Higgs field is different than the color, electromagnetic, and weak fields in a number of ways but plays a crucial role in the Standard Model: the interaction between a fermion field or gauge field with the Higgs field makes the excitations of those fields behave like massive particles. If it were not for the Higgs field, all particles in the Standard Model would be massless, like the photon or gluons! We will discuss the Higgs field and describe exactly what sense it "gives particles mass" in Chapters 26 and 27.

I mentioned earlier that all fields have "particle" and "antiparticle" excitations. For example, an antielectron e^+ (also called a **positron**) has the same mass and spin as the electron but a charge of $+1$ instead of -1. Similarly, the antidown ($\bar{d}$) has the same mass and spin as the down but a charge of $+\frac{1}{3}$ instead of $-\frac{1}{3}$. The symbol for an antiparticle is the symbol for the corresponding particle but with a line over it (except for the charged leptons, whose antiparticle symbols are more usually written simply as e^+, μ^+, and τ^+). The table only lists the "particle" excitation of each field, but it is pretty easy to infer an antiparticle's properties from the table entry for the particle.

Some fields with electrically neutral "particle" excitations have antiparticle excitations that are the *same* as the particle excitations: this is the case for the photon field, the Z field, and the Higgs field. It also turns out that the $\overline{W^+} = W^-$ and $\overline{W^-} = W^+$. Antiparticle excitations of the gluon and neutrino fields have the same charge (0) as their corresponding particle excitations, but have other properties (not listed on the table) that make them physically distinct from their corresponding particles. We will discuss these distinguishing properties in context in later chapters.

The gravitational interaction is *not* handled in the Standard Model and (for reasons we will discuss in Chapter 21) is actually extremely difficult to integrate with relativistic quantum field theory. The development of a quantum theory of gravity and integration of that theory with the Standard Model is an important but currently unsolved problem in theoretical physics.

1.4 Hadrons

We are quite familiar with the electron and the photon, and some of the other fundamental particles listed on the table (particularly the muon and the neutrinos) are quite common in the universe. But no one has ever observed any of the listed quarks as isolated particles. Instead, we commonly see protons and neutrons in the universe (not listed on the table), and high-energy interactions can generate many other kinds of particles also not listed on the table. How do these particles fit into the Standard Model?

The color (gluon) fields are different from the electromagnetic field in that they interact with each other (in contrast, the electromagnetic field is electrically neutral and so cannot interact with itself). This leads to self-feedback effects in the color field that, among other things, make the color interaction so strong that if we try to separate two bound quarks by pulling them apart (and do work on them that ends up in the color field), the field energy eventually becomes so large that it is energetically advantageous to convert some of it to the rest energy of new quarks that bind to the ones we were trying to separate (see Figure 1.3). This is why we never observe solitary quarks in nature; we instead only observe specific combinations of quarks that are collectively **color-neutral** and thus are unable to interact with the color field. Such particles *can* be separated. For similar reasons, we never observe isolated gluons (though people are currently searching for color-neutral combinations of gluons called "glueballs," which current particle colliders might be able to create).

Color-neutral quark-combinations are called **hadrons**. For reasons we will explore in Chapter 30, hadrons mostly fall into two subclasses: **mesons**, which consist of a quark-antiquark pair; and **baryons**, which consist of three quarks or three antiquarks. The proton ($p = uud$) and the neutron ($n = ddu$) are the two most common baryons in the universe. For reasons we will discuss in Chapter 27, all mesons are unstable, but they can still be observed in collider experiments. Table 1.1 lists some of the more commonly observed hadrons.

The fact that quarks are never observed alone makes measuring their masses difficult: in actual hadrons, it is difficult to detangle the quarks' masses from their kinetic energies and the gluon field energy. (For a review of how mass and energy are related and how field energies contribute to a bound system's mass, see Box 1.2.) The quark masses displayed in Figure 1.2 are masses that the quarks seem to display in the high-energy limit, where (for reasons we will discuss in Chapter 30) the quarks behave like free particles.

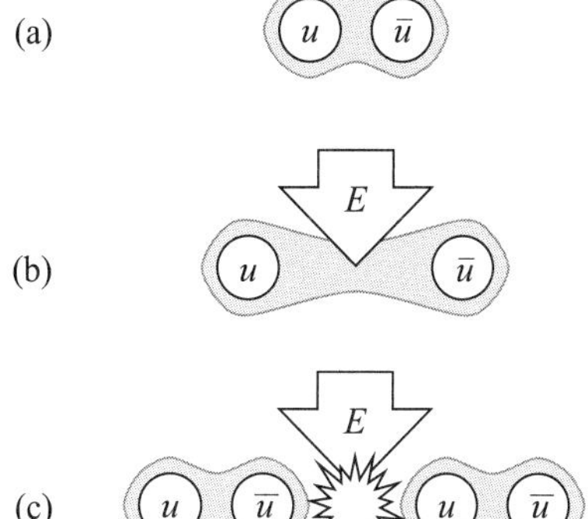

FIGURE 1.3 To separate two bound quarks, one must pour energy into the color field (qualitatively indicated by the gray enclosure) that binds them. Eventually the field energy becomes so large, the color field can actually reduce its energy by creating a new quark-antiquark pair.

Selected Mesons

Symbol	Anti	Spin	Quarks	mc^2 (GeV)	Lifetime (s)
π^+	π^-	0	$u\overline{d}, d\overline{u}$	0.13957	2.60×10^{-8}
π^0	(self)	0	$u\overline{u} + d\overline{d}$ §	0.13498	8.52×10^{-17}
K^+	K^-	0	$u\overline{s}, s\overline{u}$	0.49368	1.24×10^{-8}
K^0_L	(self)	0	$d\overline{s} + s\overline{d}$ §	0.49761	5.18×10^{-8}
K^0_S	(self)	0	$d\overline{s} - s\overline{d}$ §	0.49761	0.89×10^{-10}
η	(self)	0	mostly $s\overline{s}$ §	0.54786	$5.0 \ \times 10^{-19}$
η'	(self)	0	$u\overline{u} + d\overline{d} + s\overline{s}$ §	0.95778	$3.2 \ \times 10^{-21}$
ρ^+	ρ^-	1	$u\overline{d}, d\overline{u}$	0.77526	4.41×10^{-24}
ρ^0	(self)	1	$u\overline{u} + d\overline{d}$ §	0.77526	4.41×10^{-24}
ϕ	(self)	1	$s\overline{s}$	0.77526	$4.1 \ \times 10^{-24}$
D^+	D^-	0	$c\overline{d}, d\overline{c}$	1.8967	1.06×10^{-12}
D^0	$\overline{D}^0$	0	$c\overline{u}, u\overline{c}$	1.8648	4.10×10^{-13}
J/ψ	(self)	1	$c\overline{c}$	3.0969	$7.1 \ \times 10^{-21}$
B^+	B^-	0	$u\overline{b}, b\overline{u}$	5.2793	1.64×10^{-12}
B^0	$\overline{B}^0$	0	$b\overline{d}, d\overline{b}$	5.2797	1.52×10^{-12}
Υ	(self)	1	$b\overline{b}$	9.4603	$1.2 \ \times 10^{-20}$

§. Low-mass neutral mesons are quantum superpositions of quark types.

Selected Baryons

Symbol	Spin	Quarks	mc^2 (GeV)	Lifetime (s)
p	$\frac{1}{2}$	uud	0.93827	∞
n	$\frac{1}{2}$	ddu	0.93957	900
Λ	$\frac{1}{2}$	uds	1.11568	2.63×10^{-10}
Σ^+	$\frac{1}{2}$	uus	1.18937	0.80×10^{-10}
Σ^0	$\frac{1}{2}$	uds	1.19264	$6 \ \ \times 10^{-20}$
Σ^-	$\frac{1}{2}$	dds	1.19974	1.48×10^{-10}
Δ^{++}	$\frac{3}{2}$	uuu	1.232	$6.6 \ \times 10^{-24}$
Δ^+	$\frac{3}{2}$	uud	1.232	$6.6 \ \times 10^{-24}$
Δ^0	$\frac{3}{2}$	udd	1.232	$6.6 \ \times 10^{-24}$
Δ^0	$\frac{3}{2}$	ddd	1.232	$6.6 \ \times 10^{-24}$
Ξ^0	$\frac{1}{2}$	uss	1.31486	2.90×10^{-10}
Ξ^-	$\frac{1}{2}$	dss	1.32171	1.64×10^{-10}
Ω^-	$\frac{3}{2}$	sss	1.6724	$8.2 \ \times 10^{-11}$

TABLE 1.1   This table displays the properties of selected mesons and baryons. Values for mc^2 and mean lifetimes are from the Particle Data Group (http://pdg.lbl.gov/2020/tables/contents_tables.html), where one can find much more complete lists of known particles and their experimentally observed properties.

Hadrons also have antiparticles. One can see from Table 1.1 that the π^+ and π^- are antiparticles of each other, as are the K^+ and K^- and the K^0 and $\overline{K}^0$ (for example, $\overline{\pi^-} = \overline{d\overline{u}} = \overline{d}u = \pi^+$ because the antiparticle of an antiquark is the original quark). The π^0, η, and η' are their own antiparticles. The baryons, in contrast, have antiparticles *not* listed on the table: the antiproton p^- has quark composition $\overline{u}\,\overline{u}\,\overline{d}$, the antineutron $\overline{n}$ has quark composition $\overline{d}\,\overline{d}\,\overline{u}$, and so on. (Note that the neutron is *not* its own antiparticle.)

The practical uses of the Standard Model are to predict the masses and other properties of these quark combinations, their decay rates and decay products, as well as possible outcomes and their probabilities when we smash these particles together in particle physics experiments. Even though making such predictions can be pretty difficult (for reasons we will understand better as we go on), the Standard Model has proved remarkably successful in making such predictions when the calculations are tractable.

1.5 Symmetries

But the deeper purpose of the Standard Model is to illuminate the patterns we observe: the structure of the periodic table in Figure 1.2, the features and properties of the fundamental fermion fields, and the characteristics of the gauge fields. Why do quarks and leptons appear in pairs? Why is there only one electromagnetic field but three weak fields and eight color fields?

One of the core features of the Standard Model is that it arrives at many of its predictions by applying **symmetry principles**. A "symmetry principle" declares that the laws of physics are unchanged when we transform in a certain way the system to which those laws are applied. Requiring that the laws of physics be unchanged (for example) by our choice of inertial reference frame (so that all such frames are equivalent) puts strong constraints on the shape of those laws. Indeed, the entire theory of special relativity is simply an unraveling of this symmetry principle's consequences. (To see a simple example of how a rotation symmetry principle can shape Newtonian physics in a way that has profound and non-obvious implications, see Box 1.4.)

The Standard Model is constructed to respect the basic symmetry principles that the laws of physics are unchanged by rotations in space, one's choice of space or time origin, and one's choice of inertial reference frame. But it also recognizes new, non-obvious symmetry principles as the reasons that the electromagnetic, weak, and color fields exist and have the characteristics they do. Unlike the previously mentioned symmetries, these new symmetries involve postulating that the laws of physics are unchanged by "rotations" in various kinds of abstract spaces, even if that rotation depends on spatial location.

For example, we can multiply a quantum wave function by a phase (a complex number with magnitude 1) without changing the laws of quantum field theory. This symmetry, called a **U(1)** symmetry, is equivalent to a rotation of the wavefunction's value in the abstract complex plane. If we require that the laws of quantum field theory be invariant even if we vary that rotation angle from place to place in space, then we find (see Chapter 19) that we must add to our theory a new field that obeys Maxwell's equations. This new field is an electromagnetic field. So electromagnetism is an implication of the fact that quantum field theory respects a **local** (position-dependent) U(1) symmetry principle.

In a similar way, we will see in Chapter 30 that we can rotate a triplet of quark fields in an abstract three-dimensional complex vector space—a so-called **SU(3)** symmetry— without changing the laws of physics. A general rotation in such a space requires eight angles to specify it completely. If we require that the laws of physics be invariant under a *local* SU(3) rotation (meaning that the eight angles can vary from place to place), we must add eight new fields that obey more complicated versions of Maxwell's equations— equations that completely describe these fields' behavior just as Maxwell's equations completely describe the physics of the electromagnetic field. The excitations of these new "color" fields are the eight different gluons.

Finally, the reason that quarks and leptons appear in pairs is that the Standard Model assumes that the laws of physics are invariant under a rotation of the fields in each pair in an abstract two-dimensional complex vector space—a so-called **SU(2)** symmetry. A general rotation in such a space is specified by three angles. If we require that the laws of physics be invariant under a *local* SU(2) rotation, we must add three new "weak interaction" fields, whose excitations the W^+, the W^-, and the Z vector bosons (see Chapters 23–25).

This symmetry is complicated by the fact that interactions with the Higgs field "break" the underlying SU(2) symmetry in such a way that the quark and lepton fields in each pair end up having particles with different masses and charges, and so do *not* actually behave equivalently, as the unbroken SU(2) symmetry principle would require. This is discussed in Chapters 26 and 27.

So, according to the Standard Model, the structure of the universe at the quantum scale is as simple as 1, 2, 3—or (more technically) U(1), SU(2), and SU(3). This is an approach to physics that firmly grounds the Standard Model on simple but profound conceptual and mathematical foundations.

1.6 The Flow of the Argument

Table 1.2 describes how the rest of the book presents the details of the Standard Model. The model is founded on the **principle of relativity** (the symmetry principle requiring that physical laws be independent of one's choice of inertial reference frame). Chapters 2

	1. Introduction
Relativity	2. Special Relativity
	3. Index Notation
Classical Fields	4. A Classical Scalar Field
	5. The Klein-Gordon Equation
	6. Noether's Theorem
Quantum Mechanics	7. Basic Quantum Mechanics
	8. Operators and Time Evolution
	9. Angular Momentum
(Shortcut to 15)	10. Feynman Diagram Quick Start*
Quantum Field Theory	11. Creation and Annihilation Operators
	12. The Quantized Scalar Field
	13. Interactions and Feynman Diagrams
Connection to Experiment	14. Fermi's Golden Rules
	15. Toy Universe Predictions
Spin-$\frac{1}{2}$ Particles	16. The Dirac Equation
	17. Transforming Dirac Spinors
	18. Quantizing the Dirac Field*
The Electroweak Theory	19. Electromagnetism from U(1)
	20. Quantum Electrodynamics
	21. Renormalization
	22. Applications of QED
	23. The Weak Interaction from SU(2)
	24. The Weak Interaction's Leftist Bias
	25. The Unified Electroweak Theory
Particle Masses	26. The Higgs Mechanism
	27. Fermion Masses and Mixing
	28. Neutrino Masses and Mixing*
	29. Applications of Electroweak Theory*
Chromodynamics	30. The Color Interaction
	31. Applications of Chromodynamics*
And Beyond!	32. Beyond the Standard Model*

TABLE 1.2 This table displays the organization of chapters in this text. Chapters marked with an asterisk may be omitted without affecting continuity.

and 3 review the theory of special relativity and introduce mathematics that allows us to write equations in a way that automatically ensures they obey that symmetry principle.

Chapters 4–13 present the rudiments of quantum field theory, sufficient to see how the theory works without getting bogged down by details better handled in a graduate-level course. Chapter 13 closes by showing how to use Feynman diagrams to calculate amplitudes for various experimental outcomes. Chapters 14 and 15 examine the concepts of a **decay rate** and an interaction **cross section**, and show how to connect these experimentally measurable quantities to the amplitudes depicted by Feynman diagrams. (For those not interested in the details of quantum field theory, Chapter 10 provides a sufficient overview to allow the reader to leap ahead to Chapter 15 and over Chapter 18.)

To describe particles with spin $\frac{1}{2}$, we need a relativistic field equation involving spinors rather than the simpler scalar field equation presented earlier. We explore this new equation in Chapters 16–18.

We are then ready to understand the core of the Standard Model. Chapters 19 and 20 derive electromagnetism from the local U(1) symmetry principle and show how to calculate electromagnetic cross sections. Chapter 21 discusses how we can resolve certain mathematical difficulties that arise by applying a concept known as **renormalization**. (This important chapter also lays the foundations for Chapter 30.) Optional (but recommended) Chapter 22 explores some experimental tests and applications of quantum electrodynamics. Chapters 23–25 derive the weak interaction by applying a local SU(2) symmetry to quark and lepton pairs and show how the electromagnetic and weak fields are but two different manifestations of a unified **electroweak** set of fields.

However, without the Higgs field, all the particles in the Standard Model would be massless, which contradicts reality. Chapters 26 and 27 describe how the Higgs mechanism solves this problem and how mass mixing resolves some other issues. Optional Chapter 28 also discusses the problem of neutrino masses, which are an experimental reality but whose origin is not yet understood. Optional Chapter 29 closes this section with an exploration of some of the experimental tests of electroweak theory.

Chapter 30 and (optional) Chapter 31 finish our exploration of the Standard Model by describing how the color interaction emerges from an SU(3) symmetry, as well as the experimental tests that support that theory.

Powerful as it is, the Standard Model does not provide answers to every question about particle physics. It tells us why quarks and leptons come in pairs but not why there are *three* pairs. It tells us why particles have mass, but not the values of those masses. It does not tell us why the electron's charge or the analogous weak and color coupling constants have the values they do. It does not tell us how gravity, dark matter, and dark energy fit in.

These questions and others presumably have answers in some theory beyond the Standard Model. Optional Chapter 32 looks at some ideas for what might come next and the ongoing experiments to test those ideas.

We have a map; let's begin the journey!

BOX 1.1   Particles as Field Excitations

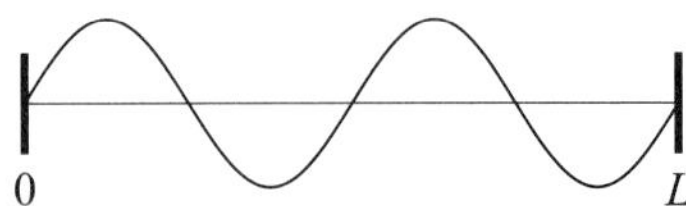

FIGURE 1.4   A normal mode of an electromagnetic field trapped between two perfectly reflective mirrors (vertical bars). The curve shows the magnitude of the electric field vector.

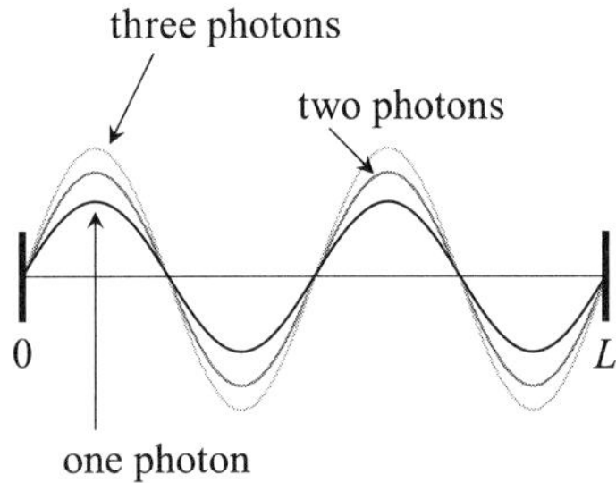

FIGURE 1.5   The amplitude of the wave in a given normal mode increases as the number of photons in that mode increases.

What does it mean to describe a particle as a "quantized excitation of a field?" In this box, we will describe this idea in more depth.

For the sake of simplicity, imagine an electromagnetic field in one dimension in a region between perfectly reflective boundaries a distance L apart. This is the electromagnetic analogue of the "particle-in-a-box" problem in basic quantum mechanics. The electromagnetic field must go to zero at a perfectly reflective boundary, so the only electromagnetic fields that are possible are oscillating fields that fit an integer number of half-wavelengths between the boundaries. Classically, we would call an oscillation corresponding to a given number of half-wavelengths one of the field's **normal modes**. Figure 1.4 shows the normal mode with four half-wavelengths fit between the boundary.

The energy in an oscillation is proportional to the square of the oscillation's amplitude. Classically, we could put as much or as little energy into a given normal mode as we want: the oscillation's amplitude can have any value. But for reasons no one really understands, each normal mode of a field can actually only store quantized amounts of energy. For an electromagnetic field, this energy is given by $E_n = nhc/\lambda$, where h is Planck's constant, c is the speed of light, and λ is the wavelength of the standing wave in the normal mode. For the given mode shown in Figure 1.4, $\lambda = L/2$, so the possible energies of the normal mode are $0, 2hc/L, 4hc/L, \ldots$, going up in steps of $2hc/L$.

A "photon" represents one step's worth of energy in the mode. So, if the mode contains an energy of $2hc/L$, we say that the mode contains one photon with wavelength $\lambda = L/2$. If the mode has energy $6hc/L$, then it contains three photons with that wavelength. If it has no energy, it contains no photons. Since the amplitude of the wave is proportional to the square root of the energy, that amplitude of the oscillation in the mode increases as $\sqrt{n}A_1$ (where A_1 is the wave amplitude corresponding to one photon) as the number n of photons in the mode increases. Figure 1.5 illustrates the basic idea.

In quantum mechanics, we call the normal modes of the field the possible **energy eigenstates** that a photon might have. The n for a given mode would thus be the number of photons sharing that eigenstate. (Because photons are bosons, any number can share a given energy state.) Photons in such states have definite energy but no specific location in the box. We can construct more localized (and so more particle-like) photon states by appropriately superposing a series of energy eigenstates, much as we can construct a pulse wave by adding appropriately weighted Fourier series of sine waves.

In classical electromagnetism, we ignore the quantization of the energies in a normal mode. For everyday-sized boxes, this is an excellent approximation.

Exercise 1.1.1   Calculate the smallest increment of energy that one could add to an electromagnetic field in a one-dimensional box of length $L = 0.2$ m. (*Hint:* $hc = 1240$ eV nm, and 1 eV $= 1.60 \times 10^{-19}$ J.)

BOX 1.2 Mass and Energy

One of the core implications of relativity theory is that an object's mass is a form of energy that can be freely converted into other forms of energy, and other forms of energy can be converted to mass. We will review in Chapter 2 how this follows from the principle of relativity, but for now, let's accept this as a fact. The energy equivalent of a mass m is given by Einstein's famous formula $E_0 = mc^2$. This energy is called a particle's **rest energy** because this is the energy that an individual particle has when it is at rest (and therefore has no additional kinetic energy due to its motion).

The cleanest example of converting energy from mass to other forms is matter-antimatter annihilation. If a positron essentially at rest meets an electron essentially at rest, the pair annihilate, producing two photons ($e^+ + e^- \rightarrow \gamma + \gamma$). The two photons must emerge from this annihilation process going in opposite directions with equal energies to conserve momentum (as we will see in Chapter 2). To conserve energy, the sum of the photon energies must be the same as the sum of the mass energies of the original electron and positron. If we let E be the energy of each photon, then $2E = 2mc^2 \Rightarrow E = mc^2 = 0.511$ MeV. Conversely, two photons, each with energy $E = 0.511$ MeV can collide and create an electron-positron pair at rest, thus converting the photons' energy into the mass-energy of the pair of particles.

Exercise 1.2.1 The "short-lifetime" neutral kaon K_S^0 in Table 1.1 typically decays into two pions, which can be either $\pi^0 + \pi^0$ or $\pi^+ + \pi^-$. If we assume that the decay is to neutral pions π^0 and that the kaon is at rest when it decays, what are the total energies E and kinetic energies $K \equiv E - m_\pi c^2$ of the outgoing pions?

Now, the rest energy of a composite system, like a meson, is the energy of the masses of its constituent quarks, plus the kinetic energy of the quarks' motion (as they orbit each other), plus any potential energy associated with their interaction. The quarks in a meson (or baryon) are bound together by the color interaction. Any attractive interaction that binds particles together will have a negative potential energy and therefore will contribute negatively to the system's total mass.

BOX 1.2 Mass and Energy (cont.)

Consider, for example, the π^+ meson. According to Table 1.1, the quark composition of this meson is $u\bar{d}$. Now, the u and d quark masses given in Figure 1.2 are tiny compared to the mass of even the π^+, so we can approximate them in any hadron as massless excitations of their respective quark fields with the same quantized energy $E = hc/\lambda$ as describes the massless photon. The color interaction basically confines these quark fields to a box of length $L \approx 1\,\text{fm} = 1 \times 10^{-15}\,\text{m} = 1 \times 10^{-6}\,\text{nm}$, which we can model as we did in Box 1.1 as a region bounded by mirrors a distance L apart.

Exercise 1.2.2   Estimate the minimum total energy that two massless quarks must have in a box that size. From the observed rest energy of the π^+, estimate the potential energy of the color field.

Technically, we should also include the electrostatic potential energy. Because the u and $\bar{d}$ both have positive charge, the electrostatic potential energy should be positive. If the quarks are separated by $r \approx L$, the effective electrostatic potential is therefore

$$V_e = \frac{kq_u q_{\bar{d}}}{r} = \frac{k(\frac{2}{3}e)(\frac{1}{3}e)}{r},$$

where k is the Coulomb constant.

Exercise 1.2.3   Calculate the electrostatic contribution to the rest energy of π^+. (*Hint: $ke^2 \approx 1.44$ eV nm*, where e is the proton charge.)

BOX 1.3 Conservation Laws for Particle Decays

Before the Standard Model, physicists found that they could explain *many* of the patterns observed in particle interactions using a few basic principles and some conservation laws. One of the most important principles were some rules for recognizing what kind of interaction was responsible for a decay:

1. Color interactions only rearrange quarks and/or create or destroy quark-antiquark pairs. Color-based decays are very rapid (lifetimes $\sim 10^{-22}$ s).

2. Electromagnetic interactions can rearrange quarks, and/or emit or absorb photons, and/or create or destroy particle-antiparticle pairs. Particles that decay electromagnetically have lifetimes $\sim 10^{-16}$ s.

3. Weak interactions can also create or destroy particle-antiparticle pairs, but weak interactions *alone* can change quark or lepton types (for example, change a $s \to d$ or a $\mu^- \to \nu_\mu$). In every weak interaction, exactly four fundamental particles are involved, with at least one going in and at least one going out. Only weak interactions can absorb, emit, or scatter neutrinos. Particles that decay weakly last $\sim 10^{-8}$ s.

4. When multiple types of interactions are possible, the fastest will dominate: color before electromagnetic before weak.

In addition, every interaction must obey the following conservation laws:

1. Energy and momentum must be conserved.

2. Total charge must be conserved.

3. Quark numbers must be conserved, where every quark has quark number $+1$ and every antiquark has quark number -1.

4. Lepton numbers for each lepton family must be *individually* conserved. Each lepton has lepton number $+1$, each antilepton lepton number -1.

For example, a muon can only decay to an electron, an electron-antineutrino, and a muon neutrino ($\mu^- \to e^- + \bar{\nu}_e + \nu_\mu$). This is because the electron is the only particle lighter than a muon, so only such a decay could conserve energy. But if the electron number (the lepton number or the electron family) was zero before the interaction, it must be zero afterward, which is why we must emit an electron-antineutrino to cancel the electron's lepton number (emitting a positron would not conserve charge). We also must conserve muon number, which was $+1$ initially, so we must have an outgoing muon neutrino to carry away that number. Because neutrinos are involved, this must be a weak-interaction decay, so we would predict a lifetime on the order of 10^{-8} s. (Experimentally, the lifetime is 2.2×10^{-6} s, even longer than is typical.)

Exercise 1.3.1 Why is the proton stable?

BOX 1.4   Rotational Symmetry and Angular Momentum

The goal of this box is to present a simple Newtonian example of how a symmetry principle can constrain a theory and lead to surprising consequences.

Consider an isolated system of two interacting particles that are at positions $\vec{r}_1$ and $\vec{r}_2$ (respectively) relative to some origin O. Newton's third law requires that the forces that these particles exert on each other must be equal in magnitude and opposite in direction: $\vec{F}_2 = -\vec{F}_1$. But Newton's laws do not specify the *directions* that $\vec{F}_1$ and $\vec{F}_2$ point, so long as they are opposite. In principle, $\vec{F}_1$ could point in any crazy direction, so long as $\vec{F}_2$ points in the opposite direction (see Figure 1.6a).

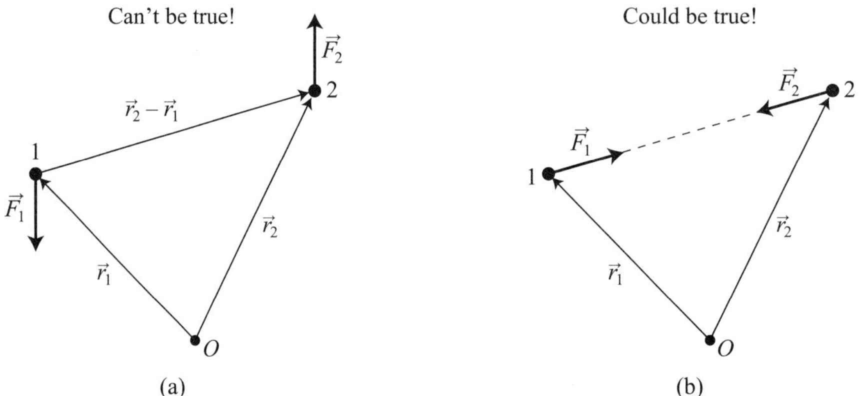

FIGURE 1.6   (a) Newton's laws allow this. (b) But the principle of rotational symmetry additionally requires that the forces point directly toward or away from the other particle along the axis connecting the particles.

But the laws of physics should be independent of one's orientation in space, meaning that all directions in space are physically equivalent. This means that we should be able to rotate the system around the line connecting the two particles without changing anything. But such a rotation *will change* the directions of $\vec{F}_1$ and $\vec{F}_2$, contrary to the principle of rotational symmetry, *unless* the forces both point *along the line connecting the particles*, as shown in Figure 1.6b. (The forces could point either directly toward the other particle—the case shown—or directly away from the other particle.) To put it a different way, if $\vec{F}_1$ were to lean away from that line, how would the interaction choose the direction in which it leans? All directions around the line are equivalent. So choosing a direction would privilege a particular direction in space over others, contrary to the principle.

So we see that the principle of rotational symmetry places an additional constraint on the laws of physics that is not present in Newton's laws: forces between particles must be parallel to the line connecting them to preserve the principle of rotational symmetry.

But this is not all! These constraints have *consequences*. Noether's theorem (which we will study in Chapter 6) states that for every symmetry principle that applies to the laws of physics, there is an associated conserved quantity. In this particular case, the conserved quantity is **angular momentum**. Recall that **torque** $\vec{\tau}$ is defined to be the rate of change of angular momentum: $\vec{\tau} = d\vec{L}/dt$. Figure 1.6b directly implies that the net torque that the interaction's two forces exert on the system around the arbitrary origin O is zero, implying that the net angular momentum of this system is conserved.

BOX 1.4 Rotational Symmetry and Angular Momentum (cont.)

Exercise 1.4.1 Prove that the sum of the torques $\vec{\tau}_1 = \vec{r}_1 \times \vec{F}_1$ and $\vec{\tau}_2 = \vec{r}_2 \times \vec{F}_2$ must be zero, independent of the location of the origin O. (*Hints:* The cross product of two vectors is distributive. Also, the cross product between two parallel or two opposite vectors is zero. The solution involves only a few steps: it should not take all the space provided in this box.)

Homework Problems

P1.1 In Compton scattering, a photon strikes an electron and both then move away with different energies and momenta. Draw a Feynman diagram for this process, using the electron-scattering diagram in Figure 1.1b as an analogy. What particle forms the horizontal line in this case?

P1.2 Draw the 10 remaining 4-vertex Feynman diagrams for which Figure 1.1c is one example. (Remember that each vertex must connect two electron lines with one photon line.)

P1.3 A deuterium nucleus consists of a proton and a neutron, and has a measured mass of 1875.6 MeV. What is the potential energy of the color interaction that binds the proton and the neutron? (Assume that the kinetic energies of the proton and neutron are negligible.)

P1.4 A K^- meson typically decays to a muon and something else, a pair of pions, or a triplet of pions.
a. Using the conservation laws listed in Box 1.3, determine the charges of any particles emitted in all three decay cases (and, in the first case, what the "something else" must be).
b. What is the total kinetic energy of the emitted particles in each case?
c. Why isn't a decay to four pions possible?

P1.5 The Λ baryon typically decays to $p + \pi^-$ or $n + \pi^0$. Considering the rules discussed in Box 1.3, what happens in the decay of the Λ to get each of the two possible sets of decay products? In particular, what happens to the s quark in each case, and what quark-antiquark pair is created? (Note that pair creation *can* create a superposition of $u\bar{u}$ and $d\bar{d}$ pairs.)

P1.6 Consider the decays of the Σ^0 and Σ^- in light of the rules discussed in Box 1.3. Typically $\Sigma^0 \to \gamma + \Lambda$ very quickly, while $\Sigma^- \to n + \pi^-$ much more slowly.
a. Why can the Σ^0 decay through a fast EM decay process rather than the slower weak process required by the decay of the Σ^-?
b. Why can't the Σ^0 decay by a still-more-rapid color process?

P1.7 Considering the rules described in Box 1.3, supply the missing particles in the following decays and explain why.
a. $\Lambda \to p + e^- + ?$ (This is possible but rare.)
b. $K_L^0 \to \pi^+ + \bar{\nu}_e + ?$
c. $\Delta^{++} \to p + ?$
d. $\rho^0 \to \pi^+ + ?$

P1.8 Considering the rules described in Box 1.3, supply the missing particles in the following interactions and explain why. If multiple possibilities exist, choose the possibility that would be most probable (color > EM > weak) and/or would require the least initial energy (which might be anything).
a. $\nu_e + n \to p + ?$
b. $\gamma + e^+ \to e^+ + e^- + ?$
c. $\pi^+ + n \to \Lambda + ?$
d. $p + p \to p + \Lambda + ?$

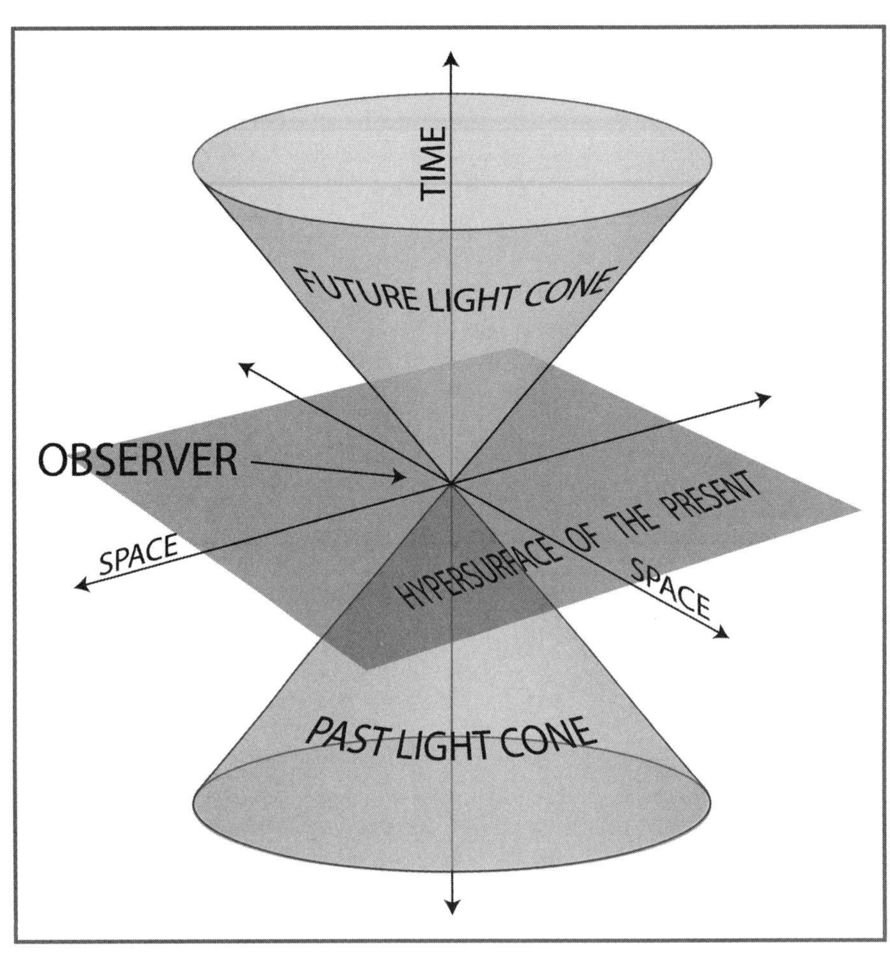

TIME
FUTURE LIGHT CONE
OBSERVER
SPACE
HYPERSURFACE OF THE PRESENT
SPACE
PAST LIGHT CONE

2. SPECIAL RELATIVITY

As we saw in the last chapter, the Standard Model is fundamentally founded on the concept of recognizing that various **symmetry principles** apply to the laws of physics. We also saw that such symmetry principles have physical consequences, and that recognizing the right principles and determining their consequences is what the Standard Model is all about.

Perhaps the most fundamental of these deep principles is the **principle of relativity**, the symmetry principle that lies at the heart of the theory of special relativity. This principle is indeed so fundamental that we are going to construct our whole mathematical approach so that any equations we write *automatically* satisfy this principle. This mathematical approach will be the main topic of Chapter 3. The purpose of this chapter is to review the principle of relativity and its most important consequences.

2.1 The Principle of Relativity

The principle of relativity states:

> The laws of physics are the same in every inertial reference frame.

Expressed in the form of a symmetry principle, this statement requires that the laws of physics be unchanged when one moves from one inertial frame to another.

To understand the full import of this statement, we first need to define some terms. An **event** is some physical process (for example, the explosion of a firecracker, or the collision of two particles) that marks a specific point in space and an instant of time. An *event* plays the same role in the four-dimensional combination of space and time that physicists call **spacetime** that a *point* does in three-dimensional space. A **reference frame** is a mechanism (real or imagined) that allows one to assign **spacetime coordinates** (three spatial coordinates and a time coordinate) to any event. An **observer** collects and interprets the data generated by the frame.

For example, we might construct a reference frame using a radar method. Imagine a system that sends out pulses of radar waves. When we receive a reflection, we can quantify the reflection event's direction by noting the horizontal and vertical angles in the sky from which the reflection comes. We can quantify its distance by multiplying half the time it took the pulse to travel from the transmitter to the reflection event and back by the speed of light. Using these three numbers, we can calculate the event's position in space by converting from spherical to Cartesian coordinates. The event's time coordinate is midway between when the pulse was sent and when the reflection was received.

We often consider reference frames attached to specific objects, and sometimes refer to a reference frame obliquely, as in "an observer riding along with the particle finds that . . . " Such a statement presumes that the observer is using a frame attached to the particle.

Depending on how a frame (or the object to which it is attached) moves through spacetime, a given reference frame may be either *inertial* or *noninertial*. An **inertial reference frame** in special relativity is a frame in which we observe a **free object** (an object that does not interact with other objects) anywhere in the frame to move at a constant velocity (as required by Newton's first law). All other reference frames are **noninertial reference frames**. Note that an observer in any frame can determine (by testing Newton's first law) whether a frame is inertial without referring to anything outside the frame. One can also prove (see Box 2.1) that a reference frame will be inertial *if and only if* it moves at a constant velocity and does not rotate relative to an established inertial reference frame.

The principle of relativity implies that all inertial reference frames are physically equivalent: no physical experiment performed entirely within the frame can distinguish it from any other inertial reference frame. This does *not* mean that physical quantities (such as a particle's energy or momentum) necessarily have the same *values* in different inertial reference frames; only that all observers find that these values obey the same laws.

On the other hand, Maxwell's equations state that light waves move in a given inertial frame with a certain speed c that appears as a fixed constant in the equations (a constant linked to experimental measurements of the strengths of static electric and magnetic interactions). Since this constant's value is *part* of the laws of electromagnetism, and since those laws are laws of physics that must obey the principle of relativity, it follows that the value of c and thus the speed of light must be the same in all inertial reference frames. Einstein was the first to fully understand this implication of the principle; his entire theory of special relativity is merely an exploration of its consequences. (Note that "radar method" for constructing a reference frame described earlier implicitly *assumes* that c is the same in every frame in the way that the method calculates distances from the flight-times of radar signals.)

2.2 Natural Units

Because the speed of light c is frame-independent, we can connect space and time units in a frame-independent way. In particular (in the spirit of the radar method), we can describe distances in terms of time units, where 1 s of distance is (by definition) the distance that light travels in 1 s of time, which is 299,792,458 m. Light moves at a speed of 1 s/s = 1 by definition in this unit system, and all other speeds are expressed by unitless fractions of the speed of light. This implies that mass m, momentum (which must have the same units as mv), and energy (which must have the same units as $\frac{1}{2}mv^2$) all have the same units. In particle physics, it is both convenient and conventional to express all of these quantities in terms of energies in GeV (where $1\ \mathrm{GeV} = 1 \times 10^9\ \mathrm{eV} = 1.60 \times 10^{-10}\ \mathrm{J}$). In these units, for example, the mass of a proton is 0.938 GeV.

Indeed, in particle physics, we go a step further. In the world of quantum mechanics, $\hbar = 1.055 \times 10^{-34}\ \mathrm{J\,s}$ (Planck's constant over 2π) plays a central role, and just as working in units where $c = 1$ simplifies equations in relativity, working in units where $\hbar = 1$ simplifies equations in quantum mechanics. You can see that Planck's constant makes a connection between energy and time units, and since c connects time and length units, we can express time, distance, energy, momentum, and mass all in terms of energy units. To bring charge into the mix, we also define the vacuum permittivity $\varepsilon_0 = 1$, which makes charge unitless. We call the unit system in which $\hbar = 1$, $c = 1$, and $\varepsilon_0 = 1$ **natural units.**

What can it possibly mean to express distance in *energy* units? In SI units, a photon with energy E has a wavelength

$$\lambda = \frac{hc}{E} = \frac{2\pi\hbar c}{E}. \tag{2.1}$$

Note that in natural units, the above equation becomes $\lambda/2\pi = 1/E$, so $1\ \mathrm{GeV}^{-1}$ of length corresponds to $1/2\pi$ times the wavelength (that is, length corresponding to a radian's worth of oscillation) of a photon whose energy is 1 GeV. Similarly, $1\ \mathrm{GeV}^{-1}$ of time corresponds to $1/2\pi$ times the oscillation period (that is, the time required for a radian's worth of oscillation) of such a photon. We see that in natural units, both distance and time have units of *inverse* energy. Table 2.1 shows SI units expressed in terms of GeV.

SI Unit	Natural Units
1 m	$5.07 \times 10^{15}\ \mathrm{GeV}^{-1}$
1/m	$1.973 \times 10^{-16}\ \mathrm{GeV}$
1/fm	$0.1973\ \mathrm{GeV}$
1 s	$1.520 \times 10^{24}\ \mathrm{GeV}^{-1}$
1/s	$6.58 \times 10^{-25}\ \mathrm{GeV}$
1/ys	$0.658\ \mathrm{GeV}$
1 kg	$5.63 \times 10^{26}\ \mathrm{GeV}$

Quantity	Natural Units
E_{Pl}	$1.221 \times 10^{19}\ \mathrm{GeV}$
e	0.3029

TABLE 2.1   This table lists the natural-unit equivalents of SI units and various quantities. ($1\ \mathrm{ys} = 10^{-24}\ \mathrm{s}$, E_{Pl} is the Planck energy, and e is the proton charge.)

Light of a given wavelength (or a beam of particles with a given de Broglie wavelength) is only able to resolve features whose dimensions are larger than the wavelength. This means, for example, if we want to resolve where the quarks in a proton are located, we need to illuminate the proton with particles whose wavelength is very small. This is why we need particle accelerators that are capable of producing particle beams with very large energies (small wavelengths) to see the subnuclear world clearly.

We can express any quantity in natural units. For example, energy densities have units of $\mathrm{J/m^3} \to \mathrm{GeV}/(\mathrm{GeV^{-1}})^3 = \mathrm{GeV^4}$, angular momentum is unitless (the same as $\hbar$), force has units of $\mathrm{kg\, m/s^2} \to \mathrm{GeV} \cdot \mathrm{GeV^{-1}}/(\mathrm{GeV^{-1}})^2 = \mathrm{GeV^2}$, and so on. The gravitational constant G has units of $\mathrm{N\, m^2/kg^2} \to \mathrm{GeV^2}(\mathrm{GeV^{-1}})^2/\mathrm{GeV^2} = \mathrm{GeV^{-2}}$. You will show in Box 2.2 that $E_{\mathrm{Pl}} = G^{-1/2} = 1.22 \times 10^{19}\,\mathrm{GeV}$. This **Planck energy** is roughly the particle energy where we'd expect quantum corrections to gravity to become significant.

2.3 Lorentz Transformations

Consider two inertial reference frames S and S' whose spatial coordinate axes are aligned, and suppose that frame S' moves along the x axis relative to S with x-velocity β. (We say that such frames are in **standard orientation**.) Define $t = 0$ in both frames to be the instant that the frames' spatial origins coincide. The principle of relativity then implies (see Box 2.3) that if a given event's coordinates are t, x, y, and z in frame S, the coordinates of that same event in frame S' are given by the matrix equation

$$
\begin{bmatrix} t' \\ x' \\ y' \\ z' \end{bmatrix} =
\begin{bmatrix} \gamma & -\gamma\beta & 0 & 0 \\ -\gamma\beta & \gamma & 0 & 0 \\ 0 & 0 & 1 & 0 \\ 0 & 0 & 0 & 1 \end{bmatrix}
\begin{bmatrix} t \\ x \\ y \\ z \end{bmatrix},
\quad \text{where } \gamma \equiv \frac{1}{\sqrt{1-\beta^2}}. \tag{2.2}
$$

Physicists call this the **Lorentz transformation equation**, and the square matrix the **Lorentz transformation matrix** or simply the **Lorentz transformation**. We can generate the **inverse Lorentz transformation** from the primed to unprimed coordinates simply by moving the primes from the components of the vector on the left to the components of the vector on the right and replacing β by $-\beta$ (see Box 2.3).

I want to be clear that these equations provide the transformation between two inertial frames only in a certain very special case (where the frames' spatial axes point in the same directions and their relative motion is along the common x axis). Physicists call this transformation a **boost** along the x direction. A more general Lorentz transformation equation might allow for a boost in an arbitrary direction and/or a rotation of the spatial coordinate axes as we go from one frame to the other. But because of another fundamental symmetry principle (the **isotropy of space**), the laws of physics are the same in all *rotated* coordinate systems. This means that we can always, without any loss of physical meaning, choose our spatial coordinate systems to be oriented so that their axes coincide and their relative motion is along their common x axis. Therefore, we really do not learn anything else from more general transformation equations.

Because the transformation is linear, the same transformation law applies to coordinate *differences* Δt, Δx, Δy, and Δz between any *pair* of events:

$$
\begin{bmatrix} \Delta t' \\ \Delta x' \\ \Delta y' \\ \Delta z' \end{bmatrix} =
\begin{bmatrix} \gamma & -\gamma\beta & 0 & 0 \\ -\gamma\beta & \gamma & 0 & 0 \\ 0 & 0 & 1 & 0 \\ 0 & 0 & 0 & 1 \end{bmatrix}
\begin{bmatrix} \Delta t \\ \Delta x \\ \Delta y \\ \Delta z \end{bmatrix}. \tag{2.3}
$$

Again, we can find the inverse transformation for such coordinate differences simply by replacing β by $-\beta$ and by swapping the primed and unprimed vectors.

2.4 The Spacetime Interval

The **spacetime interval** Δs is defined by the **metric equation**

$$\Delta s^2 = \Delta t^2 - \Delta x^2 - \Delta y^2 - \Delta z^2, \tag{2.4}$$

where Δt, Δx, Δy, and Δz are the differences between the spacetime coordinates of two events. The squared spacetime interval Δs^2 corresponds to the squared time interval between the events, as measured in an inertial frame where they occur at the *same place* ($\Delta x = \Delta y = \Delta z = 0$) or negative the squared distance between the events, as measured in an inertial frame where they occur at the same time ($\Delta t = 0$). The spacetime interval is important because it is a *frame-independent* measure of the events' separation: observers in all inertial frames agree on the value of Δs between a given pair of events (see Box 2.4). The spacetime interval and the metric equation are to spacetime what the *distance* between two points and the Pythagorean theorem are to ordinary plane geometry. This is therefore a *central* equation in relativity theory.

(The choice of the overall sign in the definition of Δs^2 in Equation 2.4 is conventional in the particle physics community, but opposite to the choice used by most of the general relativity community.)

Spacetime intervals are called

$$
\begin{aligned}
\textit{timelike} \quad &\text{if } \Delta s^2 > 0 \;\Rightarrow\; \Delta t^2 > \Delta x^2 + \Delta y^2 + \Delta z^2, \\
\textit{lightlike} \quad &\text{if } \Delta s^2 = 0 \;\Rightarrow\; \Delta t^2 = \Delta x^2 + \Delta y^2 + \Delta z^2, \\
\textit{spacelike} \quad &\text{if } \Delta s^2 < 0 \;\Rightarrow\; \Delta t^2 < \Delta x^2 + \Delta y^2 + \Delta z^2.
\end{aligned}
$$

These distinctions arise only because of the minus signs in the metric equation. Since there are no minus signs in the corresponding Pythagorean theorem for ordinary space, there is only one kind of distance in such a space. Because observers in all inertial frames agree on the value of Δs^2 (and thus its sign), all will agree on how to classify the interval between a given pair of events.

These classifications are important because events separated by a spacelike interval *cannot* be causally connected: because which event occurs after the other depends on one's choice of frame (see Box 2.5), one cannot be the cause of the other. This in turn means that no particle, message, or other causal influence can travel faster than the speed of light. Figure 2.1 illustrates the frame-independent causal structure of spacetime implied by this fact.

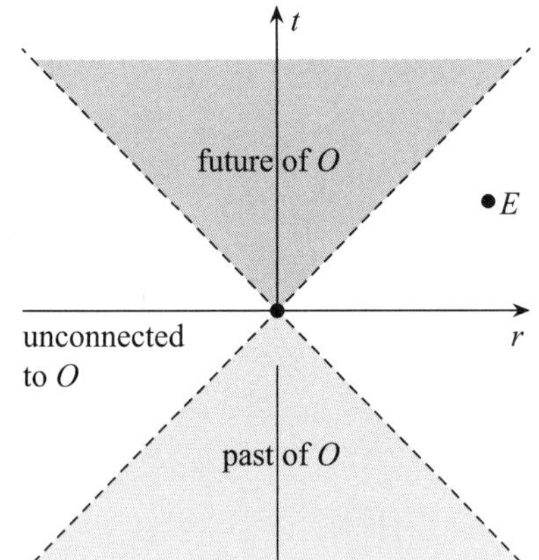

FIGURE 2.1   Events in the light gray region can influence event O (they are in O's **past**), and event O can influence events in the darker gray region (they are in O's **future**), but events in the white regions (such as event E) cannot be causally connected to event O. Here $r^2 \equiv x^2 + y^2 + z^2$.

2.5 Proper Time

A particle's **worldline** through spacetime is the sequence of events that occur along the particle's trajectory. Even if a clock is accelerating, we can calculate the **proper time** $\Delta \tau$ that it measures between two events A and B along its worldline using a method analogous to how we would calculate the path length along a curve in ordinary space. (1) We first divide the clock's worldline into a sequence of infinitesimal segments. (2) If the ith segment is short enough so that the clock's velocity is nearly constant, then the time $\Delta \tau_i$ it measures between the events delimiting that segment should be essentially the same as the time read by an imaginary *inertial* clock moving at a constant velocity between those events. Since such a clock reads the spacetime interval between the events delimiting the segment by definition, $\Delta \tau_i \approx \Delta s_i$ for that segment. The approximation becomes exact as we take the limit $\Delta \tau_i \to 0$. (3) We then sum over all segments, which, when we take the limit that $\Delta \tau_i \to 0$, becomes the integral

$$\Delta\tau = \lim_{\Delta\tau_i \to 0} \sum_i \sqrt{\Delta\tau_i^2} = \int \sqrt{ds^2} = \int \sqrt{dt^2 - dx^2 - dy^2 - dz^2}$$

$$= \int \sqrt{1 - \left(\frac{dx}{dt}\right)^2 - \left(\frac{dy}{dt}\right)^2 - \left(\frac{dz}{dt}\right)^2}\, dt = \int_{t_A}^{t_B} \sqrt{1 - v^2}\, dt, \qquad (2.5)$$

where in the last step, we are imagining we know the clock's speed v as a function of time t in whatever (arbitrary) inertial frame we are using to track the clock's motion.

Because every observer can see what the face of the clock reads between events A and B as it follows a given worldline, the proper time it registers (and therefore the number calculated using the procedure in Equation 2.5) is a *frame-independent* quantity.

2.6 Four-Velocity

We can apply the same procedure to the worldline of any particle to calculate the proper time that a hypothetical clock *would* measure if it followed the particle's worldline. The proper time τ along a worldline represents a frame-independent quantity that we can conveniently use as the parameter in the four functions $t(\tau)$, $x(\tau)$, $y(\tau)$, and $z(\tau)$ that describe a particle's trajectory through space and time in a given inertial frame.

With this in mind, imagine an infinitesimal step $d\tau$ in proper time along a given particle's worldline, with the step centered on an event E. We define an object's **four-displacement** ds during this infinitesimal time interval and the particle's **four-velocity** u at event E to be the four-component vectors

$$ds = \begin{bmatrix} dt \\ dx \\ dy \\ dz \end{bmatrix} \quad \text{and} \quad u \equiv \frac{ds}{d\tau} \equiv \begin{bmatrix} dt/d\tau \\ dx/d\tau \\ dy/d\tau \\ dz/d\tau \end{bmatrix} \equiv \begin{bmatrix} u^t \\ u^x \\ u^y \\ u^z \end{bmatrix}. \qquad (2.6)$$

Note that you should read u^t as "the time component of the four-velocity u," not as "u raised to the power t." (The next chapter will explain why we use superscripted instead of subscripted component labels.)

We define the four-velocity this way because if we know the components of u in one inertial frame, we can compute its components in another inertial frame using the Lorentz transformation. Let's see why. Assuming our primed and unprimed frames are in standard orientation, we know the components of the four-displacement ds transform according to the Lorentz transformation:

$$\begin{bmatrix} dt' \\ dx' \\ dy' \\ dz' \end{bmatrix} = \begin{bmatrix} \gamma & -\gamma\beta & 0 & 0 \\ -\gamma\beta & \gamma & 0 & 0 \\ 0 & 0 & 1 & 0 \\ 0 & 0 & 0 & 1 \end{bmatrix} \begin{bmatrix} dt \\ dx \\ dy \\ dz \end{bmatrix}. \qquad (2.7)$$

Because $d\tau$ is frame-independent, the components of an object's four-velocity u must then transform as follows:

$$\begin{bmatrix} u'^t \\ u'^x \\ u'^y \\ u'^z \end{bmatrix} = \frac{1}{d\tau} \begin{bmatrix} dt' \\ dx' \\ dy' \\ dz' \end{bmatrix} = \frac{1}{d\tau} \begin{bmatrix} \gamma & -\gamma\beta & 0 & 0 \\ -\gamma\beta & \gamma & 0 & 0 \\ 0 & 0 & 1 & 0 \\ 0 & 0 & 0 & 1 \end{bmatrix} \begin{bmatrix} dt \\ dx \\ dy \\ dz \end{bmatrix}$$

$$= \begin{bmatrix} \gamma & -\gamma\beta & 0 & 0 \\ -\gamma\beta & \gamma & 0 & 0 \\ 0 & 0 & 1 & 0 \\ 0 & 0 & 0 & 1 \end{bmatrix} \begin{bmatrix} u^t \\ u^x \\ u^y \\ u^z \end{bmatrix}. \qquad (2.8)$$

We define a **four-vector** to be any four-component quantity whose components transform according to the Lorentz transformation when we change inertial reference frames. In particle physics, one conventionally represents four-vectors with simple italic symbols (for example, u, ds, a). One (sadly) must infer from context whether a given symbol refers to a four-vector or a scalar. We will also use $\vec{a}$ to represent the spatial components of a four-vector a.

We define the **dot product** of two arbitrary four-vectors a and b as follows:

$$a \cdot b \equiv a^t b^t - a^x b^x - a^y b^y - a^z b^z = a^t b^t - \vec{a} \cdot \vec{b}, \tag{2.9}$$

and the squared magnitude of an arbitrary four-vector a to be

$$a^2 \equiv a \cdot a = (a^t)^2 - (a^x)^2 - (a^y)^2 - (a^z)^2 = (a^t)^2 - |\vec{a}|^2. \tag{2.10}$$

These quantities are interesting and important because their values are *frame-independent*, just as the squared spacetime interval between the events at the two ends of a four-displacement

$$\Delta s^2 = \Delta t^2 - \Delta x^2 - \Delta y^2 - \Delta z^2 \tag{2.11}$$

is frame-independent. The squared magnitude of a particle's four-velocity is

$$u^2 = (u^t)^2 - (u^x)^2 - (u^y)^2 - (u^z)^2 = (u^t)^2 - |\vec{u}|^2 = 1 \tag{2.12}$$

no matter how fast the object is moving through space (see Box 2.6). You will also show in that box that in a given inertial frame

$$\begin{bmatrix} u^t \\ u^x \\ u^y \\ u^z \end{bmatrix} = \begin{bmatrix} 1/\sqrt{1-v^2} \\ v_x/\sqrt{1-v^2} \\ v_y/\sqrt{1-v^2} \\ v_z/\sqrt{1-v^2} \end{bmatrix} \approx \begin{bmatrix} 1 \\ v_x \\ v_y \\ v_z \end{bmatrix} \quad \text{when } v \ll 1, \tag{2.13}$$

where $v_x \equiv dx/dt$, $v_y \equiv dy/dt$, and $v_z \equiv dz/dt$ are the components of the particle's ordinary (Newtonian) velocity and v its ordinary speed in that frame.

2.7 Four-Momentum

In particle physics, we are especially interested in a particle's **four momentum** p, which (for a particle of mass m) we define to be $p \equiv mu$, with components

$$\begin{bmatrix} p^t \\ p^x \\ p^y \\ p^z \end{bmatrix} \equiv \begin{bmatrix} mu^t \\ mu^x \\ mu^y \\ mu^z \end{bmatrix} = \begin{bmatrix} m/\sqrt{1-v^2} \\ mv_x/\sqrt{1-v^2} \\ mv_y/\sqrt{1-v^2} \\ mv_z/\sqrt{1-v^2} \end{bmatrix} \approx \begin{bmatrix} m \\ mv_x \\ mv_y \\ mv_z \end{bmatrix} \quad \text{when } v \ll 1. \tag{2.14}$$

In modern physics, we consider a particle's mass m to be a frame-independent constant. Therefore, the four components of p transform according to the Lorentz transformation when we change frames (just multiply Equation 2.8 through by m to see that this is true), which in turn means that p is a four-vector. The frame-independent squared magnitude of this four-vector is

$$p^2 \equiv p \cdot p = (mu) \cdot (mu) = m^2(u \cdot u) = m^2. \tag{2.15}$$

At sufficiently small speeds, Equation 2.14 implies that the spatial components of p reduce to the components of the particle's Newtonian momentum. In Newtonian physics, we assume that an isolated system's total Newtonian momentum is conserved. But laws of physics must be consistent with the principle of relativity, and Box 2.7 shows that a law of conservation of Newtonian momentum is *not* consistent with the principle of relativity, while a law of conservation of four-momentum *is* consistent with that principle.

We conventionally define a particle's **relativistic momentum** $\vec{p}$ to be the three-dimensional vector whose components are the *spatial* components of p, and we define $|\vec{p}|^2 = (p^x)^2 + (p^y)^2 + (p^z)^2$ to be the magnitude of this spatial vector (note how we distinguish this from $p^2 \equiv p \cdot p$). But p^t is also a conserved quantity. We call that the particle's **relativistic energy** E because

$$E \equiv p^t = \frac{m}{\sqrt{1 - v^2}} = m + \tfrac{1}{2}mv^2 + \tfrac{3}{8}mv^4 + \cdots, \qquad (2.16)$$

where in the last step, I expanded $(1 - v^2)^{-1/2}$ in a power series. We see that when $v \ll 1$, $E \approx m + \tfrac{1}{2}mv^2$, the particle's mass plus its Newtonian kinetic energy. In relativistic interactions, the *sum* of these is conserved, not either one alone, so (as Einstein was the first to realize) processes might exist that convert a portion of the energy associated with a particle's mass to that associated with its kinetic energy and vice versa. We now know many such processes.

Equations 2.10 and 2.15 and these definitions imply

$$m^2 = p^2 = (p^t)^2 - (p^x)^2 - (p^y)^2 - (p^z)^2 = E^2 - |\vec{p}\,|^2. \qquad (2.17)$$

This is a very important equation that we will use often in this text.

Now, a photon of light carries energy and so should have an associated four-momentum vector. But the proper time measured along a photon worldline is zero (see Equation 2.5 and note that $v = 1$ for a photon), so we cannot define that four-momentum using Equation 2.14. But note that the latter equation implies that for particles in general

$$p = \begin{bmatrix} E \\ E v_x \\ E v_y \\ E v_z \end{bmatrix}. \qquad (2.18)$$

Since a photon's energy and velocity are well-defined in a given inertial frame, we can use this to calculate a photon's four-momentum; indeed, we will take this to be the *definition* of that four-momentum. Since a photon's speed $v = 1$ in all frames, Equation 2.17 implies that a photon's mass must be

$$m^2 = E^2 - |\vec{p}\,|^2 = E^2 - E^2 v^2 = 0. \qquad (2.19)$$

We can think of a photon's energy as being purely kinetic energy with no associated mass energy.

Note that Equation 2.18 implies the following useful trick for determining a particle's ordinary velocity from its energy and relativistic momentum:

$$\vec{v} = \frac{\vec{p}}{E}. \qquad (2.20)$$

We will find this very handy in this course.

BOX 2.1 Inertial Frames Have Constant Relative Velocities

Our goal in this box is to prove that a rigid frame S' is inertial *if and only if* it moves at a constant velocity and does not rotate relative to an established inertial frame S (assuming that both frames can assign coordinates to events in a common region of spacetime). We start by proving the "if" clause: we will *assume* that S' moves at a constant velocity and *prove* that it is inertial.

Suppose there is a free particle that in S is observed to be at rest. Since S is inertial by hypothesis, Newton's first law implies that such a particle will remain at rest relative to S as time passes. Now consider observing this particle in frame S'. First, note that *all* observers agree about whether a particle is "free" or not: if the particle is uncharged, not magnetic, not touching anything, and not near a large gravitating object, then it cannot be participating in any kinds of interactions. Therefore, the observer in S' will agree that the particle in question is free. But S moves at a constant velocity relative to S' by hypothesis, and because both frames are rigid and not rotating relative to each other by hypothesis, *every point* in S moves at a constant velocity relative to every point in S'. So, if the particle is at rest at a point in frame S, the observer in S' must observe this free particle to move with a constant velocity (so long as the particle is in the region of spacetime that both frames can observe). Therefore, frame S' is inertial by definition.

Exercise 2.1.1 In the space below, prove the "only if" clause: Assume that S' is inertial and show that it must move at a constant velocity relative to S. Start by considering a free particle at rest in frame S'.

BOX 2.2   The Planck Energy

The Planck energy E_{Pl} is the unique combination of the fundamental constants G, $\hbar$, and c that has units of energy. By the usual rules of dimensional analysis, we would expect this quantity to define the characteristic energy scale that would appear in equations describing a theory that involves gravity, quantum mechanics, and relativity; that is, a quantum theory of gravity. (As we will discuss later in the book, there is an analogous quantity that defines the energy scale at which electrodynamics merges with the weak interaction.)

Exercise 2.2.1   Assume that $E_{\mathrm{Pl}} = G^i \hbar^j c^k$, where i, j, and k are integers or simple fractions. Show that there is a unique solution for i, j, and k that gives E_{Pl} the correct SI units for energy. Then use the resulting expression to evaluate the energy, expressing the result in GeV.

BOX 2.3   The Lorentz Transformation

There are many ways to derive the Lorentz transformation from the principle of relativity. This box outlines one particularly nice method.

We begin by considering two frames S and S' in standard orientation (their spatial axes point in the same directions, their relative motions are along the common x direction, and their origins coincide at $t = t' = 0$). First, we note that the principle of relativity requires that $y' = y$ and $z' = z$, as we will prove by contradiction. Suppose this were *not* true, and that distances perpendicular to the line of relative motion are (say) contracted (in the frame that we consider to be moving) by some factor that depends on that frame's speed β relative to the frame we consider to be at rest. Suppose that in each frame there is a length of pipe centered on the x axis that has a diameter of 1.00 m, as measured in its own frame, and say that the pipe in frame S is red while the pipe in frame S' is green. In frame S, the moving green pipe is contracted, so it should fit inside the red pipe as the frames pass. But in frame S' the moving red pipe is contracted, so it should fit inside the green pipe. But observers in *every* frame can determine whether red or green photons are reflected from the pipes' outermost surface when the pipes pass; this result must be *frame-independent*, not frame-dependent as just argued. The only way to avoid this contradiction is to reject the original assumption that the moving pipe is contracted. The same problem occurs if we imagine the moving pipe to be expanded. So we must conclude that the pipes' diameters must be the same in both frames (and that instead of one pipe passing inside the other, the pipes are observed in all frames to crash at a relativistic relative speed and vaporize).

The transformation between two inertial frames using Cartesian coordinates should plausibly also be *linear*, as a linear transformation will map a free particle's constant-velocity trajectory in one frame to a constant-velocity trajectory in the other frame, ensuring that both frames are inertial. In addition to the principle of relativity, we will assume this linearity as well as the isotropy and homogenenity of spacetime in what follows.

The transformation equations should therefore have the form

$$t' = At + Bx + Cy + Dz \tag{2.21a}$$

$$x' = Et + Fx + Gy + Hz \tag{2.21b}$$

$$y' = y \tag{2.21c}$$

$$z' = z, \tag{2.21d}$$

where the letters A through H are coefficients that can depend only on the frames' relative x-velocity β (because that is all that distinguishes the frames).

Exercise 2.3.1   Argue that C, D, G, and H must be zero for the transformation to be rotationally invariant around the common x axis. (*Hint:* Suppose we replace y by $-y$. Should this change t? Why doesn't the same argument apply to B?)

BOX 2.3 The Lorentz Transformation (cont.)

With $G = H = 0$, Equation 2.21b becomes $x' = Et + Fx$. Now, in frame S, the origin of frame S' (defined to be $x' = 0$) moves with x-velocity β along the x axis, and thus along a worldline where $x = \beta t$. The coefficients E and F (which cannot depend on anything but β) therefore must be such that any event lying along the worldline $x = \beta t$ must yield $x' = 0$.

Exercise 2.3.2   Argue that this requires $E = -\beta F \Rightarrow x' = F(-\beta t + x)$.

The only difference between the two frames is the sign of β: an observer in frame S sees frame S' moving with x-velocity β (by definition of β), while the observer in frame S' sees frame S moving with x-velocity $-\beta$ (note that if the speeds were not the same, we could physically distinguish the frames in contradiction to the principle of relativity). This means that the transformation law to x from x' and t' must be $x = F(+\beta t' + x')$ (the value of F must be the same in both cases by the same argument). Now, a light flash moving in the $+x$ direction must follow a worldline such that $x = t$ in the S frame *and* $x' = t'$ in the S' frame, because its x-velocity must be 1 in both cases.

Exercise 2.3.3   Substitute $x = t$ into $x' = F(-\beta t + x)$ to eliminate t and $x' = t'$ into $x = F(+\beta t' + x')$ to eliminate t'. Then note that if we transform from x to x' and then back to x, we should get simply $x = x$. Use this to determine the value of F.

BOX 2.3   The Lorentz Transformation (cont.)

Again, since the coefficients in the transformation equation cannot depend on anything but β (in particular, the coefficients cannot depend on whether we are looking at a light flash moving in the $+x$ direction or not), our transformation equation to x' for *any* event must be

$$x' = \gamma(-\beta t + x), \quad \text{where } \gamma = \frac{1}{\sqrt{1 - \beta^2}}. \tag{2.22}$$

Now consider the transformation equation for t'. Its coefficients A and B must be such that a light flash emitted from the origin in the $+x$ direction at $t = t' = 0$ must follow a worldline that in the S frame is described by $x = t$ and in the S' frame is described by $x' = t'$. For such a flash, we have earlier seen that $x' = \gamma(-\beta t + x) = \gamma(1 - \beta)x$ and we also have $t' = At + Bx = (A + B)x$. We see that for $x'/t' = 1$, we must have

$$A + B = \gamma(1 - \beta). \tag{2.23}$$

Similarly, we must have $x = -t$ and $x' = -t'$ for a light flash emitted from the origin at $t = t' = 0$ in the $-x$ direction.

Exercise 2.3.4   Use $x' = -t'$ and $x = -t$ to find the analogue to Equation 2.23 for this case, and then use the Equation 2.23 and the new equation you've just found to solve for the two unknowns A and B in terms of γ and β. Check that $t' = At + Bx$ is the same as the top row of Equation 2.2.

BOX 2.4   The Invariance of Δs^2

The Lorentz transformation equations for coordinate differences (Equation 2.3 expressed in equation form) are

$$\Delta t' = \gamma \, \Delta t - \gamma \beta \, \Delta x \qquad (2.24a)$$

$$\Delta x' = -\gamma \beta \, \Delta t + \gamma \, \Delta x \qquad (2.24b)$$

$$\Delta y' = \Delta y \qquad (2.24c)$$

$$\Delta z' = \Delta z. \qquad (2.24d)$$

Exercise 2.4.1   Use the above equations to verify that

$$(\Delta s')^2 \equiv (\Delta t')^2 - (\Delta x')^2 - (\Delta y')^2 - (\Delta z')^2$$
$$= \Delta s^2 \equiv \Delta t^2 - \Delta x^2 - \Delta y^2 - \Delta z^2. \qquad (2.25)$$

Does your proof require you to make any assumptions about whether Δs^2 is positive, negative, or zero?

BOX 2.5 Events with a Spacelike Separation

Consider two arbitrary events A and B that have a spacelike separation. Let us use our freedom to orient both frames S and S' so that both events occur somewhere along the frames' common x direction, and that event B (which we take to be the event that occurs *later* in the S frame) also occurs farther in the $+x$ direction than event A, so that both $\Delta x \equiv x_B - x_A$ and $\Delta t \equiv t_B - t_A$ are positive in that frame. For $\Delta s^2 \equiv \Delta t^2 - \Delta x^2 - \Delta y^2 - \Delta z^2 < 0$ in that frame, we must have $\Delta x > \Delta t$ (since $\Delta y = \Delta z = 0$ by construction). In the exercise below, you will show that in such a case, we can always find a value of β for a primed frame where $\Delta t' = t'_B - t'_A < 0$, meaning that event B occurs *before* event A in that frame. This means that A cannot cause B or vice versa, because we can always find a reference frame where the time order of these events is different, and an effect cannot precede its cause in *any* reference frame.

On the other hand, if the separation between the events is *timelike* (meaning that $\Delta x < \Delta t$), then if event A occurs before event B in frame S, it will occur before B in *all* frames, implying that event A *could* cause event B.

Exercise 2.5.1 Use the Lorentz transformation Equation 2.24a to prove that we can choose a sufficiently large value of β that is still less than 1 such that $\Delta t' = t'_B - t'_A < 0$. Also, show that this *cannot* be done if the separation between the events is timelike.

BOX 2.6   Four-Velocity

According to Equation 2.6, the components of a particle's four-velocity u are $u^t \equiv dt/d\tau$, $u^x \equiv dx/d\tau$, $u^y \equiv dy/d\tau$, and $u^z \equiv dz/d\tau$. Note also that for an infinitesimal four-displacement

$$d\tau^2 = ds^2 = dt^2 - dx^2 - dy^2 - dz^2. \tag{2.26}$$

Exercise 2.6.1   Use these results to prove that the squared magnitude of the four-velocity is *always*

$$u^2 \equiv u \cdot u \equiv (u^t)^2 - (u^x)^2 - (u^y)^2 - (u^z)^2 = 1, \tag{2.27}$$

no matter how fast the particle is moving.

Exercise 2.6.2   Use Equation 2.5 to show that the components of the four-velocity in a given frame can be written

$$u^t = \frac{1}{\sqrt{1 - v^2}}, \quad u^x = \frac{v_x}{\sqrt{1 - v^2}}, \quad u^y = \frac{v_y}{\sqrt{1 - v^2}}, \quad u^z = \frac{v_z}{\sqrt{1 - v^2}}, \tag{2.28}$$

where $v_x \equiv dx/dt$, $v_y \equiv dy/dt$, and $v_z \equiv dz/dt$ are the components of particle's ordinary velocity in that frame.

BOX 2.7 Conservation of Momentum

Consider a simple Newtonian decay event, where particle 1 with mass m at rest in frame S decays to two particles 2 and 3, both with mass $\frac{1}{2}m$ that move in the $+x$ and $-x$ directions, respectively, with the same speed v. In this frame, this decay event clearly conserves Newtonian x-momentum: the initial x-momentum is zero (because particle A is at rest), and the final Newtonian x-momentum is $\frac{1}{2}mv - \frac{1}{2}mv = 0$. (We can also conserve energy if we assume that some of particle 1's internal energy is converted to the outgoing particles' kinetic energies, but in Newtonian physics, this is a separate issue.)

Now consider looking at this situation in a frame S' that moves in the $+x$ direction with $\beta = v$ relative to S. In this frame, both particle 1 and the frame S in which it is at rest must move *backward* with speed v, and because frame S' and particle 2 move together when seen in frame S, the observer in frame S' must observe particle 2 to be at rest. But what about particle 3? The Lorentz transformation equations tell us that if any particle's x-velocity in S is $v_x \equiv dx/dt$, its x-velocity in frame S' will be

$$v'_x = \frac{dx'}{dt'} = \frac{\gamma(-\beta dt + dx)}{\gamma(dt - \beta\, dx)} = \frac{-\beta + dx/dt}{1 - \beta\, dx/dt} = \frac{v_x - \beta}{1 - \beta v_x}. \tag{2.29}$$

Now, particle 3 was moving with $v_{3x} = -v$ in frame S, and frame S' moves with $\beta = +v$ relative to frame S, so for particle 3,

$$v'_{3x} = \frac{v_{3x} - \beta}{1 - \beta v_{3x}} = \frac{-2v}{1 + v^2}. \tag{2.30}$$

Exercise 2.7.1 Show that Newtonian x-momentum is *not* conserved in reference frame S', meaning that a law of conservation of *Newtonian* momentum cannot be consistent with the Principle of Relativity. (*Hint:* Remember that particle masses are frame-independent.)

BOX 2.7 Conservation of Momentum (cont.)

Now consider the original decay from a relativistic perspective. First, consider conservation of energy. If particle 1 has mass m and is at rest in frame S, its relativistic energy is $E_1 = m/\sqrt{1-0} = m$. Particles 2 and 3 must then have masses of $m_2 = m_3 = \frac{1}{2}m\sqrt{1-v^2}$ so that the system's final energy is

$$E_2 + E_3 \equiv \frac{m_2}{\sqrt{1-v_2^2}} + \frac{m_3}{\sqrt{1-v_3^2}} = \frac{\frac{1}{2}m\sqrt{1-v^2}}{\sqrt{1-v^2}} + \frac{\frac{1}{2}m\sqrt{1-v^2}}{\sqrt{1-v^2}} = m. \quad (2.31)$$

This adjustment of the masses is required if conservation of relativistic energy is to be true in frame S. We see that the particle masses are *not* conserved in relativistic dynamics: the outgoing particles' kinetic energy here is coming at the expense of the original particle's mass energy (note that $m_2 = m_3 < \frac{1}{2}m$).

Exercise 2.7.2 Show that, given these masses, relativistic energy is conserved in frame S' and relativistic x-momentum *is* conserved in both reference frames, meaning that a law of conservation of four-momentum *is* consistent with the principle of relativity. [*Hint:* First show that $\sqrt{1-v_3'^2} = (1-v^2)/(1+v^2)$.]

Homework Problems

P2.1 Express the following quantities in natural units (that is, in units of GeV to some power).
a. 1 angstrom
b. 1 minute
c. 1 light-year
d. The density of water (≈ 1000 kg/m^3)
e. The sun's luminosity ($\approx 3.8 \times 10^{26}$ W)

P2.2 Consider a particle whose four-momentum components in a certain inertial frame S are $p^t = 5m$, $p^x = -4m$, $p^y = 0$, $p^z = 0$, where m is some constant.
a. What is the particle's mass M (in units of m)?
b. What is its velocity?
c. What is its kinetic energy $K = E - M$ (in units of m)?
d. Now, consider transforming to a frame S' moving at $\beta = \frac{4}{5}$ in the $+x$ direction. What are the components of p' in this frame?
e. What is the particle's mass in this frame?
f. What is its velocity? Compare with the Galilean result, which is $-\frac{8}{5}$.
g. What is its kinetic energy?

P2.3 This problem illustrates how one might take advantage of the frame-independent nature of the dot product of two four-vectors to make a collision calculation easier. Suppose that in the laboratory frame, a photon strikes an electron at rest with sufficient energy to create an electron-positron pair (destroying the photon). What is the minimum photon energy E required to do this? The answer is not simply $E = 2m$ (where m is the mass of an electron) because the three final particles must recoil to conserve momentum, so some of the photon's original energy must be converted to kinetic energy in the outgoing particles.

But consider this process in the **center of mass frame** (CM frame), where the photon's and original electron's spatial momenta add to zero. In this frame, a photon with lowest possible energy would create an electron-positron pair at rest alongside the original electron at rest: these particles will have a spatial momentum of zero (required to conserve momentum) and the minimum possible total energy of $3m$. So the minimum squared magnitude of the total final four-momentum is

$$p_f \cdot p_f \equiv (p_f^t)^2 - (p_f^x)^2 - (p_f^y)^2 - (p_f^z)^2 = (3m)^2. \quad (2.32)$$

However, note that conservation of momentum requires that $p_1 + p_2 = p_f$ (where p_1 and p_2 are the original photon and electron's four-momenta, respectively). Therefore, in the CM frame, $(p_1 + p_2) \cdot (p_1 + p_2) = (3m)^2 = 9m^2$. But this must be the value of the invariant dot product in *any* frame.
a. Write p_1 and p_2 in the lab frame as four-component column vectors, expressing each component in terms of the lab-frame photon energy E and/or the electron mass m. Assume that the photon moves in the $+x$ direction in that frame.

b. Express $(p_1 + p_2) \cdot (p_1 + p_2)$ in terms of E and m. (*Hint:* The product is equivalent to $p_1 \cdot p_1 + 2p_1 \cdot p_2 + p_2 \cdot p_2$.)
c. Set this equal to its minimum frame-independent value of $9m^2$ and solve for the minimum photon energy E in terms of m.

P2.4 Consider a positive pion at rest. It most often decays to an antimuon and a muon neutrino: $\pi^+ \to \mu^+ + \nu_\mu$. Assume that the direction that the muon neutrino travels after the decay defines the $+x$ direction.
a. The mass of the muon neutrino is so small that in this context, we can treat it as if it has zero mass; that is, the same as we would treat a photon. Suppose that this neutrino has energy E. What are the components of its four-momentum in terms of E?
b. Let the energy of the antimuon be E_μ. Use conservation of four-momentum to calculate how E_μ is related to E and the pion mass m_π. Also, calculate the spatial components of the antimuon's four-momentum in terms of E.
c. We can also connect E_μ to the antimuon's spatial momentum components and its mass m_μ. How?
d. Use the results of the previous parts to calculate both E_μ and neutrino energy E symbolically in terms of m_π and m_μ, and then numerically in GeV.
e. What is the velocity (magnitude and direction) of the departing antimuon? (*Hint:* Consider Equation 2.20.)

P2.5 Consider a "short-lived" neutral kaon (K_S^0) at rest. This version of the neutral kaon most often decays to a pair of charged pions: $K_S^0 \to \pi^+\pi^-$. Use conservation of four-momentum and data from Table 1.1 to calculate
a. the energy E,
b. the spatial momentum $|\vec{p}|$, and
c. the speed
of each departing pion. (*Hint:* Use Equation 2.20 in the last step.)

P2.6 About a third of the time, The η meson decays into three pions: $\eta \to 3\pi^0$. Suppose that an η initially at rest decays this way, and that one of the pions is at rest after the decay. Use conservation of four-momentum and data from Table 1.1 to calculate
a. the energy E,
b. the spatial momentum $|\vec{p}|$, and
c. the speed
of each of the *moving* pions. (*Hint:* Use Equation 2.20 in the last step.)

P2.7 Suppose that a Z boson produced at the Large Hadron Collider decays into a $\tau^+\tau^-$ pair. In the particle detector, the relativistic spatial momentum of the τ^+ is measured to be $\vec{p}_1 = [110\,\text{GeV}, -45\,\text{GeV}, 0]$, and that for the τ^- measured to be $\vec{p}_2 = [39\,\text{GeV}, 40\,\text{GeV}, 0]$. The measured mass of the tau is $1.7769\,\text{GeV}$. Use conservation of four-momentum to determine the mass of the Z boson that is implied by these results.

P2.8 The Large Hadron Collider (LHC) is able to accelerate protons so that they have a total relativistic energy of 6.5 TeV. What is the speed of those protons? (*Hint:* Your calculator will have to keep a lot of digits to distinguish between the proton speed and 1. To get a more accurate result, define δ such that $v = 1 - \delta$, and use the binomial expansion to calculate $1 - v^2$ in terms of δ. Use this result, the energy given, and the proton mass to calculate δ. Then express your final result for the proton speed as $v = 1 - \delta$.)

P2.9 Most modern particle accelerators (including the Large Hadron Collider (LHC), currently the most powerful particle accelerator in the world) operate by using powerful magnetic fields to constrain a batch of particles to travel around a circular path. Each time the particles circulate, electric fields give them an additional boost of energy. The formula for the radius of a charged particle's motion in a magnetic field of magnitude $|\vec{B}|$ is given (in SI units) by the expression

$$R = \frac{|\vec{p}|}{q|\vec{B}|}, \tag{2.33}$$

where $|\vec{p}|$ is the particle's relativistic momentum and q is the particle's charge. If protons in the LHC ultimately have an energy of 6.5 TeV, what is the magnitude of the magnetic field that the accelerator's magnets must provide to keep the protons orbiting in their circular path? The LHC has a *circumference* of 27 km. (*Hint:* Calculate each proton's momentum $|\vec{p}|$ in GeV, then convert to SI units.)

P2.10 Conventional particle physics notation does not distinguish between four-vectors and simple numbers. In many cases, this is not a huge issue: p almost always refers to four-momentum, E and m to energy and mass, and so on. Even when the symbols themselves are not conventional, context can often help you figure out what a symbol represents. For example, in the expression $a \cdot b = g$, a and b must be four-vectors (since they appear in a four-vector dot product) but g must be a scalar (since a dot product produces a scalar).

However, perhaps the most difficult symbol to interpret is x, which (in various different contexts) can be a four-vector, a simple number, or a vector component. In each of the following contexts, describe whether x refers to a four-vector, simple number, or vector component and briefly explain why. Also, indicate the one case that is completely ambiguous.

a. $x' = -\gamma \beta t + \gamma x$.
b. $\int_0^\infty x^2 e^{-a^2 x^2}\, dx = \sqrt{\pi}/4a^3$.
c. ... where particle 1 moves in the $+x$ direction.
d. $F(x) = A \sin(-p \cdot x + \theta_0)$.
e. $F(x) = A \sin(ax + \theta_0)$.
f. $\partial E_x/\partial x + \partial E_y/\partial y + \partial E_z/\partial z = \rho$.
g. $F(x) = -a/x$.
h. $G(x, y) = (x + y) \cdot (x + y)/x^2 y^2$.
i. $G(x, y) = (x + y)^2/x^2 y^2$.
j. $L_z = x p_y - y p_x$.

$R_{ik} = 0$

3. INDEX NOTATION

One way we can ensure that the laws of physics are consistent with a deep symmetry, such as the principle of relativity, is to develop a mathematical formalism that is *automatically* consistent with that principle. As we will see, the constraints that such a formalism places on the laws of physics also teach us about the consequences and implications of such a symmetry principle.

In the previous chapter, we learned about **four-vectors**, such as the four-displacement Δs, the four-velocity u, and the four-momentum p. By definition, the components a^t, a^x, a^y, and a^z of an arbitrary four-vector a transform according to the Lorentz transformation (like spacetime coordinates do) when we change inertial reference frames:

$$\begin{bmatrix} a'^t \\ a'^x \\ a'^y \\ a'^z \end{bmatrix} = \begin{bmatrix} \gamma & -\gamma\beta & 0 & 0 \\ -\gamma\beta & \gamma & 0 & 0 \\ 0 & 0 & 1 & 0 \\ 0 & 0 & 0 & 1 \end{bmatrix} \begin{bmatrix} a^t \\ a^x \\ a^y \\ a^z \end{bmatrix}, \tag{3.1}$$

assuming the frames' axes are aligned, the frames' relative motion is along their common x axis, β is the x-velocity of the primed frame relative to the unprimed frame, and $\gamma \equiv 1/\sqrt{1 - \beta^2}$. (We say that two such frames are in **standard orientation**.) We also encountered quantities such as the dot product $a \cdot b$ of two four-vectors and the squared magnitude $a^2 = a \cdot a$ of a four-vector, which are simple numbers whose values are independent of one's choice of inertial frame. We call such numbers (relativistic) **scalars**.

In this chapter, we will see that four-vectors and scalars are special cases of mathematical objects called *tensors*. A **tensor** is an object having 4^n components (where $n = 0, 1, 2, \ldots$) that transform in a well-defined way (that we will shortly define) when we change inertial reference frames. We will also develop a very powerful formalism for writing tensor equations compactly. If we write equations describing physical laws following the rules of the tensor formalism, we can be *certain* that our laws are consistent with (and therefore express the consequences of) the principle of relativity.

3.1 Index Notation

Even for relatively simple cases, the matrix notation used in Equation 3.1 can involve a lot of writing, and we will be using equations like this all the time. Our goal in this section is to develop a notation for such equations that is even more compact, elegant, and flexible.

Instead of writing a^t, a^x, a^y, a^z each time we want to refer to the components of a four-vector a, we can simply write a^μ, where the Greek superscript (we can use any Greek letter) stands for any *one* of the coordinates t, x, y, z. So,

$$a^\mu \quad \text{stands for } a^t, a^x, a^y, \text{ or } a^z. \tag{3.2}$$

We think of a^μ as representing "the μth component of a," with μ unspecified.

Similarly, we can represent any *one* of the Lorentz transformation matrix components with the symbol $\Lambda^\mu{}_\nu$, where the abstract indices μ and ν can *independently* stand for

$t, x, y,$ or z. Here is how we assign the index values:

$$\begin{bmatrix} \Lambda^t{}_t & \Lambda^t{}_x & \Lambda^t{}_y & \Lambda^t{}_z \\ \Lambda^x{}_t & \Lambda^x{}_x & \Lambda^x{}_y & \Lambda^x{}_z \\ \Lambda^y{}_t & \Lambda^y{}_x & \Lambda^y{}_y & \Lambda^y{}_z \\ \Lambda^z{}_t & \Lambda^z{}_x & \Lambda^z{}_y & \Lambda^z{}_z \end{bmatrix} \equiv \begin{bmatrix} \gamma & -\gamma\beta & 0 & 0 \\ -\gamma\beta & \gamma & 0 & 0 \\ 0 & 0 & 1 & 0 \\ 0 & 0 & 0 & 1 \end{bmatrix}. \tag{3.3}$$

Note that the first and second indices in $\Lambda^\mu{}_\nu$ refer to the matrix row and column, respectively. (It happens here that the row index is superscripted and the column index is subscripted, but that will not always be the case.)

Using this notation, we can more compactly write the four-vector transformation law given by Equation 3.1 as follows:

$$a'^\mu = \sum_{\nu=t,x,y,z} \Lambda^\mu{}_\nu a^\nu \equiv \Lambda^\mu{}_t a^t + \Lambda^\mu{}_x a^x + \Lambda^\mu{}_y a^y + \Lambda^\mu{}_z a^z. \tag{3.4}$$

The summation instruction tells us to assign the values t, x, y, z to the index ν and sum the four terms that result. The value of μ is left unspecified: if $\mu = t$, this equation corresponds to the first row of Equation 3.1; if $\mu = x$, the equation corresponds to the second row of Equation 3.1, and so on. Therefore Equation 3.4 implicitly represents all four rows of Equation 3.1.

We can make the notation even *more* compact if we adopt the rule generally known as the **Einstein summation convention**:

> If the same Greek index appears exactly once in a superscript *and* exactly once as a subscript in any single *term* of an equation, we will sum that term over the possible index values t, x, y, z.

(To clarify the terminology, in an equation $a = b + cd - eg/h$, where a through g are symbols that may or may not have indices, a, b, cd, and eg/h are different **terms** in that equation, while e, g and $1/h$ are **factors** in the term eg/h.)

Using this convention, we can write Equation 3.4 simply as

$$a'^\mu = \Lambda^\mu{}_\nu a^\nu. \tag{3.5}$$

Since the index ν appears once as a superscript and once as a subscript in the term $\Lambda^\mu{}_\nu a^\nu$, we should perform the summation shown explicitly in Equation 3.4. In contrast, there is no sum over the index μ, because it appears only once in each term in this equation, and only as a superscript.

This convention explains *part* of the reason I have written some component labels as superscripts and some as subscripts: the index positions specify exactly *how* we perform the sums. For example, since the *column* index for the matrix component $\Lambda^\mu{}_\nu$ appears in the subscript position, we see that the sum implied in Equation 3.5 is *along the row* of the Lorentz transformation matrix (and down the column of the four-vector), consistent with the implied sum in the matrix Equation 3.1. (The index position has another very important meaning that we will get to shortly.)

3.2 The Metric Tensor

We can write the metric equation using this notation if we define a square matrix (called the **metric tensor**) whose components are

$$\begin{bmatrix} g_{tt} & g_{tx} & g_{ty} & g_{tz} \\ g_{xt} & g_{xx} & g_{xy} & g_{xz} \\ g_{yt} & g_{yx} & g_{yy} & g_{yz} \\ g_{zt} & g_{zx} & g_{zy} & g_{zz} \end{bmatrix} \equiv \begin{bmatrix} +1 & 0 & 0 & 0 \\ 0 & -1 & 0 & 0 \\ 0 & 0 & -1 & 0 \\ 0 & 0 & 0 & -1 \end{bmatrix}. \tag{3.6}$$

Here, we write both indices as subscripts (unlike the Lorentz transformation matrix). With this matrix, we can write the metric equation as follows:

$$\Delta s^2 = \Delta t^2 - \Delta x^2 - \Delta y^2 - \Delta z^2 = g_{\mu\nu}\Delta x^\mu \Delta x^\nu \tag{3.7}$$

The last expression takes advantage of index notation and another convention where we define $x^t \equiv t$, $x^x \equiv x$, $x^y \equiv y$, and $x^z \equiv z$ so that Δx^μ stands for Δt, Δx, Δy, or Δz depending on the value of the index μ. Note that the Einstein summation convention applied to the rightmost expression in Equation 3.7 actually implies a sum over 16 terms:

$$\begin{aligned}
g_{\mu\nu}\Delta x^\mu \Delta x^\nu = \; & g_{tt}\Delta t^2 + g_{tx}\Delta t \Delta x + g_{ty}\Delta t \Delta y + g_{tz}\Delta t \Delta z \\
& + g_{xt}\Delta x \Delta t + g_{xx}\Delta x^2 + g_{xy}\Delta x \Delta y + g_{xz}\Delta x \Delta z \\
& + g_{yt}\Delta y \Delta t + g_{yx}\Delta y \Delta x + g_{yy}\Delta y^2 + g_{yz}\Delta y \Delta z \\
& + g_{zt}\Delta z \Delta t + g_{zx}\Delta z \Delta x + g_{zy}\Delta z \Delta y + g_{zz}\Delta z^2,
\end{aligned} \tag{3.8}$$

but since all the off-diagonal terms in the metric tensor are zero, the only nonzero terms are the four that appear in Equation 3.7. The purpose of the metric tensor is to supply the minus signs that appear in the metric equation and associate them with the correct terms.

Now, Equation 3.7 tells us how to calculate the spacetime interval, which is a frame-independent scalar, so we should have

$$\Delta s^2 = g_{\mu\nu}\Delta x^\mu \Delta x^\nu = g'_{\alpha\beta}\Delta x'^\alpha \Delta x'^\beta, \tag{3.9}$$

where $g'_{\alpha\beta}$ and $\Delta x'^\alpha$ represent the components of the metric tensor g and the four-displacement Δs as evaluated in the primed coordinate system. But we know that the components of the latter transform as $\Delta x'^\alpha = \Lambda^\alpha{}_\sigma \Delta x^\sigma$ (see Equation 3.5 and remember that the index symbols can be any Greek letter), so Equation 3.9 becomes

$$\Delta s^2 = g_{\mu\nu}\Delta x^\mu \Delta x^\nu = g'_{\alpha\beta}(\Lambda^\alpha{}_\sigma \Delta x^\sigma)(\Lambda^\beta{}_\eta \Delta x^\eta). \tag{3.10}$$

Since the components of Δs can have any value, the most general way to *ensure* that Δs^2 is frame-independent is to require that

$$g'_{\alpha\beta} = (\Lambda^{-1})^\mu{}_\alpha (\Lambda^{-1})^\nu{}_\beta \, g_{\mu\nu}, \tag{3.11}$$

where $(\Lambda^{-1})^\mu{}_\alpha$ represents components of the **inverse Lorentz transformation** matrix; that is, the matrix that takes a four-vector from the primed frame back to the unprimed frame. Since the only difference between the frames is the sign of the relative velocity, we have (for frames in standard orientation)

$$\begin{bmatrix}
(\Lambda^{-1})^t{}_t & (\Lambda^{-1})^t{}_x & (\Lambda^{-1})^t{}_y & (\Lambda^{-1})^t{}_z \\
(\Lambda^{-1})^x{}_t & (\Lambda^{-1})^x{}_x & (\Lambda^{-1})^x{}_y & (\Lambda^{-1})^x{}_z \\
(\Lambda^{-1})^y{}_t & (\Lambda^{-1})^y{}_x & (\Lambda^{-1})^y{}_y & (\Lambda^{-1})^y{}_z \\
(\Lambda^{-1})^z{}_t & (\Lambda^{-1})^z{}_x & (\Lambda^{-1})^z{}_y & (\Lambda^{-1})^z{}_z
\end{bmatrix} = \begin{bmatrix}
\gamma & +\gamma\beta & 0 & 0 \\
+\gamma\beta & \gamma & 0 & 0 \\
0 & 0 & 1 & 0 \\
0 & 0 & 0 & 1
\end{bmatrix}. \tag{3.12}$$

Since the matrices $[\Lambda]$ and $[\Lambda^{-1}]$ are the matrix inverses of each other (as you will show in Box 3.1), we have

$$\left[\Lambda^{-1}\right]\left[\Lambda\right] = \left[I\right] = \left[\Lambda\right]\left[\Lambda^{-1}\right], \tag{3.13}$$

where $[I]$ is the identity matrix. In index notation, this equation reads

$$(\Lambda^{-1})^\mu{}_\alpha \Lambda^\alpha{}_\nu = \delta^\mu{}_\nu = \Lambda^\mu{}_\alpha (\Lambda^{-1})^\alpha{}_\nu, \tag{3.14}$$

where $\delta^\mu{}_\nu$ is the **Kronecker delta**, a symbol whose value is 1 if $\mu = \nu$ and zero otherwise, making its 16 components equivalent to those of the 4×4 identity matrix.

If we substitute Equation 3.11 into Equation 3.10 and use Equation 3.14, we see that

$$
\begin{aligned}
(\Delta s')^2 &\equiv g'_{\alpha\beta}\Delta x'^{\alpha}\Delta x'^{\beta} \\
&= [(\Lambda^{-1})^{\mu}{}_{\alpha}(\Lambda^{-1})^{\nu}{}_{\beta}g_{\mu\nu}][\Lambda^{\alpha}{}_{\sigma}\Delta x^{\sigma}][\Lambda^{\beta}{}_{\eta}\Delta x^{\eta}] \\
&= g_{\mu\nu}(\Lambda^{-1})^{\mu}{}_{\alpha}\Lambda^{\alpha}{}_{\sigma}\Delta x^{\sigma}(\Lambda^{-1})^{\nu}{}_{\beta}\Lambda^{\beta}{}_{\eta}\Delta x^{\eta} \\
&= g_{\mu\nu}\delta^{\mu}{}_{\sigma}\Delta x^{\sigma}\delta^{\nu}{}_{\eta}\Delta x^{\eta} \\
&= g_{\mu\nu}\Delta x^{\mu}\Delta x^{\nu} = \Delta s^2,
\end{aligned}
\tag{3.15}
$$

meaning that the spacetime interval really is frame-independent.

Look carefully at the derivation above line by line. The first line is the definition of the spacetime interval in the primed frame. The second line uses the transformation rules given in Equations 3.5 and 3.11. The third line takes advantage of a lovely feature of index notation: since all sums are explicitly specified by paired indices, the order of factors is not important (unlike in matrix notation, where order of matrices implicitly specifies which rows and columns one sums over). I have therefore moved the inverse Lorentz transformation factors next to the Lorentz transformations with which they are summed. This makes it easy to apply Equation 3.14, which is what I do in the next step. Now note that in a factor like $\delta^{\mu}{}_{\sigma}\Delta x^{\sigma}$, the sum over σ yields four terms, but the only term that is nonzero is the one where $\sigma = \mu$; therefore, $\delta^{\mu}{}_{\sigma}\Delta x^{\sigma} = 1 \cdot \Delta x^{\mu} + 0 = \Delta x^{\mu}$. Similarly, $\delta^{\nu}{}_{\eta}\Delta x^{\eta} = \Delta x^{\nu}$.

What has basically happened is that the two inverse transformation matrices in the metric transformation equation (Equation 3.11) cancel the transformation matrices coming from four-vector transformations of the two Δx^{μ} four-vectors. This yields two identity matrices that drop out of the equation.

Of course, the metric tensor (at least if we stick to Cartesian coordinates) must have the same components ($+1, -1, -1, -1$ along the diagonal and zero off the diagonal, as given in Equation 3.6) in *every* reference frame, meaning that $g'_{\alpha\beta} = (\Lambda^{-1})^{\mu}{}_{\alpha}(\Lambda^{-1})^{\nu}{}_{\beta}\,g_{\mu\nu}$ must yield the same set of components as $g_{\mu\nu}$ in the unprimed frame. You will show that this is true (for frames in standard orientation) in Box 3.2.

3.3 Covectors

Consider now the quantity

$$
a_{\mu} \equiv g_{\mu\nu}a^{\nu}.
\tag{3.16}
$$

This quantity has four components (corresponding to the four possible values of μ). But is it a four-vector? Using a calculation similar to the one shown in Equation 3.15, one can show (see Box 3.3) that

$$
a'_{\mu} = (\Lambda^{-1})^{\nu}{}_{\mu}\,a_{\nu},
\tag{3.17}
$$

This is *not* the same as the transformation equation for a four-vector (see Equation 3.5), so a_{μ} is *not* a four-vector. However, its transformation law is *analogous*. Physicists call such a quantity a **covector**, though some authors use **one-form** or **covariant vector** to describe the same concept.

Note that for every four-vector a^{ν} there is a corresponding covector $a_{\mu} \equiv g_{\mu\nu}a^{\nu}$ whose components are the same as the four-vector's components, except that the spatial components are negated. Note also that the dot product of two four-vectors a and b is equivalent to the *simple sum* of one vector's components multiplied by the corresponding covector component of the other vector:

$$
a \cdot b \equiv a^{\mu}g_{\mu\nu}b^{\nu} = a^{\mu}b_{\mu} = a_{\nu}b^{\nu}.
\tag{3.18}
$$

The next-to-last equality is a simple application of Equation 3.16 (applied to b instead of a), but the last equality follows from that definition as well, because

$$a^{\mu} g_{\mu\nu} = g_{\mu\nu} a^{\mu} = g_{\nu\mu} a^{\mu} \equiv a_{\nu}, \tag{3.19}$$

where in the first step, I used the fact that the factors commute, and in the second step, I used the fact that the metric tensor is **symmetric**, meaning that it remains the same if we exchange rows and columns ($g_{\mu\nu} = g_{\nu\mu}$). The last step follows because the names of the indices are irrelevant so long as the nature of the implied sum is the same.

Covectors aren't always just vectors disguised by multiplying by the metric. Certain physical quantities are *naturally* covectors. For example, consider the four-gradient $\partial_{\mu} \Phi \equiv \partial \Phi / \partial x^{\mu}$ of a scalar function $\Phi(t, x, y, z)$. You can show (see Box 3.4) that

$$\partial_{\mu}' \Phi \equiv \frac{\partial \Phi}{\partial x'^{\mu}} = (\Lambda^{-1})^{\nu}{}_{\mu} \frac{\partial \Phi}{\partial x^{\mu}} \equiv (\Lambda^{-1})^{\nu}{}_{\mu} \partial_{\mu} \Phi. \tag{3.20}$$

This transformation law identifies the four-gradient as being a natural covector.

Just as we can convert vectors to covectors, we can convert covectors to vectors using the **inverse metric tensor** $g^{\mu\nu}$:

$$a^{\mu} = g^{\mu\nu} a_{\nu} \qquad \partial^{\mu} \Phi \equiv g^{\mu\nu} \partial_{\nu} \Phi. \tag{3.21}$$

The inverse metric tensor is defined as the matrix inverse of the metric tensor. But note that the metric for Cartesian components multiplied by itself is the identity matrix:

$$\begin{bmatrix} 1 & 0 & 0 & 0 \\ 0 & -1 & 0 & 0 \\ 0 & 0 & -1 & 0 \\ 0 & 0 & 0 & -1 \end{bmatrix} \begin{bmatrix} 1 & 0 & 0 & 0 \\ 0 & -1 & 0 & 0 \\ 0 & 0 & -1 & 0 \\ 0 & 0 & 0 & -1 \end{bmatrix} = \begin{bmatrix} 1 & 0 & 0 & 0 \\ 0 & 1 & 0 & 0 \\ 0 & 0 & 1 & 0 \\ 0 & 0 & 0 & 1 \end{bmatrix}, \tag{3.22}$$

(so long as we stick to Cartesian coordinates). Therefore, the metric is self-inverse, meaning that the inverse metric has the same components as the metric.

3.4 Tensors

We *define* a **tensor** to be a quantity with zero or more indices that transforms when one changes inertial reference frames in such a way that each upper index gets a factor of the Lorentz transformation matrix $\Lambda^{\mu}{}_{\nu}$ (summed over that index), and each lower index gets a factor of the inverse Lorentz transformation matrix $(\Lambda^{-1})^{\mu}{}_{\nu}$ (summed over that index). So, for example, a tensor with three indices, one upper and two lower, would transform according to

$$T'^{\alpha}{}_{\beta\mu} = \Lambda^{\alpha}{}_{\nu} (\Lambda^{-1})^{\sigma}{}_{\beta} (\Lambda^{-1})^{\eta}{}_{\mu} T^{\nu}{}_{\sigma\eta}. \tag{3.23}$$

Since each of this tensor's three indices has four possible values, the tensor T here has $4^3 = 64$ components. We say that a tensor with n indices is an **nth-rank tensor**; such a tensor has 4^n components.

We now understand the full implications of an index's position as a superscript or subscript: *the positions of a tensor's indices tell us how the tensor transforms*. This will be crucial in what follows.

Many of the quantities we have studied so far in this chapter are tensors. A *four-vector* is a first-rank tensor with one upper index. A *covector* is a first-rank tensor with one lower index. A scalar is a zeroth-rank tensor with no indices (which therefore does not transform at all when we change reference frames). Equation 3.11 tells us that the metric is a second-rank tensor with two lower indices. The inverse metric and the Kronecker delta also turn out to be tensors (see Box 3.5). Another tensor we will find useful in the future is the **electromagnetic field tensor** F (also known as the **Faraday tensor**): in upper-index form, it has the components

$$F^{\mu\nu} = \begin{bmatrix} F^{tt} & F^{tx} & F^{ty} & F^{tz} \\ F^{xt} & F^{xx} & F^{xy} & F^{xz} \\ F^{yt} & F^{yx} & F^{yy} & F^{yz} \\ F^{zt} & F^{zx} & F^{zy} & F^{zz} \end{bmatrix} = \begin{bmatrix} 0 & -E_x & -E_y & -E_z \\ E_x & 0 & -B_z & B_y \\ E_y & B_z & 0 & -B_x \\ E_z & -B_y & B_x & 0 \end{bmatrix}, \tag{3.24}$$

where E_x, E_y, E_z are the components of the ordinary electric field vector $\vec{E}$ in a given inertial frame, and B_x, B_y, B_z are the components of the ordinary magnetic field vector $\vec{B}$ in the same frame. Note that this tensor is **antisymmetric**: $F^{\mu\nu} = -F^{\nu\mu}$, meaning that transposing the matrix (exchanging rows and columns) preserves the magnitude of each component but flips its sign.

However, the transformation matrix $\Lambda^{\mu}{}_{\nu}$ and the inverse transformation matrix $(\Lambda^{-1})^{\mu}{}_{\nu}$ are *not* tensors. It makes no sense to even talk about how these matrices "transform," since they represent the transformation itself.

3.5 Tensor Equations and Operations

A **tensor equation** declares that two tensors are equal, for example,

$$A^{\mu}{}_{\nu}{}^{\alpha} = B^{\mu}{}_{\nu}{}^{\alpha}, \tag{3.25}$$

which declares that the corresponding components of the two third-rank (64-component) tensors A and B are all equal, meaning that the tensors themselves are equal. The beauty of a tensor equation is that if it is true in any one inertial reference frame, then it must be true in *every* inertial reference frame, because the two tensors (by the definition of a tensor) transform the same way when we change frames and so remain equal (component by component) in the new frame:

$$A'^{\beta}{}_{\sigma}{}^{\eta} \equiv \Lambda^{\beta}{}_{\mu}(\Lambda^{-1})^{\nu}{}_{\sigma}\Lambda^{\eta}{}_{\alpha}A^{\mu}{}_{\nu}{}^{\alpha} = \Lambda^{\beta}{}_{\mu}(\Lambda^{-1})^{\nu}{}_{\sigma}\Lambda^{\eta}{}_{\alpha}B^{\mu}{}_{\nu}{}^{\alpha} \equiv B'^{\beta}{}_{\sigma}{}^{\eta}. \tag{3.26}$$

Tensor equations therefore mathematically *embody* the symmetry expressed by the principle of relativity: any law of physics written in the form of a tensor equation will *automatically* satisfy the principle.

There are some simple rules that tensor equations must obey. Note that it makes no sense to equate quantities that have different numbers of components, so (1) the number of non-summed indices must be the same on both sides of a tensor equation. Moreover, (2) the vertical positions of those indices must be the same, or they will not transform the same way when we change reference frames. Finally, (3) the symbols used for the corresponding indices must be the same so that it is clear what you are equating. For example, equating two four-vectors a and b *should* say that their corresponding components are the same, meaning that $a^t = b^t$, $a^x = b^x$, $a^y = b^y$, and $a^z = b^z$, or (equivalently) $a^{\mu} = b^{\mu}$ for all four choices of the value of μ. On the other hand, the meaning of a statement $a^{\mu} = b^{\nu}$ (where the index symbols are different) is unclear, and might be taken to mean that *each* of the components of a is equal to *every* component of b, since we would expect the equation to be true for every possible choice of values for μ and ν.

Some technical terms will help us state things more clearly. We use the term **bound index** for an index that is summed over because it appears once as a superscript and once as a subscript in a given term (for example, the indices μ, ν, and α in Equation 3.26). We call the other indices (β, σ, and η in Equation 3.26) **free indices**. The previous paragraph states that the number of free indices, the vertical positions of those indices, and the symbols used for corresponding indices must be the same on both sides of a tensor equation.

To construct tensor equations that describe actual physical laws, it helps to understand the kinds of operations we can perform with tensors that produce tensors. The core operations are as follows.

1. **Tensor Sum.** *Example:* $p_1^\mu + p_2^\mu = p_{tot}^\mu$. *Result:* Tensor of the same rank. *Limitations:* The sum is defined between tensors having the same units, rank, and index positions.

2. **Tensor Product.** *Example:* $A^\mu B_\nu = C^\mu{}_\nu$. *Result:* Tensor of rank $n_1 + n_2$, where n_1 and n_2 are the ranks of the multiplied tensors. *Comments:* The components of the product tensor are the products of all possible pairings of the multiplied tensors' components.

3. **Contraction** (summing over an upper and lower index). *Example:* $R^\alpha{}_{\mu\alpha\nu} = R_{\mu\nu}$. *Result:* Tensor of rank $n - 2$, where n is the rank of the original tensor. *Limitations:* One can only sum over exactly one upper and one lower index, or the result is not a tensor.

4. **Gradient.** *Example:* $\partial_\alpha F^{\mu\nu}$. *Result:* Tensor of rank $n + 1$, where n is the rank of the original tensor. *Limitation:* This is only a tensor operation when we are using Cartesian coordinates.

5. **Lowering/Raising an Index.** *Examples:* $u_\mu = g_{\mu\nu} u^\nu$ or $\partial^\mu \Phi = g^{\mu\nu} \partial_\nu \Phi$. *Result:* Tensor of the same rank, but with a different index position. *Comment:* This is really just a tensor product with the metric tensor followed by a contraction over one of the metric's indices.

You will prove that the first three operations actually produce a tensor in Box 3.6 and the fourth in Box 3.4; the final case is a particularly useful special case of the product and contraction operations.

As an example, consider the tensor version of the **Lorentz force law** equation of electromagnetism, which in classical mechanics states that $\vec{F} = q(\vec{E} + \vec{v} \times \vec{B})$, where $\vec{F}$ is the force on a charged particle, q is the particle's charge, $\vec{v}$ is its ordinary velocity, and $\vec{E}$ and $\vec{B}$ are the electric and magnetic fields at the particle's location. The tensor version of this physical law is

$$\frac{dp^\mu}{d\tau} = q F^{\mu\nu} u_\nu, \tag{3.27}$$

where p^μ is the particle's four-momentum, τ its proper time, u_ν the covector version of its four-velocity, and $F^{\mu\nu}$ the Faraday tensor. The right side involves a tensor product of a scalar, second-rank tensor, and first-rank tensor, followed by a contraction over one upper and lower index. The left side involves an infinitesimal tensor difference multiplied by a scalar ($1/d\tau$) and a limit taken. Since all operations on both sides of the equation are tensor operations, Equation 3.27 is a tensor equation automatically valid in all inertial frames. You will see in Box 3.7 that it is equivalent to the classical version.

3.6 Index Renaming

The symbols we use for indices indicate how *free* indices on one side of an equation correspond to free indices on the other side, and which pairs of *bound* indices to sum. We also use Greek letters when the indices range over the four dimensions of spacetime. Otherwise, the symbol choices are arbitrary. For example, $a^\nu = b^\nu$ means the same thing as $a^\alpha = b^\alpha$, and $a_\alpha b^\alpha = a_\beta b^\beta$.

Sometimes it can really help to change index symbols. For example, consider the tensor equation $a = b^\nu u_\nu + b^\alpha w_\alpha$. Since the index symbols don't matter (so long as what is being summed with what is clearly indicated), and since addition of numbers is associative and addition and multiplication are distributive, we can change $\alpha \to \mu$ in the second term and factor out the common factor of b^μ:

$$a = b^\mu u_\mu + b^\alpha w_\alpha = b^\mu u_\mu + b^\mu w_\mu = b^\mu (u_\mu + w_\mu). \tag{3.28}$$

However, following certain rules will enable you to change index symbols (a process physicists call **renaming indices**) without turning a meaningful equation into nonsense.

1. **Rename every instance.** Whether the index is free or bound, you must change *every appearance* of a given symbol to the new symbol. **Examples:** Renaming only the index on one side of an equation, such as renaming $a^\nu = b^\nu$ to $a^\alpha = b^\nu$, creates nonsense. Similarly, renaming only one of a bound-index pair breaks the sum: $a_\beta b^\beta \neq a_\alpha b^\beta$ (the former is a scalar, the latter is a second-rank tensor).

2. **Avoid collisions.** You should never rename an index so that it becomes the same as a bound or free index already in the same term. **Example:** Consider the equation $a_\mu = g_{\mu\nu} a^\nu$. If we were to rename $\nu \to \mu$, we would get $a_\mu = g_{\mu\mu} a^\mu$. It is not clear what the latter equation even means: Is μ a free index or a bound index? Do we do a sum or not? One might possibly interpret the latter equation as meaning that $a_t = g_{tt} a^t$, $a_x = g_{xx} a^x$, $a_y = g_{yy} a^y$, $a_z = g_{zz} a^z$ (assigning all four possible values to μ and not doing a sum). But these are not the same as the four components of $a_\mu = g_{\mu\nu} a^\nu$, so the collision has changed the equation's meaning.

3. **When in doubt, write it out.** If you are ever in doubt about what a renaming operation might do, write out any implicit sums before and after the operation and make sure they are the same.

Box 3.8 exercises your ability to recognize invalid tensor equations and illegal index renamings.

3.7 Latin Index Symbols

In this text and most others, using a Latin letter as an index symbol indicates that we should assume the index ranges over only the three *spatial* index values x, y, z, not the four spacetime values. For example, we might write the ordinary three-dimensional gradient of a potential function ϕ as $\partial_i \phi$, or the ordinary dot product of two three-dimensional vectors as $\vec{a} \cdot \vec{b} = -a_i b^i$. Note in the last case that the minus sign is necessary because $a_i \equiv g_{ij} a^j$ means that each spatial covector component a_i has the same magnitude as the corresponding vector component a^i but the opposite sign, since $g_{xx} = g_{yy} = g_{zz} = -1$.

Note also that summing over only the spatial indices is not a valid tensor operation. For example, $\vec{a} \cdot \vec{b} = -a_i b^i$ is not a relativistic scalar (though it is invariant under the *subset* of Lorentz transformations that are simply rotations with zero relative frame velocity).

BOX 3.1   Inverse Lorentz Transformation

According to Equations 3.3 and 3.12, we have

$$\Lambda^{\mu}{}_{\nu} = \begin{bmatrix} \gamma & -\gamma\beta & 0 & 0 \\ -\gamma\beta & \gamma & 0 & 0 \\ 0 & 0 & 1 & 0 \\ 0 & 0 & 0 & 1 \end{bmatrix} \qquad (\Lambda^{-1})^{\mu}{}_{\nu} = \begin{bmatrix} \gamma & +\gamma\beta & 0 & 0 \\ +\gamma\beta & \gamma & 0 & 0 \\ 0 & 0 & 1 & 0 \\ 0 & 0 & 0 & 1 \end{bmatrix} \qquad (3.29)$$

(where $\gamma \equiv 1/\sqrt{1-\beta^2}$) for inertial frames in standard orientation. The forward transformation $\Lambda^{\mu}{}_{\nu}$ takes four-vector components from the unprimed to the primed frame, and $(\Lambda^{-1})^{\mu}{}_{\nu}$ is supposed to transform those components from the primed frame back to the unprimed frame. We suspect that the latter is the correct inverse transformation because the only difference between the two frames is that the primed frame moves in the $+x$ direction with speed $|\beta|$ relative to the unprimed frame, while the unprimed frame moves in the $-x$ direction with the same speed relative to the primed frame.

Exercise 3.1.1 Show that multiplying these matrices in either order yields the identity matrix, implying that they are indeed inverses of each other.

Exercise 3.1.2 Argue, therefore, that $(\Lambda^{-1})^{\mu}{}_{\nu}\Lambda^{\nu}{}_{\alpha} = (\Lambda^{-1})^{\nu}{}_{\alpha}\Lambda^{\mu}{}_{\nu} = \delta^{\mu}_{\alpha}$. (*Hint:* Consider how the implied sums correspond to those implied by the matrix multiplications in the previous exercise.)

BOX 3.2   How the Metric Transforms

Equation 3.11 claims that the metric tensor transforms according to

$$g'_{\alpha\beta} = (\Lambda^{-1})^{\mu}{}_{\alpha}(\Lambda^{-1})^{\nu}{}_{\beta}\, g_{\mu\nu}. \tag{3.30}$$

However, the metric tensor must have the components

$$g_{\mu\nu} = \begin{bmatrix} 1 & 0 & 0 & 0 \\ 0 & -1 & 0 & 0 \\ 0 & 0 & -1 & 0 \\ 0 & 0 & 0 & -1 \end{bmatrix} \tag{3.31}$$

in *every* frame for $\Delta s^2 = g_{\mu\nu}\Delta x^{\mu}\Delta x^{\nu}$ to be $\Delta t^2 - \Delta x^2 - \Delta y^2 - \Delta z^2$ in every reference frame. Are these requirements consistent with each other?

Exercise 3.2.1   Show that Equation 3.30 implies that $g'_{\mu\nu}$ does indeed have the components listed in Equation 3.31 so long as $g_{\mu\nu}$ does. (*Hint*: Either calculate the 16 components of Equation 3.30, or consider the sums implied in that equation and reexpress the equation as a matrix equation. In the latter case, note that one of the matrices will have to be transposed.)

BOX 3.3 How a Covector Transforms

In this box, we are considering the transformation properties of the quantity $a_\mu \equiv g_{\mu\nu}a^\nu$. We know that the transformation equations for $g_{\mu\nu}$ and a^ν (according to Equations 3.11 and 3.5) are

$$g'_{\alpha\beta} = (\Lambda^{-1})^\mu{}_\alpha (\Lambda^{-1})^\nu{}_\beta \, g_{\mu\nu} \tag{3.32}$$

$$a'^\beta = \Lambda^\beta{}_\sigma a^\sigma. \tag{3.33}$$

Therefore, we must have

$$a'_\alpha \equiv g'_{\alpha\beta} a'^\beta$$

$$= \left((\Lambda^{-1})^\mu{}_\alpha (\Lambda^{-1})^\nu{}_\beta \, g_{\mu\nu}\right)\left(\Lambda^\beta{}_\sigma a^\sigma\right). \tag{3.34}$$

Exercise 3.3.1 Argue that this simplifies to $a'_\alpha = (\Lambda^{-1})^\mu{}_\alpha a_\mu$, consistent with the claim made in Equation 3.17.

Please note that *whenever* we sum over an upper and lower tensor index, the transformations associated with the two indices cancel in exactly the manner illustrated in this calculation. This is why contraction over an upper and lower index produces a tensor of smaller rank, and this is also why we *never* sum indices that are either both upper or both lower indices: in such a case, the sum would not erase the transformations associated with the summed indices, meaning that the result would not transform as a tensor should.

BOX 3.4 The Gradient Is a Tensor

In this box, we will show that taking the gradient of a **tensor field** (a tensor that is a function of spacetime coordinates) transforms like a tensor with one additional lower index; for example,

$$\partial'_\mu T'^\alpha{}_{\beta\nu} \equiv \frac{\partial}{\partial x'^\mu} T'^\alpha{}_{\beta\nu}$$

$$= \left((\Lambda^{-1})^\sigma{}_\mu \partial_\sigma \right)\left(\Lambda^\alpha{}_\eta (\Lambda^{-1})^\rho{}_\beta (\Lambda^{-1})^\gamma{}_\nu T^\eta{}_{\rho\gamma} \right)$$

$$= (\Lambda^{-1})^\sigma{}_\mu \Lambda^\alpha{}_\eta (\Lambda^{-1})^\rho{}_\beta (\Lambda^{-1})^\gamma{}_\nu \left(\partial_\sigma T^\eta{}_{\rho\gamma} \right). \tag{3.35}$$

In particular, this means that the gradient of a scalar field $\Phi(t, x, y, z)$ transforms like a covector:

$$\partial'_\mu \Phi = (\Lambda^{-1})^\sigma{}_\mu \partial_\sigma \Phi. \tag{3.36}$$

This was the claim made in Equation 3.20. The first step is to note that the chain rule of partial differential calculus implies that for any function $f(t, x, y, z)$,

$$\frac{\partial f}{\partial x'^\mu} = \sum_{\text{all } \nu} \frac{\partial x^\nu}{\partial x'^\mu} \frac{\partial f}{\partial x^\nu}, \tag{3.37}$$

where we evaluate $\partial x^\nu / \partial x'^\mu$ by calculating derivatives of the transformation functions $x^\nu(t', x', y', z')$ with respect to the μth primed coordinate x'^μ.

Exercise 3.4.1 Argue that for the particular case of changing inertial frames, $\partial x^\nu / \partial x'^\mu = (\Lambda^{-1})^\nu{}_\mu$. (*Hint:* Note that the inverse Lorentz transformation states that $t = \gamma t' + \gamma\beta x'$, $x = \gamma\beta t' + \gamma x'$, $y = y'$, and $z = z'$.)

Exercise 3.4.2 Use this to prove Equation 3.35. (Be aware that in the second line, the derivative acts on the components of Λ, (Λ^{-1}), *and T*.)

BOX 3.5 The Inverse Metric and the Kronecker Delta

If the inverse metric $g^{\mu\nu}$ and the Kronecker delta δ^{μ}_{ν} are really tensors, then they should obey the tensor transformation rules, meaning that

$$g'^{\mu\nu} = \Lambda^{\mu}{}_{\alpha}\Lambda^{\nu}{}_{\beta}g^{\alpha\beta} \quad \text{and} \quad \delta'^{\mu}_{\nu} = \Lambda^{\mu}{}_{\alpha}(\Lambda^{-1})^{\beta}{}_{\nu}\delta^{\alpha}_{\beta}. \tag{3.38}$$

But (assuming Cartesian coordinates) these tensors have components

$$g^{\mu\nu} = \begin{bmatrix} 1 & 0 & 0 & 0 \\ 0 & -1 & 0 & 0 \\ 0 & 0 & -1 & 0 \\ 0 & 0 & 0 & -1 \end{bmatrix} \quad \text{and} \quad \delta^{\mu}_{\nu} = \begin{bmatrix} 1 & 0 & 0 & 0 \\ 0 & 1 & 0 & 0 \\ 0 & 0 & 1 & 0 \\ 0 & 0 & 0 & 1 \end{bmatrix} \tag{3.39}$$

in *all* inertial reference frames (because the Kronecker delta is *defined* to diagonal components equal to 1, and $g^{\mu\nu}$ is the matrix inverse of $g_{\mu\nu}$, which has the same components in all inertial reference frames). The question we want to resolve is whether these two requirements are consistent with each other.

Exercise 3.5.1 Prove that the transformation relations given in Equation 3.38 yield matrices for $g'^{\mu\nu}$ and δ'^{μ}_{ν} that have the components shown in Equation 3.39 so long as we assume that $g^{\alpha\beta}$ and δ^{α}_{β} have those components. (*Hint:* Start with the proof for the Kronecker delta, as it is easier. For the case of the inverse metric, you can use either of the methods described in Box 3.2 for proving the analogous equation for the metric.)

BOX 3.6 Tensor Operations

Our goal in this box is to prove that tensor addition (defined as the addition of the tensors' corresponding components), the tensor product, and contraction (summing over an upper and lower index) produce tensors.

As an example, consider tensor addition. Suppose we know that $A^\mu{}_\nu$ and $B^\mu{}_\nu$ are tensors. Does $C^\mu{}_\nu \equiv A^\mu{}_\nu + B^\mu{}_\nu$ transform like a tensor when we change frames? Because we know that $A^\mu{}_\nu$ and $B^\mu{}_\nu$ are tensors, we know that

$$C'^\mu{}_\nu \equiv A'^\mu{}_\nu + B'^\mu{}_\nu$$

$$= \Lambda^\mu{}_\alpha (\Lambda^{-1})^\beta{}_\nu A^\alpha{}_\beta + \Lambda^\mu{}_\sigma (\Lambda^{-1})^\eta{}_\nu B^\sigma{}_\eta. \tag{3.40}$$

We are free to rename the bound indices $\sigma \to \alpha$ and $\eta \to \beta$ in the final term because that won't cause any collisions with existing index symbols in that term. That allows us to factor out the now-common transformation symbols:

$$C'^\mu{}_\nu = \Lambda^\mu{}_\alpha (\Lambda^{-1})^\beta{}_\nu (A^\alpha{}_\beta + B^\alpha{}_\beta) \equiv \Lambda^\mu{}_\alpha (\Lambda^{-1})^\beta{}_\nu C^\alpha{}_\beta \tag{3.41}$$

by the definition of $C^\mu{}_\nu$ in the unprimed frame. So, we see that $C^\mu{}_\nu$ really does transform as a tensor having one upper index and one lower index should. You can probably see how one might generalize this argument to any number of indices in any position.

Exercise 3.6.1 In a similar manner, prove that $C^\alpha{}_{\beta\gamma} \equiv A^\alpha{}_\beta B_\gamma$ transforms like a third-rank tensor with one upper and two lower indices, assuming that $A^\alpha{}_\beta$ and B_γ are tensors. (*Hint:* This is simpler than the proof above.)

Exercise 3.6.2 Prove that $B_\gamma \equiv C^\alpha{}_{\alpha\gamma}$ transforms like a covector, assuming that $C^\alpha{}_{\beta\gamma}$ is a tensor (showing that contraction is a tensor operation).

BOX 3.7   The Lorentz Force Law

The Lorentz force law in ordinary vector form states that $\vec{F} = q(\vec{E} + \vec{v} \times \vec{B})$, where $\vec{F}$ is the electromagnetic force acting on a particle with charge q at a point where the electric and magnetic field vectors are $\vec{E}$ and $\vec{B}$, respectively. Assuming that the electromagnetic force alone acts on the particle, then the force $\vec{F} = d\vec{p}/dt$, where $\vec{p}$ is the particle's momentum and t is the coordinate time in our given reference frame. In this box, we will show that if $\vec{p}$ is the particle's *relativistic* momentum, this non-tensor law is actually frame-independent!

We begin by noting that in index notation

$$F_i \equiv \frac{dp^i}{dt} = \frac{dp^i}{d\tau}\frac{d\tau}{dt} = \frac{dp^i}{d\tau}\frac{1}{u^t} \quad \text{and} \quad v_i \equiv \frac{dx^i}{dt} = \frac{dx^i}{d\tau}\frac{d\tau}{dt} = \frac{u^i}{u^t}, \quad (3.42)$$

where F_i, p^i, and v_i represent the ith component of the vectors $\vec{F}$, $\vec{p}$, and $\vec{v}$, respectively, and u^t and u^i represent the time component and the spatial components of the particle's four-velocity u, respectively (remember that Latin indices represent *spatial* components in our notation).

Exercise 3.7.1   By writing out the implicit sum, prove that in *any* frame, taking the spatial components of the tensor expression $dp^\mu/d\tau = qF^{\mu\nu}u_\nu$ (where $F^{\mu\nu}$ is the Faraday tensor given in Equation 3.24) and dividing both sides by u^t evaluated in that frame yields the Lorentz force law. (Since a tensor law is valid in every inertial frame, so is the Lorentz force law.) Also find the classical form of the time component of $dp^\mu/d\tau = qF^{\mu\nu}u_\nu$ and interpret its physical meaning. (*Hint:* Remember that $u_\mu \equiv g_{\mu\nu}u^\nu$ means that $u_t = u^t$ and $u_i = -u^i$. But because the components of the ordinary three-dimensional vectors $\vec{F}$, $\vec{v}$, $\vec{E}$, and $\vec{B}$ *don't* transform as spatial components of a four-vector, whether we use superscripts or subscripts on these quantities is irrelevant: the meaning is the same either way.)

BOX 3.8 Legal and Illegal Tensor Equations

The following are all valid tensor equations in index notation:

1. $g_{\mu\nu}\dfrac{du^{\mu}}{d\tau}u^{\nu} = 0.$

2. $g_{\mu\nu}(\Lambda^{-1})^{\nu}{}_{\alpha} = g_{\mu\beta}\Lambda^{\beta}{}_{\alpha}.$

3. $\dfrac{dp^{\mu}}{d\tau} = 0$ for a free particle.

4. $g_{\alpha\mu}g_{\beta\nu}\partial_{\sigma}F^{\mu\nu} + g_{\sigma\mu}g_{\alpha\nu}\partial_{\beta}F^{\mu\nu} + g_{\beta\mu}g_{\sigma\nu}\partial_{\alpha}F^{\mu\nu} = 0.$

5. $\partial_{\mu}F^{\mu\nu} = 4\pi k J^{\nu}.$

6. $g_{\mu\nu}g^{\mu\nu} = 4.$

Exercise 3.8.1 Circle all the free indices in the equations above (or write "none"). How many component equations does each represent?

The following equations may or may not have index errors.

1. $p^{\mu} = mu^{\alpha}\delta^{\mu}_{\nu}.$

2. $p^{\mu}g_{\mu\nu}p^{\nu} - m^{2} = 0.$

3. $dp^{\mu}/d\tau = qF^{\mu\mu}u_{\mu}.$

4. $\partial_{\mu}F^{\alpha\beta} = 4\pi J^{\beta}.$

5. $\Lambda^{\mu}{}_{\alpha}\Lambda^{\mu}{}_{\beta}A^{\alpha\beta} = B^{\mu}.$

6. $p^{\mu} + m = 0.$

Exercise 3.8.2 Identify the equations above as being "OK" or "Not OK." If the latter, describe what the problem is.

Consider the following examples of renaming an index.

1. $\Lambda^{\alpha}{}_{\beta}B^{\beta} = B'^{\alpha}$ renamed to $\Lambda^{\alpha}{}_{\beta}B^{\beta} = B'^{\mu}.$

2. $g_{\mu\nu}p^{\mu}p^{\nu} = m^{2}$ renamed to $g_{\mu\mu}p^{\mu}p^{\mu} = m^{2}.$

3. $dp^{\mu}/d\tau = qF^{\mu\nu}u_{\nu}$ renamed to $dp^{\mu}/d\tau = qF^{\mu\beta}u_{\beta}.$

4. $\partial_{\alpha}F^{\alpha\beta} - 4\pi J^{\beta} = 0.$ renamed to $\partial_{\alpha}F^{\alpha\beta} - 4\pi J^{\mu} = 0.$

5. $g_{\mu\nu}p_1^{\mu}p_1^{\nu} + g_{\alpha\sigma}p_1^{\alpha}p_2^{\sigma} = 0$ renamed to $g_{\mu\nu}p_1^{\mu}p_1^{\nu} + g_{\mu\nu}p_1^{\mu}p_2^{\nu} = 0.$

6. $\delta^{\alpha}_{\mu}\delta^{\beta}_{\nu}F^{\mu\nu} = F^{\alpha\beta}$ renamed to $\delta^{\mu}_{\mu}\delta^{\nu}_{\nu}F^{\mu\nu} = F^{\mu\nu}.$

Exercise 3.8.3 Identify the index-renaming processes above as being "OK" or "not OK."

Homework Problems

P3.1 We can consider the components of an accelerating particle's four-velocity to be functions of the particle's proper time τ: $u = u(\tau)$.

a. With the help of index notation, prove that $d(u \cdot u)/d\tau = 2u \cdot (du/d\tau)$, as one might anticipate.

b. Argue that $a \cdot u = 0$, where $a \equiv du/d\tau$. (*Hint:* What is the frame-independent value of $u \cdot u$?)

P3.2 Consider a four-vector field $A(t, x, y, z)$ whose components are

$$A^t = a(t^2 - x^2),\ A^x = bx^2,\ A^y = cx,\ A^z = 0, \quad (3.44)$$

where a, b, and c are constants.

a. Calculate the components of A_μ.
b. Calculate the components of $\partial_\mu A_\nu$.
c. Calculate the value of $\partial_\mu A^\mu$.
d. Calculate the components of $\partial^\mu A^\nu$.
e. Calculate the value of $\partial^\mu A_\mu$.

P3.3 Suppose we define a scalar field $\phi(t, x, y, z) \equiv A \cos(p \cdot x)$, where A is a constant, x is the position four-vector with components $[t, x, y, z]$, and p is a four-vector with constant components $[p^t, p^x, p^y, p^z]$. Evaluate the gradient $\partial_\mu \phi$ of this scalar field. (*Hint:* The result must be a *covector*, since $\partial_\mu \phi$ is a covector. Explain how this happens even though p is a *vector*. Don't forget that the dot product includes the metric!)

P3.4 Suppose $a^\mu = b^\mu + \partial_\nu F^{\nu\mu}$.

a. Would it be acceptable to rename $\nu \to \mu$?

b. Write $a \cdot a$ in index form, carefully avoiding index collisions.

P3.5 Use the transformation law for the second-rank Faraday tensor $F^{\mu\nu}$ to derive transformation laws describing how the electric and magnetic fields $\vec{E}$ and $\vec{B}$ change when one goes from the unprimed inertial frame to the primed inertial frame.

P3.6 Since it is a tensor expression with no free indices, the quantity $F^{\mu\nu} F_{\mu\nu}$ must be a *scalar*, meaning that its value is independent of one's choice of reference frame. What is this frame-independent quantity in terms of the components of $\vec{E}$ and $\vec{B}$?

P3.7 (Strongly recommended for those students who have not yet seen Maxwell's equations in differential form.) In introductory physics, you may have studied Maxwell's equations for the electric and magnetic fields in integral form: $\oint \vec{E} \cdot d\vec{A} = Q/\varepsilon_0$, $\oint \vec{E} \cdot d\vec{S} = -d\Phi_B/dt$, and so on. More useful in particle physics are the so-called differential versions of these equations, which apply at every point in spacetime. In natural units, these are

$$\vec{\nabla} \cdot \vec{E} = \rho \qquad \text{(Gauss's Law)} \qquad (3.45a)$$

$$\vec{\nabla} \times \vec{B} - \frac{\partial \vec{E}}{\partial t} = \vec{J} \qquad \text{(Ampere's Law)} \qquad (3.45b)$$

$$\vec{\nabla} \cdot \vec{B} = 0 \qquad \text{(Gauss's Law for } \vec{B}) \quad (3.45c)$$

$$\vec{\nabla} \times \vec{E} + \frac{\partial \vec{B}}{\partial t} = 0 \qquad \text{(Faraday's Law)}, \qquad (3.45d)$$

where ρ is the charge density, $\vec{J} = \rho \vec{v}$ is the **current density** (where $\vec{v}$ is the velocity of charges at the point in question), and

$$\vec{\nabla} \cdot \vec{F} \equiv \frac{\partial F_x}{\partial x} + \frac{\partial F_y}{\partial y} + \frac{\partial F_z}{\partial z} \qquad (3.46a)$$

$$\vec{\nabla} \times \vec{F} \equiv \begin{bmatrix} \partial F_z/\partial y - \partial F_y/\partial z \\ \partial F_x/\partial z - \partial F_z/\partial x \\ \partial F_y/\partial x - \partial F_x/\partial y \end{bmatrix}, \qquad (3.46b)$$

where $\vec{F}$ stands for either $\vec{E}$ or $\vec{B}$. These equations are mathematically equivalent to the integral versions of the equations (as one can prove using Gauss's theorem and Stoke's theorem), but have the advantage of applying *locally* at every single point in space and time rather than requiring integration over specified surfaces or curves. (If you are familiar with these equations in SI units, note also that in units where $\varepsilon_0 = 1$ and $c = 1$, the magnetic permeability constant $\mu_0 = 1/\varepsilon_0 c^2 = 1$ as well.)

But how do we know that these equations are valid laws of physics that apply in all reference frames? If we can express them in tensor form, we are then *assured* that they are consistent with the principle of relativity. Show in fact that these equations can be written in the form

$$\partial_\mu F^{\mu\nu} = J^\nu \text{ and } \partial^\mu F^{\nu\alpha} + \partial^\nu F^{\alpha\mu} + \partial^\alpha F^{\mu\nu} = 0, \quad (3.47)$$

where $F^{\mu\nu}$ is the Faraday tensor and J^μ is a four-vector called the **four-current**, whose components are $[\rho, \rho v_x, \rho v_y, \rho v_z]$. The fact that the tensor forms are simpler than those in Equation 3.45 shows not only that the laws are consistent with the principle of relativity but also that mathematics consistent with that principle provides a more elegant expression of those laws: we are finally using the right mathematics for the job.

(*Hints:* First, show that the four components of $\partial_\mu F^{\mu\nu} = J^\nu$ yield Gauss's law and the three components of Ampere's law. The second equation in Equation 3.47 has three free indices, so it represents 64 different component equations. First show that if any of the two indices are the same, that equation reduces to the trivial $0 = 0$. There are $4 \times 3 \times 2 = 24$ choices for index values that are not the same. Show that the choice xyz for $\mu\nu\alpha$ yields Gauss's law for $\vec{B}$, while the choices txy, tyz, and txz yield the three components of Faraday's law. Remember that raising a spatial index changes its sign! The other nontrivial choices are just repetitions of these four. It is sufficient to argue that the six permutations xyz, xzy, yzx, yxz, zxy, and zyx of the first choice all yield the same thing: the other cases are analogous.)

P3.8 Consider a tensor $S^{\mu\nu}$ that is symmetric ($S^{\mu\nu} = S^{\nu\mu}$) in a given inertial frame, and a different tensor $M^{\mu\nu}$ that is antisymmetric ($M^{\mu\nu} = -M^{\nu\mu}$) in that same frame.

a. Prove that $S^{\mu\nu}$ is symmetric in *all* inertial frames.

b. Prove that $M^{\mu\nu}$ is antisymmetric in *all* inertial frames.

c. Prove that $S_{\mu\nu} \equiv g_{\mu\alpha}g_{\nu\beta}S^{\alpha\beta}$ is symmetric in all frames if $S^{\mu\nu}$ is.

d. Is $S^{\mu}{}_{\nu} \equiv g_{\nu\alpha}S^{\mu\alpha}$ symmetric ($S^{\mu}{}_{\nu} = S^{\nu}{}_{\mu}$)? Argue that this tensor is not symmetric in *any* frame if at least one off-diagonal element is nonzero. (Lesson: Symmetry or anti-symmetry of a tensor is a frame-independent concept only if the swapped indices have the same vertical position.)

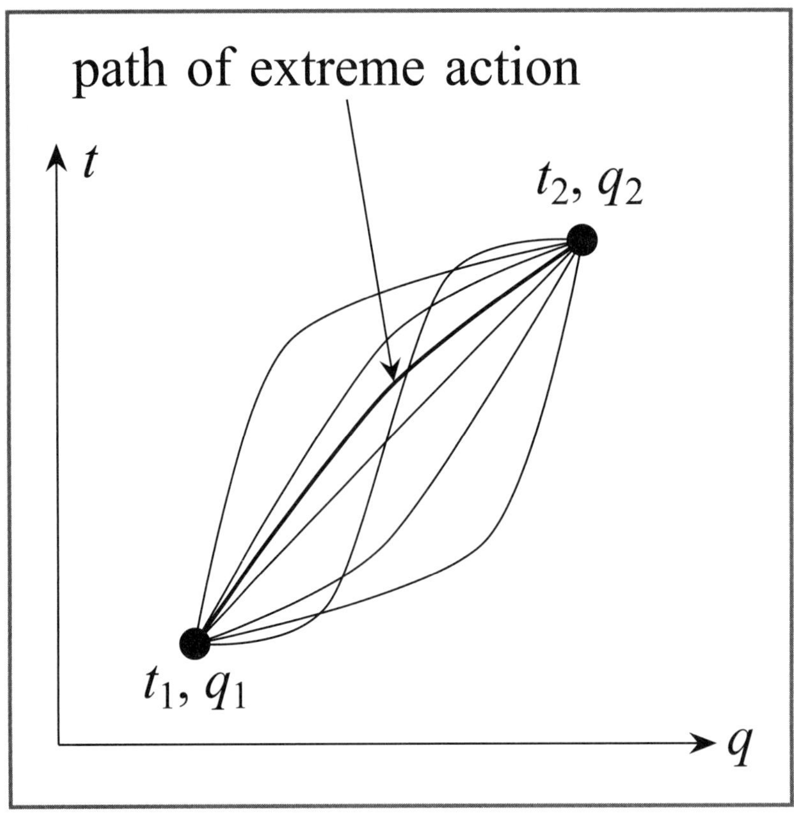

path of extreme action
t
t_2, q_2
t_1, q_1
q

4. A CLASSICAL SCALAR FIELD

The fundamental problem in developing any relativistic physical theory is that causal influences cannot travel faster than light for the reasons we explored in Chapter 2. For example, consider Coulomb's law $\vec{F}_e = (kq_1q_2/r^2)\hat{r}$, where $\vec{F}_e$ is the force that a particle with charge q_1 exerts on a particle with charge q_2 that is separated from particle 1 by a distance r in the $\hat{r}$ direction. This cannot be a *relativistically* valid law of physics because it states that particle 1 affects particle 2 instantaneously across the distance separating them: if we wiggle the position of particle 1, particle 2 instantaneously feels the effects.

In the case of electrodynamics, the solution to this problem was to instead describe the interaction between the particles in terms of a mediating electromagnetic field. Instead of particle 1 instantaneously affecting particle 2, wiggling particle 1 instead creates a local disturbance in the electromagnetic field surrounding that particle, and the disturbance ripples away from particle 1's location at the speed of light, eventually reaching particle 2. Making this picture fully consistent with relativity leads to Maxwell's equations, which describe in detail how the electromagnetic field evolves in time.

We will find that a similar approach provides the solution to the analogous problem in quantum mechanics. We will indeed find that a relativistic theory of quantum mechanics not only requires that we describe the interactions between particles in terms of fields, but that we also describe the "particles" themselves in terms of fields—hence the term "relativistic quantum field theory."

The field descriptions of even the simplest realistic particles (such as the electron) involve complications that will require some painstaking development to fully understand. We will start this journey by first considering a **scalar field**, which assigns a frame-independent number to every point in space and time. Though no low-mass (and thus physically common) elementary particle is described by such a field, the mathematics of scalar fields is much simpler than that for more realistic quantum fields, and this will help us more easily understand a number of issues common to all quantum fields. Moreover, when we consider the Higgs boson at the end of the text, we will find that it (alone of all the truly elementary particles of nature) *is* described by a scalar field, so our work on this field will be directly applicable at that point.

The behavior of both classical and quantum scalar fields are best described using Lagrangians instead of Newton's second law, so this chapter begins with a review of basic Lagrangian mechanics before moving on to explore the Lagrangian of a classical relativistic scalar field.

4.1 The Principle of Least Action

Historically, the path leading to Lagrangian mechanics arguably starts with **Fermat's Principle** (articulated by Pierre de Fermat in 1662), which states:

> The path taken by a light beam traveling between two points minimizes the travel time between those points.

From this principle, one can derive results such as Snell's law of refraction and the principle that the angle of reflection is equal to the angle of incidence when a light ray bounces off a mirror (see Box 4.1.) This formulation is strikingly different from Newtonian mechanics in that it does not look at the forces acting on the ray at each instant, but rather looks at the trajectory between the two points in its entirety and asks which of many possible trajectories minimizes the travel time, as if the universe had some purpose in making the light-ray's travel as efficient as possible.

Finding trajectories that minimized travel times in various circumstances exercised many thinkers during the 1700s. In a 1746 paper entitled *Les Loix du Mouvement et du Repos, déduites d'un Principe Metaphysique*, Pierre-Louis Moreau de Maupertuis stated a more generalized principle of minimizing something he called "action":

> This is the principle of least action, a principle so wise and so worthy of the supreme Being, and intrinsic to all natural phenomena; one observes it at work not only in every change, but also in every constancy that Nature exhibits. In the collision of bodies, motion is distributed such that the quantity of action is as small as possible, given that the collision occurs. At equilibrium, the bodies are arranged such that, if they were to undergo a small movement, the quantity of action would be smallest.

(Translation from `https://en.wikiquote.org/wiki/Principle_of_least_action`.)

In the following decades, Leonhard Euler and his student Joseph Louis Lagrange developed the calculus of variations, which made it possible to calculate trajectories that minimized various formulations of what "action" might really be. The formulation that ultimately proved successful was that due to William Hamilton in 1834, which defined the **action** to be

$$S \equiv \int L(q_i, \dot{q}_i, t)\, dt, \tag{4.1}$$

where q_i is the ith member of the set of all coordinates needed to express that system's degrees of freedom, $\dot{q}_i \equiv dq_i/dt$, t is time, and L is some function of *all* of these variables ($L(q_i, \dot{q}_i, t) = L(q_1, q_2, \ldots, \dot{q}_1, \dot{q}_2, \ldots, t)$). **Hamilton's principle** states that we can determine $q_i(t)$ (that is, how the coordinates evolve in time) by *extremizing* (not necessarily minimizing) S between fixed initial and final points. Therefore, the particular function L, which Hamilton called the **Lagrangian** in Lagrange's honor, expresses the physics of the system.

4.2 The Euler-Lagrange Equations

In principle, one might find the trajectory $q_i(t)$ that minimizes S by trial and error, but Lagrange's tremendous contribution was to figure out how to do this systematically. Pretend that we know the true trajectory $q_i(t)$ between two fixed points $q_i(t_A)$ and $q_i(t_B)$ in the system's coordinate space, and consider an infinitesimal perturbation $q_i(t) + \delta q_i(t)$ of that path such that $\delta q_i(t) = 0$ at both t_A and t_B. To first order in the perturbation, the change in the Lagrangian is

$$\delta L = \frac{\partial L}{\partial q_i}\delta q_i + \frac{\partial L}{\partial \dot{q}_i}\delta \dot{q}_i \quad \text{where } \delta \dot{q}_i = \frac{d(\delta q_i)}{dt}, \tag{4.2}$$

and where (following a convention similar to the one in relativity) we take this notation to imply an implicit sum over the repeated indexes i. But if $q_i(t)$ is truly the path that makes the action extreme, then small variations about that path should lead to essentially no change in the action:

$$0 \approx \delta S \approx \int_{t_A}^{t_B} \left[\frac{\partial L}{\partial q_i}\delta q_i + \frac{\partial L}{\partial \dot{q}_i}\delta \dot{q}_i \right] dt. \tag{4.3}$$

This approximation should become exact as the size of the perturbation goes to zero. Since $\delta \dot{q}_i = d(\delta q_i)/dt$, we can integrate the second term in the expression above by parts (see Box 4.2 for a review of this technique) to get

$$0 \approx \delta S \approx \int_{t_A}^{t_B} \left[\frac{\partial L}{\partial q_i} - \frac{d}{dt} \left(\frac{\partial L}{\partial \dot{q}_i} \right) \right] \delta q_i \, dt \; + \; \delta q_i \frac{\partial L}{\partial \dot{q}_i} \bigg|_{t_A}^{t_B}. \tag{4.4}$$

But we have *defined* our perturbation function $\delta q_i(t)$ to be zero at t_A and t_B, so the final term evaluates to zero. Finally, note that the only way the remaining integral can be zero for *all possible* path perturbations δq_i is for the quantity in brackets to be zero at all times between t_A and t_B, which are arbitrary anyway. So, the trajectory that extremizes the action will be the trajectory that solves the **Euler-Lagrange equations**

$$0 = \frac{d}{dt} \left(\frac{\partial L}{\partial \dot{q}_i} \right) - \frac{\partial L}{\partial q_i}. \tag{4.5}$$

Note that we have a *separate* equation for each coordinate q_i.

To see how this works, let's apply it to a basic Newtonian system. The simple harmonic oscillator is going to be our constant companion in this text, so it makes sense to use that as an example here. For Newtonian systems involving a single particle interacting with a potential energy function U, the Lagrangian that delivers Newton's laws of motion is

$$L = K - U, \tag{4.6}$$

where K is the particle's kinetic energy. For a simple harmonic oscillator moving in one dimension, a single coordinate suffices to express the system's degree of freedom (the particle's position), so $q = x$. The potential energy function is $U(q) = \frac{1}{2}k_s x^2 = \frac{1}{2}k_s q^2$ (where k_s is the spring constant). The particle's kinetic energy is $K = \frac{1}{2}mv^2 = \frac{1}{2}m\dot{q}^2$ (where m is the particle's mass). So,

$$\frac{\partial L}{\partial q} = \frac{\partial}{\partial q} \left(\tfrac{1}{2}m\dot{q}^2 - \tfrac{1}{2}k_s q^2 \right) = 0 - k_s q \tag{4.7a}$$

$$\frac{\partial L}{\partial \dot{q}} = \frac{\partial}{\partial \dot{q}} \left(\tfrac{1}{2}m\dot{q}^2 - \tfrac{1}{2}k_s q^2 \right) = m\dot{q} + 0. \tag{4.7b}$$

(Note that we treat the q and $\dot{q}$ as *independent variables* when evaluating the partial derivatives.) So the Euler-Lagrange equation for this system is

$$0 = \frac{d}{dt} \left(\frac{\partial L}{\partial \dot{q}} \right) - \frac{\partial L}{\partial q} = m\ddot{q} + k_s q \;\; \Rightarrow \;\; m\ddot{x} = -k_s x, \tag{4.8}$$

which is precisely Newton's second law for a one-dimensional oscillator.

Part of the beauty and power of the Lagrangian approach is that our coordinates do not have to be Cartesian spatial coordinates; they can be any kind of coordinate that conveniently captures the system's degrees of freedom. For example, consider a system whose particle is free to move in *two* dimensions in response to the radial potential energy function $U(r)$, where r is the distance from the origin. It is convenient in this case to use polar coordinates $q_1 = r$ and $q_2 = \theta$ to express the system's degrees of freedom. You can show (see Box 4.3) that in this case, the Euler-Lagrange equations imply that

$$\ell \equiv mr^2\dot{\theta} = \text{constant}, \qquad m\ddot{r} = -\frac{\partial U_{\text{eff}}}{\partial r} \;\; \text{where} \;\; U_{\text{eff}} \equiv U + \frac{\ell^2}{2mr^2}, \tag{4.9}$$

where ℓ is the particle's angular momentum. These results are much more difficult to derive using Newton's second law in Cartesian coordinates.

4.3 A Lagrangian for a String

Consider now a string modeled as N identical one-dimensional harmonic oscillators (each comprised of a particle of mass m connected to a spring with spring constant k_s) that are also connected to adjacent particles by springs with spring constant k_a (see Figure 4.1). The masses are constrained to move only vertically. In this case, the ith coordinate $\phi_i(t)$ expresses the vertical displacement of the ith mass from the position where its spring is relaxed. The system's total kinetic energy and potential energies are

$$K = \tfrac{1}{2}\sum_i m\dot{\phi}_i^2 \quad \text{and} \quad U = \tfrac{1}{2}k_s\phi_1^2 + \sum_{i=2}^{N}\left(\tfrac{1}{2}k_s\phi_i^2 + \tfrac{1}{2}k_a\big[(\phi_i - \phi_{i-1})^2 + \Delta x^2\big]\right), \quad (4.10)$$

where the last term in U follows from the Pythagorean theorem. This system's total Lagrangian is therefore

$$L = \tfrac{1}{2}m\dot{\phi}_1^2 - \tfrac{1}{2}k_s\phi_1^2 + \sum_{i=2}^{N}\left(\tfrac{1}{2}m\dot{\phi}_i^2 - \tfrac{1}{2}k_s\phi_i^2 - \tfrac{1}{2}k_a(\phi_i - \phi_{i-1})^2\right), \quad (4.11)$$

where I have dropped the term involving $\tfrac{1}{2}k_a\Delta x^2$ because it is a constant, and the Euler-Lagrange equations involve only derivatives of the Lagrangian.

Ignoring the boundary oscillators at $i = 1$ and $i = N$, you can show (see Box 4.4) that the Euler-Lagrange equations for this system imply that for any interior oscillator

$$m\ddot{\phi}_i - k_a(\phi_{i+1} - 2\phi_i + \phi_{i-1}) = -k_s\phi_i, \quad (4.12)$$

where the k_a terms come from the springs connecting adjacent oscillators.

We go to the continuum limit by taking the number of oscillators to infinity and their separation Δx to zero. In such a case, $\phi_i \equiv \phi(x_i)$ becomes a continuous function $\phi(x)$. Let's divide the equations of motion above by m, and define $\rho \equiv m/\Delta x$ to be the string's linear mass density, $T \equiv k_a\Delta x$ to be the minimum tension force exerted by the connecting springs, and $\sigma \equiv (k_s/m)^{1/2}$ to be a mass's "solo" oscillation angular frequency (due to its own vertical spring if it weren't connected to others). Noting that

$$\lim_{\Delta x \to 0} \frac{\phi_{i+1} - 2\phi_i + \phi_{i-1}}{\Delta x^2} = \lim_{\Delta x \to 0} \frac{1}{\Delta x}\left[\frac{\phi_{i+1} - \phi_i}{\Delta x} - \frac{\phi_i - \phi_{i-1}}{\Delta x}\right] = \frac{\partial^2 \phi}{\partial x^2}, \quad (4.13)$$

we see that the equations of motion become

$$\frac{\partial^2 \phi}{\partial t^2} - \frac{T}{\rho}\frac{\partial^2 \phi}{\partial x^2} = -\sigma^2\phi \quad \text{since} \quad \frac{k_a}{m} = \frac{T/\Delta x}{\rho\Delta x} = \frac{T}{\rho\Delta x^2}. \quad (4.14)$$

If $\sigma = 0$, then this would be the standard wave equation for a string under tension T (search for "string vibration" on Wikipedia). With nonzero σ, the equation's solution (see Box 4.5) is still a superposition of (co)sinusuoidal waves

$$\phi = A\cos(\omega t - kx + \theta), \quad \text{where} \quad \omega^2 = (T/\rho)k^2 + \sigma^2 \quad (4.15)$$

and where θ is a constant. (The vertical springs contribute the term σ^2 in the latter expression. Then we will see in the next chapter why we want that term.)

Now, multiplying or dividing the Lagrangian by a constant factor affects neither the trajectory that (for given initial conditions) minimizes its integral nor the Euler-Lagrange equations that determine that trajectory. Therefore, we can simplify the Lagrangian by multiplying through by $\Delta x/m$ and taking the continuous limit. Using the same definitions of ρ, T, and σ, we can write the above Lagrangian in the form

$$L = \int \left(\frac{1}{2}\dot{\phi}^2 - \frac{1}{2}\frac{T}{\rho}\left(\frac{\partial\phi}{\partial x}\right)^2 - \frac{1}{2}\sigma^2\phi^2\right)\,dx, \quad (4.16)$$

where the integral has replaced the sum and the derivative the difference in Equation 4.11.

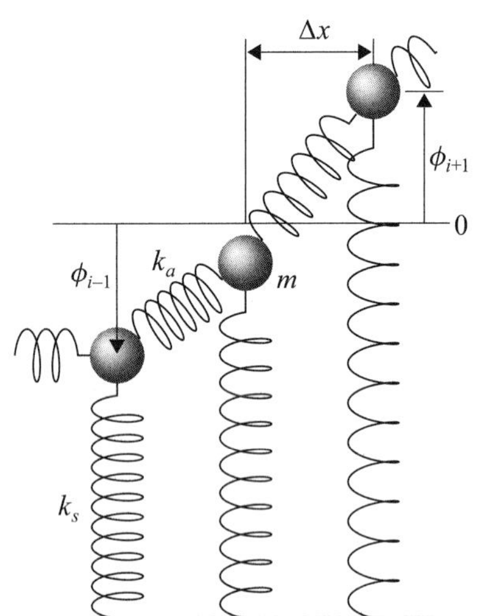

FIGURE 4.1 This figure shows three masses along the string, centered on the ith mass. A vertical spring would be relaxed if its mass were on the line marked "0."

The action for this field is therefore

$$S = \int \left(\frac{1}{2}\dot{\phi}^2 - \frac{1}{2}\frac{T}{\rho} \left(\frac{\partial \phi}{\partial x} \right)^2 - \frac{1}{2}\sigma^2\phi^2 \right) dx\, dt. \tag{4.17}$$

4.4 From String to Field

We can easily generalize our one-dimensional string model to three dimensions. In our string model, ϕ represented a vertical physical displacement, but in a three-dimensional classical scalar field, ϕ typically does not quantify a *physical displacement*, but rather something like a fluid's pressure or temperature at a given point.

The plausible generalization of the action in three dimensions becomes

$$S = \int \left(\frac{1}{2}\dot{\phi}^2 - \frac{1}{2}v_0^2 \, (\nabla\phi)^2 - \frac{1}{2}\sigma^2\phi^2 \right) d^3x\, dt, \tag{4.18}$$

where v_0 would be the speed of waves in the field if there was nothing corresponding to the oscillators' internal springs (the vertical springs in Figure 4.1), d^3x is the differential volume element, and $(\nabla\phi)^2$ is shorthand for

$$\vec{\nabla}\phi \cdot \vec{\nabla}\phi = \left(\frac{\partial \phi}{\partial x} \right)^2 + \left(\frac{\partial \phi}{\partial y} \right)^2 + \left(\frac{\partial \phi}{\partial z} \right)^2. \tag{4.19}$$

Now, note that $d^3x\, dt = d^4x$, which is the differential volume element in four-dimensional spacetime. Moreover, if v_0 were to be the speed of light ($= 1$ in our units), then the first two terms would simplify to

$$\dot{\phi}^2 - (\nabla\phi)^2 = \left(\frac{\partial \phi}{\partial t} \right)^2 - \left(\frac{\partial \phi}{\partial x} \right)^2 - \left(\frac{\partial \phi}{\partial y} \right)^2 - \left(\frac{\partial \phi}{\partial z} \right)^2$$

$$= g^{\mu\nu}\partial_\nu\phi \, \partial_\mu\phi = \partial^\mu\phi \, \partial_\mu\phi, \tag{4.20}$$

and our action becomes a beautiful tensor expression

$$S = \int \mathcal{L}\, d^4x, \quad \text{where } \mathcal{L} \equiv \tfrac{1}{2}\partial^\mu\phi \, \partial_\mu\phi - \tfrac{1}{2}\sigma^2\phi^2. \tag{4.21}$$

Here, $\mathcal{L}$ is a relativistic scalar function we call the **Lagrangian density**. Because this is a tensor equation, it is a possible frame-independent law of physics describing a three-dimensional relativistic wave.

What is the Euler-Lagrange equation for this kind of field? While the number of system "coordinates" ϕ_i has gone to infinity, the argument is otherwise very similar to that presented in Equations 4.2 to 4.5; the main difference is that we now treat time and space coordinates on an even footing. As you will show in Box 4.6, we get

$$0 = \partial_\mu \left(\frac{\partial \mathcal{L}}{\partial(\partial_\mu\phi)} \right) - \frac{\partial \mathcal{L}}{\partial \phi}, \tag{4.22}$$

which is an implicit sum over the "upper" and "lower" indices μ (the general rule is that we treat an upper index in a derivative denominator to be a lower index and vice versa).

The derivative in parentheses looks mind-boggling at first glance. What can we possibly mean by "the partial derivative of $\mathcal{L}$ with respect to $(\partial_\mu\phi)$"? But we have always treated coordinates q_i and their time-derivatives $\dot{q}_i$ as if they were independent functions in Equation 4.5. The field values $\phi(t, x, y, z)$ and their derivatives $\partial_\mu\phi$ are playing the analogous roles in Equation 4.22. If we define $\psi_\mu \equiv \partial_\mu\phi$, then we could write our Lagrangian density and the Euler-Lagrange equations for our field as

$$\mathcal{L} = \tfrac{1}{2}g^{\mu\nu}\psi_\nu\psi_\mu - \tfrac{1}{2}\sigma^2\phi^2 \quad \text{and} \quad 0 = \partial_\mu \left(\frac{\partial \mathcal{L}}{\partial \psi_\mu} \right) - \frac{\partial \mathcal{L}}{\partial \phi}. \tag{4.23}$$

We are considering the Lagrangian density $\mathcal{L}$ to be a function of the five independent fields ψ_t, ψ_x, ψ_y, ψ_z, and ϕ, and the Euler-Lagrange equations simply relate the derivatives of $\mathcal{L}$ with respect to these fields. I hope this makes the meaning of the notation in Equation 4.22 somewhat clearer.

To see how we can use this Euler-Lagrange equation to derive an equation of motion for the field $\phi(t, x, y, z)$, let's apply Equation 4.22 to our proposed Lagrangian density $\mathcal{L} = \frac{1}{2}\partial^\mu\phi\,\partial_\mu\phi - \frac{1}{2}\sigma^2\phi^2$. Evaluating the second term in Equation 4.22 is not too hard:

$$\frac{\partial\mathcal{L}}{\partial\phi} = \frac{\partial}{\partial\phi}\left(-\frac{1}{2}\sigma^2\phi^2\right) = -\sigma^2\phi. \tag{4.24}$$

The partial derivative in parentheses, however, is very tricky on a first encounter; the result (as you will show in Box 4.7) turns out to be

$$\frac{\partial\mathcal{L}}{\partial(\partial_\mu\phi)} = \partial^\mu\phi. \tag{4.25}$$

(This is consistent with the rule I have just stated about index positions in the denominator of derivatives being opposite to what they seem to be. You will understand better *why* this rule is true after you do the box.) Therefore, the Euler-Lagrange equation for our relativistic field is

$$0 = \partial_\mu\partial^\mu\phi + \sigma^2\phi = \frac{\partial^2\phi}{\partial t^2} - \frac{\partial^2\phi}{\partial x^2} - \frac{\partial^2\phi}{\partial y^2} - \frac{\partial^2\phi}{\partial z^2} + \sigma^2\phi. \tag{4.26}$$

Solutions to this equation, which you can check by direct substitution (see Box 4.8), are linear superpositions of plane waves

$$\phi = A\cos(\omega t + k_x x + k_y y + k_z z + \theta) = A\cos(k_\mu x^\mu + \theta) \tag{4.27}$$

(where k_μ is a covector of constants, A is an amplitude, and θ is a phase constant specifying the wave's phase at the origin event $t = x = y = z = 0$) so long as

$$\sigma^2 = \omega^2 - k_x^2 - k_y^2 - k_z^2 = k^\mu k_\mu. \tag{4.28}$$

The spatial components of k^μ define a vector $\vec{k} = (k^x, k^y, k^z) = (-k_x, -k_y, -k_z)$ whose direction indicates the direction of the wave's motion and whose magnitude is $2\pi/\lambda$, where λ is the wave's wavelength, as discussed in Box 4.8.

Let's summarize where we have come. We started with a simple model of masses connected by springs, took the continuum limit to get a model of a one-dimensional string, and then generalized to three dimensions. Our formulation at that point could describe classical waves in any three-dimensional medium. But to create a fully relativistic Lagrangian, we also had to take the wave's intrinsic speed (when $\sigma = 0$) to be 1, the speed of light. This makes sense, as any speed *other than* the speed of light would be frame-dependent, and so could not be a part of a physical law consistent with the principle of relativity. If we make this choice, we can then construct a Lagrangian whose associated Euler-Lagrange equations describe the behavior of a field ϕ that is the fully relativistic analogue of those simpler mechanical fields. These are the kinds of equations we will need to describe the fields corresponding to relativistic particles.

But how might we connect such fields to particles? We will begin to consider that question in the next chapter.

(If you are interested in reading more about the background and philosophy behind the "principle of least action," you may find these resources interesting:

- H. Margenau, *The Nature of Physical Reality*, McGraw-Hill, 1950.

- Feynman's lecture on least action, which you can find online at `https://www.feynmanlectures.caltech.edu/II_19.html`.)

BOX 4.1   Fermat's Principle

Fermat's principle is a powerful idea that one can use to derive many results in basic optics. We will look at two cases. First, consider points A and B that are distances a and b from a mirror and a distance d apart in the direction parallel to the mirror, as shown in Figure 4.2. Consider a light ray that travels from A to B by reflecting off the mirror. Fermat's principle implies that the position x where the ray reflects will be such that the light-travel time from A to x to B is minimized, which will be where $r_A + r_B$ is minimized.

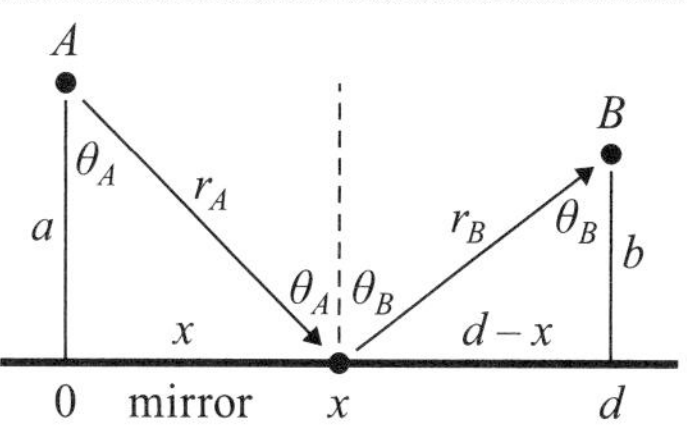

FIGURE 4.2   The setup for proving that Fermat's principle implies that $\theta_A = \theta_B$. We will vary x to find the path of minimal time between points A and B.

Exercise 4.1.1   Use calculus to show that the value of x that yields the minimum value of $r_A + r_B$ is such that

$$\frac{x}{r_A} = \frac{d-x}{r_B} \quad \Rightarrow \quad \sin\theta_A = \sin\theta_B \quad \Rightarrow \quad \theta_A = \theta_B.$$

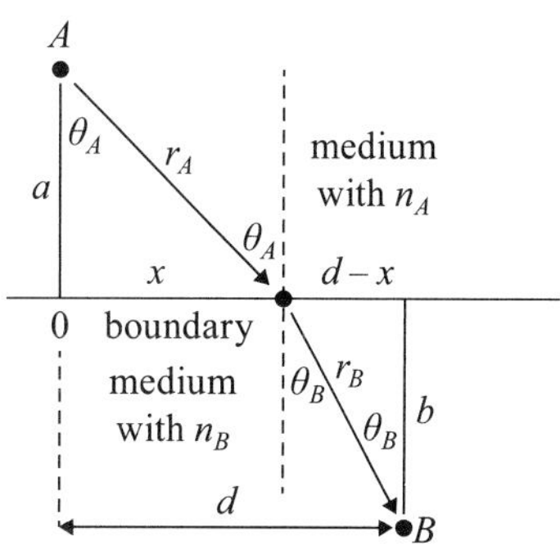

FIGURE 4.3   The setup for proving that Fermat's principle implies Snell's Law. We will vary x to find the path of minimal time between points A and B.

Now consider points A and B in media with indices of refraction n_A and n_B, respectively. Suppose points A and B are perpendicular distances a and b from the planar boundary between the media, respectively, and are d apart in the direction parallel to that boundary. Fermat's principle tells us that the position x of the point where a light ray traveling from A to B crosses that boundary (see Figure 4.3) will be whichever value minimizes the light-travel time $T \equiv r_A/c_A + r_B/c_B$ between A and B, where $c_A = c/n_A$ and $c_B = c/n_B$ are the speeds of light in media A and B, respectively.

Exercise 4.1.2   Show that the value of x that yields the minimum value of T will be such that $n_A \sin\theta_A = n_B \sin\theta_B$, as stated by Snell's law.

BOX 4.2   Review of Integration by Parts

In this box, we will review the process of integration by parts, which is a valuable technique in many areas of physics.

Consider functions $u(t)$ and $v(t)$. Note that the product rule of differential calculus implies that

$$\frac{d}{dt}(uv) = \frac{du}{dt}v + u\frac{dv}{dt}. \tag{4.29}$$

Exercise 4.2.1   Take the definite integral of both sides of the equation above from t_A to t_B and rearrange the result to prove the statement that

$$\int_{t_A}^{t_B} u\frac{dv}{dt}\,dt = uv\Big|_{t_A}^{t_B} - \int_{t_A}^{t_B} \frac{du}{dt}v\,dt. \tag{4.30}$$

(*Hint:* Don't overthink this—this is not meant to be hard.)

Note that "integration by parts" is therefore just an implication of the product rule for derivatives. The rule is often stated in the more compact form

$$\int u\,dv = uv - \int v\,du. \tag{4.31}$$

But this is really the same thing as what Equation 4.30 says, except that the integration parameter t has been eliminated and the limits are left unspecified. But Equation 4.30 is closer to the form we most often use in physics.

BOX 4.3   A Two-Dimensional Problem

In this box, we will consider a system consisting of a particle free to move in *two* dimensions in response to a radial potential energy function $U(r)$, where r is the distance from the origin. Because the potential energy only depends on r, it is convenient in this case to use polar coordinates $q_1 = r$ and $q_2 = \theta$ to express the system's degrees of freedom. The Lagrangian in this case is

$$L = K - U = \tfrac{1}{2}m\left(v_r^2 + v_\perp^2\right) - U(r), \tag{4.32}$$

where v_r is the particle's velocity component in the radial direction and $v_\perp$ is the component perpendicular to that. We can express these components in terms of polar coordinates as follows:

$$v_r = \frac{dr}{dt} \equiv \dot{r} \quad \text{and} \quad v_\perp = r\frac{d\theta}{dt} \equiv r\dot{\theta}. \tag{4.33}$$

So, the Lagrangian becomes

$$L = \tfrac{1}{2}m\dot{r}^2 + \tfrac{1}{2}mr^2\dot{\theta}^2 - U(r). \tag{4.34}$$

Because we have two coordinates here, we have two separate Euler-Lagrange equations to solve, one for each coordinate:

$$0 = \frac{d}{dt}\left(\frac{\partial L}{\partial \dot{r}}\right) - \frac{\partial L}{\partial r} \quad \text{and} \quad 0 = \frac{d}{dt}\left(\frac{\partial L}{\partial \dot{\theta}}\right) - \frac{\partial L}{\partial \theta}. \tag{4.35}$$

Since L does not depend on θ, the second equation becomes simply

$$0 = \frac{d}{dt}\left(0 + \frac{\partial}{\partial \dot{\theta}}\left[\tfrac{1}{2}mr^2\dot{\theta}^2\right] + 0\right) = \frac{d}{dt}\left(mr^2\dot{\theta}\right), \tag{4.36}$$

implying that $\ell \equiv mr^2\dot{\theta} = \text{constant}$.

Exercise 4.3.1   Now evaluate the Euler-Lagrange equation for the r coordinate, and show that one can rewrite it in the form

$$m\ddot{r} = -\frac{\partial U_{\text{eff}}}{\partial r} \quad \text{where} \quad U_{\text{eff}} \equiv U(r) + \frac{\ell^2}{2mr^2}. \tag{4.37}$$

BOX 4.4   The Wave Equation for a String

In this box, we want to evaluate the Euler-Lagrange equations for the string model Lagrangian given in Equation 4.11:

$$L = \sum_i \left(\tfrac{1}{2} m \dot{\phi}_i^2 - \tfrac{1}{2} k_s \phi_i^2 - \tfrac{1}{2} k_a (\phi_i - \phi_{i-1})^2 \right)$$

$$= \sum_i \left(\tfrac{1}{2} m \dot{\phi}_i^2 - \tfrac{1}{2} k_s \phi_i^2 - \tfrac{1}{2} k_a (\phi_i^2 - 2\phi_i \phi_{i-1} + \phi_{i-1}^2) \right). \tag{4.38}$$

Note that there is one Euler-Lagrange equation for each coordinate ϕ_i:

$$0 = \frac{d}{dt} \left(\frac{\partial L}{\partial \dot{\phi}_i} \right) - \frac{\partial L}{\partial \phi_i}. \tag{4.39}$$

Now $\partial L / \partial \dot{\phi}_i$ is not hard to evaluate, but $\partial L / \partial \phi_i$ is trickier, because terms involving a given ϕ_i appear both when the sum index is equal to i and when it is equal to $i + 1$ in Equation 4.38. The terms with nonzero ϕ_i-derivatives in that equation will be

$$\frac{\partial L}{\partial \phi_i} = \frac{\partial}{\partial \phi_i} \Big(\cdots - \tfrac{1}{2} k_s \phi_i^2 - \tfrac{1}{2} k_a \phi_i^2 + k_a \phi_i \phi_{i-1} + \cdots$$

$$+ k_a \phi_{i+1} \phi_i - \tfrac{1}{2} k_a \phi_i^2 + \cdots \Big), \tag{4.40}$$

where the terms on the first and second lines are those that involve ϕ_i when the sum index is i and when the sum index is $i + 1$, respectively.

Exercise 4.4.1   With this in mind, evaluate $\partial L / \partial \dot{\phi}_i$ and $\partial L / \partial \phi_i$ and show that the equations of motion for the string are equivalent to Equation 4.12.

BOX 4.5 Solving the String Wave Equation

The final equation of motion for our continuous string model (see Equation 4.14) is given by

$$\frac{\partial^2 \phi}{\partial t^2} - \frac{T}{\rho}\frac{\partial^2 \phi}{\partial x^2} = -\sigma^2 \phi, \tag{4.41}$$

where T/ρ and σ are constants.

Exercise 4.5.1 By direct substitution, verify that

$$\phi = A \cos\left(\omega t - kx + \theta\right) \tag{4.42}$$

is a solution to the string wave equation, where A, ω, k, and θ are constants such that $\omega^2 = (T/\rho)k^2 + \sigma^2$.

Exercise 4.5.2 Prove that if $\phi_1(t, x)$ and $\phi_2(t, x)$ satisfy Equation 4.41, then so does $a\phi_1 + b\phi_2$, where a and b are arbitrary constants. (This means that the *general* solution to Equation 4.41 will be an arbitrary *superposition* of waves of the type given in Equation 4.42.)

BOX 4.6   Finding the Euler-Lagrange Equations for a Field

In this box, we will see that the field $\phi(t, x, y, z)$ that extremizes the action $S = \int \mathcal{L}\, d^4x$ (where $\mathcal{L}$ is a Lagrangian density that depends on ϕ and $\partial_\mu \phi$) must obey the Euler-Lagrange equation given in Equation 4.22:

$$0 = \partial_\mu \left(\frac{\partial \mathcal{L}}{\partial(\partial_\mu \phi)} \right) - \frac{\partial \mathcal{L}}{\partial \phi} = \partial_\mu \left(\frac{\partial \mathcal{L}}{\partial \psi_\mu} \right) - \frac{\partial \mathcal{L}}{\partial \phi} \quad \text{where } \psi_\mu \equiv \partial_\mu \phi. \tag{4.43}$$

As we did in going from Equations 4.2 to 4.5, we first suppose that we already *know* the field $\phi(t, x, y, z)$ that extremizes the action and imagine adding a tiny perturbation $\delta\phi(t, x, y, z)$ that goes to zero on the boundary hypersurface B of a hypervolume V in spacetime (for example, a hypercylinder with an extremely large spatial radius and with caps at arbitrary past and future times t_1 and t_2). If we really are perturbing around the correct $\phi(t, x, y, z)$, then

$$0 = \delta S = \int \left[\frac{\partial \mathcal{L}}{\partial \phi} \delta\phi + \frac{\partial \mathcal{L}}{\partial \psi_\mu} \delta\psi_\mu \right] d^4x = \int \left[\frac{\partial \mathcal{L}}{\partial \phi} \delta\phi + \frac{\partial \mathcal{L}}{\partial \psi_\mu} \partial_\mu(\delta\phi) \right] d^4x, \tag{4.44}$$

since $\delta\psi_\mu = \partial_\mu(\phi + \delta\phi) - \partial_\mu\phi = \partial_\mu(\delta\phi)$. Now, note that by the product rule,

$$\partial_\mu \left(\frac{\partial \mathcal{L}}{\partial \psi_\mu} \delta\phi \right) = \left(\partial_\mu \frac{\partial \mathcal{L}}{\partial \psi_\mu} \right) \delta\phi + \frac{\partial \mathcal{L}}{\partial \psi_\mu} \partial_\mu(\delta\phi). \tag{4.45}$$

The analogue to the fundamental theorem of calculus for spacetime (the 4D version of the divergence theorem) states that for any four-vector field $f^\mu(t, x, y, z)$,

$$\int_V (\partial_\mu f^\mu)\, d^4x = \int_B n_\mu f^\mu \, d^3x, \tag{4.46}$$

where n_ν is a covector normal to the boundary B. Note that $(\partial \mathcal{L}/\partial \psi_\mu)\delta\phi$ is a function f^μ because the subscript index in the denominator of a derivative is equivalent to a raised index.

Exercise 4.6.1   Use Equation 4.45 to eliminate $(\partial \mathcal{L}/\partial \psi_\mu)\partial_\mu(\delta\phi)$ in Equation 4.44, and then use Equation 4.46 and the fact that $\delta\phi$ is arbitrary (except that $\delta\phi \to 0$ on the boundary) to show that ϕ must satisfy Equation 4.43.

BOX 4.7   Evaluating a Field's Euler-Lagrange Equation

Our goal in this box is to show that when $\mathcal{L} = \frac{1}{2}\partial^\mu\phi\,\partial_\mu\phi - \frac{1}{2}\sigma^2\phi^2$, we have

$$\frac{\partial\mathcal{L}}{\partial(\partial_\mu\phi)} = \partial^\mu\phi, \tag{4.47}$$

as claimed in Equation 4.25.

Let's simplify the notation by writing the Lagrangian in terms of $\psi_\mu \equiv \partial_\mu\phi$:

$$\mathcal{L} = \frac{1}{2}\psi^\alpha\psi_\alpha - \frac{1}{2}\sigma^2\phi^2 = \frac{1}{2}g^{\alpha\beta}\psi_\alpha\psi_\beta - \frac{1}{2}\sigma^2\phi^2, \tag{4.48}$$

where I have renamed the bound index $\mu \to \alpha$ in the first term of $\mathcal{L}$ to avoid colliding with the derivative index μ in Equation 4.47 (remember we can rename bound indices at will). In this notation, we want to find $\partial\mathcal{L}/\partial\psi_\mu$. Note that

$$\frac{\partial\psi_\alpha}{\partial\psi_\mu} = \delta^\mu_\alpha \tag{4.49}$$

because the four fields ψ_t, ψ_x, ψ_y, and ψ_z are *independent*, meaning that (for example) $\partial\psi_x/\partial\psi_y = 0$ and similarly for other index choices where $\alpha \neq \mu$. On the other hand, the derivative of any field with respect to itself is 1. (The index positions are consistent with the rule that we treat a lower index in the denominator of a derivative the same as an upper index.)

Exercise 4.7.1   From here, argue that Equation 4.47 must be correct. (*Hint:* Use the product rule, and note that $\partial^\mu\phi = g^{\mu\nu}\partial_\nu\phi = g^{\mu\nu}\psi_\nu \equiv \psi^\mu$. Alternatively, you could simply write out the sums and calculate the derivatives for each value of μ without using index notation.)

BOX 4.8 Solving the Field's Euler-Lagrange Equations

Our goal in this box is to show that solutions to the scalar field's Euler-Lagrange equation (Equation 4.26)

$$0 = \partial^\mu \partial_\mu \phi + \sigma^2 \phi \qquad (4.50)$$

are $\phi = A \cos (k_\nu x^\nu + \theta)$, where A and θ are arbitrary constants and k_μ is a covector of values satisfying $\sigma^2 = k^\mu k_\mu$. We will also discuss the physical meaning of the k^μ four-vector.

Note that by the chain rule,

$$\partial_\mu \phi = \frac{\partial}{\partial x^\mu} A \cos (k_\nu x^\nu + \theta) = -A \sin (k_\nu x^\nu + \theta) k_\nu \delta^\nu_\mu \qquad (4.51)$$

because $\partial x^\nu / \partial x^\mu$ is equal to 1 if $\nu = \mu$ and zero otherwise. Doing the sum over ν in $k_\nu \delta^\nu_\mu$ yields k_μ. So, taking the derivative pulls out a factor of k_μ.

Exercise 4.8.1 Use similar steps to show that $\phi = A \cos (k_\nu x^\nu + \theta)$ satisfies Equation 4.50 so long as $\sigma^2 = k^\mu k_\mu$. (*Hint:* Remember that $\partial^\mu \equiv g^{\mu\alpha} \partial_\alpha$.)

Since this equation of motion is linear, its general solution is a linear combination of such cosine-wave solutions with arbitrary values of A, θ, and k_μ.

Now, we often split the covector components (k_t, k_x, k_y, k_z) into temporal and spatial parts $\omega \equiv k_t$ and $\vec{k}$, where the latter has components $(k^x, k^y, k^z) = (-k_x, -k_y, -k_z)$ (remember that raising or lowering a spatial index changes its sign). What do these quantities represent physically? Note that ω describes the rate at which the wave oscillates at a fixed spatial point (x, y, z). Note also that points on a wave crest will be located at positions $\vec{r}$ where

$$2n\pi = \omega t + k_x x + k_y y + k_z z + \theta = \omega t - k^x x - k^y y - k^z z + \theta$$

$$= \omega t - \vec{k} \cdot \vec{r} + \theta, \qquad (4.52)$$

where n is an integer and the dot product here is the ordinary dot product for three-vectors. Now, if $\vec{r}_0$ is the position of a point that lies on this crest at a given instant, then so is $\vec{r}_0 + \vec{r}_1$ if $\vec{k} \cdot \vec{r}_1 = 0$. This says that the set of all points that lie on the crest at that instant (corresponding to all possible values of $\vec{r}_1$) lie on a plane that is perpendicular to $\vec{k}$. So, our solution describes a plane wave.

Note that a wave crest must move in a direction *perpendicular* to its plane (how could we measure movement of a planar crest in a direction in its own plane?), and thus in the direction of $\vec{k}$.

Exercise 4.8.2   Take the time derivative of $2\pi n = \omega t - \vec{k} \cdot \vec{r} + \theta$ (Equation 4.52) to show that the crest must move with a speed $v_p = \omega / |\vec{k}|$. This speed is the wave's *phase speed*. (*Hint:* In line with the comment above, assume that the velocity $v_p \equiv d\vec{r}/dt$ we are interested in is parallel to $\vec{k}$, and remember the basic definition of the dot product in terms of vector magnitudes.)

Finally, note that at a given instant of time, the distance λ between wave crests is the magnitude of the displacement $\Delta\vec{r}$ in the $\vec{k}$ direction that increases the value of $\vec{k} \cdot \vec{r}$ by 2π.

Exercise 4.8.3   Argue, then, that the wave solution's wavelength is

$$\lambda = \frac{2\pi}{|\vec{k}|}. \tag{4.53}$$

Homework Problems

P4.1 Consider a mass m connected to a spring with spring constant k. This system therefore oscillates with angular oscillation frequency $\omega = \sqrt{k/m}$ (meaning that $k = m\omega^2$). Suppose we release the mass at rest at $x(0) = A$; we know its subsequent motion will be described by

$$x(t) = A \cos(\omega t). \tag{4.54}$$

Though we hardly ever do this in practice, in this problem, we will explore what happens when we perturb this solution in the manner described in Section 4.2. We will then show that the solution given by the equation above really does make the action S extreme.

a. Show that the action S of the unperturbed solution for a complete cycle of the unperturbed oscillation ($t = 0$ to $t = 2\pi/\omega$) is zero.

b. Suppose that we perturb this solution by adding

$$\delta x(t) = \epsilon \sin\left(\tfrac{1}{2}\omega t\right). \tag{4.55}$$

You can see that this perturbation will go to zero at $t = 0$ and also at $t = 2\pi/\omega$, which is the endpoint of one complete oscillation. The perturbed Lagrangian becomes

$$L = \tfrac{1}{2}m(\dot{x} + \delta\dot{x})^2 - \tfrac{1}{2}m\omega^2(x + \delta x)^2. \tag{4.56}$$

Integrate this from $t = 0$ to $t = 2\pi/\omega$ to find a nonzero (negative) result, but show that the result is extremized when $\epsilon \to 0$. (*Hint:* Use the following trigonometric identities: $\sin^2\theta = \tfrac{1}{2}(1 - \cos 2\theta)$, $\cos^2\theta = \tfrac{1}{2}(1 + \cos 2\theta)$, and $2\sin\theta\cos\phi = \sin(\theta + \phi) - \sin(\theta - \phi)$.)

P4.2 Consider a situation where an object of mass m swings at the end of a string of length ℓ in the earth's gravitational field. Assume that this object is *not* confined to swinging on one plane, but can swing in any direction it likes (physicists call this a **conical pendulum**). Part of the beauty of the Lagrangian approach is that we can use whatever coordinates we like to describe the problem. In this case, it is probably easiest to use a coordinate θ to describe the angle that the string makes with the vertical, and an angle ϕ that describes the angle that a line from the vertical axis to the object makes in a horizontal plane around that axis.

a. Write down the Lagrangian for this situation.

b. What are the two Euler-Lagrange equations in this case?

c. Do we have a conserved quantity? If so, what is it?

d. In an introductory physics class, you may have learned that Newton's second law leads to the following equation of motion for a simple pendulum swinging in a plane:

$$m\ddot{\theta} = -\frac{mg}{\ell} \sin\theta. \tag{4.57}$$

If we set up initial conditions so that $\dot{\phi}(0) = 0$ initially, argue that the Euler-Lagrange equation for ϕ requires that ϕ

be constant, and the other Euler-Lagrange equation reduces to Equation 4.57.

P4.3 Consider a situation where a particle of mass m moves in one dimension along the x axis and is subject to a potential energy $U = bx$, where b is a constant with units of force.

a. Write down the Lagrangian for this situation.

b. What is the Euler-Lagrange equation in this case?

c. Solve the Euler-Lagrange equation for $x(t)$.

P4.4 Note that we generally define the action in spacetime

$$S \equiv \int \mathcal{L}\, d^4x \tag{4.58}$$

to be a unitless scalar.

a. What are the units of $\mathcal{L}$ in our energy-based unit system?

b. If the Lagrangian density is $\mathcal{L} = \tfrac{1}{2}\partial^\mu\phi\,\partial_\mu\phi - \tfrac{1}{2}\sigma^2\phi^2$, what are the units of ϕ? The units of σ?

P4.5 Consider the Lagrangian density for a scalar field for which $\sigma = 0$:

$$\mathcal{L} = \tfrac{1}{2}\partial^\mu\phi\,\partial_\mu\phi. \tag{4.59}$$

a. What are the Euler-Lagrange equations for ϕ in this case?

b. Show that $\phi = A \cos(k_\nu x^\nu)$ solves this equation subject to the requirement that $k_\mu k^\mu = 0$.

c. This does *not* mean that $\phi = $ constant (which would correspond to no wave). Suppose $k_x = k_y = 0$, but $k_z = \omega \neq 0$. Show that by choosing an appropriate nonzero value for k_t, we can satisfy $k_\mu k^\mu = 0$ and still have a nontrivial wave solution.

d. A crest of the wave will be when $k_\mu x^\mu = 2\pi n$, where n is a fixed integer. Still assuming that $k_x = k_y = 0$, solve this expression for the position $z(t)$ of a given crest, and then take the time derivative to prove that the crest must move at the speed of light.

P4.6 (This problem illustrates an approximation procedure that we will later use to handle fields that describe interacting particles.) Consider an object with mass m oscillating in one dimension at the end of a spring with spring constant k. We know that if the spring is ideal, its potential energy is $U = \tfrac{1}{2}kx^2$ (where x is the object's position relative to the equilibrium point) and, if we release the object from position A at time $t = 0$, the solution to the equation of motion is $x(t) = A \cos(\omega_0 t)$, where $\omega_0^2 = k/m$.

But suppose the potential energy of a nonideal spring is actually given by $U = \tfrac{1}{2}kx^2 + bx^4$ (this is the potential energy for what is usually called an "anharmonic oscillator"). This is like a true harmonic oscillator, except that the spring gets increasingly "stiffer" than expected when it is stretched far away from equilibrium. We would expect this system to oscillate in a similar way, but perhaps at a slightly different frequency and maybe with a not-quite cosinusoidal shape, since the spring will push back harder when the mass is far from the origin.

a. Write the exact Euler-Lagrange equation for a particle of mass m moving subject to the potential energy function $U = \frac{1}{2}kx^2 + bx^4$.

b. Let us *assume* that the oscillation is small enough so that even at the maximum amplitude, $bA^4 \ll \frac{1}{2}kA^2$, or equivalently $\varepsilon \equiv 4bA^2/k = 4bA^2/m\omega_0^2 \ll 1$ (the factor of four is for convenience, as we will see shortly). Show that we can write the Euler-Lagrange equation for this system in the form

$$\ddot{x} + \omega_0^2 x = -\varepsilon(\omega_0^2/A^2)x^3. \tag{4.60}$$

c. For small oscillations, we would expect that corrections to both the frequency and amplitude of the solution $A\cos(\omega_0 t)$ will be small and go to zero as $\varepsilon \to 0$. The general trick in such cases is to assume a zeroth-order solution that is either the same as or very much like the solution to the case where $\varepsilon = 0$, and assume that corrections to this solution are small (of order ε). We then substitute this assumption into the equation of motion, ignore any terms involving ε^2 and smaller, and use this to find an equation for the correction term.

In this particular case, it proves advantageous to try a solution of the form $x(u) = x_0(u) + \varepsilon x_1(u)$, where $x_0 \equiv A\cos u$, $u \equiv \omega t$, and $\omega \equiv \omega_0 + \varepsilon\omega_1$, where ω_1 is a correction to the original frequency ω_0. Also, define $x'' \equiv d^2x/du^2$ and note that $\ddot{x} = \omega^2 x''$ by the chain rule. Show that if we substitute these expressions into Equation 4.60 and keep only terms to order ε, we get

$$x_1'' + x_1 = 2\left(\frac{\omega_1}{\omega_0}\right)x_0 - \frac{1}{A^2}x_0^3. \tag{4.61}$$

d. This equation describes a driven harmonic oscillator. The first driving term on the right drives the oscillator its own resonant frequency ω. Like a person pushing a child on a swing, this will increase the oscillation amplitude without limit. This does not make physical sense, since we would expect the oscillation to remain stable. Use the trigonometric identity $\cos^3 u = \frac{3}{4}\cos u + \frac{1}{4}\cos 3u$ to show that we can cancel the $\cos u$ driving term if

$$\omega_1 = \tfrac{3}{8}\omega_0, \tag{4.62}$$

and that our resulting equation for x_1 with this substitution is

$$x_1'' + x_1 = -\tfrac{1}{4}A\cos 3u. \tag{4.63}$$

Equation 4.62 determines the correction to the frequency. As $\omega = \omega_0 + \varepsilon\omega_1 = (1 + \frac{3}{8}\varepsilon)\,\omega_0$, the anharmonic oscillator's frequency is a bit faster than the harmonic oscillator's, which makes sense.

e. The usual trick for solving a harmonic oscillator equation with an oscillating driving term is to attempt a solution with the same time dependence. Attempt a solution for Equation 4.63 in the form $x_1 = B\cos 3u$ and use the equation to find B in terms of A.

f. But we also need to add to $B\cos 3u$ the general solution to the homogeneous equation $x_1'' + x_1 = 0$, which is $C\cos u + D\sin u$. Choose values for C and D to satisfy the initial conditions $x_1'(0) = 0 = x_1(0)$. These initial conditions ensure that the full corrected solution is still consistent with releasing the object at rest with $x(0) = A$.

g. Finally, write down the complete corrected solution in terms of b.

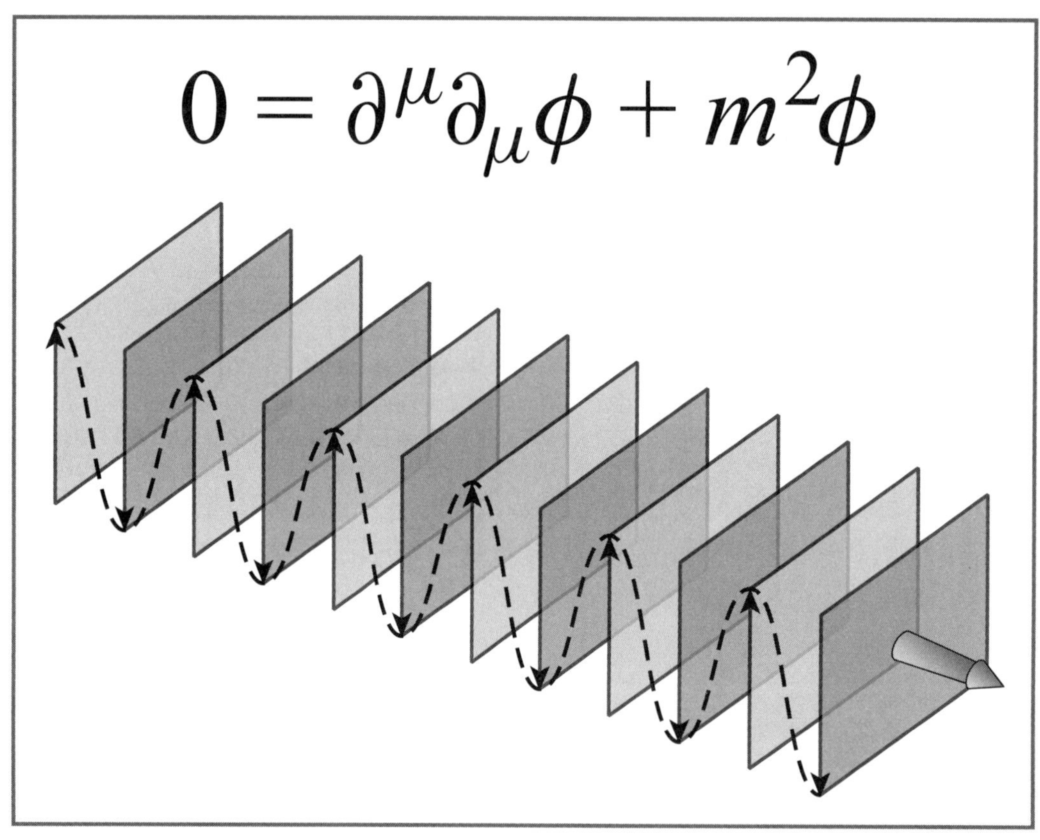

$0 = \partial^\mu \partial_\mu \phi + m^2 \phi$

5. THE KLEIN-GORDON EQUATION

In the previous chapter, we developed a Lagrangian-based description for a scalar field ϕ that generalized to three dimensions the continuum limit of a set of identical harmonic oscillators that were also connected to adjacent oscillators by springs. The Lagrangian density for this system was

$$\mathcal{L} = \tfrac{1}{2}\partial^{\mu}\phi \, \partial_{\mu}\phi - \tfrac{1}{2}\sigma^2\phi^2, \tag{5.1}$$

where $\phi(t, x, y, z)$ was the field variable and σ was related to the strength of the oscillators' *intrinsic* springs (as opposed to the springs that connect adjacent oscillators). The Euler-Lagrange equation for this system was

$$\partial^{\mu}\partial_{\mu}\phi + \sigma^2\phi = 0, \tag{5.2}$$

whose solutions could be written as superpositions of plane waves of the form

$$\phi(t, x, y, z) = A\cos(k_{\mu}x^{\mu} + \theta_0) = A\cos\left(\omega t - \vec{k}\cdot\vec{r} + \theta_0\right), \tag{5.3}$$

where A, ω, θ_0 and $\vec{k}$ are constants, with ω and $\vec{k}$ satisfying the constraint

$$k_{\mu}k^{\mu} = \omega^2 - |\vec{k}|^2 = \sigma^2. \tag{5.4}$$

We also saw that $\vec{k}$ was related to the plane-wave wavelength by the relation

$$|\vec{k}| = \frac{2\pi}{\lambda}. \tag{5.5}$$

In this chapter, we will take the first steps toward connecting this scalar field model to a relativistic quantum particle.

5.1 The Klein-Gordon Equation

The foundation on which the rest of quantum mechanics is built was laid by Louis de Broglie in 1924. In his doctoral thesis, published that year, he asserted that all free quantum particles should, like photons, be described by plane waves such that

$$E = hf = \hbar\omega \quad \text{and} \quad |\vec{p}| = \frac{h}{\lambda} = \hbar\frac{2\pi}{\lambda} = \hbar|\vec{k}|, \tag{5.6}$$

where h is Planck's constant, $\hbar = h/2\pi$, $f = \omega/2\pi$ is the wave's oscillation frequency, and $\vec{p}$ is the particle's *relativistic* momentum (the spatial components of its four-momentum). The relation $|\vec{p}| = h/\lambda$ was the key new element. Today we call that equation the **de Broglie relation.**

In tensor notation, we can express both of Equations 5.6 compactly as

$$p^{\mu} = \hbar k^{\mu}. \tag{5.7}$$

Note that this implies that

$$m^2 = p^{\mu}p_{\mu} = \hbar^2 k^{\mu}k_{\mu} = \hbar^2(\omega^2 - |\vec{k}|^2). \tag{5.8}$$

Now, this is exactly the same as the equation $k_{\mu}k^{\mu} = \omega^2 - |\vec{k}|^2 = \sigma^2$ we have for our field model (Equation 5.4) if we identify $m = \hbar\sigma$. Therefore, the Lagrangian density constructed in the last chapter describes a field whose plane-wave solutions completely satisfy de Broglie's requirements for the wave describing a relativistic quantum particle! We also now see why our model included intrinsic springs leading to term $\sigma^2 \phi^2$ in the Lagrangian density: it is this term that gives the particle a nonzero mass.

Therefore, if our field is to describe a quantum particle with a given mass m, we want to replace the constant σ with $m/\hbar$ wherever it appears. But also recall that we are working in units where $\hbar = 1$! So our Lagrangian density for the field whose excitations (waves) correspond to particles with mass m is

$$\mathcal{L} = \tfrac{1}{2}\partial^\mu \phi \, \partial_\mu \phi - \tfrac{1}{2}m^2\phi^2, \tag{5.9}$$

and the corresponding Euler-Lagrange equation is

$$\partial^\mu \partial_\mu \phi + m^2\phi = 0. \tag{5.10}$$

This last equation is called the **Klein-Gordon equation**, after Oskar Klein and Walter Gordon, who (along with a number of other people at about the same time) in 1926 proposed this equation as the relativistic generalization of the Schrödinger equation. This proposal did not gather support at the time because some of its solutions seemed to have negative energy, which would be physically impossible. But eventually, people found a different way to interpret these seemingly unphysical solutions, an approach that provided the foundation for quantum field theory.

5.2 A Quick Review of Complex Numbers

As we will see in the next few chapters, part of the key to this reinterpretation was to suppose that the field $\phi(t, x, y, z)$ attaches not a real number but rather a *complex* number to every point in spacetime. Since complex numbers will prove to be a crucial feature of the mathematics we will develop, a very brief review of the mathematics of complex numbers may be helpful.

A **complex number** c can be written in the form $a + ib$, where a and b are real and $i \equiv \sqrt{-1}$. Addition and multiplication of complex numbers are defined in sensible ways as follows:

$$c_1 + c_2 \equiv a_1 + a_2 + ib_1 + ib_2 = (a_1 + a_2) + i(b_1 + b_2) \tag{5.11a}$$
$$c_1 c_2 \equiv (a_1 + ib_1)(a_2 + ib_2) = a_1 a_2 + i(a_1 b_2 + a_2 b_1) - b_1 b_2, \tag{5.11b}$$

since $i^2 = -1$ (note that we just multiply $(a_1 + ib_1)(a_2 + ib_2)$ as if they were ordinary binomials). We define a complex number's **complex conjugate** c^* and **absolute square** $|c|^2$ to be

$$c^* = a - ib \quad (\text{if } c = a + ib) \tag{5.12a}$$
$$|c|^2 = c^* c = (a - ib)(a + ib) = a^2 + b^2. \tag{5.12b}$$

Note that $|c|^2$ is always nonnegative and real.

Euler's formula *defines* the function $e^{i\theta}$ as

$$e^{i\theta} \equiv \cos\theta + i\sin\theta. \tag{5.13}$$

We will find this function *extraordinarily useful* in this text. The logic behind this definition is that the function $e^{i\theta}$ has many properties analogous to the ordinary exponential function (see Box 5.3), except that $|e^{i\theta}|^2 = 1$ always. We can think of $e^{i\theta}$ as representing an arrow of length 1 in a two-dimensional space, where the horizontal axis is the real part of the function and the vertical axis is the imaginary part (see Figure 5.1).

We can represent any complex number $c = a + ib = Re^{i\theta}$, where $R^2 = a^2 + b^2$ and θ is such that $a = R\cos\theta$ and $b = R\sin\theta$. The multiplicative inverse of any complex number is therefore

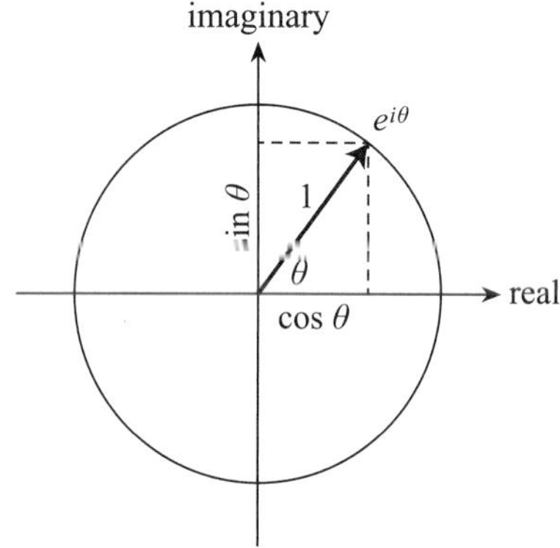

FIGURE 5.1 We can visualize the function $e^{i\theta}$ as a unit-length arrow in a two-dimensional, real-imaginary space.

$$\frac{1}{c} \equiv \frac{e^{-i\theta}}{R} \quad \text{because } c\left(\frac{1}{c}\right) = \frac{R}{R}e^{-i\theta}e^{i\theta} = 1, \tag{5.14}$$

since $e^{i\theta}e^{-i\theta} = e^{i(\theta-\theta)} = e^0 = 1$.

In Box 5.3, you will also prove these results, which you should *memorize*:

$$(e^{i\theta})^* = e^{-i\theta} \tag{5.15a}$$

$$e^{\pm i\pi/2} = \pm i, \qquad e^{\pm i\pi} = -1 \tag{5.15b}$$

$$\tfrac{1}{2}(e^{i\theta} + e^{-i\theta}) = \cos\theta, \qquad \tfrac{1}{2}(e^{i\theta} - e^{-i\theta}) = i\sin\theta. \tag{5.15c}$$

5.3 Complex Fields

As a first step toward understanding the utility of complex numbers in field theory, consider the following Lagrangian density for *two* real scalar fields:

$$\mathcal{L} = \tfrac{1}{2}\partial^\mu\phi_1\,\partial_\mu\phi_1 - \tfrac{1}{2}m_1^2\phi_1^2 + \tfrac{1}{2}\partial^\mu\phi_2\,\partial_\mu\phi_2 - \tfrac{1}{2}m_2^2\phi_2^2. \tag{5.16}$$

The Euler-Lagrange equations for these fields are

$$\partial^\mu\partial_\mu\phi_1 + m_1^2\phi_1 = 0 \tag{5.17a}$$

$$\partial^\mu\partial_\mu\phi_2 + m_2^2\phi_2 = 0. \tag{5.17b}$$

These equations are totally independent, so they describe two separate kinds of fields (type "1" and type "2") that do not interact with each other.

Now, suppose the particles that represent excitations of these fields have the same mass: $m_1 = m_2 = m$. The mass term in the Lagrangian density then becomes simply $\tfrac{1}{2}m(\phi_1^2 + \phi_2^2)$. But this case leads us to a surprising observation that will be quite important in the future. Suppose we define two new fields ϕ_a and ϕ_b such that

$$\phi_a = \cos\theta\,\phi_1 + \sin\theta\,\phi_2 \tag{5.18a}$$

$$\phi_b = -\sin\theta\,\phi_1 + \cos\theta\,\phi_2, \tag{5.18b}$$

where θ is an arbitrary angle. Here, we are essentially "rotating" the field definitions in the two-dimensional space spanned by ϕ_1 and ϕ_2 without changing their magnitudes (see Figure 5.2). You can show (see Box 5.1) that this does nothing to the Lagrangian densities:

$$\mathcal{L} = \tfrac{1}{2}\partial^\mu\phi_1\,\partial_\mu\phi_1 + \tfrac{1}{2}\partial^\mu\phi_2\,\partial_\mu\phi_2 - \tfrac{1}{2}m^2(\phi_1^2 + \phi_2^2)$$

$$= \tfrac{1}{2}\partial^\mu\phi_a\,\partial_\mu\phi_a + \tfrac{1}{2}\partial^\mu\phi_b\,\partial_\mu\phi_b - \tfrac{1}{2}m^2(\phi_a^2 + \phi_b^2). \tag{5.19}$$

Remember, though, that the fields "*a*" and "*b*" are not the same as the fields "1" and "2" but instead are weighted combinations (the quantum mechanical term would be **superpositions**) of those field types. In this case, then, determining which fields and thus which particles we are talking about *depends on making an arbitrary mathematical choice*.

Choosing which fields to work with can sometimes be tricky. We will study a somewhat more complicated set of fields in Chapter 28 that describe neutrinos. In Chapter 1, we saw that there are three types of neutrinos in the Standard Model: an electron neutrino, a muon neutrino, and a tau neutrino. A weak interaction that converts an electron to a neutrino always produces an electron neutrino, an interaction that converts a muon to a neutrino always produces a muon neutrino, and so on. But it turns out that electron, muon, and tau neutrinos are superpositions of three different neutrino fields with (slightly) different masses m_1, m_2, and m_3. Now, we *usually* classify particles according to their mass, but if we did that in this case, converting an electron to a neutrino would sometimes produce a neutrino of type 1, type 2, or type 3. Is that easier or harder to

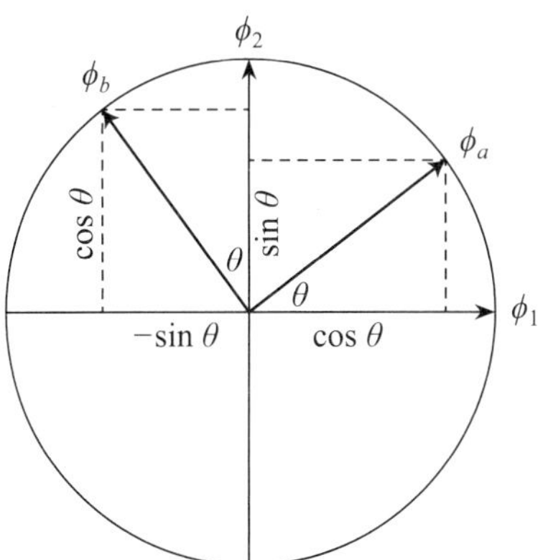

FIGURE 5.2   Rotating a pair of fields without changing the fields' magnitudes.

work with? Fortunately, we aren't forced to make any particular choice: we can "rotate" the three neutrino fields into whichever superpositions we find handy for calculation, just as we have rotated the fields above.

But returning to the case where the particle masses are identical, we can take advantage of our ability to freely "rotate" the fields to reexpress our two-field Lagrangian density in terms of a single *complex* scalar field, where

$$\phi = \left(\frac{\phi_1 + \phi_2}{2}\right) + i\left(\frac{\phi_1 - \phi_2}{2}\right). \tag{5.20}$$

In Box 5.2, you will show that Equation 5.16 reduces in this case to

$$\mathcal{L} = \partial^\mu \phi^* \, \partial_\mu \phi - m^2 \phi^* \phi. \tag{5.21}$$

Because there are really two particle fields involved here, we treat ϕ and ϕ^* as separate fields for the purposes of deriving the Euler-Lagrange equations. So,

$$0 = \partial_\mu \left(\frac{\partial \mathcal{L}}{\partial(\partial_\mu \phi^*)}\right) - \frac{\partial \mathcal{L}}{\partial \phi^*} = \partial_\mu \partial^\mu \phi + m^2 \phi \tag{5.22a}$$

$$0 = \partial_\mu \left(\frac{\partial \mathcal{L}}{\partial(\partial_\mu \phi)}\right) - \frac{\partial \mathcal{L}}{\partial \phi} = \partial_\mu \partial^\mu \phi^* + m^2 \phi^*. \tag{5.22b}$$

Evaluating the first derivative in Equation 5.22a may seem a bit tricky (because $\mathcal{L}$ depends on $\partial^\mu \phi^*$, not $\partial_\mu \phi^*$), but remember that $\partial^\alpha \phi^* \equiv g^{\alpha\beta} \partial_\beta \phi^*$, so

$$\frac{\partial \mathcal{L}}{\partial(\partial_\mu \phi^*)} = \frac{\partial}{\partial(\partial_\mu \phi^*)}\left(g^{\alpha\beta} \partial_\beta \phi^* \partial_\alpha \phi\right) = g^{\alpha\beta} \partial_\alpha \phi \, \delta^\mu_\beta = \partial^\mu \phi, \tag{5.23}$$

as claimed. Note that I have renamed indices to avoid index-symbol collisions.

We see that both fields evolve according to separate Klein-Gordon equations, as we would expect. But it is also true that the equations $0 = \partial_\mu \partial^\mu \phi + m^2 \phi$ (Equation 5.22a) and $0 = \partial_\mu \partial^\mu \phi^* + m^2 \phi^*$ (Equation 5.22b) are simply complex conjugates of each other, so really, we need only solve one equation or the other. This makes the process of working with the pair of fields much easier.

The general solution to either of these equations will be superpositions of plane-wave solutions of the form

$$\phi = A e^{\pm i k_\mu x^\mu} = A e^{\pm i p_\mu x^\mu} = A e^{\pm i p \cdot x}, \tag{5.24}$$

where A is a complex amplitude, and I have taken advantage of the fact that $\hbar k_\mu = p_\mu$ and we are working in units where $\hbar = 1$.

This situation provides a striking illustration of how complex numbers are useful in physics. We certainly could have continued to work with the complicated Lagrangian density given by Equation 5.16, with its two separate Euler-Lagrange equations. But using complex numbers combines the two fields into a single package, so we only have to deal with *one* Lagrangian density and one equation. We can even "rotate" the fields (if we want to) by simply multiplying the complex ϕ by $e^{i\theta}$ (see Box 5.4). We have not changed any of the physics involved, but we have made it easier to do calculations.

Similarly, in any case where complex numbers appear in classical physics or quantum physics, we *could* do all the calculations we need to using purely real numbers. But we would have to deal with twice as many numbers and many pairs of related equations. Complex numbers just provide a simpler and more natural language for doing certain kinds of calculations that seem to occur especially often in quantum mechanics.

In this particular situation, we have encapsulated two separate fields whose particles have the same mass into a nice package that we can represent with a single Lagrangian density and a single equation of motion. As we will see, successfully interpreting the Klein-Gordon equation's solutions will involve identifying these two distinct but connected fields as fields whose excitations are *antiparticles* of each other. (Note that antiparticles therefore have the same mass by construction!) Introducing complex fields into the Klein-Gordon equation is therefore a crucial step toward this successful interpretation.

5.4 The Dirac Delta

In this final section of the chapter, I want to discuss another mathematical tool that will become essential very soon. This tool is the **Dirac delta** (sometimes called the **Dirac delta function**, though it is technically not actually a function but rather what mathematicians call a **distribution**). We *define* the Dirac delta by its action in an integral when it multiplies another continuous function $f(x)$:

$$\int_{-\infty}^{+\infty} f(x)\, \delta(x - x_0)\, dx \equiv f(x_0).$$
(5.25)

By setting $f(x) = 1$, one can see that an immediate consequence of this definition is that

$$\int_{-\infty}^{+\infty} \delta(x - x_0)\, dx = 1,$$
(5.26)

meaning that the area under the Dirac delta "function" is 1.

The Dirac delta does for integrals what the Kronecker delta does for sums. In the sum of $\delta_\nu^\mu p^\nu$ over ν, the Kronecker delta picks out the component of the four-vector p with index μ. Similarly, in the integral of $f(x)\, \delta(x - x_0)$ over x, the Dirac delta picks out the value of $f(x)$ at x_0.

Part of the reason that the Dirac delta is not a function is that we can imagine many functional representations of $\delta(x - x_0)$ that satisfy the definition in Equation 5.25. Perhaps the easiest to visualize is a "function" that is zero everywhere except for (in the limit that $\Delta x \to 0$) a rectangular bar of width Δx and height $1/\Delta x$ centered at $x = x_0$ (see Figure 5.3). This "function" is zero except for in an infinitesimal range around x_0. So long as $f(x)$ is continuous in that range, its value at all points in that range will be approximately $f(x_0)$, and that approximation becomes exact in the limit. So, we can treat the value of $f(x)$ over the bar's infinitesimal width as being the constant value $f(x_0)$, and the integral over $f(x)\, \delta(x - x_0)$ becomes

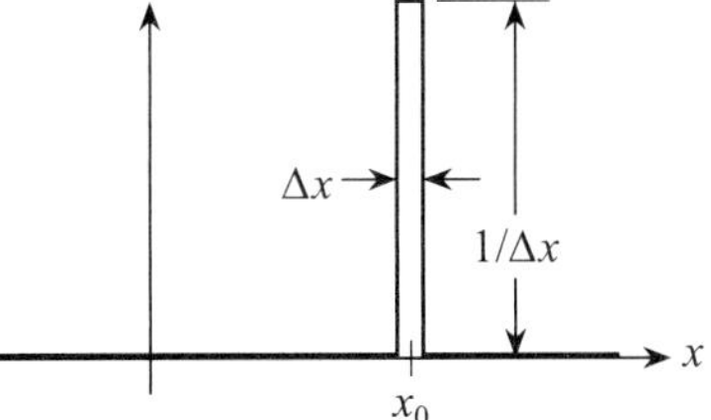

FIGURE 5.3   One can visualize the Dirac delta as being (in the limit that $\Delta x \to 0$) a thin rectangular spike of unit area centered at $x = x_0$.

$$\int_{-\infty}^{+\infty} f(x)\, \delta(x - x_0)\, dx = \lim_{\Delta x \to 0} \int_{x_0 - \frac{1}{2}\Delta x}^{x_0 + \frac{1}{2}\Delta x} f(x) \left(\frac{1}{\Delta x}\right) dx$$

$$= \lim_{\Delta x \to 0} \frac{f(x_0)}{\Delta x} \int_{x_0 - \frac{1}{2}\Delta x}^{x_0 + \frac{1}{2}\Delta x} dx = \lim_{\Delta x \to 0} f(x_0) \frac{\Delta x}{\Delta x} = f(x_0).$$
(5.27)

So, we see that this "thin bar" representation of the Dirac delta satisfies its definition. But a number of other representations work as well. We will see another important representation at the end of this section.

One can prove a number of useful properties of the Dirac delta straight from its definition (meaning that these properties are independent of our choice of representation). In Box 5.5, you will prove that (so long as the Dirac deltas in question appear in integrals)

$$\delta(a(x - x_0)) = \frac{1}{|a|}\delta(x - x_0)$$
(5.28a)

$$\delta(x_0 - x) = \delta(x - x_0)$$
(5.28b)

$$\delta(x^2 - x_0^2) = \frac{1}{2\,|x_0|}\Big[\delta(x - x_0) + \delta(x + x_0)\Big].$$
(5.28c)

These results actually all follow directly from the following very general identity:

when the function $g(x)$ is infinitely differentiable and has zeros at various points x_i, then (as you will also show in Box 5.5)

$$\delta[g(x)] = \sum_{\text{all } x_i} \frac{\delta(x - x_i)}{|dg/dx|_{x=x_i}}.$$

(5.29)

Finally (as you will see in the same box), we can prove the following identities regarding derivatives and integrals of the Dirac delta:

$$\int_{-\infty}^{+\infty} f(x) \left[\frac{d}{dx} \delta(x - x_0) \right] dx = - \left[\frac{df}{dx} \right]_{x=x_0}$$

(5.30a)

$$\int_{-\infty}^{x} \delta(x' - x_0) \, dx' = \begin{cases} 0 & x < x_0 \\ 1 & x > x_0 \end{cases} \equiv \Theta(x - x_0)$$

(5.30b)

$$\frac{d}{dx} \Theta(x - x_0) = \delta(x - x_0),$$

(5.30c)

where $\Theta(x - x_0)$ is called the **Heaviside step function**.

Note that we can construct higher-dimensional Dirac deltas out of products of one-dimensional Dirac deltas:

$$\int f(x, y, z) \, \delta(x - x_0) \, \delta(y - y_0) \, \delta(z - z_0) \, dx \, dy \, dz$$

$$\equiv \int f(\vec{r}) \, \delta^3(\vec{r} - \vec{r}_0) \, d^3x = f(\vec{r}_0)$$

(5.31a)

$$\int f(t, x, y, z) \, \delta(t - t_0) \, \delta(x - x_0) \, \delta(y - y_0) \, \delta(z - z_0) \, dt \, dx \, dy \, dz$$

$$\equiv \int f(x) \, \delta^4(x - x_0) \, d^4x = f(x_0),$$

(5.31b)

where the integrals are over all space or spacetime, respectively. In the last case, I am using the notational convention mentioned earlier where x and x_0 represent four-vectors (the clues that these are not scalars are the $\delta^4(x - x_0)$ and d^4x appearing in the integral).

Finally, in Box 5.5 you will prove an identity that will become *extremely* valuable to us as we go on:

$$\int_{-\infty}^{+\infty} e^{\pm i p(x - x_0)} \, dp = 2\pi \delta(x - x_0),$$

(5.32a)

or in an equivalent form that is equally useful,

$$\int_{-\infty}^{+\infty} e^{\pm i(p - p_0)x} \, dx = 2\pi \delta(p - p_0).$$

(5.32b)

Again, these equations have higher-dimensional versions. For example (again taking the integrals are over all space or spacetime),

$$\int e^{\pm i(\vec{p} - \vec{p}_0)\cdot\vec{r}} \, d^3x = (2\pi)^3 \delta^3(\vec{p} - \vec{p}_0)$$

(5.33a)

$$\int e^{\pm i(p - p_0)\cdot x} \, d^4x = (2\pi)^4 \delta^4(p - p_0),$$

(5.33b)

where in the last equation, p, p_0, and x in the exponent and in the final delta function are again four-vectors. These equations are important because they link the plane-wave solutions of the Klein-Gordon equation to the Dirac delta and provide a crucial tool for simplifying equations we will encounter later that involve superpositions of such solutions.

BOX 5.1 "Rotating" Fields in a Lagrangian Density

Our goal in this box is to show that when we take the Lagrangian density for two non-interacting fields with the same mass m (see Equation 5.16)

$$\mathcal{L} = \tfrac{1}{2}\partial^\mu \phi_1\, \partial_\mu \phi_1 - \tfrac{1}{2}m^2\phi_1^2 + \tfrac{1}{2}\partial^\mu \phi_2\, \partial_\mu \phi_2 - \tfrac{1}{2}m^2\phi_2^2 \qquad (5.34)$$

and "rotate" the fields according to Equations 5.18

$$\phi_a = \cos\theta\; \phi_1 + \sin\theta\; \phi_2 \qquad (5.35a)$$

$$\phi_b = -\sin\theta\; \phi_1 + \cos\theta\; \phi_2, \qquad (5.35b)$$

the Lagrangian density is unchanged except for the field labels:

$$\mathcal{L} = \tfrac{1}{2}\partial^\mu \phi_a\, \partial_\mu \phi_a - \tfrac{1}{2}m^2\phi_a^2 + \tfrac{1}{2}\partial^\mu \phi_b\, \partial_\mu \phi_b - \tfrac{1}{2}m^2\phi_b^2. \qquad (5.36)$$

We can see this for the terms involving the squares of the fields as follows:

$$
\begin{aligned}
\tfrac{1}{2}m\phi_a^2 + \tfrac{1}{2}m\phi_b^2 &= \tfrac{1}{2}m(\cos\theta\; \phi_1 + \sin\theta\; \phi_2)^2 + \tfrac{1}{2}m(-\sin\theta\; \phi_1 + \cos\theta\; \phi_2)^2 \\
&= \tfrac{1}{2}m\Big(\cos^2\theta\; \phi_1^2 + 2\cos\theta\sin\theta\; \phi_1\phi_2 + \sin^2\theta\; \phi_2^2 \\
&\qquad + \sin^2\theta\; \phi_1^2 - 2\sin\theta\cos\theta\; \phi_1\phi_2 + \cos^2\theta\; \phi_2^2 \Big) \\
&= \tfrac{1}{2}m\phi_1^2 + \tfrac{1}{2}m\phi_2^2. \qquad (5.37)
\end{aligned}
$$

Exercise 5.1.1 Show that the terms involving the field derivatives work out in a similar way. (*Hint:* Remember that $\partial^\mu = g^{\mu\nu}\partial_\nu$. This will be a bit more work, but the process is basically the same.)

BOX 5.2   The Lagrangian Density for a Complex Field

In this box, we will see that if we take the two-field Lagrangian density given in Equation 5.16

$$\mathcal{L} = \tfrac{1}{2}\partial^\mu \phi_1 \, \partial_\mu \phi_1 - \tfrac{1}{2}m^2\phi_1^2 + \tfrac{1}{2}\partial^\mu \phi_2 \, \partial_\mu \phi_2 - \tfrac{1}{2}m^2\phi_2^2 \tag{5.38}$$

and redefine the fields as stated in Equation 5.20

$$\phi = \left(\frac{\phi_1 + \phi_2}{2}\right) + i\left(\frac{\phi_1 - \phi_2}{2}\right) = \frac{1}{2}\Big[(1+i)\phi_1 + (1-i)\phi_2\Big], \tag{5.39}$$

we end up with the Lagrangian density given in Equation 5.21:

$$\mathcal{L} = \partial^\mu \phi^* \, \partial_\mu \phi - m^2 \phi^* \phi. \tag{5.40}$$

Exercise 5.2.1   Show this. (*Hint:* As in the last box, work backwards.)

BOX 5.3   The Euler Formula

The reason we describe $\cos\theta + i\sin\theta$ by the symbol $e^{i\theta}$ is that the function has most of the properties we would expect from an exponential function:

$$e^{i\theta_1}e^{i\theta_2} = e^{i(\theta_1+\theta_2)}, \tag{5.41a}$$

$$e^{i\theta}e^{-i\theta} = 1, \tag{5.41b}$$

$$\frac{d}{d\theta}e^{i\theta} = ie^{i\theta}, \qquad \int e^{i\theta}d\theta = \frac{e^{i\theta}}{i} \quad \text{where } \frac{1}{i} = -i. \tag{5.41c}$$

(Note that $e^{-i\theta} \equiv \cos(-\theta) + i\sin(-\theta) = \cos\theta - i\sin\theta$, since $\cos(-\theta) = \cos\theta$ and $\sin(-\theta) = -\sin\theta$. To see that $1/i = -i$, multiply top and bottom by i/i.) We see that we can treat $e^{i\theta}$ in most situations as if it were an ordinary exponential, so the notation serves as mnemonic for how the function behaves. However, $e^{i\theta}$ also has the following unique properties that you should *memorize*:

$$(e^{i\theta})^* = e^{-i\theta}, \tag{5.41d}$$

$$e^{\pm i\pi/2} = \pm i, \qquad e^{\pm i\pi} = -1, \tag{5.41e}$$

$$\tfrac{1}{2}(e^{i\theta} + e^{-i\theta}) = \cos\theta, \qquad \tfrac{1}{2}(e^{i\theta} - e^{-i\theta}) = i\sin\theta. \tag{5.41f}$$

Exercise 5.3.1 Prove that these equations all follow from Euler's formula $e^{i\theta} \equiv \cos\theta + i\sin\theta$. (*Hint:* For the first case, you will need the trigonometric identities for the addition of angles: $\cos(\theta_1 + \theta_2) = \cos\theta_1\cos\theta_2 - \sin\theta_1\sin\theta_2$ and $\sin(\theta_1 + \theta_2) = \sin\theta_1\cos\theta_2 + \cos\theta_1\sin\theta_2$. Note that Equation 5.41a is therefore mathematically equivalent to these identities, but is a lot easier to remember!)

BOX 5.4   Rotating the Complex Fields

In this box, we will see that multiplying our complex scalar field ϕ by $e^{i\theta}$ is equivalent to doing the rotation from real fields ϕ_1 and ϕ_2 to the fields ϕ_a and ϕ_b given by Equations 5.18, which I am repeating here for your convenience:

$$\phi_a = \cos\theta\, \phi_1 + \sin\theta\, \phi_2 \tag{5.42a}$$

$$\phi_b = -\sin\theta\, \phi_1 + \cos\theta\, \phi_2. \tag{5.42b}$$

Note also that Equation 5.20 defines our complex field as

$$\phi = \tfrac{1}{2}(\phi_1 + \phi_2) + \tfrac{1}{2}i(\phi_1 - \phi_2). \tag{5.43}$$

Exercise 5.4.1   Prove that

$$e^{i\theta}\phi = \tfrac{1}{2}(\phi_a + \phi_b) + \tfrac{1}{2}i(\phi_a - \phi_b), \tag{5.44}$$

exactly as if we had replaced ϕ_1 and ϕ_2 by the "rotated" fields ϕ_a and ϕ_b in Equation 5.20. This makes it a lot easier to "rotate" the two encapsulated fields if this is required for some reason.

BOX 5.5 Properties of the Dirac Delta

In this box, you will prove properties of the Dirac delta from its definition

$$\int_{-\infty}^{+\infty} f(x)\,\delta(x - x_0)\,dx \equiv f(x_0). \tag{5.45}$$

Exercise 5.5.1 Prove that (so long as we are in an integral),

$$\delta(a(x - x_0)) = \frac{1}{|a|}\delta(x - x_0). \tag{5.46}$$

(*Hint:* In the integral, change variables to $u \equiv ax$, and separately consider the cases where $a > 0$ and $a < 0$.) Note that $\delta(x_0 - x) = \delta(x - x_0)$ immediately follows from this (simply choose $a = -1$).

Exercise 5.5.2 Prove that (as long as we are in an integral)

$$\delta(x^2 - x_0^2) = \frac{1}{2\,|x_0|}\Big[\,\delta(x - x_0) + \delta(x + x_0)\,\Big]. \tag{5.47}$$

(*Hints:* Note that $\delta(x^2 - x_0^2) = \delta((x - x_0)(x + x_0))$, which spikes at both $x = x_0$ and $x = -x_0$, because the Dirac delta spikes whenever its argument is equal to zero (and is zero everywhere else). So, at the spike at $x = x_0$, for example, we have $\delta((x - x_0)(x + x_0)) = \delta((x - x_0)2x_0)$. Continue from there.)

BOX 5.5 Properties of the Dirac Delta (cont.)

Now consider the case where $g(x)$ has zeros at various points x_i.

Exercise 5.5.3 Prove that (so long as we are in an integral)

$$\delta[g(x)] = \sum_{\text{all } x_i} \frac{\delta(x - x_i)}{|dg/dx|_{x=x_i}}. \tag{5.48}$$

(*Hint:* Expand in a Taylor series around each point x_i where $g(x_i) = 0$, and argue why we can drop terms proportional to $(x - x_i)^n$ for $n \geq 2$. You will likely not need all of the space provided.)

BOX 5.5 Properties of the Dirac Delta (cont.)

Exercise 5.5.4 Show that

$$\int_{-\infty}^{+\infty} f(x) \left(\frac{d}{dx} \delta(x - x_0) \right) dx = - \left[\frac{df}{dx} \right]_{x=x_0} . \qquad (5.49)$$

(*Hint:* Integrate by parts.)

Exercise 5.5.5 Argue that

$$\int_{-\infty}^{x} \delta(x' - x_0)\, dx' = \begin{cases} 0 & x < x_0 \\ 1 & x > x_0 \end{cases} \equiv \Theta(x - x_0), \qquad (5.50)$$

$$\text{and that} \quad \frac{d\Theta}{dx} = \delta(x - x_0). \qquad (5.51)$$

BOX 5.5 Properties of the Dirac Delta (cont.)

Exercise 5.5.6 Show that

$$\frac{\sin(ax)}{x} = \frac{1}{2} \int_{-a}^{a} e^{ikx}\, dk.$$

$$(5.52)$$

(*Hint:* Just do the integral.)

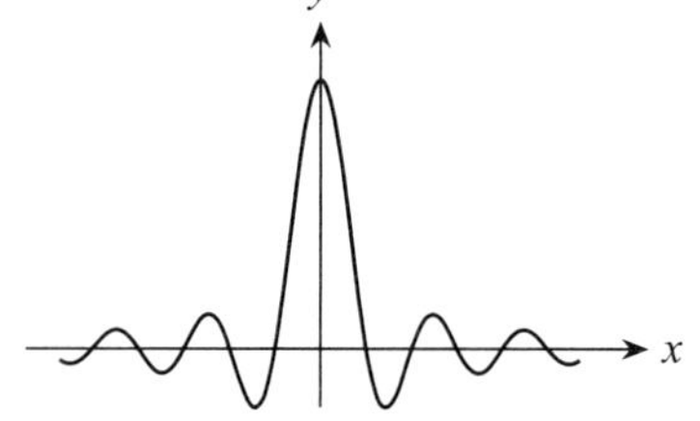

FIGURE 5.4 A graph of the function $\sin(ax)/x$.

A plot of the function $\sin(ax)/x$ appears in Figure 5.4. The width of this function's central peak is about $2\pi/a$, because the nearest values to $x = 0$ where the sine function is zero are at $ax = \pm\pi$. The height of the central peak is equal to a. So, as $a \to \infty$, the height of the central peak will grow and its width will narrow, and the curve will average to zero outside of the central peak even if $1/x$ is not all that small. Therefore, this function can be used as a representation of the Dirac delta in that limit.

Exercise 5.5.7 Use WolframAlpha or some other mathematical tool to show that the integral of $\sin(ax)/x$ over all x is π, independent of a (assuming a is positive). Then use the above and Equation 5.52 to argue that

$$\delta(x - x_0) = \frac{1}{2\pi} \int_{-\infty}^{\infty} e^{ip(x-x_0)}\, dp,$$

$$(5.53)$$

as claimed in Equation 5.32a.

Homework Problems

P5.1 Use the relation $e^{i(\theta_1 - \theta_2)} = e^{i\theta_1} e^{-i\theta_2}$ to derive the difference-of-angles trigonometric identities for $\cos(\theta_1 - \theta_2)$ and $\sin(\theta_1 - \theta_2)$.

P5.2 Given $c \equiv a + ib$, write $1/c$ in terms of a and b, and verify that your expression satisfies $c(1/c) = 1$. (*Hint:* Multiply $1/c$ by $1 = c^*/c^*$.)

P5.3 Consider complex numbers $c_1 = 2 - i$ and $c_2 = 1 + i$.
a. Calculate c_1/c_2 and c_2/c_1 in the form $a + ib$. Are they inverses? (*Hint:* See the hint to the previous problem.)
b. Calculate $|c_1/c_2|$ and $|c_2/c_1|$. Are they inverses?
c. Write c_1 in the form $Re^{i\theta}$.

P5.4 Consider a plane-wave solution $\phi = Ae^{-ik \cdot x}$ to the Klein-Gordon equation, where $k \cdot x \equiv k_\mu x^\mu \equiv g_{\nu\mu} k^\nu x^\mu = \omega t - k^x x - k^y y - k^z z$, k^μ is a four-vector whose components satisfy the constraints imposed by the Klein-Gordon equation, x^μ is the position four-vector, and A is a constant.
a. Calculate the wave's phase speed $v_p \equiv \omega/|\vec{k}|$. (This is the speed at which a crest of the wave moves.) Verify that this is faster than light.
b. Calculate the wave's group speed $v_g \equiv d\omega/d|\vec{k}|$, which is the speed at which wave *modulations* travel (and thus the speed that information travels). Verify that this is slower than light.

P5.5 Consider this Lagrangian density for a real field ϕ:

$$\mathcal{L} = \tfrac{1}{2}\partial^\mu \phi\, \partial_\mu \phi + a + b\phi - \tfrac{1}{2}m^2\phi^2, \qquad (5.54)$$

where m, a, and b are constants. In this problem, we will show that adding the a and b terms do not make the Euler-Lagrange equation for this Lagrangian density materially different from the Klein-Gordon equation. Terms that affect the physics of the Euler-Lagrange equation must therefore be at least quadratic in the field variable ϕ.
a. Argue that the a term has no effect on the Euler-Lagrange equation corresponding to this Lagrangian density.
b. Argue that if we redefine the field $\phi' = \phi + d$, where d is an appropriately chosen constant, we can make the b term disappear in the Euler-Lagrange equation for this Lagrangian density (that is, the Euler-Lagrange equation for ϕ' is just the Klein-Gordon equation).

P5.6 Argue that another possible representation of the Dirac delta is

$$\delta(x) = \lim_{\varepsilon \to 0} \frac{1}{\pi} \frac{\varepsilon}{x^2 + \varepsilon^2} \qquad (5.55)$$

by proving each of the following statements.
a. The width of the spike (defined, say, by the region where the function value is at least $\tfrac{1}{10}$ of its maximum value) goes to zero as $\varepsilon \to 0$.
b. The value of the function on the right in the stated limit is zero, unless $x = 0$.
c. The area under the curve of the spike is 1.
d. Therefore, the integral of this function times any continuous function $f(x)$ yields $f(0)$. (*Hint:* Since the width of the region where this function is nonzero is essentially zero in the limit where $\varepsilon \to 0$, we can replace the function by the constant value $f(0)$ in that infinitesimal region without introducing any errors.)

6. NOETHER'S THEOREM

The simplest and most powerful laws of physics are **conservation laws**. Quantities like energy, momentum, angular momentum, and electrical charge are useful concepts because their total value in a closed system remains *fixed*. Such laws are simple to state, easy to apply, and yet powerfully rich in implications.

Historically, physicists developed conservation laws mostly by either finding them to be a mathematical consequence of laws considered to be more fundamental (for example, deriving conservation of momentum from Newton's third law), by empirical observation (for example, conservation of charge), or through a combination of both processes (for example, conservation of energy). Though some physicists (notably Laplace and Hamilton) had glimmers of insight about how conservation laws might be connected to symmetries, Amalie "Emmy" Noether (1882–1935; see image opposite, taken c. 1910) was the first to fully understand this connection. In 1918, she provided the community with this crucial insight by proving the theorem that now bears her name. Noether's theorem showed that *each* (seemingly arbitrary) *conserved quantity is connected to a certain fundamental symmetry of nature*. For example, conservation of energy, momentum, and angular momentum are consequences of the fact that the laws of physics are independent of time, position, and orientation, respectively. More crucially, Noether also showed how one could *deduce* the conservation law that a given symmetry implies.

Because the Standard Model is fundamentally based on exploring the implications of symmetries, and those implications (in the context of quantum fields) are sufficiently obscure, we really need Noether's theorem to light our path. In this chapter, therefore, we will explore in detail how to use the theorem to derive the conserved quantity connected with a given symmetry.

6.1 Noether's Theorem Applied to Classical Particles

Consider a classical system described by a set of coordinates q_j and a Lagrangian $L(q_j, \dot{q}_j)$ (where j ranges over the number of coordinates needed to describe the system). Suppose we transform the coordinates according to some rule $q'_j = f(q_j, \epsilon)$ described by a single continuous parameter ϵ such that $\epsilon = 0$ corresponds to *no* transformation: $f(q_j, 0) = q_j$. According to the chain rule, the effect of this transformation on the value of L, evaluated at $\epsilon = 0$, is

$$\left[\frac{dL}{d\epsilon}\right]_0 = \sum_j \left[\frac{\partial L}{\partial q'_j}\frac{dq'_j}{d\epsilon} + \frac{\partial L}{\partial \dot{q}'_j}\frac{d\dot{q}'_j}{d\epsilon}\right]_0 = \sum_j \left[\frac{\partial L}{\partial q_j}\left[\frac{dq'_j}{d\epsilon}\right]_0 + \frac{\partial L}{\partial \dot{q}_j}\left[\frac{d\dot{q}'_j}{d\epsilon}\right]_0\right], \quad (6.1)$$

because when we evaluate the partial derivatives at $\epsilon = 0$, $q'_j = f(q_j, 0)$ is the same as q_j (I have kept the prime in the derivative because only q'_j depends on ϵ). Now, the Euler-Lagrange equation for each coordinate asserts that

$$0 = \frac{d}{dt}\left(\frac{\partial L}{\partial \dot{q}_j}\right) - \frac{\partial L}{\partial q_j} \quad \Rightarrow \quad \frac{d}{dt}\left(\frac{\partial L}{\partial \dot{q}_j}\right) = \frac{\partial L}{\partial q_j}. \quad (6.2)$$

So, if we require that the unmodified coordinates $q_j(t)$ solve these equations of motion, then we can substitute the above into Equation 6.1 to get

$$\left[\frac{dL}{d\epsilon}\right]_0 = \sum_j \left[\frac{d}{dt}\left(\frac{\partial L}{\partial \dot{q}_j}\right)\left[\frac{dq'_j}{d\epsilon}\right]_0 + \frac{\partial L}{\partial \dot{q}_j}\left[\frac{d\dot{q}'_j}{d\epsilon}\right]_0\right] = \sum_j \frac{d}{dt}\left[\frac{\partial L}{\partial \dot{q}_j}\frac{dq'_j}{d\epsilon}\bigg|_0\right] \quad (6.3)$$

(as you can check by applying the product rule to the quantity on the right, and noting that the order in which we apply derivatives with respect to t and ϵ does not matter). Pulling the time derivative out of the sum yields

$$\frac{d}{dt}\left[\sum_j \frac{\partial L}{\partial \dot{q}_j}\left[\frac{dq'_j}{d\epsilon}\right]_0\right] = \left[\frac{dL}{d\epsilon}\right]_0. \quad (6.4)$$

Now, if the Lagrangian L is *unchanged* by the transformation $q'_j = f(q_j, \epsilon)$ (which is what we usually mean when we say that the system's physics is "symmetric" under that transformation), then $dL/d\epsilon = 0$ and

$$\sum_i \frac{\partial L}{\partial \dot{q}_i}\left[\frac{dq'_i}{d\epsilon}\right]_0 = \text{constant}. \quad (6.5)$$

We therefore see that *such a symmetry has an associated conserved quantity* given by Equation 6.5. Again, we evaluate the derivatives at $\epsilon = 0$ so that our quantity is conserved for the *unmodified* coordinate trajectories $q_j(t)$ that solve the equation of motion and so describe how our system actually evolves.

Let's see how this works in a specific situation. Consider a particle of mass m moving in a uniform gravitational field of magnitude g. Assume the horizontal and vertical coordinates are x and z. The Lagrangian for this situation is

$$L = \tfrac{1}{2}m\dot{x}^2 + \tfrac{1}{2}m\dot{z}^2 - mgz. \quad (6.6)$$

Now consider a horizontal translation $x' = x + \epsilon$. Note that $\dot{x}' = \dot{x}$, and the undotted coordinate x does not appear in the Lagrangian, so $dL/d\epsilon = 0$. Therefore, we have a conserved quantity

$$\frac{\partial L}{\partial \dot{x}}\left[\frac{dx'}{d\epsilon}\right]_0 = (m\dot{x})(1) = m\dot{x}, \quad (6.7)$$

since $dx'/d\epsilon = 1$ here, and we have no other terms in the sum in Equation 6.5 because $dz/d\epsilon = 0$. We recognize the conserved quantity here as being the x component of momentum, which is indeed conserved for an ideal projectile.

In Box 6.1, you will in a similar way prove that conservation of angular momentum follows from rotational invariance.

Now consider a time translation $t' = t + \epsilon$. In this case, there is a complication. While a Lagrangian's *equation* rarely depends on time explicitly, its *value* usually changes as the particle it describes moves, so we cannot assume that $dL/dt = 0$ and thus that $(dL/d\epsilon)_0 = (dL/dt)(dt'/d\epsilon)_0 = (dL/dt)(1) = 0$. But in this very special case, Equation 6.4 implies that

$$\frac{d}{dt}\left[\sum_j \frac{\partial L}{\partial \dot{q}_j}\left[\frac{dq'_j}{dt'}\frac{dt'}{d\epsilon}\right]_0\right] = \left[\frac{dL}{d\epsilon}\right]_0 = \frac{dL}{dt} \Rightarrow \frac{d}{dt}\left[\sum_j \frac{\partial L}{\partial \dot{q}_j}\dot{q}_j - L\right] = 0. \quad (6.8)$$

You may recognize the quantity in the square brackets as being a classical system's **Hamiltonian**, which corresponds to its conserved energy.

So, we see that Noether's theorem tells us that the three great classical conservation laws (momentum, angular momentum, and energy) do indeed correspond to three fundamental symmetries: invariance under translations in space, rotation about an axis, and translations in time, respectively.

6.2 Currents

Classical conservation laws like those we just described are simple and obvious: the value of some defined quantity does not change with time. But when we talk about relativistic fields, things become more complicated. First, note that talking about a quantity being conserved in *time* is going to be problematic in any relativistic context: we need to treat time and space evenhandedly. Secondly, fields are distributed in time and space and (as they evolve) are flowing from place to place. We need to think more in terms of analogies from fluid flow rather than the way we think in simple particle mechanics.

The trick to expressing conservation laws in a fluid-like context is to work with **currents**. Suppose that a specific volume of a fluid contains a certain amount of some quantity Q that is a relativistic scalar (that is, all observers agree about the value of Q for the fluid in a given box, such as the number of fluid particles the box contains or the total charge of those particles). As you will see in Box 6.2, the density ρ of such a quantity transforms like the time component of a four-vector we can call J^μ, whose spatial components in a given frame are $\vec{J} = \rho\vec{v}$, where $\vec{v}$ is the ordinary local fluid velocity measured in that frame. In introductory physics, we in fact saw that the **charge density** ρ and **current density** $\vec{J} = \rho\vec{v}$ comprised the sources of the electromagnetic field in Maxwell's equations, so (though you did not know it) you have already worked with the current four-vector J^μ corresponding to electric charge Q.

The Lorentz-invariant way to express conservation of the quantity Q in terms of its current J^μ is the simple equation

$$0 = \partial_\mu J^\mu = \frac{\partial \rho}{\partial t} + \frac{\partial J^x}{\partial x} + \frac{\partial J^y}{\partial y} + \frac{\partial J^z}{\partial z} = \frac{\partial \rho}{\partial t} + \vec{\nabla} \cdot \vec{J}. \tag{6.9}$$

This nice equation is obviously Lorentz-invariant, but how do we know that it expresses conservation of Q? The quickest path to see this is to use the divergence theorem. Suppose that in a given reference frame, we integrate Equation 6.9 over a certain volume of space:

$$0 = \int \left(\frac{\partial \rho}{\partial t} + \vec{\nabla} \cdot \vec{J} \right) d^3x \ \Rightarrow \ \int \frac{\partial \rho}{\partial t} d^3x = -\int \vec{\nabla} \cdot \vec{J} \, d^3x. \tag{6.10}$$

But the left integral in the last equation is the same as $(d/dt) \int \rho \, d^3x = dQ/dt$, the rate at which the quantity Q in the volume is changing with time. By the divergence theorem of vector calculus, the right side of the equation is an integral of $\vec{J} \cdot d\vec{A}$ over the volume's surface at a given instant of time, so

$$\frac{dQ}{dt} = -\oint \vec{J} \cdot d\vec{A}. \tag{6.11}$$

Thus, the rate at which Q in the volume decreases is equal to the rate at which Q flows through the surface (see Box 6.3). This is only true if Q is conserved.

What if the quantity in the box is not a frame-independent scalar Q, but a four-vector P^μ? In such a case, the corresponding current has four components for each component of P^μ, so it must be a 16-component tensor $T^{\nu\mu}$ satisfying

$$0 = \partial_\nu T^{\nu\mu}. \tag{6.12}$$

The bottom line is that in the context of field theory, we will be looking for conservation laws in the form of $0 = \partial_\mu J^\mu$ or $0 = \partial_\nu T^{\nu\mu}$.

6.3 Noether's Theorem for Fields

Now we are ready to explore what Noether's theorem tells us about symmetries involving fields. Suppose we have a system involving an intertwined set of fields ϕ_n (where n is an index ranging over the number of fields). Following a similar kind of argument, we consider a transformation $\phi'_n = f_n(\phi_n, \epsilon)$ such that ϵ is a continuous parameter and $f_n(\phi_n, 0) = \phi_n$. If we follow steps analogous to those in Section 6.1 (see Box 6.4), then we find that

$$\frac{d\mathcal{L}}{d\epsilon}\bigg|_0 = \partial_\mu \left[\sum_n \frac{\partial \mathcal{L}}{\partial(\partial_\mu \phi_n)} \frac{d\phi'_n}{d\epsilon}\bigg|_0 \right]. \tag{6.13}$$

If the transformation does not affect the *value* of $\mathcal{L}$, then $d\mathcal{L}/d\epsilon = 0$ and the quantity in brackets is a current J^μ that obeys the conservation law $\partial_\mu J^\mu = 0$.

If the transformation involves the coordinates x^μ, however, it will almost certainly affect the value of $\mathcal{L}$, because at its most basic level, $\mathcal{L}$ is simply a scalar function of space and time. In such a case, we have (by the chain rule)

$$\left[\frac{d\mathcal{L}}{d\epsilon}\right]_0 = \frac{\partial \mathcal{L}}{\partial x^\mu}\left[\frac{dx'^\mu}{d\epsilon}\right]_0 = (\partial_\mu \mathcal{L})\frac{dx'^\mu}{d\epsilon}\bigg|_0 = \partial_\mu\left(\mathcal{L}\left[\frac{dx'^\mu}{d\epsilon}\right]_0\right) - \mathcal{L}\left[\partial_\mu\left(\frac{dx'^\mu}{d\epsilon}\right)\right]_0. \tag{6.14}$$

where the last step follows from the product rule. If $\partial_\mu(dx^\mu/d\epsilon) = 0$, as is very often the case for transformations of interest, then the last term is zero and Equation 6.13 becomes

$$0 = \partial_\mu\left[\sum_n \frac{\partial \mathcal{L}}{\partial(\partial_\mu \phi_n)}\left[\frac{d\phi'_n}{d\epsilon}\right]_0 - \mathcal{L}\left[\frac{dx'^\mu}{d\epsilon}\right]_0\right]. \tag{6.15}$$

So, the conservation law $\partial_\mu J^\mu = 0$ applies to the current

$$J^\mu \equiv \left[\sum_n \frac{\partial \mathcal{L}}{\partial(\partial_\mu \phi_n)}\left[\frac{d\phi'_n}{d\epsilon}\right]_0\right] - \mathcal{L}\left[\frac{dx^\mu}{d\epsilon}\right]_0 \quad \text{when } \partial_\mu\left(\frac{dx^\mu}{d\epsilon}\right) = 0. \tag{6.16}$$

This proves (for our purposes) to be a sufficiently general expression for finding conservation laws from symmetry principles involving fields.

6.4 The Stress-Energy Tensor

As an example, consider a coordinate transformation involving an arbitrary displacement in spacetime:

$$x'^\nu = x^\nu + \epsilon^\nu, \tag{6.17}$$

where x'^ν is the transformed coordinate (which we should think of as being a function of ϵ^ν), x^ν is the original coordinate, and ϵ^ν compactly represents four independent transformation parameters (representing independent displacements in the t, x, y, and z directions). This transformation in no way affects the *form* of the Lagrangian densities we have seen so far, because none of them depend directly on the coordinates, but rather only indirectly through the fields. However, since this is a coordinate transformation, it will affect the *value* of $\mathcal{L}$. Note that $dx^\mu/d\epsilon^\nu = \delta^\mu_\nu$, so

$$\partial_\mu \frac{dx^\mu}{d\epsilon^\nu} = \partial_\mu \delta^\mu_\nu = 0, \tag{6.18}$$

meaning this transformation satisfies the requirement in Equation 6.16. We can determine the effect of this transformation on the fields using the chain rule:

$$\left[\frac{d\phi'_n}{d\epsilon^\nu}\right]_0 = \frac{\partial \phi_n}{\partial x^\mu}\left[\frac{dx'^\mu}{d\epsilon^\nu}\right]_0 = \frac{\partial \phi_n}{\partial x^\mu}\delta^\mu_\nu = \frac{\partial \phi_n}{\partial x^\nu} = \partial_\nu \phi_n. \tag{6.19}$$

Substituting these results into Equation 6.16 yields the following definition of a conserved current:

$$T^\mu_\nu = \left[\sum_n \frac{\partial \mathcal{L}}{\partial(\partial_\mu \phi_n)}\partial_\nu \phi_n\right] - \delta^\mu_\nu \mathcal{L}. \tag{6.20}$$

This current's components are easiest to interpret if we raise the ν index:

$$T^{\mu\nu} = \left[\sum_n \frac{\partial\mathcal{L}}{\partial(\partial_\mu\phi_n)}\partial^\nu\phi_n\right] - g^{\mu\nu}\mathcal{L}, \tag{6.21}$$

which obeys the conservation law $0 = \partial_\mu T^{\mu\nu}$. This is actually the current for *four* conserved quantities (one for each value of ν) that comprise a conserved four-vector we can call P^ν. Since ϵ^t represents a translation in time, we (by analogy to the particle case) identify the corresponding conserved quantity P^t as the field's *energy*, and the conserved quantities P^i corresponding to the spatial translations ϵ^i as the field's *momentum*. We call $T^{\mu\nu}$ the field's **stress-energy tensor**, and interpret its components as follows:

- T^{tt} = density of P^t = the field's energy density
- T^{ti} = density of P^i = the field's i-momentum density
- T^{it} = flow rate of energy in the ith direction
- T^{ij} = flow rate of i-momentum in the jth direction.

Each "flow rate" above is technically a **flux density** (that is, the amount of whatever quantity per unit area per unit time is flowing) through an infinitesimal area perpendicular to whichever direction (see Box 6.3). Remember that *force* is defined as the rate at which spatial momentum flows, so the stress-energy's spatial components T^{ij} express the ith component of the force per unit area acting on a plane perpendicular to the jth direction. The diagonal components T^{ii} therefore correspond to *pressure*; that is, the component of force per unit area that acts perpendicular to a given surface. The other components are shearing forces that have components parallel to the surface of interest, such as the forces your feet exert on the ground when you start to run. In classical mechanics, the spatial components T^{ij} comprise what people call the "stress tensor," which is why people call $T^{\mu\nu}$ the "stress-energy tensor." But we will primarily focus on the $T^{t\mu}$ components of this tensor.

For the following Lagrangian density for a real scalar field

$$\mathcal{L} = \tfrac{1}{2}\partial^\mu\phi\partial_\mu\phi - \tfrac{1}{2}m^2\phi^2, \tag{6.22}$$

the field's energy density is

$$\begin{aligned}
\rho = T^{tt} &= \frac{\partial\mathcal{L}}{\partial(\partial_t\phi)}\partial^t\phi - g^{tt}\mathcal{L} = (\partial^t\phi)\partial^t\phi - \mathcal{L} \\
&= (\partial_t\phi)^2 - \tfrac{1}{2}\left[(\partial_t\phi)^2 - (\partial_x\phi)^2 - (\partial_y\phi)^2 - (\partial_z\phi)^2 - m^2\phi^2\right] \\
&= \tfrac{1}{2}\left[\dot\phi^2 + (\vec\nabla\phi)\cdot(\vec\nabla\phi) + m^2\phi^2\right].
\end{aligned} \tag{6.23}$$

[Remember that $\partial\mathcal{L}/\partial(\partial_\mu\phi) = \partial^\mu\phi$ when $\mathcal{L} = \tfrac{1}{2}\partial^\mu\phi\partial_\mu\phi + f(\phi)$.] I have abandoned tensor notation in the last expression because energy density is not a frame-independent scalar, but rather (because $\rho = T^{tt}$) transforms as the tt component of a second-rank tensor.

Similarly, the momentum density component π^i in the ith direction is

$$\pi^i = T^{ti} = \frac{\partial\mathcal{L}}{\partial(\partial_t\phi)}\partial^i\phi - g^{ti}\mathcal{L} = (\partial^t\phi)\partial^i\phi + 0 = \dot\phi\partial^i\phi. \tag{6.24}$$

In Box 6.5, you will show that the symmetry of the Lagrangian density in Equation 6.22 with respect to rotations around the z axis yields a current for a conserved quantity whose density is

$$x(\dot\phi\partial^y\phi) - y(\dot\phi\,\partial^x\phi) = x\pi^y - y\pi^x = (\vec r \times \vec\pi)_z. \tag{6.25}$$

This is plausibly the density of the z component of angular momentum.

6.5 Conservation Laws for a Complex Field

For the following Lagrangian density for a complex scalar field

$$\mathcal{L} = \partial^\mu \phi^* \, \partial_\mu \phi - m^2 \phi^* \phi, \tag{6.26}$$

the energy density and i-momentum density are

$$\rho = T^{tt} = \dot{\phi}^* \dot{\phi} + (\vec{\nabla}\phi^*) \cdot (\vec{\nabla}\phi) + m^2 \phi^* \phi \tag{6.27a}$$

$$\text{and} \quad \pi^i = \dot{\phi} \, \partial^i \phi^* + \dot{\phi}^* \partial^i \phi \tag{6.27b}$$

(see Box 6.6). These results are plausible modifications of the results in Equations 6.23 and 6.24, especially when one considers that energy density and momentum density must be real numbers.

So far, we have considered symmetries with regard to coordinate transformations, but many of the most interesting symmetries in the Standard Model are what we call **internal symmetries**: transformations of the *fields* that leave the Lagrangian density unchanged.

A very important example is the following. The Lagrangian density in Equation 6.26 is unchanged (in both form and value) by the field transformation

$$\phi' = \phi \, e^{-iq\theta}, \tag{6.28}$$

where θ is a parameter (playing the role of the transformation parameter ϵ considered earlier) and q is an arbitrary real constant. As you will show in Box 6.7, this symmetry implies a conserved current

$$J^\mu = -iq(\phi \, \partial^\mu \phi^* - \phi^* \partial^\mu \phi). \tag{6.29}$$

What is the conserved quantity corresponding to this current? In Chapter 11, we will interpret solutions to the Klein-Gordon equation in a way that suggests the conserved quantity might be *electrical charge*. This will be confirmed in Chapter 19, when we will derive all of electromagnetism from this symmetry! This is your first glimpse into what is one of the most amazing results from the Standard Model: *electromagnetism is a consequence of the phase-invariance of complex fields.*

BOX 6.1   Conservation of Angular Momentum

In this box, we will consider the Lagrangian for an object of mass m orbiting a very massive object of mass M in two dimensions. Because we are using a Lagrangian, there is no cost to using polar coordinates, so we will assume two coordinates, $q_1 \equiv r$ and $q_2 \equiv \theta$. The system's Lagrangian in these coordinates is

$$L = \tfrac{1}{2}m\dot{r}^2 + \tfrac{1}{2}m(r\dot{\theta})^2 + \frac{GMm}{r}. \tag{6.30}$$

Equation 6.5 tells us that for a transformation involving a single parameter ϵ that does not affect the value of L,

$$\sum_i \frac{\partial L}{\partial \dot{q}_i}\left[\frac{dq_i'}{d\epsilon}\right]_0 = \text{constant}. \tag{6.31}$$

Exercise 6.1.1   Consider the rotation transformation $\theta' = \theta + \epsilon$. This does not affect the Lagrangian because the Lagrangian does not depend explicitly on θ. Calculate the corresponding conserved quantity and argue that it is the object's orbital angular momentum.

BOX 6.2   The Four-Current J^μ

In this box, we will argue that the *density* of a Lorentz-invariant quantity Q transforms like the time component of a four-vector. Consider a classical fluid of particles we can count, and suppose that in a certain reference frame the average velocity of those particles is zero: this will be the fluid's rest frame. In this frame, we define a tiny box of volume V centered on a given point. The number density of the particles in that box is therefore $n = N/V$, where N is the number of particles in the box. Now, N is an example of a Lorentz scalar since every observer will agree on how many particles are in the box. If each particle has a Lorentz-invariant property like charge q, this means that the total charge $Q = Nq$ in the box is also Lorentz-invariant.

However, neither the number density n nor the charge density $\rho = Q/V$ is a Lorentz scalar, because V is not. In an inertial frame (the primed frame) moving with x-velocity β relative to the fluid rest frame, the box will be Lorentz-contracted in the x direction by the factor $\sqrt{1 - \beta^2}$, so the box's volume in the new frame is smaller by that factor: $V' = V\sqrt{1 - \beta^2}$. This means that the number density in the new frame is

$$n' = \frac{N}{V'} = \frac{N}{V\sqrt{1 - \beta^2}} = \gamma n \;\Rightarrow\; \rho' = qn' = \gamma qn = \gamma \rho. \qquad (6.32)$$

Now, suppose that we define a four-vector J^μ whose four components are $[\rho,\ \rho v_x,\ \rho v_y,\ \rho v_z]$, where $\vec{v}$ is the average fluid velocity at the point in question. In the fluid's rest frame, $\vec{v} = 0$.

Exercise 6.2.1   Use the Lorentz transformation equations to show that in the primed frame, $J'^\mu = [\rho',\ \rho' v'_x,\ \rho' v'_y,\ \rho' v'_z]$. This shows that the density is the time component of the current four-vector so defined.

BOX 6.3 Spatial Current Components Are Flux Densities

In this box, we will show that $\vec{J} \cdot d\vec{A}$ for a current $J^\mu = (\rho, \vec{J}) = (\rho, \rho\vec{v})$ (where ρ is the density of the associated conserved quantity Q) describes the **flux** (amount per time) of Q flowing through the infinitesimal surface described by the area element $d\vec{A}$. This also implies that if the surface is perpendicular to an axis (say, the x^i axis), so that $d\vec{A}$ is parallel to that axis, then $J^i = \vec{J} \cdot d\vec{A}/dA$ is the **flux density** (flux per unit area) flowing through such a surface.

We saw in the Section 6.2 that the current's spatial components had the form $\rho\vec{v}$, where ρ was the density of the associated conserved quantity. If that quantity is a Lorentz scalar (such as charge), then the corresponding current is a four-vector J^μ and the density is that current's time component J^t. If the quantity is a component of a four-vector (such as four-momentum), then the current is a second-rank tensor $T^{\mu\nu}$ and the density is one of the four components $T^{t\mu}$.

In this box, let us use ρ to stand for either J^t or $T^{t\mu}$ and Q to stand for the associated conserved quantity, whether it is a charge or a component of the four-momentum. (It won't matter in the argument that follows.)

Now consider the diagram shown in Figure 6.1. This figure shows a tiny surface of area dA represented by the vector $d\vec{A}$. If the fluid particles have an average velocity component $v_\perp \equiv \vec{v} \cdot (d\vec{A}/dA)$ perpendicular to the surface, then (on average) all the fluid particles in the box shown in the diagram will move through that surface during the tiny time interval dt (any extra particles that get through because they are moving faster than average will be canceled by those that don't make it because they are slower than average). If the box is small enough so that ρ and $v_\perp$ are approximately the same in all parts of the box, then the edges of the box will be well-defined and the total amount of our conserved quantity Q that flows through the surface in time dt will be equal to the density ρ of that quantity times the box's volume $dA\, v_\perp dt$.

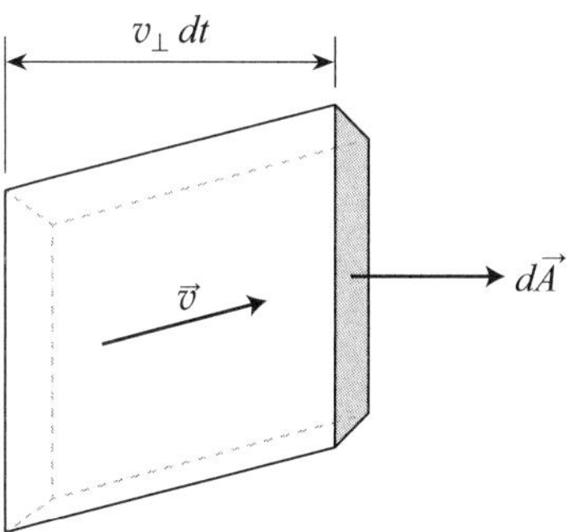

FIGURE 6.1 All of the fluid in the box will flow through the shaded surface during the time interval dt. (For simplicity, the drawing assumes that the fluid moves in the plane of the figure.)

Exercise 6.3.1 Show that $\vec{J} \cdot d\vec{A}$ is indeed the amount of Q flowing per unit time $= dQ/dt$ through the surface, and J^i is the flux per unit area through a surface perpendicular to the x^i axis. (*Hint:* Use the equations above.)

BOX 6.4   Noether Currents for a Classical Field

In this box, you will repeat the argument outlined in Equations 6.1 to 6.4 but for a Lagrangian density $\mathcal{L}$ involving fields ϕ_n instead of a Lagrangian L involving coordinates q_i. Remember that the Euler-Lagrange equations for fields are

$$0 = \partial_\mu \left(\frac{\partial \mathcal{L}}{\partial(\partial_\mu \phi_n)} \right) - \frac{\partial \mathcal{L}}{\partial \phi_n}. \tag{6.33}$$

We want to end up with Equation 6.13, repeated here:

$$\left[\frac{d\mathcal{L}}{d\epsilon} \right]_0 = \partial_\mu \left[\sum_n \frac{\partial \mathcal{L}}{\partial(\partial_\mu \phi_n)} \left[\frac{d\phi'_n}{d\epsilon} \right]_0 \right]. \tag{6.34}$$

which is for fields what Equation 6.4 was for coordinates.

Exercise 6.4.1   Write out the steps for fields that are analogous to Equations 6.1–6.4. (*Hint:* Remember that time derivatives in the coordinate case become ∂_μ in the field case. Also remember that we are considering infinitesimal transformations of the coordinates and/or fields characterized by a single continuous parameter ϵ such that $\epsilon = 0$ corresponds to no transformation.)

BOX 6.5 Conservation of Angular Momentum for a Field

In this box, we will calculate the density $\rho = J^t$ of the quantity that is conserved when a field's Lagrangian density is independent of rotations around the z axis.

Remember that Equation 6.16 says that any transformation characterized by a continuous parameter ϵ that leaves the form of a Lagrangian density $\mathcal{L}$ unchanged has an associated conserved quantity whose current is

$$J^\mu = \left[\sum_n \frac{\partial \mathcal{L}}{\partial(\partial_\mu \phi_n)} \left[\frac{d\phi'_n}{d\epsilon} \right]_0 \right] - \mathcal{L} \left[\frac{dx'^\mu}{d\epsilon} \right]_0, \tag{6.35}$$

where the sum is over the number of fields involved. Assume that the Lagrangian density

$$\mathcal{L} = \tfrac{1}{2} \partial^\mu \phi\, \partial_\mu \phi - \tfrac{1}{2} m^2 \phi^2 \tag{6.36}$$

for a real scalar field is invariant under a rotation of coordinates such that

$$x' = (\cos \epsilon)x + (\sin \epsilon)y \qquad y' = -(\sin \epsilon)x + (\cos \epsilon)y. \tag{6.37a}$$

This implies that

$$\left[\frac{dx'}{d\epsilon} \right]_0 = \left[(-\sin \epsilon)x + (\cos \epsilon)y \right]_0 = y \quad \text{and} \tag{6.37b}$$

$$\left[\frac{dy}{d\epsilon} \right]_0 = \left[(-\cos \epsilon)x + (-\sin \epsilon)y \right]_0 = -x, \tag{6.37c}$$

and by the chain rule for partial derivatives

$$\left[\frac{d\phi'}{d\epsilon} \right]_0 = \frac{\partial \phi}{\partial x} \left[\frac{dx'}{d\epsilon} \right]_0 + \frac{\partial \phi}{\partial y} \left[\frac{dy}{d\epsilon} \right]_0 = (\partial_x \phi)y - (\partial_y \phi)x. \tag{6.38}$$

Exercise 6.5.1 From here, find an expression for the density $\rho = J^t$ of the conserved quantity associated with this symmetry. (This will be the density of the z component of angular momentum.) Compare with Equation 6.25.

BOX 6.6   Energy and Momentum Densities for a Complex Field

In this box, we will find expressions for the energy and momentum densities for the Lagrangian density for a complex scalar field given in Equation 6.26:

$$\mathcal{L} = \partial^{\alpha}\phi^* \, \partial_{\alpha}\phi - m^2\phi^*\phi. \tag{6.39}$$

Recall that the energy and momentum densities are the values of the $T^{t\nu}$ components of the stress-energy tensor given in Equation 6.21, repeated here:

$$T^{\mu\nu} = \left[\sum_n \frac{\partial \mathcal{L}}{\partial(\partial_\mu \phi_n)} \partial^\nu \phi_n \right] - g^{\mu\nu}\mathcal{L}. \tag{6.40}$$

Exercise 6.6.1   Find expressions for the energy and momentum densities for this Lagrangian density. (*Hints:* Remember that we treat ϕ and ϕ^* as two independent fields, and that $\partial^{\alpha}\phi^* \, \partial_{\alpha}\phi = g^{\alpha\beta}\partial_{\beta}\phi^* \, \partial_{\alpha}\phi$.)

BOX 6.7   Current for a Phase Transformation

In this box, we will show that a phase transformation $\phi' = \phi\, e^{-iq\theta}$ applied to the standard Lagrangian density for a complex field yields the current for an associated conserved quantity given in Equation 6.29, repeated here:

$$J^{\mu} = -iq(\phi\, \partial^{\mu}\phi^{*} - \phi^{*}\partial^{\mu}\phi). \tag{6.41}$$

We start with the general equation for the current associated with a symmetry under a one-parameter transformation (Equation 6.16). In this case, the parameter is θ, so simply replacing the parameter ϵ by θ in that equation yields

$$J^{\mu} = \left[\sum_{n} \frac{\partial \mathcal{L}}{\partial(\partial_{\mu}\phi_{n})}\left[\frac{d\phi'_{n}}{d\theta}\right]_{0}\right] - \mathcal{L}\left[\frac{dx'^{\mu}}{d\theta}\right]_{0}. \tag{6.42}$$

This transformation changes the fields, not the coordinates, so $dx'^{\mu}/d\theta = 0$. Remember also that in the case of the complex Lagrangian density, we have exactly *two* fields, ϕ and ϕ^{*}.

Exercise 6.7.1   Find the conserved current associated with this transformation. (*Hints:* Remember that we evaluate the θ-derivative at $\theta = 0$. You also may find results from the previous box useful.)

Homework Problems

P6.1 Calculate the Hamiltonian for the Lagrangian

$$L = \tfrac{1}{2}m\dot{x}^2 + \tfrac{1}{2}m\dot{z}^2 - mgz. \tag{6.43}$$

Explain why your result makes sense if we interpret the Hamiltonian to describe the system's conserved energy.

P6.2 Consider two real scalar fields ϕ_1 and ϕ_2 subject to the Lagrangian density

$$\begin{aligned}
\mathcal{L} = {}&\tfrac{1}{2}\partial^\mu\phi_1\,\partial_\mu\phi_1 - \tfrac{1}{2}m_1^2\phi_1^2 + \tfrac{1}{2}\partial^\mu\phi_2\,\partial_\mu\phi_2 \\
&- \tfrac{1}{2}m_2^2\phi_2^2 - \lambda\phi_1^2\phi_2^2,
\end{aligned} \tag{6.44}$$

with $\lambda > 0$. The term $-\lambda\phi_1^2\phi_2^2$ describes an *interaction* between the fields.

a. What are the units of λ?

b. Write down an expression for the energy density for this $\mathcal{L}$.

c. Note that we have terms like $m_1^2\phi_1^2 + m_2^2\phi_2^2$ in the energy density, which suggests that ϕ_1^2 is the number density of the particles associated with field 1, and ϕ_2^2 is the number density of the particles associated with field 2 (as then the terms $m_1^2\phi_1^2 + m_2^2\phi_2^2$ will correspond to density of rest energy). With this in mind, does the interaction term describe an attractive or repulsive interaction? (*Hint:* How does the value of this term change as the density of the two types of particles increases at a given location?)

P6.3 Compute all four components of $T^{x\mu}$ for the complex-field Lagrangian density $\mathcal{L} = \partial^\mu\phi^*\,\partial_\mu\phi - m^2\phi^*\phi$.

P6.4 Prove that the stress-energy tensor for the Lagrangian density $\mathcal{L} = \partial^\mu\phi^*\,\partial_\mu\phi - m^2\phi^*\phi$ is symmetrical: $T^{\mu\nu} = T^{\nu\mu}$.

P6.5 Using an approach similar to the methods we employed in Box 6.5, find the expression for the conserved angular momentum for the complex scalar field whose Lagrangian density is $\mathcal{L} = \partial^\mu\phi^*\,\partial_\mu\phi - m^2\phi^*\phi$.

P6.6 Later in the book, we will encounter Lagrangian densities involving multiplets of fields, such as

$$\Phi = \begin{bmatrix} \phi_1 \\ \phi_2 \end{bmatrix}, \tag{6.45}$$

where ϕ_1 and ϕ_2 are real scalar fields. Consider the Lagrangian density

$$\mathcal{L} = \tfrac{1}{2}\partial^\mu\Phi^T\partial_\mu\Phi - \tfrac{1}{2}m^2\Phi^T\Phi, \tag{6.46}$$

which uses matrix notation to write the Lagrangian density for the field in a compact way. Consider "rotating" the fields according to the following transformation:

$$\Phi' = \begin{bmatrix} \phi_1' \\ \phi_2' \end{bmatrix} = \begin{bmatrix} \cos\theta & \sin\theta \\ -\sin\theta & \cos\theta \end{bmatrix} \begin{bmatrix} \phi_1 \\ \phi_2 \end{bmatrix} \equiv \mathbf{R}\Phi. \tag{6.47}$$

a. What is the corresponding transformation equation for $\Phi^T \equiv [\phi_1, \phi_2]$?

b. Show that neither term in the Lagrangian density is changed by this transformation.

c. What is the conserved current corresponding this symmetry? Express your answer both in terms of Φ and in terms of ϕ_1 and ϕ_2. (*Hints:* Treat the multiplet Φ and its transpose Φ^T as separate fields, as we do with a complex field and its complex conjugate. Write the derivative of the transformation evaluated at $\theta = 0$ in the form of $i\mathbf{X}\Phi$, where $\mathbf{X}$ is a square matrix.)

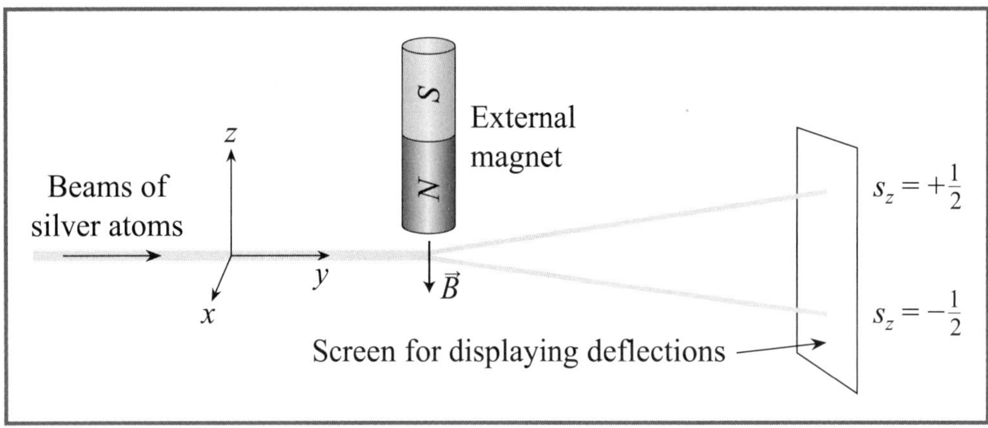

z
x
y
Beams of silver atoms
S
N
External magnet
$\vec{B}$
Screen for displaying deflections
$s_z = +\frac{1}{2}$
$s_z = -\frac{1}{2}$

In Chapters 4–6, we considered the behavior of classical scalar fields. The next step is to "quantize" those fields so that they describe the particles we encounter in quantum mechanics. But before taking that step, it will help us to review standard non-relativistic quantum mechanics so that we can more clearly see the path forward, and also so that we can see how quantum field theory differs from non-relativistic quantum mechanics.

In what follows, I will assume a background in quantum mechanics similar to the basic wave mechanics one might see in a typical "Modern Physics" course.

7.1 The Phenomenology of Spin

As you may know, many quantum particles (for example, electrons) have an intrinsic and invariant property called "spin." We will start our review of quantum mechanics by considering the quantum mechanics of spin not only because both the concept and mathematics of spin will prove central to quantum field theory but also because the structure of ordinary quantum mechanics appears in its simplest and clearest form when applied to spin.

Physicists at the turn of the twentieth century had observed that atoms in strong magnetic fields exhibited behavior that might be explained if the atoms' electrons acted like tiny permanent magnets. This could be explained by imagining an electron to be a tiny, spinning charged ball. The spin would make the electron's charge a circulating current analogous to the current flowing in an electromagnet (see Figure 7.1).

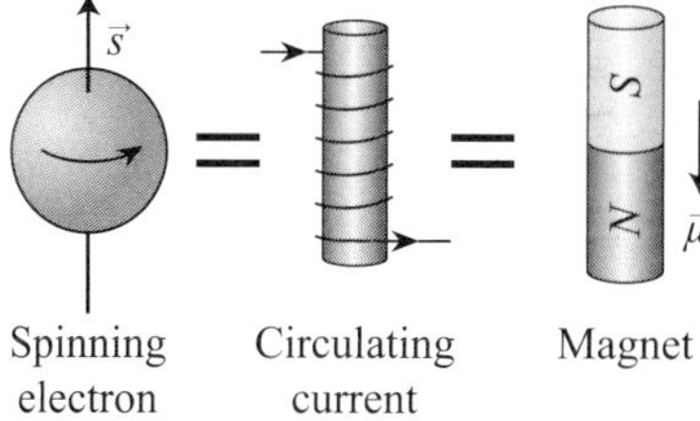

FIGURE 7.1   A spinning charged ball behaves as an electromagnet.

Specifically, an *electron* spinning counterclockwise (whose spin angular momentum $\vec{s}$ thus points *upward* by the usual right-hand rule for angular momentum) corresponds to a conventional current flowing *clockwise* because the electron's charge is negative. The equivalent electromagnet's **magnetic moment** vector $\vec{\mu}$, whose direction specifies the direction of that electromagnet's north pole relative to its south pole, is indicated by your right thumb when your right fingers curl along with the circulating conventional current. Applying the rule here shows that the electron's magnetic moment points downward, *opposite* to that of its angular momentum vector, as shown in the figure.

In 1922, Walther Gerlach executed an experiment proposed by his colleague Otto Stern that was able to use this magnetic effect to measure the projection of a given electron's spin along a certain axis. (The fact that Stern and Gerlach thought they were measuring something else, and that it took three years to clear up the confusion, is a story for another time.) The physics behind this experiment (though it is fascinating— see problem P7.1) is not relevant to our present needs: suffice it to say that when we send a beam of electrons into a "Stern-Gerlach" device, the electrons are deflected along a certain axis according to their spin projection along that axis.

Since one would expect the electrons in a beam to be randomly oriented, one would expect the electrons to experience a continuous range of deflections, whose limits would correspond to an electron's spin being completely aligned or completely anti-aligned with the axis. However, contrary to this expectation, Gerlach observed only *two* projections, corresponding quantitatively to the projection along the axis being $\pm\frac{1}{2}$ ($\pm\frac{1}{2}\hbar$ in SI units). The projection of an electron's spin angular momentum on a given axis is therefore *quantized.*

This result is independent of one's choice of axis. How the same beam of electrons can be simultaneously aligned or anti-aligned with, say, the z direction and also aligned

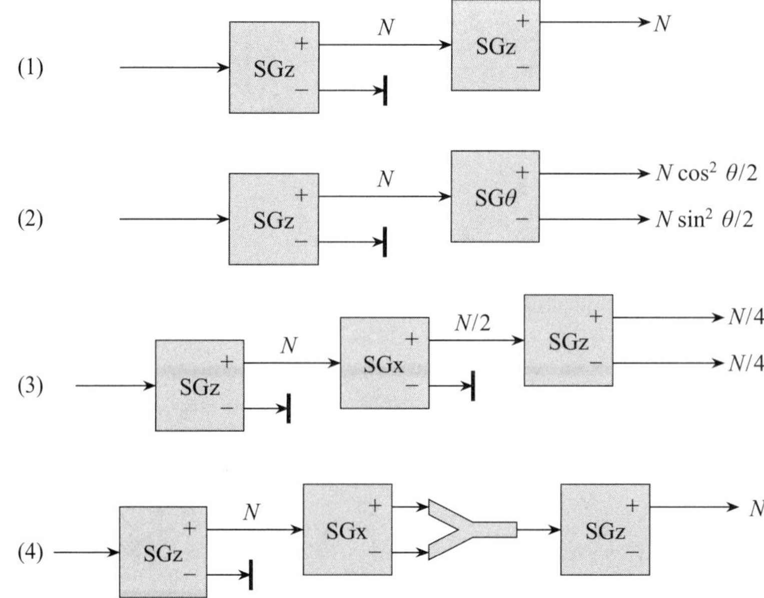

FIGURE 7.2   These four basic experiments display the quantum behavior of spin.

or anti-aligned with the x direction is an intrinsically quantum behavior that does not make classical sense.

There are other things about a classical spinning-ball model for an electron that do not make sense. An electron having the observed amount of spin would have an equatorial speed much greater than that of light (see Problem P7.2). If the electron were a simple spinning ball, it would also make sense that collisions between electrons might make some spin faster and some slower. But we *never* observe the projection of an electron's angular momentum on an axis to be anything other than $\pm\frac{1}{2}$. We will see as this course progresses that even though electron spin is genuine angular momentum that is conserved along with other forms of angular momentum, it is an intrinsic quantum property of the electron that has nothing to do with the electron actually rotating.

Figure 7.2 shows four experiments that illustrate the quantum behavior of spin. In this schematic diagram, a box labeled SGx, SGz, or SGθ represents an idealized Stern-Gerlach device, where the final letter indicates the axis along which the device is measuring the electron spin's projection ("θ" means that the axis makes an angle of θ with the z axis in the xz plane). Arrows show the flow of electrons. Electrons emerging from the device's "+" and "−" channels have spin projections of $\pm\frac{1}{2}$ on that axis, respectively. A vertical dark bar after a channel means that any electrons emerging from that channel are blocked.

Experiment 1 illustrates the *self-consistency* of a Stern-Gerlach measurement: if we measure N electrons to have spin up in the z direction and then immediately measure s_z again, we find that all still have spin up. However, Experiment 2 illustrates that if we rotate the magnetic field axis of the second SG device by an angle of θ relative to the first, the probabilities of measuring the spin of a given electron (which we had previously measured to be up in the z direction) will be up or down in the new direction are $\cos^2\frac{1}{2}\theta$ and $\sin^2\frac{1}{2}\theta$, respectively. This illustrates the probabilistic nature of quantum measurements: we can only predict *statistically* how identically prepared electrons will behave when we perform a measurement that is *not* a repeat of the first measurement.

In particular, if $\theta = 90°$, corresponding to the magnetic field direction being along the x axis, the probabilities are $\cos^2 45° = \sin^2 45° = \frac{1}{2}$. This makes some kind of sense: if an electron whose spin is in the $+z$ is in some sense "forced" by the second device to choose between being aligned with the $+x$ direction or the $-x$ direction, the probabilities must be equal by symmetry.

Experiment 3 seems to support this model. In this situation, electrons that had been measured to have spins in the $+z$ are forced by the second device to choose between being aligned with the $+x$ or $-x$ directions, and half choose the former. If we select the $+x$ set and then force them to choose between the $+z$ and $-z$ axes, half must make each choice by symmetry. But note that *all* the electrons had been originally measured to have spins in the $+z$ direction! This makes it clear that a Stern-Gerlach device does not simply *reveal* a pre-existing spin projection, the way that uncovering a flipped coin reveals whether it is heads or tails. The SGx measurement *does something* to the electron's state: it effectively *erases* the knowledge we gained from the first SGz measurement, as if that experiment had never been performed! We therefore cannot assign meaningful values to the quantities s_x and s_z simultaneously: measuring one effectively destroys any knowledge we might have of the other. This is the spin version of the Heisenberg uncertainty principle.

Experiment 4 undermines the simplistic picture that a Stern-Gerlach measurement "forces an electron to choose" a spin orientation. This experiment is just like the previous one, except that instead of throwing away the electrons with $s_x = -\frac{1}{2}$, we recombine the electrons emerging from the SGx device in such a way that we cannot tell even in principle which particular electron emerged from which channel of that device. Then we see that all electrons emerge from the final SGz device as though the middle device were not there at all! So, the middle device is not actually forcing the electrons to choose: something more subtle is going on. This is the spin manifestation of the concept of quantum interference.

These experiments clearly illustrate that any classical understanding of electron spin is inadequate. We need a fundamentally new kind of mechanics to explain and help us predict spin phenomena.

7.2 The Rules of Quantum Mechanics

The basic mathematical tool this new mechanics employs is a certain kind of vector space called a **Hilbert space**, whose denizens are N-dimensional column vectors with complex components that we denote abstractly by a symbol $|\psi\rangle$ we call a **ket**. For every ket in the space, there is a corresponding **bra** (pronounced with the "a" as in "cat," not as in "ah") $\langle\psi|$ that denotes a row vector whose components are the complex conjugates of the components of the ket $|\psi\rangle$:

$$|\psi\rangle = \begin{bmatrix} \psi_1 \\ \psi_2 \\ \vdots \end{bmatrix}, \qquad \langle\psi| = [\,\psi_1^*, \quad \psi_2^*, \quad \dots\,], \tag{7.1}$$

where $\psi_1, \psi_2, \dots$ are complex numbers. We say that $|\psi\rangle$ and $\langle\psi|$ are **adjoints** of each other. If kets $|u\rangle$ and $|w\rangle$ have the same number of components, we define their **inner product** by the bra(c)ket

$$\langle u|w\rangle = \begin{bmatrix} u_1^*, & u_2^*, & \dots \end{bmatrix} \begin{bmatrix} w_1 \\ w_2 \\ \vdots \end{bmatrix} = u_1^* w_1 + u_2^* w_2 + \cdots, \tag{7.2}$$

which is just a complex number. We say that the two kets are **orthogonal** if their inner product (bra(c)ket) $\langle u|w\rangle = 0$ and describe a ket $|\psi\rangle$ as being **normalized** if its **squared magnitude** $\langle\psi|\psi\rangle$ is 1; that is, if

$$\langle\psi|\psi\rangle = \psi_1^*\psi_1 + \psi_2^*\psi_2 + \cdots = |\psi_1|^2 + |\psi_2|^2 + \cdots = 1. \tag{7.3}$$

In quantum mechanics, we also distinguish between a quantum particle's **intrinsic properties** (such as mass, charge, and spin magnitude) that never change, and **observables** (such as energy, position, momentum, and spin projection on an axis) that might change. The fundamental goal of quantum mechanics is to link *preparation* to *observation*: given a set of particles "prepared" in a certain way, we seek to predict what we will observe in subsequent measurements of observable values.

We can express the logical structure of quantum mechanics in terms of seven basic rules that link the mathematical structure just described to empirical reality. Some of these rules may seem arbitrary or even crazy, but they are simple and easy to use, and also quite reliably predict the behavior of non-relativistic quantum particles in a context where the particles are neither created nor destroyed.

Rule 1: The State Rule

A suitable preparation puts a particle in a well-defined **quantum state** we can represent mathematically by a *normalized ket* $|\psi\rangle$ in an appropriately chosen Hilbert space.

Commentary: In Newtonian mechanics, we would describe an electron's spin state by specifying the components s_x, s_y, s_z of its spin angular momentum. In quantum mechanics, we specify an electron's spin state using a ket in what we will find to be a two-dimensional Hilbert space. The two complex components of this ket do *not* specify the values of the spin components, but rather (as we will see) encoded probability information about possible experimental outcomes.

Rule 2: The Eigenvector Rule

For every possible numerical value a given observable can have, there is an associated normalized Hilbert-space ket we call that value's **eigenvector**. (The value associated with that ket is called the ket's **eigenvalue**.) Any given observable's eigenvalues are mutually orthogonal.

Corollary: Our kets must have as many components as there are possible observable values.

Commentary: *Eigen* is a German root meaning "characteristic." The connection between this language and the eigenvectors and values of a matrix will become clearer in the next section. The corollary follows from the mathematical fact that having N mutually orthogonal vectors requires an N-dimensional space. Since the observables s_x, s_y, and s_z associated with an electron's spin each have two possible values ($\pm\frac{1}{2}$), describing these observables' eigenvectors requires a two-dimensional Hilbert space. Figure 7.3 lists those eigenvectors when we make the conventional choice of choosing the orthogonal eigenvectors for the s_z observable as the basis vectors of our space.

	s_x	s_y	s_z	s_θ
$+\frac{1}{2}$	$\|+x\rangle = \begin{bmatrix} \sqrt{\frac{1}{2}} \\ \sqrt{\frac{1}{2}} \end{bmatrix}$	$\|+y\rangle = \begin{bmatrix} \sqrt{\frac{1}{2}} \\ i\sqrt{\frac{1}{2}} \end{bmatrix}$	$\|+z\rangle = \begin{bmatrix} 1 \\ 0 \end{bmatrix}$	$\|+\theta\rangle = \begin{bmatrix} \cos\frac{1}{2}\theta \\ \sin\frac{1}{2}\theta \end{bmatrix}$
$-\frac{1}{2}$	$\|-x\rangle = \begin{bmatrix} \sqrt{\frac{1}{2}} \\ -\sqrt{\frac{1}{2}} \end{bmatrix}$	$\|-y\rangle = \begin{bmatrix} \sqrt{\frac{1}{2}} \\ -i\sqrt{\frac{1}{2}} \end{bmatrix}$	$\|-z\rangle = \begin{bmatrix} 0 \\ 1 \end{bmatrix}$	$\|-\theta\rangle = \begin{bmatrix} \sin\frac{1}{2}\theta \\ -\cos\frac{1}{2}\theta \end{bmatrix}$

FIGURE 7.3   This table lists the eigenvectors for spin observables. (The observable s_θ is the projection along an axis that makes an angle of θ with the z axis in the xz plane.)

> ### Rule 3: The Outcome Probability Rule
>
> In an experiment that measures the value of an observable a for a particle prepared in state $|\psi\rangle$, the **quantum amplitude** for getting a particular result a_n from the set of N possible eigenvalues $a_1, a_2, \ldots, a_N$ is the inner product $\langle a_n|\psi\rangle$ with the initial state on the right and the eigenvector corresponding to the value a_n on the left. The *probability* of actually getting that result is the amplitude's absolute square $|\langle a_n|\psi\rangle|^2$.
>
> *Corollary:* The kets $|\psi\rangle$ and $e^{i\phi}|\psi\rangle$ are physically equivalent for any ϕ.

Commentary: This rule implies that quantum mechanics is intrinsically probablistic: we can make only *statistical* predictions about the results of experiments performed on a set of particles even though they have all been prepared identically. (See Box 7.1 for a discussion of the corollary, and how the requirement that both states and eigenvectors be *normalized* ensures that $|\langle a_n|\psi\rangle|^2$ yields a meaningful probability.)

> ### Rule 4: The Collapse Rule
>
> Measuring the value of an observable "collapses" the particle's state to the eigenvector corresponding to the observed result.
>
> *Corollary:* We can prepare a particle in a well-defined quantum state by determining the value of one of its observables.

Commentary: This rule is simple to both state and use, but presents vexing theoretical and philosophical problems that have eluded satisfactory resolution for almost a century. Exactly when and how does this collapse occur? No one knows. But this rule is essential for ensuring that measurement results are self-consistent (see Box 7.3).

> ### Rule 5: The Sequence Rule
>
> If we prepare a particle in state $|\psi\rangle$, measure observable a and obtain the result a_n, and then immediately measure observable b, the amplitude for obtaining a particular eigenvalue b_j is $\langle b_j|a_n\rangle\langle a_n|\psi\rangle$.

> ### Rule 6: The Superposition Rule
>
> In an experiment where a particle can take either one of two paths from its initial preparation to a final observation, and we cannot know (even in principle) which a given particle actually takes, the quantum amplitude for any given final result is the *sum of the amplitudes* for each path.

Commentary: These rules are needed to explain phenomena involving "quantum interference," such as the result of Experiment 4 described previously.

In Box 7.3, you will check that these six rules are sufficient to predict the results of Experiments 1–4 described in the previous section. We will discuss the seventh and final rule in Chapter 8.

7.3 Position and Momentum Observables

The spin observables we have considered so far have had just two eigenvalues and so could be represented by two-dimensional vectors. But as far as we know, a particle's position (even in one spatial dimension) can be any real number x, and that particle's momentum could also be any real number p_x. These observables thus have an infinite number of eigenvalues (and corresponding eigenvectors), so we need an infinite-dimensional Hilbert space to distinguish all those eigenvectors.

How can we work with infinite vectors? Suppose as a first step we confine our particle in a box where its range of positions is $0 \leq x \leq L$, and divide this region of space into discrete positions $x_1, x_2, x_3, \ldots, x_N$ separated by equal steps $\Delta x = L/N$, starting at $x_1 = \frac{1}{2}\Delta x$. Each discrete position x_k collectively represents position outcomes within the range $x_k \pm \frac{1}{2}\Delta x$. If Δx is very small, this should decently approximate reality. Then the position eigenstates will reside in a large but finite N-dimensional space. Just as we conventionally choose the s_z eigenvectors to provide the basis for electron spin states, we conventionally choose the basis vectors for states describing a particle's position or momentum (or observables related to these) to be position eigenvectors:

$$|x_1\rangle \equiv \begin{bmatrix} 1 \\ 0 \\ 0 \\ \vdots \\ 0 \end{bmatrix}, \quad |x_2\rangle \equiv \begin{bmatrix} 0 \\ 1 \\ 0 \\ \vdots \\ 0 \end{bmatrix}, \quad |x_3\rangle \equiv \begin{bmatrix} 0 \\ 0 \\ 1 \\ \vdots \\ 0 \end{bmatrix}, \quad \ldots \; |x_N\rangle \equiv \begin{bmatrix} 0 \\ 0 \\ 0 \\ \vdots \\ 1 \end{bmatrix}. \quad (7.4)$$

A general quantum state $|\psi\rangle$ is then a column vector of components ψ_1, ψ_2, ψ_3, and so on, each of which is the quantum amplitude for measuring the particle to be at the corresponding position: for example,

$$\langle x_3|\psi\rangle = \begin{bmatrix} 0 & 0 & 1 & \ldots & 0 \end{bmatrix} \begin{bmatrix} \psi_1 \\ \psi_2 \\ \psi_3 \\ \vdots \\ \psi_N \end{bmatrix} = \psi_3. \quad (7.5)$$

The inner product $\langle u|w\rangle$ of vectors $|u\rangle$ and $|w\rangle$ is

$$\langle u|w\rangle = \begin{bmatrix} u_1^* & u_2^* & u_3^* & \cdots & u_N^* \end{bmatrix} \begin{bmatrix} w_1 \\ w_2 \\ w_3 \\ \vdots \\ w_N \end{bmatrix} = \sum_{k=1}^{N} u_i^* w_k. \quad (7.6)$$

Now, what happens as we make $N \to \infty$ (and thus $\Delta x \to 0$)? A general state vector $|\psi\rangle$ becomes a longer and denser list of components $\psi_k \equiv \langle x_k|\psi\rangle$. Note that we can index this list by the value of the corresponding position x_k instead of an integer index k. In the limit, this list of numbers indexed by position becomes a *function* $\psi(x_k) \to \psi(x)$ that we call the particle's **wavefunction**.

In the same limit (and as we widen our box to include all space), we can convert the sum in Equation 7.6 to an integral:

$$\langle u|w\rangle = \sum_{k=1}^{N} u_k^* w_k \longrightarrow \int_{-\infty}^{+\infty} u^*(x)\, w(x)\, dx. \quad (7.7)$$

The normalization condition for kets becomes the integral

$$1 = \langle \psi|\psi\rangle = \int_{-\infty}^{+\infty} \psi^*(x)\, \psi(x)\, dx = \int_{-\infty}^{+\infty} |\psi(x)|^2\, dx. \quad (7.8)$$

Technically, this conversion from sums to integrals actually requires us to rescale our position eigenvector's components to supply the extra factor of dx inside the integral. After rescaling, the position eigenvectors (which we should now call **eigenfunctions**) become Dirac deltas:

$$\psi_{x_0}(x) = \langle x|x_0\rangle = \delta(x - x_0). \quad (7.9)$$

The probability for finding the particle's position to be within a range of width dx centered on x is not $|\langle x|\psi\rangle|^2$, but rather

$$\Pr(x)\,dx = |\langle x|\psi\rangle|^2 dx = |\psi(x)|^2 dx. \tag{7.10}$$

Box 7.4 spells out the details.

How about momentum eigenfunctions? The de Broglie relation $|p_x| = 2\pi/\lambda$ ($|p_x| = h/\lambda$ in SI units), where λ is the particle's wavelength, tells us what the momentum eigenfunctions $\psi_{p_0}(x)$ must be. To have a definite momentum, the particle's wavefunction must be a wave with a definite wavelength. Possible examples of such waves are

$$\psi_{p_0}(x) = A\cos\left(\frac{2\pi x}{\lambda_0}\right) \quad\text{or}\quad A\sin\left(\frac{2\pi x}{\lambda_0}\right), \tag{7.11}$$

where A is some normalization constant and $\lambda_0 = 2\pi/p_0$ is the wavelength associated with momentum magnitude $|p_{0x}| = p_0$. Note that A cannot depend on position because the Fourier theorem tells us that the wave must have a constant amplitude over all space to have a precisely defined wavelength.

But both of these proposed waves have the problem that the amplitude for finding the particle at a position varies with position. In particular, we would *never* find the particle to be at $x = \frac{1}{4}\lambda_0, \frac{3}{4}\lambda_0, \frac{5}{4}\lambda_0$, and so on in the first case or at $x = \frac{1}{2}\lambda_0, \lambda_0, \frac{3}{2}\lambda_0$, and so on in the second case. But a momentum eigenfunction should not pick out any special places in space: all points in space should be equivalent to a free particle with a given momentum.

However, suppose $\psi_{p_{0x}}(x) = Ae^{\pm i2\pi x/\lambda_0}$. The probability of measuring the particle to have position in a range of width dx centered on any x in this case is simply $|\psi_{p_{0x}}(x)|^2 dx = |A|^2 dx$, which is the same for all x and therefore satisfies our requirement. The momentum eigenfunctions should therefore be

$$\psi_{p_{0x}}(x) = Ae^{\pm i2\pi x/\lambda_0} = Ae^{ip_{0x}x}, \tag{7.12}$$

where in the middle step I used the de Broglie relation $p_0 = 2\pi/\lambda_0$ ($p_0 = h/\lambda_0$ in SI units) to eliminate λ_0. The orthonormality condition that is conventional in ordinary quantum mechanics for wavefunctions whose eigenvalues a are continuous is $\langle\psi_{a_1}|\psi_{a_2}\rangle = \delta(a_1 - a_2)$, as in Equation 7.9. Therefore, we define A so that

$$\psi_{p_{0x}}(x) = \sqrt{\frac{1}{2\pi}}\,e^{ip_{0x}x}, \quad\text{because then we have}$$

$$\langle\psi_{p_{1x}}|\psi_{p_{2x}}\rangle = \frac{1}{2\pi}\int_{-\infty}^{+\infty} e^{i(p_{1x}-p_{2x})x}\,dx = \delta(p_{1x} - p_{2x}), \tag{7.13}$$

(see Equation 5.32b).

From this point, the generalization to three spatial dimensions is straightforward. Wavefunctions representing quantum states become three-dimensional functions $\psi(x, y, z) \equiv \psi(\vec{x})$. The probability of finding a particle whose state is $|\psi\rangle$ to be within a box with volume $dx\,dy\,dz \equiv d^3x$ centered on position $\vec{x}$ is

$$\Pr(\vec{x})\,d^3x = |\psi(\vec{x})|^2\,d^3x. \tag{7.14}$$

Inner products become three-dimensional integrals

$$\langle u|w\rangle = \iiint [u(x, y, z)]^*\,w(x, y, z)\,dx\,dy\,dz = \int [u(\vec{x})]^*\,w(\vec{x})\,d^3x, \tag{7.15}$$

where the integral is over all space. Position eigenfunctions become

$$|\vec{x}_0\rangle = \psi_{\vec{x}_0}(\vec{x}) = \delta(x - x_0)\delta(y - y_0)\delta(z - z_0) = \delta^3(\vec{x} - \vec{x}_0). \tag{7.16}$$

Momentum eigenfunctions become

$$|\vec{p}_0\rangle = \psi_{\vec{p}_0}(\vec{x}) = \left(\frac{1}{2\pi}\right)^{3/2} e^{ip_{0x}x} e^{ip_{0y}y} e^{ip_{0z}z} = \left(\frac{1}{2\pi}\right)^{3/2} e^{i\vec{p}_0 \cdot \vec{x}}. \tag{7.17}$$

We normalize wavefunctions according to

$$1 = \langle\psi|\psi\rangle = \int |\psi(\vec{x})|^2 \, d^3x.$$

Eigenfunctions $|a_n\rangle = \psi_{a_n}(x)$ for an observable A having (even an infinite number of) discrete eigenvalues a_n are orthonormal:

$$\langle a_k|a_n\rangle = 1 \quad \text{if } k = n, \quad \text{and 0 otherwise.} \tag{7.18}$$

If the eigenfunctions correspond to eigenvalues that are continuous (for example, momentum), then the orthonormality condition is instead

$$\langle\vec{p}_1|\vec{p}_2\rangle = \delta^3(\vec{p}_1 - \vec{p}_2). \tag{7.19}$$

The point is that even in infinite-dimensional Hilbert spaces, we have all of the tools we need to describe states and calculate amplitudes and probabilities according to the six rules we have seen so far. The logical structure is the same, even if the mathematical tools are not.

See Box 7.5 for an illustration of how we can apply the rules to understand the classic two-slit interferometer.

BOX 7.1 Probabilities

Suppose a particle has state $|\psi\rangle$. According to the Outcome Probability Rule, the probability of getting a result a_n for an arbitrary observable a is $|\langle a_n|\psi\rangle|^2$, where $\langle a_n|$ is the bra form of the eigenvector $|a_n\rangle$ corresponding to a_n.

Exercise 7.1.1 Show that if the state were $e^{i\phi}|\psi\rangle$ instead of $|\psi\rangle$, the probabilities are exactly the same, independent of ϕ. Since a is arbitrary, this means that no experiment we can perform would distinguish $e^{i\phi}|\psi\rangle$ from $|\psi\rangle$: these states are physically equivalent. (*Hint:* The inner product is linear.)

The **Cauchy-Schwarz inequality** applies to both real and complex vector spaces (see many proofs online). It tells us that $|\langle a_n|\psi\rangle|^2 \leq \langle a_n|a_n\rangle\langle\psi|\psi\rangle = 1$ if both $|a_n\rangle$ and $|\psi\rangle$ are normalized. Since $|\langle a_n|\psi\rangle|^2$ is an absolute square, it is real and ≥ 0. So, $0 \leq |\langle a_n|\psi\rangle|^2 \leq 1$, the basic condition for a probability. Now note that N mutually orthogonal eigenvectors $|a_n\rangle$ provide a complete basis for an N-dimensional complex space, meaning that we should be able to expand any ket $|\psi\rangle$ in the space as a linear combination of these basis vectors:

$$|\psi\rangle = |a_1\rangle\langle a_1|\psi\rangle + |a_2\rangle\langle a_2|\psi\rangle + \cdots = \left[\sum_n |a_n\rangle\langle a_n|\right]|\psi\rangle$$

$$\Rightarrow \sum_n |a_n\rangle\langle a_n| = I, \tag{7.20}$$

an expression we call the **completeness relation.**

Exercise 7.1.2 Show that the completeness relation implies that the total probability for all possible outcomes is $\sum_n |\langle a_n|\psi\rangle|^2 = 1$, as it must be. (*Hint:* First show that $\langle w|u\rangle = \langle u|w\rangle^*$ for any kets $|u\rangle$ and $|w\rangle$.)

BOX 7.2   The Spin Eigenvectors

In this box, we will examine the eigenvectors listed in Figure 7.3.

Exercise 7.2.1   Show that each vector is normalized, and that each pair of vectors for a given observable are orthogonal, as required by the rules.

BOX 7.3   The Rules Are Sufficient

In this box, we will show that the six rules stated in this chapter are sufficient to explain the results of Experiments 1–4 in Figure 7.2.

First, consider Experiment 1.

Exercise 7.3.1   Use the rules to argue that if we measure any observable a, obtain the result a_n, and then immediately measure a again, we get the same result with a probability of 1, and that all other probabilities must be zero.

Exercise 7.3.2   Now consider Experiment 2. Use the rules and the eigenvectors listed in Figure 7.3 to explain the results of this experiment.

BOX 7.3 The Rules Are Sufficient (cont.)

Exercise 7.3.3 Now consider Experiment 3. Use the rules and the eigenvectors in Figure 7.3 to explain this experiment's results.

Exercise 7.3.4 Finally, consider Experiment 4. Use the rules and the eigenvectors in Figure 7.3 to explain this experiment's results. Also, show that if we were to add the *probabilities* for the two paths (as we would do in classical mechanics), we would get a different (and thus incorrect) result.

BOX 7.4   Normalization in Infinite-Dimensional Spaces

In this box, we will discuss why we need to change the way we scale vector components to make the shift from sums to integrals as we move from a finite number of dimensions to an infinite number of dimensions.

Consider again the case where we put our particle in a box to limit its position to the range $0 \leq x \leq L$, and then divide that range into N discrete positions $x_1, x_2, x_3, \ldots, x_N$ that are $\Delta x = L/N$ apart starting at $x_1 = \frac{1}{2}\Delta x$. In this situation, our state vectors are N-dimensional vectors. The inner product between vectors $|u\rangle$ and $|w\rangle$ is

$$\langle u|w\rangle = \sum_{k=0}^{N} u_k^* \, w_k = \sum_{k=0}^{N} \langle u|x_k\rangle \langle x_k|w\rangle \qquad (7.21)$$

since $w_k \equiv \langle x_k|w\rangle$ and $u_k^* \equiv \langle x_k|u\rangle^* = \langle u|x_k\rangle$. We need this sum to look like

$$\langle u|w\rangle = \sum_{k=0}^{N} [\, u_k'\,]^* \, w_k' \, \Delta x \qquad (7.22)$$

so that when we take limits $N \to \infty$ and $\Delta x \to 0$, the sum becomes an integral. We want the numerical value of $\langle u|w\rangle$ to be the same, so the components u_k' and w_k' in Equation 7.22 can't be the same as u_k and w_k in Equation 7.21.

But we can get things to work out if we simply rescale the position eigenvectors. Define new position eigenvectors $|x\rangle_k' = |x_k\rangle/\sqrt{\Delta x}$. Then

$$\sum_{k=0}^{N} [\, u_k'\,]^* \, w_k' \, \Delta x = \sum_{k=0}^{N} \langle u|x_k'\rangle \langle x_k'|w\rangle \, \Delta x = \sum_{k=0}^{N} \frac{\langle u|x_k\rangle}{\sqrt{\Delta x}} \frac{\langle x_k|w\rangle}{\sqrt{\Delta x}} \Delta x$$

$$= \sum_{k=0}^{N} u_k^* w_k = \langle u|w\rangle. \qquad (7.23)$$

The quantities $u_k' \equiv \langle x_k'|u\rangle$ and $w_k' \equiv \langle x_k'|w\rangle$ in the integral-like sum in Equation 7.22 are what become the wavefunctions $u(x)$ and $w(x)$ that represent the states $|u\rangle$ and $|w\rangle$ when we take $N \to \infty$, $\Delta x \to 0$.

Exercise 7.4.1   What is the probability that a particle in the state $|\psi\rangle$ will be determined to have position x_k expressed in terms of the wavefunction value $\psi(x_k) \equiv \langle x_k'|\psi\rangle$ before we take the limits $N \to \infty$, $\Delta x \to 0$? After?

BOX 7.4 Normalization in Infinite-Dimensional Spaces (cont.)

If you think about it, the position probability being $|\psi(x)|^2\, dx$ instead of $|\langle x|\psi\rangle|^2$ makes good practical sense. Since the total probability of being somewhere (anywhere) must be 1, the probability of being at any *specific* position x must go to zero as $N \to \infty$. The rescaling converts $|\psi(x)|^2$ into a linear *probability density* (probability per unit length). For a range of positions dx small enough so that the probability density $|\psi(x)|^2$ is approximately constant, the probability of being in that range is simply the probability per unit length times the length. This rescaling also has the salutory effect of keeping magnitude of the function $\psi(x) \equiv \langle x'|\psi\rangle$ finite even though $\langle x|\psi\rangle \to 0$ as $N \to \infty$.

Now consider the wavefunction $\psi_{x_n}(x) \equiv \langle x'|x'_n\rangle$ that will represent the rescaled position eigenfunction. Before we take the limits $N \to \infty$, $\Delta x \to 0$, the components of $|x'_k\rangle$ are all zero except for the entry corresponding to x_k, which now has a value of $1/\sqrt{\Delta x}$ instead of 1. The quantity $\psi_{x_n}(x_k) \equiv \langle x'_k|x'_n\rangle$ that will become the wavefunction in the limit is therefore

$$\psi_{x_n}(x_k) \equiv \langle x'_k|x'_n\rangle = \frac{1}{\Delta x} \quad \text{when } k = n, \text{ and 0 otherwise.} \tag{7.24}$$

One can view this "function" as being a bar of height $1/\Delta x$ over the range $x_k \pm \frac{1}{2}\Delta x$ and zero elsewhere.

Exercise 7.4.2   Argue that when we take the limits $N \to \infty$, $\Delta x = L/N \to 0$, the function becomes a Dirac delta:

$$\psi_{x_n}(x) = \delta(x - x_n), \tag{7.25}$$

as claimed in Equation 7.9. (*Hint:* Review the beginning of section 5.4 about how the Dirac delta is defined. You needn't repeat any derivations in that section: simply describe how they apply in this case.)

BOX 7.5 The Two-Slit Interferometer

Figure 7.4 shows a simplified setup for a basic two-slit interferometer. Given the assumptions shown, if the distance between slit S_1 and the detector D is D_1, then that between slit S_2 and D is $D_1 + d \sin\theta$, where d is the slit separation. We will assume that the photons are free particles with the same energy and thus the same fixed momentum magnitude p_0.

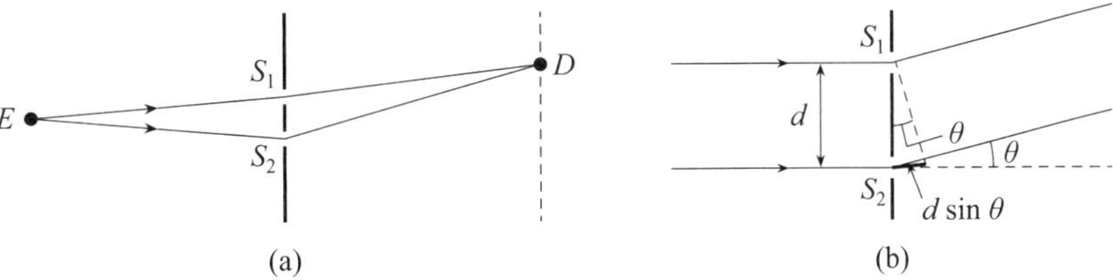

(a)   (b)

FIGURE 7.4   **(a)** The basic setup for a two-slit interferometer (with the distance between slits greatly exaggerated). Photons from an emitter E illuminate slits S_1 and S_2 and are ultimately detected by detector D somewhere along the dotted line. **(b)** A close-up diagram of the slits, which we assume are so close together that the two incoming lines from E are essentially parallel, and the outgoing lines to D are also essentially parallel.

Suppose that the amplitudes for a photon from emitter E to go through slits S_1 and S_2 are $\langle\psi_1|\psi_E\rangle$ and $\langle\psi_2|\psi_E\rangle$, respectively, and that we have arranged things symmetrically so that $\langle\psi_1|\psi_E\rangle = \langle\psi_2|\psi_E\rangle$. Since the photons are free particles, we will assume that they are in momentum eigenstates so that beyond slit $n = 1, 2$, a photon's wavefunction has the form $\langle\vec{x}|\psi_n\rangle = \psi_{p_0}(r_n) = Ae^{ip_0 r_n}$, where r_n is the straight-line distance from slit S_n to a given position $\vec{x}$. We will ignore complexities about how the waves evolve in time and how the amplitude A depends on r and θ and simply assume that θ and the difference in r are both small enough that A is the same for both paths. Finally, suppose we send one photon at a time through this apparatus.

Exercise 7.5.1 Use the rules of quantum mechanics to show that the probability the detector registers that photon is a function of θ that looks like an interference pattern.

Homework Problems

P7.1 In the Stern-Gerlach experiment, Gerlach sent a beam of silver atoms with a reasonably well-defined speed into a region permeated by a vertically downward magnetic field whose magnitude increases upward. A simplified diagram of the experiment appears in Figure 7.5.

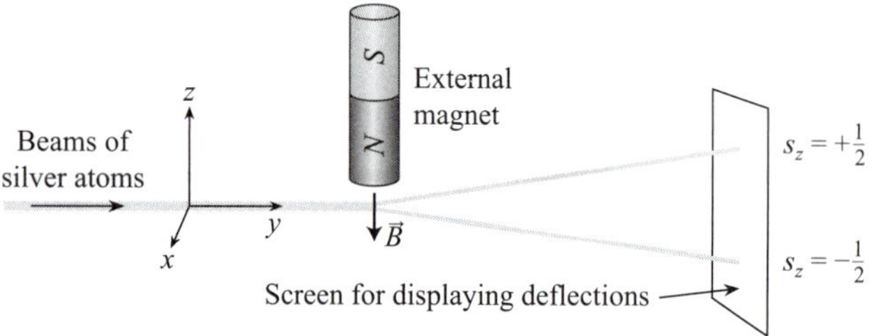

FIGURE 7.5   A simplified diagram of a Stern-Gerlach experiment. A beam of silver atoms passes just below the north pole of an external magnet, which creates a vertical magnetic field whose strength increases as one approaches the pole. Such a field will deflect a silver atom vertically depending on the projection of that atom's outer electron's magnetic moment (and thus its spin) on the field direction. One can calculate that spin projection by observing the deflection.

a. The magnetic moment $\vec{\mu}$ of a circular loop carrying current I in a circle of area A has a direction perpendicular to the circle's plane and a magnitude $|\vec{\mu}| = I A$. Imagine a spinning ball consisting of tiny particles, the ith of which has mass m_i and charge q_i, and is a distance r_i from the rotation axis. Assume each tiny particle's charge is proportional to its mass, so that $q_i/m_i = Q/M$, where Q and M are the ball's total charge and total mass, and assume that the ball rotates rigidly with some angular velocity $\vec{\omega}$. Show then that $\vec{\mu} = (Q/2M)\vec{s}$, where $\vec{s}$ is the ball's total spin angular momentum.

b. An external magnetic field $\vec{B}$ exerts a torque $\vec{\tau} = \vec{\mu} \times \vec{B}$ on a magnet with magnetic moment $\vec{\mu}$ that seeks to align the magnet with the field, as a compass needle with no initial angular momentum would. Why don't all the electrons in the silver atoms simply align themselves with the external field? Show that classical mechanics would predict that, since the electrons have spin angular momentum $\vec{s}$ and torque is defined to be $\vec{\tau} = d\vec{s}/dt$, the electrons will precess in a way that preserves the component of $\vec{s}$ in the field direction.

c. In a vertical external magnetic field $\vec{B}$ whose strength increases in the upward direction, a magnet will experience a net vertical force. The dipole model of a magnet asserts that a magnet behaves like north and south poles with "magnetic charge" $+q_m$ and $-q_m$, respectively, separated by a distance ℓ such that $|\vec{\mu}| = q_m\ell$. The force on each pole is $\vec{F}_m = q_m\vec{B}$. Use this model to show that the net force is vertical with magnitude $\mu_z(\partial B_z/\partial z)$.

d. In a silver atom, the magnetic moments of all electrons but one cancel out, so the deflecting force on a silver atom will be the same as that on a single electron. But why not use bare electrons instead of silver atoms? Estimate the magnitude of the direct magnetic force on a moving charged electron and compare it to the deflection force. How does using a silver atom avoid this problem?

P7.2 The radius R of an electron is known experimentally to be smaller than 10^{-18} m. If we assume the electron is a classical spinning solid sphere whose moment of inertia is $I = \frac{2}{5}mR^2$ and whose mass is $m = 9.11 \times 10^{-31}$ kg, show that an electron with spin angular momentum $I\omega = \frac{1}{2}\hbar$ in SI units (where ω is the angular rotation rate in radians/s) has an equatorial speed much greater than the speed of light.

P7.3 Calculate the outcome probabilities if the final Stern-Gerlach device in Experiment 3 were SGθ instead of SGz.

P7.4 Consider an experiment like Experiment 3, except that the middle device is an SGθ and the final device is an SGx device. What angle θ maximizes the probability that an electron going through the experiment emerges from the $+$ channel of that final device?

P7.5 This problem further examines Experiment 4.
a. Using the same method as in Exercise 7.3.4, explicitly calculate the final outcome probabilities if the middle device were SGθ instead of SGx. (*Hint:* The results should still be 1 and 0, for all θ.)
b. Indeed, use the facts (1) that the eigenvectors $|\pm n\rangle$ for the results of a measurement of the projection of an electron's spin along any possible direction $\vec{n}$ provide a complete basis for our spin Hilbert space, and (2) that, for a complete basis, $|+n\rangle\langle+n| + |-n\rangle\langle-n| = I$ according to the completeness relation in Equation 7.20, to argue that the results would be the same if we put *any* Stern-Gerlach device in the middle of Experiment 4.

P7.6 Suppose that we prepare electrons in an unknown quantum state $|\psi\rangle$. If we send such electrons through an SGz device, we find that the probabilities of obtaining $\pm\frac{1}{2}$ are $\frac{9}{25}$ and $\frac{16}{25}$, respectively. The probabilities if we send such electrons through an SGx device are $\frac{1}{2}$ for each result, and the probabilities if we send them through an SGy device are $\frac{1}{50}$ and $\frac{49}{50}$, respectively. What are the components of the state? (*Hints:* Assume that $|\psi\rangle$ has possibly complex components a and b, and use the experimental data in the order provided to successively constrain the values. Remember that if $|c|^2 = d$, where d is real, then $c = e^{i\theta}\sqrt{d}$, where θ could be anything. Finally, remember that probability information can only determine a state vector up to an *overall* phase factor $e^{i\phi}$—the same phase factor applied to both components. But the two components might have a *relative* phase difference.

P7.7 Why must quantum mechanics involve complex numbers? If we prepare electrons in a state $|+z\rangle$ and measure s_x, the text argued that the probabilities of both outcomes $\pm\frac{1}{2}$ must be $\frac{1}{2}$ by symmetry, since the axes are $90°$ apart. For the same reason, if we were to measure s_y, the probabilities should also be $\frac{1}{2}$. But if we prepare our electron in state $|+x\rangle$ and measure s_y, both probabilities should *still* be $\frac{1}{2}$, again by symmetry. Show that we cannot satisfy all of these constraints unless we use complex numbers. (*Hint:* Begin by supposing we choose the kets $|\pm z\rangle$ to be the basis vectors for our space. That is, they are the vectors given in Figure 7.3. Then assume that the components of $|\pm x\rangle$ and $|\pm y\rangle$ are all real. Show that we cannot satisfy all the probability constraints given this assumption. Then show that the eigenvectors in Figure 7.3 *do* satisfy all the constraints.)

P7.8 Consider an interference experiment like that discussed in Box 7.5, but with three slits, where both side slits are the same distance d from the center slit. What is the probability that the detector will register a given photon in this case as a function of θ? Submit a computer-drawn graph of this function.

P7.9 In the two-slit interference experiment, if we somehow can determine which slit the photon goes through, the interference pattern disappears. Use the rules to explain why.

P7.10 Suppose that a particle moving in one spatial dimension has a quantum state described by the wavefunction $\psi(x) = Ae^{ikx}e^{-ax^2}$, where A, a, and k are positive and real constants.

a. Use the rules of quantum mechanics to calculate the probability that, if we measure the particle's momentum, we get the result p_x. Express your result as a function of p_x, A, a, and k. (*Hint:* You may find the following integral helpful:

$$\int_{-\infty}^{+\infty} e^{-ax^2+bx} = e^{b^2/4a}\sqrt{\frac{\pi}{a}}.)$$

b. What is the most probable value of p_x?

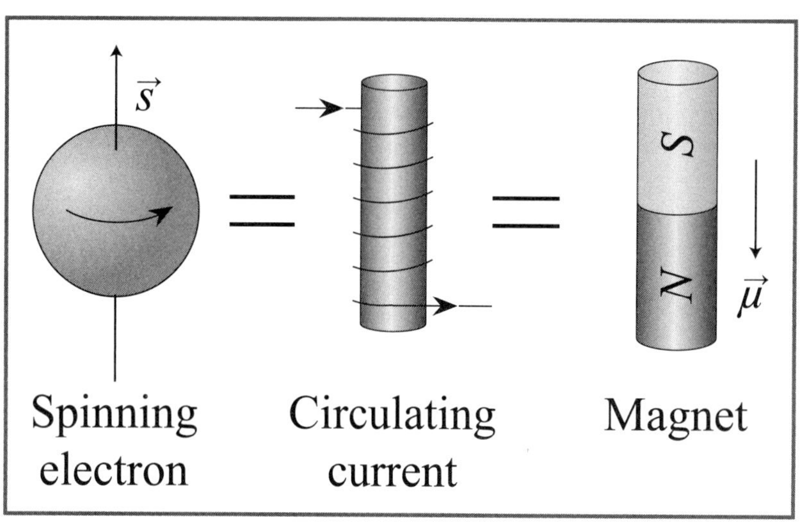

$\vec{s}$
Spinning electron
Circulating current
S
N
$\vec{\mu}$
Magnet

8. OPERATORS AND TIME EVOLUTION

In the last chapter, we discussed six rules that connect the quantum Hilbert-space formalism with physical reality. In this chapter, we finish our review of basic quantum mechanics by examining its seventh and final rule. However, to most concisely state that rule, we need to define what we mean by an *operator* and explore the roles that operators play in quantum mechanics. Operators play even more important roles in the Standard Model, so our efforts here will provide an important foundation for future chapters.

8.1 Operators and Observables

An **operator** O is a mathematical object that converts one ket to another:

$$|\psi_2\rangle = O|\psi_1\rangle. \tag{8.1}$$

In Hilbert space, we represent a (linear) operator by a square matrix with possibly complex components.

In quantum mechanics, people often use a "hat" ($\hat{O}$) to distinguish an operator from other kinds of quantities. In this text, I will follow the particle physics convention, which is to avoid the hat. In most cases, this will not be a problem, as either context or the symbol's definition itself will make the meaning clear. I will also often use a capital letter to distinguish an operator from a quantity designated by a small letter.

One reason that operators are useful in quantum mechanics is because *we can use an operator to represent an observable quantity*. The Eigenvector Rule requires that for every observable quantity (for example, the projection of a particle's spin on the x axis), we have a list of eigenvalues (possible measurement results $\pm\frac{1}{2}$) and a set of corresponding orthogonal and normalized eigenvectors (see Figure 7.3). Suppose we use these vectors and values to construct the following operator:

$$
\begin{aligned}
S_x &\equiv (+\tfrac{1}{2})|+x\rangle\langle+x| \;+\; (-\tfrac{1}{2})|-x\rangle\langle-x| \\[4pt]
&= \tfrac{1}{2}\begin{bmatrix} \sqrt{\tfrac{1}{2}} \\ \sqrt{\tfrac{1}{2}} \end{bmatrix}\begin{bmatrix} \sqrt{\tfrac{1}{2}} & \sqrt{\tfrac{1}{2}} \end{bmatrix} - \tfrac{1}{2}\begin{bmatrix} \sqrt{\tfrac{1}{2}} \\ -\sqrt{\tfrac{1}{2}} \end{bmatrix}\begin{bmatrix} \sqrt{\tfrac{1}{2}} & -\sqrt{\tfrac{1}{2}} \end{bmatrix} \\[4pt]
&= \tfrac{1}{2}\begin{pmatrix} \tfrac{1}{2} & \tfrac{1}{2} \\ \tfrac{1}{2} & \tfrac{1}{2} \end{pmatrix} - \tfrac{1}{2}\begin{pmatrix} \tfrac{1}{2} & -\tfrac{1}{2} \\ -\tfrac{1}{2} & \tfrac{1}{2} \end{pmatrix} = \tfrac{1}{2}\begin{pmatrix} 0 & 1 \\ 1 & 0 \end{pmatrix}.
\end{aligned}
\tag{8.2}
$$

(Note that while the "inner product" $\langle\phi|\psi\rangle$ of a bra and ket yields a complex number, the "outer product" $|\phi\rangle\langle\psi|$ yields a matrix.) Now, you may have learned in a linear algebra or math methods class that $|\psi\rangle$ is an "eigenvector" of a certain operator O if $|\psi\rangle$ satisfies the "eigenvector equation"

$$O|\psi\rangle = \lambda|\psi\rangle \tag{8.3}$$

for some number λ that we call that eigenvector's eigenvalue. You can see that as defined by this equation, the $|+x\rangle$ and $|-x\rangle$ kets are eigenvectors of the S_x operator with eigenvalues $+\frac{1}{2}$ and $-\frac{1}{2}$, respectively:

$$S_x|{+}x\rangle = \frac{1}{2}\begin{pmatrix} 0 & 1 \\ 1 & 0 \end{pmatrix}\begin{bmatrix} \sqrt{\frac{1}{2}} \\ \sqrt{\frac{1}{2}} \end{bmatrix} = \frac{1}{2}\begin{bmatrix} \sqrt{\frac{1}{2}} \\ \sqrt{\frac{1}{2}} \end{bmatrix} = \frac{1}{2}|{+}x\rangle, \tag{8.4a}$$

$$S_x|{-}x\rangle = \frac{1}{2}\begin{pmatrix} 0 & 1 \\ 1 & 0 \end{pmatrix}\begin{bmatrix} \sqrt{\frac{1}{2}} \\ -\sqrt{\frac{1}{2}} \end{bmatrix} = \frac{1}{2}\begin{bmatrix} -\sqrt{\frac{1}{2}} \\ \sqrt{\frac{1}{2}} \end{bmatrix}$$

$$= -\frac{1}{2}\begin{bmatrix} \sqrt{\frac{1}{2}} \\ -\sqrt{\frac{1}{2}} \end{bmatrix} = -\frac{1}{2}|{-}x\rangle. \tag{8.4b}$$

Since a 2×2 matrix can have at most two eigenvalues and associated eigenvectors, these are the *only* eigenvectors for this matrix.

We can similarly construct an operator whose eigenvectors and associated eigenvalues (the solutions to the eigenvector equation) are *any* given set of orthogonal kets, each paired with a numerical value. Now, solving the eigenvector equation for a square matrix is relatively straightforward. So, instead of giving someone a list of eigenvectors and associated eigenvalues for an observable, one can simply give that person an operator that contains *exactly the same information* in a nicely wrapped package.

Only certain kinds of operators can represent an observable this way. The **adjoint** $O^\dagger$ of an operator O is its **conjugate transpose**; that is, the matrix you get if you flip the original matrix across its diagonal and take the complex conjugate of its components. For a 2×2 matrix,

$$O^\dagger = \begin{pmatrix} a & b \\ c & d \end{pmatrix}^\dagger = \left[\begin{pmatrix} a & b \\ c & d \end{pmatrix}^T\right]^* = \begin{pmatrix} a^* & c^* \\ b^* & d^* \end{pmatrix}. \tag{8.5}$$

A **Hermitian** operator is **self-adjoint**: $O^\dagger = O$. You can see that the operator S_x we constructed above to represent the observable "spin projection on the x axis" s_x is Hermitian. Indeed, in Box 8.2, you will show that any operator that we construct out of orthogonal eigenvectors and real eigenvalues in a manner analogous to the way we constructed the S_x operator in Equation 8.2 is automatically Hermitian. You will also show the converse: any Hermitian operator has real eigenvalues and the corresponding eigenvectors are orthogonal.

We can therefore restate the second rule of quantum mechanics this way:

Rule 2: The Operator Rule

For every observable a, there is a $N \times N$ Hermitian operator A whose N eigenvalues are the results we might obtain when measuring this observable. The eigenvector $|a_n\rangle$ associated with each eigenvalue a_n provides the key to calculating the probability of that result.

The value of using operators to represent observables goes beyond providing a nice package for containing the eigenvalues and eigenvectors that we need. In many cases, we can also *anticipate* what the operator might be for an observable. For example, suppose we have an experimental apparatus that is able to measure the squared magnitude of an electron's spin. Since we know that $|\vec{s}|^2 = s_x^2 + s_y^2 + s_z^2$ in classical physics, might the operator S^2 representing that observable be equal to $S_x^2 + S_y^2 + S_z^2$? In Box 8.3, you will construct that operator and use it to predict that an electron in *any* spin state will be observed by such an apparatus to have $|\vec{s}|^2 = \frac{3}{4}$ (not $\frac{1}{4}$ as you might expect). This prediction turns out experimentally to be correct!

8.2 The Time Evolution Rule

Recall that the most basic definition of an operator is an object that converts a vector into another vector:

$$|\psi_2\rangle = O|\psi_1\rangle. \tag{8.6}$$

We can always represent such an operator by a square Hilbert-space matrix so long as the operator is *linear;* that is,

$$O\big(a|\psi_1\rangle + b|\psi_3\rangle\big) = aO|\psi_1\rangle + bO|\psi_3\rangle, \tag{8.7}$$

where a and b are (possibly complex) numbers. In what follows, we will assume that all of the operators we discuss are linear.

One operation that we would very much like to perform is evolving a particle's state forward in time. We ought to be able to find a "time-evolution operator" $U(t)$ that in a given situation converts the particle's state at time 0 to what we would predict that state to be at time t:

$$|\psi(t)\rangle = U(t)|\psi(0)\rangle. \tag{8.8}$$

Since U in this equation converts one ket to another, it satisfies the definition of an operator, and we should be able to represent it with a matrix (though unlike the operators we have seen so far, this matrix will depend on time).

How might we determine what this operator is in a given situation? We begin by putting some constraints on how this operator must behave. Since a particle's state must always be normalized, the operator U must preserve the ket's magnitude. Now, the equations

$$|\psi_2\rangle = O|\psi_1\rangle \quad \Leftrightarrow \quad \langle\psi_2| = \langle\psi_1|O^\dagger \tag{8.9}$$

are adjoints of each other (see Box 8.1). So, preserving a quantum state's normalization as it evolves requires that

$$1 = \langle\psi(t)|\psi(t)\rangle = \big[\langle\psi(0)|U^\dagger\big]U|\psi(0)\rangle = \langle\psi(0)|\psi(0)\rangle = 1. \tag{8.10}$$

This will be true in general only if $U^\dagger U = I$, where I is the identity matrix. We call an operator that satisfies this condition a **unitary** operator.

Now consider taking just an infinitesimal step dt forward in time. In such a case, U should be almost equal to the identity operator I (the state is barely changed if dt is infinitesimal), so we can write it as I plus a small correction:

$$U(dt) = I - iH\,dt, \tag{8.11}$$

where H is an operator that we call the **Hamiltonian** and also the **generator** of the U operator (the factor of $-i$ is conventional for reasons that will become clearer as we go on). To ensure that U is unitary, we must have

$$U^\dagger U = \big(I + iH^\dagger dt\big)\,(I - iH\,dt) = I + i(H^\dagger - H)\,dt = I \tag{8.12}$$

since $i^\dagger = i^* = -i$ and, in the limit that $dt \to 0$, we can ignore the dt^2 term. We see that to make this equation work to first order in dt, we must have $H^\dagger = H$; that is, H must be Hermitian. (The point of the factor of $-i$ is to ensure this. We will see why we use $-i$ as opposed to i shortly.)

Now, suppose we apply the infinitesimal operator at time t to move the state ahead in time to $t + dt$. Note that

$$\frac{d}{dt}|\psi(t)\rangle \equiv \lim_{dt \to 0} \frac{|\psi(t+dt)\rangle - |\psi(t)\rangle}{dt}$$

$$= \lim_{dt \to 0} \frac{1}{dt}\left[(I - iH\,dt) - I\right]|\psi(t)\rangle = -iH|\psi(t)\rangle \quad \text{or}$$

$$i\frac{d}{dt}|\psi(t)\rangle = H|\psi(t)\rangle \quad \text{or} \quad i\hbar\frac{d}{dt}|\psi(t)\rangle = H|\psi(t)\rangle \text{ in SI units.} \tag{8.13}$$

We call this equation the **time-dependent Schrödinger equation**. It is to quantum mechanics what Newton's second law is to classical mechanics. (Note how the $-i$ in the definition of H makes this equation nicely positive.)

Because H is Hermitian, it might correspond to an observable quantity for our particle. To see what this observable might be, suppose $|h_n\rangle$ is an eigenvector of H with eigenvalue h_n, and suppose the particle's state is $|h_n\rangle$ at time $t = t_0$. In such a case, Equation 8.13 becomes

$$\frac{d}{dt}|h_n\rangle = -ih_n|h_n\rangle. \tag{8.14}$$

If H is constant in time (so that $|h_n\rangle$ always *remains* an eigenvector of H with eigenvalue h_n), then we can integrate both sides of this expression (remember that the components of $|h_n\rangle$ are just numbers):

$$|h_n(t)\rangle = e^{-ih_n t}|h_n\rangle. \tag{8.15}$$

However, we saw in the last chapter that $e^{i\phi}|\psi\rangle$ is physically equivalent to $|\psi\rangle$ for all ϕ and $|\psi\rangle$. Therefore, we see that an eigenvector of H does not *physically* evolve in time. This in turn means that if we measure the observable corresponding to H at time $t = 0$ and get the result h_n, and we then measure that observable again at time t, we will get the same value h_n with probability 1. Therefore, the value of h_n *is conserved in time.*

So, we see that if Equation 8.13 is unchanged by a translation in time (including the matrix definition of H), we have an associated conserved quantity. Generally, we call the conserved quantity associated with time symmetry *energy*. On this basis, we identify H as the operator that represents a particle's energy. (The seemingly gratuitous factor of $\hbar$ in Equation 8.13 ensures that H also has the right units: $\hbar$ has units of kg m^2/s = energy $\cdot$ time.)

This result provides the basis for the seventh rule of quantum mechanics.

> **Rule 7: The Time-Evolution Rule**
>
> In between measurements, a particle's state evolves in time according to the time-dependent Schrödinger equation (TDSE):
>
> $$i\frac{d}{dt}|\psi\rangle = H|\psi\rangle \quad \left(\text{or } i\hbar\frac{d}{dt}|\psi\rangle = H|\psi\rangle \text{ in SI units}\right),$$
>
> where H is the operator corresponding to the particle's *energy*.

8.3 An Electron in a Magnetic Field

As an example, consider an electron in a vertical magnetic field. Classically, the potential energy of a magnetic moment $\vec{\mu}$ in such a magnetic field $\vec{B}$ is

$$V = -\vec{\mu} \cdot \vec{B} = -\mu_z B_z. \tag{8.16}$$

(This formula basically says that the magnet's energy is proportional to the projection of its magnetic moment on the field direction and is lowest when the magnet is aligned with the field.) Now, the magnetic moment of an electron happens to be $\vec{\mu} = -(e/m)\vec{s}$, where $-e$ and m are the charge and mass of an electron, respectively. Therefore,

$$V = \frac{e}{m} s_z B_z \equiv \omega s_z \quad \text{where } \omega \equiv \frac{e B_z}{m}. \tag{8.17}$$

Now, if the electron is isolated and motionless so that it has no other form of energy, then we might reasonably expect the energy operator (the Hamiltonian) in this case to be

$$H = \omega S_z = \frac{\omega}{2} \begin{pmatrix} 1 & 0 \\ 0 & -1 \end{pmatrix}. \tag{8.18}$$

Suppose the electron's state at time $t = 0$ is $|+x\rangle$. You will show in Box 8.4 that the electron's state at time t is

$$|\psi(t)\rangle = \begin{bmatrix} e^{-i\omega t/2}\sqrt{\tfrac{1}{2}} \\ e^{+i\omega t/2}\sqrt{\tfrac{1}{2}} \end{bmatrix} = \begin{pmatrix} e^{-i\omega t/2} & 0 \\ 0 & e^{+i\omega t/2} \end{pmatrix} \begin{bmatrix} \sqrt{\tfrac{1}{2}} \\ \sqrt{\tfrac{1}{2}} \end{bmatrix} \tag{8.19}$$

and that this is consistent with the electron's spin alignment rotating with angular frequency ω in the xy plane. This is exactly the behavior we would classically expect (and which has been experimentally verified).

8.4 Operator Functions

We can write time-evolution operator U in this case as

$$U(t) = \begin{pmatrix} e^{-i\omega t/2} & 0 \\ 0 & e^{+i\omega t/2} \end{pmatrix}. \tag{8.20}$$

Note that unlike the operators we have been using to represent observables, the components of this operator depend on time.

In Box 8.5, we will see that we can write this in the form

$$U(t) = e^{-i\omega S_z t}. \tag{8.21}$$

But what can it possibly mean to take the exponential of a matrix?

We really have three mathematical things we can do with or to $N \times N$ square matrices. We can add or subtract two matrices, we can multiply two matrices, and we can multiply a matrix by a number. Now, if x in e^x is a number, then we could write the exponential as the power series

$$e^x = 1 + x + \frac{1}{2!}x^2 + \frac{1}{3!}x^3 + \cdots. \tag{8.22}$$

When x is a number, this is not the *only* way we could define what e^x is, but it is certainly a valid way.

By analogy, we can *define* the exponential of a matrix A as follows:

$$e^A \equiv I + A + \frac{1}{2!}A^2 + \frac{1}{3!}A^3 + \cdots. \tag{8.23}$$

Note that all the mathematical operations on the right side of this expression are perfectly well-defined, so the matrix that results is perfectly well-defined.

Indeed, we define *any* function $f(A)$ of a matrix to be the power series we get if we substitute A for the number x in a valid power series expansion for $f(x)$: *we define matrix functions in terms of power series.* The notation $f(A)$ is therefore simply a shorthand expression for the power series.

In the chapters that follow, we will find the exponential of a matrix to be an especially useful matrix function. This is because the exponential also happens to be the correct expression for the limit of a series of infinitesimal unitary operations. For example, suppose H is the generator of the time-evolution operator $U(t/N) = I - iH(t/N)$, where N is very large so that t/N is very small. To get to a finite value of t, we would take N such time steps forward, which amounts to applying the operator $U(t/N)$ to our original state N times. But applying $U(t/N)$ operator N times is the same as applying the operator $[\,U(t/N)\,]^N$ once. Of course, to make t/N truly infinitesimal, we should take the limit that $N \to \infty$. As you will see in Box 8.6, it happens that

$$\lim_{N \to \infty} \left(I - iH\frac{t}{N} \right)^N = e^{-iHt}. \tag{8.24}$$

However, it is very important to note that this only applies if the generator H does *not* depend on time, because the argument that applying $U(t/N)$ successively N times is equivalent to applying $[\,U(t/N)\,]^N$ once depends on the value of H being the same at each successive instant when $U(t/N)$ is applied.

But if H *is* constant in time (as is commonly the case), we can simply write

$$|\psi(t)\rangle = e^{-iHt}|\psi(0)\rangle \tag{8.25}$$

instead of integrating the time-dependent Schrödinger equation. This is often quite a bit easier.

8.5 Infinite-Dimensional Operators

When our observables are position $\vec{x}$, momentum $\vec{p}$, or observables related to these (such as potential energy or kinetic energy), our Hilbert space has infinite dimensions, and using square matrices to represent an operator is not practical. But the basic definition of an operator is an object that linearly converts one quantum state to another. In an infinite-dimensional Hilbert space, we represent states by functions $\psi(x)$ instead of column vectors, so an operator in this context is an object that linearly converts one *function* to another.

In particular, we will see in Box 8.7 that the position and momentum operators (for a particle moving in one spatial dimension) are

$$X = x, \qquad P_x = -i\frac{d}{dx} \left(= -i\hbar\frac{d}{dx} \text{ in SI units} \right). \tag{8.26}$$

Note that $X\psi(x) = x\psi(x)$ and $P_x\psi(x) = -id\psi/dx$ are certainly different functions than $\psi(x)$. These operators are also clearly linear: for example,

$$x\big[a\psi_1(x) + b\psi_2(x)\big] = ax\psi(x) + bx\psi(x), \tag{8.27}$$

since multiplication by ordinary numbers is linear.

We can construct operators for another observable out of these basic operators by taking the classical expressions for the observable quantity in terms of x and p_x and converting $x \to X$ and $p_x \to P_x$. For example, a particle's total energy when it moves in one dimension in a region where its potential energy is $V(x)$ is $E = p_x^2/2m + V(x)$, so the Hamiltonian operator should be

$$H = \frac{P_x^2}{2m} + V(X) = -\frac{1}{2m}\frac{d^2}{dx^2} + V(x), \tag{8.28}$$

assuming that we can write $V(x)$ as a power series. To find the energy eigenfunctions and eigenvalues for a particle in this situation, we solve this operator's "eigenvector" equation:

$$H\psi(x) = -\frac{\hbar^2}{2m}\frac{d^2\psi}{dx^2} + V(x)\psi(x) = E\psi(x) \text{ (in SI units)}. \tag{8.29}$$

You may recognize this equation as the **time-independent Schrödinger equation**. The time-dependent Schrödinger equation (in SI units) becomes

$$i\hbar\frac{\partial\psi}{\partial t} = H\psi(t, x) = -\frac{\hbar^2}{2m}\frac{\partial^2\psi}{\partial x^2} + V(x)\psi(t, x) \tag{8.30}$$

(we need partial derivatives because we are now thinking of the particle's wavefunction as being a function of both position and time). We see that in both cases, the operator equations that were matrix equations in the spin context become *differential equations* in this context.

The generalization to three spatial directions is again reasonably straightforward. We have three position component operators

$$X = x, \qquad Y = y, \qquad Z = z \tag{8.31a}$$

and three momentum component operators

$$P_x = -i\frac{\partial}{\partial x}, \qquad P_y = -i\frac{\partial}{\partial y}, \qquad P_z = -i\frac{\partial}{\partial z} \tag{8.31b}$$

or, in a more compact notation

$$\vec{X} = \vec{x}, \qquad \vec{P} = -i\vec{\nabla}. \tag{8.31c}$$

The two Schrödinger equations become (in SI units)

$$-\frac{\hbar^2}{2m}\nabla^2\psi(x, y, z) + V(x, y, z)\psi(x, y, z) = E\psi(x, y, z) \quad \text{and} \tag{8.32a}$$

$$-\frac{\hbar^2}{2m}\nabla^2\psi(t, x, y, z) + V(x, y, z)\psi(t, x, y, z) = i\hbar\frac{\partial\psi}{\partial t}, \tag{8.32b}$$

where $\nabla^2\psi(t, x, y, z) \equiv \vec{\nabla}\cdot\vec{\nabla}\psi = \partial^2\psi/\partial x^2 + \partial^2\psi/\partial y^2 + \partial^2\psi/\partial z^2$.

You have now seen the basic structure of quantum mechanics! The main purpose of a course in quantum mechanics is to develop techniques for actually *solving* and applying these fairly complicated partial differential equations in realistic contexts. (Thankfully, that is not our purpose here.)

BOX 8.1   Adjoint Rules

In this box, we will prove some useful theorems about adjoints.

The adjoint of a matrix is its conjugate transpose: we exchange rows for columns and take the complex conjugate of each component. This means that the adjoint of a ket is a bra and vice versa:

$$(|\psi\rangle)^\dagger = \left(\begin{bmatrix} \psi_1 + a \\ \psi_2 \\ \vdots \end{bmatrix}\right)^\dagger = \begin{bmatrix} \psi_1 & \psi_2 & \cdots \end{bmatrix}^* = \begin{bmatrix} \psi_1^* & \psi_2^* & \cdots \end{bmatrix} \equiv \langle\psi|, \quad (8.33a)$$

$$(\langle\psi|)^\dagger = \left(\begin{bmatrix} \psi_1 & \psi_2 & \cdots \end{bmatrix}\right)^\dagger = \begin{bmatrix} \psi_1^* \\ \psi_2^* \\ \vdots \end{bmatrix}^* = \begin{bmatrix} \psi_1 \\ \psi_2 \\ \vdots \end{bmatrix} \equiv |\psi\rangle. \quad (8.33b)$$

The adjoint of a complex number is simply its complex conjugate, since a number has no rows or columns to interchange.

Exercise 8.1.1 Show that $(O|\psi\rangle)^\dagger = \langle\psi|O^\dagger$ and that $(AB)^\dagger = B^\dagger A^\dagger$. (*Hints:* Assume a two-dimensional complex space: the extension to N dimensions is straightforward. You may calculate each side and show that they are equal.)

Similarly, one can show that $(\langle u|w\rangle)^\dagger = \langle w|u\rangle = \langle u|w\rangle^*$, and that $(|u\rangle\langle w|)^\dagger = |w\rangle\langle u|$ for arbitrary kets $|u\rangle$ and $|w\rangle$. The basic idea is that the adjoint of a product is the *reversed* product of the adjoints.

However, the adjoint of a sum is the sum of the adjoints. For example,

$$(A + B)^\dagger = \left[\begin{pmatrix} a & b \\ c & d \end{pmatrix} + \begin{pmatrix} e & f \\ g & h \end{pmatrix}\right]^\dagger = \left[\begin{pmatrix} a+e & b+f \\ c+g & d+h \end{pmatrix}\right]^\dagger \quad (8.34)$$

$$= \begin{pmatrix} a^*+e^* & c^*+g^* \\ b^*+f^* & d^*+h^* \end{pmatrix} = \begin{pmatrix} a^* & c^* \\ b^* & d^* \end{pmatrix} + \begin{pmatrix} e^* & g^* \\ f^* & h^* \end{pmatrix} = A^\dagger + B^\dagger.$$

BOX 8.2   Hermitian Matrices

In this box, we will prove some useful theorems about Hermitian matrices. In particular, we will show that the eigenvalues of a Hermitian matrix are real and that the eigenvectors for distinct eigenvalues are orthogonal, making such matrices suitable for representing observables.

Exercise 8.2.1   Consider a matrix A of the form $A = \sum_n a_n |a_n\rangle\langle a_n|$, where a_n are a set of real eigenvalues and $|a_n\rangle$ are the corresponding eigenvectors. Use the results discussed in Box 8.1 to show that such a matrix is Hermitian.

Exercise 8.2.2   The adjoint of the eigenvector equation $O|\psi\rangle = \lambda|\psi\rangle$ is given by $\langle\psi|O^\dagger = \langle\psi|\lambda^*$. Use this to show that any eigenvalue λ of a Hermitian matrix must be real. (*Hint:* Compare $\langle\psi|\big[O|\psi\rangle\big] = \langle\psi|O|\psi\rangle = \big[\langle\psi|O^\dagger\big]|\psi\rangle$.)

Exercise 8.2.3   Suppose $|\psi_1\rangle$ and $|\psi_2\rangle$ are distinct eigenvectors of a Hermitian operator O with different eigenvalues λ_1 and λ_2. Show that $|\psi_1\rangle$ and $|\psi)_2\rangle$ must be orthogonal. (*Hint:* Use a similar approach to the above except with $\langle\psi_2|O|\psi_1\rangle$.)

BOX 8.3 The Squared-Spin Operator

Exercise 8.3.1 Use the approach illustrated in Equation 8.2 to construct the matrices representing the observables s_y and s_z.

Exercise 8.3.2 Construct the operator $S^2 \equiv S_x^2 + S_y^2 + S_z^2$ and argue that *all* two-dimensional vectors are eigenvectors of this operator with eigenvalue $\frac{3}{4}$. (We don't have just a pair of orthogonal eigenvectors in this case because the eigenvalues for different eigenvectors are not distinct: see the assumptions of the proof in Exercise 8.2.3.)

BOX 8.4 Spin Precession

In this box, we will solve the time-dependent Schrödinger equation for the situation where we have an electron in a vertical magnetic field. We saw in Equation 8.18 that the Hamiltonian operator in this case is

$$H = \omega S_z = \frac{\omega}{2} \begin{pmatrix} 1 & 0 \\ 0 & -1 \end{pmatrix}. \tag{8.35}$$

Let the components of our state vector $|\psi\rangle$ be a and b. The time-dependent Schrödinger equation $i\, d|\psi\rangle/dt = H|\psi\rangle$ in this case tells us that

$$i \frac{d}{dt} \begin{bmatrix} a \\ b \end{bmatrix} = \frac{\omega}{2} \begin{pmatrix} 1 & 0 \\ 0 & -1 \end{pmatrix} \begin{bmatrix} a \\ b \end{bmatrix} = \frac{\omega}{2} \begin{bmatrix} a \\ -b \end{bmatrix}$$

$$\Rightarrow \begin{bmatrix} da/dt \\ db/dt \end{bmatrix} = \begin{bmatrix} (-i\omega/2)a \\ (+i\omega/2)b \end{bmatrix}, \tag{8.36}$$

where in the final step, I have multiplied both sides by $-i$.

Exercise 8.4.1 Solve these differential equations for $a(t)$ and $b(t)$, assuming that $a(0) = \sqrt{\tfrac{1}{2}} = b(0)$.

BOX 8.4   Spin Precession (cont.)

We have calculated that the quantum state of an electron in a vertical magnetic field, if the electron's initial state is $|\psi(0)\rangle = |+x\rangle$, is given by

$$|\psi(t)\rangle = \begin{bmatrix} e^{-i\omega t/2}\sqrt{\tfrac{1}{2}} \\ e^{+i\omega t/2}\sqrt{\tfrac{1}{2}} \end{bmatrix}. \tag{8.37}$$

But what does this mean physically?

The physical meaning is a bit easier to see if we pull out an overall phase:

$$|\psi(t)\rangle = e^{-i\omega t/2}\begin{bmatrix} \sqrt{\tfrac{1}{2}} \\ e^{+i\omega t}\sqrt{\tfrac{1}{2}} \end{bmatrix}. \tag{8.38}$$

We have seen that an overall phase factor has no physical meaning. Therefore, we can throw it away: the previous state is physically equivalent to

$$|\psi(t)\rangle = \begin{bmatrix} \sqrt{\tfrac{1}{2}} \\ e^{+i\omega t}\sqrt{\tfrac{1}{2}} \end{bmatrix}. \tag{8.39}$$

Exercise 8.4.2   At what time is this state the same as $|-x\rangle$? $|+y\rangle$? $|-y\rangle$? When does it return to being $|+x\rangle$? What do the answers to these questions suggest the electron's spin state is doing as time passes?

BOX 8.5 An Operator Function Example

In this box, we will see why

$$e^{-i\omega S_z t} = \begin{bmatrix} e^{-i\omega t/2} & 0 \\ 0 & e^{+i\omega t/2} \end{bmatrix}. \tag{8.40}$$

First, note that

$$S_z = \tfrac{1}{2}\begin{pmatrix} 1 & 0 \\ 0 & -1 \end{pmatrix}, \qquad S_z^2 = \left(\tfrac{1}{2}\right)^2 \begin{pmatrix} 1 & 0 \\ 0 & 1 \end{pmatrix},$$

$$S_z^3 = \left(\tfrac{1}{2}\right)^3 \begin{pmatrix} 1 & 0 \\ 0 & -1 \end{pmatrix}, \qquad S_z^4 = \left(\tfrac{1}{2}\right)^4 \begin{pmatrix} 1 & 0 \\ 0 & 1 \end{pmatrix}, \quad \text{and so on.} \tag{8.41}$$

We see that even powers of S_z look like the identity matrix times that power of $\tfrac{1}{2}$, while odd powers look like S_z times the appropriate power of $\tfrac{1}{2}$.

Exercise 8.5.1 With this in mind, expand $e^{-i\omega S_z t}$ as a power series and add the resulting matrices to get Equation 8.40.

BOX 8.6   An Exponential as the Product of Small Steps

In this box, we will see why

$$\lim_{N \to \infty} \left(I - iH\frac{t}{N} \right)^N = e^{-iHt}, \tag{8.42}$$

as claimed in Equation 8.24. First, note that

$$(1 + x)^N = 1 + Nx + \frac{N(N-1)}{2!}x^2 + \frac{N(N-1)(N-2)}{3!}x^3 + \cdots, \tag{8.43}$$

as you can easily find from various online sources. This means that

$$\left(1 + \frac{x}{N} \right)^N = 1 + N\frac{x}{N} + \frac{N(N-1)}{2!}\left(\frac{x}{N}\right)^2 + \frac{N(N-1)(N-2)}{3!}\left(\frac{x}{N}\right)^3 + \cdots$$

$$= 1 + x + \frac{x^2}{2!}\frac{N(N-1)}{N^2} + \frac{x^3}{3!}\frac{N(N-1)(N-2)}{N^3} + \cdots. \tag{8.44}$$

Exercise 8.6.1   Argue that in the limit that $N \to \infty$, this will become

$$\lim_{N \to \infty} \left(1 + \frac{x}{N} \right)^N = 1 + x + \frac{x^2}{2!} + \frac{x^3}{3!} + \cdots = e^x. \tag{8.45}$$

(*Hint:* You needn't strive for a mathematically rigorous proof that worries about convergence or extremely high-order terms. Go for a "physicist's proof" that focuses on terms involving finite powers of x.)

Exercise 8.6.2   How does this imply the validity of Equation 8.42? (*Hint:* No math needed! A couple of sentences describing the connection is sufficient.)

BOX 8.7 The Operators for x-Position and x-Momentum

In this box, we will argue that the operators $X = x$ and $P_x = -i\, d/dx$ have eigenfunctions and corresponding eigenvalues that we derived in other ways in Chapter 7 (see Equations 7.9 and 7.12). As we saw in the last chapter, an operator describes an observable if and only if its eigenvalues and the corresponding eigenfunctions are consistent with what one can infer for that observable from experiment or other arguments. First, consider the operator $X = x$.

Exercise 8.7.1 Show that the function $\psi_{x_0}(x) = \delta(x - x_0)$ is an eigenfunction of the X operator with eigenvalue x_0.

Exercise 8.7.2 Show that the function $\psi_{p_{0x}}(x) = A e^{i p_{0x} x}$ (where A is the appropriate normalization constant) is an eigenfunction of the P_x operator with eigenvalue p_{0x}.

Homework Problems

P8.1 Construct the matrix that represents the observable s_θ.

P8.2 Construct the matrix $e^{-iS_x\theta}$ for a spin-$\frac{1}{2}$ particle.

P8.3 Construct the matrix $e^{-iS_y\theta}$ for a spin-$\frac{1}{2}$ particle.

P8.4 Use properties of the exponential function to argue that any matrix of the form $U = e^{-iM\theta}$ is unitary so long as M is Hermitian.

P8.5 Show that the eigenvalues of a unitary matrix U must be of the form $e^{i\phi}$ for some real ϕ, no matter what the matrix's eigenvectors might be. (*Hint:* Any unitary matrix must satisfy the requirement that $U^\dagger U = I$.)

P8.6 For the case of an electron in a vertical magnetic field, suppose the electron's initial state is $|\psi(0)\rangle = |+z\rangle$.
a. What is the electron's state $|\psi(t)\rangle$ at time t?
b. Does this answer make physical sense? Explain. (*Hint:* The electron's spin axis should precess around the field direction.)

P8.7 For the case of an electron in a vertical magnetic field, suppose the electron's initial state is $|\psi(0)\rangle = |+\theta\rangle$.
a. What is the electron's state $|\psi(t)\rangle$ at time t?
b. Does this answer make physical sense? Explain. (*Hint:* The electron's spin axis should precess around the field direction.)
c. What is the angular frequency of this precession? Justify your response. (*Hint:* When does the state return to its initial value?)

P8.8 Consider the case of an electron in a magnetic field pointing in the $+x$ direction, and suppose the electron's initial state is $|\psi(0)\rangle = |+z\rangle$.
a. What is the electron's state $|\psi(t)\rangle$ at time t? (*Hint:* This problem is easier when paired with Problem P8.2.)
b. Does this answer make physical sense? Explain. (*Hint:* The electron's spin axis should precess around the field direction.)
c. What is the angular frequency of this precession? Justify your response. (*Hint:* When does the state return to its initial value?)

P8.9 In an infinite-dimensional Hilbert space, how can we tell whether an operator is Hermitian (since we have no matrix to transpose)? Note that if $|\psi_2\rangle = O|\psi_1\rangle$, then

$$\langle\psi_3|\psi_2\rangle = \langle\psi_3|O|\psi_1\rangle = \langle\psi_2|\psi_3\rangle^* = \langle\psi_1|O^\dagger|\psi_3\rangle^*. \quad (8.46)$$

This means that $\langle\psi_3|O|\psi_1\rangle = \langle\psi_1|O|\psi_3\rangle^*$ for arbitrary kets $|\psi_1\rangle$ and $|\psi_3\rangle$ if and only if O is Hermitian.
a. Show that the $X = x$ operator is Hermitian.
b. Show that the $P_x = -i\,d/dx$ operator is Hermitian. (*Hint:* This is a bit tougher. Use integration by parts

$$\int_{-\infty}^{+\infty} u\left(\frac{dv}{dx}\right)\,dx = uv\Big|_{-\infty}^{+\infty} - \int_{-\infty}^{+\infty}\left(\frac{du}{dx}\right)v\,dx \quad (8.47)$$

and the fact that normalizable functions go to zero at $\pm\infty$.)

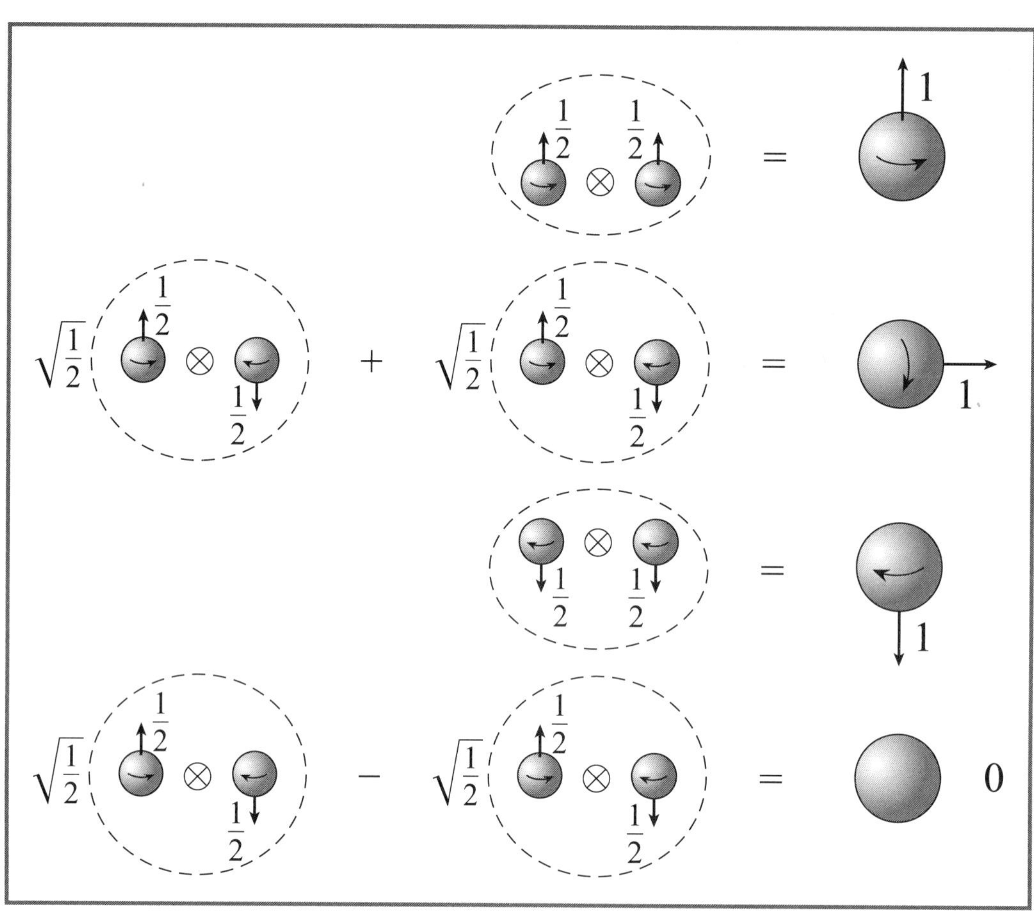

9. ANGULAR MOMENTUM

Symmetry under rotations is one of the most basic symmetries of the universe. We know from Noether's theorem that rotational symmetry is associated with conservation of angular momentum, a conservation law that is important in the Standard Model. Understanding angular momentum is also crucial for understanding particle spin, another central aspect of the Standard Model. Finally, rotation serves as a model for other symmetries in the Standard Model that are similar but not due to rotations in physical space.

This chapter's goal, therefore, is to explore the mathematics of rotation in order to provide a background for these important applications.

9.1 Rotation Operators for Electron Spin

We saw in the last chapter that the time-evolution operator $U(t) = e^{-i\omega S_z t}$ has the effect of rotating an electron's effective spin axis counterclockwise around the $+z$ axis at a rate of ωt. In fact, as Box 9.1 demonstrates, the operators

$$R_x(\theta) = e^{-iS_x\theta}, \qquad R_y(\theta) = e^{-iS_y\theta}, \qquad R_z(\theta) = e^{-iS_z\theta} \tag{9.1}$$

rotate a spin state by an angle θ around the $+x$, $+y$, and $+z$ axes, respectively.

We can think about this in the reverse direction. Just as the generator H of the time-evolution operator e^{-iHt} turns out to be the operator for energy (the observable that is conserved when the physics of the system does not depend on time), the generators S_x, S_y, and S_z of rotations around the x, y, and z axes should be the operators corresponding to the components of angular momentum along those axes (that is, the observables that are conserved when the physics of the system does not depend on its orientation in space). This logic provides an alternative means to *deriving* these operators instead of simply appealing to experiment. Knowing these operators would have allowed us to *predict* what the eigenvalues and eigenvectors of angular momentum should be.

In fact, if we know the matrix form of the operator $R_y(\theta)$, then

$$\frac{dR_y(\theta)}{d\theta} = -iS_y e^{-iS_y\theta} \quad \Rightarrow \quad S_y = i\left[\frac{dR_y(\theta)}{d\theta}\right]_{\theta=0} \tag{9.2}$$

allows us to directly calculate S_y. This is precisely the approach that we will use in future chapters to derive the generators of various other kinds of transformations.

9.2 Rotation Operators in General

Equation 9.1 gives the rotation operators in the situation where we are rotating electron spin quantum states. How might we find rotation operators in a more general situation?

We start by noting something very fundamental about rotation: *rotations around different axes do not commute.* You can see this for yourself: lay a book flat on a table, face up. First, rotate the book 90° clockwise around the vertical axis, then 90° clockwise around a horizontal axis facing you. You should see the book standing on its original top edge, with its front facing to your right. Now restore the book to its original position and perform the two rotation operations in reverse order. You should now see the book standing on what was its original right edge, with its back facing you; not the same result at all. (I am deliberately not including a picture because I want you to actually *do* this experiment to see the result vividly for yourself.)

No matter what kind of mathematical object we are rotating in space (electron spin quantum states, ordinary three-vectors in space, or whatever), this basic lack of commutativity must appear. In fact, the common, precise way that infinitesimal rotations *fail* to commute turns out to be sufficient to determine what the rotation matrices must be in any context, as we are about to see.

We can always define an infinitesimal rotation around an axis as the identity operator plus a small matrix correction:

$$R_x(d\theta) = I - i J_x \, d\theta, \qquad R_y(d\theta) = I - i J_y \, d\theta, \qquad R_z(d\theta) = I - i J_z \, d\theta, \quad (9.3)$$

where $d\theta$ is the infinitesimal rotation angle, the negative sign is conventional, and the i ensures that J_z is Hermitian. (I am using J here instead of S because we are no longer necessarily talking about *spin* angular momentum: J_x, J_y, and J_z might instead represent components of an electron's *orbital* angular momentum $\vec{L}$. J conventionally represents both possibilities. Note also that in natural units, J_x, J_y, and J_z all have the same units as $\hbar = 1$, and so are unitless.)

As we saw in the last chapter, we can construct the matrix representing a rotation through a finite angle by successively applying infinitesimal rotations

$$R_x(\theta) = \lim_{N \to \infty} \left(I - i J_x \frac{\theta}{N} \right)^N = e^{-i J_x \theta}, \qquad (9.4)$$

and similarly for rotations around other axes.

In Box 9.2, you will apply this mathematics to the infinitesimal rotation of ordinary three-dimensional displacement vectors in actual space (where one can see more intuitively what the rotation must be) to show that the lack of commutativity of rotation implies that the generators J_x, J_y, and J_z obey the following **commutation relations:**

$$[J_x, J_y] = i J_z, \qquad [J_y, J_z] = i J_x, \qquad [J_z, J_x] = i J_y, \qquad (9.5)$$

where $[A, B] \equiv AB - BA$. (A useful mnemonic for the relations above is to note that the subscripts always go in alphabetical order, and when you get to z, you wrap back around to x.) As these relations express the fundamental lack of commutativity of rotations, they must apply to the generators of rotations of whatever kind of mathematical object we are considering, not just ordinary displacement vectors in real space.

9.3 Quantum Angular Momentum Eigenstates

In this chapter, we are looking for operators that spatially rotate particle quantum states in N-dimensional Hilbert space. The generators J_x, J_y, and J_z of these rotations will represent angular momentum observables in the quantum-mechanical context.

In Box 9.3, you will see that if and only if observable operators commute, they have common eigenvectors. Since the J_x, J_y, and J_z don't commute, we cannot specify a particle or system's angular momentum quantum state (the way we would classically) by specifying the numerical values of the corresponding observables j_x, j_y, and j_z, because the eigenvectors corresponding to definite values of these observables are not the same. However, you will see in Box 9.4 that $J^2 \equiv J_x^2 + J_y^2 + J_z^2$ does commute with all three. The best we can do for angular momentum quantum states is to label them by the eigenvalues of the J^2 operator and the eigenvalues of any one of the spin projection operators. We conventionally select the second operator to be J_z:

$$J^2 |\lambda, m\rangle = \lambda |\lambda, m\rangle, \qquad J_z |\lambda, m\rangle = m |\lambda, m\rangle. \qquad (9.6)$$

We need to do this in general because knowing only the projection of the angular momentum on a single axis doesn't tell us anything about the *magnitude* of that angular momentum, whereas simultaneously specifying the eigenvalues λ and m of J^2 and J_z, respectively, gives us necessary information about both the magnitude and projection of an object's angular momentum. We didn't worry about the magnitude when talking about electron spin states because electron spins always have the *same* magnitude. But in this more general context, we need this extra information to completely specify the quantum state.

9.4 Angular Momentum Eigenvalues

To determine what the possible values of λ and m might be, it helps to define two new operators

$$J_+ \equiv J_x + i J_y \quad \text{and} \quad J_- \equiv J_x - i J_y. \tag{9.7}$$

As you will show in Box 9.5, these operators have the effect of converting an angular momentum eigenstate into one with a higher or lower value of m: if we define $|\lambda, m_+\rangle \equiv J_+|\lambda, m\rangle$ and $|\lambda, m_-\rangle \equiv J_-|\lambda, m\rangle$, then

$$J_z|\lambda, m_+\rangle = (m + 1)|\lambda, m_+\rangle \quad \text{and} \quad J_z|\lambda, m_-\rangle = (m - 1)|\lambda, m_-\rangle, \tag{9.8}$$

meaning that

$$|\lambda, m_+\rangle = c_+|\lambda, m + 1\rangle \quad \text{and} \quad |\lambda, m_-\rangle = c_-|\lambda, m - 1\rangle, \tag{9.9}$$

where c_+ and c_- are constants yet to be specified. (Note that the eigenvector equation $A|a_n\rangle = a_n|a_n\rangle$ only specifies the eigenvector $|a_n\rangle$ to an overall constant, since $c|a_n\rangle$ satisfies that equation as well.)

However, we can't raise or lower the value of m without limit, because the absolute value of the maximum projection of an object's angular momentum along any axis cannot be greater than that angular momentum's magnitude. In our context, this means that $|m| \leq \sqrt{\lambda}$, or (more conveniently)

$$m^2 \leq \lambda \tag{9.10}$$

(since the eigenvalue λ of the J^2 operator must be positive if it is to represent the squared magnitude of an object's angular momentum). Define j to be the maximum value of m and k to be the minimum value of m for a given value of λ. This means that we must have

$$J_+|\lambda, j\rangle = 0 \quad \text{and} \quad J_-|\lambda, k\rangle = 0 \tag{9.11}$$

because if this were not true, we could use $J_\pm$ to go on raising or lowering the value of a state's m indefinitely. But, as you will see in Box 9.6, the equations above require that

$$\lambda = j(j + 1) \quad \text{and} \quad k = -j \tag{9.12}$$

(one might suspect the latter result by symmetry).

Finally, if we start with the ket $|\lambda, -j\rangle$ and increment its value of m by successive applications of J_+, we must eventually get to $c|\lambda, j\rangle$ (where c is some constant) because otherwise we violate our assumption that j is the maximum possible value of m. But this means that there must exist an integer number of possible steps between $-j$ and j. Since the size of each step is 1, the possible values of j must be such that $2j = $ some integer, or

$$j = 0, \tfrac{1}{2}, 1, \tfrac{3}{2}, 2, \tfrac{5}{2}, \ldots . \tag{9.13}$$

For any given value of j, there are $2j + 1$ possible values of m:

$$m = -j, -j + 1, \ldots, j - 1, j. \tag{9.14}$$

Since j has nice, easy-to-memorize values, while the values of $\lambda = j(j+1)$ are not so memorable, we will (following convention) switch the labels on our angular momentum states from $|\lambda, m\rangle$ to $|j, m\rangle$ from now on. This switch has no consequences because, if we know the value of j, we can calculate the value of λ and vice versa.

The fact that an electron has two spin projection eigenstates therefore means that $j = \frac{1}{2}$ for an electron. We therefore say that an electron is a "spin-$\frac{1}{2}$ particle." In $|j, m\rangle$ notation, our old basis state vectors are written

$$|+z\rangle = \left|\tfrac{1}{2}, +\tfrac{1}{2}\right\rangle \quad \text{and} \quad |-z\rangle = \left|\tfrac{1}{2}, -\tfrac{1}{2}\right\rangle. \tag{9.15}$$

If we measure the squared angular momentum for such a particle, we always find that $J^2 = \frac{1}{2}(\frac{1}{2}+1) = \frac{3}{4}$.

Later in the course, we will encounter particles with spin $j = 0$ and spin $j = 1$. If we measure J^2 or J_z for a spin-0 particle, we always get zero. If we measure J^2 for a spin-1 particle, we get 2, with possible spin projections of -1, 0, and $+1$. While particles that are constructed of elementary particles can have spins with other values of j, the spins of all *elementary* particles in the Standard Model are either 0, $\frac{1}{2}$, or 1.

The *orbital* angular momenta of particles in a system must have integer values of j, ultimately because a particle's wavefunction $\psi(\vec{x})$ must connect with itself after a rotation of 2π around any axis (see Problem P9.5 for more discussion). We will see why this does *not* apply to particle spin in later chapters.

9.5 Matrices for J_x, J_y, and J_z

Since we now know the eigenvectors and eigenvalues for the J^2 and J_z operators, we can construct matrices for these operators quite easily if we use (as is conventional) the $|j, m\rangle$ quantum states as a basis. For example, for a spin-1 particle, the basis vectors are (by convention)

$$|1, 1\rangle \equiv \begin{bmatrix} 1 \\ 0 \\ 0 \end{bmatrix}, \qquad |1, 0\rangle \equiv \begin{bmatrix} 0 \\ 1 \\ 0 \end{bmatrix}, \qquad |1, -1\rangle \equiv \begin{bmatrix} 0 \\ 0 \\ 1 \end{bmatrix}, \tag{9.16}$$

and the matrix for the J_z operator is

$$J_z = +|1, 1\rangle\langle 1, 1| + 0|1, 0\rangle\langle 1, 0| - |1, -1\rangle\langle 1, -1|$$

$$= \begin{bmatrix} 1 \\ 0 \\ 0 \end{bmatrix} \begin{bmatrix} 1 & 0 & 0 \end{bmatrix} + 0 - \begin{bmatrix} 0 \\ 0 \\ 1 \end{bmatrix} \begin{bmatrix} 0 & 0 & 1 \end{bmatrix}$$

$$= \begin{pmatrix} 1 & 0 & 0 \\ 0 & 0 & 0 \\ 0 & 0 & 0 \end{pmatrix} - \begin{pmatrix} 0 & 0 & 0 \\ 0 & 0 & 0 \\ 0 & 0 & 1 \end{pmatrix} = \begin{pmatrix} 1 & 0 & 0 \\ 0 & 0 & 0 \\ 0 & 0 & -1 \end{pmatrix}. \tag{9.17}$$

Finding the matrices for the J_x and J_y operators is a bit trickier. It is best to go through the J_+ and J_- operators. In Box 9.7, you will show that we can define the values of c_+ and c_- in Equation 9.9 to be such that

$$J_+|j, m\rangle = \sqrt{j(j+1) - m(m+1)}\,|j, m+1\rangle \quad \text{and} \tag{9.18a}$$

$$J_-|j, m\rangle = \sqrt{j(j+1) - m(m-1)}\,|j, m-1\rangle. \tag{9.18b}$$

From these relations, one can construct the matrices for J_+ and J_-, and from there, find the matrices for

$$J_x = \tfrac{1}{2}(J_+ + J_-) \quad \text{and} \quad J_y = i\tfrac{1}{2}(J_- - J_+). \tag{9.19}$$

See Box 9.8 for details.

9.6 Addition of Angular Momentum

Now consider a system consisting of two spin-$\frac{1}{2}$ particles, perhaps an electron and a positron in a bound state. From a purely classical perspective, we might consider the magnitude of the total angular momentum of two particles with spin projections of $\pm\frac{1}{2}$ on any axis to be as large as 1 (if the spins are aligned) or as small as zero (if the spins are opposite). Since quantum mechanics only allows total j to be integers or half-integers, we might expect total j values for this system to be 0, $\frac{1}{2}$, or 1.

To be sure, we need to do a more careful analysis. In quantum mechanics, we can describe the system's angular momentum state using a basis comprised of what we call **direct product states** $|j_1, m_1\rangle \otimes |j_2, m_2\rangle$ (often written simply as $|j_1, m_1\rangle|j_2, m_2\rangle$), where the subscripts indicate which particles we have arbitrarily chosen to be particles 1 and 2. For a pair of spin-$\frac{1}{2}$ particles, there are four possible direct-product states corresponding to all possible combinations of the two states available to each particle:

$$\left|\tfrac{1}{2}, +\tfrac{1}{2}\right\rangle \otimes \left|\tfrac{1}{2}, +\tfrac{1}{2}\right\rangle, \qquad \left|\tfrac{1}{2}, +\tfrac{1}{2}\right\rangle \otimes \left|\tfrac{1}{2}, -\tfrac{1}{2}\right\rangle,$$

$$\left|\tfrac{1}{2}, -\tfrac{1}{2}\right\rangle \otimes \left|\tfrac{1}{2}, +\tfrac{1}{2}\right\rangle, \quad \text{and} \quad \left|\tfrac{1}{2}, -\tfrac{1}{2}\right\rangle \otimes \left|\tfrac{1}{2}, -\tfrac{1}{2}\right\rangle. \tag{9.20}$$

Since these states exhaust all of the possibilities for spin alignment with the z axis, we expect these states to completely span the angular-momentum Hilbert space for this system: it should be possible to write *any* of the system's total angular momentum states as a linear combination of the states above.

Now, the operator corresponding to the total projection of the system's spin angular momentum on the z axis is $J_z = J_{1z} \otimes I + I \otimes J_{2z}$, where the first operator in each product acts on the first ket in each operator product and the second operator acts on the second ket. For example,

$$
\begin{aligned}
J_z\left|\tfrac{1}{2}, +\tfrac{1}{2}\right\rangle \otimes \left|\tfrac{1}{2}, -\tfrac{1}{2}\right\rangle &= (J_{1z} \otimes I + I \otimes J_{2z})\left|\tfrac{1}{2}, +\tfrac{1}{2}\right\rangle \otimes \left|\tfrac{1}{2}, -\tfrac{1}{2}\right\rangle \\
&= J_{1z} \otimes I \left|\tfrac{1}{2}, +\tfrac{1}{2}\right\rangle \otimes \left|\tfrac{1}{2}, -\tfrac{1}{2}\right\rangle + I \otimes J_{2z} \left|\tfrac{1}{2}, +\tfrac{1}{2}\right\rangle \otimes \left|\tfrac{1}{2}, -\tfrac{1}{2}\right\rangle \\
&= \tfrac{1}{2} \cdot 1 \left|\tfrac{1}{2}, +\tfrac{1}{2}\right\rangle \otimes \left|\tfrac{1}{2}, -\tfrac{1}{2}\right\rangle + 1 \cdot \left(-\tfrac{1}{2}\right)\left|\tfrac{1}{2}, +\tfrac{1}{2}\right\rangle \otimes \left|\tfrac{1}{2}, -\tfrac{1}{2}\right\rangle \\
&= \left(\tfrac{1}{2} - \tfrac{1}{2}\right)\left|\tfrac{1}{2}, +\tfrac{1}{2}\right\rangle \otimes \left|\tfrac{1}{2}, -\tfrac{1}{2}\right\rangle = 0\left|\tfrac{1}{2}, +\tfrac{1}{2}\right\rangle \otimes \left|\tfrac{1}{2}, -\tfrac{1}{2}\right\rangle, \quad (9.21)
\end{aligned}
$$

meaning that the ket $\left|\tfrac{1}{2}, +\tfrac{1}{2}\right\rangle \otimes \left|\tfrac{1}{2}, -\tfrac{1}{2}\right\rangle$ is an eigenvector of the total angular momentum projection operator J_z with eigenvalue 0. In a similar way, you can perhaps convince yourself that the $\left|\tfrac{1}{2}, +\tfrac{1}{2}\right\rangle \otimes \left|\tfrac{1}{2}, +\tfrac{1}{2}\right\rangle, \left|\tfrac{1}{2}, -\tfrac{1}{2}\right\rangle \otimes \left|\tfrac{1}{2}, +\tfrac{1}{2}\right\rangle$, and $\left|\tfrac{1}{2}, -\tfrac{1}{2}\right\rangle \otimes \left|\tfrac{1}{2}, -\tfrac{1}{2}\right\rangle$ kets are eigenvectors of that operator with eigenvalues 1, 0, and -1, respectively.

What are the total-j values for these states? The ket $\left|\tfrac{1}{2}, +\tfrac{1}{2}\right\rangle \otimes \left|\tfrac{1}{2}, +\tfrac{1}{2}\right\rangle$ must have a total-j of 1, since the J_z eigenvalue for this state is $m = 1$, and m cannot be greater than j. Similarly, since the state $\left|\tfrac{1}{2}, -\tfrac{1}{2}\right\rangle \otimes \left|\tfrac{1}{2}, -\tfrac{1}{2}\right\rangle$ has the eigenvalue $m = -1$, it must also have a total-j value of 1. So, these states must be the same as $|j, m\rangle = |1, 1\rangle$ and $|1, -1\rangle$, respectively.

But the other two states (which have $m = 0$) are ambiguous: the system's total angular momentum could be either $j = 0$ or $j = 1$ and still have a projection of 0 on the z axis.

However, we can actually *construct* the correct $|1, 0\rangle$ state by taking the $|1, 1\rangle = \left|\tfrac{1}{2}, +\tfrac{1}{2}\right\rangle \otimes \left|\tfrac{1}{2}, +\tfrac{1}{2}\right\rangle$ ket and applying the total-angular-momentum lowering operator $J_- = J_{1-} \otimes I + I \otimes J_{2-}$. Note that

$$
\begin{aligned}
J_{1-} &\otimes I \left|\tfrac{1}{2}, +\tfrac{1}{2}\right\rangle \otimes \left|\tfrac{1}{2}, +\tfrac{1}{2}\right\rangle \\
&= \sqrt{\tfrac{1}{2}(\tfrac{1}{2} + 1) - \tfrac{1}{2}(\tfrac{1}{2} - 1)}\left|\tfrac{1}{2}, -\tfrac{1}{2}\right\rangle \otimes \left|\tfrac{1}{2}, +\tfrac{1}{2}\right\rangle \\
&= \sqrt{\tfrac{3}{4} + \tfrac{1}{4}}\left|\tfrac{1}{2}, -\tfrac{1}{2}\right\rangle \otimes \left|\tfrac{1}{2}, +\tfrac{1}{2}\right\rangle = \left|\tfrac{1}{2}, -\tfrac{1}{2}\right\rangle \otimes \left|\tfrac{1}{2}, +\tfrac{1}{2}\right\rangle \tag{9.22}
\end{aligned}
$$

and an analogous result applies for the $I \otimes J_{2-}$ operator. Therefore,

$$J_- \left|\tfrac{1}{2}, +\tfrac{1}{2}\right\rangle \otimes \left|\tfrac{1}{2}, +\tfrac{1}{2}\right\rangle = (J_{1-} \otimes I + I \otimes J_{2-}) \left|\tfrac{1}{2}, +\tfrac{1}{2}\right\rangle \otimes \left|\tfrac{1}{2}, +\tfrac{1}{2}\right\rangle$$

$$= \left|\tfrac{1}{2}, -\tfrac{1}{2}\right\rangle \otimes \left|\tfrac{1}{2}, +\tfrac{1}{2}\right\rangle + \left|\tfrac{1}{2}, +\tfrac{1}{2}\right\rangle \otimes \left|\tfrac{1}{2}, -\tfrac{1}{2}\right\rangle. \qquad (9.23)$$

But this is supposed to be equal to

$$J_- |1, 1\rangle = \sqrt{1(1+1) - 1(1-1)} \, |1, 0\rangle = \sqrt{2} \, |1, 0\rangle. \qquad (9.24)$$

Equating these expressions and dividing through by $\sqrt{2}$ yields

$$|1, 0\rangle = \sqrt{\tfrac{1}{2}} \left|\tfrac{1}{2}, -\tfrac{1}{2}\right\rangle \otimes \left|\tfrac{1}{2}, +\tfrac{1}{2}\right\rangle + \sqrt{\tfrac{1}{2}} \left|\tfrac{1}{2}, +\tfrac{1}{2}\right\rangle \otimes \left|\tfrac{1}{2}, -\tfrac{1}{2}\right\rangle. \qquad (9.25)$$

So, this particular combination of our original basis states corresponds to the situation that we might picture classically as the spins being aligned but perpendicular to the z axis.

Since eigenstates of the J^2 operator should be mutually orthogonal, the $|0, 0\rangle$ state must be orthogonal to the $|1, 0\rangle$ ket we have just constructed. It must also be constructed of the same two states out of which we constructed the $|1, 0\rangle$, since only these states have $m = 0$. Up to an overall phase factor, this state must therefore be

$$|0, 0\rangle = \sqrt{\tfrac{1}{2}} \left|\tfrac{1}{2}, +\tfrac{1}{2}\right\rangle \otimes \left|\tfrac{1}{2}, -\tfrac{1}{2}\right\rangle - \sqrt{\tfrac{1}{2}} \left|\tfrac{1}{2}, -\tfrac{1}{2}\right\rangle \otimes \left|\tfrac{1}{2}, +\tfrac{1}{2}\right\rangle. \qquad (9.26)$$

You can easily check that this is the only normalized possibility (again, up to a phase factor): see Problem P9.8. This is the case where the spins cancel.

This is an interesting and important result. Our four direct product states map to four total angular momentum states, the so-called **triplet** $|1, 1\rangle$, $|1, 0\rangle$, and $|1, -1\rangle$, and the **singlet** $|0, 0\rangle$. We also see that any total spin-$\tfrac{1}{2}$ states are excluded: these four states completely cover our 4D Hilbert space.

We will see situations like this in the future.

BOX 9.1   Spin Rotation Operators

If the operator $U(t) = e^{-i\omega S_z t}$ has the effect of rotating an electron's effective spin axis counterclockwise around the $+z$ axis at a rate of ω, then it would make sense that $R_z(\theta) = e^{-i S_z \theta}$ rotates a state by θ around that axis: this simply corresponds to evaluating the U matrix at an instant when $\theta = \omega t$. There is also nothing special about the z axis, so $R_x(\theta) = e^{-i S_x \theta}$ and $R_y(\theta) = e^{-i S_y \theta}$ certainly *should* correspond to rotations around their respective axes as well.

Still, it doesn't hurt to see how this works out. We saw in Equation 8.2 that

$$S_x = \tfrac{1}{2} \begin{pmatrix} 0 & 1 \\ 1 & 0 \end{pmatrix}. \tag{9.27}$$

Note that

$$S_x^2 = \left(\frac{1}{2}\right)^2 \begin{pmatrix} 0 & 1 \\ 1 & 0 \end{pmatrix} \begin{pmatrix} 0 & 1 \\ 1 & 0 \end{pmatrix} = \left(\frac{1}{2}\right)^2 \begin{pmatrix} 1 & 0 \\ 0 & 1 \end{pmatrix}, \tag{9.28}$$

which is just a multiple of the identity operator. This means that

$$(i S_x \theta)^n = i(-1)^{(n-1)/2} \left(\frac{\theta}{2}\right)^n \begin{pmatrix} 0 & 1 \\ 1 & 0 \end{pmatrix} \quad \text{if } n \text{ is odd, and} \tag{9.29a}$$

$$(i S_x \theta)^n = (-1)^{n/2} \left(\frac{\theta}{2}\right)^n \begin{pmatrix} 1 & 0 \\ 0 & 1 \end{pmatrix} \quad \text{if } n \text{ is even,} \tag{9.29b}$$

since $i^n = (-1)^{n/2}$ if n is even.

Exercise 9.1.1   Use this result and the power series

$$\cos\theta = 1 - \frac{\theta^2}{2!} + \frac{\theta^4}{4!} + \cdots, \qquad \sin\theta = \theta - \frac{\theta^3}{3!} + \frac{\theta^5}{5!} + \cdots \tag{9.30}$$

to construct the matrix for $R_x(\theta)$ in terms of $\cos\theta$ and $\sin\theta$.

BOX 9.1   Spin Rotation Operators (cont.)

You should have found in the previous exercise that

$$R_x(\theta) = \begin{pmatrix} \cos \frac{1}{2}\theta & -i \sin \frac{1}{2}\theta \\ -i \sin \frac{1}{2}\theta & \cos \frac{1}{2}\theta \end{pmatrix}.$$

Suppose that we use this matrix to act on the initial state $|+z\rangle$.

Exercise 9.1.2   If $R_x(\theta)$ really rotates a state by an angle θ counterclockwise around the $+x$ axis (that is, counterclockwise when that axis is pointing toward us), then, intuitively, what angle should take us from $|+z\rangle$ to $|-z\rangle$? To $|-y\rangle$? To $|+y\rangle$? Verify that this matrix does the job you intuitively expect it should.

Exercise 9.1.3   What if the initial state were $|+x\rangle$? What would you expect to happen if we rotate through an arbitrary angle θ? What does happen?

BOX 9.2 Commutation Relations for J_x, J_y, J_z

Consider an ordinary three-dimensional physical space, and consider rotating a vector counterclockwise around the x axis. The matrix for such a rotation in physical space is

$$R_x(\theta) = \begin{pmatrix} 1 & 0 & 0 \\ 0 & \cos\theta & -\sin\theta \\ 0 & \sin\theta & \cos\theta \end{pmatrix}. \tag{9.31}$$

You can see that this is correct by considering what this vector does to the unit vectors $[\,1, 0, 0\,]$, $[\,0, 1, 0\,]$, and $[\,0, 0, 1\,]$ that are the basis vectors in the x, y, and z directions, respectively. The $[\,1, 0, 0\,]$ vector rotates to itself, the $[\,0, 1, 0\,]$ vector rotates to $[\,\cos\theta, \sin\theta, 0\,]$, and the $[\,0, 0, 1\,]$ vector rotates to $[\,-\sin\theta, \cos\theta, 0\,]$, which are all correct (see Figure 9.1a). If the matrix correctly rotates the basis vectors for a space, it correctly rotates any vector, since any vector is just a linear combination of basis vectors.

Equation 9.2 tells us that the generator J_x for this rotation is

$$J_x = i\left[\frac{dR_x}{d\theta}\right]_{\theta=0} = i\begin{pmatrix} 0 & 0 & 0 \\ 0 & -\sin 0 & -\cos 0 \\ 0 & \cos 0 & -\sin 0 \end{pmatrix} = \begin{pmatrix} 0 & 0 & 0 \\ 0 & 0 & -i \\ 0 & i & 0 \end{pmatrix}. \tag{9.32}$$

Similarly, from Figure 9.1b,c we can see that

$$R_y(\theta) = \begin{pmatrix} \cos\theta & 0 & \sin\theta \\ 0 & 1 & 0 \\ -\sin\theta & 0 & \cos\theta \end{pmatrix}, \qquad R_z(\theta) = \begin{pmatrix} \cos\theta & -\sin\theta & 0 \\ \sin\theta & \cos\theta & 0 \\ 0 & 0 & 1 \end{pmatrix}. \tag{9.33}$$

Exercise 9.2.1   What are the matrices for the generators J_y and J_z?

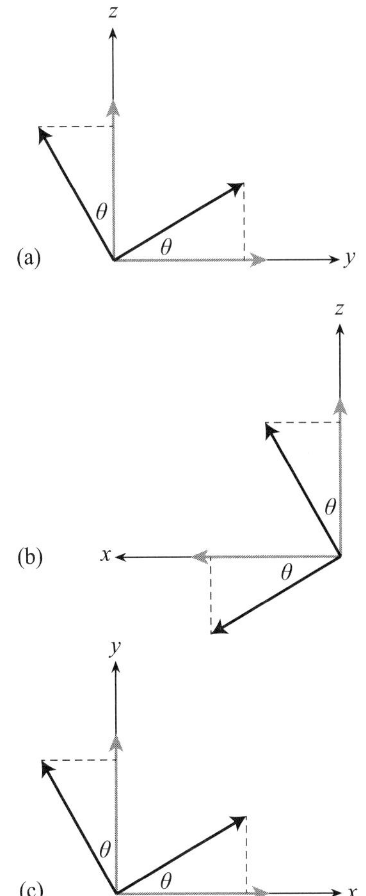

FIGURE 9.1   Rotating the gray basis vectors by an angle θ counterclockwise around either the x, y, or z axis—as shown in diagrams (a), (b), and (c), respectively—leads to the pair of black vectors displayed in each case. (The rotation axis points toward the viewer in each case.)

BOX 9.2 Commutation Relations for J_x, J_y, J_z (cont.)

You should have found in the previous exercise that

$$J_x = \begin{pmatrix} 0 & 0 & 0 \\ 0 & 0 & -i \\ 0 & i & 0 \end{pmatrix}, \quad J_y = \begin{pmatrix} 0 & 0 & i \\ 0 & 0 & 0 \\ -i & 0 & 0 \end{pmatrix}, \quad J_z = \begin{pmatrix} 0 & -i & 0 \\ i & 0 & 0 \\ 0 & 0 & 0 \end{pmatrix}. \quad (9.34)$$

Exercise 9.2.2 Verify that $[J_x, J_y] = i J_z$, $[J_y, J_z] = i J_x$, and $[J_z, J_x] = i J_y$, as claimed in Equation 9.5. If the generators of rotation obey these in *this* context, they will in *every* context, since they are basically describing the way that rotations around different axes do not commute.

BOX 9.3 Operators That Commute Have Common Eigenvectors

In this box, we will show that any pair of operators that commute will have eigenvectors in common.

Suppose we have an operator A whose normalized eigenvectors $|a_n\rangle$ all have *different* eigenvalues a_n. Also, suppose we have another operator B that commutes with A: $0 = [\,A,\,B\,] = AB - BA \Rightarrow AB = BA$.

Exercise 9.3.1 Start with the eigenvector equation for A:

$$A|a_n\rangle = a_n|a_n\rangle. \tag{9.35}$$

Multiply both sides on the left by B. Use the commutation relation and the fact that *only* $|a_n\rangle$ has eigenvalue a_n to argue that $B|a_n\rangle = b_k|a_n\rangle$, where b_k is some number. This means that $|a_n\rangle$ is also an eigenvector of B.

Now, note that if A has two different eigenvectors $|a_{n1}\rangle$ and $|a_{n2}\rangle$ with the *same* eigenvalue a_n, then the vector $|a_{n3}\rangle = c_1|a_{n1}\rangle + c_2|a_{n2}\rangle$ for any coefficients c_1 and c_2 is *also* an eigenvector of A with eigenvalue a_n. In such a case, we don't know whether $B|a_{n1}\rangle$ is $|a_{n1}\rangle$ or $|a_{n2}\rangle$ or any linear combination of those vectors. But we can typically find *some* linear combination of those vectors that *is* an eigenvector of B with a unique eigenvalue b_n. Since A doesn't care what linear combination we use, we can label the unique state that is both an eigenvector of A with eigenvalue a_n *and* an eigenvalue of B with eigenvalue b_k as $|b_k, a_n\rangle$.

This is precisely what we are doing with the commuting operators J^2 and J_z. J_z cannot tell us anything about the angular momentum magnitude, because angular momenta with different magnitudes can still have the same projection on a given axis. But J^2 cannot tell us anything about the orientation of the angular momentum in space. Together, however, these commuting operators can specify unique quantum states $|j, m\rangle$ that specify everything we can know about a particle's or a system's angular momentum.

BOX 9.4   The J^2 Operator

The J^2 operator is defined as

$$J^2 = J_x^2 + J_y^2 + J_z^2.\tag{9.36}$$

Remember that $[\,J_x, J_y\,] = i\,J_z$, $[\,J_y, J_z\,] = i\,J_x$, and $[\,J_z, J_x\,] = i\,J_y$. Note also that because $[\,A, B\,] \equiv AB - BA$,

$$[\,B, A\,] = BA - AB = -(AB - BA) = -[\,A, B\,]\tag{9.37a}$$

$$BA = AB - [\,A, B\,].\tag{9.37b}$$

Exercise 9.4.1   Use the above to show that $[\,J^2, J_x\,] = 0$. (Since J_x, J_y, and J_z appear symmetrically in J^2, this implies that $[\,J^2, J_y\,] = [\,J^2, J_z\,] = 0$ also.)

BOX 9.5 Raising and Lowering Operators

In Equation 9.7, we defined

$$J_+ \equiv J_x + i J_y \quad \text{and} \quad J_- \equiv J_x - i J_y. \tag{9.38}$$

These operators have the interesting property that

$$[J_z, J_\pm] = \pm J_\pm. \tag{9.39}$$

Exercise 9.5.1 Prove this.

Exercise 9.5.2 Use this to show that $J_\pm|j, m\rangle$ is an eigenvector of J_z with eigenvalue $(m \pm 1)$. This means that $J_+|j, m\rangle = c_+|j, m + 1\rangle$ and $J_-|j, m\rangle = c_-|j, m - 1\rangle$, respectively, where c_+ and c_- are (as yet) unknown numbers.

BOX 9.6   The Possible Values of λ

In this box, we will determine the possible values for the eigenvalue λ of the J^2 operator. Equation 9.11 tells us that if j and k are the maximum and minimum values of m, respectively, then we must have

$$J_+|\lambda, j\rangle = 0 \quad \text{and} \quad J_-|\lambda, k\rangle = 0. \tag{9.40}$$

Exercise 9.6.1   First, show that

$$J_-J_+ = J^2 - J_z^2 - J_z \quad \text{and} \quad J_+J_- = J^2 - J_z^2 + J_z. \tag{9.41}$$

Exercise 9.6.2   If $J_+|\lambda, j\rangle = 0$, then $J_-J_+|\lambda, j\rangle = 0$ also. Use this to show that $\lambda = j(j + 1)$. In a similar way, use $J_+J_-|\lambda, k\rangle = 0$ to show that $k = -j$. (*Hint:* You may use WolframAlpha to solve the quadratic equation for k.)

BOX 9.7 What $J_\pm$ Do to States

Our goal in this box is to justify Equations 9.18a and 9.18b:

$$J_+|j, m\rangle = \sqrt{j(j+1) - m(m+1)}\,|j, m+1\rangle \quad \text{and} \qquad (9.42a)$$

$$J_-|j, m\rangle = \sqrt{j(j+1) - m(m-1)}\,|j, m-1\rangle. \qquad (9.42b)$$

We already know from Equation 9.9 that $J_+|j, m\rangle = c_+|j, m+1\rangle$ and that $J_-|j, m\rangle = c_-|j, m-1\rangle$, so what exactly are we doing by determining the values of c_+ and c_-? We simply want to ensure that if $|j, m\rangle$ is normalized, then the vectors $|j, m+1\rangle$ and $|j, m-1\rangle$ that we are creating are also normalized.

It will help us to note that $J_\pm$ are adjoints of each other:

$$J_+^\dagger = (J_x + iJ_y)^\dagger = J_x^\dagger - iJ_y^\dagger = J_x - iJ_y = J_- \qquad (9.43a)$$

$$J_-^\dagger = (J_x - iJ_y)^\dagger = J_x^\dagger + iJ_y^\dagger = J_x + iJ_y = J_+ \qquad (9.43b)$$

since both J_x and J_y are Hermitian. Therefore, taking the adjoints of both sides of $J_+|j, m\rangle = c_+|j, m+1\rangle$ and $J_-|j, m\rangle = c_-|j, m-1\rangle$ yields

$$\langle j, m|J_+^\dagger = \langle j, m|J_- = \langle j, m+1|c_+^*, \qquad (9.44a)$$

$$\langle j, m|J_-^\dagger = \langle j, m|J_+ = \langle j, m-1|c_-^*. \qquad (9.44b)$$

Exercise 9.7.1 Use the results above as well as Equations 9.41 to evaluate $\langle j, m|J_-J_+|j, m\rangle$ and $\langle j, m|J_+J_-|j, m\rangle$ and so determine $|c_+|^2$ and $|c_-|^2$. Note: since kets with different overall phases are physically equivalent, we can *choose* the phase of each of the raised or lowered states so that $c_\pm$ is real.

BOX 9.8 The Matrices for J_x and J_y

In this box, we will determine the matrices that represent J_x and J_x for the quantum spin states of a spin-1 particle. We begin by determining the matrix representations of J_+ and J_-. If (as is conventional) we choose our basis states to be the J_z eigenstates, then for a spin-1 particle the basis states are

$$|1, 1\rangle = \begin{bmatrix} 1 \\ 0 \\ 0 \end{bmatrix}, \qquad |1, 0\rangle = \begin{bmatrix} 0 \\ 1 \\ 0 \end{bmatrix}, \qquad |1, -1\rangle = \begin{bmatrix} 1 \\ 0 \\ 0 \end{bmatrix}. \qquad (9.45)$$

Now, note that

$$J_+|1, 1\rangle = 0, \tag{9.46a}$$

$$J_+|1, 0\rangle = \sqrt{1(1+1) - 0(0+1)}\,|1, 1\rangle = \sqrt{2}|1, 1\rangle, \tag{9.46b}$$

$$J_+|1, -1\rangle = \sqrt{1(1+1) + 1(-1+1)}\,|1, 0\rangle = \sqrt{2}|1, 0\rangle. \tag{9.46c}$$

A trick for evaluating matrices is that the first column of the matrix is the vector that results from the matrix operating on the first basis vector, the second column is the vector that results from the matrix operating on the second basis vector, and so on. The equations above then imply that

$$J_+ = \sqrt{2} \begin{pmatrix} 0 & 1 & 0 \\ 0 & 0 & 1 \\ 0 & 0 & 0 \end{pmatrix}. \tag{9.47}$$

You can easily check that this transforms each basis vector in a manner consistent with what Equation 9.46 requires.

Exercise 9.8.1 Find the matrix representing J_- in this context.

BOX 9.8 The Matrices for J_x and J_y (cont.)

We can now construct the matrices for J_x and $\hat{J}_y$ using Equations 9.19:

$$J_x = \tfrac{1}{2}(J_+ + J_-) \quad \text{and} \quad J_y = i\tfrac{1}{2}(J_- - J_+). \tag{9.48}$$

Exercise 9.8.2 Find the matrices representing J_x and J_y in this context, and check that they are Hermitian.

Homework Problems

P9.1 How can we be *certain* that the eigenvalue λ of the J^2 operator is nonnegative? Note that $J_x|\lambda, m\rangle = c\,|\psi\rangle$, where $|\psi\rangle$ is some normalized ket. Use this equation and its adjoint to prove that $\langle\lambda, m|J_x^2|\lambda, m\rangle \geq 0$, and from that argue that $\lambda \geq 0$.

P9.2 Verify the following commutator identities:
a. $[A, B + C] = [A, B] + [A, C]$
    and $[A + B, C] = [A, C] + [B, C]$,
b. $[A, BC] = [A, B]C + B[A, C]$
    and $[AB, C] = [A, C]B + A[B, C]$.
(Note the similarity to the sum and product rules for derivatives.)

P9.3 Show that if two operators A and B have a *complete* set of eigenvectors $|a, b\rangle$ in common (complete in the sense that we can write *any* ket $|\psi\rangle$ as a linear combination of the eigenvectors $|a, b\rangle$), then $[A, B] = 0$.

P9.4 By expanding to second order, show that $e^{A+B} = e^A e^B$ *only* if $[A, B] = 0$.

P9.5 Consider the orbital angular momentum associated with a particle moving through space around another particle, such as an electron orbiting a proton. We describe the angular momentum quantum eigenstate $|j, m\rangle$ of such a particle with a wavefunction $\psi_{j,m}(\vec{x})$. Now, any $\psi(\vec{x})$ must be *unchanged* as we rotate it through an angle of 2π around an arbitrary axis through the origin, because otherwise the function is a poorly defined function of $\vec{x}$, having multiple values of $\psi(\vec{x})$ assigned to the same spatial point $\vec{x}$. Show that

$$R_z(2\pi)\,|j, m\rangle = e^{-iJ_z 2\pi}|j, m\rangle = -|j, m\rangle \quad (9.49)$$

if j is not equal to an integer. This is acceptable if $|j, m\rangle$ is a *vector* (because $-|j, m\rangle$ and $|j, m\rangle$ are physically equivalent as vectors), but is not acceptable for a wavefunction.

P9.6 Use the methods described in this chapter to determine the matrices for J_x, J_y, and J_z for a spin-$\frac{1}{2}$ particle, using the kets $|+z\rangle = |\frac{1}{2}, +\frac{1}{2}\rangle$ and $|-z\rangle = |\frac{1}{2}, -\frac{1}{2}\rangle$ as the basis vectors. Do you get the same matrices as we found experimentally for S_x, S_y, and S_z in Chapter 8?

P9.7 Determine the normalized eigenvectors of the J_x operator you found in Box 9.8, assuming that its eigenvalues are 1, 0, and -1. (*Hint:* Assume that the components of the eigenvector are $[\,a, b, c\,]$ in a column, and use the eigenvector equation to link two of these components to the third. Then use the normalization condition to determine that third component. Remember that we can determine these vectors only up to an overall phase.

P9.8 In Equation 9.25, we saw that when adding the angular momentum of two spin-$\frac{1}{2}$ particles, the state with total $j = 1$ and total $m = 0$ was

$$|1, 0\rangle = \sqrt{\tfrac{1}{2}}\,|\tfrac{1}{2}, -\tfrac{1}{2}\rangle \otimes |\tfrac{1}{2}, +\tfrac{1}{2}\rangle + \sqrt{\tfrac{1}{2}}\,|\tfrac{1}{2}, +\tfrac{1}{2}\rangle \otimes |\tfrac{1}{2}, -\tfrac{1}{2}\rangle.$$

$$(9.50)$$

We argued that the $|0, 0\rangle$ state must also be constructed out of the same direct-product states $|\frac{1}{2}, -\frac{1}{2}\rangle \otimes |\frac{1}{2}, +\frac{1}{2}\rangle$ and $|\frac{1}{2}, +\frac{1}{2}\rangle \otimes |\frac{1}{2}, -\frac{1}{2}\rangle$, but must be orthogonal to the state above (as well as being normalized). Find the most general normalized combination of these direct-product states that is orthogonal to $|1, 0\rangle$. Then take advantage of our freedom to choose a state's overall phase to choose the coefficient of the first state to be real and positive. Do you get the result in Equation 9.26?

P9.9 Simply by counting states, argue that the total spin quantum number j for a system of two spin-1 particles must be 2, 1, or 0.

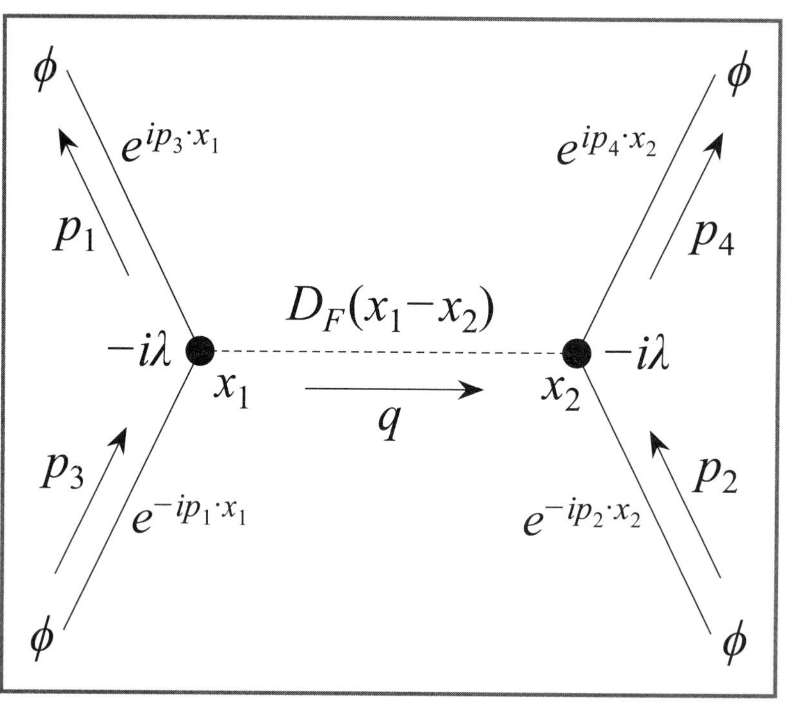

ϕ
$e^{ip_3 \cdot x_1}$
p_1
$-i\lambda$
x_1
p_3
$e^{-ip_1 \cdot x_1}$
ϕ
$D_F(x_1 - x_2)$
q
ϕ
$e^{ip_4 \cdot x_2}$
p_4
$-i\lambda$
x_2
p_2
$e^{-ip_2 \cdot x_2}$
ϕ

10. FEYNMAN DIAGRAM QUICK START

In Chapters 4–6, we constructed a relativistic but classical theory of fields based on Lagrangian densities. In Chapters 7–9, we reviewed the basic features of non-relativistic quantum mechanics (NRQM). On these twin foundations, one can construct a fully relativistic version of quantum mechanics called **relativistic quantum field theory** (QFT). While quantum field theory is built on non-relativistic quantum mechanics and is consistent with it (in that it reduces to NRQM in the non-relativistic limit), quantum field theory is also quite different conceptually and mathematically.

This book offers two tracks for exploring quantum field theory. Chapters 11–14 explore in some detail how one can marry the principles of quantum mechanics and classical field theory to yield a complete and reasonable relativistic quantum field theory. Alternatively, this chapter provides an overview of how one calculates the experimental predictions of quantum field theory without peering too deeply into the mechanisms behind those calculations. Reading this chapter allows one to leap directly to Chapter 15, and therefore more expeditiously get to exploring the implications of the Standard Model (though at the expense of one's depth of understanding).

10.1 Transition Rates

The basic problem with non-relativistic quantum mechanics is that it *assumes* that the particles it describes exist forever. The time-dependent Schrödinger equation describes how a particle's quantum state evolves with time, but is incapable of describing how such a particle might be created or destroyed. This makes non-relativistic quantum mechanics structurally incapable of addressing the physics that we observe in high-energy particle experiments, where particles are routinely created and/or decay to other particles.

In relativistic quantum field theory, a "quantum state" does not describe the state of a particle but rather the state of an *entire system* by describing (1) how many particles of each type the system contains and (2) what four-momentum each particle has. It provides tools for calculating a **transition rate** Γ, the *rate* at which an initial system quantum state $| i \rangle$ evolves to a final system quantum state $| f \rangle$, defined as

$$\Gamma = \frac{\text{probability of } i \to f \text{ during time } T}{T} = \frac{|\langle f | U(t_0 + T, t_0)|i\rangle|^2}{T}, \quad (10.1)$$

where $U(t_f, t_i)$ is a time-evolution operator that acts on QFT states and T is the duration of an experiment.

Commonly, though, the final situation of interest involves a whole *set* of system quantum states. For example, we may be interested in the rate at which a particle decays to final particles having *all possible momenta*, not evolving to a *single* final system state where the outgoing particles have specific momenta. Therefore, we generally define the transition rate to be the total rate for which a given set of final particles appears without regard to their momenta:

$$\Gamma \equiv \frac{\sum_f |\langle f | U(t_0 + T, t_0)| i \rangle|^2}{T}, \quad (10.2)$$

where the sum is over states corresponding all possible outgoing particle momenta. If a decay or scattering process might produce different sets a, b, ... of particles, then the total transition rate is the sum of the rates for each outcome:

$$\Gamma_{\text{tot}} = \Gamma_a + \Gamma_b + \cdots. \tag{10.3}$$

10.2 Decay Rates and Cross Sections

In particle physics experiments, people often measure *decay rates* of unstable particles and what are called *scattering cross sections* for collisions between particles. We would like to use quantum field theory to calculate values for these quantities.

In the case of particle decay, if we have N unstable particles initially and the transition rate to all possible outgoing particles is Γ, then the number of particles that decay in a differential time interval dt is Γdt. As you will show in Box 10.1, the number of particles $N(t)$ that remain undecayed as a function of time t and the **average lifetime** τ of a given initial particle are

$$N(t) = N_0 e^{-\Gamma t} \quad \text{and} \quad \tau = \frac{1}{\Gamma}. \tag{10.4}$$

In many particle physics experiments, physicists either shoot a stream of high-energy particles at other stationary particles, like arrows at a target, or (in more modern colliders such as the Large Hadron Collider) shoot two beams of high-energy particles directly at each other. In either case, we are interested in the probability that a projectile particle "hits" a target particle. Of course, "hitting" could mean any number of things. We might ask in a scattering experiment what is the probability that the particles are deflected by a certain angle. We might also ask about the relative probabilities of certain types of collision outcomes; for example, the probability that colliding electrons produce a pair of muons compared to the probability that they simply scatter.

Figure 10.1 shows two cylindrical bunches of particles containing N_1 and N_2 particles that are approaching each other with relative velocity $|\vec{v}_1 - \vec{v}_2|$. Suppose that each bunch has length ℓ and cross-sectional area A. Let us arbitrarily consider bunch 1 to be "projectiles" and bunch 2 to be the "targets." Consider a given target particle that sits on a plane perpendicular to the cylinders' common axis. All projectile particles pass through that plane during a time $T = \ell/|\vec{v}_1 - \vec{v}_2|$. The probability that a given projectile particle interacts with that target particle to yield a desired outcome is the transition rate Γ for that outcome times the duration of the experiment T. If we multiply this by the beam's cross-sectional area A, we get a quantity σ with units of area that we call the **cross section** σ for that outcome:

$$\sigma \equiv \Gamma T A. \tag{10.5}$$

You can think of σ as being the effective cross-sectional area that the target particle presents to the projectile particles as they stream in: the probability ΓT of the desired outcome occurring is the fraction of the beam area that the target particle effectively "presents" to that incoming projectile: $\Gamma T = \sigma / A$.

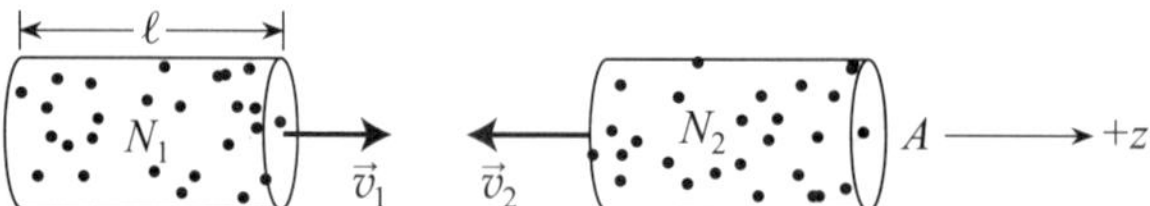

FIGURE 10.1   A pair of particle bunches approaching each other in a particle accelerator experiment. (Note that it is conventional in particle physics experiments to choose the $+z$ axis to be along the direction of $\vec{v}_1 - \vec{v}_2$.)

What we actually *measure* is the total rate $\Gamma_{\text{tot}} = N_1 N_2 \Gamma$ of the desired outcomes during the experiment. This is related to the cross section as follows:

$$\sigma = \Gamma\, T A = \frac{N_1 N_2 \Gamma T A}{N_1 N_2} = \frac{N_1 N_2 \Gamma}{N_1 N_2 / (AT)} = \frac{\Gamma_{\text{tot}}}{\mathcal{L}}, \qquad (10.6)$$

where $\mathcal{L} \equiv N_1 N_2 / AT$ is the particle accelerator's **luminosity**, defined as the number of target particles N_2 times the number of projectile particles N_1 the accelerator sends through a unit area per unit time in the target particles' frame. Dividing the observed desired outcome rate by the accelerator's measured luminosity therefore gives one the cross section σ for that outcome.

10.3 Fermi's Golden Rules

We connect the measurable decay rates and cross sections to quantities we can theoretically calculate as follows: **Fermi's Golden Rule for Decays** states that if particle 1 decays to particles 2 through N, the decay rate is

$$\Gamma = \frac{S}{2E_1} \int \left[\left(\frac{d^3 p_2}{(2\pi)^3\, 2E_2} \right) \left(\frac{d^3 p_3}{(2\pi)^3\, 2E_3} \right) \cdots \left(\frac{d^3 p_N}{(2\pi)^3\, 2E_N} \right) \right.$$

$$\left. \times\, (2\pi)^4 \delta^4(p_1 - p_2 - p_3 \cdots - p_N)\, |\mathcal{M}|^2 \right], \qquad (10.7)$$

where $E_1, E_2, \ldots$ is shorthand for $E_{\vec{p}_1}, E_{\vec{p}_2}, \ldots$ and S is a "degeneracy factor" that is the product of $1/n!$ for every group of n identical particles in the set of outgoing particles, and $\mathcal{M}$ is a quantity called the **scattering amplitude**.

Much of what this rule claims makes sense. We credibly want to integrate over all possible outcome particle momenta $\vec{p}_2$ to $\vec{p}_N$ to get the total decay rate. The factors of E_2 through E_N in the denominators turn out to be necessary to ensure that the integrals are Lorentz-invariant. The factor of $\delta^4(p_1 - p_2 \cdots - p_N)$ ensures that the process conserves four-momentum while we are integrating over all the momenta. We need the degeneracy factor S because integrating over all outgoing particle momenta overcounts the possibilities when the particles are identical, since the system quantum state where the set of outgoing identical particles have momenta $\vec{p}_2$, $\vec{p}_3$, and so on is physically equivalent to a state where those particles have any permutation of those momenta.

The main purpose of the rule is actually to remove elements common to *all* decays from the definition of $\mathcal{M}$, which will therefore express the physics that distinguishes a *particular* decay process from other decays. Our focus in what follows will be to calculate $\mathcal{M}$ in various situations of interest.

Equation 10.7 handles *decay* processes. In *scattering* processes where particles 1 and 2 collide and produce outgoing particles 3 through N, **Fermi's Golden Rule for Scattering** declares that the interaction cross section is

$$\sigma = \frac{S}{4\sqrt{(p_1 \cdot p_2)^2 - (m_1 m_2)^2}} \int \left[\left(\frac{d^3 p_3}{(2\pi)^3\, 2E_3} \right) \left(\frac{d^3 p_4}{(2\pi)^3\, 2E_4} \right) \cdots \left(\frac{d^3 p_N}{(2\pi)^3\, 2E_N} \right) \right.$$

$$\left. \times\, (2\pi)^4 \delta^4(p_1 + p_2 - p_3 - p_4 \cdots - p_N)\, |\mathcal{M}|^2 \right], \qquad (10.8)$$

where m_1 and m_2 are the incoming particles' masses. This rule again removes the elements common to all scattering processes from the definition of $\mathcal{M}$.

Though I have argued that these formulas' overall structure makes sense, I have not justified the denominator of the factor in front of the integral in each expression. Chapter 14 provides a justification for each denominator, but those justifications are part of the complexities we are skipping over in this chapter.

10.4 Feynman Diagrams

So, how do we calculate $\mathcal{M}$? The details of such a calculation intricately depend on the logical structure of quantum field theory, and even the treatment in Chapters 11–14 does not unravel all of the complexities involved. But it turns out that one can express the calculation of $\mathcal{M}$ as an infinite series of terms, like terms in a power series, and Richard Feynman showed that it was possible to represent each term with a map that people now call a **Feynman diagram**. Given the map for a particular term, one can apply **Feynman Rules** to directly read off the mathematical factors that must appear in that term.

To see how this works, let's start with the simplest possible quantum field theory that involves interacting fields. Imagine a "toy universe" consisting of two real fields $\phi(t, x, y, z)$ and $\eta(t, x, y, z)$ whose physics is expressed by the following Lagrangian density:

$$\mathcal{L} = \tfrac{1}{2}\partial_\mu \phi \, \partial^{\,\mu}\phi - \tfrac{1}{2}m_\phi^2\phi^2 + \tfrac{1}{2}\partial_\mu \eta \, \partial^{\,\mu}\eta - \tfrac{1}{2}m_\eta^2\eta^2 - \lambda\phi^2\eta. \tag{10.9}$$

If this Lagrangian density did not have the final term, then its Euler-Lagrange equations would be independent Klein-Gordon equations for each of the fields ϕ and η (see Box 10.2.) In quantum field theory, we interpret excitations of these fields as being particles. Let's call the particles associated with the ϕ and η fields **phions** and **etaons**, respectively. The Klein-Gordon equations for these fields imply that phions have mass m_ϕ and etaons have mass m_η. We could think of phions and etaons as being somewhat analogous to electrons and photons (particularly if we were to choose $m_\eta = 0$), but without the complicating aspects of electron and photon spins.

If the final term in the Lagrangian density were not present, these fields would be free and independent: the solutions to their Euler-Lagrange equations would be superpositions of plane waves moving past each other. But the final term describes an *interaction* between the ϕ and η fields. This is where the interesting physics is. (To see what this interaction term implies for classical fields in the non-relativitic limit, see Appendix A.)

The nonlinear nature of even this simple interaction term makes it impossible to solve the resulting Euler-Lagrange equations analytically, so quantum field theory adopts a perturbative approach, where we start with the plane-wave solutions for the non-interacting fields and use an iterative approach that effectively writes the solution in terms of a power series involving successive amounts of modification by the interaction term.

In a Feynman diagram, we represent the interaction term by a **vertex** where two phion lines meet with a single etaon line. The lines represent the unperturbed plane wave solutions of the fields, and the vertex represents the effect of the interaction in perturbing those solutions.

Figure 10.2 shows the six qualitatively distinct ways that two phion lines could meet an etaon line. Interpreted as a spacetime diagram, the first diagram represents a phion absorbing an etaon and afterward moving on with a new momentum, the second as a phion emitting an etaon and then moving away with a new momentum, and so on. Quantum field theory requires that four-momentum be conserved at each vertex, so the final two diagrams (where two phions and an etaon are created from or destroyed to nothing) are forbidden.

One constructs a complete Feynman diagram by wiring together a set of vertices selected from the four allowed vertex diagrams in Figure 10.2. Which diagrams are possible depends on which particles come in and which particles come out of a given process. For example, suppose we are considering *phion scattering*, where two incoming phions interact and emerge with different momenta. The simplest diagrams we can construct where two phions go in and two phions come out are the three two-vertex diagrams shown in Figure 10.3. Note that Figures 15.1a and b are the same except that

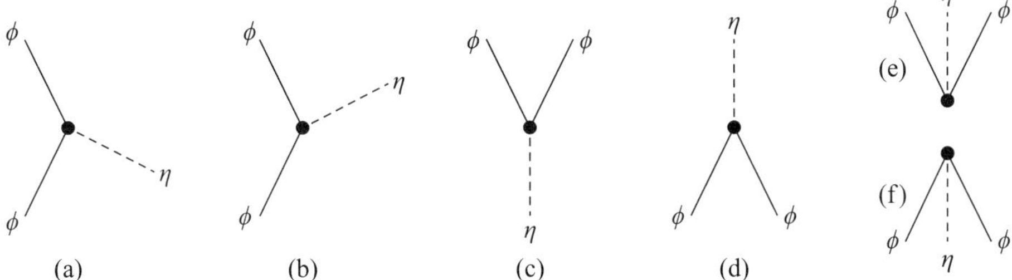

FIGURE 10.2 Six possible vertices that show two phion lines meeting an etaon line. Time is vertically upward in these spacetime diagrams. Diagrams **(e)** and **(f)** manifestly violate conservation of four-momentum and so are forbidden. (The first two vertices, if isolated, would also fail to conserve four-momentum, but they can be part of a larger diagram: see Box 10.3.)

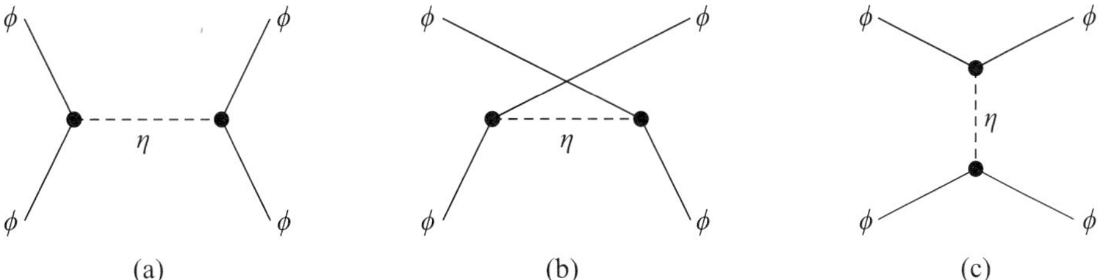

FIGURE 10.3 The three possible two-vertex diagrams that describe a phion scattering process $\phi + \phi \rightarrow \phi + \phi$. (For reasons unknown to me, the processes depicted by these common second-order diagram shapes are called **t-channel, u-channel,** and **s-channel** processes, respectively.)

outgoing phions are switched, but we treat them as distinct diagrams to represent that we have no way of telling which outgoing phion is connected to which incoming phion.

Remember, however, that these diagrams are maps to help us do calculations, not actual spacetime diagrams of actual physical processes. The slopes of particle lines on the diagrams are irrelevant except to qualitatively indicate which particles are incoming, which are outgoing, and which are internal. With internal particles, it is not even relevant to indicate which vertex emits the particle and which receives it: the internal line simply represents a factor in the calculation that is the same in both cases. We usually draw internal lines as either horizontal (as in Figures 10.3a and 10.3b) or vertical (as in Figure 10.3c).

10.5 Feynman Rules for Our Toy Theory

The entire point of a Feynman diagram is to describe how to calculate the scattering amplitude $\mathcal{M}$ for a process. A set of **Feynman Rules** for a given theory specify how to translate the diagram to a formula of $\mathcal{M}$. Here are the rules for our toy universe.

1. **Draw diagrams**. Draw worldlines for the incoming and outgoing particles and connect them with lines representing internal particles. These lines must intersect at vertices that connect the fields appearing in the Lagrangian density's interaction term. (In our toy theory, that term is $-\lambda\eta\phi^2$, so we will have two phion lines connecting to an etaon line.)

2. **Label lines.** On your diagram, draw an arrow next to each particle worldline that indicates the direction of the particle's momentum. (For internal lines, this direction is arbitrary, but *choosing* a direction is important for subsequent steps.) Label the arrows for incoming or outgoing particles with four-momenta $p_1, p_2, \ldots$ and those for internal particles with four-momenta $q_1, q_2, \ldots$.

3. **Vertex factors.** Each vertex in our toy theory contributes a **vertex factor** of $-i\lambda$, where λ is the coupling constant in the interaction term.

4. **Propagators.** Each internal scalar field line in the diagram contributes a **propagator** factor of $i/(q_j^2 - m_j^2)$, where q_j is the four-momentum associated with the jth internal line and m_j is the mass of the particle associated with that line. (Other types of fields have different propagators.)

5. **Conserve four-momentum at each vertex.** Four-momentum must be conserved at each vertex. For diagrams having a single internal particle line, this requirement will determine that particle's four-momentum q. (In more complex diagrams, we must integrate over any undetermined internal particle four-momenta, as we will discuss in a future chapter.)

6. **The result is** $-i\mathcal{M}$. Multiply by i to get the scattering amplitude $\mathcal{M}$.

7. **Add the amplitudes.** To get the total amplitude $\mathcal{M}_{\mathrm{tot}}$ to a given order n, *add* the amplitudes for all possible diagrams having n or fewer vertices. This yields the value of $\mathcal{M}_{\mathrm{tot}}$ accurate to that order of approximation.

There is a logic behind some of these rules that one can appreciate even without delving deeply into quantum field theory, as Box 10.3 discusses.

The last rule reminds us that each Feynman diagram represents a term in a power-series expansion for $\mathcal{M}_{\mathrm{tot}}$. If the interaction coupling constant λ is "small" (in a sense yet to be precisely defined), then higher-order terms will be smaller than lower-order terms, and we can get a decent approximation for $\mathcal{M}_{\mathrm{tot}}$ by adding only the first few terms. Also remember that in quantum field theory (as in quantum mechanics) we *add* all relevant amplitudes *before* taking the absolute square to find probabilities.

Let's see how this works for phion-phion scattering. Figure 10.3 implements the first step in the process by illustrating all of the lowest-order diagrams for that process. Consider the first of the three diagrams shown (the "t-channel" diagram). The second step is to draw arrows on the diagram that show the direction of each particle's four-momentum. Figure 10.4 illustrates that diagram with the second step implemented (when an internal line is horizontal, it is conventional to choose the direction of its four-momentum to be rightward). Applying rules 3–6 to this labeled diagram then yields

$$-i\mathcal{M}_t = (-i\lambda)\frac{i}{q^2 - m_\eta^2}(-i\lambda) \;\Rightarrow\; \mathcal{M}_t = \frac{\lambda^2}{(p_1 - p_3)^2 - m_\eta^2}, \qquad (10.10)$$

since conservation of four-momentum at the left vertex requires that $q = p_1 - p_3$. In Box 10.4, you will apply the rules to calculate the scattering amplitudes for the u-channel and s-channel diagrams. By the last rule, the total lowest-order scattering amplitude $\mathcal{M}_{\mathrm{tot}} = \mathcal{M}_t + \mathcal{M}_u + \mathcal{M}_s$.

To get a more accurate result, we should include higher-order diagrams. Figure 10.5 illustrates one higher-order diagram that illustrates some of the kinds of extra features that a higher-order diagram might include.

10.6 Interpreting Feynman Diagrams

Feynman diagrams provide a wonderfully easy method to determine and calculate amplitudes for experimentally relevant processes without delving too deeply into the nuts and bolts of quantum field theory. But I need to warn you against overinterpreting what these diagrams mean. The story that each diagram tells is so clear and credible as a physical process that it is natural to assume that they describe what is physically happening in each interaction (for example, the t-channel diagram seems to tell the story that one of the incoming phions emits an etaon and recoils, while the other incoming phion receives that etaon and absorbs its momentum). But I urge you to resist the temptation of thinking that these diagrams describe things that are actually happening.

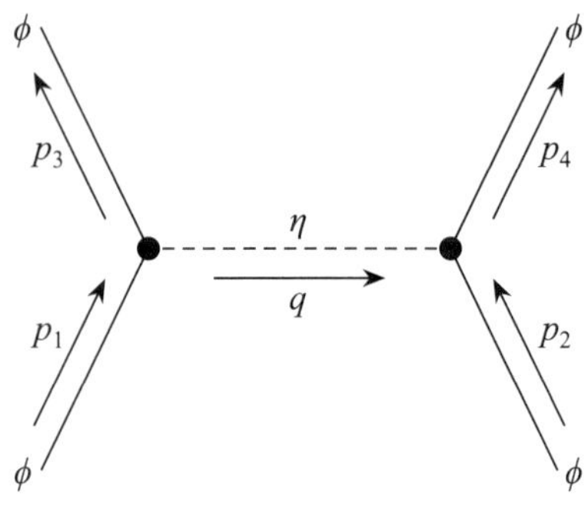

FIGURE 10.4   This Feynman diagram describes one of three second-order phion scattering processes.

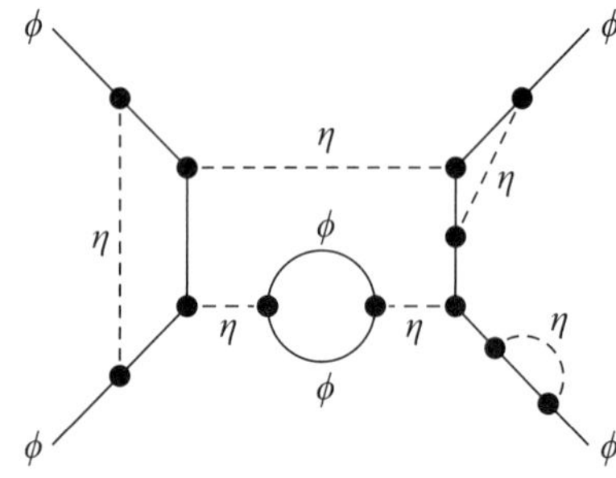

FIGURE 10.5   This is one of many possible 12th-order Feynman diagrams for phion scattering in our toy theory. In drawing higher-order diagrams, we can put one or more phion "bubbles" along any etaon line, one or more etaon "half-bubbles" along any phion line, and/or we can bridge any vertex with an etaon line as well as adding large-scale structures, as illustrated by this diagram.

These diagrams are tools that help us calculate terms in a power series, nothing more. For a number of reasons, interpreting any given diagram—or even the set of all diagrams—as depicting what is "really" going on is highly problematic. First of all, remember that our incoming and outgoing system states describe infinite incoming and outgoing plane waves, which are smeared out over all space, not the classical-particle-like trajectories that appear in a diagram. Though the diagram pretends that the interaction between these plane waves happens at a single event in space and time, the plane waves actually interact *all the time* and *everywhere* (and the deep mathematics behind the diagram involves integrating over all space and time).

Perhaps even more importantly, the real physical process is represented by the sum of an infinite number of terms in a power series, while a diagram only represents one term, a *tiny piece* of the whole process. As an analogy, we can fruitfully describe a pulse wave mathematically as an infinite superposition of sinusoidal waves. But is it helpful to consider that the pulse "really is" (at the fundamental level) a bunch of sine waves, or especially that any single sine wave is something that is "really happening" in the wave? No! Fourier decomposition is a wonderful mathematical tool for understanding how general waves work, but ascribing reality to the individual sine waves in the decomposition does not make any sense, particularly if one keeps in mind that I could have just as easily described the pulse as a superposition of Dirac deltas (which comprises an alternative orthonormal basis for writing functions).

All that being said, let us duly praise and thank Richard Feynman for developing this amazing tool. Were it not for his diagrams, the mathematics involved in quantum field theory would be much more difficult: Feynman diagrams really make quantum field theory practical!

BOX 10.1   The Mathematics of Particle Decay

We have defined the transition rate Γ to be the probability per unit time that an initial particle evolves to a final QFT system state involving a certain set of output particles, summed over all possible outgoing particles and all possible particle momenta. So long as the duration of our experiment is very long compared to whatever dynamics might be involved in the decay process itself, the fact that the laws of physics are independent of time implies that the decay rate Γ for a given decay process should be independent of time. Therefore, a given single particle should have a fixed probability $\Gamma\,dt$ of decaying during an infinitesimal time interval dt (but with dt still long compared to the time required for the decay process itself). This means that the change dN in the number N of undecayed particles N during the interval dt is

$$dN = -N\Gamma\,dt \qquad (10.11)$$

(negative because the number of undecayed particles is *decreasing*).

Exercise 10.1.1   Integrate this from time $t = 0$ to time t to prove that the number of particles that remain undecayed at time t is $N(t) = N(0)e^{-\Gamma t}$.

BOX 10.1   The Mathematics of Particle Decay (cont.)

Now we would like to show that the average lifetime of a particle is $\tau = 1/\Gamma$. But what do we mean by the concept "average lifetime?" Suppose that we have N_0 systems at time $t = 0$. The number $|dN(t)|$ that decay during a tiny time interval of length dt centered on t is given by

$$|dN(t)| = \left|\frac{dN}{dt}\right| dt = \left|-\Gamma N_0\, e^{-\Gamma t}\right| dt = +\Gamma N_0\, e^{-\Gamma t} dt. \qquad (10.12)$$

(We need the absolute value, because $dN < 0$.) The systems that decay during this interval have lasted for time t, so that is these particular systems' lifetime. To get the average lifetime, we have to add up the results for all time intervals and divide by the total initial number of systems. In the limit that N_0 becomes large enough that $|dN(t)|$ is large even for time intervals $dt \ll 1/\Gamma$, we can approximate the sum by the integral

$$\tau \approx \frac{1}{N_0} \int_0^\infty t\, [\Gamma N_0 e^{-\Gamma t}]\, dt. \qquad (10.13)$$

Exercise 10.1.2   This is actually an easy integral to do by parts. First, change variables to $x \equiv \Gamma t$. The usual statement of integration by parts is that $\int u\, dv = uv - \int v\, du$, where u and v are functions of x, $dv = (dv/dx)\, dx$, and so on (as we saw in Chapter 4). Define $u \equiv x$ and $v \equiv -e^{-x}$ in this case to calculate the integral.

BOX 10.2 Euler-Lagrange Equations for Our Toy Theory

In this box, we seek the Euler-Lagrange equations for the toy universe described by the Lagrangian density given in Equation 10.9 and repeated here.

$$\mathcal{L} = \tfrac{1}{2}\partial_\mu\phi\,\partial^\mu\phi - \tfrac{1}{2}m_\phi^2\phi^2 + \tfrac{1}{2}\partial_\mu\eta\,\partial^\mu\eta - \tfrac{1}{2}m_\eta^2\eta^2 - \lambda\phi^2\eta. \qquad (10.14)$$

Exercise 10.2.1 Use the methods discussed in Chapters 4–5 to find the Euler-Lagrange equations for this Lagrangian density, and show that they reduce to independent Klein-Gordon equations when $\lambda = 0$. (*Hints:* Remember that the two distinct fields ϕ, η in this case are real, not complex. Also, I recommend rewriting $\partial^\mu\phi\,\partial_\mu\phi$ as $g^{\alpha\beta}\partial_\alpha\phi\,\partial_\beta\phi$ to make evaluating $\partial\mathcal{L}/\partial(\partial_\mu\phi)$ easier. Do something similar with the analogous η term.)

BOX 10.3 Behind the Rules

The purpose of this box is to discuss justifications for the Feynman rules that one can understand without delving deeply into quantum field theory.

We begin by considering external particles. Consider a plane-wave solution $e^{-ip\cdot x} = e^{-iE_{\vec{p}}t}e^{+i\vec{p}\cdot\vec{r}}$ to the Klein-Gordon equation. We have seen that in quantum mechanics, this is the time-dependent amplitude $\langle \vec{r}, t \,|\, \vec{p} \,\rangle$ that an incoming particle with definite momentum $\vec{p}$ will be found at spatial position $\vec{r}$ at time t. When we consider particles that are *outgoing* particles, we are interested in the amplitude $\langle \vec{p} \,|\, \vec{r}, t \,\rangle$ that a particle starting from position $\vec{r}$ at time t comes out with definite momentum $\vec{p}$. That is given by $\langle \vec{p} \,|\, \vec{r}, t \,\rangle = \langle \vec{r}, t|\vec{p} \,\rangle^* = e^{+ip\cdot x}$.

Now, these phion fields might interact with the etaon field at any four-position x_1 in space and time. Suppose that at x_1, the interaction excites the eta field, producing an etaon at that four-position. This etaon then travels to a different four-position x_2, where the interaction between the fields de-excites the etaon field. The problem with this picture is that it is impossible for the etaon's four-momentum q in this instance to satisfy the requirement that $q^2 = m_\eta^2$. We describe the etaon in this case as being "off the mass shell" (imagining m_η as the radius of a spherical surface in four-dimensional momentum space), or **off-shell** for short. This means that the etaon cannot be a real particle (whose four-momentum *must* be "on-shell"), but rather is what physicists call a **virtual particle**: a field excitation that has all the characteristics of a real particle except for satisfying this core requirement.

How do we know that the etaon is off-shell? Consider the left vertex of Figure 10.4, as shown in Figure 10.6. Suppose we are in the initial phion's rest frame, so that it has initial energy m_ϕ and no spatial momentum. Suppose the final phion has spatial momentum $|\vec{p}\,|$ in the $-x$ direction. The final phion is a real outgoing particle, so it must be on-shell, meaning that its final energy must be $E_3 = (|\vec{p}\,|^2 + m_\phi^2)^{1/2}$. But one of the rules is that we must conserve four-momentum at each vertex.

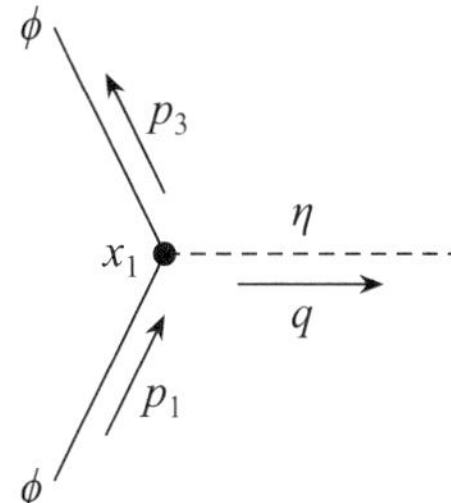

FIGURE 10.6   Half of a process containing an internal particle.

Exercise 10.3.1 Calculate the components of the etaon four-momentum q and the value of q^2 in the rest frame of the initial phion. Noting that the value of q^2 is frame-independent, explain how we *know* the etaon must be off-shell.

BOX 10.3   Behind the Rules (cont.)

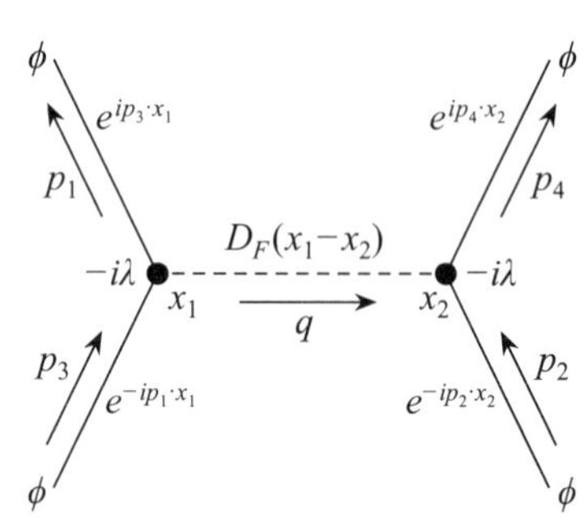

FIGURE 10.7   Amplitude factors that go with each of the features of the phion-scattering Feynman diagram in Figure 10.4.

The amplitude that an off-shell etaon created at x_1 arrives at x_2 turns out to be given by the **Feynman propagator**

$$D_F(x_1 - x_2) = \int \frac{d^4q}{(2\pi)^4} \frac{i e^{iq\cdot(x_1 - x_2)}}{q^2 - m_\eta^2}, \tag{10.15}$$

where q is the etaon's four-momentum. We can interpret this formula as follows. As we discussed at the beginning of the box, a factor of $e^{+iq\cdot x_1}$ is the amplitude that the etaon leaving four-position x_1 has the (off-shell) four-momentum q. The factor of $e^{-iq\cdot x_2}$ is the amplitude that an etaon with four-momentum q arrives at four-position x_2. The denominator "weights" the product of these amplitudes by how close to being "real" the etaon is: the closer that q^2 is to m_η^2, the larger the weighted amplitude. Finally, because in quantum mechanics the total amplitude is the sum of the amplitudes for all possibilities, we integrate over all four-momenta q.

Finally, we associate with each vertex a factor of $-i\lambda$, where λ is the coupling constant that appears in the term that connects the phion and etaon fields in the Lagrangian density. While we cannot justify the factor of $-i$ without delving into the details of quantum field theory, it makes sense that the total amplitude for the process at each vertex should be proportional to the magnitude of λ, because this constant describes how strongly the fields interact.

Figure 10.7 shows all of the elements of the t-channel Feynman diagram for phion scattering, and the contributions to the total amplitudes that go with each. If we multiply all of these amplitude factors together, we get

$$\mathcal{A}(x_1, x_2) = e^{-ip_1\cdot x_1}(-i\lambda)e^{+ip_3\cdot x_1} \int \frac{d^4q}{(2\pi)^4} \frac{i e^{iq\cdot(x_1 - x_2)}}{q^2 - m_\eta^2} e^{-ip_2\cdot x_2}(-i\lambda)e^{+ip_4\cdot x_2},$$

where $\mathcal{A}(x_1, x_2)$ is the complete amplitude for the process assuming that the etaon moves between the four-positions x_1 and x_2. But this exchange process could happen between *any* pair of such positions, so to get the *total* amplitude for the phion scattering process, we need to integrate over all possible positions:

$$\mathcal{A} = -i\lambda^2 \int d^4x_1\, d^4x_2 \frac{d^4q}{(2\pi)^4} \frac{e^{-i(p_1 - p_3 - q)\cdot x_1} e^{-i(p_2 - p_4 + q)\cdot x_2}}{q^2 - m_\eta^2},$$

where I have also gathered together all the exponentials that involve x_1 and all the exponentials that involve x_2.

We can now simplify this expression. We first use Equation 5.33b to do the integral over x_1, treating the other quantities as constants. Since

$$\int d^4x_1\, e^{-i(p_1 - p_3 - q)\cdot x_1} = (2\pi)^4 \delta^4(p_1 - p_3 - q),$$

the expression for the amplitude becomes

$$\mathcal{A} = -i\lambda^2 \int d^4x_2\, d^4q\, \frac{e^{-i(p_2 - p_4 + q)\cdot x_2}}{q^2 - m_\eta^2} \delta^4(p_1 - p_3 - q).$$

If we now integrate over d^4q, the Dirac delta sets $q = p_1 - p_3$ everywhere else in the expression, thereby enforcing conservation of four-momentum at the left vertex, as I have previously claimed is necessary. The result is

$$\mathcal{A} = -i\lambda^2 \int d^4x_2\, \frac{e^{-i(p_2 - p_4 + p_1 - p_4)\cdot x_2}}{(p_1 - p_3)^2 - m_\eta^2}.$$

Exercise 10.3.2   Now do the integration over d^4x_2. You should find that the result proportional to $(2\pi)^4\delta^4(p_1 + p_2 - p_3 - p_4)$ times the scattering amplitude $\mathcal{M}$ that we found by applying the Feynman rules (see Equation 10.10). (*Hint:* Note that everything other than x_2 is a constant in this integration.)

We see now where the overall Dirac delta comes from in Fermi's Golden Rules: the integrals over d^4x_1 and d^4x_2 end up yielding Dirac deltas that have the effect of conserving four-momentum at each vertex, and the combination of those Dirac deltas yields a single Dirac delta, which (when we do the integrals in Fermi's Golden Rule for whatever range of final four-momenta we consider relevant in our experiment) enforces conservation of four-momentum for the entire process. But since this Dirac delta is going to be a feature common to all processes, we extract it from the definition of the scattering amplitude $\mathcal{M}$.

I hope you can now better see how the Feynman Rules simply express the results of a longer calculation of the type discussed in this box (as well as some results from quantum field theory that we are simply taking as given). The rules therefore provide a quick way to calculate $\mathcal{M}$ without doing all of the mathematics we did in this box.

BOX 10.4 The Other Second-Order Phion-Scattering Processes

The Feynman diagrams for the other second-order phion scattering processes are shown in Figure 10.8. The u-channel process is the same as the t-channel process discussed in the text, except that the outgoing phions are exchanged (since we can't tell them apart, we don't know which outgoing phion was connected with which incoming phion, so we must include both processes). The s-channel process is a completely different way of getting from two initial phions to two final phions.

In this box, we will apply the Feynman rules (briefly summarized below for easy reference) to calculate scattering amplitudes for these processes.

1. **Draw the Diagrams.** (See Figure 10.8.)

2. **Label Lines.** (This has already been done in the diagrams.)

3. **Vertex Factors.** Each vertex contributes $-i\lambda$.

4. **Propagator.** Each internal line contributes $i/(q^2 - m^2)$.

5. **Conserve four-momentum at each vertex.**

6. **The result is $-i\mathcal{M}$.**

Exercise 10.4.1 Apply the Feynman rules to the diagrams to calculate the corresponding scattering amplitudes $\mathcal{M}_u$ and $\mathcal{M}_s$. (By Rule 8, the total second-order amplitude is $\mathcal{M}_{\text{tot}} = \mathcal{M}_t + \mathcal{M}_u + \mathcal{M}_s$.)

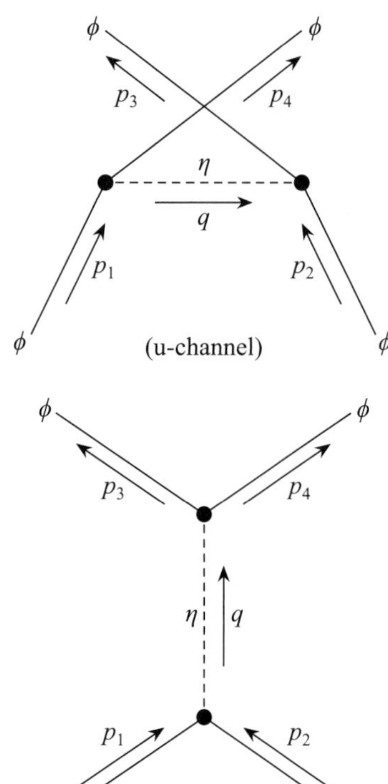

FIGURE 10.8 The Feynman diagram for the other two second-order phion scattering processes.

Homework Problems

P10.1 Suppose the decay rate for a certain particle is $\Gamma = 10$ MeV.
a. What is the average lifetime of this particle in seconds?
b. We sometimes describe how long an unstable particle lasts by stating its **half-life**, the time required for half of the particles that exist at time $t = 0$ to decay. What is this particle's half-life?

P10.2 In particle physics experiments, physicists often express cross sections in barns, where 1 barn = 10^{-28} m^2 = 100 fm^2. Express this unit as an appropriate power of GeV. (The barn is roughly the cross section for a neutron to cause a uranium nucleus to fission. This is as big as "the broad side of a barn" for most subatomic particle interactions. For a re-counting of the origin of this name, see Holloway and Baker, "How the barn was born," *Phys Today* **25**, 7, p. 9 [1972].)

P10.3 The Large Hadron Collider accelerates bunches of protons toward each other at virtually the speed of light. There are about 115 billion protons in each bunch, the length of each bunch is about 0.4 m, and the collider's luminosity is about 2×10^{36} cm^{-2}s^{-1}.
a. If we model each bunch as a cylinder, what is the area of the cylinder's circular end cap?
b. The cross section for elastic scattering of protons at all angles is roughly 10 mb (see Problem P10.2 for the definition of a barn). What would be the event rate for elastic scattering during the time the bunches are interacting?

P10.4 Not all interaction terms that one might dream up are physically meaningful. Consider the Lagrangian density

$$\mathcal{L} = \tfrac{1}{2}\partial^\mu \phi\, \partial_\mu \phi - \tfrac{1}{2}m_\phi^2\phi^2$$
$$+ \tfrac{1}{2}\partial^\mu \eta\, \partial_\mu \eta - \tfrac{1}{2}m_\eta^2\eta^2 - \lambda\eta\phi. \tag{10.16}$$

It looks like this Lagrangian density has an interaction term $-\lambda\eta\phi$. But suppose we define new rotated fields

$$\phi' = \cos\theta\,\phi + \sin\theta\,\eta \quad \text{and}$$
$$\eta' = -\sin\theta\,\phi + \cos\theta\,\eta. \tag{10.17}$$

Show that the Lagrangian density

$$\mathcal{L} = \tfrac{1}{2}\partial^\mu \phi'\, \partial_\mu \phi' - \tfrac{1}{2}m_1^2\phi'^2$$
$$+ \tfrac{1}{2}\partial^\mu \eta'\, \partial_\mu \eta' - \tfrac{1}{2}m_2^2\eta'^2, \tag{10.18}$$

with *no* interaction term (but with different masses for the ϕ' and η' fields), is physically equivalent to original Lagrangian density with an appropriate choice of θ and show how to calculate θ, m_1^2, and m_2^2 given λ, m_ϕ^2, and m_η^2 (you may find some double-angle trig identities helpful). The point is that the original Lagrangian density does *not* describe two fields that interact but rather two free (but different) fields in a hidden guise.

P10.5 If $m_\eta > 2m_\phi$, then an etaon can decay to two phions.
a. Draw the lowest-order Feynman diagram for this process.
b. Use the Feynman Rules to calculate $\mathcal{M}$ for that diagram.
c. Draw the four possible third-order diagrams for this process. (*Hint:* Study Figure 10.5 for inspiration.)
d. If $m_\phi > 2m_\eta$, is it possible for a phion to decay to two etaons? If so, draw the lowest-order Feynman diagram and calculate $\mathcal{M}$ for that diagram. If not, explain why not.

P10.6 Consider the scattering process $\eta + \phi \to \eta + \phi$.
a. Draw the possible Feynman diagrams with the lowest non-trivial order for this process. (*Hint:* There are two.)
b. Compare your diagrams to diagrams we have seen in this chapter to determine the channel names for each subprocess shown. Why is the third channel not present in this case?
c. Are there diagrams with three vertices? If so, draw them. If not, explain why not.
d. Draw the diagrams with four vertices. There are six. (*Hint:* Study Figure 10.5 for inspiration.)

P10.7 In our toy universe, each quanton is its own anti-quanton. Consider the phion-phion "annihilation" process $\phi + \phi \to \eta + \eta$ and assume in this case that the etaon mass m_η is smaller than the phion mass m_ϕ, so that the annihilation is possible for phions at rest.
a. Draw the possible Feynman diagrams with the lowest non-trivial order for this process. (*Hint:* There are two.)
b. Compare your diagrams to diagrams we have seen in this chapter to determine the channel names for each subprocess shown. Why is the third channel not present in this case?
c. Use the Feynman rules to calculate the scattering amplitude $\mathcal{M}$ for each diagram.

P10.8 Consider the bottom vertex in the s-channel diagram for phion-phion scattering. Let's look at this in the center-of-mass frame where the incoming phions have opposite spatial momenta. Do an analysis similar to that in Box 10.3 to show that, even though the virtual etaon's four-momentum q in this case is positive, we still have $q^2 \neq m_\eta^2$ in all frames unless the phion energies in the center-of-mass frame happen to be such that $2E_1 = 2E_2 = m_\eta$. (In that special case, note that $\mathcal{M}_s$ calculated in Box 10.4 does blow up, meaning that this process will dominate over the other processes. Getting the energies *exactly* right is extremely unlikely, but you will get a peak in the cross section as you get close to this energy. This is an example of what particle physicists call a *resonance*.)

P10.9 The Feynman propagator (see Equation 10.15) is a curious beast. Suppose we consider it to be a function of x_2. The only part of the propagator that depends on x_2 is the plane-wave factor $e^{-iq \cdot x_2}$ in the numerator; everything else is a constant. But since $q^2 \neq m_\eta^2$, this wave is *not* a solution to the Klein-Gordon equation, so one would not think that the propagator could describe a particle moving through space at all. However, show that the propagator as a whole (after one does the integral over d^4q) *is* a solution to the Klein-Gordon equation for all values of x_2 except at $x_2 = x_1$. (It is not very surprising that a quantity that is supposed to be the amplitude for a particle going from x_1 to x_2 might blow up when $x_1 = x_2$.) (*Hints:* If we are considering $D_F(x_1 - x_2)$ to be a function of x_2, then the partial derivatives in the Klein-Gordon equation $(\partial^\alpha \partial_\alpha + m_\eta^2) D_F(x_1 - x_2)$ will be with respect to x_2. You will find Equation 5.33b helpful at the end.)

P10.10 We considered the Feynman propagator (see Equation 10.15) to be the amplitude that a particle created at four-position x_1 is found at four-position x_2. But show that the propagator for a particle created at x_2 arriving at x_1 is exactly the same (in other words, the propagator is symmetric under the exchange $x_1 \leftrightarrow x_2$), and discuss, therefore, whether talking about the direction the virtual particle actually travels is meaningful. (*Hint:* It is relevant that we are integrating over all possible four-momenta q.)

$a^{\dagger}_{\vec{p}}$
$\vec{p}$
poof!
poof!
$a_{\vec{p}}$

11. CREATION AND ANNIHILATION OPERATORS

Chapters 7–9 have outlined the basic structure of non-relativistic quantum mechanics (NRQM). A key aspect of NRQM is that **observables** (such as angular momentum, position, energy, and so on) are represented mathematically by **operators**. **Commutation relations** (for example, $[\hat{J}_x, \hat{J}_y] = i\hbar\hat{J}_z$) between these operators tell us much about the physical behavior of their corresponding observables; in particular, how that behavior differs from classical mechanics (where observables quantities are ordinary numbers that commute).

Our goal in the next few chapters is to build from this foundation a self-consistent *relativistic* version of quantum mechanics called **relativistic quantum field theory** (QFT). While QFT is built on non-relativistic quantum mechanics and is consistent with it (in that QFT reduces to NRQM in the non-relativistic limit), QFT is also quite different both conceptually and mathematically. So, be prepared for a change in mindset!

11.1 Fock Space

One of the most important differences between non-relativistic quantum mechanics and quantum field theory is that the states $|\psi\rangle$ in NRQM describe particles whose existence we assume is permanent. We prepare a particle in a certain state and then measure its observables, but we have no way of describing a particle's appearance or destruction in a physical process. However, as we have seen, particles in relativistic systems can be created and destroyed, so we need a new mathematical approach that can handle such events.

The solution to this problem involves changing our focus. Instead of focusing on an individual *particle*, we instead focus on a *system* of interacting particles. Instead of asking "What is this particle's state?" we ask "How many particles in our system occupy such-and-such a state?"

In quantum field theory, we consider the states that a particle might occupy to be free-particle momentum eigenstates $|\vec{p}\,\rangle$. To make things easier at first, let's assume that these momentum states are discrete: $|\vec{p}_k\rangle$ with an integer index k. This would be the case if our particles were confined to move in one dimension around a ring: since a particle's wavefunction must oscillate an integer number k times as we go around the ring (so that it matches itself when we return to where we started), that particle's momentum in a ring of radius R must be

$$p_k = \frac{h}{\lambda} = \frac{(2\pi\hbar)}{(2\pi R/k)} = \frac{k\hbar}{R} = \frac{k}{R}, \tag{11.1}$$

where k is any positive or negative integer (and remember that we are working in units where $\hbar = 1$). There are an infinite number of such states (from $k = -\infty$ to $k = +\infty$, excluding 0), but at least the states are discrete.

Now, the states $|\vec{p}_k\rangle$ here are just the states in an ordinary Hilbert space that we might use in ordinary quantum mechanics to describe a single particle. But from our particle *system* perspective, this state represents an abstract possibility: it is a state that a particle *might* have. Moreover, more than one particle in our system might have the *same* momentum state $|\vec{p}_k\rangle$. Our focus is not on a particle that has a certain state, but rather on a state that might have a certain number of particles sharing that state.

The states we will use to describe our *system* are therefore states in a different kind of space. The basis states of this new space have an infinite list of slots (one slot for each possible particle momentum state $|\vec{p}_k\rangle$), and in each slot, we specify how many particles share or **occupy** that momentum state. For example, the system state

$$|\ldots, n_{k-1}, n_k, n_{k+1}, \ldots\rangle = |\ldots, 2, 0, 1, \ldots\rangle \qquad (11.2)$$

is the system basis state describing a situation where we have two particles with the (ordinary Hilbert-space) momentum state $|\vec{p}_{k-1}\rangle$, no particles with the particle state $|\vec{p}_k\rangle$, one with the particle state $|\vec{p}_{k+1}\rangle$, and so on. In what follows, we will assume that we are talking about *bosons*, where n_k can be any integer ≥ 0.

We call the space whose basis kets have the form $|\ldots, n_{k-1}, n_k, n_{k+1}, \ldots\rangle$ a **Fock space**. A Fock space is a kind of "super"-Hilbert space, whose system states are vast collections of Hilbert-space states of the type ordinarily used in quantum mechanics. The beauty of a Fock space, however, is that it has the same mathematical structure no matter how many particles our system contains or whether those particles are bosons or fermions (for fermions, we need only to restrict n_k to being 0 or 1). In particular, Fock space very naturally handles states involving sets of identical particles, which are *very* awkward to describe using normal Hilbert-space states.

A particularly important Fock-space state is the **vacuum state**

$$|0\rangle = |\ldots, 0, 0, 0, \ldots\rangle \qquad (11.3)$$

describing a system that contains no particles in any state.

11.2 Creation and Annihilation Operators

An operator, by definition, operates on a state to convert it to another state. We *define* the **annihilation operator** a_k as taking a Fock-space ket and lowering the number in its kth slot (annihilating a particle with momentum $\vec{p}_k$):

$$a_k |\ldots, n_{k-1}, n_k, n_{k+1}, \ldots\rangle \equiv c_{n_k} |\ldots, n_{k-1}, (n_k - 1), n_{k+1}, \ldots\rangle, \quad (11.4a)$$

where c_{n_k} is an as yet unspecified (possibly complex) number. To make the notation for Fock-space kets a bit cleaner, let's focus on the single slot that is modified, writing the previous equation as simply

$$a_k |, n_k, \rangle \equiv c_{n_k} |, n_k - 1, \rangle, \qquad (11.4b)$$

where I am using the commas to remind us that this is a Fock-space ket with slots other than the kth slot (and not just a normal Hilbert space ket). As the value in a given slot cannot be a negative number, we cannot further lower a state where n_k is already zero, so we must have

$$a_k |, 0, \rangle = 0 \quad \Rightarrow \quad c_0 = 0. \qquad (11.5)$$

But, of course, we want $c_{n_k} \neq 0$ if $n_k \geq 1$.

Now let's explore what the adjoint $a_k^\dagger$ of this operator does. Note that by definition of the adjoint,

$$\langle, n_k', |a_k^\dagger|, n_k, \rangle = \langle, n_k, |a_k|, n_k', \rangle^*$$

$$= \langle, n_k, |c_{n_k'}|, n_k' - 1, \rangle^*$$

$$= c_{n_k'}^* \langle, n_k, |, n_k' - 1, \rangle^*$$

$$= c_{n_k'}^* \quad \text{if } n_k = n_k' - 1 \quad \text{(otherwise, 0)}, \qquad (11.6)$$

assuming that the basic kets of our Fock space are orthonormal. This means that $n'_k = n_k + 1$ to yield a nontrivial result. Therefore, the action of $a^\dagger_k$ on the ket on the *right* side of the original bra-ket must be

$$a^\dagger_k \, |, n_k, \rangle = c^*_{n_k+1} \, |, n_k + 1, \rangle \tag{11.7}$$

to yield the nonzero result given in Equation 11.6. So, we see that $a^\dagger_k$ acts as a **creation operator**: it increases the number in the kth slot (which amounts to *creating* a particle with momentum $\vec{p}_k$).

Now consider a new operator $N_k \equiv a^\dagger_k a_k$. What does this operator do to a Fock-space ket? Note that the a_k lowers the value in the kth slot by one, then the $a^\dagger_k$ raises that value back to the original value, meaning that we get the original state back again times a constant:

$$\begin{aligned}
N_k \, |, n_k, \rangle &\equiv a^\dagger_k a_k \, |, n_k, \rangle \\
&= a^\dagger_k \big[\, c_{n_k} \, |, n_k - 1, \rangle \big] \\
&= c^*_{(n_k-1)+1} c_{n_k} \, |, n_k, \rangle \\
&= |c_{n_k}|^2 \, |, n_k, \rangle.
\end{aligned} \tag{11.8}$$

So, we see that Fock-space basis kets are eigenstates of the N_k operator with a real and nonnegative eigenvalue $|c_{n_k}|^2$. Since we can freely choose the overall phase of the eigenstates generated from the vacuum state by the creation operator $a^\dagger_k$, we can choose c_{n_k} to be real for all n_k. If we also choose

$$c_{n_k} \equiv \sqrt{n_k} \quad \text{so that} \tag{11.9a}$$

$$a_k \, |, n_k, \rangle = \sqrt{n_k} \, |, n_k - 1, \rangle \tag{11.9b}$$

$$a^\dagger_k \, |, n_k, \rangle = \sqrt{n_k + 1} \, |, n_k + 1, \rangle \tag{11.9c}$$

(note that this choice is consistent with the requirement that $c_0 = 0$), then we have completely defined the action of both a_k and $a^\dagger_k$, and we see that

$$N_k \, |, n_k, \rangle = n_k \, |, n_k, \rangle. \tag{11.10}$$

So, N_k acting on a basis ket simply returns the number in the kth slot; that is, the number of particles sharing the particle momentum state $|\vec{p}_k\rangle$. We therefore call $N_k \equiv a^\dagger_k a_k$ the **number operator** for the kth momentum state.

You may recognize some similarities between creation and annihilation operators and the "raising and lowering" operators $J_\pm$ for angular momentum states in non-relativistic quantum mechanics. The basic mathematics is indeed pretty similar. However, the interpretation is somewhat different: in ordinary quantum mechanics, we change a single-particle state $|j, m\rangle$ to a different single-particle state $|j, m \pm 1\rangle$ with a different value of the quantum number m, but here we are actually increasing or decreasing the number of particles sharing a *given* particle momentum state, which amounts to creating or destroying a particle. Don't worry! The mathematics we develop will ensure that energy is conserved in these creation or annihilation events. But these operators make it possible to model physical processes that actually do create and destroy particles, something impossible to model in ordinary quantum mechanics.

11.3 Commutation Relations

These definitions require that a_k and $a^\dagger_k$ obey the commutation relations

$$[\, a_j, a^\dagger_k] = \delta_{jk} \quad \text{and} \quad [\, a_j, a_k] = 0 \quad \text{and} \quad [\, a^\dagger_j, a^\dagger_k] = 0. \tag{11.11}$$

Let me illustrate how to prove the relation $[a_j, a_k^\dagger] = \delta_{j,k}$ in the case where $j = k$:

$$
\begin{aligned}
[a_k, a_k^\dagger]\,|, n_k, \rangle &\equiv (a_k a_k^\dagger - a_k^\dagger a_k)\,|, n_k, \rangle \\
&= a_k a_k^\dagger\,|, n_k, \rangle - N_k\,|, n_k, \rangle \\
&= a_k \sqrt{n_k + 1}\,|, n_k + 1, \rangle - n_k\,|, n_k, \rangle \\
&= \sqrt{n_k + 1}\sqrt{n_k + 1}\,|, n_k, \rangle - n_k\,|, n_k, \rangle \\
&= (n_k + 1 - n_k)\,|, n_k, \rangle = |, n_k, \rangle,
\end{aligned}
\tag{11.12}
$$

meaning that $[a_k, a_k^\dagger]$ acts like the identity operator on any Fock-space basis ket. In Box 11.1, you will verify the other relations yourself.

One can also do this math in the other direction: one can *start* with the commutation relations and show that $N_k \equiv a_k^\dagger a_k$ is a number operator, a_k is an annihilation operator, and so on. Box 11.2 explores this approach. The point is that the commutation relations actually tell you *everything* you need to know about the operators!

We can construct an *arbitrary* Fock-space state from the vacuum state by successively applying appropriate creation operators. For example, to generate the state with three particles with momentum p_j and one with momentum p_k, we apply three $a_j^\dagger$ operators and one $a_k^\dagger$ operator to the vacuum state:

$$
|\ldots 3 \ldots 1 \ldots\rangle = \frac{a_j^\dagger}{\sqrt 3}\frac{a_j^\dagger}{\sqrt 2}\frac{a_j^\dagger}{\sqrt 1}\frac{a_k^\dagger}{\sqrt 1}|0\rangle = \frac{(a_j^\dagger)^3}{\sqrt{3!}}\frac{(a_k^\dagger)^1}{\sqrt{1!}}|0\rangle
\tag{11.13}
$$

(where in the left ket, the 3 is in the jth slot and the 1 is in the kth slot). Since all creation operators commute with each other, the order in which we apply the operators does not matter. But also note that in order to ensure that the ket is *normalized*, we must divide by $\sqrt{n_k}$ (where n_k is the number of particles in the kth slot after each creation operation on that slot) because the creation operator acting by itself *multiplies* the ket by $\sqrt{n_k}$ according to Equation 11.9c.

The general expression for constructing an arbitrary Fock-space ket is

$$
|\ldots, n_{k-1}, n_k, n_{k+1}, \ldots\rangle = \prod_k \frac{(a_k^\dagger)^{n_k}}{\sqrt{n_k!}}|0\rangle.
\tag{11.14}
$$

See Equation 11.13 for justification of the factorial. This expression takes advantage of the fact that the creation operators all commute to apply them in a nice sequential order.

11.4 The Continuum Limit

However, one-dimensional rings are rare in physics. We generally consider both position space and momentum space to be three-dimensional, infinite, and continuous. This means that the momentum eigenvalues we use to label the momentum states we are using as the foundation of our Fock-space are continuous rather than having discrete values. How can we handle this?

In Chapter 7, we saw that we can represent a particle momentum state $|\vec p\rangle$ in an infinite three-dimensional space using a wavefunction

$$
\psi_{\vec p}(x, y, z) = \left(\frac{1}{2\pi}\right)^{3/2} e^{-i\vec p \cdot \vec x},
\tag{11.15}
$$

where the components of $\vec p$ can be any set of three real numbers (see Equation 7.17). The constant in front of this wavefunction has been chosen so that the states satisfy the normalization condition given in Equation 7.19:

$$
\langle \vec p\,'|\vec p\,\rangle = \delta^3(\vec p\,' - \vec p) \quad \text{(in NRQM)}.
\tag{11.16}
$$

This is the convention in non-relativistic quantum mechanics, but in quantum field theory, we conventionally normalize particle momentum states so that

$$\langle \vec{p}\,'|\vec{p}\,\rangle = (2\pi)^3 \delta^3(\vec{p}\,' - \vec{p}) \quad \text{(in QFT)}. \tag{11.17}$$

This has the effect of making a number of other important equations simpler.

How we normalize particle momentum states in quantum field theory is relevant for two reasons. First, remember that particle momentum states provide the foundation on which we build our Fock space: our Fock-space states count the number of particles occupying each particle momentum state. Second (and more directly to the point), knowing how momentum states are normalized in the continuum limit tells us what the commutation relations for creation and annihilation operators are in that limit. The argument follows.

In Fock space, the vacuum state $|0\rangle$ is unique and well-defined. Assume that we normalize it to have unit magnitude $\langle 0|0\rangle = 1$. We can raise the vacuum state to create a Fock-space state describing a single particle with momentum $\vec{p}$, which is really the same thing as the Hilbert space state $|\vec{p}\,\rangle$. Therefore,

$$a_{\vec{p}}^\dagger |0\rangle = \sqrt{1}\,|,\vec{p}, \rangle = |\vec{p}\,\rangle. \tag{11.18}$$

In Box 11.3, you will show that the normalization condition in Equation 11.17 requires that

$$[\, a_{\vec{p}'},\ a_{\vec{p}}^\dagger\,] = (2\pi)^3\, \delta^3(\vec{p}\,' - \vec{p}\,). \tag{11.19a}$$

This is the only tricky commutator, as the ones that were zero are still zero:

$$[\, a_{\vec{p}'},\ a_{\vec{p}}\,] = 0 \quad \text{and} \quad [\, a_{\vec{p}'}^\dagger,\ a_{\vec{p}}^\dagger\,] = 0. \tag{11.19b}$$

These equations will be *very important* in the next chapter.

In the continuum limit, our Fock-space kets no longer have discrete "slots" into which we put occupation numbers, but this does not really matter. We can still construct an arbitrary state by applying the appropriate number of creation operators to the vacuum state, and this will turn out to be sufficient.

11.5 Ensuring Lorentz Invariance of Integrals

In quantum field theory, we will be doing lots of integrals of the form

$$q = \int \frac{d^3 p}{(2\pi)^3} f(\vec{p}\,), \tag{11.20}$$

where $f(\vec{p})$ is some function of a possible particle momentum $\vec{p}$ and we want q to be a Lorentz-invariant scalar. This is problematic because the volume element $d^3 p$ is *not* Lorentz-invariant.

How can we ensure the Lorentz invariance of such integrals? We begin by noting that if $h(p^\mu)$ is a Lorentz-invariant scalar, then so is

$$q \equiv \int d^4 p\, h(p^\mu), \tag{11.21}$$

because $d^4 p$ *is* a Lorentz-invariant volume (see Box 11.4).

We normally do integrals only over the *spatial* momentum components because if you know a particle's spatial momentum $\vec{p}$, then you know its four-momentum time component $p^t = E_{\vec{p}} = (|\vec{p}\,|^2 + m^2)^{1/2}$. So, how do we constrain a four-dimensional integral to include only those points in four-momentum space where $p^t = (|\vec{p}\,|^2 + m^2)^{1/2}$?

We can write the three-dimensional integral in Equation 11.20 as a four-dimensional integral if we use a Dirac delta:

$$q = \int d^4p\, \delta(p^2 - m^2)\, \Theta(p^t)\, h(p^\mu), \tag{11.22}$$

where $p^2 = p \cdot p = (p^t)^2 - |\vec{p}\,|^2$. The Dirac delta is a Lorentz scalar because its argument is a Lorentz scalar, and it constrains p^t to be $\pm\sqrt{|\vec{p}\,|^2 + m^2}$. To ensure we get only the positive value, I have included the $\Theta(p^t)$ function, which is 1 if $p^t > 0$ and 0 otherwise. Even though it does not look like this is a Lorentz scalar (because it depends on a *component* of a four-vector), no Lorentz transformation can change the sign of p^t in a timelike four-momentum (and a real quanton's four-momentum is always timelike), so all observers will agree on whether p^t is positive or not. Therefore, the entire integrand is a Lorentz scalar, meaning that the integral is also a Lorentz scalar.

Now, if you perform the integral over p^t and use properties of the Dirac delta derived in Chapter 4, you can show (see Box 11.5) that this becomes

$$q = \int \frac{d^3p}{2E_{\vec{p}}}\, h(E_{\vec{p}}, \vec{p}\,) \tag{11.23}$$

(where the notation for $h(p^\mu)$ is meant to suggest that we have now forced p^t to be equal to $E_{\vec{p}}$). Therefore, $d^3p/2E_{\vec{p}}$ is the correct volume element to use to yield a Lorentz-invariant result subject to our constraint.

The remaining trouble is that the momentum kets $|\vec{p}\,\rangle$ we are using as the foundation of our Fock space are not Lorentz-invariant either. We have normalized them so that

$$\int \frac{d^3p}{(2\pi)^3} \langle \vec{p} \mid \vec{p}\,' \rangle = \int \frac{d^3p}{(2\pi)^3} (2\pi)^3 \delta^3(\vec{p} - \vec{p}\,') = 1. \tag{11.24}$$

But because d^3p is not Lorentz-invariant, the $|\vec{p}\rangle$ and $|\vec{p}\,'\rangle$ states cannot be Lorentz-invariant if the integral yields the Lorentz-invariant result 1. However, we can define new explicitly Lorentz-invariant momentum states $|p\rangle$ such that

$$\int \frac{d^3p}{(2\pi)^3} \frac{1}{2E_{\vec{p}}} \langle p \mid p' \rangle = 1. \tag{11.25}$$

If we compare this equation with the previous equation, we see that

$$|p\rangle = \sqrt{2E_{\vec{p}}}\,|\vec{p}\,\rangle. \tag{11.26}$$

Finally, let's consider the creation and annihilation operators. Since every inertial observer will agree that the Fock-space vacuum state $|0\rangle$ is the vacuum state, it must therefore be a Lorentz-invariant state. We can define a Lorentz-invariant creation operator $\alpha_p^\dagger$ that acts on the Lorentz-invariant vacuum state to create a Lorentz-invariant particle momentum state $|p\rangle$:

$$|p\rangle = \alpha_p^\dagger |0\rangle. \tag{11.27}$$

If we compare this to $|\vec{p}\,\rangle = a_{\vec{p}}^\dagger |0\rangle$ (as stated in Equation 11.18), we see that the Lorentz-invariant raising operator is related to $a_{\vec{p}}^\dagger$ as follows:

$$\alpha_{\vec{p}}^\dagger = \sqrt{2E_{\vec{p}}}\, a_{\vec{p}}^\dagger. \tag{11.28}$$

We now have the tools we need to construct a quantum field theory for free, non-interacting bosons. We will take up that task in the next chapter.

BOX 11.1 Proving the Commutation Relations

In this box, we will prove the commutation relations

$$[a_j, a_k^\dagger] = 0 \quad \text{for } j \neq k \tag{11.29a}$$

$$[a_j, a_k] = 0 \quad \text{and} \quad [a_j^\dagger, a_k^\dagger] = 0 \quad \text{for all } j, k \tag{11.29b}$$

in the situation where we have *discrete* one-dimensional momentum states with eigenvalues p_k (where k is an integer index).

Exercise 11.1.1 Prove these relations. (*Hints:* Separate the $j = k$ and $j \neq k$ cases. The two $j = k$ cases in Equation 11.29b are relatively easy. When $j \neq k$, examine the operators' action on a Fock-space ket $|\ldots n_j \ldots n_k \ldots\rangle$. Also, handle the $n_j = 0$ case separately whenever a commutator involves a_j.)

BOX 11.2 Starting with the Commutation Relations

In this box, we will see that in the situation where we have discrete momentum states labeled by eigenvalues p_k (where k is an integer index), the relations

$$[\, a_j, a_k^\dagger \,] = \delta_{jk}, \quad [\, a_j, a_k \,] = 0, \quad \text{and} \quad [\, a_j^\dagger, a_k^\dagger \,] = 0 \qquad (11.30)$$

are sufficient to show that $a_k^\dagger$ is a creation operator, a_k is an annihilation operator, and $N_k = a_k^\dagger a_k$ is a number operator for Fock-space kets.

Exercise 11.2.1 Directly from the commutation relations, prove that

$$[\, N_k, a_k^\dagger \,] = a_k^\dagger \quad \text{and} \quad [\, N_k, a_k \,] = -a_k. \qquad (11.31)$$

(*Hint:* Remember the basic identity $[AB, C\,] = A[\, B, C\,] + [\, A, C\,]B$.)

Now, every operator has eigenstates with corresponding eigenvalues. Let $|, \eta_k, \rangle$ be the (as yet undetermined) system eigenstate in Fock space for the operator N_k corresponding to the eigenvalue η_k:

$$N_k \,|, \eta_k, \rangle = \eta_k \,|, \eta_k, \rangle. \qquad (11.32)$$

Now consider the ket $|, \psi_+, \rangle \equiv a_k^\dagger \,|, \eta_k, \rangle$.

Exercise 11.2.2 Show that $|, \psi_+, \rangle$ is also an eigenstate of N_k, but with eigenvalue $\eta_k + 1$. (*Hint:* First use the result of the previous exercise to show that $N_k a_k^\dagger = a_k^\dagger + a_k^\dagger N_k$.)

BOX 11.2 Starting with the Commutation Relations (cont.)

That $|, \psi_+, \rangle \equiv a_k^\dagger |, \eta_k, \rangle$ is an eigenstate of N_k with eigenvalue $\eta_k + 1$ means that the operator $a_k^\dagger$ serves as a creation operator for eigenstates of N_k: given any initial eigenvector, we can create eigenstates with incrementally higher eigenvalues by successive applications of the $a_k^\dagger$ operator.

Exercise 11.2.3 In a similar way, show that the state $|, \psi_-, \rangle \equiv a_k |, \eta_k, \rangle$ is an eigenstate of N_k with eigenvalue $\eta_k - 1$, making a_k an annihilation operator.

Exercise 11.2.4 Argue that $\langle, \psi_-, |, \psi_-, \rangle = \eta_k \langle, \eta_k, |, \eta_k, \rangle$ and that therefore $\eta_k \geq 0$. (*Hint:* The norms of kets must be nonnegative.)

BOX 11.2 Starting with the Commutation Relations (cont.)

In the last exercise, we found that the eigenvalues of the N_k operator must be nonnegative. But we can also lower those eigenvalues indefinitely by applying the annihilation operator a_k. This seems contradictory. The only way out of the contradiction is to require that there is some lowest eigenstate $|\eta_{k,\min}\rangle$ such that $a_k |\eta_{k,\min}\rangle = 0$. Then we will not be able to lower the state any further.

Exercise 11.2.5 By applying N_k to the state $|,\eta_{k,\min},\rangle$, argue that $\eta_{k,\min} = 0$.

We see, therefore, that the eigenvalues of the N_k operators are nonnegative integers, starting with 0, so N_k has the characteristics of a number operator.

Now let's determine how the eigenstates of N_k are related to the basis kets of our Fock space.

Exercise 11.2.6 Show that the basic commutation relations given in Equation 11.30 imply that $[\, N_j,\, N_k \,] = 0$ if $j \neq k$.

But this means that the number operators with different indices all commute, which means that these operators have eigenstates in common. This means that we can describe these eigenstates as Fock-space states $|\ldots n_k \ldots\rangle$ with one slot for each possible different value of the index k.

None of the above requires that these slots describe the number of particles in a given *momentum* state, and indeed, we could construct a Fock space based on any complete sets of fundamental particle states that we care to use. For free particles, however, momentum states are convenient.

BOX 11.3   Commutators in the Continuum Limit

In this box, we will show that if we require that the normalization

$$\langle \vec{p}\,'|\vec{p}\,\rangle = (2\pi)^3\,\delta^3(\vec{p}\,' - \vec{p}\,) \tag{11.33}$$

in the continuum limit applies to the Fock-space single-particle state

$$|\vec{p}\,\rangle \equiv a_{\vec{p}}^{\dagger}\,|\,0\,\rangle, \tag{11.34}$$

then the commutator $[\,a_{\vec{p}\,'},\ a_{\vec{p}}^{\dagger}\,]$ in that same limit must be

$$[\,a_{\vec{p}\,'},\ a_{\vec{p}}^{\dagger}\,] = (2\pi)^3\,\delta^3(\vec{p}\,' - \vec{p}\,). \tag{11.35}$$

Exercise 11.3.1   Show this. (*Hint:* It is acceptable to work backwards: assume the commutation relation and show that we get Equation 11.33.)

BOX 11.4 The Volume of d^4p Is Invariant

Determining how the volume element in a multiple integral transforms when one changes variables is not trivial. For example, consider the transformation from Cartesian coordinates x, y to polar coordinates r, θ. Since $x = r \cos \theta$ and $y = r \sin \theta$, one might naively think that

$$dx\, dy = \left(\frac{\partial x}{\partial r} dr + \frac{\partial x}{\partial \theta} d\theta \right) \left(\frac{\partial y}{\partial r} dr + \frac{\partial y}{\partial \theta} d\theta \right)$$

$$= (\cos \theta\, dr - r \sin \theta\, d\theta)\, (\sin \theta\, dr + r \cos \theta\, d\theta)$$

$$= (\cos \theta \sin \theta)(dr^2 - r^2\, d\theta^2) + (\cos^2 \theta - \sin^2 \theta) r\, dr\, d\theta, \qquad (11.36)$$

which is not close to being the correct transformation $dx\, dy \to r\, dr\, d\theta$.

This approach is wrong because it does not address the right question. What we *really* want to know is the *area* in x, y coordinates of the polar-coordinate area element, whose edges are tiny displacements: one (call it $d\vec{r}$) in the direction of increasing r (while holding θ constant) spanned by the coordinate difference dr, and the other (call it $d\vec{\theta}$) in the direction of increasing θ (while holding r constant) spanned by $d\theta$. The components of $d\vec{r}$ and $d\vec{\theta}$ in the x, y coordinate system are

$$dr_x = \frac{\partial x}{\partial r} dr,\ dr_y = \frac{\partial y}{\partial r} dr \quad \text{and} \quad d\theta_x = \frac{\partial x}{\partial \theta} d\theta,\ d\theta_y = \frac{\partial y}{\partial \theta} d\theta, \qquad (11.37)$$

since in the first case $d\theta = 0$ and in the second $dr = 0$. Now, the area of the parallelogram spanned by two vectors is equal to the magnitude of those vectors' *cross product*. So, the equivalent area of our polar-coordinate area element is

$$\text{area} = |d\vec{r} \times d\vec{\theta}| = |\, dr_x\, d\theta_y - dr_y\, d\theta_x\, | = \left| \frac{\partial x}{\partial r} \frac{\partial y}{\partial \theta} - \frac{\partial y}{\partial r} \frac{\partial x}{\partial \theta} \right| dr\, d\theta$$

$$= |\cos \theta (r \cos \theta) + \sin \theta (r \sin \theta)|\, dr\, d\theta = r\, dr\, d\theta. \qquad (11.38)$$

Abstractly, when we transform coordinates p, q to coordinates u, v, the integration area element transforms according to

$$dp\, dq \to \left| \frac{\partial p}{\partial u} \frac{\partial q}{\partial v} - \frac{\partial q}{\partial u} \frac{\partial p}{\partial v} \right| du\, dv. \qquad (11.39)$$

Because it is easier to generalize to more dimensions, people usually write the cross product as the determinant of the matrix of the partials:

$$dp\, dq \to \begin{vmatrix} \dfrac{\partial p}{\partial u} & \dfrac{\partial q}{\partial u} \\[2mm] \dfrac{\partial p}{\partial v} & \dfrac{\partial q}{\partial v} \end{vmatrix} du\, dv. \qquad (11.40)$$

(Note that enclosing a matrix with vertical bars rather than parentheses indicates that matrix's determinant, not its absolute value. But to relate the *areas,* one should generally *also* take the absolute value of that determinant, as indicated in Equation 11.39.) Mathematicians call the matrix (and sometimes also the matrix's determinant) the **Jacobian** for the transformation.

A similar argument shows that the 3D volume element transforms as

$$dp\, dq\, dr \to \begin{vmatrix} \dfrac{\partial p}{\partial u} & \dfrac{\partial q}{\partial u} & \dfrac{\partial r}{\partial u} \\[2mm] \dfrac{\partial p}{\partial v} & \dfrac{\partial q}{\partial v} & \dfrac{\partial r}{\partial v} \\[2mm] \dfrac{\partial p}{\partial w} & \dfrac{\partial q}{\partial w} & \dfrac{\partial r}{\partial w} \end{vmatrix} du\, dv\, dw. \qquad (11.41)$$

BOX 11.4 The Volume of d^4p Is Invariant (cont.)

The generalization to four and more dimensions is analogous.

Now, our fundamental goal in this box is to prove that the area of the four-dimensional volume element d^4p in momentum space is Lorentz-invariant. Assuming that the primed and unprimed frames are in standard orientation, the Lorentz transformation of four-momentum components is

$$p'^t = \gamma p^t - \gamma \beta p^x \tag{11.42a}$$

$$p'^x = -\gamma \beta p^t + \gamma p^x \tag{11.42b}$$

$$p'^y = p^y \tag{11.42c}$$

$$p'^z = p^z. \tag{11.42d}$$

Exercise 11.4.1   Using the methods described in this section, show that $d^4p' = d^4p$. (*Hint:* If you don't remember how to evaluate the determinant of a 4×4 matrix, ask WolframAlpha for the determinant of the matrix we get.)

BOX 11.5 Lorentz-Invariant Integrals Over d^3p

In this box, we are going to integrate Equation 11.22

$$q = \int d^4p \, \delta(p^2 - m^2) \, \Theta(p^t) \, h(p^\mu) \tag{11.43}$$

with respect to p^t, and use properties of the Dirac delta and the Heaviside step function to arrive at the simpler expression in Equation 11.23:

$$q = \int \frac{d^3p}{2E_{\vec{p}}} \, h(E_{\vec{p}}, \vec{p}\,). \tag{11.44}$$

Exercise 11.5.1 Do this integration. (*Hint:* The key is to recognize that the Dirac delta is zero for *two* values of p^t: when $p^t = \pm(|\vec{p}\,|^2 + m^2)^{1/2}$. Use Equation 4.25 in Chapter 4, repeated below for easy reference.)

$$\delta[\,g(x)\,] = \sum_{\text{all}x_i} \frac{\delta(x - x_i)}{|dg/dx|_{x=x_i}} \quad \text{for all } x_i \text{ such that } g(x_i) = 0. \tag{11.45}$$

Homework Problems

P11.1 In the context of a Fock space built on discrete particle momentum states $|\vec{p}_k\rangle$, show directly from the commutation relations involving the creation and annihilation operators that:

a. $[\,a_k,\,(a_k^\dagger)^n\,] = n(a_k^\dagger)^{n-1}$. (*Hint:* Use induction.)

b. $\langle\,0\,|(a_k)^n(a_k^\dagger)^m|\,0\,\rangle = n!\,\delta_{nm}$.

P11.2 In the context of a Fock space built on discrete particle momentum states $|\vec{p}_k\rangle$, what is the value of $\langle 0|a_k a_k a_k^\dagger a_k^\dagger|0\rangle$?

P11.3 In the context of a Fock space built on discrete particle momentum states $|\vec{p}_k\rangle$, suppose we have a sequence of four raising and/or lowering operators sandwiched between two vacuum states; for example, $\langle 0|a_k^\dagger a_j a_n a_k^\dagger|0\rangle$, where the subscripts may or may not be the same and there may not be an equal number of raising and lowering operators. Specify as completely as possible what conditions must apply for this quantity to be nonzero. Do the same for a sequence of five raising and/or lowering operators.

P11.4 In the context of a Fock space built on discrete particle momentum states $|\vec{p}_k\rangle$, prove that Equation 11.14

$$|\ldots, n_{k-1}, n_k, n_{k+1}, \ldots\rangle = \prod_k \frac{(a_k^\dagger)^{n_k}}{\sqrt{n_k!}}|0\rangle \qquad (11.46)$$

correctly specifies how to construct a general Fock-space state. (*Hint:* Use induction to show this for a single given value of the index k. Then argue that the product over all values of k is correct.)

P11.5 Consider the two-particle Fock state

$$|\vec{q}\,\vec{p}\,\rangle = a_{\vec{q}}^\dagger\, a_{\vec{p}}^\dagger|\,0\,\rangle \qquad (11.47)$$

(where $\vec{p}$ and $\vec{q}$ are momenta) in the continuum limit. Evaluate the inner product $\langle\,\vec{p}\,'\,\vec{q}\,'\,|\,\vec{q}\,\vec{p}\,\rangle$ in that limit. (*Hint:* The result will involve a sum of products of Dirac deltas.)

P11.6 Use the Jacobian method to find the correct volume element for a volume integral over in three-dimensional physical space when we are using spherical coordinates r, θ, and ϕ.

P11.7 Show, as suggested in Box 11.4, that under a Lorentz transformation $d^3p \to d^3p' = \gamma\,d^3p$.

P11.8 Using the Jacobian method, argue that under a Lorentz transformation

$$\frac{d^3p}{2E_{\vec{p}}} = \frac{d^3p'}{2E_{\vec{p}'}}. \qquad (11.48)$$

(*Hint:* Use the fact that $E_{\vec{p}} = [(p^x)^2 + (p^y)^2 + (p^z)^2 + m^2]^{1/2}$ to show that $\partial E_{\vec{p}}/\partial p^x = p^x/E_{\vec{p}}$ and similarly for p^y and p^z.)

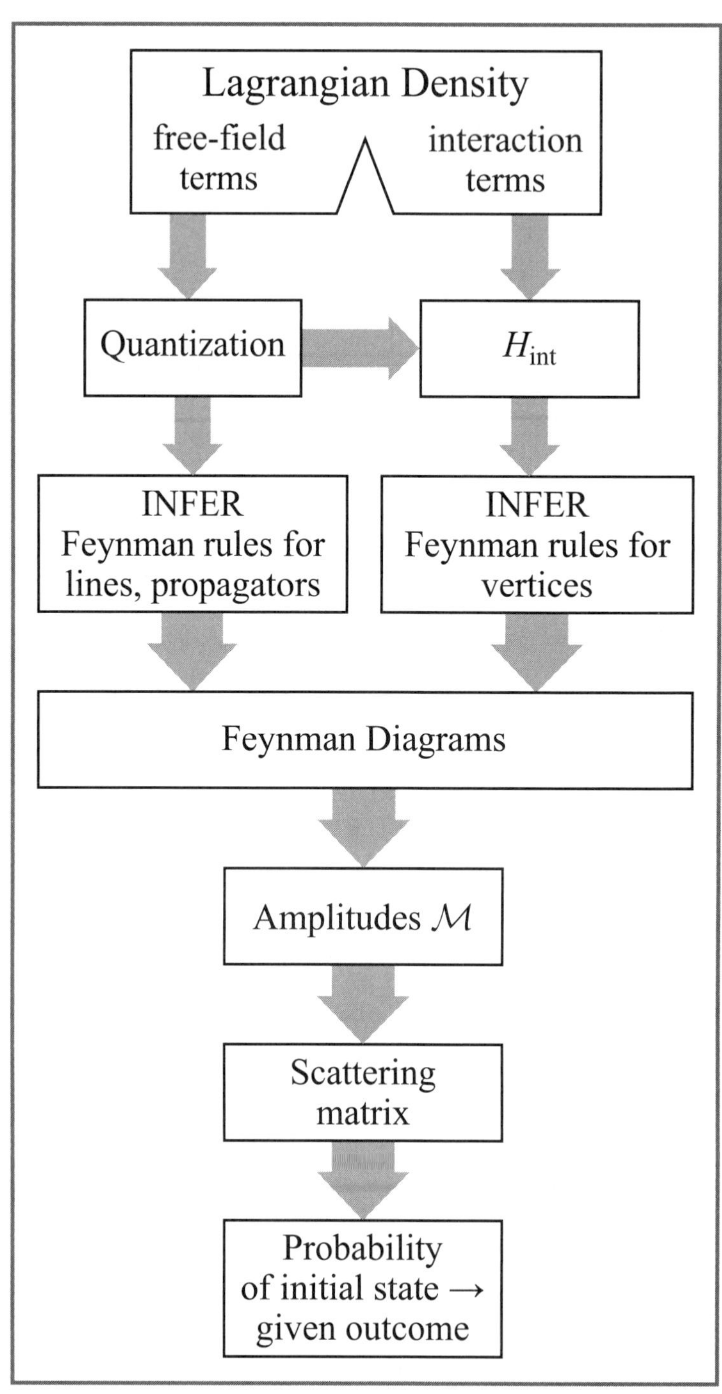

Lagrangian Density
free-field terms
interaction terms
Quantization
H_{int}
INFER
Feynman rules for lines, propagators
INFER
Feynman rules for vertices
Feynman Diagrams
Amplitudes $\mathcal{M}$
Scattering matrix
Probability of initial state → given outcome

In Chapter 5, we discussed the Lagrangian density for a classical field $\phi(t, x, y, z)$ whose equation of motion is the fully relativistic Klein-Gordon equation. In Chapter 6, we saw that we could define a position-dependent energy density for this field that could correspond to something like a density of particles of a given type. But we have not yet seen how to make $\phi(t, x, y, z)$ a *quantum* field or what that might mean.

In the previous chapter, we explored how to use a Fock space to describe the state of quantum systems that involve a variable number of particles, and defined creation and annihilation operators that act on that space to raise or lower the number of particles sharing a specific state of definite momentum. Connecting our field to this mathematical backdrop is the next crucial step in constructing a *quantum* field theory.

12.1 Quantizing the Field

The key to connecting the ϕ-field to our Fock space is to make it an *operator* field, that is, a field that assigns a Fock-space *operator* $\phi(x)$ to every point $x \equiv [t, x, y, z]$ in space and time instead of a complex number. (Note that I am using x here to represent an entire position four-vector instead of just its x component: this is potentially confusing but is conventional notation in particle physics). The *operator* $\phi(x)$ can change one Fock-space state describing a certain number of particles with some arrangement of momenta to another Fock-space state describing a different set of particles with a different arrangement of momenta, which is exactly what we want a theory to do. This is analogous to the move we made in ordinary quantum mechanics where we took observables (such as spin, energy, and momentum) that are represented by numbers in classical mechanics and converted them to operators. This move is in fact the main thing that distinguishes classical mechanics from quantum mechanics. Similarly, converting our complex scalar field to an *operator* field is what distinguishes classical field theory from quantum field theory.

To connect with the Fock space, our operator field should be some linear combination of raising and lowering operators. The field's whole purpose is to remove particles from certain initial particle momentum states and place a possibly different number of particles into different final particle momentum states as some kind of interaction proceeds. This is going to involve annihilating particles in the initial momentum states and creating particles in the final momentum states.

As a base case, consider a system of a certain specific type of non-interacting free particles, which we will call phions. For this particular case, the correct linear combination of creation and annihilation operators turns out to be

$$\phi(x) = \int \frac{d^3p}{(2\pi)^3} \frac{1}{2E_{\vec{p}}} \left(\alpha_{\vec{p}}\, e^{-ip\cdot x} + \beta_{\vec{p}}^{\dagger}\, e^{+ip\cdot x} \right)$$

$$= \int \frac{d^3p}{(2\pi)^3} \frac{1}{\sqrt{2E_{\vec{p}}}} \left(a_{\vec{p}}\, e^{-ip\cdot x} + b_{\vec{p}}^{\dagger}\, e^{+ip\cdot x} \right), \tag{12.1}$$

where $p \cdot x$ is our usual shorthand for $g_{\mu\nu} p^{\mu} x^{\nu}$, $E_{\vec{p}} = (|\vec{p}|^2 + m^2)^{1/2}$ (where m is the phion mass), and $\alpha_{\vec{p}} = (2E_{\vec{p}})^{1/2}\, a_{\vec{p}}$ is the Lorentz-invariant version of the annihilation operator $a_{\vec{p}}$ for phions that have momentum $\vec{p}$ (see Equation 11.28). We interpret the other operator $\beta_{\vec{p}}^{\dagger} \equiv (2E_{\vec{p}})^{1/2}\, b_{\vec{p}}^{\dagger}$ as the Lorentz-invariant version of a creation operator

$b_{\vec{p}}^{\dagger}$ for **antiphions**. Since the plane waves that we attach to the phions and antiphions both are solutions to the same Klein-Gordon equation with the same mass term m, phions and antiphions must have the same mass.

The first form of this equation makes it clear that the integral's result (though it is an operator) transforms as if it were a Lorentz scalar: the quantity in parentheses transforms like a Lorentz scalar function of particle momentum, and the factor of $d^3p/2E_{\vec{p}}$ is the Lorentz-invariant volume element (as discussed in the last chapter). The second line simply expresses the same linear combination in terms of the normal forms of the annihilation and creation operators.

How people dreamed up this particular superposition is a long story. Superficially, it looks like just a general superposition of plane-wave solutions $e^{\pm ip\cdot x}$ of the Klein-Gordon equation $\partial^{\mu}\partial_{\mu}\phi + m^2\phi = 0$, with the annihilation and creation operators replacing what would be momentum-dependent amplitudes $A_{\vec{p}}$ and $B_{\vec{p}}$ in a truly general superposition. Indeed, one can start with a general superposition including those amplitudes and then show that various additional requirements such as Lorentz-invariance constrain those amplitudes in ways that eventually whittle them away entirely. The argument is long and not particularly illuminating, so let's simply accept Equation 12.1 as a starting point and instead show that it *makes sense*.

12.2 Total Energy and Total Charge

We'll begin by calculating the energy density associated with this field. In Chapter 6 (Equation 6.27), we saw that the energy density for a classical but complex scalar field was given by

$$\rho = T^{tt} = \dot{\phi}^*\dot{\phi} + (\vec{\nabla}\phi^*)\cdot(\vec{\nabla}\phi) + m^2\phi^*\phi. \tag{12.2}$$

Now that our field is an operator, we should replace all complex conjugates by adjoints, but otherwise the equation remains the same. Now, by the chain rule,

$$\partial_{\mu}e^{\pm ip\cdot x} = \partial_{\mu}e^{\pm i(p_{\alpha}x^{\alpha})} = e^{\pm i(p_{\alpha}x^{\alpha})}(\pm ip_{\alpha}\,\delta^{\alpha}_{\mu}) = \pm ip_{\mu}e^{\pm ip\cdot x}, \tag{12.3}$$

where $p_{\mu} \equiv g_{\mu\nu}p^{\nu}$. We also have

$$\phi^{\dagger}(x) = \int \frac{d^3p}{(2\pi)^3}\frac{1}{(2E_{\vec{p}})^{1/2}}\left[a_{\vec{p}}^{\dagger}\,e^{+ip\cdot x} + b_{\vec{p}}\,e^{-ip\cdot x}\right] \tag{12.4}$$

and, using Equation 12.3, we have

$$\partial_{\mu}\phi = \int \frac{d^3p}{(2\pi)^3}\frac{-ip_{\mu}}{(2E_{\vec{p}})^{1/2}}\left[a_{\vec{p}}\,e^{-ip\cdot x} - b_{\vec{p}}^{\dagger}\,e^{+ip\cdot x}\right] \tag{12.5a}$$

$$\partial_{\mu}\phi^{\dagger} = \int \frac{d^3p}{(2\pi)^3}\frac{ip_{\mu}}{(2E_{\vec{p}})^{1/2}}\left[a_{\vec{p}}^{\dagger}\,e^{+ip\cdot x} - b_{\vec{p}}\,e^{-ip\cdot x}\right]. \tag{12.5b}$$

(For later reference, recall that raising the index on p_{μ} does nothing if it is a time index and changes its sign if it is a spatial index: $p^j = -p_j$.)

With these results in mind, let us calculate the first term in the expression for the energy density given in Equation 12.2, which is

$$\dot{\phi}^{\dagger}\dot{\phi} = \left(\int \frac{d^3p}{(2\pi)^3}\frac{ip_t}{(2E_{\vec{p}})^{1/2}}\left[a_{\vec{p}}^{\dagger}\,e^{+ip\cdot x} - b_{\vec{p}}\,e^{-ip\cdot x}\right]\right)$$

$$\times \left(\int \frac{d^3q}{(2\pi)^3}\frac{-iq_t}{(2E_{\vec{q}})^{1/2}}\left[a_{\vec{q}}\,e^{-iq\cdot x} - b_{\vec{q}}^{\dagger}\,e^{+iq\cdot x}\right]\right). \tag{12.6}$$

Note that $\dot{\psi}$ and $\dot{\psi}^\dagger$ involve *independent* sums over the plane wave solutions, so to distinguish the sums, I have used q to denote the four-momentum in the second integral (this is analogous to avoiding index-name collisions when we are using index notation for tensors). Keeping the integration variables separate allows us to combine these separate sums into one gigantic integral, a generalization of the operation $(a + b)(c + d) = ac + ad + bc + bd$. Doing this, while also noting that $p_t = E_{\vec{p}}$ and $q_t = E_{\vec{q}}$, yields

$$\dot{\phi}^\dagger\dot{\phi} = \int \frac{d^3p\, d^3q}{(2\pi)^6} \frac{-i^2 E_{\vec{p}} E_{\vec{q}}}{2(E_{\vec{p}} E_{\vec{q}})^{1/2}} \left[a_{\vec{p}}^\dagger e^{+ip\cdot x} - b_{\vec{p}} e^{-ip\cdot x} \right]\left[a_{\vec{q}} e^{-iq\cdot x} - b_{\vec{q}}^\dagger e^{+iq\cdot x} \right]$$

$$= \int \frac{d^3p\, d^3q}{(2\pi)^6} \frac{(E_{\vec{p}} E_{\vec{q}})^{1/2}}{2} \left[a_{\vec{p}}^\dagger a_{\vec{q}}\, e^{i(p-q)\cdot x} - a_{\vec{p}}^\dagger b_{\vec{q}}^\dagger\, e^{i(p+q)\cdot x} \right.$$

$$\left. - b_{\vec{p}} a_{\vec{q}}\, e^{-i(p+q)\cdot x} + b_{\vec{p}} b_{\vec{q}}^\dagger\, e^{i(q-p)\cdot x} \right]. \tag{12.7}$$

Now, in order to evaluate this, note that what we are really interested in is the total energy, that is, the energy density integrated over all space. We also know from Chapter 4 that

$$\int d^3x\, e^{-i(\vec{p}\pm\vec{q})\cdot\vec{r}} = \int d^3x\, e^{+i(\vec{p}\pm\vec{q})\cdot\vec{r}} = (2\pi)^3\, \delta^3(\vec{p}\pm\vec{q}), \tag{12.8}$$

which (since $p \cdot x = p^t t - p^x x - p^y y - p^z z = E_{\vec{p}} t - \vec{p}\cdot\vec{r}$) means that

$$\int d^3x\, e^{-i(p\pm q)\cdot x} = \int d^3x\, e^{-i(E_{\vec{p}}\pm E_{\vec{q}})t} e^{+i(\vec{p}\pm\vec{q})\cdot\vec{r}} = e^{-i(E_{\vec{p}}\pm E_{\vec{q}})t}(2\pi)^3\, \delta^3(\vec{p}\pm\vec{q}),$$

and likewise $\quad \int d^3x\, e^{+i(p\pm q)\cdot x} = e^{+i(E_{\vec{p}}\pm E_{\vec{q}})t}(2\pi)^3\, \delta^3(\vec{p}\pm\vec{q}). \tag{12.9}$

This result implies that

$$\int d^3x\, \dot{\phi}^\dagger\dot{\phi} = \int \frac{d^3p\, d^3q}{(2\pi)^6}\, d^3x\, \frac{(E_{\vec{p}} E_{\vec{q}})^{1/2}}{2}$$

$$\times \left[a_{\vec{p}}^\dagger a_{\vec{q}}\, e^{i(p-q)\cdot x} - a_{\vec{p}}^\dagger b_{\vec{q}}^\dagger\, e^{i(p+q)\cdot x} - b_{\vec{p}} a_{\vec{q}}\, e^{-i(p+q)\cdot x} + b_{\vec{p}} b_{\vec{q}}^\dagger\, e^{i(q-p)\cdot x} \right]$$

$$= \int \frac{d^3p\, d^3q}{(2\pi)^3} \frac{(E_{\vec{p}} E_{\vec{q}})^{1/2}}{2} \left[a_{\vec{p}}^\dagger a_{\vec{q}}\, \delta^3(\vec{p}-\vec{q})\, e^{i(E_{\vec{p}}-E_{\vec{q}})t} - a_{\vec{p}}^\dagger b_{\vec{q}}^\dagger\, \delta^3(\vec{p}+\vec{q})\, e^{i(E_{\vec{p}}+E_{\vec{q}})t} \right.$$

$$\left. - b_{\vec{p}} a_{\vec{q}}\, \delta^3(\vec{p}+\vec{q})\, e^{-i(E_{\vec{p}}+E_{\vec{q}})t} + b_{\vec{p}} b_{\vec{q}}^\dagger\, \delta^3(\vec{p}-\vec{q})\, e^{i(E_{\vec{q}}-E_{\vec{p}})t} \right], \tag{12.10}$$

where I have performed the integral over all space using Equation 12.9. But now we can do the integral over d^3q: when we integrate a term involving $\delta^3(\vec{p}-\vec{q})$, it will set $\vec{q} = \vec{p}$ in the rest of that term; and when we integrate a term involving $\delta^3(\vec{p}+\vec{q})$, it will set $\vec{q} = -\vec{p}$ in the rest of that term. Also, note that $E_{\vec{p}} = E_{-\vec{p}}$, so all these integrations will set $E_{\vec{q}} = E_{\vec{p}}$. The final result is therefore

$$\int d^3x\, \dot{\phi}^\dagger\dot{\phi} = \int \frac{d^3p}{(2\pi)^3} \frac{E_{\vec{p}}}{2} \left[a_{\vec{p}}^\dagger a_{\vec{p}} - a_{\vec{p}}^\dagger b_{-\vec{p}}^\dagger\, e^{i2E_{\vec{p}}t} - b_{\vec{p}} a_{-\vec{p}}\, e^{-i2E_{\vec{p}}t} + b_{\vec{p}} b_{\vec{p}}^\dagger \right]. \tag{12.11}$$

In Box 12.1, you will similarly calculate the other two terms in the energy density. When the dust settles, you will find that the *total* energy associated with the field is

$$E = \int d^3x\, \rho = \int \frac{d^3p}{(2\pi)^3} E_{\vec{p}} \left(a_{\vec{p}}^\dagger a_{\vec{p}} + b_{\vec{p}} b_{\vec{p}}^\dagger \right). \tag{12.12}$$

In an entirely similar manner, you can show that the time component

$$J^t = -iq \left(\phi \dot{\phi}^\dagger - \phi^\dagger \dot{\phi} \right) \tag{12.13}$$

of the current for the conserved quantity associated with the Lagrangian density's symmetry under the transformation $\phi' = e^{-iq\theta}\phi$, when integrated over all space, becomes a conserved scalar quantity

$$Q = \int d^3x\, J^t = q \int \frac{d^3p}{(2\pi)^3} \frac{1}{2}\left[a_{\vec{p}}\, a_{\vec{p}}^\dagger + a_{\vec{p}}^\dagger a_{\vec{p}} - b_{\vec{p}}^\dagger b_{\vec{p}} - b_{\vec{p}}\, b_{\vec{p}}^\dagger \right]. \quad (12.14)$$

See Box 12.2 for the details.

12.3 Normal Ordering

Now, we can simplify the expression for Q by using the commutation relations for creation and annihilation operators:

$$[\, a_{\vec{p}}\, ,\ a_{\vec{p}}^\dagger\,] \equiv a_{\vec{p}}\, a_{\vec{p}}^\dagger - a_{\vec{p}}^\dagger a_{\vec{p}} = (2\pi)^3\, \delta^3(\vec{p} - \vec{p}\,)$$

$$\Rightarrow\ a_{\vec{p}}\, a_{\vec{p}}^\dagger = a_{\vec{p}}^\dagger a_{\vec{p}} + (2\pi)^3\, \delta^3(\vec{p} - \vec{p}\,), \quad (12.15)$$

and a similar relation for $b_{\vec{p}}$. If we substitute these back into Equation 12.14, the Dirac deltas cancel and we get

$$Q = q \int \frac{d^3p}{(2\pi)^3}\left[a_{\vec{p}}^\dagger a_{\vec{p}} - b_{\vec{p}}^\dagger b_{\vec{p}} \right]. \quad (12.16)$$

As we saw in the last chapter, when $a_{\vec{p}}^\dagger a_{\vec{p}}$ acts on a Fock-space state, it returns the number of phions having momentum $\vec{p}$. Similarly, $b_{\vec{p}}^\dagger b_{\vec{p}}$ yields the number of antiphons with that momentum. If we integrate over all momenta, we get the total numbers of each type of particle:

$$N_a = \int \frac{d^3p}{(2\pi)^3} a_{\vec{p}}^\dagger a_{\vec{p}}, \qquad N_b = \int \frac{d^3p}{(2\pi)^3} b_{\vec{p}}^\dagger b_{\vec{p}}. \quad (12.17)$$

If we interpret q as the charge of a phion and Q as total charge, then Equation 12.16 makes total sense: $Q = qN_a + (-q)N_b$ because phions and antiphions have opposite charges.

Because we have not yet introduced electromagnetism in any form into our theory, the connection between Q and electric charge is at this point only speculative. Later in the course, however, we will see that the same symmetry that leads to this conserved Q also produces Maxwell's equations, which will fully justify our identification of Q with charge.

If we apply the same procedure to the expression for E, we get

$$E = \int \frac{d^3p}{(2\pi)^3} E_{\vec{p}} \left[a_{\vec{p}}^\dagger a_{\vec{p}} + b_{\vec{p}}^\dagger b_{\vec{p}} + (2\pi)^3 \delta^3(\vec{p} - \vec{p}\,) \right]. \quad (12.18)$$

The first two terms are fine: if we multiply the total number of particles of each type with momentum $\vec{p}$ by the energy $E_{\vec{p}}$ that each particle with that momentum has and integrate over all momenta, we *should* get the total energy E. (Note that both particles and antiparticles have *positive* energy.) But the third term looks like a total disaster, because $\delta^3(\vec{p} - \vec{p}\,) = \infty$! Not only that, we are integrating this infinity over $d^3p\, E_{\vec{p}}$, making this term infinitely infinite!

This is just the first of many infinities that we will encounter and dispose of in our journey through quantum field theory. Here, one can argue quite reasonably that this infinity is only an apparent problem, not a true problem with the theory. The issue arises because we are taking a classical field theory (where all factors commute) and converting it into an operator field theory (where factors don't commute). Note, for example that $\phi^*\phi$ (one of the terms in the classical expression for the energy) is the same as $\phi\phi^*$, but $\phi^\dagger\phi$ is *not* the same as $\phi\phi^\dagger$. Thus, we have an ordering ambiguity in the classical field theory that we must resolve when we move to quantum field theory. One should resolve this ambiguity in a way that makes physical sense in the new context. Therefore, when doing calculations like the energy calculation we are doing here, the standard practice is to simply arrange all products of creation and annihilation operators in **normal order** (that is, with all annihilation operators on the right, creation operators on the left) in the first place, without worrying about how these products arrived at whatever ordering they seem to have in the equation. This avoids generating the infinite term altogether. (Since all annihilation operators commute with one another, and all creation operators commute with one another, the ordering within these subgroups is not an issue.) The majority opinion among physicists is that this simply resolves the classical ordering ambiguity in a way that makes physical sense (that is, a way that does not involve infinite terms).

Not everyone buys this argument (see Klauber, *Student Friendly Quantum Field Theory*, 2nd edition, Sandtrove Press, 2013, page 61), but there is no question that this *works* in the sense that the theory with normal ordering included is very successful about making accurate predictions about the universe.

12.4 Reviewing the Process

We now have a complete quantum field theory for a certain type of free, non-interacting particle and its antiparticle. Let's review the process by which we have arrived at this theory so that we can see the general pattern we need to follow in creating more realistic theories in the future.

- **Step 1.** Write down a classical Lagrangian density involving classical fields. This is where a lot of creativity enters the process, because many Lagrangian densities are possible, and one must choose one that successfully explains what one observes in nature.

- **Step 2.** Use Noether's theorem to calculate the energy density and any other current densities that might be interesting.

- **Step 3.** Develop creation and annihilation operators that act on a Fock space to create and destroy particles and that obey certain commutation relations that constrain their behavior.

- **Step 4.** "Quantize" the field by converting it into an *operator* field that is a linear combination of plane-wave solutions of the Lagrangian density's Euler-Lagrange equation, but with annihilation and creation operators in place of the plane-wave amplitudes.

- **Step 5.** Reorder any operator products in that result to make them consistent with "normal order" (annihilation operators to the right), ignoring the commutation conditions that would give rise to infinities.

In the particular case of the quantum field theory we have developed so far, we did **Step 1** in Chapter 4, postulating the Lagrangian density

$$\mathcal{L} = \partial^\mu \phi^* \partial_\mu \phi - m^2 \phi^* \phi \tag{12.19}$$

representing two fields ϕ and ϕ^*, whose Euler-Lagrange equations are

$$0 = \partial_\mu \partial^\mu \phi + m^2 \phi \tag{12.20a}$$

$$0 = \partial_\mu \partial^\mu \phi^* + m^2 \phi^*. \tag{12.20b}$$

We did **Step 2** in Chapter 5, where we used Noether's theorem to derive the conserved currents corresponding to the Lagrangian density's symmetries under spacetime translations and the field's complex phase. In particular, we found the energy density of the field and the charge density of the field to be

$$T^{tt} = \dot{\phi}^*\dot{\phi} + (\vec{\nabla}\phi^*)\cdot(\vec{\nabla}\phi) + m^2\phi^*\phi \tag{12.21a}$$

$$J^t = -iq\left(\phi\dot{\phi}^* - \dot{\phi}\phi^*\right). \tag{12.21b}$$

We did **Step 3** in the last chapter, where we developed creation and annihilation operators and discovered the commutation relations that described and constrained their behavior. We performed **Step 4** in this chapter, where we defined the operator field to be the superposition

$$\phi(x) = \int \frac{d^3p}{(2\pi)^3}\frac{1}{\sqrt{2E_{\vec{p}}}}\left[a_{\vec{p}}\,e^{-ip\cdot x} + b_{\vec{p}}^\dagger\,e^{ip\cdot x}\right], \tag{12.22}$$

substituted it (with $\phi^* \to \phi^\dagger$) into the equations above, and (preserving the original operator ordering) got (after a lot of work)

$$E = \int d^3x\, T^{tt} = \int \frac{d^3p}{(2\pi)^3}E_{\vec{p}}\left[a_{\vec{p}}^\dagger a_{\vec{p}} + b_{\vec{p}}\,b_{\vec{p}}^\dagger\right], \tag{12.23a}$$

$$Q = \int d^3x\, J^t = q\int \frac{d^3p}{(2\pi)^3}\frac{1}{2}\left[a_{\vec{p}}^\dagger a_{\vec{p}} + a_{\vec{p}}\,a_{\vec{p}}^\dagger - b_{\vec{p}}^\dagger b_{\vec{p}} - b_{\vec{p}}\,b_{\vec{p}}^\dagger\right]. \tag{12.23b}$$

Finally, we performed **Step 5** in this chapter. Putting the operator products in the expressions above into normal order yields equations

$$E = \int d^3x\, T^{tt} = \int \frac{d^3p}{(2\pi)^3}E_{\vec{p}}\left[a_{\vec{p}}^\dagger a_{\vec{p}} + b_{\vec{p}}^\dagger b_{\vec{p}}\right], \tag{12.24a}$$

$$Q = \int d^3x\, J^t = q\int \frac{d^3p}{(2\pi)^3}\left[a_{\vec{p}}^\dagger a_{\vec{p}} - b_{\vec{p}}^\dagger b_{\vec{p}}\right], \tag{12.24b}$$

which now make sense physically: since the $a_{\vec{p}}^\dagger a_{\vec{p}}$ and $b_{\vec{p}}^\dagger b_{\vec{p}}$ operators acting on a Fock-space state yield the number of phions and antiphions that have momentum $\vec{p}$, respectively, the integrals over all momenta do indeed yield the correct total positive energy and the correct net charge if phions and antiphions have opposite charge.

This is how we interpret the operator version of a *complex* scalar field. The operator analogy of a *real* scalar field is a field whose adjoint is equal to itself. To get this, the superposition in Equation 12.22 must become

$$\phi(x) = \int \frac{d^3p}{(2\pi)^3}\frac{1}{\sqrt{2E_{\vec{p}}}}\left[a_{\vec{p}}\,e^{-ip\cdot x} + a_{\vec{p}}^\dagger\,e^{ip\cdot x}\right]. \tag{12.25}$$

By the interpretation we have developed, such a field describes a particle that is its own antiparticle and whose charge must therefore be zero.

BOX 12.1   Energy Terms

In the body of this chapter, we took the trial field expression given in Equation 12.1

$$\phi(x) = \int \frac{d^3 p}{(2\pi)^3} \frac{1}{(2E_{\vec{p}})^{1/2}} \left[a_{\vec{p}}\, e^{-ip\cdot x} + b_{\vec{p}}^{\dagger}\, e^{+ip\cdot x} \right] \qquad (12.26)$$

and calculated the first term in the expression for the field's energy density

$$\rho = T^{tt} = \dot{\phi}^{\dagger}\dot{\phi} + (\vec{\nabla}\phi^{\dagger})\cdot(\vec{\nabla}\phi) + m^2\phi^{\dagger}\phi \qquad (12.27)$$

(as given in Equation 12.2) in terms of $a_{\vec{p}}$ and $b_{\vec{p}}^{\dagger}$. Our goal in this box is to calculate the other two terms in Equation 12.27 and combine them with the first term to arrive at the final result quoted in Equation 12.12.

Exercise 12.1.1   Follow the same steps as in the text to show that the third term in Equation 12.27, integrated over all space, becomes

$$\int d^3 x\, m^2 \phi^{\dagger}\phi \qquad (12.28)$$

$$= \int \frac{d^3 p}{(2\pi)^3} \frac{m^2}{2E_{\vec{p}}} \left[a_{\vec{p}}^{\dagger} a_{\vec{p}} + a_{\vec{p}}^{\dagger} b_{-\vec{p}}^{\dagger}\, e^{i2E_{\vec{p}}t} + b_{\vec{p}}\, a_{-\vec{q}}\, e^{-i2E_{\vec{p}}t} + b_{\vec{p}}\, b_{\vec{p}}^{\dagger} \right].$$

(*Hint:* Use Equation 12.9 to simplify things after integrating over all space.)

BOX 12.1 Energy Terms (cont.)

Now let us look at the second term. Remember that Equation 12.3 says that

$$\partial_\mu e^{\pm ip\cdot x} = \pm ip_\mu e^{\pm ip\cdot x}. \tag{12.29}$$

Also, note that the way to write $(\vec{\nabla}\phi^\dagger)\cdot(\vec{\nabla}\phi)$ in index notation is

$$(\vec{\nabla}\phi^\dagger)\cdot(\vec{\nabla}\phi) = -g^{jk}\partial_j\phi^\dagger\,\partial_k\phi \quad \text{since } g^{jk} = -1 \text{ when } j = k. \tag{12.30}$$

Exercise 12.1.2 Using similar methods as in the last exercise (but remembering that $q_j = -p_j$ for the cross terms after integrating over d^3x), show that

$$\int d^3x\,(\vec{\nabla}\phi^\dagger)\cdot(\vec{\nabla}\phi) \tag{12.31}$$

$$= \int \frac{d^3p}{(2\pi)^3}\frac{|\vec{p}|^2}{2E_{\vec{p}}}\left[a_{\vec{p}}^\dagger a_{\vec{p}} + a_{\vec{p}}^\dagger b_{-\vec{p}}^\dagger e^{i2E_{\vec{p}}t} + b_{\vec{p}} a_{-\vec{p}} e^{-i2E_{\vec{p}}t} + b_{\vec{p}} b_{\vec{p}}^\dagger \right].$$

BOX 12.1   Energy Terms (cont.)

Exercise 12.1.3   Finally, put together Equations 12.11, 12.31, and 12.28 to get Equation 12.12, repeated here:

$$\int d^3x \,\rho = \int \frac{d^3p}{(2\pi)^3} E_{\vec{p}} \left(a_{\vec{p}}^{\dagger} a_{\vec{p}} + b_{\vec{p}} b_{\vec{p}}^{\dagger} \right). \tag{12.32}$$

(*Hint:* Note that $E_{\vec{p}}^2 - |\vec{p}|^2 = m^2$ for solutions of the Klein-Gordon equation.)

BOX 12.2 Calculating the Total Charge

Our goal in this box is to compute the total charge associated with the current whose time component is given by Equation 12.13

$$J^t = -iq\left(\phi\dot{\phi}^\dagger - \phi^\dagger\dot{\phi}\right) \tag{12.33}$$

in terms of the operators $a_{\vec{p}}$ and $b_{\vec{p}}^\dagger$ appearing in Equation 12.1.

Exercise 12.2.1   Using the methods illustrated in the text, show that

$$\int d^3x\, \phi\, \dot{\phi}^\dagger = \frac{i}{2} \int \frac{d^3p}{(2\pi)^3}\left[a_{\vec{p}}\, a_{\vec{p}}^\dagger - b_{\vec{p}}^\dagger\, b_{\vec{p}} - a_{\vec{p}}\, b_{-\vec{p}}\, e^{-i2E_{\vec{p}}t} + b_{\vec{p}}^\dagger\, a_{-\vec{p}}^\dagger\, e^{+i2E_{\vec{p}}t} \right].$$

$$\tag{12.34}$$

(*Hints:* Starting with Equation 12.1, calculate ϕ, $\phi^\dagger$, $\dot{\phi}$, and $\dot{\phi}^\dagger$ first. Also, remember that $\partial_t\, e^{\pm ip\cdot x} = \pm i E_{\vec{p}}\, e^{\pm ip\cdot x}$.)

BOX 12.2 Calculating the Total Charge (cont.)

Exercise 12.2.2 Similarly, calculate $\int d^3x \, \phi^\dagger \dot{\phi}$.

Exercise 12.2.3 Finally, verify that, as claimed in Equation 12.14,

$$Q \equiv -iq \int d^3x \left(\phi \, \dot{\phi}^\dagger - \phi^\dagger \dot{\phi} \right) = q \int \frac{d^3p}{(2\pi)^3} \frac{1}{2} \left(a_{\vec{p}} \, a_{\vec{p}}^\dagger + a_{\vec{p}}^\dagger \, a_{\vec{p}} - b_{\vec{p}}^\dagger \, b_{\vec{p}} - b_{\vec{p}} \, b_{\vec{p}}^\dagger \right).$$

$$(12.35)$$

(*Hint:* To show that the cross terms cancel, separate out the integrals involving the cross terms and change the integration variable on half of them to $\vec{p}\,' = -\vec{p}$. Also, two annihilation operators commute, as do two creation operators.)

Homework Problems

P12.1 Assume that we expand a scalar operator field $\phi(x)$ as in Equation 12.1:

$$\phi(x) = \int \frac{d^3p}{(2\pi)^3} \frac{1}{(2E_{\vec{p}})^{1/2}} \left[a_{\vec{p}}\, e^{-ip\cdot x} + b_{\vec{p}}^\dagger\, e^{+ip\cdot x} \right].$$

$$(12.36)$$

a. Show that the expansion of $\dot{\phi}\, \partial^j\phi^\dagger$ (where j is a spatial index) is

$$\dot{\phi}\, \partial^j\phi^\dagger = \int \frac{d^3p}{(2\pi)^3} \frac{d^3q}{(2\pi)^3} \frac{E_{\vec{p}}\, q^j}{2(E_{\vec{p}}E_{\vec{q}})^{1/2}}$$

$$\times \left[a_{\vec{p}}\, a_{\vec{q}}^\dagger\, e^{-i(p-q)\cdot x} - a_{\vec{p}}\, b_{\vec{q}}\, e^{-i(p+q)\cdot x} \right. \quad (12.37)$$

$$\left. - b_{\vec{p}}^\dagger\, a_{\vec{q}}^\dagger\, e^{+i(p+q)\cdot x} + b_{\vec{p}}^\dagger\, b_{\vec{q}}\, e^{+i(p-q)\cdot x} \right].$$

b. In a similar way, calculate the expansion of $\dot{\phi}^\dagger\, \partial^j\phi$.

P12.2 Consider the results of Problem P12.1.
a. Use those results to calculate the total momentum component $P^j = \int d^3x\, T^{tj}$ for a scalar operator field $\phi(x)$.
b. In light of the interpretation of the coefficients $a_{\vec{p}}^\dagger\, a_{\vec{p}}$ and $b_{\vec{p}}^\dagger\, b_{\vec{p}}$ that we have developed in this chapter, does your result make sense? Explain why it does (or does not).

P12.3 Consider a hypothetical real *four-vector* field B^α whose Lagrangian density is given by

$$\mathcal{L} = \partial_\alpha B^\beta \partial_\beta B^\alpha - m^2 B_\alpha B^\alpha. \qquad (12.38)$$

(This Lagrangian density is not physically useful, but let's pretend.) Note that I have chosen index names for the bound indices that avoid μ, since ∂_μ will appear in the Euler-Lagrange equations for this Lagrangian density.
a. Noting that $B_\alpha B^\alpha = g_{\alpha\beta} B^\beta B^\alpha$, show that $\partial\mathcal{L}/\partial B^\nu = -2m^2 B_\nu$. (*Hint:* First, argue that $\partial B^\alpha/\partial B^\nu = \delta^\alpha_\nu$, and use the product rule.)
b. Show that $\partial\mathcal{L}/\partial(\partial_\mu B^\nu) = 2\partial_\nu B^\mu$. (*Hint:* First, argue that we must have $\partial(\partial_\alpha B^\beta)/\partial(\partial_\mu B^\nu) = \delta^\mu_\alpha \delta^\beta_\nu$.)
c. Use the results of the previous part to find the Euler-Lagrange equations for our Lagrangian density.
d. Assume that we "quantize" this field by converting it to an operator field described by the following superposition:

$$B^\mu(x) = \int \frac{d^3p}{(2\pi)^3} \frac{\epsilon^\mu}{(2E_{\vec{p}})^{1/2}} \left[a_{\vec{p}}\, e^{-ip\cdot x} + a_{\vec{p}}^\dagger\, e^{+ip\cdot x} \right],$$

$$(12.39)$$

where ϵ^μ is a real four-vector of components we will call the field's "polarization" four-vector. Use the Euler-Lagrange equation to put a condition on how the polariza-

tion vector, the momentum vector p^μ, and the mass m are interrelated.
e. What does this relationship imply physically if $m = 0$?
f. Express the energy density of this field in terms of B^μ and its partial derivatives. Do you see anything that might indicate why this is not a physically useful Lagrangian density?

P12.4 One of the standard difficulties everyone encounters when learning about quantum field theory for the first time is that it seems so completely different from standard quantum theory that it is hard to see how the theories are connected. In this problem, we will see that quantum field theory *contains* ordinary quantum mechanics.

In ordinary quantum mechanics, we describe quantum states by wavefunctions $\psi(x) \equiv \psi(t, \vec{r})$. So far, our QFT can only handle free particles where energy eigenfunctions are the same as momentum eigenfunctions, so we will stick to such wavefunctions. The wavefunction for such an energy eigenfunction (including its time dependence) is

$$\psi_{\vec{p}}(t, \vec{r}) = \psi_{\vec{p}}(\vec{r})e^{-iE_{\vec{p}}t} = e^{i\vec{p}\cdot\vec{r}}e^{-iE_{\vec{p}}t} = e^{-ip\cdot x} \quad (12.40)$$

(where I am using QFT normalization). We can see that the phion (as opposed to the antiphion) part of our operator field $\phi(x)$ is a superposition of *all* momentum phion eigenfunctions multiplied by the corresponding annihilation operators $a_{\vec{p}}$:

$$\phi_+(x^\mu) = \int \frac{d^3p}{(2\pi)^3} e^{-ip\cdot x} a_{\vec{p}} = \int \frac{d^3p}{(2\pi)^3} \psi_{\vec{p}}(t, \vec{r})\, a_{\vec{p}}.$$

$$(12.41)$$

Now, our field operates on Fock-space states, but ordinary quantum mechanics deals only with single-particle states. We can construct a Fock-space single-particle state from the Fock-space vacuum as follows:

$$|\vec{p}\rangle = a_{\vec{p}}^\dagger|0\rangle. \qquad (12.42)$$

a. Show that $\langle 0|\phi_+|\vec{p}\,'\rangle = \psi_{\vec{p}\,'}(t, \vec{r})$. This means that our field operator ϕ_+ delivers the correct wavefunction in a context when there is just one phion. (Note that I am using a prime to distinguish $\vec{p}\,'$ from the momentum $\vec{p}$ that we are integrating over in Equation 12.41.)
b. The Hamiltonian operator H that acts in Fock space is constructed from the free-particle quantum mechanics Hamiltonian (which is $H_{\text{QM}} = -\nabla^2/2m$) as follows:

$$H = \int d^3x\, \phi_+^\dagger(t, \vec{r})\, H_{\text{QM}}\, \phi_+(t, \vec{r})$$

$$= \int d^3x\, \phi_+^\dagger(t, \vec{r}) \left[-\frac{\nabla^2}{2m} \right] \phi_+(t, \vec{r}). \quad (12.43)$$

Show that this simplifies to

$$H = \int \frac{d^3p}{(2\pi)^3} E_{\vec{p}}\, a_{\vec{p}}^{\dagger} a_{\vec{p}}. \qquad (12.44)$$

c. Show that when this acts on the Fock-space single-particle state $|\vec{p}\,'\rangle$, we get the energy of that state back:

$$H|\vec{p}\,'\rangle = E_{\vec{p}\,'}|\vec{p}\,'\rangle. \qquad (12.45)$$

The point is that our operator field embeds ordinary free-particle quantum mechanics within a larger structure.

P12.5 An interesting historical note is that Schrödinger actually briefly considered the Klein-Gordon equation as a possibility for describing the evolution of the ordinary quantum mechanical wavefunction before devising the non-relativistic equation we now call the time-dependent Schrödinger equation. He abandoned the Klein-Gordon equation because (as we saw in Chapter 8), the energy operator H is the same as $i\hbar\, d/dt$ (see the Time-Evolution Rule in Chapter 8). This is $i\, d/dt$ in our units.

a. Show that both $e^{-ip\cdot x}$ and $e^{+ip\cdot x}$ are plane-wave solutions to the Klein-Gordon equation $\partial^{\mu}\partial_{\mu}\phi + m^2\phi = 0$.

b. Show that when you apply the Hamiltonian operator $H = i\, d/dt$ to the $e^{+ip\cdot x}$ solution, you find that it is an eigenfunction of the Hamiltonian, but with negative energy. (*Hint*: In the exponential, p_t specifies the plane wave's angular frequency, which is positive by definition.)

Negative energies would be a huge problem, as quantum systems could radiate energy continuously without ever reaching a ground state to settle into. One also cannot simply reject the $e^{+ip\cdot x}$ solutions as being "unphysical" without getting into other problems. This led Schrödinger to retreat to the non-relativistic treatment that led to his eponymous equation. The discovery (by Richard Feynman and Ernst Stueckelberg) that one could successfully treat the "negative-energy solutions" as instead representing *positive-energy antiparticles* in a Fock-space context (as we have seen in this chapter) was historically the crucial step forward.

P12.6 Though interpreting what the operator $\phi(t, \vec{r})$ *does* is not crucial for using it (as we will see), here is one way to arrive at an *approximate* description. (To avoid some mathematical complexity, I am going to focus on what this operator does at $t = 0$.)

a. Suppose that we allow the field $\phi(0, \vec{r})$ to act on the vacuum state. Argue that the result is

$$\phi(0, \vec{r})|\,0\,\rangle = \int \frac{d^3p}{(2\pi)^3}\frac{e^{-i\vec{p}\cdot\vec{r}}}{\sqrt{2E_{\vec{p}}}}|\vec{p}\,\rangle, \qquad (12.46)$$

where $|\vec{p}\,\rangle$ is the particle momentum state for the antiphion. (*Hint:* Note that $a_{\vec{p}}$ annihilates a phion, but $b_{\vec{p}}^{\dagger}$ creates an antiphion.)

b. This is a superposition of particle momentum states. Now, note that one of the definitions of the Dirac delta implies that

$$\int \frac{d^3p}{(2\pi)^3}e^{-i\vec{p}\cdot\vec{r}} = \delta(\vec{r}). \qquad (12.47)$$

This means that an unweighted superposition of plane waves $e^{-i\vec{p}\cdot\vec{r}}$ integrated over all momentum space delivers a spike at the position $\vec{r}$. This suggests by analogy that the *state* of a particle having a definite position is given by

$$|\vec{r}\,\rangle = \int \frac{d^3p}{(2\pi)^3}\, e^{-i\vec{p}\cdot\vec{r}}|\vec{p}\,\rangle \qquad (12.48)$$

(as one can verify using methods from ordinary quantum mechanics that I don't want to get into here). Comparing this last equation to Equation 12.46 suggests that we might interpret $\phi(0, \vec{r})$ as acting on the vacuum to create an antiparticle in the single-particle definite position state $|\vec{r}\,\rangle$. Indeed, some field theory books say just that. But this lovely interpretation is spoiled by the extra $1/\sqrt{2E_{\vec{p}}}$ factor in Equation 12.46, which is required for Lorentz invariance. This factor suppresses the integrand at small wavelengths (high energies), which has the effect of smearing out the spike. So, we can interpret $\phi(0, \vec{r})$ acting on the vacuum as creating an antiparticle *approximately* at position $\vec{r}$ at time $t = 0$.

With this in mind, describe what $\phi^{\dagger}(0, \vec{r})$ does when it acts on the vacuum state. Give both a mathematical result and a conceptual interpretation. [*Hint:* Note that $\delta^3(-\vec{r}) = \delta^3(\vec{r})$.]

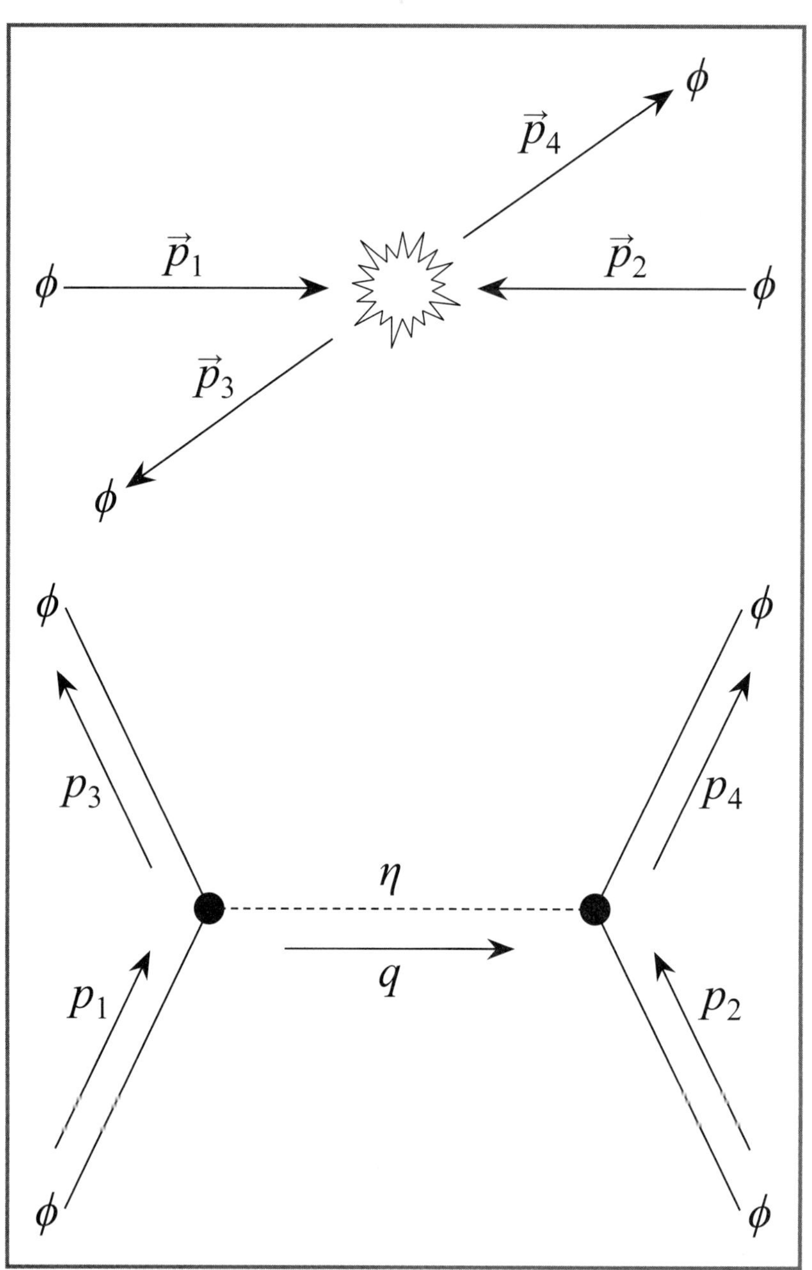

φ
φ
φ
φ
$\vec{p}_1$
$\vec{p}_2$
$\vec{p}_3$
$\vec{p}_4$
φ
φ
φ
φ
p_1
p_2
p_3
p_4
η
q

13. INTERACTIONS AND FEYNMAN DIAGRAMS

In the last chapter, we worked out a complete and consistent quantum field theory for free, noninteracting particles. A universe of free, noninteracting particles is, however, a very dull place. Particles serenely glide along straight paths at constant velocities, passing through each other like ghosts when their paths happen to intersect. Nothing ever happens.

What makes the universe interesting is that particles *interact*. In this chapter, we will explore how to add interactions to our theory and how to calculate the effects of those interactions. (I will, however, deliberately brush past certain mathematical details in order to make the big picture as clear as possible.)

13.1 Adding Interactions

In one sense, adding interactions to our theory is very easy: we simply add a term to the Lagrangian density that multiplies the fields whose excitations are different particles. For example, consider a toy universe that contains just two kinds of particles, "phions" of mass m_ϕ, described by a Hermitian operator field $\phi(x)$; and "etaons" with mass m_η, described by a Hermitian operator field $\eta(x)$. (As mentioned at the end of the last chapter, a Hermitian field describes an uncharged particle that is its own antiparticle, like the photon.) Let the Lagrangian density for this universe be

$$\mathcal{L} = \tfrac{1}{2}\partial^\mu\phi\,\partial_\mu\phi - \tfrac{1}{2}m_\phi^2\phi^2 + \tfrac{1}{2}\partial^\mu\eta\,\partial_\mu\eta - \tfrac{1}{2}m_\eta^2\eta^2 - \lambda\eta\phi^2. \tag{13.1}$$

The first four terms simply describe two separate free real scalar classical fields, each obeying the Klein-Gordon equation. The new feature is the final $\mathcal{L}_{\text{int}} \equiv -\lambda\eta\phi^2$ term, which describes an *interaction* between the two fields. The term in the classical expression for the *energy* due to this interaction is

$$E_{\text{int}} = +\lambda \int d^3x\, \eta(x)\,\phi(x)\,\phi(x). \tag{13.2}$$

This term thus amounts to a potential energy of interaction between the fields.

In a classical context, such a term attaches a potential energy to positions where the η and ϕ fields are both nonzero. This would mean that if a bump in the ϕ field were to move toward a positive bump in the η field, they would repel each other as the bumps began to overlap. On the other hand, a bump in the ϕ field would be attracted to a negative bump in the η field. (For more about the classical meaning of such an interaction, see Appendix A.)

Now, this is not the only kind of interaction term we might add. Simpler terms are often physically meaningless (see Problem P13.1), but terms such as $+\lambda\phi^4$ (which would imply mutual repulsion of bumps in the ϕ field) are certainly possible. We will see, however, that many of the interaction terms in Standard Model Lagrangian densities look like $-\lambda\eta\phi^2$, so this is a good place to begin.

After we quantize the field, the classical energy becomes the quantum Hamiltonian operator H. The interaction term in this Hamiltonian is

$$H_{\text{int}} = +\lambda \int d^3x\, \eta(x)\,\phi(x)\,\phi(x). \tag{13.3}$$

The operator expansion for a Hermitian scalar field is

$$\phi(x) = \int \frac{d^3p}{(2\pi)^3} \frac{1}{\sqrt{2E_{\vec{p}}}} \left[a_{\vec{p}}\, e^{-ip\cdot x} + a_{\vec{p}}^{\dagger}\, e^{+ip\cdot x} \right], \tag{13.4}$$

so the interaction Hamiltonian is

$$\begin{aligned}
H_{\text{int}} = \lambda \int d^3x \, \frac{d^3p_1}{(2\pi)^3} \frac{d^3p_2}{(2\pi)^3} \frac{d^3p_3}{(2\pi)^3} \frac{1}{\sqrt{8E_{\vec{p}_1} E_{\vec{p}_2} E_{\vec{p}_3}}} \left[c_{\vec{p}_1}\, e^{-ip_1\cdot x} + c_{\vec{p}_1}^{\dagger}\, e^{+ip_1\cdot x} \right] \\
\times \left[a_{\vec{p}_2}\, e^{-ip_2\cdot x} + a_{\vec{p}_2}^{\dagger}\, e^{+ip_2\cdot x} \right] \left[a_{\vec{p}_3}\, e^{-ip_3\cdot x} + a_{\vec{p}_3}^{\dagger}\, e^{+ip_3\cdot x} \right],
\end{aligned} \tag{13.5}$$

where the $c^{\dagger}$ and c operators are creation and annihilation operators for etaons and the $a^{\dagger}$ and a operators are the same for phions. (We also must normal-order the creation and annihilation operators when we work out the products.)

13.2 The S-Matrix

Just as in non-relativistic quantum mechanics, the Hamiltonian operator in quantum field theory tells us how system states evolve in time:

$$i\frac{d}{dt}|\psi(t)\rangle = H(t)|\psi(t)\rangle = \left[H_0 + H_{\text{int}}(t) \right]|\psi(t)\rangle \tag{13.6}$$

(see the Time-Evolution Rule in Chapter 8). Here, H_0 is the Hamiltonian for free particles, $H_{\text{int}}(t)$ is the new interaction term (which depends on time because the fields depend on time) and the kets here are Fock-space kets for the entire system. Now, we have actually already built the effect of the free-particle Hamiltonian into the time-dependence of the fields if we use Equation 13.4 to define the fields, so in quantum field theory, the equation above reduces to

$$i\frac{d}{dt}|\psi(t)\rangle = H_{\text{int}}(t)|\psi(t)\rangle. \tag{13.7}$$

One tricky part of this is that even a simple interaction term like $\lambda\eta\phi^2$ is nonlinear, and nonlinear differential equations are notoriously difficult to solve analytically. (Numerical solutions are always possible, but analytic solutions give more insight.) Another complication is that, unlike most of the cases one considers in non-relativistic quantum mechanics, H_{int} depends on time. Abstractly, we could integrate both sides of the above expression to get

$$|\psi(t)\rangle = |\psi(t_0)\rangle - i \int_{t_0}^{t} dt'\, H_{\text{int}}(t')|\psi(t')\rangle. \tag{13.8}$$

But to actually do the integral, we would have to know how $|\psi(t)\rangle$ depends on time, which is what we are trying to find.

A time-honored tactic that physicists often use when confronted with otherwise intractable problems is an iterative approach to finding a solution. In this iterative approach, we calculate $|\psi_1(t)\rangle$ on the left side assuming on the right side that $|\psi(t')\rangle = |\psi(0)\rangle$. Then we insert $|\psi_1(t')\rangle$ on the right side to get a better approximation $|\psi_2(t)\rangle$ on the left, and so on, until we get as accurate a solution as we desire. The result, as you will show in Box 13.1, is

$$|\psi(t)\rangle = U(t, t_0)|\psi(t_0)\rangle, \quad \text{where}$$

$$U(t, t_0) = I - i \int_{t_0}^{t} dt' \, H_{\mathrm{int}}(t') + (-i)^2 \int_{t_0}^{t} dt' \, H_{\mathrm{int}}(t') \left(\int_{t_0}^{t'} dt'' \, H_{\mathrm{int}}(t'') \right)$$

$$+ (-i)^3 \int_{t_0}^{t} dt' \, H_{\mathrm{int}}(t') \left(\int_{t_0}^{t'} dt'' \, H_{\mathrm{int}}(t'') \left(\int_{t_0}^{t''} dt''' \, H_{\mathrm{int}}(t''') \right) \right) + \cdots. \quad (13.9)$$

Now, the assumption behind this approach is that the interaction Hamiltonian H_{int} is in some sense "small," so that we only need to calculate a few terms in this power series to get a decent result. This assumption can fail at many levels. It can fail because the interaction is *not* small, as in the case of the color interaction (where a different approach is needed to make the calculations work). It can also fail if individual higher-order terms diverge, as can happen even in electromagnetic and weak interactions (we will later use the technique of "renormalization" to deal with these infinities). In many cases, though, keeping only a handful of terms *does* yield excellent agreement with reality.

So, let's assume that we can actually do this. We'll also assume that the system's state is essentially some pure Fock-space state $|i\rangle$ long *before* the interaction, and that it is some *different* pure Fock-space state $|f\rangle$ sufficiently long after the interaction. (This assumption also raises a number of mathematical issues beyond our scope here.) What we would like to do is to calculate the "scattering matrix" or **S-matrix**, defined as the quantum *amplitude* that an initial state $|i\rangle$ ultimately evolves to the final state $|f\rangle$:

$$S_{fi} \equiv \lim_{\substack{t \to +\infty \\ t_0 \to -\infty}} \langle f \, | U(t, t_0)| \, i \rangle \equiv \langle f \, | U(+\infty, -\infty)| \, i \rangle, \quad |f\rangle \neq |i\rangle. \quad (13.10)$$

Note that for a given initial state, there may be many possible final states, so the system's actual final state $|f_{\mathrm{tot}}\rangle$ will be the following superposition:

$$|f_{\mathrm{tot}}\rangle = \sum_{f} |f\rangle\langle f \, | U(+\infty, -\infty)| \, i \rangle, \quad (13.11)$$

and the probability that the outcome is a particular state $|f\rangle \neq |i\rangle$ is

$$\frac{|\langle f \, | U(+\infty, -\infty)| \, i \rangle|^2}{\langle f \, | f \rangle\langle i \, | i \rangle} = \frac{|S_{fi}|^2}{\langle f \, | f \rangle\langle i \, | i \rangle} \quad (13.12)$$

(where the factors in the denominator are there so that we do not have to assume any particular normalization for the initial and final states).

Even though, in non-relativistic quantum mechanics, we can find analytic solutions in several useful cases (the simple harmonic oscillator, the hydrogen atom, etc.), in quantum field theory, we have an analytic solution for exactly *one* case: the free particle. The method outlined above cleverly leverages our knowledge of this solution to enable us to solve many other cases (at least approximately).

13.3 The S-Matrix for a Decay Process

Let's see how this might work in a specific case. Consider a hypothetical decay of an etaon into two phions.

$$\eta \to \phi + \phi. \quad (13.13)$$

Our goal in this section is to calculate the first-order contribution $S_{fi}^{(1)}$ to S_{fi} in the power series expansion of the time-evolution operator given in Equation 13.9. In this case, we can generate the initial (relativistically invariant) Fock-space state by creating the initial incoming etaon state from the vacuum:

$$|i\rangle = \sqrt{2E_{\vec{p}_1}}\, c_{\vec{p}_1}^{\dagger}|0\rangle. \quad (13.14)$$

This will yield the plane-wave state for an incoming etaon with momentum $\vec{p}_1$. Similarly, we can generate the outgoing two-phion state by applying the phion creation operator twice to the vacuum:

$$|f\rangle = \sqrt{4E_{\vec{p}_2} E_{\vec{p}_3}}\, a^{\dagger}_{\vec{p}_2} a^{\dagger}_{\vec{p}_3} |0\rangle. \tag{13.15}$$

This final state describes two phions with momenta $\vec{p}_2$ and $\vec{p}_3$. We can get the bra form of this by taking the adjoint:

$$\langle f| = \langle 0| a_{\vec{p}_2} a_{\vec{p}_3} \sqrt{4E_{\vec{p}_2} E_{\vec{p}_3}}. \tag{13.16}$$

Now we need to calculate H_{int}. Expanding out the products in Equation 13.5 yields eight terms with creation and annihilation operators in various orders. But as you will see in Box 13.2, the only term that yields something nonzero is the one that annihilates the incoming etaon and creates the outgoing phions:

$$H_{\text{int}} = \lambda \int d^3x \, \frac{d^3q_1}{(2\pi)^3} \frac{d^3q_2}{(2\pi)^3} \frac{d^3q_3}{(2\pi)^3} \frac{1}{\sqrt{8E_{\vec{q}_1} E_{\vec{q}_2} E_{\vec{q}_3}}} a^{\dagger}_{\vec{q}_3} a^{\dagger}_{\vec{q}_2} c_{\vec{q}_1}\, e^{-i(q_1 - q_2 - q_3)\cdot x}. \tag{13.17}$$

(I have used $\vec{q}_1$, $\vec{q}_2$, and $\vec{q}_3$ to distinguish the integrated momenta from the incoming and outgoing particles' actual momenta $\vec{p}_1$, $\vec{p}_2$, and $\vec{p}_3$.)

To get the scattering matrix, we put everything together: to lowest order

$$S^{(1)}_{fi} = \langle f|U(+\infty, -\infty)|i\rangle \approx -i \int_{-\infty}^{\infty} dt'\, \langle f|H_{\text{int}}(t')|i\rangle$$

$$= \sqrt{8E_{\vec{p}_1} E_{\vec{p}_2} E_{\vec{p}_3}}\, \langle 0| \left[\int d^4x \, \frac{d^3q_1}{(2\pi)^3} \frac{d^3q_2}{(2\pi)^3} \frac{d^3q_3}{(2\pi)^3} \frac{-i\lambda}{\sqrt{8E_{\vec{q}_1} E_{\vec{q}_2} E_{\vec{q}_3}}} \right.$$

$$\left. \times e^{-i(q_1 - q_2 - q_3)\cdot x} \left(a_{\vec{p}_3} a_{\vec{p}_2} a^{\dagger}_{\vec{q}_3} a^{\dagger}_{\vec{q}_2} c_{\vec{q}_1} c^{\dagger}_{\vec{p}_1} \right) \right] |0\rangle. \tag{13.18}$$

(Note that H_{int} contains an integral over d^3x and we integrate that over time, yielding the integral over d^4x in the expression above.)

A bit of work (see Box 13.3) shows that all this reduces to

$$S^{(1)}_{fi} = \int d^4x \, (-i\lambda) e^{-ip_1 \cdot x} e^{ip_2 \cdot x} e^{ip_3 \cdot x}. \tag{13.19}$$

Now, we can visually represent this expression by a schematic drawing called a **Feynman diagram.** The $e^{-ip_1 \cdot x}$ factor is the amplitude $\langle x|\psi_{p_1}\rangle$ that the incoming etaon will be found at four-position x, and the $e^{+ip_2 \cdot x}$ and $e^{+ip_3 \cdot x}$ are the amplitudes $\langle \psi_{p_2}|x\rangle$ and $\langle \psi_{p_3}|x\rangle$ that the outgoing phions created at four-position x will have four-momenta p_2 and p_3. In the diagram (which we construct to resemble a spacetime diagram), we metaphorically depict these factors as the straight worldlines that the corresponding free particles would follow if they were classical, and draw arrows next to each that indicate the particle's four-momentum. The net effect of the single annihilation and two creation operators in the interaction Hamiltonian was to annihilate the etaon and create two phions: we represent this process as a **vertex** where the three lines meet. The interaction Hamiltonian brought in with it a factor of $-i\lambda$: we can consider this factor to be "attached" to the vertex. The resulting diagram appears in Figure 13.1.

Now, because the initial and final particles actually have wavefunctions that are plane waves, and so have the same probability of being at any given point in space and time, the vertex could be anywhere. To get the total amplitude, therefore, we integrate over all space and time. The identity $\int d^4x \, e^{-i(p_1 - p_2 - p_3)\cdot x} = (2\pi)^4 \delta^4(p_1 - p_2 - p_3)$ means that the final scattering amplitude is

$$S^{(1)}_{fi} = -i\lambda(2\pi)^4 \, \delta^4(p_1 - p_2 - p_3), \tag{13.20}$$

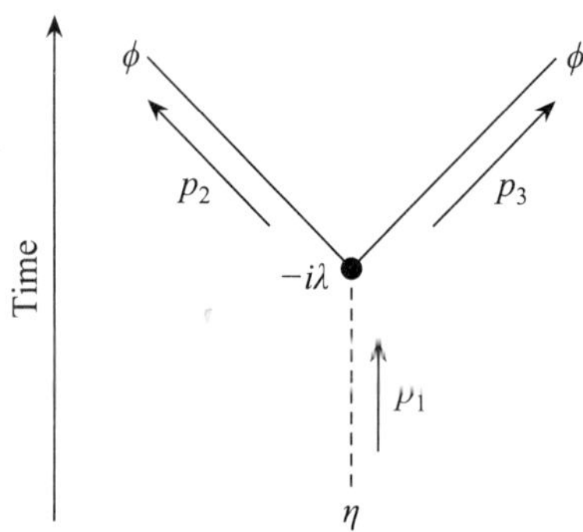

FIGURE 13.1   This Feynman diagram visually represents the lowest-order term in the interaction Hamiltonian for η-decay in our toy universe.

where I am again using a conventional simplified notation where

$$\delta^4(p_1 - p_2 - p_3)$$
$$\equiv \delta(p_1^t - p_2^t - p_3^t)\,\delta(p_1^x - p_2^x - p_3^x)\,\delta(p_1^y - p_2^y - p_3^y)\,\delta(p_1^z - p_2^z - p_3^z). \quad (13.21)$$

Equation 13.20 shows that the scattering matrix is zero for any combination of initial and final states where $p_1^\mu \neq p_2^\mu + p_3^\mu$, so the Dirac delta enforces conservation of four-momentum for this decay process.

13.4 Higher-Order Terms

However, this is only the first-order approximation for the S-matrix. What about higher-order terms? The second-order term in Equation 13.9 involves a product of two H_{int} operators. In Box 13.4, you will see that among the possible products of creation and annihilation operators that result, no terms yield a nonzero result for the η-decay process. But we might have seen this less mathematically by attempting to construct the Feynman diagram. We simply cannot construct a diagram with two vertices (where two phion lines meet an etaon line) that connects one incoming etaon line with two outgoing phion lines. Trying (and failing) to construct such a diagram allows one to arrive at the zero-amplitude result without examining the 64 different possible combinations of creation and annihilation operators in this case.

The decay process does have third-order terms (see Problem P13.2 for diagrams). But to see what happens with higher-order terms, it will be much easier to consider a process that has nonzero *second-order* terms. One such process is *phion scattering,* where we have two phions entering with four-momenta p_1 and p_2 and two phions departing with different four-momenta p_3 and p_4 (see Figure 13.2). In this case, we have no *first*-order term because the one etaon operator in H_{int} must either annihilate or create an etaon, and we have no initial etaon to destroy nor any final etaon left. But we do have three possible second-order processes, one of which is shown in Figure 13.3. In this process, we have a vertex where an internal etaon is created and another where it is annihilated.

The second-order contribution $S_{fi,1}^{(2)}$ for this process 1 (of three) is

$$S_{fi,1}^{(2)} = \frac{(-i\lambda)^2}{2!}\sqrt{16 E_{\vec{p}_1} E_{\vec{p}_2} E_{\vec{p}_3} E_{\vec{p}_4}}\,\langle 0|a_{\vec{p}_3} a_{\vec{p}_4}$$

$$\times T\left(\left[\int d^4x_2 \frac{d^3k_4}{(2\pi)^3}\frac{d^3k_2}{(2\pi)^3}\frac{d^3q_2}{(2\pi)^3}\left(a_{\vec{k}_4}^\dagger a_{\vec{k}_2} c_{\vec{q}_2}\right)\frac{e^{-i(k_2+q_2-k_4)\cdot x_2}}{\sqrt{8 E_{\vec{k}_2} E_{\vec{q}_2} E_{\vec{k}_4}}}\right]\right.$$

$$\left.\times\left[\int d^4x_1 \frac{d^3k_3}{(2\pi)^3}\frac{d^3k_1}{(2\pi)^3}\frac{d^3q_1}{(2\pi)^3}\left(a_{\vec{k}_3}^\dagger a_{\vec{k}_1} c_{\vec{q}_1}^\dagger\right)\frac{e^{-i(k_1-q_1-k_3)\cdot x_1}}{\sqrt{8 E_{\vec{k}_1} E_{\vec{q}_1} E_{\vec{k}_3}}}\right]\right)$$

$$\times a_{\vec{p}_1}^\dagger a_{\vec{p}_2}^\dagger|0\rangle, \quad (13.22)$$

where k_1 through k_4 are the integrated momenta linked to the phion creation/annihilation operators in H_{int}, q_1 and q_2 are the same for the etaon operators, and the $T()$ indicates that we should technically time-order the integrals so that the etaon's creation always occurs before its destruction. As happened in Box 13.3, reordering the creation and annihilation operators yields Dirac deltas, and integrating over those deltas yields the simpler expression

$$S_{fi,1}^{(2)} = (-i\lambda)^2 \int d^4x_1\, d^4x_2\, e^{-ip_1\cdot x_1}\, e^{+ip_2\cdot x_1} e^{-ip_3\cdot x_2} e^{+ip_4\cdot x_2}$$

$$\times \int \frac{d^3q}{(2\pi)^3}\frac{e^{-iq\cdot(x_2-x_1)}}{2E_{\vec{q}}}. \quad (13.23)$$

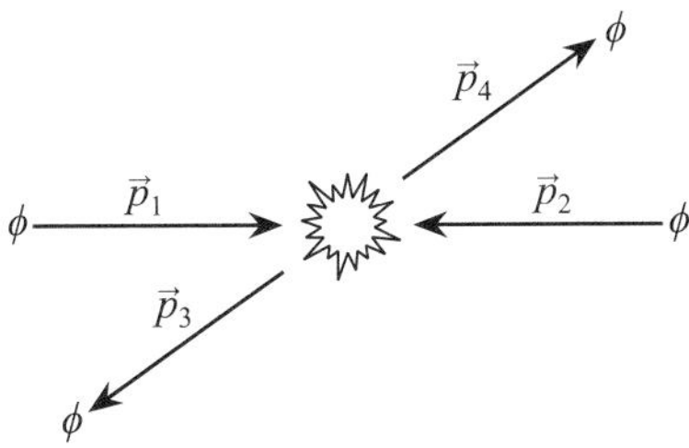

FIGURE 13.2 This diagram shows a phion scattering as viewed in the center-of-mass frame.

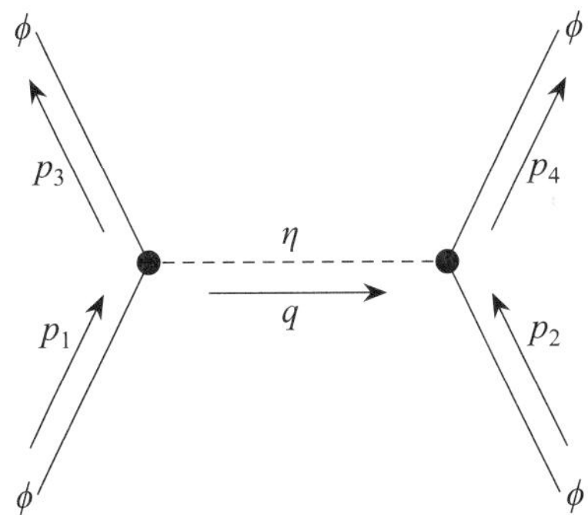

FIGURE 13.3 This Feynman diagram shows one of three second-order phion scattering processes.

Many parts of this equation are analogous to Equation 13.19 in the one-vertex decay process. We have an $e^{-ip\cdot x}$ factor for each incoming phion and an $e^{+ip\cdot x}$ for each outgoing phion, represented (as usual) by straight lines on the diagram. We have two vertices, each of which contributes a factor of $-i\lambda$. We integrate *each* vertex's position over all space and time. The one new feature is the factor on the second line that describes the internal etaon. This factor specifies the quantum amplitude that an etaon created at one vertex makes it to the other vertex as a function of the spacetime vertex positions x_1 and x_2.

When we do actual calculations involving this factor, we will find it easier to replace this expression with the following mathematically equivalent factor:

$$\int \frac{d^3q}{(2\pi)^3} \frac{e^{-iq\cdot(x_2-x_1)}}{2E_{\vec{q}}} = \int \frac{d^4q}{(2\pi)^4} \frac{ie^{-iq\cdot(x_2-x_1)}}{q^2 - m_\eta^2}. \tag{13.24}$$

The proof that these expressions are equivalent involves complicated math that is not really illuminating, so let's simply accept the result. But one might rightly be skeptical: isn't the squared magnitude q^2 of the etaon's four-momentum q *equal* to the squared etaon mass m_η^2 (making the integral infinite)?

Actually, it is *not* infinite. Gathering the various exponential terms yields $e^{-i(p_1-p_3-q)\cdot x_1}e^{-i(p_2+q-p_4)\cdot x_2}$, and integrating these over all possible vertex positions yield Dirac deltas $(2\pi)^4\delta^4(p_1 - p_3 - q)(2\pi)^4\delta^4(p_2 + q - p_4)$ that enforce conservation of four-momentum at each vertex. But we cannot conserve four-momentum *and simultaneously* satisfy the constraint that $q^2 = m_\eta^2$ for the internal etaon (see Box 13.5). We call the internal etaon a **virtual particle** whose four-momentum is **off-shell** ($q^2 \neq m_\eta^2$). This not only means that the etaon can never be detected as a real particle (a *real* particle *must* have $q^2 = m_\eta^2$), but also that the quantum amplitude for it getting from one vertex to the other is *quite credibly* inversely proportional to just how illegal it is (that is, how large $q^2 - m_\eta^2$ is), as the formula implies.

Substituting Equation 13.24 into Equation 13.23 thus yields

$$S_{fi,1}^{(2)} = -i \int d^4x_1\, d^4x_2\, \frac{d^4q}{(2\pi)^4}\, e^{-i(p_1-p_3-q)\cdot x_1}e^{-i(p_2+q-p_4)\cdot x_2}\frac{\lambda^2}{q^2 - m_\eta^2}$$

$$= -i \int \frac{d^4q}{(2\pi)^4}(2\pi)^8\delta^4(p_1 - p_3 - q)\,\delta^4(p_2 - p_4 + q)\frac{\lambda^2}{q^2 - m_\eta^2}$$

$$= -i(2\pi)^4\delta^4(p_1 + p_2 - p_3 - p_4)\left[\frac{\lambda^2}{(p_1 - p_3)^2 - m_\eta^2}\right], \tag{13.25}$$

since (in going from the middle to final step) integrating the $\delta^4(p_1 - p_3 - q)$ over d^4q sets $q = p_1 - p_3$ everywhere else.

Now, if we compare this expression to that in Equation 13.20 for etaon decay, we see that both have a $-i(2\pi)^4\delta^4(\Delta p)$ factor that enforces conservation of energy and momentum. Such a factor *always* shows up in calculations of the S-matrix, so the physically distinctive (and thus interesting) part is the *other* part: the λ in the etaon decay and the factor in square brackets above for phion scattering. We call this part the process's **scattering amplitude** $\mathcal{M}$.

13.5 The Feynman Rules for Scalar Fields

Now that we see how these calculations work, we can go directly from a process description to its scattering amplitude $\mathcal{M}$ by applying these simple rules:

1. **Draw Diagrams.** Draw worldlines for the incoming and outgoing particles and connect them with lines for internal (virtual) particles. These lines must intersect at vertices that connect the fields appearing in the Lagrangian density's interaction

term. (In our toy theory, that term is $-\lambda\eta\phi^2$, so we will have two phion lines connecting with an etaon line.)

2. **Label Lines.** On your diagram, draw an arrow next to each particle worldline that indicates the direction of the particle's momentum. (For internal lines, this direction is arbitrary, but *choosing* a direction is important for subsequent steps.) Label the arrows for incoming or outgoing particles with four-momenta p_1, p_2, . . . and those for internal particles with four-momenta q_1, q_2, ...

3. **Vertex factors.** Each vertex in our toy theory contributes a **vertex factor** of $-i\lambda$ (other kinds of fields and/or interaction terms will contribute different vertex factors).

4. **Propagators.** Each internal scalar field line in the diagram contributes a **propagator** factor of $i/(q_j^2 - m_j^2)$, where q_j is the four-momentum associated with the jth internal line and m_j is the mass of the particle associated with that line. (Other types of fields have different propagators.)

5. **Conserve Four-Momentum at Each Vertex.** We have seen that calculations lead to Dirac deltas that conserve four-momentum at each vertex. This always happens, so rather than go through the mathematics every time, we are simply going to assume this. For most of the simple diagrams we will construct, this will allow us to determine the four-momentum q_j of any internal particle. (We'll talk about how to handle closed loops of internal particles in Chapter 21.)

6. **The Result is** $-i\mathcal{M}$. Multiply by i to get the scattering amplitude $\mathcal{M}$.

7. **Add the Amplitudes.** To get the total amplitude $\mathcal{M}_{\text{tot}}$ to any given order (n), *add* the amplitudes for all possible diagrams up to that order. The scattering matrix to that order is $S_{fi}^{(n)} = -i(2\pi)^4\delta^4(\Delta p)\mathcal{M}$.

Rules 3–5 replace all the complicated calculations we have done in this chapter. This is a *huge* step forward. (In Box 13.6, you will practice using the rules to calculate the amplitudes of the other two second-order scattering processes.)

13.6 Interpreting Feynman Diagrams

The point is that Feynman diagrams represent a brilliant way to visually represent the physical implications of the complicated sequences of creation and annihilation factors that result from calculating products of an interaction Hamiltonian. It is also much easier to construct the relevant diagrams for any order of approximation rather than multiply out all the creation and annihilation operators and interpret the results. We will find, as we go on, that we can in fact use Feynman diagrams to rapidly calculate the results that we *would* have gotten from a lengthy explicit calculation of the type described above.

However, one should remember that the total scattering matrix is the *sum* of the scattering matrices associated with an infinite number of such diagrams. This whole process will be pointless if we cannot stop at some point and have a reasonable approximation. Since there is one factor of the coupling constant λ associated with each vertex, if λ is in some sense "small," diagrams with a sufficiently large number of vertices (even if there are many of them) will contribute little to the sum, so we can ignore them. (We will see later how to decide how "small" this constant is and thus how many diagrams to include.)

I also need to warn you against overinterpreting what these diagrams mean. The story that each diagram tells is so clear and credible as a physical process that it is natural to assume that they describe what is physically happening in each interaction. But I urge you to resist this temptation.

These diagrams are tools that help us calculate terms in a power series, nothing more. A number of reasons make interpreting any given diagram, or even the set of all diagrams, as depicting what is "really" going on highly problematic. First of all, remember that our incoming and outgoing Fock-space states describe infinite incoming and outgoing plane waves, which are smeared out over all space, not the classical-particle-like trajectories that appear in a diagram. Though the diagram pretends that a vertex involving creation and annihilation occurs at a specific event in spacetime, the mathematics behind the diagram involves integrating the vertex positions over all space. The diagrams also pretend that the creation and annihilation operators actually represent the physical creation and annihilation of real particles, rather than just representing a general mathematical basis for describing operators that allow us to evolve one Fock-space state to another.

Perhaps even more importantly, the real physical process is represented by the sum of an infinite number of terms in a power series, while a diagram only represents one term, a *tiny piece* of the whole process. As an analogy, we can fruitfully describe a pulse wave mathematically as an infinite superposition of sinusoidal waves, but is it helpful to consider that the pulse "really is" (at the fundamental level) a bunch of sine waves, or especially that any single sine wave is something that is "really happening" in the wave? No! Fourier decomposition is a wonderful mathematical tool for understanding how general waves work, but ascribing reality to the individual sine waves in the decomposition does not make any sense, particularly if one keeps in mind that I could have just as easily described the pulse as a superposition of Dirac deltas (which comprises an alternative orthonormal basis for writing functions).

All that being said, let us duly praise and thank Richard Feynman for developing this amazing tool. Were it not for his diagrams, the mathematics involved in quantum field theory would be much more difficult: Feynman diagrams really make quantum field theory practical!

We have come a long way in this chapter in seeing how to use the abstract mathematics of quantum field theory to connect with physical reality. But to connect fully with experimental physics, we need to be able to quantitatively predict experimentally measurable quantities such as decay lifetimes, scattering cross sections, and so on. We will begin exploring how to connect scattering amplitudes with experiments starting in the next chapter.

BOX 13.1   Iterating the Interaction Hamiltonian

In this box, we will apply the iterative method for solving the integral Equation 13.8 (repeated here for convenience):

$$|\psi(t)\rangle = |\psi(t_0)\rangle - i \int_{t_0}^{t} dt' \, H_{\text{int}}(t') |\psi(t')\rangle. \tag{13.26}$$

The idea is to substitute $|\psi_0(t')\rangle \equiv |\psi(t_0)\rangle$ for $|\psi(t')\rangle$ to calculate $|\psi_1(t')\rangle$ on the left. Then we substitute $|\psi_1(t')\rangle$ for $|\psi(t')\rangle$ on the right to get $|\psi_2(t')\rangle$ on the right, and so on for as many steps as desired. The first two steps are thus

$$|\psi_1(t)\rangle = |\psi(t_0)\rangle - i \int_{t_0}^{t} dt' \, H_{\text{int}}(t') |\psi(t_0)\rangle = \left(I - i \int_{t_0}^{t} dt' \, H_{\text{int}}(t') \right) |\psi(t_0)\rangle,$$

$$\tag{13.27}$$

$$|\psi_2(t)\rangle = |\psi(t_0)\rangle - i \int_{t_0}^{t} dt' \, H_{\text{int}}(t') |\psi_1(t')\rangle$$

$$= |\psi(t_0)\rangle - i \int_{t_0}^{t} dt' \, H_{\text{int}}(t') \left(I - i \int_{t_0}^{t'} dt'' \, H_{\text{int}}(t'') \right) |\psi(t_0)\rangle$$

$$= \left(I - i \int_{t_0}^{t} dt' \, H_{\text{int}}(t') \left(I - i \int_{t_0}^{t'} dt'' \, H_{\text{int}}(t'') \right) \right) |\psi(t_0)\rangle, \tag{13.28}$$

where I have relabeled the integration variable in the right-most integral to keep that integral distinct from the integral to its left.

Exercise 13.1.1   Do the next step and then multiply out the terms to get the power series in Equation 13.9.

BOX 13.2   The Hamiltonian for a Decay Process

Equation 13.5, repeated below, gives the interaction Hamiltonian for our toy universe:

$$H_{\text{int}} = \lambda \int d^3x \, \frac{d^3q_1}{(2\pi)^3} \frac{d^3q_2}{(2\pi)^3} \frac{d^3q_3}{(2\pi)^3} \frac{1}{\sqrt{8E_{\vec{q}_1} E_{\vec{q}_2} E_{\vec{q}_3}}} \left[c_{\vec{q}_1} e^{-iq_1 \cdot x} + c_{\vec{q}_1}^{\dagger} e^{+iq_1 \cdot x} \right]$$

$$\times \left[a_{\vec{q}_2} e^{-iq_2 \cdot x} + a_{\vec{q}_2}^{\dagger} e^{+iq_2 \cdot x} \right] \left[a_{\vec{q}_3} e^{-iq_3 \cdot x} + a_{\vec{q}_3}^{\dagger} e^{+iq_3 \cdot x} \right]. \qquad (13.29)$$

I have written the internal momenta as $\vec{q}_1, \vec{q}_2, \vec{q}_3$ here to avoid confusing these momenta with the momenta $\vec{p}_1, \vec{p}_2, \vec{p}_3$ of the actual particles in the $\eta \to \phi + \phi$ decay process. Our goal in this box is to find any terms that contribute something nonzero to the scattering matrix element

$$\langle f | H | i \rangle = \sqrt{8E_{\vec{p}_1} E_{\vec{p}_2} E_{\vec{p}_2}} \langle 0 | a_{\vec{p}_3} a_{\vec{p}_2} H_{\text{int}} c_{\vec{p}_1}^{\dagger} | 0 \rangle. \qquad (13.30)$$

Exercise 13.2.1   Focusing entirely on the creation and annihilation operators (ignoring all other numerical factors), write out all the operator products implied by Equation 13.29, being careful to put each operator product in normal order (that is, with all creation operators for a given type of particle to the left). Also, for ease in comparison, put all the c-operators on the right (all c-operators commute with all a-operators, so this is entirely legal).

BOX 13.2   The Hamiltonian for a Decay Process (cont.)

You should have found that one of the eight terms in this expansion was $a_{\vec{q}_2} a_{\vec{q}_3} c_{\vec{q}_1}$. If we put this term in the middle of the "braket" in Equation 13.30, we get (ignoring all the numerical factors)

$$\langle 0 | a_{\vec{p}_3} a_{\vec{p}_2} a_{\vec{q}_2} a_{\vec{q}_3} c_{\vec{q}_1} c_{\vec{p}_1}^\dagger | 0 \rangle. \tag{13.31}$$

Now, etaons are completely independent of phions, so we might write this expression as

$$\langle 0_\phi | a_{\vec{p}_3} a_{\vec{p}_2} a_{\vec{q}_2} a_{\vec{q}_3} | 0_\phi \rangle \langle 0_\eta | c_{\vec{q}_1} c_{\vec{p}_1}^\dagger | 0_\eta \rangle, \tag{13.32}$$

where $| 0_\phi \rangle$ is the vacuum for phions (where no phions exist) and $| 0_\eta \rangle$ is the vacuum for etaons. Now, the right-hand factor might be nonzero (if $\vec{q}_1 = \vec{p}_1$), because the $c_{\vec{p}_1}^\dagger$ creates an etaon, but the $c_{\vec{q}_1}$ destroys it, restoring the etaon vacuum state (which would be necessary, because we are calculating the inner product with $\langle 0_\eta |$). But the phion factor is clearly zero, because when the first $a_{\vec{q}_3}$ hits the vacuum state, we get zero. But even more abstractly, if we do not have an equal number of creation and annihilation operators between the vacuum states, the inner product will be zero, since any non-vacuum Fock-space state that might be created will be orthogonal to the vacuum bra on the left.

Exercise 13.2.2   Similarly analyze the other seven terms, and for each term, briefly explain why it can or cannot be nonzero.

BOX 13.3 Calculating the Decay S-matrix

Equation 13.18, repeated here,

$$
S_{fi}^{(1)} = -i\lambda \sqrt{8 E_{\vec{p}_1} E_{\vec{p}_2} E_{\vec{p}_3}} \, \langle 0 | \left[\int d^4 x \, \frac{d^3 q_1}{(2\pi)^3} \frac{d^3 q_2}{(2\pi)^3} \frac{d^3 q_3}{(2\pi)^3} \frac{1}{\sqrt{8 E_{\vec{q}_1} E_{\vec{q}_2} E_{\vec{q}_3}}} \right.
$$

$$
\left. \times \, e^{-i(q_1 - q_2 - q_3)\cdot x} \left(a_{\vec{p}_3} a_{\vec{p}_2} a_{\vec{q}_3}^\dagger a_{\vec{q}_2}^\dagger c_{\vec{q}_1} c_{\vec{p}_1}^\dagger \right) \right] | 0 \rangle \qquad (13.33)
$$

tells us how to calculate the lowest-order estimate for the S-matrix for the $\eta \to \phi + \phi$ decay process. In this box, you will evaluate all the integrals in this expression to get the much simpler result in Equation 13.19:

$$
S_{fi}^{(1)} = \int d^4 x \, (-i\lambda) e^{-ip_1 \cdot x} e^{ip_2 \cdot x} e^{ip_3 \cdot x}. \qquad (13.34)
$$

First, note that the basic commutation relation for c-operators implies that

$$
(2\pi)^3 \delta^3(\vec{q}_1 - \vec{p}_1) = [\, c_{\vec{q}_1}, \, c_{\vec{p}_1}^\dagger \,] = c_{\vec{q}_1} c_{\vec{p}_1}^\dagger - c_{\vec{p}_1}^\dagger c_{\vec{q}_1}
$$

$$
\Rightarrow \quad c_{\vec{q}_1} c_{\vec{p}_1}^\dagger = (2\pi)^3 \delta^3(\vec{q}_1 - \vec{p}_1) + c_{\vec{p}_1}^\dagger c_{\vec{q}_1}
$$

$$
\Rightarrow \quad c_{\vec{q}_1} c_{\vec{p}_1}^\dagger | 0 \rangle = (2\pi)^3 \delta^3(\vec{q}_1 - \vec{p}_1) | 0 \rangle + 0, \qquad (13.35)
$$

because $c_{\vec{q}_1} | 0 \rangle = 0$. We can do the same kind of thing with the phion operators. Start with the $a_{\vec{p}_2} a_{\vec{q}_3}^\dagger$ pair and work outward:

$$
a_{\vec{p}_3} (a_{\vec{p}_2} a_{\vec{q}_3}^\dagger) a_{\vec{q}_2}^\dagger | 0 \rangle = a_{\vec{p}_3} \left(a_{\vec{q}_3}^\dagger a_{\vec{p}_2} + (2\pi)^3 \delta^3(\vec{p}_2 - \vec{q}_3) \right) a_{\vec{q}_2}^\dagger | 0 \rangle
$$

$$
= a_{\vec{p}_3} a_{\vec{q}_3}^\dagger a_{\vec{p}_2} a_{\vec{q}_2}^\dagger | 0 \rangle + a_{\vec{p}_3} a_{\vec{q}_2}^\dagger (2\pi)^3 \delta^3(\vec{p}_2 - \vec{q}_3) | 0 \rangle = \cdots
$$

$$
= (2\pi)^6 [\, \delta^3(\vec{p}_3 - \vec{q}_3) \delta^3(\vec{p}_2 - \vec{q}_2) + \delta^3(\vec{p}_3 - \vec{q}_2) \delta^3(\vec{p}_2 - \vec{q}_3) \,] | 0 \rangle. \quad (13.36)
$$

Exercise 13.3.1   Fill in the missing steps corresponding to the $\cdots$ above.

BOX 13.3 Calculating the Decay S-matrix (cont.)

Our complete result is therefore

$$a_{\vec{p}_3}(a_{\vec{p}_2}a^{\dagger}_{\vec{q}_3})\, a^{\dagger}_{\vec{q}_2} c_{\vec{q}_1} c^{\dagger}_{\vec{p}_1}|0\rangle$$

$$= (2\pi)^6[\, \delta^3(\vec{p}_3 - \vec{q}_3)\delta^3(\vec{p}_2 - \vec{q}_2) + \delta^3(\vec{p}_3 - \vec{q}_2)\delta^3(\vec{p}_2 - \vec{q}_3)\,]$$

$$\times (2\pi)^3\delta^3(\vec{q}_1 - \vec{p}_1)|0\rangle. \tag{13.37}$$

Since the outgoing phions are identical, it is ambiguous which connects to which. So, are these separate cases, or are they really the same thing? This gets into the mathematical weeds that I am trying to avoid. It is enough to know for present that we can pick either *one* of these cases and go on. Let's pick the term connecting $\vec{p}_2$ to $\vec{q}_2$ and $\vec{p}_3$ to $\vec{q}_3$.

Exercise 13.3.2 Show that if we now substitute Equation 13.37 back into Equation 13.33 and integrate over the momenta, we get Equation 13.34. (*Hints:* Note that $\langle 0 | 0 \rangle = 1$. Note also that when a Dirac delta sets, say, $\vec{p}_1 = \vec{q}_1$, then it is also setting $p_1^{\mu} = q_1^{\mu}$ (or $p_1 = q_1$ in our compact notation) because the time component of each four-momentum is determined by its spatial components.)

BOX 13.4 Second-Order Contributions to the S-matrix

In Box 13.2, we calculated the eight operator products that appear in H_{int}, each of which involves two a-operators and one c-operator with different arrangements of daggers. The operator products that appear in the second-order term $H_{\text{int}}(t_2)H_{\text{int}}(t_1)$ will be this eight-term operator polynomial times the same thing with different momentum subscripts. The result contains terms such as

$$a_{\vec{q}_2}a_{\vec{q}_3}a_{\vec{q}_5}a_{\vec{q}_6}c_{\vec{q}_1}c_{\vec{q}_4} + a_{\vec{q}_2}a_{\vec{q}_3}a_{\vec{q}_5}^{\dagger}a_{\vec{q}_6}c_{\vec{q}_1}c_{\vec{q}_4} + a_{\vec{q}_2}a_{\vec{q}_3}a_{\vec{q}_5}^{\dagger}a_{\vec{q}_6}^{\dagger}c_{\vec{q}_4}c_{\vec{q}_1} + \cdots \tag{13.38}$$

integrated over all six momenta (note that I have moved all etaon operators to the right). Fortunately, we need not multiply out all 64 combinations to see why *each* term yields zero when inserted in place of X inside $\langle 0|a_{\vec{p}_3}a_{\vec{p}_2}Xc_{\vec{p}_1}^{\dagger}|0\rangle$.

Exercise 13.4.1 Explain why no term could possibly yield a nonzero result. (*Hint:* How many etaon creation and annihilation operators appear in each term X? Why is this a problem?)

BOX 13.5 Internal Particles Must Be Off-Shell

Consider the process shown in Figure 13.4. This is the left vertex in the Feynman diagram shown in Figure 13.3. The goal of this box is to show why the etaon must be off-shell.

Suppose we are in the initial phion's rest frame, so that it has initial energy m_ϕ and no spatial momentum. Suppose that the final phion has spatial momentum $|\vec{p}|$ in the $-x$ direction. Since it is a detectable outgoing particle, it must be on-shell, so its final energy must be $E_3 = (|\vec{p}|^2 + m_\phi^2)^{1/2}$. But we also have a Dirac delta that enforces the conservation of the components of the total four-momentum, so we must have $q^x = +|\vec{p}|$, $q_y = 0$, $q_z = 0$, and $q^t = m_\phi - E_3 = m_\phi - (|\vec{p}|^2 + m_\phi^2)^{1/2}$, which is negative. Now, this is not the "energy" of the etaon: since it is not a real particle, it doesn't really have either energy or momentum. You should think of quantities q^t, q^x, q^y, and q_z instead as simply being parameters that appear in the wavefunction $e^{-iq\cdot(x_2-x_1)}$ that describes this line. But it is clear already, just from the fact that q^t is negative in this particular frame, that the etaon can't be a real particle.

Now, q^t might be positive in some other reference frame if q is a spacelike four-vector. Of course, if q *is* spacelike, that completely proves that the etaon cannot be a real particle, since any real particle must have a timelike four-momentum.

We can show that the four-vector q is indeed spacelike as follows. Start with

$$q^t = m_\phi - \sqrt{|\vec{p}|^2 + m_\phi^2}. \tag{13.39}$$

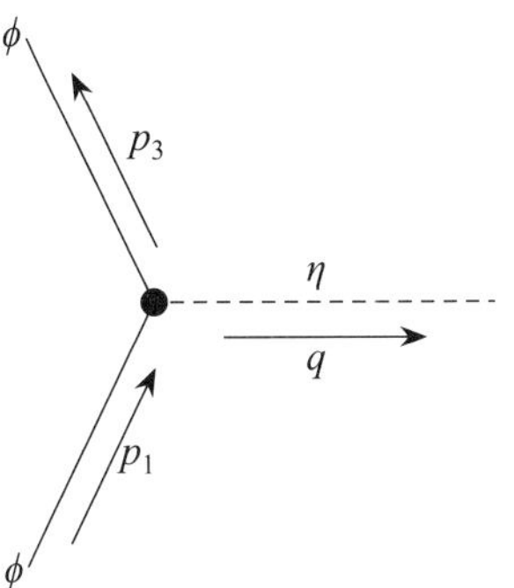

FIGURE 13.4 This diagram shows half of a process containing an internal particle.

Exercise 13.5.1 Isolate the square root, cancel some terms, and then calculate $q^2 = (q^t)^2 - (q^x)^2 - (q^y)^2 - (q^z)^2$. You should find that in the rest frame of the initial phion, q^2 is negative. But q^2 is a Lorentz invariant, so this means that the vector is indeed spacelike in all reference frames. Even more to the point, it is off-shell: $q^2 \neq m_\eta^2$ since the latter is a positive number.

BOX 13.6 The Other Second-Order Phion-Scattering Processes

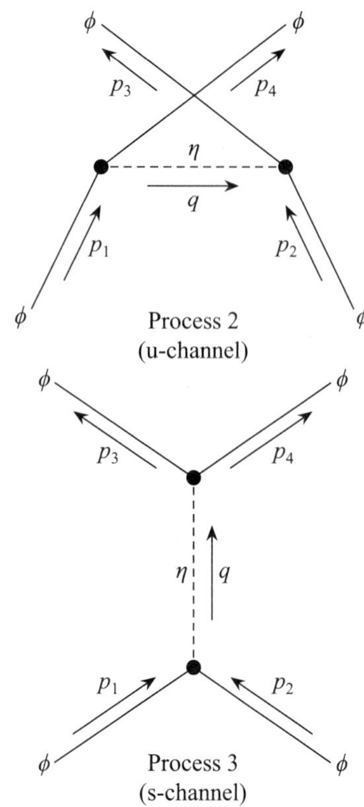

FIGURE 13.5 These Feynman diagrams represent the other two second-order phion-scattering processes.

The Feynman diagrams for the other second-order phion scattering processes are shown in Figure 13.5. Process 2 is the same as process 1 discussed in the text but with the outgoing phions exchanged (since we can't tell them apart, we don't know which outgoing phion was connected with which incoming phion, so we must include both processes). Process 3 is a completely different way of getting from two initial phions to two final phions. By the way, the processes 1, 2, and 3 represented by these very common diagrams are called (for no reason known to me) **t-channel, u-channel,** and **s-channel** processes, respectively.

In this box, we will apply the Feynman rules (briefly summarized below for easy reference) to calculate scattering amplitudes for these processes.

1. **Draw the Diagrams.** (See Figure 13.5.)

2. **Label Lines.** (This has already been done in the diagrams.)

3. **Vertex Factors.** Each vertex contributes $-i\lambda$.

4. **Propagator.** Each internal line contributes $i/(q_j^2 - m_j^2)$.

5. **Conserve four-momentum at each vertex.**

6. **The result is** $-i\mathcal{M}$.

Exercise 13.6.1 Apply the Feynman rules to the diagrams to calculate the corresponding scattering amplitudes $\mathcal{M}_2$ and $\mathcal{M}_3$. (By Rule 8, the total second-order amplitude is $\mathcal{M}_{\text{tot}} = \mathcal{M}_1 + \mathcal{M}_2 + \mathcal{M}_3$.)

Homework Problems

P13.1 Not all interaction terms that one might dream up are physically meaningful. Consider the Lagrangian density

$$\mathcal{L} = \tfrac{1}{2}\partial^\mu\phi\,\partial_\mu\phi - \tfrac{1}{2}m_\phi^2\phi^2 + \tfrac{1}{2}\partial^\mu\eta\,\partial_\mu\eta - \tfrac{1}{2}m_\eta^2\eta^2 - \lambda\eta\phi. \tag{13.40}$$

It looks like this Lagrangian density has an interaction term $-\lambda\phi\eta$. But suppose we define new rotated fields

$$\phi' = \cos\theta\,\phi + \sin\theta\,\eta \quad \text{and}$$
$$\eta' = -\sin\theta\,\phi + \cos\theta\,\eta. \tag{13.41}$$

Show that the Lagrangian density

$$\mathcal{L} = \tfrac{1}{2}\partial^\mu\phi'\,\partial_\mu\phi' - \tfrac{1}{2}m_1^2\phi'^2 + \tfrac{1}{2}\partial^\mu\eta'\,\partial_\mu\eta' - \tfrac{1}{2}m_2^2\eta'^2 \tag{13.42}$$

with *no* interaction term (but with different masses for the ϕ' and η' fields) is physically equivalent to original Lagrangian density with an appropriate choice of θ. Find the values of θ, m_1^2, and m_2^2 in terms of λ, m_ϕ^2, and m_η^2. The point is that the original Lagrangian density does *not* describe two fields that interact, but rather two free (but different) fields in a hidden guise.

P13.2 One possible third-order diagram for the etaon decay process is

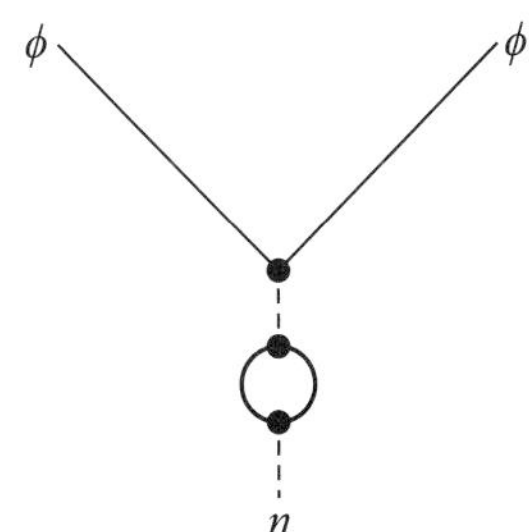

Draw the three other possible third-order diagrams.

P13.3 Consider the decay process $\eta \to \phi + \phi + \eta$ in our toy universe.
a. Show or argue from the pattern of creation and annihilation operators that no first-order contribution is nonzero. (*Hint:* First, argue that the scattering matrix has the form $\langle 0|a_{\vec{p}_4}a_{\vec{p}_3}c_{\vec{p}_2}H_{\text{int}}c_{\vec{p}_1}^\dagger|0\rangle$.)
b. Construct the simplest Feynman diagram for this process. (*Hint:* Valid Feynman diagrams must be constructed out of connected sets of vertices where two phion lines connect to an etaon line.)
c. Argue that this must correspond to a *second-order* contribution to the scattering matrix (involving two factors of H_{int}). *(continues →)*

d. Convert your diagram into a sequence of operators and argue that the entire sequence (including the operators at $t = \pm\infty$) could yield a nonzero result.
e. Argue, however, that the scattering matrix should be zero for a very basic physical reason. (This will be enforced by a Dirac delta once we have done all the integrations. I am not asking you to do those integrations, but rather to think about why this decay should be forbidden physically.)

P13.4 Consider the scattering process $\eta + \phi \to \eta + \phi$.
a. Draw the possible Feynman diagrams with the lowest nontrivial order for this process. (*Hint:* There are two.)
b. Compare your diagrams to diagrams we have seen in this chapter to determine the channel names of each subprocess shown. Why is the third channel not present in this case?
c. Are there diagrams with three vertices? If so, draw them. If not, explain why not.
d. Draw the diagrams with four vertices. There are six. (*Hint:* Some of them will have "bubbles" like the diagram shown in Problem P13.2.)

P13.5 In our toy universe, each quanton is its own anti-quanton. Consider the phion-phion "annihilation" process $\phi + \phi \to \eta + \eta$, and assume in this case that the etaon mass m_η is smaller than the phion mass m_ϕ so that the annihilation is possible for phions at rest.
a. Draw the possible Feynman diagrams with the lowest nontrivial order for this process. (*Hint:* There are two.)
b. Compare your diagrams to diagrams we have seen in this chapter to determine the channel names for each subprocess shown. Why is the third channel not present in this case?
c. Use the Feynman rules to calculate the scattering amplitude $\mathcal{M}$ for each diagram.

P13.6 Consider the bottom vertex in the second-order process 3 (the s-channel) for phion-phion scattering, and let's look at this in the center-of-mass frame where the incoming phions have opposite spatial momenta. Do an analysis similar to that in Box 13.5 to show that, even though the virtual etaon's four-momentum q in this case is timelike, we still have $q^2 \neq m_\eta^2$ in all frames unless the phion energies in the center-of-mass frame happen to be such that $2E_1 = 2E_2 = m_\eta$. (In that special case, note that $\mathcal{M}_3$ calculated in Box 13.6 does blow up, meaning that this process will dominate over the other processes. Getting the energies *exactly* right is extremely unlikely, but you will get a peak into the value of cross section as you get close to this energy. This is an example of what particle physicists call a *resonance.*)

14. FERMI'S GOLDEN RULES

In the last chapter, we learned how to calculate S-matrix values that allow us to calculate the probability that particles in a specific initial Fock-space state end up as (possibly different) particles in a specific final state. But to experimentally test these predictions, we need to connect these S-matrix values to experimentally measurable quantities. When our process involves a single initial particle, the experimentally measurable quantity is a **decay lifetime** τ. When we have two (or more) initial particles that collide, the measurable quantity is something we call a **cross section** σ. The purpose of this chapter is to define these concepts and to mathematically connect both to the S-matrix.

14.1 The Transition Rate and Decay Lifetimes

We usually call a process that starts with a single particle and ends with two or more particles a **decay process**. In previous studies, you have encountered situations where atoms or nuclei in excited states decay by emitting particles such as photons, electrons, positrons, or helium nuclei. In a similar manner, many of the particles we study in particle physics decay to particles with smaller masses. Of all the particles we can generate in particle physics experiments, only the photon, the electron, neutrinos, and (maybe?) the proton are completely stable (as in these cases, decaying would violate certain basic conservation laws).

Decay processes involving subatomic particles, like those involving atoms and nuclei, are random: we can never predict exactly when a given particle, atom, or nucleus will decay. Rather, a quantum object such as a particle, atom, or nucleus (at rest) has a certain characteristic and constant **transition rate** Γ (probability per unit time) of decaying. That this rate is *constant* is mathematically convenient but physically remarkable: it means that a radioactive nucleus (for example) that has existed for a billion years is no more or less likely to decay in the next second than the same kind of nucleus manufactured minutes ago in a nuclear physics experiment. This means that an individual quantum object does not "age" like human beings do: a 90-year-old human has a much higher probability of dying in the next year than a 20-year-old human.

Suppose that a single given quantum object has a probability Γdt of decaying during an infinitesimal time interval dt. This means that the change dN in the number N of such systems remaining in a given sample is given by

$$dN = -N\Gamma\, dt, \tag{14.1}$$

where dN is negative because the number of systems remaining is decreasing. We can integrate this immediately to find that

$$dN = -N\Gamma\, dt \;\Rightarrow\; \int_{N_0}^{N(t)} \frac{dN}{N} = -\Gamma \int_0^t dt \;\Rightarrow\; \ln N - \ln N_0 = -\Gamma t$$

$$\Rightarrow\; \frac{N(t)}{N_0} = e^{-\Gamma t} \;\Rightarrow\; N(t) = N_0 e^{-\Gamma t}. \tag{14.2}$$

This is the characteristic exponential decay pattern that may be familiar if you have studied nuclear physics. In Box 14.1, you will show that the **average lifetime** τ of the quantum object (in the limit that $N \to \infty$) is simply

$$\tau = \frac{1}{\Gamma}. \tag{14.3}$$

In nuclear physics especially, we are often interested in a sample's **half-life**: $t_{1/2}$, the time required for the number of objects remaining to decrease to half its initial value. This is related to τ and Γ as follows:

$$\frac{1}{2} = \frac{N(t_{1/2})}{N_0} = e^{-\Gamma t_{1/2}} \;\Rightarrow\; \ln \tfrac{1}{2} = -\ln 2 = -\Gamma t_{1/2}$$

$$\Rightarrow\; t_{1/2} = \frac{\ln 2}{\Gamma} = (\ln 2)\tau. \tag{14.4}$$

We see that the transition rate Γ is the key connection between the concept of probability (which quantum mechanics and, by extension, quantum field theory is designed to calculate) and experimentally measurable quantities such as a radioactive sample's average lifetime or half-life. Note that since Γdt is a unitless probability, Γ itself must have units of inverse time, which in our natural unit system is the same as units of GeV. We will see that this transition rate concept applies to more than just decays.

Before we leave this topic, though, it is important to note that many quantum objects can decay in different ways. For example, the $^{223}_{88}\mathrm{Ra}$ nucleus can decay by emitting an alpha particle (helium nucleus) or by emitting a positron. Similarly, a particle known as a pion can decay to a muon and a muon neutrino or an electron and an electron neutrino, along with a few other even less-probable modes. In such cases, using the decay rate is advantageous because the decay rates for all these modes simply add, because probabilities add in the infinitesimal limit. Therefore,

$$\Gamma_{\mathrm{tot}} = \sum_i \Gamma_i. \tag{14.5}$$

When multiple decay modes are possible, we can also experimentally determine **branching ratios**:

$$\text{Branching ratio for the } i\text{th mode} \equiv \frac{\Gamma_i}{\Gamma_{\mathrm{tot}}}. \tag{14.6}$$

This ratio tells you what fraction of the decays occur in the ith mode. If we can use quantum field theory to calculate the decay rate Γ_i for each mode, then we can calculate both the average lifetime and the branching ratios.

14.2 Counting Momentum States

The S-matrix component $S_{fi} \equiv \langle f | U(\infty, -\infty) | i \rangle$ gives us the amplitude for an interaction taking us from an initial Fock-space state $| i \rangle$ at time $t_i \to -\infty$ to a final Fock-space state $| f \rangle \neq | i \rangle$ at time $t_f \to \infty$. The probability that a transition occurs during this time is

$$P_{fi}(\infty, -\infty) = \frac{|S_{fi}|^2}{\langle i | i \rangle \langle f | f \rangle}, \tag{14.7}$$

where I have allowed for the possibility that $\langle i | i \rangle \neq 1$ and $\langle f | f \rangle \neq 1$. Let us imagine (for the moment) that $t_f - t_i \equiv T$ is a long but finite time interval: we will take the limit $T \to \infty$ eventually. Since the probability per unit time is constant, the transition *rate* (transition probability per unit time) is

$$\Gamma_{fi} = \frac{P_{fi}(+\frac{1}{2}T, -\frac{1}{2}T)}{T} = \frac{|S_{fi}|^2}{T \langle i | i \rangle \langle f | f \rangle}. \tag{14.8}$$

But this is the rate for ending up in a *specific* final state $| f \rangle$. To get the total transition rate, we need to sum the transition rates for *all possible* final states.

Suppose that our process involves two final particles with momenta $\vec{p}_2$ and $\vec{p}_3$. The Lorentz-invariant Fock-space final state in this case is $|p_2, p_3\rangle$, meaning that we have one particle with momentum p_2 and another with momentum p_3. To sum over all final states, it would make sense to integrate over all possible momentum states of this type. But how many distinct one-particle momentum states are in a differential volume d^3p of momentum space?

To answer this question, suppose that we are conducting our experiment in one-dimensional box with length L (we'll take the limit $L \to \infty$ eventually). At the box's boundaries, we will impose "periodic boundary conditions" requiring all wavefunctions to have the same value on the box's boundaries at $x = \pm\frac{1}{2}L$. This implies that we must fit an integer number of wavelengths λ within the inter-boundary distance L. The de Broglie equation then implies that

$$L = n_x \lambda = \frac{n_x h}{p_x} \quad \Rightarrow \quad p_x = \frac{n_x h}{L} = \frac{2\pi n_x \hbar}{L} = \frac{2\pi n_x}{L}, \tag{14.9}$$

where n_x is a nonzero integer (in the last step, I used the fact that in our unit system, $\hbar = 1$). In a *cubical* box whose sides have length L, the same must be true for p_y and p_z. As Figure 14.1 illustrates, the momentum state for a vector $\vec{p}$ for a specific set of possible values $p_x = 2\pi n_x/L$, $p_y = 2\pi n_y/L$, and $p_z = 2\pi n_z/L$, when plotted in momentum space, corresponds to a specific dot that is an integer number of steps of length $2\pi/L$ along each axis. The set of all such dots comprise a cubic lattice where each dot is separated from its nearest neighbors by $2\pi/L$ along each of the axis directions. This means that each possible momentum state effectively occupies a tiny cube in momentum space with a volume of $(2\pi/L)^3$; thus, the number of states in a differential volume d^3p of momentum space will be that volume divided by the momentum-space volume per state:

$$\text{number of states} = \frac{d^3p}{(2\pi/L)^3} = V\frac{d^3p}{(2\pi)^3}, \tag{14.10}$$

where $V = L^3$ is the spatial volume of our cubical box.

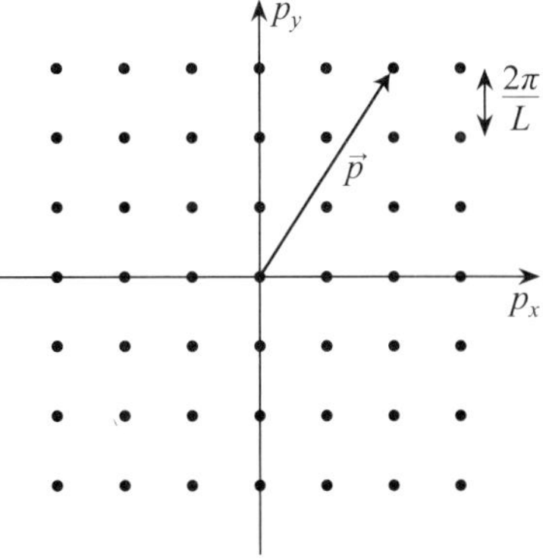
FIGURE 14.1 This figure illustrates possible momentum states as dots in the p_x-p_y plane of momentum space.

14.3 The Transition Rate for Decays

The total transition rate for any given outcome will be the total number of final momentum states consistent with the desired outcome, each multiplied by the transition rate Γ_{fi} for that particular final state $|f\rangle$. If our final states involve two outgoing particles, then this is

$$\Gamma = \int V\frac{d^3p_2}{(2\pi)^3} V\frac{d^3p_3}{(2\pi)^3}\Gamma_{fi} = \int \frac{d^3p_2}{(2\pi)^3}\frac{d^3p_3}{(2\pi)^3}V^2\frac{|S_{fi}|^2}{T\langle i\,|\,i\rangle\langle f\,|\,f\rangle}. \tag{14.11}$$

For a decay to a specific pair of final particles, the integral would be over *all* possible final momenta for that pair.

Now let's work on the normalization of the initial and final states. In Box 14.2, you will show that when we have an initial state involving only one particle (as would be the case in a particle decay),

$$\langle i\,|\,i\rangle\langle f\,|\,f\rangle = 8E_{\vec{p}_1}E_{\vec{p}_2}E_{\vec{p}_3}V^3. \tag{14.12}$$

Finally, let's deal with the S-matrix component. Remember that we define the scattering amplitude $\mathcal{M}$ so that $S_{fi} = -i(2\pi)^4\delta^4(\Delta p)\mathcal{M}$, where Δp is total initial minus the total final four-momentum (the Dirac delta therefore enforces conservation of four-momentum). This means that the absolute square of the scattering matrix must be

$$|S_{fi}|^2 = (2\pi)^8[\,\delta^4(p_1 - p_2 - p_3)\,]^2|\mathcal{M}|^2. \tag{14.13}$$

The squared Dirac delta is ugly: in an integral over the final momenta, the first delta will set $p_1 = p_2 + p_3$ and the second will then evaluate to $\delta^4(0) = \infty$. But we can apply the identity

$$(2\pi)^4 \delta^4(\Delta p) = \int d^4x \, e^{-i\Delta p \cdot x} \tag{14.14}$$

to one of the Dirac deltas. Then, as Box 14.2 discusses, the integral in the finite box and for the finite time limit will evaluate (in the limit that $\Delta p \to 0$) to VT, where V is the volume of our box and T is the duration $t_f - t_i$ of our experiment. Putting this all together yields

$$\Gamma = \int \frac{d^3p_2}{(2\pi)^3} \frac{d^3p_3}{(2\pi)^3} V^2 \frac{(2\pi)^4 \delta^4(p_1 - p_2 - p_3) \, VT \, |\mathcal{M}|^2}{T \, (8E_{\vec{p}_1} E_{\vec{p}_2} E_{\vec{p}_3} V^3)}$$

$$= \int \frac{d^3p_2}{(2\pi)^3} \frac{d^3p_3}{(2\pi)^3} \frac{(2\pi)^4 \delta^4(p_1 - p_2 - p_3) \, |\mathcal{M}|^2}{8E_{\vec{p}_1} E_{\vec{p}_2} E_{\vec{p}_3}}. \tag{14.15}$$

Since all the factors of V and T have canceled out, this will also be the result in the limit as $V \to \infty$ and $T \to \infty$.

This is spectacular! We now have a connection between the scattering amplitude $\mathcal{M}$ (which we can calculate from the Feynman diagrams for a decay process) and the observable transition rate Γ, which in this case is a *decay rate*.

This is a specific case of **Fermi's Golden Rule** for decays. The general rule states that if particle 1 decays to particles 2, 3, ... N, the decay rate is

$$\Gamma = \frac{S}{2E_1} \int \left[\left(\frac{d^3p_2}{(2\pi)^3 \, 2E_2} \right) \left(\frac{d^3p_3}{(2\pi)^3 \, 2E_3} \right) \cdots \left(\frac{d^3p_N}{(2\pi)^3 \, 2E_N} \right) \right.$$

$$\left. \times (2\pi)^4 \delta^4(p_1 - p_2 - p_3 \cdots - p_N) \, |\mathcal{M}|^2 \right], \tag{14.16}$$

where $E_1, E_2, \ldots$ is shorthand for $E_{\vec{p}_1}, E_{\vec{p}_2}, \ldots$ and S is a "degeneracy factor" that is the product of $1/n!$ for every group of n identical particles in the set of outgoing particles. (We need this factor because the integrations over all momenta basically overcount the possibilities if the particles are identical, since the state where one outgoing particle has momentum $\vec{p}_2$ and the other has momentum $\vec{p}_3$ is actually the same as the state where the momenta are switched.)

This equation implies that, other things being equal, the more momentum space available to the final states (within the constraint implied by the Dirac delta), the greater the decay rate. The final states can occupy more momentum space if the decay products have lots of kinetic energy (and thus lots of possible momenta), so decays will generally go faster if the products have small masses compared to the decaying particle.

Note that Equation 14.16 implies that *the units of $\mathcal{M}$ depend on the number of outgoing particles*. See Problem P14.1 for more discussion.

14.4 The Cross Section

In particle physics experiments, we are often either shooting a stream of high-energy particles at other stationary particles, like arrows at a target, or (in more modern colliders such as the Large Hadron Collider) shooting two beams of high-energy particles directly at each other. In either case, a quantity that would tell us something about the probability of one particle hitting another would be the *cross-sectional area σ* that one particle (the "target") presents to the other particles (the "projectiles"). Of course, "hitting" the target in a particle physics context could mean any number of things. We could ask in a scattering experiment what is the probability that the particles are deflected by a certain

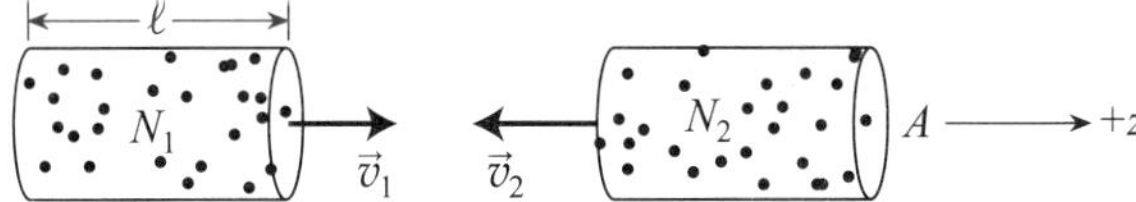

FIGURE 14.2 A pair of particle bunches approaching each other in a particle accelerator experiment. (Note that it is conventional in particle physics experiments to choose the $+z$ axis to be along the direction of $\vec{v}_1 - \vec{v}_2$.)

angle. We might also ask about the relative probabilities of certain types of collision outcomes, for example, the probability that colliding electrons produce a pair of muons compared to the probability of simply scattering.

Quite generally, the **cross section** σ for a given experimental outcome is

$$\sigma \equiv \frac{\Gamma_{\text{tot}}}{\mathcal{L}}, \tag{14.17}$$

where Γ_{tot} is the total transition rate for the particular outcome of interest and $\mathcal{L}$ is the beam **luminosity**, defined as the number of target particles N_2 times the number of projectile particles N_1 passing through a unit area per unit time. To see why this makes sense, consider the general situation shown in Figure 14.2, where we have two cylindrical bunches of particles containing N_1 and N_2 particles that are approaching each other with relative velocity $|\vec{v}_1 - \vec{v}_2|$. Suppose that each bunch has length ℓ and cross-sectional area A. We'll arbitrarily consider bunch 1 to be projectiles and bunch 2 to be the targets. Consider a given target particle that sits on a plane perpendicular to the cylinders' common axis. The probability that a given projectile particle interacts with the target as it passes through the plane is the ratio of the target's cross section σ to the bunch's total cross-sectional area A. *All* the particles in the projectile bunch will pass through that plane in a time $T = \ell / |\vec{v}_1 - \vec{v}_2|$. This means that the probability per time per target particle is indeed

$$\frac{\Gamma_{\text{tot}}}{N_2} = \frac{N_1}{T}\frac{\sigma}{A} \quad \Rightarrow \quad \sigma = \frac{\Gamma_{\text{tot}}}{N_1 N_2/(AT)} = \frac{\Gamma_{\text{tot}}}{\mathcal{L}}, \tag{14.18}$$

as claimed. Now, the total transition rate Γ_{tot} when we have a total of $N_1 N_2$ pairs of particles that might interact is just $N_1 N_2 \Gamma$, where Γ is the transition rate for one projectile hitting one target. Therefore, we can also write

$$\sigma = \frac{N_1 N_2 \Gamma}{N_1 N_2/(AT)} = \Gamma A T = \frac{\Gamma A \ell}{|\vec{v}_1 - \vec{v}_2|} = \frac{\Gamma V}{|\vec{v}_1 - \vec{v}_2|}, \tag{14.19}$$

since the duration of the experiment is $T = \ell / |\vec{v}_1 - \vec{v}_2|$. V here is the volume of the bunch.

14.5 Fermi's Golden Rule for Scattering

Now, the transition rate for a given two-particle outcome in this context is still given by Equation 14.11, but the outgoing four-momenta are p_3 and p_4:

$$\Gamma = \int V \frac{d^3 p_3}{(2\pi)^3} V \frac{d^3 p_4}{(2\pi)^3} \Gamma_{fi} = \int \frac{d^3 p_3}{(2\pi)^3} \frac{d^3 p_4}{(2\pi)^3} V^2 \frac{|S_{fi}|^2}{T \langle i \,|\, i \rangle \langle f \,|\, f \rangle}. \tag{14.20}$$

The integral in this case would be only over the final momentum states that count in the desired outcome (for example, one or the other particle ending up in a particular detector). Also in this case, we have *two* incoming particles, so

$$\langle i \,|\, i \rangle \langle f \,|\, f \rangle = 16 E_{\vec{p}_1} E_{\vec{p}_2} E_{\vec{p}_3} E_{\vec{p}_4} V^4, \tag{14.21}$$

where p_1 and p_2 are the initial ("incoming") particle momenta and p_3 and p_4 the final ("outgoing") particle momenta. This yields

$$\Gamma = \frac{1}{4E_{\vec{p}_1}E_{\vec{p}_2}V} \int \frac{d^3p_3}{(2\pi)^3}\frac{d^3p_4}{(2\pi)^3}\frac{|S_{fi}|^2}{4E_{\vec{p}_3}E_{\vec{p}_4}VT}$$

$$= \frac{1}{4E_{\vec{p}_1}E_{\vec{p}_2}V}\int \frac{d^3p_3}{(2\pi)^3}\frac{d^3p_4}{(2\pi)^3}\frac{(2\pi)^4\delta^4(p_1+p_2-p_3-p_4)|\mathcal{M}|^2 VT}{4E_{\vec{p}_3}E_{\vec{p}_4}VT}, \quad (14.22)$$

where I have again used $|S_{fi}|^2 = (2\pi)^4\delta^4(\Delta p)|\mathcal{M}|^2VT$. I have put the one remaining factor of the box volume V outside the integral (where we can see it). To get the cross section, we can use $\sigma = \Gamma V/|\vec{v}_1 - \vec{v}_2|$ (Equation 14.19):

$$\sigma = \frac{1}{4E_{\vec{p}_1}E_{\vec{p}_2}|\vec{v}_1-\vec{v}_2|}\int \frac{d^3p_2}{(2\pi)^3}\frac{d^3p_3}{(2\pi)^3}\frac{(2\pi)^4\delta^4(p_1+p_2-p_3-p_4)|\mathcal{M}|^2}{4E_{\vec{p}_3}E_{\vec{p}_4}}. \quad (14.23)$$

Here, I have canceled the factors of V, since there is no reason that we can't choose the box in which we are counting momentum states to be the same as the volume of a bunch (the latter will still be enormous compared to the wavelengths of the particles involved, and so might as well be infinity).

Now, the cross section should be a Lorentz invariant with regard to boosts parallel to the axis along which the particle bunches are moving, since it is an area perpendicular to that axis and distances perpendicular to the boost are not affected by a Lorentz transformation. The integral also has the correct form to be Lorentz-invariant (see Chapter 11). Box 14.3 shows that the clearly Lorentz-invariant expression $\sqrt{(p_1 \cdot p_2)^2 - (m_1m_2)^2}$ becomes vE_1E_2 in a frame where either initial particle is at rest and $v = |\vec{v}_1 - \vec{v}_2|$. So (again simplifying our notation by writing $E_1 \equiv E_{\vec{p}_1}$, and so on), a fully and clearly Lorentz-invariant expression for the cross section is

$$\sigma = \frac{1}{4\sqrt{(p_1 \cdot p_2)^2 - (m_1m_2)^2}}\int \frac{d^3p_3}{(2\pi)^3}\frac{d^3p_4}{(2\pi)^3}\frac{(2\pi)^4\delta^4(p_1+p_2-p_3-p_4)|\mathcal{M}|^2}{4E_3E_4}.$$

$$(14.24)$$

This is again a specific case of **Fermi's Golden Rule** for the cross section. The general rule states that if the interaction of particles 1 and 2 produces particles $3, 4, \ldots N$, then the cross section is given by the expression

$$\sigma = \frac{S}{4\sqrt{(p_1 \cdot p_2)^2 - (m_1m_2)^2}}\int \left[\left(\frac{d^3p_3}{(2\pi)^3\,2E_3}\right)\left(\frac{d^3p_4}{(2\pi)^3\,2E_4}\right)\cdots\left(\frac{d^3p_N}{(2\pi)^3\,2E_N}\right)\right.$$

$$\left.\times (2\pi)^4\delta^4(p_1+p_2-p_3-p_4\cdots-p_N)\,|\mathcal{M}|^2\right], \quad (14.25)$$

where S is the same "degeneracy factor" as before (a factor of $1/n!$ for every group of n identical outgoing particles).

BOX 14.1   Average Lifetime in a Decay Process

In this box, we will show that the average lifetime τ in an exponential decay process where $N(t) = N_0\, e^{-\Gamma t}$ is given simply by $\tau = 1/\Gamma$.

What can we mean by the concept "average lifetime?" Suppose that we have N_0 systems at time $t = 0$. The number $|dN(t)|$ that decay during a tiny time interval of length dt centered on t is given by

$$|dN(t)| = \left|\frac{dN}{dt}\right|\, dt = \left|-\Gamma N_0\, e^{-\Gamma t}\right|\, dt = +\Gamma N_0\, e^{-\Gamma t} dt. \qquad (14.26)$$

(We need the absolute value, because $dN < 0$.) The systems that decay during this interval have lasted for time t, so that is the lifetime of these particular systems. To get the average lifetime, we have to add up the results for all time intervals and divide by the total initial number of systems. In the limit that N_0 becomes large enough that $|dN(t)|$ is large even for time intervals $dt \ll 1/\Gamma$, we can approximate the sum by the integral

$$\tau \approx \frac{1}{N_0}\int_0^\infty t\,[\,\Gamma\, N_0 e^{-\Gamma t}\,]\,dt. \qquad (14.27)$$

Exercise 14.1.1 This is actually an easy integral to do by parts. First, change variables to $x \equiv \Gamma t$. The usual statement of integration by parts is that $\int u\, dv = uv - \int v\, du$, where u and v are functions of x, $dv = (dv/dx)\, dx$, and so on (as we saw in Chapter 3). Define u and v appropriately for this situation and calculate the integral.

BOX 14.2 Normalizing States in a Box

Our goal in this box is to calculate the value of $\langle i \,|\, i \rangle \langle f \,|\, f \rangle$, where the initial state $|\, i \,\rangle$ is the relativistically normalized single-particle Fock-space ket $|\, p_1 \,\rangle = \chi^\dagger_{\vec{p}_1} |\, 0 \,\rangle$ and the final state $|\, f \,\rangle$ is the similarly normalized two-particle Fock-space ket $|\, p_2, p_3 \,\rangle = \alpha^\dagger_{\vec{p}_2} \alpha^\dagger_{\vec{p}_3} |\, 0 \,\rangle$. Note that the $\alpha^\dagger$ and $\chi^\dagger$ operators here are the relativistic versions of the creation operators $a^\dagger_{\vec{p}}$ and $c^\dagger_{\vec{p}}$ that produce phions and etaons, respectively. In Section 11.5, we found that the link between $\alpha^\dagger_{\vec{p}}$ and $a^\dagger_{\vec{p}}$ was $\alpha^\dagger_{\vec{p}} = \sqrt{2E_{\vec{p}}}\, a^\dagger_{\vec{p}}$: the analogous relation applies to $\chi^\dagger_{\vec{p}}$.

Let's work on the initial state first. We have

$$\langle i \,|\, i \rangle = \langle 0 |\chi_{\vec{p}_1}\, \chi^\dagger_{\vec{p}_1}|\, 0 \,\rangle = 2E_{\vec{p}_1}\langle 0 |c_{\vec{p}_1}\, c^\dagger_{\vec{p}_1}|\, 0 \,\rangle$$
$$= 2E_{\vec{p}_1}\langle 0 |\big[c^\dagger_{\vec{p}_1}c_{\vec{p}_1} + (2\pi)^3\delta^3(\vec{p}_1 - \vec{p}_1)\big]|\, 0 \,\rangle, \tag{14.28}$$

where I have used the commutation relation for the creation operator. But the annihilation operator acting on the vacuum ket yields zero, and the Dirac delta behaves just like a number that we can pull out of the braket. So,

$$\langle i \,|\, i \rangle = 2E_{\vec{p}_1}(2\pi)^3\delta^3(0)\langle 0 |\, 0 \,\rangle = 2E_{\vec{p}_1}(2\pi)^3\delta^3(0) \tag{14.29}$$

since the vacuum state is normalized to 1.

Exercise 14.2.1 Similarly, compute $\langle f \,|\, f \rangle$. (*Hint:* Assume $\vec{p}_3 \neq \vec{p}_2$ and note that creation and annihilation operators with different subscripts commute.)

Putting everything together yields

$$\langle i \,|\, i \rangle \langle f \,|\, f \rangle = 8E_{\vec{p}_1}E_{\vec{p}_2}E_{\vec{p}_3}\big[\, (2\pi)^3\delta^3(0) \,\big]^3, \tag{14.30}$$

which is all fine and good, except that $\delta^3(0) = \infty$ normally. But what if we evaluate this in a box of finite volume V? That is what we will do when the box continues on the next page.

BOX 14.2 Normalizing States in a Box (cont.)

To go on from here, we need to go back to the definition of the Dirac delta of a momentum component in one spatial dimension:

$$2\pi \delta(p) = \lim_{L \to \infty} \int_{-L/2}^{L/2} dx \, e^{ipx}. \tag{14.31}$$

In a finite box whose sides have length L, we simply don't take the limit.

Exercise 14.2.2 Evaluate this integral at $p = 0$ but without taking the limit as $L \to \infty$. (*Hint:* This is easy.)

By extension, in a cubical box whose sides have length L, we see that

$$(2\pi)^3 \delta^3(\vec{p} = 0) \equiv (2\pi)^3 \delta(p_x = 0)\delta(p_y = 0)\delta(p_z = 0) = L^3 = V. \tag{14.32}$$

Similarly, in a region of spacetime spanning a spatial box of volume V and a time interval of $t_f - t_i = T$, we will have

$$(2\pi)^4 \delta^4(0) = VT. \tag{14.33}$$

These are the results required in the body of the chapter.

This means that for a decay of one particle to two particles in a box of finite volume V, Equation 14.30 becomes

$$\langle i \,|\, i \rangle \langle f \,|\, f \rangle = 8 E_{\vec{p}_1} E_{\vec{p}_2} E_{\vec{p}_3} \big[(2\pi)^3 \delta^3(0) \big]^3$$

$$= 8 E_{\vec{p}_1} E_{\vec{p}_2} E_{\vec{p}_3} V^3, \tag{14.34}$$

as claimed in Equation 14.12, Perhaps you can also see that in a scattering experiment where we have two initial and two final particles, we can calculate the value of $\langle i \,|\, i \rangle$ just as we calculated $\langle f \,|\, f \rangle$ in this box. So, in the case of two-particle to two-particle scattering in a finite-volume box, we have

$$\langle i \,|\, i \rangle \langle f \,|\, f \rangle = 16 E_{\vec{p}_1} E_{\vec{p}_2} E_{\vec{p}_3} E_{\vec{p}_4} \big[(2\pi)^3 \delta^3(0) \big]^4$$

$$= 16 E_{\vec{p}_1} E_{\vec{p}_2} E_{\vec{p}_3} E_{\vec{p}_4} V^4. \tag{14.35}$$

BOX 14.3 A Lorentz-Invariant Expression for vE_1E_2

Consider two particles with four-momenta p_1 and p_2 and masses m_1 and m_2, respectively. In this box, we will show that the Lorentz-invariant expression

$$\sqrt{(p_1 \cdot p_2)^2 - (m_1 m_2)^2} = vE_1E_2 \tag{14.36}$$

in a frame where either particle is at rest, where E_1 and E_2 are the energies of the two particles, and v is the speed of the moving particle in that frame.

Let's pick one of the particles (say, particle 1) to be at rest, and for the sake of simplicity, choose coordinates so that the other particle moves with speed v in the $+z$ direction, consistent with the particle accelerator convention.

Exercise 14.3.1   Write out the four-momentum components for each particle in the rest frame of particle 1, then evaluate the square root above. (*Hint:* First, argue that the moving particle's z-momentum is E_2v.)

Homework Problems

P14.1 What are the units of $\mathcal{M}$ **(a)** when we have one particle decaying to two particles and **(b)** when we have two particles scattering to two particles? (*Hint:* First, argue that $\delta^4(\Delta p)$ must have units of E^{-4}.)

P14.2 Suppose that the decay rate for a certain quantum object is $\Gamma = 10$ MeV. What is the (average) lifetime of this object in seconds? What is its half-life in seconds?

P14.3 Find an expression for the number of momentum states whose magnitude is within a differential range of $|d\vec{p}|$ of $|\vec{p}|$ if we are in a box with volume V? (*Hint:* Consider the volume of a spherical shell in momentum space.)

P14.4 In particle physics experiments, physicists often express cross sections in barns, where $1\,\text{barn} = 10^{-28}\,\text{m}^2 = 100\,\text{fm}^2$. Express this in an appropriate power of GeV. (The barn is roughly the cross section for an incident neutron to cause a uranium nucleus to fission. This is as big as "the broad side of a barn" for most subatomic particle interactions. For a recounting of the origin of this name, see Holloway and Baker, "How the barn was born," *Phys Today* **25**, 7, p.9 (1972).)

P14.5 As a classical example of a cross section, suppose we are shooting very tiny particles at an elastic sphere of radius R.

a. Suppose that a projectile particle is traveling along a line that would miss the sphere's center by a distance $b < R$. (The quantity b, which is useful in classical scattering experiments, is called the projectile's **impact parameter.**). Show that the projectile will be deflected by an angle θ (where $0 \le \theta \le \pi$) such that

$$b = R \cos\left(\tfrac{1}{2}\theta\right). \qquad (14.37)$$

(*Hint:* Draw a picture. For an elastic sphere, the angle of incidence will be the same as the angle of reflection. How does this angle relate to the total deflection angle θ?)

b. If we consider θ to be a function of b (or vice versa), then the range of angles $d\theta$ (considered positive) through which a projectile will be deflected is related to a range db of impact parameters by

$$d\theta = \left|\frac{d\theta}{db}\right| db = \frac{db}{|db/d\theta|}. \qquad (14.38)$$

From this, find a formula for the cross section that a projectile will be deflected to within a range of $d\theta$ around a given angle θ. (*Hint:* Consider a ring of radius b and thickness db around a line parallel to the beam of particles and going through the target sphere's center.)

c. If $R = 0.02$ m, find the cross section that a projectile particle will be registered by a detector centered at a deflection angle of $25°$, but that spans an angular width of $1°$ in θ and an angular width of $2°$ in the azimuthal angle ϕ around the axis parallel to which the projectiles are being fired.

d. If the luminosity of the projectile beam is 10,000 projectile particles per square meter per second (note that we have just one target here), what is the *rate* at which particles are detected by the detector in the previous part?

e. What is the total cross section for all possible nonzero deflections?

P14.6 By doing appropriate Lorentz transformations, prove that the value of $E_1 E_2 |\vec{v}_1 - \vec{v}_2|$ is indeed frame-independent for boosts $\vec{\beta}$ along the common axis on which $\vec{v}_1$ and $\vec{v}_2$ both lie. (*Hint:* Remember that $\vec{v} = \vec{p}/E$.)

P14.7 The Large Hadron Collider accelerates bunches of protons toward each other at virtually the speed of light. Each bunch has 115 billion protons and a length of about 0.4 m, and the collider's luminosity is about $2 \times 10^{36}\,\text{cm}^{-2}\text{s}^{-1}$.

a. What is the cross-sectional area of each bunch?

b. The cross section for elastic scattering of protons at all angles is roughly 10 mb (see Problem P14.4 for the definition of a barn). What would be the event rate for elastic scattering during the time the bunches are interacting?

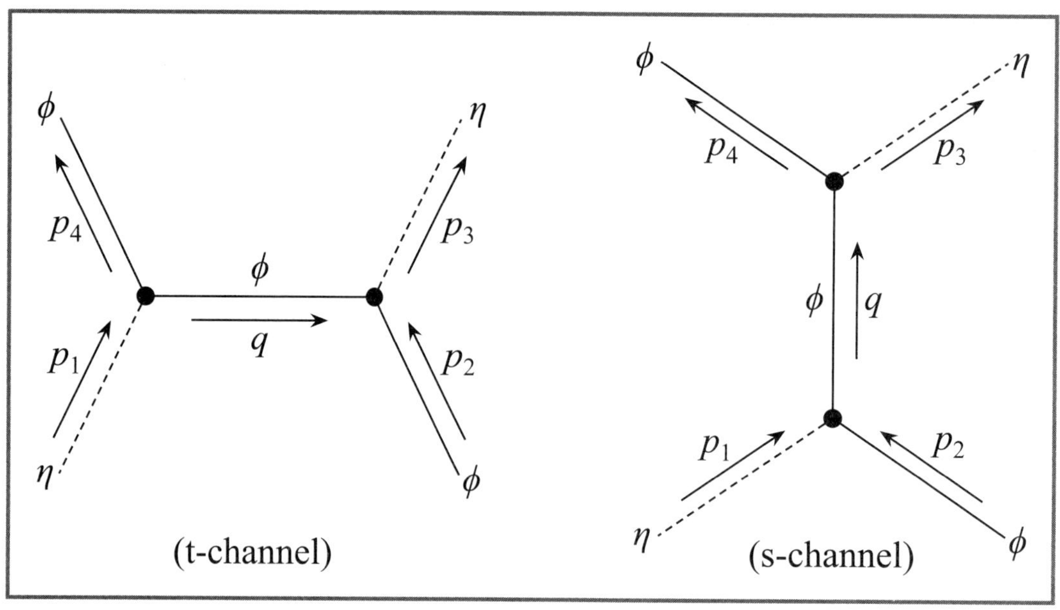

φ
p_4
η
p_3
φ
q
p_1
p_2
η
φ
(t-channel)
φ
η
p_4
p_3
φ
q
p_1
p_2
η
φ
(s-channel)

15. TOY UNIVERSE PREDICTIONS

In this chapter, we will use Fermi's Golden Rule and Feynman diagrams to calculate experimental outcomes in the context of our toy universe occupied by phions and etaons. This will give us practice with these calculations in a simple universe before we deal with the greater complexities of our universe.

15.1 Decay of One Particle to Two Particles

According to Fermi's Golden Rule, the transition rate Γ for one particle of mass M decaying to two particles is

$$\Gamma = \frac{S}{2E_1} \int \frac{d^3 p_2}{(2\pi)^3 2E_2} \frac{d^3 p_3}{(2\pi)^3 2E_3} (2\pi)^4 \delta^4(p_1 - p_2 - p_3) \, |\mathcal{M}|^2. \qquad (15.1)$$

Assume that we want to calculate the rate in the frame where the original particle is at rest, so $E_1 = M$. Equation 15.1 then implies that

$$\Gamma = \frac{S}{32\pi^2 M} \int d^3 p_2 \, d^3 p_3 \, \frac{\delta(M - E_2 - E_3) \, \delta^3(0 - \vec{p}_2 - \vec{p}_3)}{E_2 E_3} \, |\mathcal{M}|^2 . \qquad (15.2)$$

When we integrate over $d^3 p_3$, the Dirac delta sets $\vec{p}_3 = -\vec{p}_2$, yielding

$$\Gamma = \frac{S}{32\pi^2 M} \int d^3 p_2 \, \frac{\delta(M - 2E_2)}{E_2^2} \, |\mathcal{M}|^2 . \qquad (15.3)$$

Now, $E_2 = \sqrt{|\vec{p}_2|^2 + m_2^2}$ only depends on the *magnitude* of $\vec{p}_2$, which to save writing we will call $q \equiv |\vec{p}_2|$. If $|\mathcal{M}|^2$ is also independent of the direction of $\vec{p}_2$ (which it usually is for decays to two particles), we can use spherical coordinates to describe the momentum space for the remaining integral, yielding

$$\Gamma = \frac{S}{32\pi^2 M} \int 4\pi q^2 \, dq \, \frac{\delta(M - 2\sqrt{q^2 + m_2^2})}{m_2^2 + q^2} \, |\mathcal{M}|^2 , \qquad (15.4)$$

where the integral over the angular aspect of $\vec{p}_2$ has produced the factor of 4π. Using properties of the Dirac delta (see Box 15.1), one can show that the integral over the remaining Dirac delta yields the final result

$$\Gamma = \frac{S|\vec{p}_F|}{8\pi M^2} \, |\mathcal{M}|^2 , \qquad (15.5)$$

where $|\vec{p}_F|$ is the momentum magnitude of either outgoing particle allowed by conservation of energy, and $\mathcal{M}$ is evaluated at that momentum. We will find this general result for the decay of a particle at rest to two particles useful later.

15.2 The Decay of the Etaon

In the particular case of an etaon decaying to two identical phions, the degeneracy factor S (which is $1/n!$ for each set of n identical outgoing particles) is $S = 1/2! = 1/2$ because the outgoing phions are identical. Conservation of energy requires that

$$M = m_\eta = 2E_2 = 2\sqrt{|\vec{p}_F|^2 + m_\phi^2} \quad \Rightarrow \quad 4|\vec{p}_F|^2 = m_\eta^2 - 4m_\phi^2$$

$$\Rightarrow \quad |\vec{p}_F| = \tfrac{1}{2}\sqrt{m_\eta^2 - 4m_\phi^2}. \tag{15.6}$$

The Feynman rules imply that $\mathcal{M} = \lambda$ for this one-vertex decay process. So, the decay rate for the etaon must be

$$\Gamma = \frac{\lambda^2 \sqrt{m_\eta^2 - 4m_\phi^2}}{32\pi m_\eta^2}. \tag{15.7}$$

We see that the decay rate increases as the square of λ, the coupling constant in the Lagrangian density interaction term $-\lambda\eta\phi^2$. This constant basically tells us how strongly the fields interact with each other, so a more rapid decay rate if the fields interact strongly makes sense.

Let me digress a moment to talk about the units of λ. In Chapter 4, we saw that $S = \int d^4x\,\mathcal{L}$, where S is the action and $\mathcal{L}$ is the Lagrangian density. S is a unitless scalar and d^4x has units of (energy)$^{-4}$, so *all Lagrangian densities have units of* (energy)^4 (memorize this!). Now, the Lagrangian density for a free and real scalar field is

$$\mathcal{L} = \tfrac{1}{2}\partial^\mu\phi\,\partial_\mu\phi - \tfrac{1}{2}m_\phi^2\phi^2. \tag{15.8}$$

The mass term in particular makes it clear that the field ϕ must have units of energy, and the same will be true for the η field. Therefore, since the interaction term $-\lambda\eta\phi^2$ must have units of (energy)^4, λ must have units of energy.

The decay rate must have units of energy since it is probability per time and time has units of inverse energy. We see that our expression does have the right units: λ^2/m_η^2 is unitless, and the square root has units of energy.

Indeed, we see that the ratio λ/m_η provides a basic overall measure of the decay rate. The larger this ratio is, the more rapid the decay. On the other hand, very massive initial particles will decay more slowly than lighter particles, other things being equal.

The part that depends on the phion mass m_ϕ is the square root. The decay rate increases as m_ϕ becomes smaller compared to m_η because of the greater volume of momentum space available to the outgoing phions if they have lower mass. Also, note that as $2m_\phi \to m_\eta$, the decay rate goes to zero and is exactly zero for $2m_\phi \geq m_\eta$, because then the decay would violate conservation of energy.

15.3 The Cross Section for Scattering to Two Particles

It is often easiest to calculate scattering cross sections in the center-of-mass frame. This is also realistic experimentally, since the largest particle accelerators (such as the Large Hadron Collider) typically collide particles in counter-rotating beams so that the laboratory frame *is* the center-of-mass frame: this allows the maximum amount of the beam particles' kinetic energy to be converted into the mass energy of new particles (because the product particles need not carry away kinetic energy to conserve momentum).

In the center-of-mass frame, the momenta of the incoming particles are equal and opposite. Figure 15.1 shows a schematic drawing of an interaction of identical particles in the center-of-mass frame with initial momenta $\vec{p}_1$ and $\vec{p}_2 = -\vec{p}_1$ and final momenta $\vec{p}_3$ and $\vec{p}_4 = -\vec{p}_3$, and shows how we define the deflection angle θ in this case.

In Box 15.2, you will show that generally in the center-of-mass frame, where $\vec{p}_2 = -\vec{p}_1$, we have

$$\sqrt{(p_1 \cdot p_2)^2 - (m_1 m_2)^2} = (E_1 + E_2)|\vec{p}_1|. \tag{15.9}$$

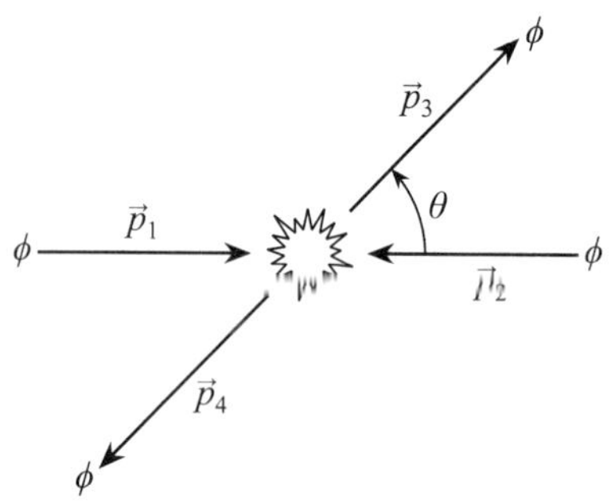

FIGURE 15.1 This diagram shows a scattering process between two particles that yields two particles, as viewed in the center-of-mass frame.

The cross section in this frame therefore becomes

$$\sigma = \frac{S}{4(E_1 + E_2)|\vec{p}_1|} \int \frac{d^3 p_3}{(2\pi)^3} \frac{d^3 p_4}{(2\pi)^3} \frac{(2\pi)^4 \delta^4(p_1 + p_2 - p_3 - p_4)|\mathcal{M}|^2}{4E_{\vec{p}_3} E_{\vec{p}_4}}. \quad (15.10)$$

One can do the integral over $d^3 p_4$ and also over all possible *magnitudes* (but not directions) of p_3. The result (see Box 15.3) is

$$d\sigma = \frac{S|\mathcal{M}|^2}{(8\pi)^2(E_1 + E_2)^2} \frac{|\vec{p}_3|}{|\vec{p}_1|} \, d\Omega, \quad (15.11)$$

where $d\Omega \equiv \sin\theta \, d\theta \, d\phi$ is the **differential solid angle** and we again evaluate $\mathcal{M}$ using the specific momenta required by conservation of energy. (I have written the result as $d\sigma$ instead of σ because we still have to do some integrating over the angular ranges for $\vec{p}_3$ that we accept as being part of the desired outcome for this experiment, which might be all possible angles or just the angular range spanned by our outgoing-particle detectors.)

This is a general result for the center-of-mass frame for any scattering process that yields two outgoing particles, and we will use it often in the future. We can't go further to complete the integral over the angular parts of $\vec{p}_3$ because $\mathcal{M}$ generally depends on the angle θ of that outgoing momentum and sometimes even on ϕ, the angle around the beam axis. But it is actually quite remarkable that we can get to the equation above without knowing anything in particular about $\mathcal{M}$.

Let me say a few words about the solid angle concept for those of you who have not seen this before or need a review. Consider a detector that is a distance R from the position where the particle bunches are colliding. The **solid angle** spanned by the detector is defined to be the area A of the detector's active face as a fraction of the area $4\pi R^2$ of the sphere of radius R that contains that face, multiplied by 4π: solid angle = A/R^2. The solid angle corresponding to *all* directions is therefore 4π. I like to think of an observer sitting at the origin looking at the sky and asking what fraction of the whole sky the detector occupies. The detector's solid angle is that fraction multiplied by 4π.

15.4 Phion-Phion Scattering

In the specific case of phion-phion scattering, we can simplify the general result given in Equation 15.11, because $E_2 = E_1$ (since the phions are identical) and because $|\vec{p}_3| = |\vec{p}_1|$ (since the outgoing phions must have the same total energy as the incoming phions to conserve energy). We also have $S = \frac{1}{2}$ because the outgoing phions are identical. So, we have

$$d\sigma = \frac{|\mathcal{M}|^2}{2(16\pi)^2 E_1^2} \sin\theta \, d\theta \, d\phi = \frac{|\mathcal{M}|^2}{2(16\pi)^2 E_1^2} d\Omega, \quad (15.12)$$

where $\mathcal{M}$ is as given in Equation 15.14 below.

We saw earlier that the scattering amplitudes for the three possible second-order processes for phion-phion scattering were

$$\mathcal{M}_n = \frac{\lambda^2}{p^2 - m_\eta^2}, \quad \text{where } p \equiv \begin{cases} p_1 - p_3 & \text{for } \mathcal{M}_t \\ p_1 - p_4 & \text{for } \mathcal{M}_u \\ p_1 + p_2 & \text{for } \mathcal{M}_s \end{cases}. \quad (15.13)$$

In Box 15.4, we will find that the total amplitude $\mathcal{M} = \mathcal{M}_t + \mathcal{M}_u + \mathcal{M}_s$ in the center-of-mass frame is

$$\mathcal{M} = \frac{-\lambda^2}{2|\vec{p}_1|^2(1 - \cos\theta) + m_\eta^2} + \frac{-\lambda^2}{2|\vec{p}_1|^2(1 + \cos\theta) + m_\eta^2}$$

$$+ \frac{\lambda^2}{4E_1^2 - m_\eta^2}. \quad (15.14)$$

The overall angular dependence of $\mathcal{M}$ in Equation 15.14 is quite complicated. But two limits are interesting. If etaon mass $m_\eta \gg E_1 > m_\phi$, then $\mathcal{M}_t \approx \mathcal{M}_u \approx \mathcal{M}_s \approx -\lambda^2/m_\eta^2$, so the differential cross section becomes simply

$$d\sigma = \frac{9\lambda^4}{2(16\pi)^2 E_1^2 m_\eta^4}\, d\Omega \quad \Rightarrow \quad \sigma(\text{all angles}) = \frac{9\lambda^4}{128\pi\, E_1^2 m_\eta^4}. \qquad (15.15)$$

(Since $d\sigma$ is independent of angle here, integrating over $d\Omega$ is the same as multiplying by 4π.) Note that the cross section gets smaller as E_1 increases (because the phions spend less time near each other) and also as m_η increases (because creating a virtual etaon so far off its mass shell is improbable).

The other interesting case is where the etaon has zero mass, making it somewhat like a photon. In this case,

$$\begin{aligned}
\mathcal{M}_t + \mathcal{M}_u &= \frac{-\lambda^2}{2|\vec{p}_1|^2(1-\cos\theta)} + \frac{-\lambda^2}{2|\vec{p}_1|^2(1+\cos\theta)} \\
&= \frac{-\lambda^2[\,(1+\cos\theta)+(1-\cos\theta)\,]}{2|\vec{p}_1|^2(1-\cos^2\theta)} = \frac{-\lambda^2}{|\vec{p}_1|^2\sin^2\theta}.
\end{aligned} \qquad (15.16)$$

If we also take the non-relativistic limit, where $|\vec{p}_1| \ll E_1 \approx m_\phi$, then the s-channel amplitude $\mathcal{M}_s \ll \mathcal{M}_t + \mathcal{M}_u$, and we end up with

$$d\sigma = \frac{\lambda^4}{2(16\pi)^2 m_\phi^2 |\vec{p}_1|^4 \sin^4\theta}\, d\Omega = \frac{1}{2}\left(\frac{\lambda^2/8\pi m_\phi^2}{4K_1 \sin^2\theta}\right)^2 d\Omega, \qquad (15.17)$$

where $K_1 = |\vec{p}_1|^2/2m_\phi$ is the phion's non-relativistic kinetic energy. This has a fairly simple angular dependence.

15.5 Phion-Etaon Scattering

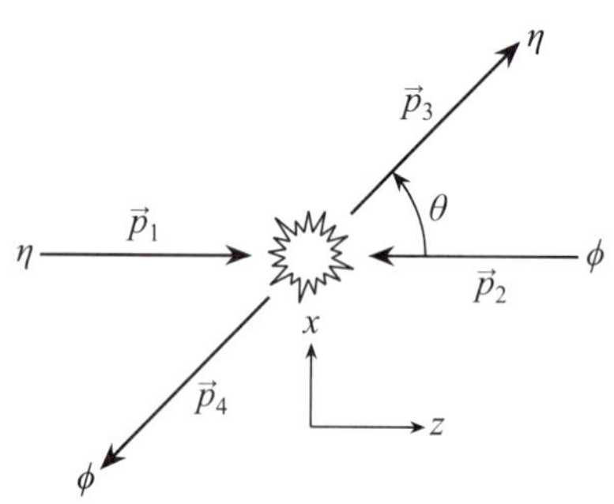

FIGURE 15.2   This diagram shows a scattering process between two particles that yields two particles, as viewed in the center-of-mass frame

Let's consider one more case before we leave our toy universe. Consider the scattering process $\eta + \phi \to \eta + \phi$ in the center-of-mass frame. Let p_1 and p_2 be the four-momenta for the incoming etaon and phion, respectively, and let p_3 and p_4 be the same for the outgoing etaon and phion, respectively. In the center-of-mass frame, the etaon and phion must have opposite spatial momenta $\vec{p}_2 = -\vec{p}_1$, and conservation of four-momentum implies that each particle must depart with the same energy it had originally ($E_1 = E_3$ and $E_2 = E_4$) and with opposite momenta $\vec{p}_4 = -\vec{p}_3$ whose magnitude must be the same as the initial etaon's momentum magnitude: $|\vec{p}_3| = |\vec{p}_1|$.

Let us define the deflection angle θ to be the angle of the departing etaon's momentum relative to that of the incoming etaon (that is, the angle between $\vec{p}_3$ and $\vec{p}_1$: see Figure 15.2). To make things easier, let us also choose coordinate axes so that the incoming etaon is moving in the $+z$ direction and the outgoing particles' momenta are in the xz plane. Then

$$p_1 = \begin{bmatrix} E_1 \\ 0 \\ 0 \\ \rho \end{bmatrix}, \quad p_2 = \begin{bmatrix} E_2 \\ 0 \\ 0 \\ -\rho \end{bmatrix}, \quad p_3 = \begin{bmatrix} E_1 \\ \rho\sin\theta \\ 0 \\ \rho\cos\theta \end{bmatrix}, \quad p_4 = \begin{bmatrix} E_2 \\ -\rho\sin\theta \\ 0 \\ -\rho\cos\theta \end{bmatrix}, \qquad (15.18)$$

where ρ is a shorthand for $|\vec{p}_1|$, $E_1 \equiv (\rho^2 + m_\eta^2)^{1/2}$, and $E_2 \equiv (\rho^2 + m_\phi^2)^{1/2}$.

In this case, because $|\vec{p}_1| = |\vec{p}_3|$, and because outgoing particles are different (so $S = 1$), Equation 15.11 becomes simply

$$\frac{d\sigma}{d\Omega} = \frac{S|\mathcal{M}|^2}{(8\pi)^2(E_1+E_2)^2}\frac{|\vec{p}_3|}{|\vec{p}_1|} = \frac{|\mathcal{M}|^2}{(8\pi)^2(E_1+E_2)^2}. \qquad (15.19)$$

Now, we only need to calculate the scattering amplitude $\mathcal{M}$.

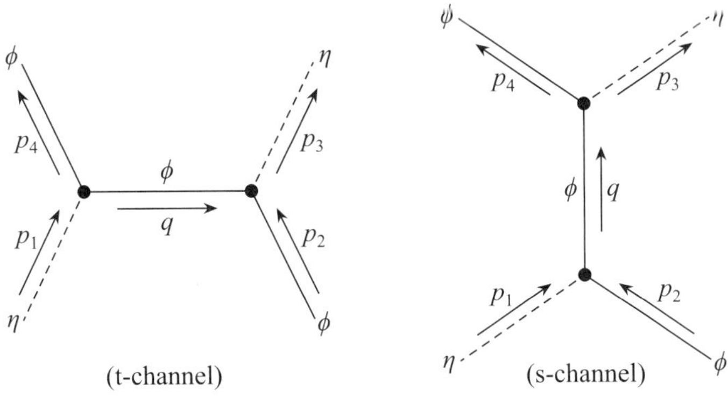

FIGURE 15.3   These are the labeled Feynman diagrams for the t-channel and s-channel subprocesses for $\phi + \eta \to \phi + \eta$ scattering.

Figure 15.3 shows the two possible second-order Feynman diagrams for this situation, using their conventional names. We do not here have a u-channel diagram as we did in the phion-scattering case because the two outgoing particles are not the same. Let's apply the Feynman rules for our toy theory to calculate the scattering amplitude for the first diagram. We have drawn and labeled the Feynman diagrams, which are the first two steps. The remaining steps are:

3. **Vertex Factors.** Each vertex contributes a factor of $-i\lambda$.

4. **Propagator.** Each internal line contributes $i/(q_j^2 - m_j^2)$. Here we have only one internal line, which is a phion with mass m_ϕ. Putting together all the items we have so far in left-to-right order yields

$$-i\mathcal{M}_t = (-i\lambda)\frac{i}{q^2 - m_\phi^2}(-i\lambda). \tag{15.20}$$

5. **Conserve four-momentum at each vertex.** Conservation of four-momentum at the left vertex yields $q = p_1 - p_4$.

6. **The Result is** $-i\mathcal{M}$. Multiplying by i yields

$$\mathcal{M}_t = \frac{\lambda^2}{(p_1 - p_4)^2 - m_\phi^2}. \tag{15.21}$$

Following the same steps for the s-channel subprocess yields

$$\mathcal{M}_s = \frac{\lambda^2}{(p_1 + p_2)^2 - m_\phi^2}. \tag{15.22}$$

We can use Equation 15.18 to calculate these factors as a function of θ and $\rho \equiv |\vec{p}_1|$ as follows:

$$(p_1 - p_4) \cdot (p_1 - p_4) - m_\phi^2 = p_1^2 + p_4^2 - 2p_1 \cdot p_4 - m_\phi^2$$

$$= m_\phi^2 + m_\eta^2 - 2E_1 E_2 + 2p_1^z p_4^z - m_\phi^2$$

$$= m_\eta^2 - 2E_1 E_2 - 2\rho^2 \cos\theta, \tag{15.23}$$

$$(p_1 + p_2) \cdot (p_1 + p_2) - m_\phi^2 = p_1^2 + p_2^2 + 2p_1 \cdot p_2 - m_\phi^2$$

$$= m_\phi^2 + m_\eta^2 + 2E_1 E_2 - 2p_1^z p_2^z - m_\phi^2$$

$$= m_\eta^2 + 2E_1 E_2 + 2\rho^2. \tag{15.24}$$

Now, $\mathcal{M} = \mathcal{M}_t + \mathcal{M}_s$ is generally pretty complicated. But an interesting subcase is where the etaon mass m_η is zero (implying that $E_1 = \rho$) and when $\rho \ll m_\phi$, implying that $E_2 = (\rho^2 + m_\phi^2)^{1/2} \approx m_\phi$. This situation is roughly analogous to Compton scattering (scattering of a photon off an electron) in our universe, with the etaon playing the role of a photon. In this case,

$$\mathcal{M} = -\frac{\lambda^2}{2m_\phi E_1 + 2E_1^2 \cos\theta} + \frac{\lambda^2}{2m_\phi E_1 + 2E_1^2}$$

$$= \frac{\lambda^2}{2E_1} \left(\frac{-m_\phi - E_1 + m_\phi + E_1 \cos\theta}{(m_\phi + E_1 \cos\theta)(m_\phi + E_1)} \right) \approx \frac{-\lambda^2(1 - \cos\theta)}{2m_\phi^2}. \tag{15.25}$$

Substituting this into Equation 15.19 yields

$$d\sigma = \frac{1}{(8\pi)^2(E_1 + E_2)^2} \left[\frac{-\lambda^2(1 - \cos\theta)}{2m_\phi^2} \right]^2 d\Omega$$

$$= \frac{\lambda^4(1 - 2\cos\theta + \cos^2\theta)}{(16\pi)^2 m_\phi^6} d\Omega \tag{15.26}$$

since $E_1 + E_2 \approx m_\phi$ in this limit. The differential cross $d\sigma$ for actual Compton scattering is proportional to just $1 + \cos^2\theta$, so our toy theory is clearly missing some important aspects of reality. (What is missing turns out to be particle spin.) We will start working on describing our real universe in the next chapter.

BOX 15.1 Integrating over the Dirac Delta

While calculating the general decay-rate formula for a particle of mass M at rest decaying to two particles of mass m, we arrived at Equation 15.4, which I have repeated here for easy reference:

$$\Gamma = \frac{S}{32\pi^2 M} \int 4\pi q^2 \, dq \, \frac{\delta(M - 2\sqrt{q^2 + m^2})}{m^2 + q^2} \, |\mathcal{M}|^2 \, . \qquad (15.27)$$

In this box, we will use properties of the Dirac delta to show that the integral over the remaining Dirac delta yields the final result

$$\Gamma = \frac{S|\vec{p}_F|}{8\pi M^2} \, |\mathcal{M}|^2 \, , \qquad (15.28)$$

where $|\vec{p}_F|$ is the particular value of $q = |\vec{p}_2|$ selected by the Dirac delta and $\mathcal{M}$ is evaluated at that momentum magnitude.

Remember from Chapter 4 that

$$\delta[\, g(x) \,] = \sum_{\text{all} x_i} \frac{\delta(x - x_i)}{|dg/dx|_{x_i}} \, , \qquad (15.29)$$

where the x_i are all values where $g(x_i) = 0$. In the particular case at hand, we have $g(q) = M - 2\sqrt{q^2 + m^2}$, which is zero when

$$M^2 = 4q_0^2 + 4m^2 \;\Rightarrow\; 2q_0 = \pm\sqrt{M^2 - 4m^2} \;\Rightarrow\; q_0 = \pm\tfrac{1}{2}\sqrt{M^2 - 4m^2}. \quad (15.30)$$

However, in our integral, $q \geq 0$ by definition, so only positive root applies.

Exercise 15.1.1 Use Equation 15.29 to transform $\delta[\, g(q) \,]$ into something times $\delta(q - q_0)$, where $q_0 \equiv \tfrac{1}{2}\sqrt{M^2 - 4m^2}$. Then do the requested integral, expressing the result in terms of $|\vec{p}_F| = q_0$.

BOX 15.2 Calculating the Scattering Factor

Consider a situation where we have scattering of two incoming particles. In this box, you will show that generally in the center-of-mass frame, where $\vec{p}_2 = -\vec{p}_1$, we have

$$\sqrt{(p_1 \cdot p_2)^2 - (m_1 m_2)^2} = (E_1 + E_2)|\,\vec{p}_1\,|. \tag{15.31}$$

We can use our freedom to choose coordinates to orient our axes so that particle 1 is moving in the $+z$ direction. Let's also save writing by defining $\rho \equiv |\,\vec{p}_1\,|$.

Exercise 15.2.1 Write out the four-momentum components for each particle in the center-of-mass frame, then evaluate the square root above. (*Hint:* Note that $E_1^2 = m_1^2 + \rho^2$ and $E_2^2 = m_2^2 + \rho^2$, because $\rho^2 = |\,\vec{p}_1\,|^2 = |\,\vec{p}_2\,|^2$.)

BOX 15.3 Calculating the Scattering Integral

Consider a two-particle to two-particle scattering situation in the center-of-mass frame. In Equation 15.10 (repeated here), we saw that

$$\sigma = \frac{S}{4(E_1 + E_2)|\,\vec{p}_1\,|} \int \frac{d^3 p_3}{(2\pi)^3} \frac{d^3 p_4}{(2\pi)^3} \frac{(2\pi)^4 \delta^4(p_1 + p_2 - p_3 - p_4)|\mathcal{M}|^2}{4 E_3 E_4}. \quad (15.32)$$

Our goal in this box is to do the integral over $d^3 p_4$ and the magnitude part of the integral over $d^3 p_3$. The first step is to note that

$$\delta^4(p_1 + p_2 - p_3 - p_4) = \delta(E_1 + E_2 - E_3 - E_4)\, \delta^3(\vec{p}_1 + \vec{p}_2 - \vec{p}_3 - \vec{p}_4). \quad (15.33)$$

If we substitute this in and do the integral over $d^3 p_4$, the result is simply to set $\vec{p}_4 = \vec{p}_1 + \vec{p}_2 - \vec{p}_3$ in all other parts of the integral (including in the scattering amplitude $|\mathcal{M}|^2$, which might depend on any or all of the momenta). Since we are in the center-of-mass frame where $\vec{p}_1 + \vec{p}_2 = 0$, this basically sets $\vec{p}_4 = -\vec{p}_3$. What we have left (after moving around some numerical factors) is

$$\sigma = \frac{S}{64\pi^2(E_1 + E_2)|\,\vec{p}_1\,|} \int d^3 p_3 \frac{\delta(E_1 + E_2 - E_3 - E_4)|\mathcal{M}|^2}{E_3 E_4}. \quad (15.34)$$

Now, E_3 and E_4 only depend on the *magnitude* of $\vec{p}_3$. To save writing, let's define $\rho \equiv |\,\vec{p}_3\,|$ so that $d^3 p = d\rho\, \rho \sin\theta\, d\theta\, \rho\, d\varphi \equiv \rho^2\, d\rho\, d\Omega$. To go on, we *could* express the energies in terms of ρ and use the identity we used in Box 15.1 to express the Dirac delta as something times $\delta(\rho - \rho_0)$, where ρ_0 is the value of ρ where $E_3 + E_4 = E_1 + E_2$. However, it turns out to be more effective in this particular case to change variables from ρ to

$$E \equiv E_3 + E_4 = \sqrt{m_3^2 + \rho^2} + \sqrt{m_4^2 + \rho^2}. \quad (15.35)$$

Exercise 15.3.1 Show that

$$dE = \frac{E\rho}{E_3 E_4}\, d\rho. \quad (15.36)$$

BOX 15.3 Calculating the Scattering Integral (cont.)

If we substitute $(\rho^2/E_3E_4)\,d\rho = \rho\,dE/E$ into Equation 15.34, we get

$$\sigma = \frac{S}{64\pi^2(E_1 + E_2)|\,\vec{p}_1\,|} \int d\Omega\,dE\,\frac{\rho}{E}\,\delta(E_1 + E_2 - E)\,|\mathcal{M}|^2, \quad (15.37)$$

where we are now considering ρ as a function of E (the inverse function of Equation 15.35). We can now do the integral over dE: integrating over the Dirac delta sets $E = E_1 + E_2$, sets the value of $\rho = |\,\vec{p}_3\,|$ to whatever value is consistent with $E = E_1 + E_2$ (conservation of energy), and sets the value of $|\mathcal{M}|^2$ to whatever value results when we set $\vec{p}_4 = -\vec{p}_3$ and $\vec{p}_3$ to the value that conserves energy. Indeed, this is exactly what we have done with $\mathcal{M}$ in the body of this chapter: we have *assumed* that four-momentum is conserved before we evaluate $\mathcal{M}$. (It is also good to note that if four-momentum *cannot* be conserved—for example, if the final particle masses are greater than $E_1 + E_2$—then the cross section is zero no matter what $\mathcal{M}$ might be.)

Note, however, that we cannot integrate with respect to the angular variables θ and φ (or equivalently with respect to solid angle $d\Omega$) without knowing exactly how $|\mathcal{M}|^2$ depends on angular variables: integrating when we *don't* have a Dirac delta requires knowing the actual function. So, the farthest we can go without knowing more is to evaluate the differential cross section

$$d\sigma = \frac{S}{64\pi^2(E_1 + E_2)|\,\vec{p}_1\,|}\left[\int dE\,\frac{\rho}{E}\,\delta(E_1 + E_2 - E)\,|\mathcal{M}|^2\right]d\Omega. \quad (15.38)$$

Exercise 15.3.2 Evaluate the integral given above and compare the result to Equation 15.11. (*Hint:* Because of the Dirac delta, this is not hard.)

BOX 15.4   The Total Phion-Phion Scattering Amplitude

The scattering amplitudes for the three possible second-order processes for phion-phion scattering were given in Equation 15.13, repeated here:

$$\mathcal{M}_t = \frac{\lambda^2}{(p_1 - p_3)^2 - m_\eta^2}, \qquad \mathcal{M}_u = \frac{\lambda^2}{(p_1 - p_4)^2 - m_\eta^2}, \quad \text{and}$$

$$\mathcal{M}_s = \frac{\lambda^2}{(p_1 + p_2)^2 - m_\eta^2}. \tag{15.39}$$

The goal of this box is to show that, in the center-of-mass frame, the total second-order scattering amplitude $\mathcal{M} = \mathcal{M}_t + \mathcal{M}_u + \mathcal{M}_s$ is

$$\mathcal{M} = \frac{-\lambda^2}{2|\vec{p}_1|^2(1 - \cos\theta) + m_\eta^2} + \frac{-\lambda^2}{2|\vec{p}_1|^2(1 + \cos\theta) + m_\eta^2} + \frac{\lambda^2}{4E_1^2 - m_\eta^2}. \tag{15.40}$$

Now, in the center-of-mass frame, we have $\vec{p}_2 = -\vec{p}_1$ and $\vec{p}_3 = -\vec{p}_4$. Moreover, because the incoming particles are the same as the outgoing particles, we have $|\vec{p}_1| = |\vec{p}_2| = |\vec{p}_3| = |\vec{p}_4|$. Also, we have defined θ to be the angle of $\vec{p}_3$ relative to $\vec{p}_1$. If we choose coordinate axes so that $\vec{p}_1$ is in the $+z$ direction and everything lies in the xz plane (as is conventional in particle physics), then the four-momenta are

$$p_1 = \begin{bmatrix} E_1 \\ 0 \\ 0 \\ |\vec{p}_1| \end{bmatrix}, \ p_2 = \begin{bmatrix} E_1 \\ 0 \\ 0 \\ -|\vec{p}_1| \end{bmatrix}, \ p_3 = \begin{bmatrix} E_1 \\ |\vec{p}_1|\sin\theta \\ 0 \\ |\vec{p}_1|\cos\theta \end{bmatrix}, \ p_4 = \begin{bmatrix} E_1 \\ -|\vec{p}_1|\sin\theta \\ 0 \\ -|\vec{p}_1|\cos\theta \end{bmatrix}.$$

$$\tag{15.41}$$

Exercise 15.4.1 Evaluate the scattering amplitudes in Equation 15.39 and show that the result is Equation 15.40. (*Hint:* $(p \pm q)^2 = p^2 \pm 2p \cdot q + q^2$.)

Homework Problems

P15.1 In our toy universe, consider the etaon decay process $\eta \rightarrow \phi + \phi$. Suppose that the η has a mass $m_\eta = 0.1$ GeV (about the mass of a muon), the ϕ has a mass $m_\phi = \frac{2}{5}M$, and λ is whatever power of (0.001 GeV) yields its correct units. Calculate the lifetime of the η in seconds.

P15.2 In this problem, we will consider the units of the amplitude $\mathcal{M}$.

a. Argue from the definition of the Dirac delta that $\delta^4(p_1 + \cdots)$ must have units of (energy)$^{-4}$.

b. Argue then from the units of Fermi's Golden Rule for decays (Equation 14.16) that $\mathcal{M}$ must have units of (energy)$^{4-N}$, where N is the total number of particles involved (incoming + outgoing).

c. Does the same rule apply for scattering processes? (If not, what is the analogous rule for scattering?)

d. Demonstrate that this rule (or your replacement rule) gives the right units in Equation 15.5 (the general formula for one particle decaying to two).

e. Demonstrate that this rule gives the right units in Equation 15.11 (the general formula for two particles scattering to two particles).

P15.3 Verify that $\int d\Omega = \int \sin\theta \, d\theta \, d\phi = 4\pi$ when we integrate over all angles for a sphere. (*Hint:* Think carefully about limits. What are the limits of the θ integral in particular? Drawing a picture might help.)

P15.4 The Moon is roughly 0.5° in diameter when viewed from the Earth. What solid angle does it span?

P15.5 Verify Equation 15.22 for the s-channel diagram for η-ϕ scattering.

P15.6 In our toy universe, the phion and etaon are both real scalars, so they are their own antiparticles. Suppose that the phions are massive but the etaon is massless ($m_\eta = 0$). Consider the process $\phi + \phi \rightarrow \eta + \eta$, where a phion annihilates another phion and produces two massless etaons as a result. (This will be analogous in our universe to a particle-antiparticle annihilation that produces two photons.)

a. Draw all possible second-order Feynman diagrams for this process and label each as describing the t-channel, u-channel, s-channel, or some other (as yet unnamed) channel.

b. Calculate the amplitude $\mathcal{M}$ for each diagram.

c. Assume that we are in the center-of-mass frame, that the phions approach each other along the z axis, and that the departing etaon with four-momentum p_3 lies in the xz plane at an angle of θ with respect to the $+z$ direction. Calculate the differential cross section $d\sigma$ in terms of λ, the phion mass m_ϕ, and the angle θ.

d. Consider the non-relativistic limit $|\vec{p}_1| \approx m_\phi v \ll m_\phi$. Calculate the total cross section in this limit.

e. If $\lambda/m_\phi = 1/1000$, $m_\phi = 0.1$ GeV, and $v = 1/1000$, calculate the total cross section σ ($d\sigma$ integrated over all angles) for this reaction in fm^2 and barns.

P15.7 For η-ϕ scattering, consider the case where $m_\eta \gg m_\phi$ and $m_\eta \gg \rho$.

a. Calculate the differential cross section $d\sigma$ and the total cross section σ integrated over all angles.

b. If $\lambda = m_\eta = 1$ GeV, calculate σ in fm^2 and barns.

$$\boxed{i\gamma^\mu \partial_\mu \psi - m\psi = 0}$$

$$\gamma^t = \begin{pmatrix} 0 & 0 & 1 & 0 \\ 0 & 0 & 0 & 1 \\ 1 & 0 & 0 & 0 \\ 0 & 1 & 0 & 0 \end{pmatrix} \qquad \gamma^x = \begin{pmatrix} 0 & 0 & 0 & 1 \\ 0 & 0 & 1 & 0 \\ 0 & -1 & 0 & 0 \\ -1 & 0 & 0 & 0 \end{pmatrix}$$

$$\gamma^y = \begin{pmatrix} 0 & 0 & 0 & -i \\ 0 & 0 & i & 0 \\ 0 & i & 0 & 0 \\ -i & 0 & 0 & 0 \end{pmatrix} \qquad \gamma^z = \begin{pmatrix} 0 & 0 & 1 & 1 \\ 0 & 0 & 0 & -1 \\ -1 & 0 & 0 & 0 \\ 0 & 1 & 0 & 0 \end{pmatrix}$$

Solutions when $\vec{p}$ is in $\pm z$ direction:

Particle:
$$\psi = e^{-ip\cdot x} \begin{bmatrix} +\sqrt{E_{\vec{p}} - p_z}\; a \\ +\sqrt{E_{\vec{p}} + p_z}\; b \\ +\sqrt{E_{\vec{p}} + p_z}\; a \\ +\sqrt{E_{\vec{p}} - p_z}\; b \end{bmatrix},$$

Antiparticle:
$$\psi = e^{+ip\cdot x} \begin{bmatrix} +\sqrt{E_{\vec{p}} - p_z}\; a \\ +\sqrt{E_{\vec{p}} + p_z}\; b \\ -\sqrt{E_{\vec{p}} + p_z}\; a \\ -\sqrt{E_{\vec{p}} - p_z}\; b \end{bmatrix},$$

normalized so that $\overline{\psi}\psi = 2m$ if $|a|^2 + |b|^2 = 1$.

16. THE DIRAC EQUATION

In the past few chapters, we have developed a method of calculating decay rates and cross sections in a toy universe. To connect these techniques with the real universe, however, we need to face the fact that (with the lone exception of the Higgs boson) scalar fields of the type we have been considering do not accurately describe *any* of the elementary particles in our universe. The fields representing all other elementary particles have *spin*, something that (as we will see) scalar fields lack.

The machinery for handling and interpreting fields with spin will take three full chapters to develop. (This is why we began by considering scalar fields!) This chapter will present an alternative to the Klein-Gordon equation for describing relativistic fields. The next chapter will show that the classical fields that are solutions to this new equation indeed have spin (and scalar fields don't). In Chapter 18, we will explore how to quantize these new fields.

16.1 Pauli Matrices

Before we begin, we must briefly review the angular momentum matrices associated with electron spin that we developed in Chapters 8 and 9. In units where $\hbar = 1$, these matrices are

$$S_x = \frac{1}{2}\begin{pmatrix} 0 & 1 \\ 1 & 0 \end{pmatrix}, \qquad S_y = \frac{1}{2}\begin{pmatrix} 0 & -i \\ i & 0 \end{pmatrix}, \qquad S_z = \frac{1}{2}\begin{pmatrix} 1 & 0 \\ 0 & -1 \end{pmatrix}. \tag{16.1}$$

These three matrices satisfy the standard angular momentum commutation relations $[S_x, S_y] = i S_z$, $[S_y, S_z] = i S_x$, and $[S_z, S_x] = i S_y$.

These matrices are useful in a wider range of contexts than you might imagine. In these other contexts, it is conventional to get rid of the factor of $\frac{1}{2}$:

$$\sigma_x = \begin{pmatrix} 0 & 1 \\ 1 & 0 \end{pmatrix}, \qquad \sigma_y = \begin{pmatrix} 0 & -i \\ i & 0 \end{pmatrix}, \qquad \sigma_z = \begin{pmatrix} 1 & 0 \\ 0 & -1 \end{pmatrix}. \tag{16.2}$$

Memorize these **Pauli matrices** before going on. From the commutation relations above, we see that $[\sigma_x, \sigma_y] = 2i\sigma_z$, $[\sigma_z, \sigma_x] = 2i\sigma_y$, and $[\sigma_y, \sigma_z] = 2i\sigma_x$, or, more compactly,

$$[\sigma_j, \sigma_k] = 2i \sum_m e_{jkm}\sigma_m, \tag{16.3}$$

where e_{jkm} is the **Levi-Civita symbol**, whose value is $+1$ if its indices $jkm = xyz, zxy,$ yzx (cyclically alphabetic) or -1 if $jkm = yxz, zyx,$ or yxz (anti-alphabetic), and 0 otherwise (any two indices are the same).

16.2 The Dirac Equation

Remember that the Klein-Gordon equation has the form

$$0 = \partial^\mu \partial_\mu \phi + m^2 \phi \equiv (\partial^2 + m^2)\phi, \tag{16.4}$$

where I am using ∂^2 as a shorthand for $\partial^\mu \partial_\mu$. In 1928, Paul Dirac was looking for an alternative to this equation with a single time derivative instead of the double time derivative that appears in the Klein-Gordon equation, as this would solve certain mathematical issues that concerned him. Specifically, he pondered whether it might be possible to factor

257

the operator $(\partial^2 + m^2)$ as follows:

$$(\partial^2 + m^2) = (\sqrt{\partial^2} + im)(\sqrt{\partial^2} - im). \tag{16.5}$$

But how can we possibly define the square root of the ∂^2 operator? Dirac's clever idea was to try to find a set of coefficients γ^μ such that

$$\partial^2 \equiv \frac{\partial^2}{\partial t^2} - \frac{\partial^2}{\partial x^2} - \frac{\partial^2}{\partial y^2} - \frac{\partial^2}{\partial z^2} = \left(\gamma^t \frac{\partial}{\partial t} + \gamma^x \frac{\partial}{\partial x} + \gamma^y \frac{\partial}{\partial y} + \gamma^z \frac{\partial}{\partial z} \right)^2. \tag{16.6}$$

The quantity in parentheses could then be taken to define $\sqrt{\partial^2}$.

In order to satisfy the above relation, we need the coefficients γ^μ to satisfy the relations

$$(\gamma^t)^2 = 1, \quad (\gamma^x)^2 = (\gamma^y)^2 = (\gamma^z)^2 = -1. \tag{16.7}$$

But to eliminate the cross terms, we also need, for all $\mu \neq \nu$,

$$\{\gamma^\mu, \gamma^\nu\} \equiv \gamma^\mu \gamma^\nu + \gamma^\nu \gamma^\mu = 0, \tag{16.8}$$

where the symbol $\{a, b\} \equiv ab + ba$ is called the **anticommutator** of a and b. We can combine these two requirements into a single relation

$$\{\gamma^\mu, \gamma^\nu\} = 2g^{\mu\nu} \tag{16.9}$$

that is valid for all choices of values for μ and ν.

The condition in Equation 16.8 in particular means that our "coefficients" cannot possibly be ordinary complex numbers. Indeed, Dirac found that the simplest possible objects that could satisfy the above requirements were 4×4 matrices. There are various possible sets of such matrices that will do the trick, but in this book we will use the **chiral representation** (also known as the **Weyl basis**) for these matrices, where

$$\gamma^t = \begin{pmatrix} 0 & 0 & 1 & 0 \\ 0 & 0 & 0 & 1 \\ 1 & 0 & 0 & 0 \\ 0 & 1 & 0 & 0 \end{pmatrix}, \qquad \gamma^x = \begin{pmatrix} 0 & 0 & 0 & 1 \\ 0 & 0 & 1 & 0 \\ 0 & -1 & 0 & 0 \\ -1 & 0 & 0 & 0 \end{pmatrix},$$

$$\gamma^y = \begin{pmatrix} 0 & 0 & 0 & -i \\ 0 & 0 & i & 0 \\ 0 & i & 0 & 0 \\ -i & 0 & 0 & 0 \end{pmatrix}, \qquad \gamma^z = \begin{pmatrix} 0 & 0 & 1 & 0 \\ 0 & 0 & 0 & -1 \\ -1 & 0 & 0 & 0 \\ 0 & 1 & 0 & 0 \end{pmatrix}. \tag{16.10}$$

Writing out 4×4 matrices gets old quickly, but we can leverage our memorization of the Pauli matrices (which you have done by now, right?) to write each of these 4×4 matrices as a 2×2 matrix of 2×2 Pauli matrices:

$$\gamma^t = \begin{pmatrix} 0 & I \\ I & 0 \end{pmatrix}, \qquad \gamma^j = \begin{pmatrix} 0 & \sigma_j \\ -\sigma_j & 0 \end{pmatrix}, \tag{16.11}$$

where $j = x, y, z$. Indeed, one can save even more writing by defining $\sigma^\mu \equiv (I, \sigma_x, \sigma_y, \sigma_z)$ and $\bar{\sigma}^\mu \equiv (I, -\sigma_x, -\sigma_y, -\sigma_z)$; then we simply have

$$\gamma^\mu = \begin{pmatrix} 0 & \sigma^\mu \\ \bar{\sigma}^\mu & 0 \end{pmatrix}. \tag{16.12}$$

However, if you ever find this super-compact notation confusing, you can always step back and write out these matrices in full. (I should note that most authors use numerical indices $\gamma^0, \gamma^1, \gamma^2, \gamma^3$ instead of $\gamma^t, \gamma^x, \gamma^y, \gamma^z$, but I thought that this would be an unnecessary layer of abstraction.)

The factor with the plus sign in Equation 16.5 becomes the **Dirac equation**

$$(\gamma^\mu \partial_\mu + im I)\psi = 0 \quad \text{or} \quad i\gamma^\mu \partial_\mu \psi - m\psi = 0, \tag{16.13}$$

where $\psi(x)$ is our field. (The latter is the more usual form, with the i associated with the derivative and the I understood.)

16.3 The Dirac Spinor

In order to make this equation work, however, $\psi(x)$ *cannot* be a simple scalar field. It must instead be some kind of four-component field

$$\psi(x) = \begin{bmatrix} \psi_1(x) \\ \psi_2(x) \\ \psi_3(x) \\ \psi_4(x) \end{bmatrix} \tag{16.14}$$

(because then the Dirac equation sets one four-component object $i\gamma^\mu \partial_\mu \psi$ equal to another four-component object $m\psi$). Indeed, people conventionally and deliberately use $\psi(x)$ for solutions of the Dirac equation and $\phi(x)$ for solutions to the Klein-Gordon equation to remind everyone that $\psi(x)$ and $\phi(x)$ are completely different kinds of objects. It is also important to know that $\psi(x)$ is not a four-vector despite having four components: its components transform under Lorentz transformations in an entirely different way (as we will see in Chapter 17).

In our compact notation, this four-component object (which we call a **Dirac spinor**) becomes a column vector of two two-component **Weyl spinors**:

$$\psi(x) = \begin{bmatrix} \psi_L \\ \psi_R \end{bmatrix} \quad \text{where } \psi_L = \begin{bmatrix} \psi_{L1} \\ \psi_{L2} \end{bmatrix} \text{ and } \psi_R = \begin{bmatrix} \psi_{R1} \\ \psi_{R2} \end{bmatrix}. \tag{16.15}$$

(Weyl is pronounced "vial." The meaning of the subscripts will become clearer later.) The Dirac equation applied to these quantities tells us that

$$\left[i \begin{pmatrix} 0 & \partial_t + \vec{\sigma} \cdot \vec{\nabla} \\ \partial_t - \vec{\sigma} \cdot \vec{\nabla} & 0 \end{pmatrix} - \begin{pmatrix} m & 0 \\ 0 & m \end{pmatrix} \right] \begin{bmatrix} \psi_L \\ \psi_R \end{bmatrix} = 0 \tag{16.16}$$

(where, again, each ∂_t is understood to be multiplied by a 2×2 I). By multiplying out the matrices, we can rewrite this as

$$i(\partial_t + \vec{\sigma} \cdot \vec{\nabla})\psi_R = m\psi_L, \tag{16.17a}$$

$$i(\partial_t - \vec{\sigma} \cdot \vec{\nabla})\psi_L = m\psi_R. \tag{16.17b}$$

Note how these equations couple the ψ_R and ψ_L portions of the full spinor. But if $m = 0$, the equations uncouple:

$$i(\partial_t + \vec{\sigma} \cdot \vec{\nabla})\psi_R = 0, \tag{16.18a}$$

$$i(\partial_t - \vec{\sigma} \cdot \vec{\nabla})\psi_L = 0. \tag{16.18b}$$

The Dirac equation basically says that massless Dirac particles come in two completely independent types: **left-handed particles** (described by a two-component field ψ_L), and **right-handed particles** (described by a two-component field ψ_R). One of the advantages of the chiral representation is that ψ_L and ψ_R are restricted to the upper two slots and the lower two slots of the Dirac spinor, respectively. Since the equations are uncoupled, a purely left-handed massless particle can never evolve to being a right-handed particle, or vice versa. Now we see the origin of the L and R subscripts, but why these types of particles are called "left-handed" and "right-handed" will be clarified in Chapter 17.

Massive particles do *not* respect this division between left- and right-handed spinors. A right-handed particle can evolve into a left-handed particle (and vice versa) as time passes.

16.4 Rest Solutions to the Dirac Equation

We first consider the case of a particle at rest, where the only nonzero component of the particle's four-momentum is the time component $p^t = E$. We look for solutions of the Dirac equation of the form

$$\psi = \begin{bmatrix} a \\ b \\ c \\ d \end{bmatrix} e^{-iEt}, \tag{16.19}$$

where a, b, c, d are constants. The Dirac equation for this trial solution becomes $i\gamma^t \partial_t \psi = m\psi$, which becomes the matrix equation

$$\begin{pmatrix} 0 & 0 & i\partial_t & 0 \\ 0 & 0 & 0 & i\partial_t \\ i\partial_t & 0 & 0 & 0 \\ 0 & i\partial_t & 0 & 0 \end{pmatrix} \begin{bmatrix} a \\ b \\ c \\ d \end{bmatrix} e^{-iEt} = m \begin{bmatrix} a \\ b \\ c \\ d \end{bmatrix} e^{-iEt}$$

$$\Rightarrow \begin{pmatrix} -m & 0 & E & 0 \\ 0 & -m & 0 & E \\ E & 0 & -m & 0 \\ 0 & E & 0 & -m \end{pmatrix} \begin{bmatrix} a \\ b \\ c \\ d \end{bmatrix} = 0. \tag{16.20}$$

This is a standard eigenvector/eigenvalue problem whose eigenvalues are $E = \pm m$, with eigenvectors $c = a$, $b = d$ for $E = +m$ and $c = -a$, $d = -b$ for $E = -m$ (see Box 16.1).

As we discussed in detail in Chapter 12, the solutions that seem to have negative energy in this context actually correspond to positive-energy *antiparticles*. In our compact notation, we write these rest solutions as

$$\text{Particle}: \quad u_0 e^{-imt} \quad \text{where } u_0 = \begin{bmatrix} (u_0)_L \\ (u_0)_R \end{bmatrix} = \sqrt{m} \begin{bmatrix} \xi \\ \xi \end{bmatrix}, \tag{16.21a}$$

$$\text{Antiparticle}: \quad v_0 e^{+imt} \quad \text{where } v_0 = \begin{bmatrix} (v_0)_L \\ (v_0)_R \end{bmatrix} = \sqrt{m} \begin{bmatrix} \eta \\ -\eta \end{bmatrix}, \tag{16.21b}$$

where u_0 is the four-component Dirac spinor amplitude; $(u_0)_L$ and $(u_0)_R$ are the two-component Weyl spinor amplitudes; ξ is an arbitrary two-component constant vector for the particle rest solution; and v_0, $(v_0)_L$, $(v_0)_R$, and η are the corresponding objects for the antiparticle rest solution. The $\sqrt{m}$ is a normalizing factor (we will see why this makes sense later). With this overall normalizing factor, we can normalize the Weyl spinors ξ and η in the usual way:

$$\xi^\dagger \xi = 1 = \eta^\dagger \eta. \tag{16.22}$$

With these normalizations, we have

$$u_0^\dagger u_0 = m \begin{bmatrix} \xi^\dagger & \xi^\dagger \end{bmatrix} \begin{bmatrix} \xi \\ \xi \end{bmatrix} = m(\xi^\dagger \xi + \xi^\dagger \xi) = 2m, \tag{16.23a}$$

$$v_0^\dagger v_0 = m \begin{bmatrix} \eta^\dagger & -\eta^\dagger \end{bmatrix} \begin{bmatrix} \eta \\ -\eta \end{bmatrix} = m(\eta^\dagger \eta + \eta^\dagger \eta) = 2m. \tag{16.23b}$$

We will see that these factors become a bit different for moving particles.

16.5 Solutions for Particles in Motion

The fastest way to generate the solutions when $\vec{p} \neq 0$ is to take the rest solutions and perform a Lorentz transformation. We will do this in the next chapter, but let me give you the results here:

$$\text{Particle}: \quad \psi = u(p)e^{-ip\cdot x} \quad \text{where } u(p) = \begin{bmatrix} \sqrt{p\cdot\sigma}\,\xi \\ \sqrt{p\cdot\overline{\sigma}}\,\xi \end{bmatrix}, \tag{16.24a}$$

$$\text{Antiparticle}: \quad \psi = v(p)e^{+ip\cdot x} \quad \text{where } v(p) = \begin{bmatrix} \sqrt{p\cdot\sigma}\,\eta \\ -\sqrt{p\cdot\overline{\sigma}}\,\eta \end{bmatrix}. \tag{16.24b}$$

Note that we generally use $u(p)$ and $v(p)$ to describe the "amplitude" part of the solution (the dependence on space and time is entirely in the $e^{\pm ip\cdot x}$ part), and these spinor amplitudes generally depend on the particle or antiparticle's four-momentum p. In the expressions $\sqrt{p\cdot\sigma}$ and $\sqrt{p\cdot\overline{\sigma}}$, we are again using the convention where $\sigma^{\mu} \equiv (I, \sigma_x, \sigma_y, \sigma_z)$ and $\overline{\sigma}^{\mu} \equiv (I, -\sigma_x, -\sigma_y, -\sigma_z)$. The notation $\sqrt{p\cdot\sigma}$ looks simple at first, but at second glance it looks crazy because the σ^{μ} are matrices. But how can we take the square root of a matrix? Box 16.2 fully discusses the notation, but the only case we will actually *use* is when the particle is moving with momentum $|\vec{p}|$ in the $+z$ direction, where

$$p\cdot\sigma \equiv p^{\mu}g_{\mu\nu}\sigma^{\mu} = E_{\vec{p}}\begin{pmatrix} 1 & 0 \\ 0 & 1 \end{pmatrix} - |\vec{p}|\begin{pmatrix} 1 & 0 \\ 0 & -1 \end{pmatrix} = \begin{pmatrix} E_{\vec{p}} - |\vec{p}| & 0 \\ 0 & E_{\vec{p}} + |\vec{p}| \end{pmatrix}$$

$$\sqrt{p\cdot\sigma} \equiv \begin{pmatrix} \sqrt{E_{\vec{p}} - |\vec{p}|} & 0 \\ 0 & \sqrt{E_{\vec{p}} + |\vec{p}|} \end{pmatrix} \tag{16.25}$$

and $\sqrt{p\cdot\overline{\sigma}}$ is the same with the signs reversed. (Unfortunately, the process for calculating $\sqrt{p\cdot\sigma}$ and $\sqrt{p\cdot\overline{\sigma}}$ in other cases is not as simple as just taking the square root of each matrix component.)

The abstract notation is useful, however, because one can take advantage of the following identities:

$$(p\cdot\sigma)(p\cdot\overline{\sigma}) = (p\cdot\overline{\sigma})(p\cdot\sigma) = m^2 I, \tag{16.26a}$$

$$p\cdot\sigma + p\cdot\overline{\sigma} = 2E_{\vec{p}}I. \tag{16.26b}$$

(See Box 16.3.) The latter implies that generally the solutions are normalized so that

$$u^{\dagger}u = \begin{bmatrix} \xi^{\dagger}\sqrt{p\cdot\sigma} & \xi^{\dagger}\sqrt{p\cdot\overline{\sigma}} \end{bmatrix} \begin{bmatrix} \sqrt{p\cdot\sigma}\,\xi \\ \sqrt{p\cdot\overline{\sigma}}\,\xi \end{bmatrix}$$

$$= \xi^{\dagger}(p\cdot\sigma)\xi + \xi^{\dagger}(p\cdot\overline{\sigma})\xi = \xi^{\dagger}(p\cdot\sigma + p\cdot\overline{\sigma})\xi = 2E_{\vec{p}}, \tag{16.27a}$$

$$v^{\dagger}v = \begin{bmatrix} \eta^{\dagger}\sqrt{p\cdot\sigma} & -\eta^{\dagger}\sqrt{p\cdot\overline{\sigma}} \end{bmatrix} \begin{bmatrix} \sqrt{p\cdot\sigma}\,\eta \\ -\sqrt{p\cdot\overline{\sigma}}\,\eta \end{bmatrix}$$

$$= \eta^{\dagger}(p\cdot\sigma)\eta + \eta^{\dagger}(p\cdot\overline{\sigma})\eta = \eta^{\dagger}(p\cdot\sigma + p\cdot\overline{\sigma})\eta = 2E_{\vec{p}}. \tag{16.27b}$$

Note that this reduces to $2m$ in the limit that $|\vec{p}| \to 0$.

16.6 The Adjoint Spinor and the Lagrangian Density

The fact that $u^{\dagger}u = v^{\dagger}v = 2E_{\vec{p}}$ is actually not good, because we would like to generate the Dirac equation from a Lagrangian density like $\mathcal{L} = \psi^{\dagger}\gamma^{\mu}\partial_{\mu}\psi - m\psi^{\dagger}\psi$. But $\mathcal{L}$ needs to be not just a scalar but a *Lorentz-invariant* scalar. Though $\psi^{\dagger}\psi = 2E_{\vec{p}}$ is a number, it is not Lorentz-invariant.

The trick to solve this problem is to define a new **adjoint spinor**

$$\overline{\psi} \equiv \psi^{\dagger}\gamma^{t}. \tag{16.28}$$

(Note that the bar means something different here than it does when $\sigma^{\mu} \to \overline{\sigma}^{\mu}$.) One can show (see Box 16.4) that

$$\overline{u}u = u^{\dagger}\gamma^{t}u = 2m \quad \text{and} \quad \overline{v}v = v^{\dagger}\gamma^{t}v = -2m. \tag{16.29}$$

These normalizations are now Lorentz-invariant. (It may also be clearer now why it is valuable to distinguish between the particle and antiparticle solutions.) We can now write the Lagrangian density for the Dirac equation

$$\mathcal{L} = i\overline{\psi}\gamma^\mu \partial_\mu \psi - m\overline{\psi}\psi. \tag{16.30}$$

The Euler-Lagrange equation for $\overline{\psi}$ yields the Dirac equation

$$0 = \partial_\mu \frac{\partial \mathcal{L}}{\partial(\partial_\mu \overline{\psi})} - \frac{\partial \mathcal{L}}{\partial \overline{\psi}} = 0 - (i\gamma^\mu \partial_\mu \psi - m\psi) \;\Rightarrow\; i\gamma^\mu \partial_\mu \psi - m\psi = 0. \tag{16.31}$$

The Euler-Lagrange equation for ψ yields the same equation in adjoint form:

$$0 = \partial_\mu \frac{\partial \mathcal{L}}{\partial(\partial_\mu \psi)} - \frac{\partial \mathcal{L}}{\partial \psi} = i\partial_\mu(\overline{\psi}\gamma^\mu) + m\overline{\psi} \;\Rightarrow\; -i\partial_\mu \overline{\psi}\gamma^\mu - m\overline{\psi} = 0. \tag{16.32}$$

16.7 Basis States for the Dirac Solutions

It is useful to define basis vectors to describe solutions for the Dirac equation. We typically define

$$\xi^+ = -\eta^- = \begin{bmatrix} 1 \\ 0 \end{bmatrix} \quad \text{and} \quad \xi^- = \eta^+ = \begin{bmatrix} 0 \\ 1 \end{bmatrix}. \tag{16.33}$$

(The reason for the backward superscripts and the odd sign choice for η^- will become clear in the next chapter.) The basis states for particles and antiparticles become (with $s = +, -$)

$$u^s(p) = \begin{bmatrix} \sqrt{p \cdot \sigma}\,\xi^s \\ \sqrt{p \cdot \overline{\sigma}}\,\xi^s \end{bmatrix} \quad \text{and} \quad v^s(p) = \begin{bmatrix} \sqrt{p \cdot \sigma}\,\eta^s \\ -\sqrt{p \cdot \overline{\sigma}}\,\eta^s \end{bmatrix}. \tag{16.34}$$

respectively. The particle states are normalized as follows:

$$\overline{u}^r(p)\,u^s(p) = +2m\delta^{rs}, \quad \text{or} \quad u^{r\dagger}(p)\,u^s(p) = +2E_{\vec{p}}\,\delta^{rs}, \tag{16.35}$$

where δ^{rs} is the usual Kronecker delta (1 if $r = s$, 0 otherwise). The antiparticle states are normalized similarly

$$\overline{v}^r(p)\,v^s(p) = -2m\delta^{rs}, \quad \text{or} \quad v^{r\dagger}(p)\,v^s(p) = +2E_{\vec{p}}\,\delta^{rs}, \tag{16.36}$$

but note the sign difference in the first case. The particle and antiparticle states are also orthogonal to each other when we use the adjoints:

$$\overline{u}^r(p)v^s(p) = \overline{v}^r(p)u^s(p) = 0 \tag{16.37}$$

(see Box 16.5), but this is *not* true of $u^{r\dagger}(p)v^s(p)$ or $v^{r\dagger}(p)u^s(p)$. However, it is useful for the future to know that

$$u^{r\dagger}(p)\,v^s(\overline{p}) = v^{r\dagger}(p)\,u^s(\overline{p}) = 0, \tag{16.38}$$

where $p = [E_{\vec{p}},\, \vec{p}\,]$ and $\overline{p} = [E_{\vec{p}},\, -\vec{p}\,]$; that is, the particle and antiparticle have opposite spatial momenta (but the same energy, since $E_{\vec{p}} = (|\vec{p}\,|^2 + m^2)^{1/2}$).

We will see that when we evaluate Feynman diagrams in the future, we will often have occasion to sum over the products of the basis states in the order $u^s(p)\overline{u}^s(p)$. In this reversed order, the product is a 4×4 matrix, not a single number. You can use the identity in Equation 16.26a to show that

$$\sum_{s=+,-} u^s(p)\,\overline{u}^s(p) = \gamma \cdot p + mI \quad \text{and} \tag{16.39a}$$

$$\sum_{s=+,-} v^s(p)\,\overline{v}^s(p) = \gamma \cdot p - mI \tag{16.39b}$$

(see Box 16.6). These are called **completeness relations** for the spinors u and v. Because the combination $\gamma \cdot p$ and similar products occur so often in Feynman diagrams, Feynman himself created a special shorthand notation for this kind of product: if a^μ is any four-vector, then

$$\rlap{/}{a} \equiv \gamma \cdot a \equiv \gamma^\mu a_\mu. \tag{16.40}$$

We will find this **Feynman slash notation** also quite handy as we go on.

16.8 But What Does It All Mean?

My father (a philosophy professor), after hearing a long exposition of some complicated technical subject, was fond of asking, "But what does it all mean?" In this chapter, we have developed a brand-new relativistic wave equation for free particles, described and cataloged its solutions, and developed a Lagrangian density that generates the equation. But why do we *need* this equation? What is the physical meaning of the basis states that we have defined? Particularly puzzling is the fact that the solutions must have four components. Why four components, and what do these components physically mean?

The terminology hints at the answers. First, we have seen that each four-component Dirac spinor divides into "right-handed" and "left-handed" parts, each a two-component Weyl spinor. The Weyl spinors, in turn, look a lot like the vectors we use to describe the spin states of spin-$\frac{1}{2}$ particles. In the next chapter, we will prove that the Weyl spinors do indeed describe the spin states of the right- and left-handed versions of the particle. So, the *entire point* of the Dirac equation is that it describes spin-$\frac{1}{2}$ particles, something that the Klein-Gordon equation cannot do (we will in fact see that the Klein-Gordon equation describes spin-0 particles). Since all quarks and leptons—the "matter" particles in the Standard Model (as opposed to "field quanta" like photons and gluons)—are spin-$\frac{1}{2}$ particles, this makes the Dirac equation central to describing the universe we live in. Somehow, the cute trick that Dirac devised (while attempting to address mathematical issues that turned out to be irrelevant) turns out to be an essential part of the machinery of the universe!

In order to prove and more deeply understand the meaning of the components, however, we need to understand how these spinors behave under rotations. We will pick up this problem in the next chapter.

BOX 16.1   Rest Solutions to the Dirac Equation

In this box, you will solve the Dirac equation for a particle at rest. According to Equation 16.20, the Dirac equation for the spinor part is simply

$$\begin{pmatrix} -m & 0 & E & 0 \\ 0 & -m & 0 & E \\ E & 0 & -m & 0 \\ 0 & E & 0 & -m \end{pmatrix} \begin{bmatrix} a \\ b \\ c \\ d \end{bmatrix} = 0, \tag{16.41}$$

where E is the particle's energy (the parameter in the e^{-iEt} oscillating part of the solution), m is the particle's mass, and a, b, c, and d are constants to be determined.

Exercise 16.1.1   Mathematically, this is a standard eigenvalue/eigenvector problem, with m playing the role of the eigenvalue. Show that the eigenvalue part implies that $m^2 = E^2$, then find the eigenvectors. (*Hints:* To evaluate the determinant, use the *cofactor method*—also known as the *minor expansion formula*—which is easy to do here. See the top of the Wikipedia article "Determinant" to see how to do this. You can use WolframAlpha only as a check. Then, since physically m must be positive and E is the variable, $E^2 = m^2$ means that $E = \pm m$. Substitute these two values into the equation to determine the values of a, b, c, d for the eigenvectors.)

BOX 16.2 Interpreting $\sqrt{p \cdot \sigma}$

The super-compressed notation $\sqrt{p \cdot \sigma}$ and $\sqrt{p \cdot \bar{\sigma}}$ is actually a shorthand for the following process:

1. Calculate the matrix quantity A inside the square root.

2. Find A's eigenvalues a_1, a_2 and normalized eigenvectors $|a_1\rangle$, $|a_2\rangle$.

3. Calculate the eigenvalues' square roots $\sqrt{a_1}$, $\sqrt{a_2}$.

4. Reconstruct the matrix: $\sqrt{A} \equiv \sqrt{a_1}|a_1\rangle\langle a_1| + \sqrt{a_2}|a_2\rangle\langle a_2|$.

Let's apply this in the case where the particle is moving with momentum $|\,\vec{p}\,|$ in the $+z$ direction. Because $\sigma^\mu \equiv (I, \sigma_x, \sigma_y, \sigma_z)$ (sorry about the confusing but conventional lower index positions on the σ-matrices), we have

$$p \cdot \sigma = p_\mu \sigma^\mu = p_t I + p_x \sigma_x + p_y \sigma_y + p_z \sigma_z = p^t I - p^x \sigma_x - p^y \sigma_y - p^z \sigma_z$$

$$= E_{\vec{p}}\begin{pmatrix} 1 & 0 \\ 0 & 1 \end{pmatrix} - |\,\vec{p}\,|\begin{pmatrix} 1 & 0 \\ 0 & -1 \end{pmatrix} = \begin{pmatrix} E_{\vec{p}} - |\,\vec{p}\,| & 0 \\ 0 & E_{\vec{p}} + |\,\vec{p}\,| \end{pmatrix}. \quad (16.42)$$

Since this matrix is diagonal, we can read the eigenvalues $E_{\vec{p}} - |\,\vec{p}\,|$ and $E_{\vec{p}} + |\,\vec{p}\,|$ directly from the diagonal. The corresponding eigenvectors are

$$|-\rangle = \begin{bmatrix} 1 \\ 0 \end{bmatrix} \quad \text{and} \quad |+\rangle = \begin{bmatrix} 0 \\ 1 \end{bmatrix}, \quad (16.43)$$

where, just to be clear, $|-\rangle$ is the eigenvector for $E_{\vec{p}} - |\,\vec{p}\,|$ and $|+\rangle$ is the eigenvector for $E_{\vec{p}} + |\,\vec{p}\,|$.

Exercise 16.2.1   From here, reconstruct the matrix $\sqrt{p \cdot \sigma}$.

BOX 16.3   Some Important Identities

The purpose of this box is to prove some useful identities.

Exercise 16.3.1   Evaluate $(p \cdot \sigma)(p \cdot \overline{\sigma})$ and $(p \cdot \overline{\sigma})(p \cdot \sigma)$. (*Hint:* Construct the $p \cdot \sigma$ and $p \cdot \overline{\sigma}$ matrices, give short names to each component to save writing, then multiply the matrices.)

Exercise 16.3.2   Evaluate $(p \cdot \sigma) + (p \cdot \overline{\sigma})$.

BOX 16.4 State Normalization

In this box, we will see that $\bar{u}u = -\bar{v}v = 2m$, where $\bar{u} \equiv u^\dagger \gamma^t$ and $\bar{v} \equiv v^\dagger \gamma^t$, where

$$u(p) = \begin{bmatrix} \sqrt{p \cdot \sigma}\, \xi \\ \sqrt{p \cdot \bar{\sigma}}\, \xi \end{bmatrix} \quad \text{and} \quad v(p) = \begin{bmatrix} \sqrt{p \cdot \sigma}\, \eta \\ -\sqrt{p \cdot \bar{\sigma}}\, \eta \end{bmatrix}, \tag{16.44}$$

and where $\xi^\dagger \xi = \eta^\dagger \eta = 1$.

Exercise 16.4.1 Calculate $\bar{u}u$ and $\bar{v}v$. (*Hint:* Work in compact notation, but take advantage of the results in the previous exercise.)

BOX 16.5   Orthogonality of Particle-Antiparticle States

In this box, we want to verify that (for any choice of r, s)

$$\overline{u}^{\,r}(p)v^s(p) = \overline{v}^{\,r}(p)u^s(p) = 0, \tag{16.45}$$

and that $u^{r\dagger}(p)v^s(p) \neq 0$ and $v^{r\dagger}(p)u^s(p) \neq 0$, but that

$$u^{r\dagger}(p)\, v^s(\overline{p}) = v^{r\dagger}(p)\, u^s(\overline{p}) = 0, \tag{16.46}$$

where the particle and antiparticle have opposite spatial momenta.

Let's show that $\overline{u}^{\,r}(p)v^s(p) = 0$. We have

$$\overline{u}^{\,r}(p)\, v^s(p) = [\, u^r(p)\,]^{\dagger}\, \gamma^t\, v^s(p) \tag{16.47}$$

$$= [\, \xi^{r\dagger} \sqrt{p \cdot \sigma} \quad \xi^{r\dagger}\sqrt{p \cdot \overline{\sigma}}\,] \begin{pmatrix} 0 & I \\ I & 0 \end{pmatrix} \begin{bmatrix} \sqrt{p \cdot \sigma}\, \eta^s \\ -\sqrt{p \cdot \overline{\sigma}}\, \eta^s \end{bmatrix} \tag{16.48}$$

$$= [\, \xi^{r\dagger} \sqrt{p \cdot \sigma} \quad \xi^{r\dagger}\sqrt{p \cdot \overline{\sigma}}\,] \begin{bmatrix} -\sqrt{p \cdot \overline{\sigma}}\, \eta^s \\ \sqrt{p \cdot \sigma}\, \eta^s \end{bmatrix} \tag{16.49}$$

$$= -\xi^{r\dagger}\sqrt{(p \cdot \sigma)(p \cdot \overline{\sigma})}\, \eta^s + \xi^{r\dagger}\sqrt{(p \cdot \overline{\sigma})(p \cdot \sigma)}\, \eta^s \tag{16.50}$$

$$= \xi^{r\dagger}(-m + m)\, \eta^s = 0, \tag{16.51}$$

no matter what s and r are.

Exercise 16.5.1   Similarly, verify equation $\overline{v}^{\,r}(p)u^s(p) = 0$.

BOX 16.5 Orthogonality of Particle-Antiparticle States (cont.)

Now, note that if we *don't* have the factor of γ^t, then

$$[\, u^r(p)\,]^\dagger\, v^s(p) = [\, \xi^{r\dagger}\sqrt{p\cdot\sigma} \quad \xi^{r\dagger}\sqrt{p\cdot\overline{\sigma}}\,]\begin{bmatrix} \sqrt{p\cdot\sigma}\,\eta^s \\ -\sqrt{p\cdot\overline{\sigma}}\,\eta^s \end{bmatrix} \tag{16.52}$$

$$= \xi^{r\dagger}(p\cdot\sigma)\,\eta^s - \xi^{r\dagger}(p\cdot\overline{\sigma})\,\eta^s = \xi^{r\dagger}(p\cdot\sigma - p\cdot\overline{\sigma})\,\eta^s. \tag{16.53}$$

Since $\sigma^t = \overline{\sigma}^t$ but $\overline{\sigma}^i = -\sigma^i$, the above becomes $-\xi^{r\dagger}(2\vec{p}\cdot\vec{\sigma})\,\eta^s \neq 0$, even if $r \neq s$ because $\vec{p}\cdot\vec{\sigma}$ is a matrix that is not necessarily the identity matrix.

Exercise 16.5.2   Show that $v^{r\dagger}(p)u^s(p) \neq 0$ also, but adapt these calculations to verify Equation 16.46. (*Hint:* For the latter, remember that if $p = [E_{\vec{p}}, \vec{p}\,]$, then $\overline{p} \equiv [E_{\vec{p}}, -\vec{p}\,]$. It will help to prove that $\overline{p}\cdot\sigma = p\cdot\overline{\sigma}$ and that $\overline{p}\cdot\overline{\sigma} = p\cdot\sigma$.)

BOX 16.6   Sums of Basis States

The purpose of this box is to show that

$$\sum_{s=+,-} u^s(p)\,\overline{u}^s(p) = \gamma \cdot p + mI \quad \text{and} \tag{16.54a}$$

$$\sum_{s=+,-} v^s(p)\,\overline{v}^s(p) = \gamma \cdot p - mI. \tag{16.54b}$$

Exercise 16.6.1   Prove the expression above. (*Hint:* Use the compact notation. Also, show that within the two-dimensional subspaces of that notation

$$\sum_{s=+,-} \xi^s \xi^{s\dagger} = \sum_{s=+,-} \eta^s \eta^{s\dagger} = \begin{pmatrix} 1 & 0 \\ 0 & 1 \end{pmatrix} = I.) \tag{16.55}$$

Homework Problems

P16.1 In this problem, we will discover some useful Pauli matrix identities.

a. Prove the following identity for the Pauli matrices:

$$(\vec{a} \cdot \vec{\sigma})(\vec{b} \cdot \vec{\sigma}) = \vec{a} \cdot \vec{b}\, I + i(\vec{a} \times \vec{b}) \cdot \vec{\sigma}. \quad (16.56)$$

(*Hint:* You will find using index notation very helpful. Also note that $(\vec{a} \times \vec{b})_m = \sum_{jk} a_j b_k \epsilon_{jkm}$.)

b. Use the result above to argue that if $\vec{n}$ is a unit vector, then

$$(\vec{n} \cdot \vec{\sigma})^{2k} = I, \qquad (\vec{n} \cdot \vec{\sigma})^{2k+1} = \vec{n} \cdot \vec{\sigma}, \quad (16.57)$$

where k is a nonnegative integer.

P16.2 Verify that γ^t and γ^x satisfy Equation 16.9.

P16.3 Verify by direct substitution that

$$\psi = u(p)\, e^{-ip\cdot x} = \begin{bmatrix} \sqrt{p \cdot \sigma}\, \xi \\ \sqrt{p \cdot \overline{\sigma}}\, \xi \end{bmatrix} e^{-ip\cdot x} \quad (16.58)$$

(where ξ is an arbitrary two-dimensional column vector with constant components) solves the Dirac equation. Assume that the particle is moving with momentum $|\vec{p}|$ in the $+z$ direction so that

$$\sqrt{p \cdot \sigma} = \begin{pmatrix} \sqrt{E_{\vec{p}} - |\vec{p}|} & 0 \\ 0 & \sqrt{E_{\vec{p}} + |\vec{p}|} \end{pmatrix}, \quad (16.59a)$$

$$\sqrt{p \cdot \overline{\sigma}} = \begin{pmatrix} \sqrt{E_{\vec{p}} + |\vec{p}|} & 0 \\ 0 & \sqrt{E_{\vec{p}} - |\vec{p}|} \end{pmatrix}. \quad (16.59b)$$

(Otherwise, this would be somewhat tricky.)

P16.4 Suppose an electron (whose mass $m = 0.511$ MeV) is traveling in the $+z$ direction with an energy $E = \frac{5}{3}m$.

a. Write the two basis states for the electron's field $\psi(x)$ as explicit four-component spinors times the appropriate exponential factors, all expressed in terms of m.

b. Similarly, write the two basis states for $\overline{\psi}(x)$.

c. Write the basis states for $\psi(x)$ if the particle is a positron.

d. Write the basis states for $\overline{\psi}(x)$ if the particle is a positron.

P16.5 What are the two basis states for a particle moving in the $+z$ direction with an energy so large that $E_{\vec{p}} - p^z \to 0$ (meaning that the particle's mass is negligible compared to its energy)? What about the basis states for an antiparticle moving in the same direction? Argue that either the right-handed or left-handed part of the Dirac spinor becomes suppressed in this negligible-mass limit.

P16.6 Find the basis states for a particle moving with momentum $|\vec{p}|$ in the $+x$ direction. (*Hint:* Follow the complete prescription outlined in Box 16.2 for how to evaluate $\sqrt{p \cdot \sigma}$ and $\sqrt{p \cdot \overline{\sigma}}$.)

P16.7 One does have some choice in defining the γ^μ matrices. In this chapter, I have chosen to use matrices that lead to the **chiral representation** (also called the **Weyl representation**) of the spinors, which has several advantages for our purposes. But other choices are possible. Show that

$$\gamma^t = \begin{pmatrix} I & 0 \\ 0 & -I \end{pmatrix} \quad \text{and} \quad \gamma^j = \begin{pmatrix} 0 & \sigma_j \\ -\sigma_j & 0 \end{pmatrix}$$

(where j is a spatial index) also satisfy the constraints described in Equation 16.9. This set of matrices leads to the **Dirac-Pauli representation** (also known as the **standard representation**) of the spinors.

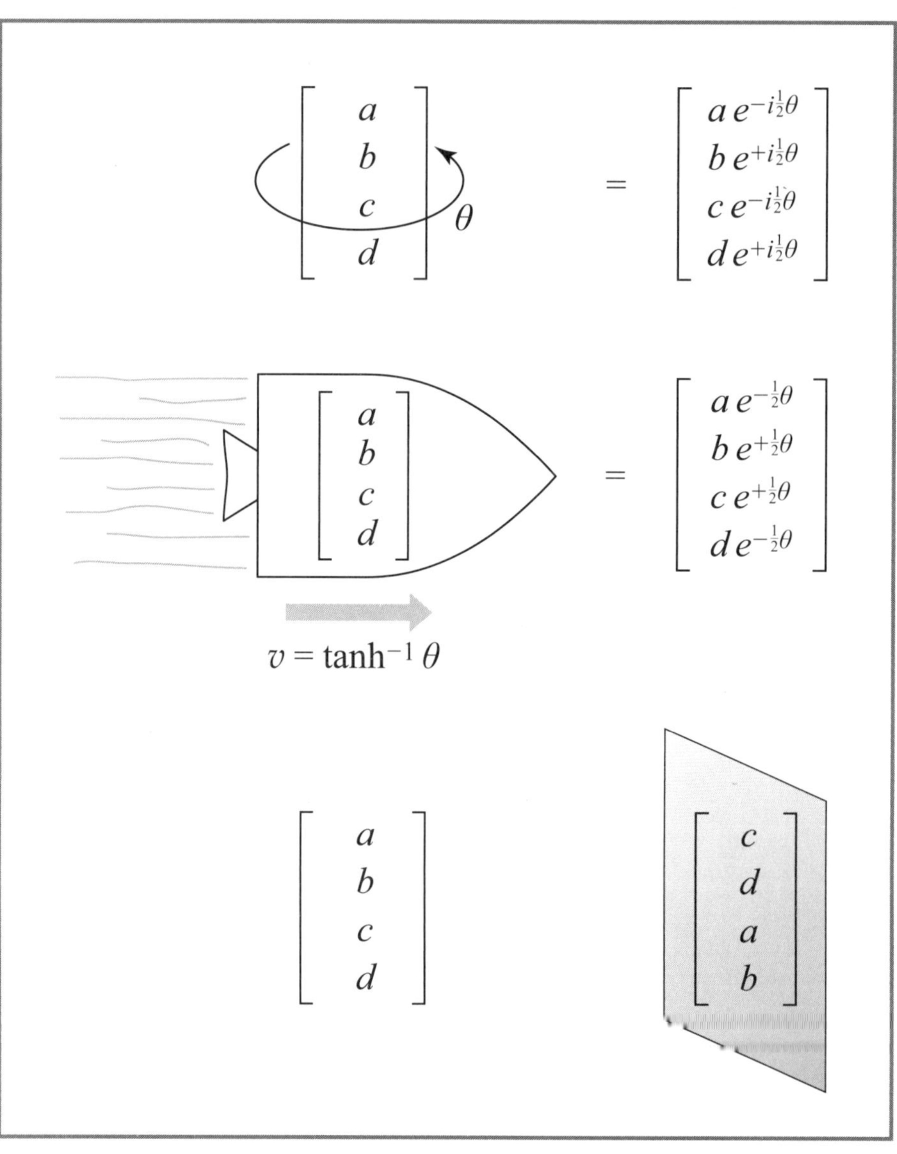

17. TRANSFORMING DIRAC SPINORS

Equation 16.30 in the last chapter introduced the Dirac Lagrangian density for a free spin-$\frac{1}{2}$ particle

$$\mathcal{L} = i\overline{\psi}\gamma^\mu\partial_\mu\psi - m\overline{\psi}\psi, \tag{17.1}$$

where $\overline{\psi} = \psi^\dagger\gamma^t$ and the 4×4 γ^μ matrices (in compact notation) are

$$\gamma^t = \begin{pmatrix} 0 & I \\ I & 0 \end{pmatrix} \quad \text{and} \quad \gamma^j = \begin{pmatrix} 0 & \sigma_j \\ -\sigma_j & 0 \end{pmatrix}, \tag{17.2}$$

where σ_j are the 2×2 Pauli matrices and $j = x, y, z$. The Euler-Lagrange equations for our Lagrangian density are the Dirac equation

$$i\gamma^\mu\partial_\mu\psi - m\psi = 0 \tag{17.3}$$

and its adjoint. The solutions to the Dirac equation are

$$\text{Particle}: \ \psi = u(p)e^{-ip\cdot x}, \ \text{where } u(p) = \begin{bmatrix} u_L(p) \\ u_R(p) \end{bmatrix} = \begin{bmatrix} \sqrt{p\cdot\sigma}\,\xi \\ \sqrt{p\cdot\overline{\sigma}}\,\xi \end{bmatrix}, \tag{17.4a}$$

$$\text{Antiparticle}: \ \psi = v(p)e^{+ip\cdot x}, \ \text{where } v(p) = \begin{bmatrix} v_L(p) \\ v_R(p) \end{bmatrix} = \begin{bmatrix} \sqrt{p\cdot\sigma}\,\eta \\ -\sqrt{p\cdot\overline{\sigma}}\,\eta \end{bmatrix}, \tag{17.4b}$$

and where $u(p)$ and $v(p)$ are four-component vectors called *Dirac spinors*. Each of the two symbols in the brackets stands for a two-component *Weyl spinor*, where $\sigma^\mu \equiv (I, \sigma_x, \sigma_y, \sigma_z)$ and $\overline{\sigma}^\mu \equiv (I, -\sigma_x, -\sigma_y, -\sigma_z)$, and where the square root of a matrix is defined in Box 16.2.

However, our treatment left us with a variety of questions. Why do the solutions have four components? What do these components mean? Why do we call the Weyl spinors "right-" and "left-handed?" How do we know that these are the solutions? We can answer all of these questions by considering how Dirac spinors transform under rotations, boosts, and other kinds of operations. Exploring these issues is the focus of this chapter.

17.1 Lorentz Transformations

In spite of having four components, Dirac spinors are *not* four-vectors: the components transform in an entirely different manner that we cannot intuit easily. Fortunately, we have the tools we need to determine exactly how these spinors transform.

Let the transformation matrix that transforms a Dirac spinor during an arbitrary Lorentz transformation be M_Λ, meaning that

$$\psi' = M_\Lambda\psi \tag{17.5}$$

by definition. Now, we know that $\overline{\psi}\psi$ must transform like a Lorentz scalar because an $m\overline{\psi}\psi$ term appears in the Lagrangian density, which must itself be a Lorentz-invariant scalar. This will be true if and only if $\overline{\psi}' = \overline{\psi}\,M_\Lambda^{-1}$ because only then will

$$\overline{\psi}'\psi' = (\overline{\psi}\,M_\Lambda^{-1})(M_\Lambda\psi) = \overline{\psi}\,I\psi = \overline{\psi}\psi. \tag{17.6}$$

We also see that for $i\overline{\psi}\gamma^\mu\partial_\mu\psi$ to transform like a Lorentz-invariant scalar, we must have

$$M_\Lambda^{-1}\gamma^\mu M_\Lambda = \Lambda^\mu{}_\nu\gamma^\nu, \tag{17.7}$$

where $\Lambda^\mu{}_\nu$ are the standard Lorentz transformation coefficients, so that the Lorentz transformation of the upper index μ on the gamma matrices cancels the inverse transformation associated with the lower index μ:

$$i\overline{\psi}'\gamma^\mu\partial_\mu'\psi' = i\overline{\psi}M^{-1}\gamma^\mu(\Lambda^{-1})^\alpha{}_\mu\partial_\alpha M_\Lambda\psi = i\overline{\psi}M^{-1}\gamma^\mu M_\Lambda(\Lambda^{-1})^\alpha{}_\mu\partial_\alpha\psi$$

$$= i\overline{\psi}\Lambda^\mu{}_\nu\gamma^\nu(\Lambda^{-1})^\alpha{}_\mu\partial_\alpha\psi = i\overline{\psi}\gamma^\nu\delta^\alpha_\nu\partial_\alpha\psi = i\overline{\psi}\gamma^\nu\partial_\nu\psi. \tag{17.8}$$

The calculation above uses the facts that the transformation coefficients $\Lambda^\mu{}_\nu$ are just numbers (not matrices in the spinor space that might not commute with M_Λ or γ^μ), and that $\Lambda^\mu{}_\nu(\Lambda^{-1})^\alpha{}_\mu = \delta^\alpha_\nu$ (see Equation 3.14).

This provides us with the clue we need to determine the matrix M_Λ for various kinds of Lorentz transformations. First, remember that general Lorentz transformations include not only **boosts** (transformations to a new frame moving at an arbitrary velocity relative to the original frame) but also rotations around an arbitrary axis. For a given boost direction or rotation axis, a single parameter (call it θ) describes how fast we are boosting or how far we are rotating. For such a transformation, we can write the M_Λ matrix in the form

$$M_\Lambda = e^{-i\theta X_\Lambda}. \tag{17.9}$$

Knowing the generator X_Λ thus allows us to calculate M_Λ as a function of the parameter θ. Note that the definition above implies that $M_\Lambda^{-1} = e^{+i\theta X_\Lambda}$ (since $e^A e^B = e^{A+B}$ as long as A and B commute, and X_Λ does commute with itself).

Let's consider first a rotation by an infinitesimal angle $\delta\theta$ around the z axis. In this small-angle limit, the transformation matrix and its inverse are

$$M_{Rz}(\delta\theta) \approx \left(I - i\delta\theta X_{Rz}\right) \quad\text{and}\quad M_{Rz}^{-1}(\delta\theta) \approx \left(I + i\delta\theta X_{Rz}\right) \tag{17.10}$$

to first order in $\delta\theta$. This means that to the same order,

$$M_{Rz}^{-1}\gamma^\mu M_{Rz} \approx \left(I + i\delta\theta X_{Rz}\right)\gamma^\mu\left(I - i\delta\theta X_{Rz}\right)$$

$$\approx \gamma^\mu + i\delta\theta\left(X_{Rz}\gamma^\mu - \gamma^\mu X_{Rz}\right) = \gamma^\mu + i\delta\theta[X_{Rz}, \gamma^\mu]. \tag{17.11}$$

We also know that since the x and y components of vectors rotate like

$$\begin{bmatrix} x' \\ y' \end{bmatrix} = \begin{pmatrix} \cos\delta\theta & -\sin\delta\theta \\ \sin\delta\theta & \cos\delta\theta \end{pmatrix}\begin{bmatrix} x \\ y \end{bmatrix} \approx \begin{pmatrix} 1 & -\delta\theta \\ \delta\theta & 1 \end{pmatrix}\begin{bmatrix} x \\ y \end{bmatrix} \tag{17.12}$$

to first order in $\delta\theta$, the gamma matrices ought to rotate like

$$\Lambda^\mu{}_\nu\gamma^\nu = \begin{bmatrix} \gamma^t \\ \gamma^x - \delta\theta\gamma^y \\ \delta\theta\gamma^x + \gamma^y \\ \gamma^z \end{bmatrix}. \tag{17.13}$$

Comparing Equations 17.13 and 17.11 tells us that

$$[X_{Rz}, \gamma^t] = 0, \quad [X_{Rz}, \gamma^x] = i\gamma^y, \quad [X_{Rz}, \gamma^y] = -i\gamma^x, \quad [X_{Rz}, \gamma^z] = 0. \tag{17.14}$$

You can show (see Box 17.1) that the matrix satisfying these constraints is

$$X_{Rz} = \frac{1}{2}\begin{pmatrix} \sigma_z & 0 \\ 0 & \sigma_z \end{pmatrix}, \tag{17.15}$$

and you might guess quite reasonably (and correctly!) that the generators for rotations around the x and y axes, respectively, are,

$$X_{Rx} = \frac{1}{2} \begin{pmatrix} \sigma_x & 0 \\ 0 & \sigma_x \end{pmatrix} \quad \text{and} \quad X_{Ry} = \frac{1}{2} \begin{pmatrix} \sigma_y & 0 \\ 0 & \sigma_y \end{pmatrix} \tag{17.16}$$

Now let's consider a Lorentz boost that increases the particle's x-velocity by a tiny amount. We can write a Lorentz transformation that boosts the primed frame in the $+x$ direction relative to the unprimed frame as

$$\begin{bmatrix} t' \\ x' \end{bmatrix} = \begin{pmatrix} \gamma & -\gamma\beta \\ -\gamma\beta & \gamma \end{pmatrix} \begin{bmatrix} t \\ x \end{bmatrix} = \begin{pmatrix} \cosh\theta & -\sinh\theta \\ -\sinh\theta & \cosh\theta \end{pmatrix} \begin{bmatrix} t \\ x \end{bmatrix} \tag{17.17}$$

if we define the primed frame's **rapidity** θ relative to the unprimed frame to be such that $\cosh\theta \equiv \gamma$ and $\sinh\theta \equiv \gamma\beta$ (see Problem P17.1). (Using the rapidity highlights the similarities between the rotations and Lorentz boosts.) But to boost the *particle's* rapidity in the $+x$ direction, we want to go to a frame that is moving faster in the $-x$ direction, so our Lorentz transformation becomes

$$\Lambda^\mu{}_\nu = \begin{pmatrix} \cosh\theta & \sinh\theta & 0 & 0 \\ \sinh\theta & \cosh\theta & 0 & 0 \\ 0 & 0 & 1 & 0 \\ 0 & 0 & 0 & 1 \end{pmatrix}. \tag{17.18}$$

For a small $\delta\theta$, $\cosh\delta\theta \to 1$ and $\sinh\delta\theta \to \delta\theta$, so the equivalent to Equation 17.13 for this transformation will be

$$\Lambda^\mu{}_\nu \gamma^\nu = \begin{bmatrix} \gamma^t + \delta\theta\gamma^x \\ \delta\theta\gamma^t + \gamma^x \\ \gamma^y \\ \gamma^z \end{bmatrix}. \tag{17.19}$$

By following similar steps to those we used for rotation (see Box 17.2), you can show that the generator for boosts of a particle's rapidity in the x direction is

$$X_{Bx} = -\frac{i}{2} \begin{pmatrix} \sigma_x & 0 \\ 0 & -\sigma_x \end{pmatrix}, \tag{17.20}$$

and boosts in the other directions are analogous. In Box 17.3, you will show how one can use $M_B = e^{-i\theta X_B}$ to boost the Dirac rest solution to generate the solutions given in Equations 17.4. But you should note that the generators of boosts, unlike the generators of rotations, are *not* Hermitian (don't be caught off-guard by this!).

17.2 Spin

As we have seen in ordinary quantum mechanics, the generator of the rotation operation is the angular momentum operator. In the context of Dirac spinors, the operator corresponding to projections of spin angular momentum on the z axis is X_{Rz}, as given in Equation 17.15. Applying this to the u_0^+ basis vector defined in the last chapter shows that for a particle at rest

$$X_{Rz}u_0^+ = \frac{1}{2} \begin{pmatrix} \sigma_z & 0 \\ 0 & \sigma_z \end{pmatrix} \begin{bmatrix} \sqrt{m}\,\xi^+ \\ \sqrt{m}\,\xi^+ \end{bmatrix} \quad \text{where} \quad \xi^+ \equiv \begin{bmatrix} 1 \\ 0 \end{bmatrix}$$

$$= \frac{\sqrt{m}}{2} \begin{pmatrix} 1 & 0 & 0 & 0 \\ 0 & -1 & 0 & 0 \\ 0 & 0 & 1 & 0 \\ 0 & 0 & 0 & -1 \end{pmatrix} \begin{bmatrix} 1 \\ 0 \\ 1 \\ 0 \end{bmatrix} = \frac{\sqrt{m}}{2} \begin{bmatrix} 1 \\ 0 \\ 1 \\ 0 \end{bmatrix} = \frac{1}{2}u_0^+. \tag{17.21}$$

For the u_0^- basis vector for a particle at rest, we have

$$X_{Rz} u_0^- = \frac{1}{2} \begin{pmatrix} \sigma_z & 0 \\ 0 & \sigma_z \end{pmatrix} \begin{bmatrix} \sqrt{m}\, \xi^- \\ \sqrt{m}\, \xi^- \end{bmatrix} \quad \text{where} \quad \xi^- \equiv \begin{bmatrix} 0 \\ 1 \end{bmatrix}$$

$$= \frac{\sqrt{m}}{2} \begin{pmatrix} 1 & 0 & 0 & 0 \\ 0 & -1 & 0 & 0 \\ 0 & 0 & 1 & 0 \\ 0 & 0 & 0 & -1 \end{pmatrix} \begin{bmatrix} 0 \\ 1 \\ 0 \\ 1 \end{bmatrix} = \frac{\sqrt{m}}{2} \begin{bmatrix} 0 \\ -1 \\ 0 \\ -1 \end{bmatrix} = -\frac{1}{2} u_0^-. \quad (17.22)$$

So we see that these basis states are eigenstates of the spin operator corresponding to the particle having a spin of $\frac{1}{2}$ in the $+z$ or $-z$ direction, respectively. Similarly (see Box 17.4), the basis states for an antiparticle at rest have the same eigenvalues: $X_{Rz} v_0^- = \frac{1}{2} v_0^-$ and $X_{Rz} v_0^+ = -\frac{1}{2} v_0^+$.

We now see clearly and unequivocally that the Dirac equation describes particles that have *intrinsic* angular momentum (= spin) whose projection on any axis (since all axis directions are equivalent) is $\pm\frac{1}{2}$ (or $\pm\frac{1}{2}\hbar$ in a unit system where $\hbar \neq 1$). This is *required* by the very nature of the Dirac equation.

(By the way, we can use the same method to conclude that a *scalar* field $\phi(x)$ solving the Klein-Gordon equation describes a particle with spin 0. A scalar's value is (by definition) not affected by rotation, so the rotation operator in the mathematical space occupied by the single component $\phi(x)$ must be $M_R = 1$ for a rotation of any θ around any axis. The generator X_R for $e^{-iX_R\theta} = M_R = 1$ must be $X_R = 0$. The projection of the scalar particle's intrinsic angular momentum on any axis is therefore $X_R\, \phi(x) = 0$. It has spin 0.)

Please note that our chosen basis states don't remain eigenstates of the X_{Rz} operator for moving particles that have momentum components in the x or y directions. This is because the factors $\sqrt{p \cdot \sigma}$ and $\sqrt{p \cdot \bar{\sigma}}$ in the general states contain σ_x and σ_y, which don't commute with σ_z. This in turn means that we cannot pull these factors out in front the way we did with the $\sqrt{m}$ in the calculations above. So simply having $\xi = \xi^+$ doesn't *necessarily* imply that a moving particle's spin is purely "up" in the z direction.

There is a separate issue with antiparticles that I must note here, though it won't make much sense until later. Even though I just showed that the X_{Rz} eigenvalues of the basis rest states v_0^- and v_0^+ are $+\frac{1}{2}$ and $-\frac{1}{2}$, respectively (contradicting their superscripts), we consider the "physical spin" of antiparticles to be "up" (in the $+z$ direction) for the v_0^+ state and "down" for the v_0^- state, consistent with the superscripts but opposite to what the eigenvalues say and opposite to our usual vector notation spinors for antiparticles:

$$\eta^+ = \begin{bmatrix} 0 \\ 1 \end{bmatrix} = \text{spin up} \quad \text{and} \quad \eta^- = -\begin{bmatrix} 1 \\ 0 \end{bmatrix} = \text{spin down.} \quad (!!) \quad (17.23)$$

This explains the superscripts. One can remember this (in a semiclassical way) by thinking that because the antiparticle solutions have negative energy (and thus negative mass when at rest), their rotation must be opposite to that of the corresponding particle to give the same angular momentum. (The real reason for this in quantum field theory is very difficult to explain, so let's simply accept Equation 17.23 and move on. It all does work out mathematically, I promise!)

17.3 Helicity and Chirality

We can measure a massive particle's spin unambiguously by going to the particle's rest frame. However, no rest frame exists for a massless particle, so even *defining* its "spin" is problematic. But we can meaningfully talk about a massless Dirac particle's **helicity**. We define the helicity operator as

$$\hat{h} \equiv \frac{\vec{p} \cdot \vec{\Sigma}}{|\vec{p}|} \quad \text{where} \quad \vec{\Sigma} \equiv \pm\frac{1}{2} \begin{pmatrix} \vec{\sigma} & 0 \\ 0 & \vec{\sigma} \end{pmatrix}, \quad (17.24)$$

where the positive sign applies to particles and the negative sign to antiparticles, consistent with the "physical spin" direction for each type (the hat distinguishes this operator from Planck's constant h). This operator essentially measures the projection of the particle's physical spin on its direction of motion. Now, it turns out that in the extreme relativistic limit where $|\vec{p}|/m \to \infty$, our basis states approach

$$u^{+}(p) \to \sqrt{2E_{\vec{p}}} \begin{bmatrix} 0 \\ 0 \\ 1 \\ 0 \end{bmatrix}, \qquad u^{-}(p) \to \sqrt{2E_{\vec{p}}} \begin{bmatrix} 0 \\ 1 \\ 0 \\ 0 \end{bmatrix}, \qquad (17.25a)$$

$$v^{-}(p) \to \sqrt{2E_{\vec{p}}} \begin{bmatrix} 0 \\ 0 \\ 1 \\ 0 \end{bmatrix}, \qquad v^{+}(p) \to \sqrt{2E_{\vec{p}}} \begin{bmatrix} 0 \\ 1 \\ 0 \\ 0 \end{bmatrix} \qquad (17.25b)$$

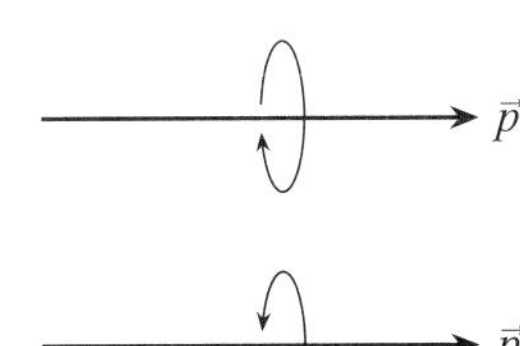

FIGURE 17.1 The top drawing shows a right-handed particle; the bottom one shows a left-handed particle.

(see Problem 16.5). In Box 17.5, you will show that the particle states $u^{+}(p)$ and $u^{-}(p)$ have helicity $+\frac{1}{2}$ and $-\frac{1}{2}$, respectively. You can also see that these states are purely right-handed and left-handed, respectively ($u_L = 0$ in the first case and $u_R = 0$ in the second). We finally understand the physical meaning of these names. As Figure 17.1 shows, a particle whose physical spin is aligned with its direction of motion will have a spin such that if your right fingers curl in the direction of the rotation, your right thumb indicates the direction of motion, so the particle is right-handed. If the particle has the opposite helicity, you must use your left hand to get your thumb to point in the direction of the momentum.

But before we get too comfortable, note that the antiparticle states $v^{-}(p)$ and $v^{+}(p)$ have helicities of $-\frac{1}{2}$ and $+\frac{1}{2}$, respectively (consistent with their superscripts), meaning that the positive-helicity antiparticle state has $v_R = 0$ and the negative-helicity antiparticle state has $v_L = 0$. This means that they are left-handed and right-handed, respectively, exactly opposite to what your hands are telling you. The right-handed and left-handed labels for the upper and lower Weyl spinors in a Dirac spinor correctly describe the helicities of the particle states, not the antiparticle states.

What the R and L subscripts actually refer to is the particle's **chirality** (from a Greek word for "hand"). We define the chirality operator as

$$\gamma^5 \equiv i\gamma^t\gamma^x\gamma^y\gamma^z = \begin{pmatrix} -I & 0 \\ 0 & I \end{pmatrix} \qquad (17.26)$$

(see Box 17.6). You can see by inspection that the states $u^{+}(p)$ and $v^{-}(p)$ both have chirality $+1$ and have $u_L = v_L = 0$: they have "right-handed chirality." The states $u^{-}(p)$ and $v^{+}(p)$ have chirality -1, which is "left-handed chirality." Positive helicity means positive chirality for massless particles, but the two are opposite for massless antiparticles. (The conventional basis states for massive particles are not generally either helicity or chirality eigenstates.)

For future reference, note that we can isolate the chirality eigenstates of any Dirac spinor using the projection operators

$$P_L = \frac{1-\gamma^5}{2} = \begin{pmatrix} I & 0 \\ 0 & 0 \end{pmatrix} \quad \text{and} \quad P_R = \frac{1+\gamma^5}{2} = \begin{pmatrix} 0 & 0 \\ 0 & I \end{pmatrix} \qquad (17.27)$$

(where the conventional "1" in each equation should be interpreted as being the four-dimensional I), because

$$P_L u = \begin{pmatrix} I & 0 \\ 0 & 0 \end{pmatrix} \begin{bmatrix} u_L \\ u_R \end{bmatrix} = \begin{bmatrix} u_L \\ 0 \end{bmatrix} \quad \text{and} \quad P_R u = \begin{pmatrix} 0 & 0 \\ 0 & I \end{pmatrix} \begin{bmatrix} u_L \\ u_R \end{bmatrix} = \begin{bmatrix} 0 \\ u_R \end{bmatrix} \qquad (17.28)$$

and similarly for v. This will be essential when we consider the weak interaction.

Note that the spinors on the right sides of Equations 17.28 are what we *mean* by chirally left-handed and chirally right-handed states.

17.4 Charge Conjugation

We can also imagine a **charge conjugation operator** C that converts particles to antiparticles and vice versa. For a complex scalar field, this operator is simply $C\phi = \phi^*$ since that exchanges particle solutions $e^{-ip\cdot x}$ for antiparticle solutions $e^{+ip\cdot x}$. The correct transformation for Dirac particles is a bit more complicated:

$$\psi^c \equiv C\psi = -i\gamma^y\psi^*. \tag{17.29}$$

In Box 17.7, you will show that (assuming our strange sign for η^-)

$$Cu^+(p) = v^+(p), \qquad Cv^+(p) = u^+(p), \tag{17.30a}$$

$$Cu^-(p) = v^-(p), \qquad Cv^-(p) = u^-(p), \tag{17.30b}$$

meaning that (for particles at rest) the operator converts a physical spin-up particle to a physical spin-up antiparticle or vice versa, and converts a spin-down particle to a physical spin-down antiparticle or vice versa. This is the point of the sign choice for η^-. Note also that $CC\psi = \psi$, as required.

17.5 Parity

We can also define a so-called **parity operator** P that transforms $\phi(t, \vec{r}) \to \phi(t, -\vec{r})$. Such a transformation reverses the direction of ordinary vectors (for example, $\vec{p} = m\, d\vec{r}/d\tau \to m\, d(-\vec{r})/d\tau = -\vec{p}$) but does *not* reverse the direction of so-called **axial vectors** (for example, $\vec{L} = \vec{r} \times \vec{p} \to -\vec{r} \times -\vec{p} = +\vec{L}$). This is equivalent to viewing the situation in a mirror. How does a Dirac spinor transform under a parity operation?

The electromagnetic interaction is experimentally parity-invariant, so we expect the Dirac equation to be parity-invariant so that its solutions can be parity-invariant. That is, if $\psi'(t, \vec{r}) = M_P\, \psi(t, -\vec{r})$ (where the 4×4 matrix M_P handles the transformation of the spinor part of ψ), we want

$$i\gamma^v\partial'_v\psi' - m\psi' = 0 \quad \longleftrightarrow \quad i\gamma^\mu\partial_\mu\psi - m\psi = 0. \tag{17.31}$$

Note that the partial is involved in this parity transformation because

$$\partial'_v \equiv \frac{\partial}{\partial x'_v} = \frac{\partial x^\mu}{\partial x'_v}\frac{\partial}{\partial x^\mu} = \frac{\partial x^\mu}{\partial x'_v}\partial_\mu \equiv \Gamma_v{}^\mu \partial_\mu, \tag{17.32}$$

where $\Gamma_v{}^\mu$ stands for one of a set of 16 numbers such that

$$\Gamma_v{}^\mu = \begin{cases} 1 & \text{if } \mu = v = t \\ -1 & \text{if } \mu = v = x, y, z \\ 0 & \text{if } \mu \neq v \end{cases}. \tag{17.33}$$

Substituting this into the above yields

$$0 = i\gamma^v\partial'_v\psi' - m\psi' = i\gamma^v\Gamma_v{}^\mu\partial_\mu M_P\psi - M_P(m\psi)$$

$$\Rightarrow \quad 0 = iM_P^{-1}\gamma^v M_P\Gamma_v{}^\mu\partial_\mu\psi - m\psi, \tag{17.34}$$

where in the last step I multiplied both sides on the left by M_P^{-1}, and noted that I can pull the M_P past the derivative and the simple number $\Gamma_v{}^\mu$. If we compare this with the unprimed Dirac equation in Equation 17.31, then we see that this equation will imply the unprimed equation so long as

$$M_P^{-1}\gamma^v M_P\Gamma_v{}^\mu = \gamma^\mu,$$

$$\text{that is,} \quad M_P^{-1}\gamma^t M_P = \gamma^t \quad \text{and} \quad M_P^{-1}\gamma^j M_P = -\gamma^j, \tag{17.35}$$

where $j = x, y, z$. Now, these relations will be satisfied if

$$[\gamma^t, M_P] = 0 \quad \rightarrow \quad M_P \gamma^t = \gamma^t M_P, \tag{17.36a}$$

$$\{\gamma^j, M_P\} = 0 \quad \Rightarrow \quad M_P \gamma^j = -\gamma^j M_P \tag{17.36b}$$

because substituting Equations 17.36 into Equations 17.35 will put the M_P next to the M_P^{-1} in each case, yielding the identity. We know from the previous chapter that $\{\gamma^\mu, \gamma^\nu\} = 2g^{\mu\nu}I$, so we see that the only possible solution is

$$M_P = b\,\gamma^t, \tag{17.37}$$

where b is a constant. But since two applications of the parity operator should take us back to where we started, and because $(\gamma^t)^2 = I$, we must have $b = \pm 1$. Setting $b = 1$ simply chooses the irrelevant overall phase of the transformed vector and is the simplest choice, so we will choose that.

We see that the action of the parity operator on the Dirac spinor is to exchange its right-handed and left-handed parts:

$$M_P \psi = \gamma^t \psi = \begin{pmatrix} 0 & I \\ I & 0 \end{pmatrix} \begin{bmatrix} \psi_L \\ \psi_R \end{bmatrix} = \begin{bmatrix} \psi_R \\ \psi_L \end{bmatrix}. \tag{17.38}$$

So any theory that is parity-invariant will treat the (chirally) left-handed and right-handed parts of the spinor identically.

Because particles and antiparticles at rest have spinors

$$u_0 = \sqrt{m}\begin{bmatrix} \xi \\ \xi \end{bmatrix} \quad \text{and} \quad v_0 = \sqrt{m}\begin{bmatrix} \eta \\ -\eta \end{bmatrix}, \tag{17.39}$$

we can see by substitution into the previous equation that Dirac particles are positive-parity eigenstates ($M_P u_0 = +u_0$) while antiparticles are negative-parity eigenstates ($M_P v_0 = -v_0$), at least with our (conventional) choice of b above. (The states of moving particles are not generally parity eigenstates.)

Parity is interesting because parity is experimentally conserved for both the color and electromagnetic interactions (meaning that if a process is possible, its mirror image is also possible and equally probable). This can help us construct appropriate (parity-invariant) Lagrangian densities for these interactions.

The concept of parity also delivers some insight into why Dirac spinors have four components instead of the two that one might think sufficient for describing a spin-$\frac{1}{2}$ particle. We have seen that for massless particles, the Dirac equation splits into two independent chiral (right- and left-handed) subtypes, each of which has the expected two-component spinor. So why does nature need both subtypes in one equation? We see that the parity transformation swaps the subtypes. Since the color and electromagnetic interactions are parity-invariant, we need both types to evolve together in the right way so that parity can be conserved. Parity-invariance therefore requires that the Dirac equation involves four components to describe the two coevolving chiral subtypes.

BOX 17.1   The Generator for Rotations

Our goal in this box is to show that

$$X_{Rz} = \frac{1}{2} \begin{pmatrix} \sigma_z & 0 \\ 0 & \sigma_z \end{pmatrix} \tag{17.40}$$

satisfies the constraints we established in Equation 17.14 (repeated here):

$$[X_{Rz}, \gamma^t] = 0, \quad [X_{Rz}, \gamma^x] = i\gamma^y, \quad [X_{Rz}, \gamma^y] = -i\gamma^x, \quad [X_{Rz}, \gamma^z] = 0 \tag{17.41}$$

for the generator of rotations around the z axis.

Exercise 17.1.1   Prove that the given X_{Rz} satisfies those constraints. (*Hint:* Remember that $[\sigma_j, \sigma_k] = 2i \sum_m \epsilon_{jkm} \sigma_m$, where j, k, and m are indices with values x, y, z, and $\epsilon_{jkm} = +1$ if the indices go cyclically in alphabetical order, -1 if they go cyclically in reverse order, and 0 if any two indices are the same.)

BOX 17.2 The Generator for Boosts

The goal of this box is to show that we can use Equation 17.19, repeated here,

$$\Lambda^{\mu}{}_{\nu}\gamma^{\nu} = \begin{bmatrix} \gamma^t + \delta\theta\gamma^x \\ \delta\theta\gamma^t + \gamma^x \\ \gamma^y \\ \gamma^z \end{bmatrix} \qquad (17.42)$$

and the analogue of Equation 17.11

$$M_{Bx}^{-1}\gamma^{\mu}M_{Bx} \approx \left(I + i\,\delta\theta\,X_{Bx}\right)\gamma^{\mu}\left(I - i\,\delta\theta\,X_{Bx}\right)$$

$$\approx \gamma^{\mu} + i\,\delta\theta\left(X_{Bx}\gamma^{\mu} - \gamma^{\mu}X_{Bx}\right) = \gamma^{\mu} + i\,\delta\theta[X_{Bx}, \gamma^{\mu}] \quad (17.43)$$

to generate a set of commutation relations that are solved by

$$X_{Bx} = -\frac{i}{2}\begin{pmatrix} \sigma_x & 0 \\ 0 & -\sigma_x \end{pmatrix}. \qquad (17.44)$$

Exercise 17.2.1 Find the four commutation relations implied by Equation 17.43 and 17.42, and verify that X_{Bx} satisfies those relations. (*Hint:* You will find the identity $\{\sigma_i, \sigma_j\} = 2\delta_{ij}I$ helpful.)

BOX 17.3   Boosting from Rest

In this box, we will see that we can Lorentz-boost the Dirac spinor of a particle at rest to get the spinor for a moving particle. We saw in Equation 7.15 that the generator for boosts in the $+z$ direction is

$$X_{Bz} = -\frac{i}{2} \begin{pmatrix} \sigma_z & 0 \\ 0 & -\sigma_z \end{pmatrix}. \tag{17.45}$$

This means that the transformation matrix is

$$M_{Bz}(\theta) = e^{-iX_{Bz}\theta} = \begin{pmatrix} e^{-\frac{1}{2}\sigma_z\theta} & 0 \\ 0 & e^{+\frac{1}{2}\sigma_z\theta} \end{pmatrix}. \tag{17.46}$$

Exercise 17.3.1   Argue that $e^{\pm\frac{1}{2}\sigma_z\theta} = \cosh(\frac{1}{2}\theta)I \pm \sinh(\frac{1}{2}\theta)\sigma_z$ and write M_{Bz} as a 4×4 matrix. (*Hints:* Look up the power series for $\cosh x$ and $\sinh x$, and to save writing, define $c \equiv \cosh\frac{1}{2}\theta$ and $s \equiv \sinh\frac{1}{2}\theta$ at the right point.)

BOX 17.3 Boosting from Rest (cont.)

Now, note that

$$\cosh(\tfrac{1}{2}\theta) \pm \sinh(\tfrac{1}{2}\theta) = e^{\pm\frac{1}{2}\theta} = \sqrt{e^{\pm\theta}} = \sqrt{\cosh\theta \pm \sinh\theta}. \quad (17.47)$$

Also, note that in an ordinary Lorentz transformation of the four-momentum of a particle at rest, we would have

$$E = p'^{\,t} = \cosh\theta\ p^t + \sinh\theta\ p^z = \cosh\theta\ m + 0, \quad (17.48a)$$

$$|\,\vec{p}\,| = p'^{\,z} = \sinh\theta\ p^t + \cosh\theta\ p^z = \sinh\theta\ m + 0, \quad (17.48b)$$

so $\cosh\theta = E/m$ and $\sinh\theta = |\,\vec{p}\,|/m$. Finally, recall that the amplitude part of the solution for a particle at rest is

$$\begin{bmatrix} u_L(p) \\ u_R(p) \end{bmatrix} = \begin{bmatrix} \sqrt{m}\ \xi \\ \sqrt{m}\ \xi \end{bmatrix}. \quad (17.49)$$

Exercise 17.3.2 Use the transformation matrix from the last exercise to find the four-component Dirac spinor after boosting, expressing that spinor in terms of E, $|\,\vec{p}\,|$, and the components a and b of the spinor ξ. Then argue that this solution is consistent with the statement that the general solution is

$$\begin{bmatrix} u_L(p) \\ u_R(p) \end{bmatrix} = \begin{bmatrix} \sqrt{p\cdot\sigma}\ \xi \\ \sqrt{p\cdot\overline{\sigma}}\ \xi \end{bmatrix}. \quad (17.50)$$

BOX 17.4   Spin Eigenvalues of Antiparticle Basis States

The goal of this box is to determine the spin eigenvalues of the basis states for an antiparticle at rest:

$$v_0^- = \begin{bmatrix} \sqrt{m}\,\eta^- \\ -\sqrt{m}\,\eta^- \end{bmatrix} \quad \text{and} \quad v_0^+ = \begin{bmatrix} \sqrt{m}\,\eta^+ \\ -\sqrt{m}\,\eta^+ \end{bmatrix}, \tag{17.51a}$$

$$\text{where} \quad \eta^- \equiv -\begin{bmatrix} 1 \\ 0 \end{bmatrix} \quad \text{and} \quad \eta^+ \equiv \begin{bmatrix} 0 \\ 1 \end{bmatrix}. \tag{17.51b}$$

Exercise 17.4.1   Using the same approach as in the body of the chapter, show that these states are eigenvectors of the generator of rotations around the z axis, and determine their eigenvalues.

BOX 17.5 Helicity Values for Dirac Basis States

The helicity operator is

$$\hat{h} \equiv \frac{\vec{p} \cdot \vec{\Sigma}}{|\vec{p}|}, \quad \text{where} \quad \vec{\Sigma} \equiv \pm \frac{1}{2} \begin{pmatrix} \vec{\sigma} & 0 \\ 0 & \vec{\sigma} \end{pmatrix} \qquad (17.52)$$

and where the positive sign applies to particles and the negative sign to antiparticles, to make the operator eigenvalues consistent with the "physical spin" direction for each type.

Exercise 17.5.1 Show that the following extremely relativistic states

$$u^+(p) = \sqrt{2E_{\vec{p}}} \begin{bmatrix} 0 \\ 0 \\ 1 \\ 0 \end{bmatrix}, \qquad u^-(p) = \sqrt{2E_{\vec{p}}} \begin{bmatrix} 0 \\ 1 \\ 0 \\ 0 \end{bmatrix}, \qquad (17.53a)$$

$$v^-(p) = \sqrt{2E_{\vec{p}}} \begin{bmatrix} 0 \\ 0 \\ 1 \\ 0 \end{bmatrix}, \quad \text{and} \quad v^+(p) = \sqrt{2E_{\vec{p}}} \begin{bmatrix} 0 \\ 1 \\ 0 \\ 0 \end{bmatrix} \qquad (17.53b)$$

are eigenstates of the helicity operator (assuming that the particle's momentum is entirely in the $+z$ direction) and determine the eigenvalues.

BOX 17.6   The Chirality Operator

Exercise 17.6.1   Show that

$$\gamma^5 \equiv i\gamma^t\gamma^x\gamma^y\gamma^z = \begin{pmatrix} -I & 0 \\ 0 & I \end{pmatrix}. \tag{17.54}$$

(*Hint:* Work in compact notation first.)

BOX 17.7 The Charge-Conjugation Operator

In this box, we will see that charge-conjugation operator C, defined by the expression

$$\psi^c \equiv C\psi = -i\gamma^y \psi^*, \tag{17.55}$$

correctly transforms a solution to the Dirac equation to the one corresponding to the antiparticle of that solution with the same physical spin. We can most conveniently do this in the particle's rest frame.

Exercise 17.7.1   Calculate what C does to the Dirac basis states $\psi = u_0^s e^{-imt}$ and $\psi = v_0^s e^{+imt}$ (where $s = +$ or $-$) for a particle at rest.

Homework Problems

P17.1 In this problem, we will explore the *rapidity* θ as an alternative way of describing Lorentz transformations.

a. Suppose we define $\theta \equiv \tanh^{-1}\beta$, where β is the relative x-velocity of two frames. The definition of the inverse hyperbolic tangent is

$$\tanh^{-1}x = \frac{1}{2}\ln\frac{1+x}{1-x}. \qquad (17.56)$$

What does θ approach as $\beta \to \pm 1$?

b. The definitions of the hyperbolic sine and cosine are

$$\sinh\theta = \tfrac{1}{2}(e^{\theta}-e^{-\theta}), \quad \cosh\theta = \tfrac{1}{2}(e^{\theta}+e^{-\theta}).(17.57)$$

(Note the similarity to the expressions for $\sin\theta$ and $\cos\theta$ in terms of $e^{\pm i\theta}$.) Show that $\sinh\theta = \gamma\beta$, and $\cosh\theta = \gamma$, where (as usual) $\gamma \equiv (1-\beta^2)^{-1/2}$.

c. Note that if frame S'' has an x-velocity of β_2 as observed in frame S', which in turn has an x-velocity of β_1 with respect to frame S, then the x-velocity of frame S'' with respect to S is

$$\beta = \frac{dx}{dt} = \frac{\gamma_1\beta_1 dt' + \gamma_1 dx'}{\gamma_1 dt' + \gamma_1\beta_1 dx'}$$
$$= \frac{\beta_1 + dx'/dt'}{1+\beta_1 dx'/dt'} = \frac{\beta_1+\beta_2}{1+\beta_1\beta_2}. \qquad (17.58)$$

(This is the famous Einstein velocity addition formula.) Use Equation 17.56 to show that $\tanh^{-1}\beta = \theta_1 + \theta_2$, meaning that the rapidities of the two frames simply *add*. This is one of the main advantages of using the rapidity. (*Hint:* Work from both ends.)

P17.2 The adjoint of the spinor transformation $\psi' = M_\Lambda\psi$ is $\overline{\psi}' = \psi'^\dagger\gamma^t = \psi^\dagger M_\Lambda^\dagger\gamma^t$. For the latter to be equivalent to $\overline{\psi}' = \overline{\psi}M_\Lambda^{-1} = \psi^\dagger\gamma^t M_\Lambda^{-1}$, we must have $M_\Lambda^\dagger\gamma^t = \gamma^t M_\Lambda^{-1}$. In this problem, we will show that this is true, which is an important consistency check. It is enough to show that this is true for the infinitesimal transformation $M_\Lambda \approx I - i\delta\theta X_\Lambda$, because we can build a finite transformation out of a sequence of such transformations.

a. The generators X_{Rx}, X_{Ry}, and X_{Rz} for rotations are all Hermitian (because the Pauli matrices are Hermitian). Show that for Hermitian generators, $M_\Lambda^\dagger\gamma^t = \gamma^t M_\Lambda^{-1}$ will be true for infinitesimal transformations so long as $[X_R, \gamma^t] = 0$, which is true by Equation 17.14.

b. However, the generators X_{Bx}, X_{By}, and X_{Bz} for boosts are all anti-Hermitian matrices ($X_B^\dagger = -X_B$). Show that for anti-Hermitian matrices, $M_\Lambda^\dagger\gamma^t = \gamma^t M_\Lambda^{-1}$ will be true for infinitesimal transformations so long as the anticommutator $\{X_B, \gamma^t\} = 0$, and verify that the latter is true for all three boost generators.

P17.3 Calculate M_{Bx} for a finite rapidity θ.

P17.4 Lagrangian densities that we will develop will include quantities such as $\overline{\psi}\gamma^\mu\psi$ and $\overline{\psi}\gamma^\mu\gamma^5\psi$, which physicists call **bilinears**. We have seen that the bilinear $\overline{\psi}\psi$ is a Lorentz scalar.

a. Show that under a Lorentz transformation, $\overline{\psi}\gamma^\mu\psi$ is a set of four simple numbers that transform like the components of a four-vector.

b. Show that the four numbers $\overline{\psi}\gamma^\mu\gamma^5\psi$ also transform like the components of a four-vector.

c. How do the spatial components of these quantities transform under the parity operation? Which would we call a "normal" vector and which would we call a "pseudo-vector" (a vector that, like angular momentum, is not changed by a parity operation)?

P17.5 In this problem, we will generate the Dirac equation by boosting the spinor for a particle at rest.

a. For a particle at rest, $u_R = u_L$. Show that this means that

$$(\gamma^t - I)u(0) = 0. \qquad (17.59)$$

b. Prove that $M_{Bz}\gamma^t M_{Bz}^{-1} = p_\mu\gamma^\mu/m$, where p^μ is the boosted particle's four-momentum.

c. Use this to show that $(p_\mu\gamma^\mu - mI)u(p) = 0$, where $u(p)$ is the boosted particle's spinor.

d. Substitute $i\partial_\mu = p_\mu$ to arrive at the Dirac equation. (Note that $i\partial_\mu e^{-ip\cdot x} = p_\mu e^{-ip\cdot x}$.)

P17.6 We will see in the next chapter that the Hamiltonian operator for a free Dirac particle is

$$H = -i\gamma^j\partial_j + m \quad \text{(where j is a *spatial* index).} \quad (17.60)$$

a. We know from ordinary quantum mechanics that the momentum operator $p^j = -i\partial_j$, so the orbital angular momentum operator is

$$L_z = (\vec{r}\times\vec{p})_z = xp_y - yp_x = -ix\partial_y + iy\partial_x. \; (17.61)$$

Calculate the commutator $[H, L_z]$. Note that these operators do *not* commute, meaning that L_z is *not* conserved for a Dirac particle.

b. Calculate the commutator $[H, \Sigma_z]$, where

$$\Sigma_z = X_{Rz} = \frac{1}{2}\begin{pmatrix} \sigma_z & 0 \\ 0 & \sigma_z \end{pmatrix} \qquad (17.62)$$

is the spin angular momentum operator derived in this chapter. Note that H and Σ_z do not commute either, meaning that spin angular momentum is also *not* conserved.

c. Show that $[H, L_z + \Sigma_z] = 0$, meaning that the z component of the *total* angular momentum of a Dirac particle is conserved.

$\left(a_{\vec{p},\text{boson}}^{\dagger}\right)^2$
$\vec{p}$
$\vec{p}$
poof!
$\left(a_{\vec{p},\text{fermion}}^{\dagger}\right)^2$
sorry, dude

18. QUANTIZING THE DIRAC FIELD

The past two chapters explored the Dirac equation, which describes free spin-$\frac{1}{2}$ particles and antiparticles. This equation is crucial because most of the particles in the universe we consider "matter" are spin-$\frac{1}{2}$ particles (the photon is the only common particle that is not). But our treatment in Chapters 16 and 17 treated the solutions to the Dirac equation as classical fields. In this chapter, we will take the crucial step of converting these fields to operator fields, following the trajectory that worked for the scalar (spin-0) field.

However, we will see that duplicating the steps that worked for the scalar field does *not* work for quantizing the Dirac field. Solving the problem will require a somewhat different approach that will imply that spin-$\frac{1}{2}$ particles must behave as *fermions* (which obey the Pauli exclusion principle) rather than *bosons* (which do not). This is the crucial "spin-statistics" theorem that lies at the heart of quantum physics in general.

Readers who have not read Chapters 11–14 should skip this chapter.

18.1 Our First Attempt at Quantization

Recall from Chapter 13 that we developed a process for quantizing a field:

Step 1. Write down a classical Lagrangian density.

Step 2. Calculate the energy density and other current densities.

Step 3. Develop creation and annihilation operators.

Step 4. Quantize the field by writing it as a linear combination of plane wave solutions with annihilation and creation operators serving as the plane wave amplitudes.

Step 5. Normal-order the resulting operators in the expressions for the current densities (ignoring the commutation relations).

Let us work through these steps in sequence.

Step 1. The Dirac field Lagrangian density (see Equation 17.1) is

$$\mathcal{L} = i\overline{\psi}\gamma^{\mu}\partial_{\mu}\psi - m\overline{\psi}\psi. \tag{18.1}$$

Step 2. We saw in Chapter 6 that we can calculate the stress-energy tensor from the Lagrangian density for a set of fields ψ_n as follows:

$$T^{\mu\nu} = \left[\sum_{n}\frac{\partial\mathcal{L}}{\partial(\partial_{\mu}\psi_n)}\partial^{\nu}\psi_n\right] - g^{\mu\nu}\mathcal{L}, \tag{18.2}$$

where the sum is over the different fields that appear in the Lagrangian density. In this case, we have two fields ($\overline{\psi}$ and ψ), but only the latter has a derivative that appears in the Lagrangian density. Therefore, in our particular case,

$$T^{\mu\nu} = \frac{\partial\mathcal{L}}{\partial(\partial_{\mu}\psi)}\partial^{\nu}\psi - g^{\mu\nu}\mathcal{L}. \tag{18.3}$$

In Box 18.1, you will show that the energy-density component is

$$T^{tt} = \overline{\psi}(-i\gamma^{j}\partial_j + m)\psi, \tag{18.4}$$

where j is a spatial index. But note that the Dirac equation itself implies that

$$0 = i\overline{\psi}\gamma^\mu \partial_\mu \psi - m\overline{\psi}\psi = i\overline{\psi}\gamma^t \partial_t \psi + i\overline{\psi}\gamma^j \partial_j \psi - m\overline{\psi}\psi$$

$$\Rightarrow \;\; i\overline{\psi}\gamma^t \partial_t \psi = -i\overline{\psi}\gamma^j \partial_j \psi + m\overline{\psi}\psi. \tag{18.5}$$

Comparing this with Equation 18.4, we see that

$$T^{tt} = i\overline{\psi}\gamma^t \partial_t \psi = i\psi^\dagger \partial_t \psi \tag{18.6}$$

since $\overline{\psi} \equiv \psi^\dagger \gamma^t$ and $(\gamma^t)^2 = I$. This will be a *lot* easier to deal with.

We will also be interested in the conserved current corresponding to the Lagrangian density's invariance under phase transformations $\psi' = e^{-iq\theta}\psi$, which we will see in the next chapter is connected to electric charge. Again, we see from Chapter 6 that the formula for this current is

$$J^\mu = \sum_n \frac{\partial \mathcal{L}}{\partial(\partial_\mu \psi_n)}\left[\frac{d\psi_n}{d\theta}\right]_0 \tag{18.7}$$

since the transformation does not involve the coordinates. Again, for the Dirac case, we have only one field that has a partial derivative in the Lagrangian density. In Box 18.1, you will show that the current in this case is simply

$$J^\mu = q\overline{\psi}\gamma^\mu \psi. \tag{18.8}$$

Step 3 involves defining creation and annihilation operators. We will assume as a first guess that they will be the same as the operators we defined in Chapter 12 for the scalar field, obeying the same commutation relations.

Step 4 involves writing the quantized field ψ as an operator field comprised of a linear combination of all possible plane wave solutions to the Dirac equation with annihilation and creation operators serving as the coefficients:

$$\psi = \int \frac{d^3p}{(2\pi)^3} \frac{1}{\sqrt{2E_{\vec{p}}}} \sum_s \left[u^s(p)\, a_{s\vec{p}}\, e^{-ip\cdot x} + v^s(p)\, b_{s\vec{p}}^\dagger\, e^{+ip\cdot x} \right]. \tag{18.9}$$

This is just like the corresponding expression for the scalar field, except for two things. First, we must of course include the Dirac amplitude spinors $u^s(p)$ and $v^s(p)$ as part of the plane wave expansion, because the resulting field must be a four-component spinor. Second, we must also sum over the two possible spin states, and the expansion coefficients will in general depend on the spin state, so s appears as a subscript on the b and c coefficients: $a_{s\vec{p}}$ will annihilate a particle with spin s and momentum $\vec{p}$, and $b_{s\vec{p}}^\dagger$ will create an antiparticle with spin s and momentum $\vec{p}$.

Step 5. If we now substitute Equation 18.9 into Equation 18.6, we find (see Box 18.2) that the total energy for our field is

$$E = \int \frac{d^3p}{(2\pi)^3} E_{\vec{p}} \sum_s \left[a_{s\vec{p}}^\dagger a_{s\vec{p}} - b_{s\vec{p}}\, b_{s\vec{p}}^\dagger \right]. \tag{18.10}$$

A very similar calculation (see Box 18.3) yields the following result for the total charge:

$$Q = q \int \frac{d^3p}{(2\pi)^3} \sum_s \left[a_{s\vec{p}}^\dagger a_{s\vec{p}} + b_{s\vec{p}}\, b_{s\vec{p}}^\dagger \right]. \tag{18.11}$$

These results are disastrous! When we did these same calculations for the scalar field Lagrangian density, we got a plus sign in the equation for the energy and a minus sign in the equation for the charge, allowing us to interpret the b-particles as antiparticles with positive energy and negative charge. But such an interpretation is not possible here. We are in deep trouble, because it seems that there is no lower bound on the energy of a system of Dirac particles.

18.2 Successfully Quantizing the Field

The way to avoid this disaster was discovered by Jordan and Wigner in 1928. What if
the Dirac creation and annihilation operators satisfied the following *anticommutation*
relations:

$$\{a_{r\vec{p}}, a^{\dagger}_{s\vec{q}}\} = \delta_{rs}(2\pi)^3\delta^3(\vec{p} - \vec{q}), \tag{18.12a}$$

$$\{a_{r\vec{p}}, a^{\dagger}_{s\vec{q}}\} = \delta_{rs}(2\pi)^3\delta^3(\vec{p} - \vec{q}), \tag{18.12b}$$

$$\{b_{r\vec{p}}, b_{s\vec{q}}\} = \{b^{\dagger}_{r\vec{p}}, b^{\dagger}_{s\vec{q}}\} = \{b_{r\vec{p}}, b_{s\vec{q}}\} = \{b^{\dagger}_{r\vec{p}}, b^{\dagger}_{s\vec{q}}\} = 0, \tag{18.12c}$$

instead of the corresponding *commutation* relations obeyed by the scalar fields? Then,
because the relevant anticommutation relation implies that

$$b_{r\vec{p}}\, b^{\dagger}_{r\vec{p}} = -b^{\dagger}_{r\vec{p}}\, b_{r\vec{p}} + \delta_{rs}(2\pi)^3\delta^3(0), \tag{18.13}$$

we could rewrite Equation 18.10 as

$$E = \int \frac{d^3p}{(2\pi)^3} E_{\vec{p}} \sum_s \left[a^{\dagger}_{s\vec{p}} a_{s\vec{p}} + b^{\dagger}_{s\vec{p}} b_{s\vec{p}} - (2\pi)^3\delta^3(0) \right]. \tag{18.14}$$

In other words, putting the operators in normal order (and throwing away the infinite
$\delta^3(0)$ that results, just as we did before) delivers precisely the positive-energy result we
need! The same trick works to fix Equation 18.11: putting the operators in normal order
and ignoring the resulting $\delta^3(0)$ yields

$$Q = q \int \frac{d^3p}{(2\pi)^3} \sum_s \left[a^{\dagger}_{s\vec{p}} a_{s\vec{p}} - b^{\dagger}_{s\vec{p}} b_{s\vec{p}} \right], \tag{18.15}$$

which allows us to interpret this as the total charge of a set of positively charged particles
and negatively charged antiparticles.

But back in Chapter 12, didn't we prove that the basic character of the creation
and annihilation operators *requires* the commutation relations? Yes, but we implicitly
assumed that one could put as many particles as one wanted into every momentum state.
The anticommutation relations also work if we relax this assumption. Let's see how this
happens.

We can begin as we did in Chapter 12 by assuming a set of discrete momentum
values $\vec{p}_k$, and a Fock space whose kets specify the number of particles having each
momentum value: $| \ldots, n_{k-1}, n_k, n_{k+1}, \ldots \rangle$. There must exist an annihilation operator
a_k that lowers the count in slot k by 1. In particular, we must be able to annihilate one
particle in the kth slot to zero:

$$a_k | \ldots, 1, \ldots \rangle = c_1 | \ldots, 0, \ldots \rangle. \tag{18.16}$$

We also can't lower the number in this slot below zero, so we must have

$$a_k | \ldots, 0, \ldots \rangle = 0. \tag{18.17}$$

To save writing, let's define

$$| \ldots, n_k = 1, \ldots \rangle = |, 1_k, \rangle \quad \text{and} \quad | \ldots, n_k = 0, \ldots \rangle = |, 0_k, \rangle. \tag{18.18}$$

Now consider the adjoint operator. Multiply Equation 18.16 on the left by $\langle, 0_k, |$
to get

$$\langle, 0_k, |a_k|, 1_k, \rangle = c_1\langle, 0_k, |, 0_k, \rangle = c_1, \tag{18.19}$$

assuming that the vacuum state is normalized. Taking the adjoint of this yields

$$\langle, 1_k, |a^{\dagger}_k|, 0_k, \rangle = c^*_1 \quad \Rightarrow \quad a^{\dagger}_k|, 0_k, \rangle = c^*_1|, 1_k, \rangle \tag{18.20}$$

(as one can see by multiplying both sides of the second equation by $\langle 1_k |$ on the left). So we see that if a_k is an annihilation operator, $a_k^\dagger$ will be a creation operator. Note also that if we want $N_k \equiv a_k^\dagger a_k$ to be a number operator, then we must normalize the annihilation operator so that

$$1 = \langle, 1_k, |N_k|, 1_k, \rangle = \langle, 1_k, |a_k^\dagger a_k|, 1_k, \rangle$$
$$= \langle, 0_k, |c_1^* c_1|, 0_k, \rangle = |c_1|^2. \tag{18.21}$$

So far we have not assumed anything about the commutation relations. But suppose that a_k and $a_k^\dagger$ obey the anticommutation relation

$$\{a_k, a_k^\dagger\} \equiv a_k a_k^\dagger + a_k^\dagger a_k = 1 \;\Rightarrow\; a_k a_k^\dagger = 1 - a_k^\dagger a_k = 1 - N_k. \tag{18.22}$$

Is this even possible? If so, under what conditions? Consider trying to create a second particle with momentum $\vec{p}_k$:

$$a_k^\dagger |, 1_k, \rangle = c_2^* |, 2_k, \rangle. \tag{18.23}$$

This implies that

$$|c_2|^2 = \langle, 1_k, |\, a_k a_k^\dagger |, 1_k, \rangle = \langle, 1_k, |(1 - N_k)|, 1_k, \rangle$$
$$= \langle, 1_k, |(1 - 1)|, 1_k, \rangle = 0. \tag{18.24}$$

This implies that we *cannot* create an additional particle in that state: we can have at most *one* particle with any particular momentum value. On the other hand, there is no contradiction with creating the first particle:

$$|c_1|^2 = \langle, 0_k, |\, a_k a_k^\dagger |, 0_k, \rangle = \langle, 0_k, |(1 - N_k)|, 0_k, \rangle$$
$$= \langle, 0_k, |(1 - 0)|, 0_k, \rangle = 1, \tag{18.25}$$

which is the value we found before for $|c_1|^2$. So having the creation and annihilation operators obey the anticommutation relation we need works just fine *if* our possible Fock-space states for momentum $\vec{p}_k$ are limited to $|, 0_k, \rangle$ and $|, 1_k, \rangle$. In particular, this means that we cannot put more than one Dirac particle into any given momentum state. This is the **Pauli exclusion principle**.

One can show that this limitation directly implies that we also must have

$$\{a_j, a_k^\dagger\} = 0 \quad \text{when } j \neq k, \tag{18.26a}$$

$$\{a_j, a_k\} = 0, \quad \text{and} \quad \{a_j^\dagger, a_k^\dagger\} = 0. \tag{18.26b}$$

(See Box 18.4 for the details.) These are exactly like the *commutation* relations for scalar (spin 0) particles, simply with anticommutators replacing the commutators. This is a good thing (it makes things easier to remember).

When we go to the continuous limit and apply the anticommutation relations to both the a and b operators for Dirac particles and antiparticles, respectively, while also accounting for spin, the relations become

$$\{a_{r\vec{p}}, a_{s\vec{q}}^\dagger\} = \delta_{rs}(2\pi)^3 \delta^3(\vec{p} - \vec{q}), \tag{18.27a}$$

$$\{b_{r\vec{p}}, b_{s\vec{q}}^\dagger\} = \delta_{rs}(2\pi)^3 \delta^3(\vec{p} - \vec{q}), \tag{18.27b}$$

$$\{a_{r\vec{p}}, a_{s\vec{q}}\} = \{a_{r\vec{p}}^\dagger, a_{s\vec{q}}^\dagger\} = \{b_{r\vec{p}}, b_{s\vec{q}}\} = \{b_{r\vec{p}}^\dagger, b_{s\vec{q}}^\dagger\} = 0, \tag{18.27c}$$

as stated previously.

18.3 The Spin-Statistics Theorem

What we have just seen is a huge result! We saw in the last chapter that Dirac fields describe spin-$\frac{1}{2}$ particles and antiparticles, and we now see that in order to construct a meaningful quantum field theory for these fields, the associated particles must obey the Pauli exclusion principle and therefore Fermi-Dirac statistics. This means that Dirac particles are **fermions**. We saw earlier that scalar fields (which describe spin-0 particles) satisfy Bose-Einstein statistics and are therefore **bosons**. Physicists call this astonishing result the **spin-statistics theorem**: it describes the necessary logical connection between a particle's spin and the statistics that the particle must obey.

The mathematics not only explains the connection between spin and statistics, it also explains why the wavefunctions of particle combinations must be symmetric or antisymmetric under particle exchange. Consider constructing a two-particle momentum state by applying two creation operators to the vacuum, creating particles with distinct momenta $\vec{p}$ and $\vec{q}$, respectively (but the same spin s):

$$| \vec{q}, \vec{p} \rangle = a^{\dagger}_{s\vec{q}} a^{\dagger}_{s\vec{p}} | 0 \rangle. \tag{18.28}$$

Now, *boson* creation operators obey the *commutation* principle $[a^{\dagger}_{\vec{q}}, a^{\dagger}_{\vec{p}}] = 0$, meaning that $a^{\dagger}_{\vec{q}} a^{\dagger}_{\vec{p}} = a^{\dagger}_{\vec{p}} a^{\dagger}_{\vec{q}}$. Therefore, this two-particle state is *symmetric* under the exchange of the particles:

$$| \vec{q}, \vec{p} \rangle = a^{\dagger}_{\vec{q}} a^{\dagger}_{\vec{p}} | 0 \rangle = +a^{\dagger}_{\vec{p}} a^{\dagger}_{\vec{q}} | 0 \rangle = +| \vec{p}, \vec{q} \rangle. \tag{18.29}$$

But *fermion* creation operators obey the *anticommutation* relation $\{ a^{\dagger}_{s\vec{q}}, a^{\dagger}_{s\vec{p}} \} = 0$, meaning that $a^{\dagger}_{s\vec{q}} a^{\dagger}_{s\vec{p}} = -a^{\dagger}_{s\vec{p}} a^{\dagger}_{s\vec{q}}$. Therefore, this two-particle state is *antisymmetric* under the exchange of the particles (when they have the same spin):

$$| \vec{q}, \vec{p} \rangle = a^{\dagger}_{s\vec{q}} a^{\dagger}_{s\vec{p}} | 0 \rangle = -a^{\dagger}_{s\vec{p}} a^{\dagger}_{s\vec{q}} | 0 \rangle = -| \vec{p}, \vec{q} \rangle. \tag{18.30}$$

This explains the requirement that in standard quantum mechanics, we must construct symmetric or antisymmetric wavefunctions to represent systems of identical bosons or fermions, respectively.

Please note, however, that we don't have to worry about this issue at all when we use Fock-space kets, which simply tell us we have one particle with each value of momentum. It is only when we effectively *label* the particles by using ordinary quantum states like $| \vec{q}, \vec{p} \rangle$, which says "particle 1 is in a state with momentum $\vec{q}$ and particle 2 is in a state with momentum $\vec{p}$," that we must explicitly make the states of identical particles symmetric or antisymmetric.

The spin-statistics theorem is one of the most consequential ideas in all of physics. We see that the mathematics of Lorentz invariance allows for two completely different kinds of solutions to the free field problem: solutions to the Klein-Gordon equation and solutions to the Dirac equation. To construct a logically self-consistent quantum field theory for solutions to the former, we must use commutation relations; to do the same for the latter requires anticommutation relations. Commutation relations imply Bose-Einstein statistics; anticommutation relations imply Fermi-Dirac statistics. Both kinds of particles exist in our universe, almost as if something allowed mathematically and logically must exist physically.

The existence of fermions is extremely consequential because the rich structure of the chemical elements express the implications of the Pauli exclusion principle. A universe whose elementary particles were bosons would be a very uninteresting place (because the ground state of every system would have all its particles sitting undifferentiated in the same physical quantum state). Chemistry would not exist, biology would not exist, and we would not exist. We would not be around to contemplate a universe whose important elementary particles were not solutions to the Dirac equation.

As consequential as the spin-statistics theorem is, we only poorly understand why all the assumptions that go into it are necessary. Feynman once brashly stated that one should be able to explain to a first-year college student any physical principle physicists truly understood. He was challenged by David Goodstein (a colleague at Caltech) to explain the spin-statistics theorem at such a level. Feynman said, "I'll prepare a [first-year] lecture on it." But only days later, Feynman admitted defeat, stating, "That means we really don't understand it." (From Goodstein and Goodstein, *Feynman's Lost Lecture*, Norton, 1996, p. 52.)

18.4 Units

A brief discussion of units is useful here. A Lagrangian density must have units of $[E]^4$ (where I am using $[E]$ as a shorthand for "units of energy"). The scalar field and Dirac field Lagrangian densities are

$$\mathcal{L} = \partial^\mu \phi^\dagger \partial_\mu \phi - m^2 \phi^\dagger \phi \quad \text{and} \quad \mathcal{L} = i\overline{\psi}\gamma^\mu \partial_\mu \psi - m\overline{\psi}\psi. \tag{18.31}$$

For the units of $\mathcal{L}$ to be $[E]^4$, the units of ϕ must be $[E]$ and the units of ψ must be $[E]^{3/2}$ (look at the mass terms). If we look at the field definitions

$$\phi = \int \frac{d^3p}{(2\pi)^3} \frac{1}{\sqrt{2E_{\vec{p}}}} \left[a_{\vec{p}}\, e^{-ip\cdot x} + b_{\vec{p}}^\dagger\, e^{+ip\cdot x} \right] \quad \text{and} \tag{18.32a}$$

$$\psi = \int \frac{d^3p}{(2\pi)^3} \frac{1}{\sqrt{2E_{\vec{p}}}} \sum_s \left[u^s(p)\, a_{s\vec{p}}\, e^{-ip\cdot x} + v^s(p)\, b_{s\vec{p}}^\dagger\, e^{+ip\cdot x} \right], \tag{18.32b}$$

we see that since $u(p)$ and $v(p)$ are normalized so that $u^\dagger u = v^\dagger v = 2E_{\vec{p}}$, it makes sense that ψ has an extra factor of $[E]^{1/2}$ compared to ϕ. But since d^3p has units of $[E]^3$ it also looks like the ϕ field should have units of $[E]^3/E^{1/2} = [E]^{5/2}$, not $[E]$. What is going on here?

The answer is that the a and b operators also have units. They obey commutation or anticommutation relations $[a_{\vec{p}}, a_{\vec{q}}^\dagger]$ or $\{a_{\vec{p}}^\dagger, a_{\vec{q}}^\dagger\} = (2\pi)^3 \delta^3(\vec{p} - \vec{q})$. Since a basic property of the Dirac delta is that

$$1 = \int d^3p\, \delta^3(\vec{p} - \vec{q}), \tag{18.33}$$

we see that the Dirac delta must have units of $[E]^{-3}$. Therefore, the creation and annihilation operators have units of $E^{-3/2}$, which leaves ϕ with units of $[E]$ and ψ with units of $[E]^{3/2}$, as required.

BOX 18.1 Important Currents

The goal in this box is to take the Dirac Lagrangian density

$$\mathcal{L} = i\overline{\psi}\gamma^{\mu}\partial_{\mu}\psi - m\overline{\psi}\psi \tag{18.34}$$

and calculate the energy density and charge current.

Exercise 18.1.1 According to Equation 18.3, we have

$$T^{\mu\nu} = \frac{\partial\mathcal{L}}{\partial(\partial_{\mu}\psi)}\partial^{\nu}\psi - g^{\mu\nu}\mathcal{L}. \tag{18.35}$$

Calculate the energy density (T^{tt}) from this.

Exercise 18.1.2 According to Equation 18.7, we have

$$J^{\mu} = \sum_{n}\frac{\partial\mathcal{L}}{\partial(\partial_{\mu}\psi_{n})}\left[\frac{d\psi_{n}}{d\theta}\right]_{0}. \tag{18.36}$$

Find the conserved current corresponding to the symmetry of the Dirac Lagrangian density under the transformation $\psi' = e^{-iq\theta}\psi$.

BOX 18.2 The Expression for the Energy

Our goal in this box is to use Equation 18.9 (repeated here)

$$\psi = \int \frac{d^3p}{(2\pi)^3} \frac{1}{\sqrt{2E_{\vec{p}}}} \sum_s \left[u^s(p)\, a_{s\vec{p}}\, e^{-ip\cdot x} + v^s(p)\, b^\dagger_{s\vec{p}}\, e^{+ip\cdot x} \right] \quad (18.37)$$

to express the energy $E = \int d^3x\, T^{tt} = \int d^3x\, i\psi^\dagger \partial_t \psi$ in terms of the annihilation operators $a_{s\vec{p}}$ and $b_{s\vec{p}}$ and their adjoint creation operators).

We begin by noting that $\partial_t\, e^{\pm ip\cdot x} = \pm i E_{\vec{p}}\, e^{\pm ip\cdot x}$. With this in mind, substituting Equation 18.37 into the energy expression yields

$$
\begin{aligned}
E = \int d^3x\, &\frac{d^3p}{(2\pi)^3} \frac{d^3q}{(2\pi)^3} \frac{i}{\sqrt{2E_{\vec{p}}}\sqrt{2E_{\vec{q}}}} \sum_{r,s} \left(\left[u^{r\dagger}(q)\, a^\dagger_{r\vec{q}}\, e^{+iq\cdot x} + v^{r\dagger}(q)\, b_{r\vec{q}}\, e^{-iq\cdot x} \right] \right. \\
&\times \left. \left[-iE_{\vec{p}}\, u^s(p)\, a_{s\vec{p}}\, e^{-ip\cdot x} + iE_{\vec{p}}\, v^s(p)\, b^\dagger_{s\vec{p}}\, e^{+ip\cdot x} \right] \right) \\
= \int d^3x\, &\frac{d^3p}{(2\pi)^3} \frac{d^3q}{(2\pi)^3} \frac{E_{\vec{p}}}{\sqrt{2E_{\vec{p}}}\sqrt{2E_{\vec{q}}}} \\
&\times \sum_{r,s} \left(u^{r\dagger}(q)\, a^\dagger_{r\vec{q}}\, u^s(p)\, a_{s\vec{p}}\, e^{-i(p-q)\cdot x} - u^{r\dagger}(q)\, a^\dagger_{r\vec{q}}\, v^s(p)\, b^\dagger_{s\vec{p}}\, e^{+i(p+q)\cdot x} \right. \\
&\left. + v^{r\dagger}(q)\, b_{r\vec{q}}\, u^s(p)\, a_{s\vec{p}}\, e^{-i(p+q)\cdot x} - v^{r\dagger}(q)\, b_{r\vec{q}}\, v^s(p)\, b^\dagger_{s\vec{p}}\, e^{+i(p-q)\cdot x} \right).
\end{aligned}
$$

Exercise 18.2.1 Now integrate over d^3x and d^3q to arrive at Equation 18.10. (*Hints:* Recall from Chapter 5 that $\int d^3x\, e^{\pm(\vec{p}-\vec{q})\cdot\vec{r}} = (2\pi)^3 \delta^3(\vec{p}-\vec{q})$. Note that $E_{\vec{q}} = E_{\vec{p}}$ when $\vec{q} = \pm\vec{p}$, and recall from Chapter 16 that

$$u^{r\dagger}(p)u^s(p) = v^{r\dagger}(p)v^s(p) = 2E_{\vec{p}}\, \delta^{rs}, \qquad u^{r\dagger}(p)v^s(\overline{p}) = v^{r\dagger}(p)u^s(\overline{p}) = 0,$$

where $p = [E_{\vec{p}}, +\vec{p}\,]$ and $\overline{p} = [E_{\vec{p}}, -\vec{p}\,]$. $\qquad\qquad$ (18.38)

BOX 18.3 The Expression for the Charge

Our goal in this box is to use the expansion in Equation 18.9 (repeated here)

$$\psi = \int \frac{d^3p}{(2\pi)^3}\frac{1}{\sqrt{2E_{\vec{p}}}}\sum_s \left[u^s(p)\, a_{s\vec{p}}\, e^{-ip\cdot x} + v^s(p)\, b^\dagger_{s\vec{p}}\, e^{+ip\cdot x} \right] \quad (18.39)$$

to express the total charge

$$Q = \int d^3x\, J^t = q \int d^3x\, \overline{\psi}\gamma^t\psi \qquad (18.40)$$

in terms of the operators $a_{s\vec{p}}$, $a^\dagger_{s\vec{p}}$, $b_{s\vec{p}}$, and $b^\dagger_{s\vec{p}}$.

Note that $\overline{\psi} \equiv \psi^\dagger \gamma^t$, so

$$\overline{\psi}\gamma^t\psi = \psi^\dagger(\gamma^t)^2\psi = \psi^\dagger\psi, \qquad (18.41)$$

because $(\gamma^t)^2 = I$. This is the same as the expression for the energy density T^{tt}, but without the $i\partial_t$ in the middle.

Exercise 18.3.1 Substitute Equations 18.39 and 18.41 into Equation 18.40 to find Q as a function of $a_{s\vec{p}}$, $a^\dagger_{s\vec{p}}$, $b_{s\vec{p}}$, and $b^\dagger_{s\vec{p}}$. (*Hint:* Look over the calculation in the previous box and see how removing the $i\partial_t$ will affect the result. Describe the changes rather than repeating the whole calculation.)

BOX 18.4 Anticommutation Relations

The goal of this box is to prove the anticommutation relations stated in Equation 18.26, repeated here:

$$\{a_j,\, a_k^\dagger\} = 0 \quad \text{when } j \neq k, \tag{18.42a}$$

$$\{a_j,\, a_k\} = 0, \quad \text{and} \quad \{a_j^\dagger,\, a_k^\dagger\} = 0, \tag{18.42b}$$

assuming discrete momentum states and that no more than one particle can occupy any given momentum state.

Note that $a_k a_k |0_k\rangle = 0$ because we can't lower the zero state. We also see that $a_k a_k |1_k\rangle = a_k(c_1|0_k\rangle) = c_1 a_k |0_k\rangle = 0$. Therefore, $a_k a_k = 0$ when acting on any possible ket. This also means that $\{a_k,\, a_k\} = a_k a_k + a_k a_k = 0$. The same argument applies to $a_k^\dagger a_k^\dagger$: we cannot raise the $|1_k\rangle$ state any further, and $a_k^\dagger a_k^\dagger |0_k\rangle = c_1^* a_k^\dagger |1_k\rangle = 0$. Therefore, $\{a_k^\dagger,\, a_k^\dagger\} = 0$ also.

To prove the relations for $j \neq k$, consider the following state:

$$\sqrt{\tfrac{1}{2}} \left(a_j^\dagger + a_k^\dagger\right)|0\rangle, \tag{18.43}$$

where $j \neq k$ and $|0\rangle$ is the vacuum state. We would naturally interpret this as representing a single particle that is in a superposition state having momentum $\vec{p}_j$ or $\vec{p}_k$ with equal probability. Therefore, we must interpret $\sqrt{\tfrac{1}{2}}\,(a_j^\dagger + a_k^\dagger)$ as being the creation operator for such a particle. Since we can't ever create two such particles in that state, we must have

$$(a_j^\dagger + a_k^\dagger)(a_j^\dagger + a_k^\dagger) = 0 \tag{18.44}$$

for all possible Fock-space states.

Exercise 18.4.1   Argue from here that $\{a_j^\dagger,\, a_k^\dagger\} = 0$ for $j \neq k$. Then take the adjoint to prove that $\{a_j,\, a_k\} = 0$ for $j \neq k$.

BOX 18.4   Anticommutation Relations (cont.)

The remaining case we must prove is that $\{a_j, a_k^\dagger\} = 0$ when $j \neq k$. We can accomplish this by looking at all relevant states that this might act on. For example,

$$(a_j a_k^\dagger + a_k^\dagger a_j)|\,0_j, 0_k, \rangle = 0 \tag{18.45}$$

because the a_j operator cannot lower the zero in the jth slot.

Exercise 18.4.2 Similarly, show that the two cases where the kth slot has a 1 and the jth slot has a 0 or a 1 both yield a zero result. (*Hint:* The argument is like the case considered above.)

The trickier case is $(a_j a_k^\dagger + a_k^\dagger a_j)|\,1_j, 0_k, \rangle$ because we might in principle be able to bump up the zero in the kth slot while lowering the one in the jth slot. However, note that because $a_j^\dagger|\,0_j, \rangle = c_1^*|\,1_j, \rangle$,

$$c_1^*(a_j a_k^\dagger + a_k^\dagger a_j)|\,1_j, 0_k\rangle = (a_j a_k^\dagger + a_k^\dagger a_j)a_j^\dagger|\,0_j, 0_k, \rangle$$

$$= (a_j a_k^\dagger a_j^\dagger + a_k^\dagger a_j a_j^\dagger)|\,0_j, 0_k, \rangle$$

$$= (-a_j a_j^\dagger a_k^\dagger + a_k^\dagger[1 - N_j])|\,0_j, 0_k, \rangle, \tag{18.46}$$

where the minus sign in the first term comes from the fact that $a_j^\dagger$ and $a_k^\dagger$ anticommute, and in the second term we have $a_j a_j^\dagger = 1 - a_j^\dagger a_j = 1 - N_j$.

Exercise 18.4.3 Do a similar thing to the $-a_j a_j^\dagger a_k^\dagger$ term and show that the result is zero. (*Hint:* This is *not* because the two terms in the factor multiplying the ket cancel, though.)

Homework Problems

P18.1 Prove the following anticommutator identities.
a. $\{A, B\} = \{B, A\}$.
b. $\{cA, B\} = c\{A, B\}$, where c is a complex number.
c. $\{A + B, C\} = \{A, C\} + \{B, C\}$.
d. $\{AB, C\} = A[B, C] + \{A, C\}B$ (the commutator in the first term on the right is not an error).

P18.2 **(a)** Prove directly from the anticommutation relations that $(a^\dagger_{s\vec{p}})^2 = 0$ and **(b)** explain physically in terms of the Pauli exclusion principle why this must be so, no matter what Fock-space state the operator acts on.

P18.3 Consider the creation and annihilation operators for a complex scalar field ϕ. The mathematics leading to Equations 12.12 and 12.14 from the basic quantum field expansion in Equation 12.1 does not depend on the commutation relations. But what happens if we were to use the anticommutation relations $\{a_{\vec{p}}, a^\dagger_{\vec{p}}\} = (2\pi)^3 \delta^3(\vec{p} - \vec{p})$ (and the analogous expression for $b_{\vec{p}}$ and $b^\dagger_{\vec{p}}$) instead of the commutation relation given in Equation 12.15? Show that disaster ensues in the expressions for Q and E instead of the nice results given in Equations 12.16 and 12.18.

P18.4 Consider the Lagrangian density for a massless Dirac particle.
a. Show that this Lagrangian is invariant under a global transformation such that $\psi' = e^{-i\gamma^5\theta}\psi$.
b. Find the corresponding conserved current.
(*Hints:* As we did in the previous chapter, consider an infinitesimal transformation $e^{-i\gamma^5\delta\theta} \approx I - i\gamma^5\delta\theta$ with $\delta\theta \ll 1$, and keep only first-order terms in the calculations. The proof is still general because we can build a finite transformation out of a sequence of infinitesimal transformations. You may find it helpful to prove that $\gamma^5\gamma^\mu = -\gamma^\mu\gamma^5$ before you begin.)

P18.5 Consider the momentum density (see Equation 18.2)

$$\pi^j = T^{tj} = \left[\sum_n \frac{\partial \mathcal{L}}{\partial(\partial_t\psi_n)}\partial^j\psi_n \right] - g^{tj}\mathcal{L}. \quad (18.47)$$

a. Evaluate this in terms of ψ for the Dirac Lagrangian.
b. Calculate the total momentum P^j by expressing the density as a linear combination of creation and annihilation operators, and then integrating over all space.
c. Does your result make sense? Explain your response.

P18.6 Show that two fermions with opposite spins can be in a state with the same momentum $\vec{p}$, but that the state is still antisymmetric under exchange of the spins. (*Hint:* Adapt the argument in Section 18.3 and contrast with the case where the fermions' spin and momenta are the same.)

And God said,

"Let U(1) be
an active local
symmetry."

and there was light.

19. ELECTROMAGNETISM FROM U(1)

In this chapter, we take an important step toward constructing a quantum field theory that actually describes our real universe. In Chapter 18, we mentioned that the Lagrangian density

$$\mathcal{L} = i\overline{\psi}\gamma^{\mu}\partial_{\mu}\psi - m\overline{\psi}\psi \tag{19.1}$$

for a complex Dirac field is symmetric under the transformation $\psi' = e^{-iq\theta}\psi$, and as a result has a conserved current

$$J^{\mu} = q\overline{\psi}\gamma^{\mu}\psi. \tag{19.2}$$

We showed that the "charge" associated with this current was

$$Q = \int d^3x\, J^t = q(N_a - N_b), \tag{19.3}$$

where N_a and N_b are the numbers of Dirac particles and antiparticles, respectively. Though I claimed that this symmetry was connected to electromagnetism and that q referred to the particles' electric charge, I was not able to further justify that connection. This chapter will justify that connection.

Physicists generally call any transformation of a quantity in a physics equation that has no physical consequences (such as the transformation $\psi' = e^{-iq\theta}\psi$ here) a **gauge transformation** (for unhelpful historical reasons). We will see in this chapter that if we convert this into what we call a "local" gauge transformation, then all of electromagnetism naturally emerges. This will be our first example of how **gauge symmetries** lead to physics in the Standard Model and serve as a prototype for our future treatment of other symmetries.

19.1 The U(1) Group

A **group** in mathematics is a set of **group elements** g_j with a defined operation $g_j \circ g_k$ that exhibits the following properties: (1) **closure** ($g_j \circ g_k$ is a group element), (2) **identity** (there is an element I such that $I \circ g_j = g_j \circ I = g_j$), (3) **inverse** (every element has an inverse such that $g_j^{-1} \circ g_j = g_j \circ g_j^{-1} = I$), and (4) **associativity** [$(g_j \circ g_k) \circ g_m = g_j \circ (g_k \circ g_m)$.] The elements can be discrete (the numbers -1 and $+1$ form a group under multiplication) or continuous (the real numbers form a group under addition).

In physics, we are interested in groups in the context of *transformations*. If a Lagrangian density is symmetric under a certain set of transformations $\psi' = M_j\psi$, and the set of transformation operators M_j (often a set of matrices that can be multiplied) form a group, then we can take advantage of group mathematics to better analyze the symmetry.

The set of transformation operators $M(\theta) = e^{-iq\theta}$, where θ is a continuous parameter, form a group under simple multiplication (see Problem P19.1). This group is called **U(1)** because its elements are **unitary** ($M^{\dagger}M = I$) and the "matrices" that represent it are 1×1 matrices (that is, numbers).

19.2 Local Gauge Invariance

The gauge transformation $\psi' = e^{-iq\theta}\psi$ we have just been considering is a **global gauge transformation** because the phase change is applied universally to the field ψ independent of position. We can make this U(1) transformation a **local gauge transformation** simply by allowing the parameter θ in the transformation to depend on position:

$$\psi'(x) = e^{-iq\theta(x)}\psi(x). \tag{19.4}$$

Under this local transformation, the Lagrangian density becomes

$$\mathcal{L}' = i\overline{\psi}'\gamma^\mu\partial_\mu\psi' - m\overline{\psi}'\psi' = i(\psi')^\dagger\gamma^t\gamma^\mu\partial_\mu\psi' - m(\psi')^\dagger\gamma^t\psi'$$

$$= i[\,e^{-iq\theta(x)}\psi\,]^\dagger\gamma^t\gamma^\mu\partial_\mu[\,e^{-iq\theta(x)}\psi(x)\,] - m[\,e^{-iq\theta(x)}\psi\,]^\dagger\gamma^t[\,e^{-iq\theta(x)}\psi\,]$$

$$= i[\,e^{+iq\theta(x)}\psi^\dagger\,]\gamma^t\gamma^\mu\partial_\mu[e^{-iq\theta(x)}\psi(x)\,] - m[\,e^{+iq\theta(x)}\psi^\dagger\,]\gamma^t[\,e^{-iq\theta(x)}\psi\,]. \tag{19.5}$$

The exponentials will cancel in the final term (leaving that term invariant), but we have trouble with the derivative terms because the parameter θ now depends on x. Specifically, we see that, by the product rule, we have

$$\partial_\mu[\,e^{-iq\theta(x)}\psi(x)\,] = -iq[\,\partial_\mu\theta(x)\,]e^{-iq\theta(x)}\psi(x) + e^{-iq\theta(x)}\partial_\mu\psi(x). \tag{19.6}$$

If we had only the last term in the derivative, we would be fine: the exponentials would cancel and we'd be left with $i\psi^\dagger\gamma^t\gamma^\mu\partial_\mu\psi = i\overline{\psi}\gamma^\mu\partial_\mu\psi$, implying that $\mathcal{L}' = \mathcal{L}$. But because of the extra term involving the derivative of the parameter $\theta(x)$, our Lagrangian density is *not* invariant under the local gauge transformation.

But what if we *made* the Lagrangian density invariant by adding a new (real) field $A_\mu(x)$ whose purpose is to cancel out the new term? Suppose that we adjust our Lagrangian density to be

$$\mathcal{L} = i\overline{\psi}\gamma^\mu\partial_\mu\psi - m\overline{\psi}\psi - J^\mu A_\mu, \tag{19.7}$$

where $J^\mu = q\overline{\psi}\gamma^\mu\psi$, as defined in Equation 19.2. As you will show in Box 19.1, this Lagrangian density *will* be invariant under the gauge transformation so long as we also require that

$$A'_\mu = A_\mu + \partial_\mu\theta(x) \tag{19.8}$$

under the same gauge transformation.

Since the problem in the original Lagrangian density was in the derivatives, a general approach to finding a gauge-invariant Lagrangian density (rather than just guessing, the way that I did in Equation 19.7) is to adjust the original derivative terms to yield the correct cancellation (assuming Equation 19.8). Suppose we define a **covariant derivative**

$$D_\mu\psi \equiv (\partial_\mu + iqA_\mu)\,\psi. \tag{19.9}$$

Under a gauge transformation, this will become

$$D_\mu\psi' = (\partial_\mu + iqA'_\mu)\,\psi' = (\partial_\mu + iq[\,A_\mu + \partial_\mu\theta\,])\,e^{-iq\theta}\psi$$

$$= e^{-iq\theta}(\partial_\mu\psi - iq[\,\partial_\mu\theta\,]\psi + iqA_\mu\psi + iq[\,\partial_\mu\theta\,]\psi)$$

$$= e^{-iq\theta}(\partial_\mu + iqA_\mu)\,\psi = e^{-iq\theta}D_\mu\psi, \tag{19.10}$$

meaning that $D_\mu\psi$ transforms under the local transformation just like $\partial_\mu\psi$ does under the global transformation. So the trick is to take the Lagrangian density that was invariant under the *global* transformation and replace its ordinary partial derivatives with these new covariant derivatives. This will introduce the required new field in just the right way to keep the Lagrangian density invariant under the *local* transformation. This will be very helpful when we consider other types of transformations in the future.

In Box 19.2, you will verify that the Lagrangian density

$$\mathcal{L} = i\overline{\psi}\gamma^{\mu}D_{\mu}\psi - m\overline{\psi}\psi \tag{19.11}$$

is equivalent to the Lagrangian density in Equation 19.7.

19.3 Completing the Lagrangian Density

We do not yet have a complete Lagrangian density, though, because any new quantum field must have terms in the Lagrangian density that describe a *free* field (in absence of other fields that might interact with it). That is what the first two terms in Equation 19.7 provide for the field $\psi(x)$. We need to provide something analogous for the field $A_{\mu}(x)$.

What can the free-field terms look like? If they are like the terms for the *scalar* field $\phi(x)$, the terms might involve squares of the field and/or squares of the first derivatives of the field. The terms must also be Lorentz scalars (for example, $A_{\mu}A^{\mu}$), because the Lagrangian density as a whole must be a scalar. Finally, we must also be careful that these terms are invariant under the local gauge transformation $A'_{\mu} = A_{\mu} + \partial_{\mu}\theta$: otherwise, we will introduce new terms in the gauge transformation that will need to be erased somehow.

This last requirement means that our free-field Lagrangian density for A_{μ} cannot involve a mass term like $M^2 A_{\mu}A^{\mu}$, because

$$M^2 A'_{\mu}A'^{\mu} = M^2(A_{\mu} + \partial_{\mu}\theta)(A^{\mu} + \partial^{\mu}\theta)$$

$$= M^2(A_{\mu}A^{\mu} + A^{\mu}\partial_{\mu}\theta + A_{\mu}\partial^{\mu}\theta + [\,\partial^{\mu}\theta\,][\,\partial_{\mu}\theta\,]), \tag{19.12}$$

so it is certainly not invariant. Therefore, the particle described by the field A_{μ} must be *massless*.

We run into similar troubles with terms like $\partial_{\mu}A^{\nu}\,\partial^{\mu}A_{\nu}$. But if we define

$$F_{\mu\nu} \equiv \partial_{\mu}A_{\nu} - \partial_{\nu}A_{\mu}, \tag{19.13}$$

then you can see that under a gauge transformation,

$$F'_{\mu\nu} = \partial_{\mu}A'_{\nu} - \partial_{\nu}A'_{\mu}$$

$$= \partial_{\mu}(A_{\nu} + \partial_{\nu}\theta) - \partial_{\nu}(A_{\mu} + \partial_{\mu}\theta)$$

$$= \partial_{\mu}A_{\nu} - \partial_{\nu}A_{\mu} + \partial_{\mu}\partial_{\nu}\theta - \partial_{\nu}\partial_{\mu}\theta = F_{\mu\nu} \tag{19.14}$$

since the order of partial derivatives is irrelevant. Therefore, $F_{\mu\nu}F^{\mu\nu}$ will combine products of the partial derivatives of A_{μ} in just the right way to yield a relativistic scalar, while also ensuring that the terms introduced by the local gauge transformation cancel out. Now we can normalize the field any way that we like because multiplying A_{μ} by a constant simply adjusts how strongly the field interacts with particles through the $J^{\mu}A_{\mu}$ term, which is the same as adjusting the value of q that appears in the expression for J^{μ}. We conventionally normalize A_{μ} so that the complete Lagrangian density becomes

$$\mathcal{L} = i\overline{\psi}\gamma^{\mu}\partial_{\mu}\psi - m\overline{\psi}\psi - J^{\mu}A_{\mu} - \tfrac{1}{4}F_{\mu\nu}F^{\mu\nu}, \tag{19.15}$$

then choose the value of q for the particle to reflect experimental results. The free-field Lagrangian density for the field represented by A_{μ} is therefore

$$\mathcal{L}_{A,\,\text{free}} = -\tfrac{1}{4}F_{\mu\nu}F^{\mu\nu}. \tag{19.16}$$

19.4 Interpreting the Lagrangian Density

The simple Lagrangian density in Equation 19.15 turns out to contain everything we already know about electromagnetism, including Maxwell's equations, the Lorentz force law, the energy density of an electromagnetic field, and so on.

To see how this works, let's write out the components of the $F_{\mu\nu}$ tensor:

$$
F_{\mu\nu} = \begin{bmatrix}
\partial_t A_t - \partial_t A_t & \partial_t A_x - \partial_x A_t & \partial_t A_y - \partial_y A_t & \partial_t A_z - \partial_z A_t \\
\partial_x A_t - \partial_t A_x & \partial_x A_x - \partial_x A_x & \partial_x A_y - \partial_y A_x & \partial_x A_z - \partial_z A_x \\
\partial_y A_t - \partial_t A_y & \partial_y A_x - \partial_x A_y & \partial_y A_y - \partial_y A_y & \partial_y A_z - \partial_z A_y \\
\partial_z A_t - \partial_t A_z & \partial_z A_x - \partial_x A_z & \partial_z A_y - \partial_y A_z & \partial_z A_z - \partial_z A_z
\end{bmatrix}
$$

$$
\equiv \begin{bmatrix}
0 & E_x & E_y & E_z \\
-E_x & 0 & -B_z & B_y \\
-E_y & B_z & 0 & -B_x \\
-E_z & -B_y & B_x & 0
\end{bmatrix}, \tag{19.17}
$$

where E_x, E_y, E_z, B_x, B_y, and B_z are (at this point) just short names for the differences that are not identically zero. Note that we have just six independent differences because each difference above the diagonal has a version with the opposite sign below the diagonal. If we recall that raising a time index does nothing but raising a spatial index changes the sign of the component, then we can see that

$$
F^{\mu\nu} = \begin{bmatrix}
0 & -E_x & -E_y & -E_z \\
E_x & 0 & -B_z & B_y \\
E_y & B_z & 0 & -B_x \\
E_z & -B_y & B_x & 0
\end{bmatrix}. \tag{19.18}
$$

Now, as I said, E_x, E_y, E_z, B_x, B_y, and B_z are at this point just names for components of the tensor. What do these components mean physically?

The definition of $F_{\mu\nu}$ in terms of A_μ directly implies (see Box 19.3) that

$$
\partial_\alpha F_{\mu\nu} + \partial_\mu F_{\nu\alpha} + \partial_\nu F_{\alpha\mu} = 0. \tag{19.19}
$$

(Note how the index symbols rotate cyclically.) In principle, this actually stands for $4 \times 4 \times 4 = 64$ equations, but most of these equations are identically zero. As you will see in Box 19.3, the four meaningful equations that emerge are the components of the source-free Maxwell's equations (Gauss's law for the magnetic field and Faraday's law) in differential form:

$$
\vec{\nabla} \cdot \vec{B} = 0 \quad \text{and} \quad \vec{\nabla} \times \vec{E} + \frac{\partial \vec{B}}{\partial t} = 0, \tag{19.20a}
$$

which you might know in the mathematically equivalent integral forms

$$
\oint \vec{B} \cdot d\vec{A} = 0 \quad \text{and} \quad \oint \vec{E} \cdot d\vec{\ell} = \frac{d}{dt} \int \vec{B} \cdot d\vec{A}. \tag{19.20b}
$$

You will see in Box 19.4 that the Euler-Lagrange equations applied to the full Lagrangian density imply that

$$
\partial_\mu F^{\mu\nu} = J^\nu, \tag{19.21}
$$

which, when expressed in terms of the six components E_x, E_y, E_z, B_x, B_y, and B_z defined above, become the other two Maxwell's equations (Gauss's law and the Ampère-Maxwell relation):

$$
\vec{\nabla} \cdot \vec{E} = \rho \quad \text{and} \quad \vec{\nabla} \times \vec{B} - \frac{\partial \vec{E}}{\partial t} = \vec{J}, \tag{19.22a}
$$

which again you might recognize in their equivalent integral guise:

$$\oint \vec{E} \cdot d\vec{A} = Q_{\text{enc}} \quad \text{and} \quad \oint \vec{B} \cdot d\vec{\ell} + \frac{d}{dt} \int \vec{E} \cdot d\vec{A} = \iota_{\text{enc}}. \qquad (19.22b)$$

The absence of constants μ_0 and ε_0 reflects our choice of field normalization, which amounts to choosing units for the field where these constants are 1.

(**Please note:** The subscripts on E_x, B_x, and so on are part of the *names* for these quantities, not relativistic indexes. They are meant to identify these quantities as the components of the three-vectors $\vec{E}$ and $\vec{B}$ that we use in classical electromagnetism. But these quantities do *not* transform like the spatial components of a four vector (rather, they transform as components of a 4×4 tensor), and one cannot "raise" these subscripts as one would raise a four-vector or tensor index. The subscript position has no meaning here.)

You can also see that because $F^{\mu\nu}$ is antisymmetric by construction

$$\partial_\nu \partial_\mu F^{\mu\nu} = \partial_\mu \partial_\nu F^{\mu\nu} = -\partial_\mu \partial_\nu F^{\nu\mu} = -\partial_\nu \partial_\mu F^{\mu\nu} = 0, \qquad (19.23)$$

where, in the next-to-last step, I renamed indices $\mu \leftrightarrow \nu$, and in the last step, I used the idea that something equal to negative itself must be zero. But since $\partial_\mu F^{\mu\nu} = J^\nu$, this means that

$$0 = \partial_\nu(\partial_\mu F^{\mu\nu}) = \partial_\nu J^\nu, \qquad (19.24)$$

meaning that the charge current is conserved, as it must be.

19.5 More about Gauge Invariance

We see that the components of the **Faraday tensor** $F_{\mu\nu}$ obey Maxwell's equations for the electromagnetic field that we are familiar with in classical physics. But we see in this context that the familiar electromagnetic field is constructed from the more fundamental field A^μ. What is this field?

If we look at the definition of the electric field components in Equation 19.17 and raise the indices on the components of A_μ, we see that

$$-E_j = \partial_j A_t - \partial_t A_j = \partial_j A^t + \partial_t A^j \;\; \Rightarrow \;\; E_j = -\partial_j A^t - \partial_t A^j \qquad (19.25)$$

since raising a spatial index of the four-vector A_μ flips its sign. For a static field, we can ignore the time derivative part, and the other part says that $\vec{E} = -\vec{\nabla} A^t$. But you may remember from introductory electrodynamics that we can write a static electric field as the negative gradient of the **electrostatic potential** Φ: $\vec{E} = -\vec{\nabla}\Phi$. We can therefore identify $A^t = \Phi$. We rarely study the spatial components of A^μ in introductory physics, but we call $\vec{A}$ the **vector potential** and A^μ as a whole the **four-potential**. You can see from how Equation 19.17 defines the components of $\vec{B}$ that the magnetic field is simply the *curl* of the vector potential:

$$B_x \equiv -\partial_y A_z + \partial_z A_y = \partial_y A^z - \partial_z A^y, \;\; B_y \equiv -\partial_z A_x + \partial_x A_z = \partial_z A^x - \partial_x A^z,$$

$$B_z \equiv -\partial_x A_y + \partial_y A_x = \partial_x A^y - \partial_y A^x \;\; \Rightarrow \;\; \vec{B} = \vec{\nabla} \times \vec{A}. \qquad (19.26)$$

The vector potential also adds a crucial term to the definition of the electric field when the fields are dynamic (time-dependent).

Now, we have *constructed* the components of the tensor $F_{\mu\nu}$ (and thus the electric and magnetic fields $\vec{E}$ and $\vec{B}$) so that they are not affected by the gauge transformation

$$A'_\mu = A_\mu + \partial_\mu \theta(x), \qquad (19.27)$$

where $\theta(x)$ is *any* function of spacetime position $x = x^\mu$. Because we will be working more with A^μ in this course, it is convenient to **choose a gauge** (that is, choose a specific function for $\theta(x)$) that makes working with A^μ easier.

For example, in Box 19.5, we will see that can choose $\theta(x)$ so that

$$\partial_\mu A^\mu = 0. \qquad (19.28)$$

Physicists call the gauge for which this is true the **Lorenz gauge** (after Ludvig Lorenz, not the Hendrik Lorentz of the Lorentz transformation). This gauge is very handy because the Euler-Lagrange equations for the field become

$$J^\nu = \partial_\mu F^{\mu\nu} = \partial_\mu (\partial^\mu A^\nu - \partial^\nu A^\mu) = \partial_\mu \partial^\mu A^\nu - \partial^\nu (\partial_\mu A^\mu)$$

$$= \partial_\mu \partial^\mu A^\nu - 0 = \frac{\partial^2 A^\nu}{\partial t^2} - \frac{\partial^2 A^\nu}{\partial x^2} - \frac{\partial^2 A^\nu}{\partial y^2} - \frac{\partial^2 A^\nu}{\partial z^2}. \tag{19.29}$$

In this gauge, the differential equations for the components of A^ν uncouple, leaving us with four simple, independent, and well-studied equations for the components of A^ν.

In the absence of a current J^ν, the equations are $\partial_\mu \partial^\mu A^\nu = 0$, which are simply four independent Klein-Gordon equations for a massless field. The solutions for the free field are therefore linear combinations of the basis functions

$$A^\nu_{\vec{p}}(x) = \epsilon^\nu_{\vec{p}} \, e^{-ip \cdot x}, \tag{19.30}$$

where ϵ^ν is a set of four amplitude constants.

We have not yet exhausted our ability to choose a gauge. As Box 19.5 shows, we can choose a gauge where the Lorenz gauge condition is not only satisfied but also where $\vec{\nabla} \cdot \vec{A} = 0$. Physicists call this gauge the **Coulomb gauge**. In empty space, we can additionally require that $A^t = 0$. This gauge makes calculating a free field's stress-energy tensor easier (see Box 19.6).

In the Coulomb gauge, the free-field basis solutions are $\vec{A}_{\vec{p}} = \vec{\epsilon}_{\vec{p}} \, e^{-ip \cdot x}$. The Lorenz gauge condition applied to such a solution tells us that

$$0 = \vec{\nabla} \cdot (\vec{\epsilon}_{\vec{p}} \, e^{-ip \cdot x}) = +i \vec{p} \cdot \vec{\epsilon}_{\vec{p}} \, e^{-ip \cdot x} \quad \Rightarrow \quad \vec{p} \cdot \vec{\epsilon}_{\vec{p}} = 0. \tag{19.31}$$

In this gauge, not only is the time component of $\epsilon^\mu_{\vec{p}}$ zero but the component parallel to the direction of wave motion is zero as well: the wave is perpendicular to the direction of motion. The free-wave basis solution can therefore be written as a superposition of two polarizations: for a wave moving in the $+z$ direction, we have (in the Coulomb gauge)

$$A^\nu_{\vec{p}}(x) = \begin{bmatrix} 0 \\ \epsilon_{1\vec{p}} \\ 0 \\ 0 \end{bmatrix} e^{-ip \cdot x} + \begin{bmatrix} 0 \\ 0 \\ \epsilon_{2\vec{p}} \\ 0 \end{bmatrix} e^{-ip \cdot x}. \tag{19.32}$$

The fact that an electromagnetic wave has two possible polarizations may not be surprising to you (since this is often discussed in introductory physics), but this approach to proving that result is probably novel.

19.6 Quantizing the Field

As mentioned above, in the Lorenz gauge, the free-field Euler-Lagrange equation for the field is simply

$$\partial_\mu \partial^\mu A^\nu = 0, \tag{19.33}$$

which represents four uncoupled massless Klein-Gordon equations for the real components of A^μ. Those of you who have gone through Chapters 11–14 know how to quantize real Klein-Gordon fields, and there is nothing substantially different here except that an excitation of the four-component A^μ field is a *single* particle we call a **photon** (not four separate scalar-like fields). The amplitude of a plane-wave excitation of this field is given by a polarization four-vector $\epsilon^\mu_{s\vec{p}}$ (where $s = 1, 2$ specifies the polarization state) that (in the Coulomb gauge) satisfies the transverse wave conditions $\epsilon^t_{s\vec{p}} = 0$ and $\vec{p} \cdot \vec{\epsilon}_{s\vec{p}} = 0$. We also know that particles that are excitations of real scalar fields are their own antiparticles, and the same applies here: because the A^μ field is real, the photon is its own antiparticle.

19.7 What Have We Done?

Let's review the three-step process we have just executed. The first step involved constructing a Lagrangian density that was symmetric under a *global* U(1) transformation. The second step was to make that symmetry *local* by allowing the transformation parameter θ to depend on position. This required introducing a new field A^μ. The third and crucial step was to make this field *active* by including a new free-field term in the Lagrangian density that allows the A^μ field to have an independent existence. Without this last step, the A^μ field would merely be a mathematical bookkeeping device.

An analogy might help clarify what I mean. Imagine that all the world agreed to use a single currency, say the euro. If later, the entire world decides to switch to the yen, but provides an immutable conversion factor for converting euros to yen, that would be a global transformation. But suppose everyone decides later that individual countries should be free to use their own currency and a position-dependent conversion factor is created to convert between the global and local currency. Currency exchange brokers passively manage these conversions at the local level by applying the position-dependent currency conversion factor (a currency exchange "field," if you like, analogous to A^μ). So far, however, we have not really *changed* the global financial system: it will behave the same way as when there was a global currency.

But what if we allow the local currencies to float in value relative to each other, and give exchange brokers the freedom to speculate on these relative values and to spend and make money. Now we have introduced a real change in the "physics" of the system: the brokers become active entities and money can flow in new and different ways. This is essentially what we are doing in the third step by including the free-field terms in the Lagrangian density: we are giving the conversion field A^μ an independent and physically active reality. We might call the whole process making the global symmetry *actively local*.

But why should the universe *require* an independent and active field for a given symmetry? Why does this three-step process *work* in producing a theory that reflects reality? That is a deep mystery. But it *does* apparently work!

19.8 Postscript

It is worth pausing to reflect on the amazing result of this chapter. Simply by requiring that our Lagrangian density have an *actively local* U(1) symmetry, we have seen that the entire theory of electromagnetism naturally emerges: we have essentially *derived* Maxwell's equations from first principles! This is huge!

You may have seen the T-shirt that says, "And God said (a list of Maxwell's equations) and there was light." An even more geeky T-shirt might say, "And God said 'Let U(1) be an active local symmetry' and there was light."

We are now set to make some real predictions about the universe we actually inhabit. We will begin this task in the next chapter.

BOX 19.1   Making the Lagrangian Density Gauge Invariant

In this box, we will show that the modified Lagrangian density

$$\mathcal{L} = i\overline{\psi}\gamma^{\mu}\partial_{\mu}\psi - m\overline{\psi}\psi - J^{\mu}A_{\mu} \tag{19.34}$$

(where the current density $J^{\mu} = q\overline{\psi}\gamma^{\mu}\psi$) is invariant under a local gauge transformation $\psi' = e^{-iq\theta(x)}\psi$ so long as

$$A'_{\mu} = A_{\mu} + \partial_{\mu}\theta(x). \tag{19.35}$$

We begin by recalling that Equation 19.6 tells us that

$$\partial_{\mu}\left[e^{-iq\theta}\psi\right] = -iq\left[\partial_{\mu}\theta\right]e^{-iq\theta}\psi + e^{-iq\theta}\partial_{\mu}\psi. \tag{19.36}$$

(For simplicity, I have removed the explicit indications of position dependence, but remember that both θ and ψ depend on x.)

Exercise 19.1.1   Use Equations 19.35 and 19.36 to show that the Lagrangian density in Equation 19.34 satisfies $\mathcal{L}' = \mathcal{L}$ under the gauge transformation.

BOX 19.2 Checking the Covariant Derivative

In this box, we will verify that the Lagrangian density

$$\mathcal{L} = i\overline{\psi}\gamma^{\mu}D_{\mu}\psi - m\overline{\psi}\psi \qquad (19.37)$$

(where $D_{\mu} \equiv \partial_{\mu} + iqA_{\mu}$) is equivalent to the Lagrangian density in Equation 19.7, repeated here:

$$\mathcal{L} = i\overline{\psi}\gamma^{\mu}\partial_{\mu}\psi - m\overline{\psi}\psi - J^{\mu}A_{\mu}, \qquad (19.38)$$

where $J^{\mu} = q\overline{\psi}\gamma^{\mu}\psi$.

Exercise 19.2.1 Verify that these are the same. (*Hint:* You will likely not need all of the space provided below.)

BOX 19.3 Maxwell's Equations I

In this box, we will show that the definition of $F_{\mu\nu}$ directly implies

$$\partial_\alpha F_{\mu\nu} + \partial_\mu F_{\nu\alpha} + \partial_\nu F_{\alpha\mu} = 0 \tag{19.39}$$

and that the nontrivial index choices reflect components of Maxwell's equations $\vec{\nabla} \cdot \vec{B} = 0$ and $\vec{\nabla} \times \vec{E} + \partial \vec{B}/\partial t = 0$.

Exercise 19.3.1   Show that Equation 19.39 follows from $F_{\mu\nu} \equiv \partial_\mu A_\nu - \partial_\nu A_\mu$.

Now, most of the 64 components of this equation are zero. If any two indices are the same, the result is trivially zero because $F_{\mu\nu}$ is antisymmetric. For example, if $\mu = \nu$, then

$$\partial_\alpha F_{\mu\mu} + \partial_\mu F_{\mu\alpha} + \partial_\mu F_{\alpha\mu} = 0 + \partial_\mu F_{\mu\alpha} - \partial_\mu F_{\mu\alpha} = 0 \tag{19.40}$$

(where no sum is implied by the repeated indexes). There are only four choices for three indexes that must not be the same: txy, txz, tyz, and xyz, and because the equations are symmetric in the three indices, permutations of these four choices will yield the same results.

Exercise 19.3.2   Show that the choice txy yields a component of Faraday's law and that the choice xyz yields $\vec{\nabla} \cdot \vec{B} = 0$.

BOX 19.4 Maxwell's Equations II

The goal of this box is to show that the Euler-Lagrange equations for the field A^μ in the full Lagrangian density

$$\mathcal{L} = i\overline{\psi}\gamma^\mu \partial_\mu \psi - m\overline{\psi}\psi - J^\mu A_\mu - \tfrac{1}{4}F^{\mu\nu}F_{\mu\nu} \quad \text{are } \partial_\mu F^{\mu\nu} = J^\nu, \quad (19.41)$$

and that the four components of the latter equation correspond to Gauss's law and the three components of the Ampère-Maxwell relation.

The Euler-Lagrange equation for A_ν requires that

$$0 = \partial_\mu \frac{\partial \mathcal{L}}{\partial\,(\partial_\mu A_\nu)} - \frac{\partial \mathcal{L}}{\partial A_\nu} = -\frac{1}{4}\partial_\mu \frac{\partial\,(F^{\alpha\beta}F_{\alpha\beta})}{\partial\,(\partial_\mu A_\nu)} + J^\nu. \quad (19.42)$$

(Note how I renamed the indices $F^{\mu\nu}F_{\mu\nu} \to F^{\alpha\beta}F_{\alpha\beta}$ to avoid index collisions.) The second term was easy. Focusing on the derivative in the first term (and being meticulous about avoiding new index collisions), we have

$$\frac{\partial\,(F^{\alpha\beta}F_{\alpha\beta})}{\partial\,(\partial_\mu A_\nu)} = \frac{\partial\,(g^{\alpha\lambda}g^{\beta\sigma}F_{\lambda\sigma}F_{\alpha\beta})}{\partial\,(\partial_\mu A_\nu)}$$

$$= \frac{\partial\,(g^{\alpha\lambda}g^{\beta\sigma}[\,\partial_\lambda A_\sigma - \partial_\sigma A_\lambda\,][\,\partial_\alpha A_\beta - \partial_\beta A_\alpha\,])}{\partial\,(\partial_\mu A_\nu)}$$

$$= \frac{\partial\,(g^{\alpha\lambda}g^{\beta\sigma}[\,\partial_\lambda A_\sigma \partial_\alpha A_\beta - \partial_\lambda A_\sigma \partial_\beta A_\alpha - \partial_\sigma A_\lambda \partial_\alpha A_\beta + \partial_\sigma A_\lambda \partial_\beta A_\alpha\,])}{\partial\,(\partial_\mu A_\nu)}. \quad (19.43)$$

Let's just look at the first term in brackets here. By the product rule,

$$\frac{\partial(\partial_\lambda A_\sigma \partial_\alpha A_\beta)}{\partial(\partial_\mu A_\nu)} = \frac{\partial(\partial_\lambda A_\sigma)}{\partial(\partial_\mu A_\nu)}\partial_\alpha A_\beta + \partial_\lambda A_\sigma \frac{\partial(\partial_\alpha A_\beta)}{\partial(\partial_\mu A_\nu)}$$

$$= \delta^\mu_\lambda \delta^\nu_\sigma \partial_\alpha A_\beta + \partial_\lambda A_\sigma \delta^\mu_\alpha \delta^\nu_\beta, \quad (19.44)$$

where in the last step, I noted that $\partial(\partial_\lambda A_\sigma)/\partial(\partial_\mu A_\nu)$ will be 1 if both $\lambda = \mu$ and $\beta = \nu$ but zero otherwise, and that we can represent this fact in index form as $\delta^\mu_\lambda \delta^\nu_\sigma$ (respecting the principle that a lower index in the denominator of a derivative is equivalent to an upper index). A similar argument applies to the second term: $\partial(\partial_\alpha A_\beta)/\partial(\partial_\mu A_\nu) = \delta^\mu_\alpha \delta^\nu_\beta$.

Exercise 19.4.1 Similarly, work through the other terms in Equation 19.43, then sum over the delta and inverse metric indices to arrive at $\partial_\mu F^{\mu\nu} = J^\nu$.

BOX 19.4 Maxwell's Equations II (cont.)

Exercise 19.4.2   Check that the t and x components of this equation yield $\vec{\nabla} \cdot \vec{E} = \rho$ and $(\vec{\nabla} \times \vec{B} - \partial \vec{E}/\partial t = \vec{J})_x$, respectively. (*Hints:* Remember that $J^t = \rho$. You likely will not need all the space provided.)

BOX 19.5 Gauge Magic

In this box, we will see that we can always choose the parameter $\theta(x)$ so that

$$\partial_\mu A^\mu = 0 \quad \text{and} \quad \vec{\nabla} \cdot \vec{A} = 0 \qquad (19.45)$$

corresponding to Lorenz and Coulomb gauges, respectively, and also that in empty space, we can choose $A^t = 0$ in the Coulomb gauge.

To see the first case, consider choosing $\theta(x)$ so that

$$0 = \partial_\mu A'^\mu = \partial_\mu (A^\mu + \partial^\mu \theta) \;\Rightarrow\; \partial_\mu \partial^\mu \theta = -\partial_\mu A^\mu. \qquad (19.46)$$

Can we do this? $\partial_\mu A^\mu$ is just some function of four-position $f(t, x, y, z)$. The differential equation $\partial_\mu \partial^\mu g = f$ is a well-studied equation that has solutions g for all continuous and differentiable functions f. So we can always choose $\theta(x)$ so that the Lorenz gauge condition $0 = \partial_\mu A'^\mu$ is true.

This, however, does not exhaust our freedom to choose $\theta(x)$. If $\theta(x)$ solves $\partial_\mu \partial^\mu \theta = -\partial_\mu A^\mu$, then so does $\theta(x) + \xi(x)$, so long as $\partial_\mu \partial^\mu \xi = 0$, because

$$\partial_\mu \partial^\mu (\theta + \xi) = \partial_\mu \partial^\mu \theta + \partial_\mu \partial^\mu \xi = -\partial_\mu A^\mu + 0 = -\partial_\mu A^\mu. \qquad (19.47)$$

So assume we have chosen $\theta(x)$ so that A'^μ satisfies the Lorenz gauge condition $0 = \partial_\mu A'^\mu$. Taking the spatial divergence and the time derivative of both sides of $A''^\mu(x) = A'^\mu(x) + \partial^\mu \xi(x)$ yields

$$\vec{\nabla} \cdot \vec{A}'' = \vec{\nabla} \cdot \vec{A}' - \vec{\nabla} \cdot \vec{\nabla} \xi \quad \text{and} \quad \frac{\partial A''^t}{\partial t} = \frac{\partial A'^t}{\partial t} + \frac{\partial^2 \xi}{\partial t^2}. \qquad (19.48)$$

We can always find a function ξ that solves $\vec{\nabla} \cdot \vec{\nabla} \xi = \nabla^2 \xi = \vec{\nabla} \cdot \vec{A}'$ (and thus sets $\vec{\nabla} \cdot \vec{A}'' = 0$): $\nabla^2 \xi = \vec{\nabla} \cdot \vec{A}'$ is just a version of Poisson's equation $\nabla^2 g = f$, which we also know always has solutions for reasonable functions $f = \vec{\nabla} \cdot \vec{A}'$. But since $\partial^\mu \partial_\mu \xi = 0$, we must have $\partial^2 \xi / \partial t^2 = \nabla^2 \xi = \vec{\nabla} \cdot \vec{A}'$. Substituting this into the time-derivative part of Equation 19.48 yields

$$\frac{\partial A''^t}{\partial t} = \frac{\partial A'^t}{\partial t} + \vec{\nabla} \cdot \vec{A}' = \partial_\mu A'^\mu = 0. \qquad (19.49)$$

So, with these choices, $\partial A''^t / \partial t = 0$. But we can't conclude that we can choose $A''^t = 0$ unless setting $\partial A''^t / \partial x = \partial A''^t / \partial y = \partial A''^t / \partial z = 0$ is also allowed.

Exercise 19.5.1 The only equation that puts any constraints on the spatial derivatives of A''^t is $J^\nu = \partial_\mu \partial^\mu A''^\nu$ (Equation 19.29). Show that this equation forbids setting $A''^t = 0$ where $\rho \neq 0$, but allows it as one solution when $\rho = 0$.

BOX 19.6   Energy Density of the EM Field

The goal of this box is to calculate the energy density of the electromagnetic field. According to our usual formula, the stress-energy tensor for the electromagnetic field is given by

$$T_{\text{em}}^{\mu\nu} = \frac{\partial \mathcal{L}_{\text{em}}}{\partial \left(\partial_\mu A_\sigma\right)} \partial^\nu A_\sigma - g^{\mu\nu} \mathcal{L}_{\text{em}}$$

$$= -\frac{1}{4} \frac{\partial \left(F^{\alpha\beta} F_{\alpha\beta}\right)}{\partial \left(\partial_\mu A_\sigma\right)} \partial^\nu A_\sigma + \frac{1}{4} g^{\mu\nu} F^{\alpha\beta} F_{\alpha\beta}. \tag{19.50}$$

From Box 19.4, we have seen that the partial derivative is simply equal to $4F^{\mu\sigma}$. So the field's stress-energy tensor is

$$T_{\text{em}}^{\mu\nu} = -F^{\mu\sigma} \partial^\nu A_\sigma + \tfrac{1}{4} g^{\mu\nu} F^{\alpha\beta} F_{\alpha\beta}. \tag{19.51}$$

In particular, the field's energy density is

$$T_{\text{em}}^{tt} = -F^{t\sigma} \partial^t A_\sigma + \tfrac{1}{4} g^{tt} F^{\alpha\beta} F_{\alpha\beta} = E_j \partial^t A_j + \tfrac{1}{4} F^{\alpha\beta} F_{\alpha\beta}, \tag{19.52}$$

where $j = x, y, z$, because $F^{tt} = 0$ and $F^{tj} = -E_j$. But note that by definition

$$E_j \equiv \partial_t A_j - \partial_j A_t \tag{19.53}$$

(see Equation 19.17) and $A_t = 0$ in the Coulomb gauge.

Exercise 19.6.1   From here, show that $T_{\text{em}}^{tt} = \tfrac{1}{2}\left(\,|\vec{E}|^2 + |\vec{B}|^2\,\right)$, the expected formula for electric field energy density. (*Hint:* Write out $F^{\alpha\beta} F_{\alpha\beta}$.)

Homework Problems

P19.1 Show that the set of all possible "matrices" of the form $M(\theta) \equiv e^{-iq\theta}$ obey all the required properties of a group under multiplication.

P19.2 Consider the set of matrices

$$M_R(\theta) = \begin{pmatrix} \cos\theta & -\sin\theta \\ \sin\theta & \cos\theta \end{pmatrix}. \qquad (19.54)$$

This group represents simple rotations of two-dimensional vectors in space.

a. Show that this set of matrices satisfies all the required properties of a group under the operation of matrix multiplication. (*Hint:* Sum-of-angles trig identities will help.)

b. Show that $M_R(\theta)M_R(\phi) = M_R(\phi)M_R(\theta)$. A group whose matrices commute is called an **Abelian group**.

c. Show that for every element $M_R(\theta)$ in this group, there is a corresponding element $M(\theta) = e^{-iq\theta}$ such that the product $M_R(\theta)M_R(\phi)$ is the matrix corresponding to the product $M(\theta)M(\phi)$ and that the M group is also Abelian. This means that these two groups are identical in structure (isomorphic). $M_R(\theta)$ and $M(\theta)$ constitute two **representations** of the same group structure.

P19.3 For a complex *scalar* field, consider

$$\mathcal{L} = \partial_\mu \phi \, \partial^\mu \phi^* - m^2 \phi^* \phi. \qquad (19.55)$$

Like the Dirac Lagrangian density considered in this chapter, this is also invariant under a global U(1) transformation $\phi' = e^{-iq\theta}\phi$.

a. To make this a *local* gauge transformation, we would write $\mathcal{L} = D_\mu \phi (D^\mu \phi)^* - m^2 \phi^* \phi$, where $D_\mu \equiv \partial_\mu + iqA_\mu$. What are the additional terms that are added to the Lagrangian density in this case? (*Hints:* Note that the conserved current J^μ implied by this symmetry for the complex scalar field is given in Equation 6.29. You will get an additional term here that does not appear in the Dirac $\mathcal{L}$.)

b. Find the equation of motion for ϕ in the limit that the field is weak (so that $A \cdot A$ is small compared to terms with just one power of A^μ).

P19.4 A Lagrangian density for a hypothetical photon with nonzero mass M might be

$$\mathcal{L} = -\tfrac{1}{4} F^{\alpha\beta} F_{\alpha\beta} + \tfrac{1}{2} M^2 A^\mu A_\mu. \qquad (19.56)$$

(This is called the **Proca Lagrangian**.)

a. What are the Euler-Lagrange equations for this field?

b. Assume a plane-wave solution $A^\mu = \epsilon^\mu e^{-ip\cdot x}$. Rewrite the Euler-Lagrange equation in terms of dot products of p^μ and ϵ^μ with themselves and each other, and argue that $p \cdot \epsilon = 0$ even if $M \neq 0$.

c. This means that for a photon moving in the z direction, the solutions in Equation 19.32 still work. But for a massive photon, there is a third solution as well. Show that a polar-

ization vector such that $\epsilon^t = p^z$, $\epsilon^x = 0$, $\epsilon^y = 0$, $\epsilon^z = E_p$ (which is clearly perpendicular to the other two) also satisfies the condition.

d. Find the fields $\vec{E}$ and $\vec{B}$ for this solution, and show that these fields disappear in the limit that the photon mass $M \to 0$. (*Hint:* Don't be alarmed if the fields come out with imaginary amplitudes. You should find the fields to be multiplied by $e^{ip\cdot x}$, and technically, we would take the real part of the combination.)

P19.5 You may already know that the photon has spin 1, but has only two spin states (= polarizations). How do we know that it has spin 1, and what happened to the expected third spin state? Considering the massive photon described in the previous problem makes this easier to understand.

a. Consider the plane-wave solution $A^\mu = \epsilon^\mu e^{-ip\cdot x}$. Because A^μ is a four-vector, ϵ^μ must be a four-vector. Under a rotation around the z axis, a four-vector will transform like

$$\epsilon'^\mu = \begin{pmatrix} 1 & 0 & 0 & 0 \\ 0 & \cos\theta & -\sin\theta & 0 \\ 0 & \sin\theta & \cos\theta & 0 \\ 0 & 0 & 0 & 1 \end{pmatrix} \begin{bmatrix} \epsilon^t \\ \epsilon^x \\ \epsilon^y \\ \epsilon^z \end{bmatrix}. \qquad (19.57)$$

Find this transformation's generator, which will be the S_z operator for this massive particle.

b. Verify that $\epsilon^\mu = (0, a, \pm ia, 0)$ (where a is arbitrary) and the ϵ^μ given in part (c) of the previous problem are orthogonal eigenvectors of the S_z operator. (It turns out there is no fourth eigenvector because of the constraint $p \cdot \epsilon = 0$.) Why do we say the photon has spin 1?

c. Which eigenvalue is suppressed in the limit that $M \to 0$?

P19.6 Consider the four-potential function

$$A^\mu(x) = \begin{bmatrix} 0 \\ A\sin(\omega t - \omega z) \\ A\sin(\omega t - \omega z) \\ 0 \end{bmatrix}, \qquad (19.58)$$

where A and ω are real constants.

a. Does this four-potential satisfy the Lorenz gauge conditions? The Coulomb gauge conditions?

b. Calculate $\vec{E}(x)$ and $\vec{B}(x)$ for this field.

c. Calculate the value of $\vec{E} \cdot \vec{B}$ for this field.

d. Calculate $\vec{E} \times \vec{B}$ for this field. (This vector, called the **Poynting vector**, gives the energy per unit time per unit area transported by a dynamic electromagnetic field.)

e. $\partial_\mu F^{\mu\nu} = J^\nu$ (Equation 19.21) describes how an electromagnetic field is affected by local charge or current. What are the components of J^ν in the region of space where this solution applies?

f. In the gauge satisfied by this particular $A^\mu(x)$, the equation $\partial^\mu \partial_\mu A^\nu = J^\nu$ (see Equation 19.29) ought to represent the same thing as $\partial_\mu F^{\mu\nu} = J^\nu$. Does this $A^\mu(x)$ satisfy this equation for the values of J^μ you found in the previous part?

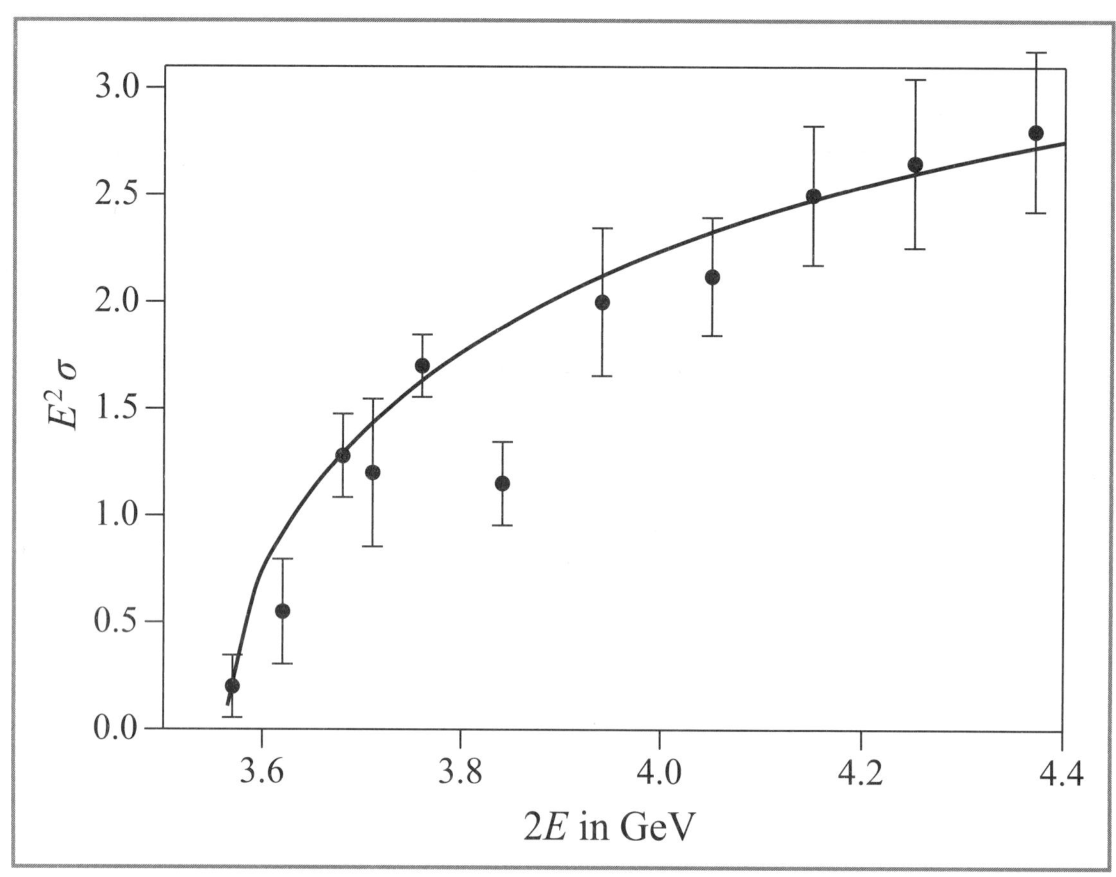

20. QUANTUM ELECTRODYNAMICS

Recent chapters have discussed a Lagrangian density that couples fermions with the electromagnetic field that emerges from a local U(1) gauge symmetry:

$$\mathcal{L} = i\overline{\psi}\gamma^{\mu}\partial_{\mu}\psi - m\overline{\psi}\psi - \tfrac{1}{4}F^{\mu\nu}F_{\mu\nu} - J^{\mu}A_{\mu}, \tag{20.1}$$

where $F^{\mu\nu} \equiv \partial^{\mu}A^{\nu} - \partial^{\nu}A^{\mu}$, $J^{\mu} \equiv q\overline{\psi}\gamma^{\mu}\psi$, and q is the charge of the particle (not the antiparticle) associated with our fermion field. The first two terms describe the Lagrangian density for a fermion with mass m and the third term describes the free-field Lagrangian density for the massless photon. The interesting part is the final term, which describes the interaction between the fermion and electromagnetic fields. This interaction term has the form $-q\overline{\psi}\gamma^{\mu}\psi A_{\mu}$, which is similar in many ways to the interaction term $-\lambda\eta\phi^{2}$ that we studied in Chapter 10 or Chapters 12 and 14. We will leverage our experience there to develop rules for constructing Feynman diagrams that express this interaction.

Studying this Lagrangian density and its implications amounts to studying what physicists call **quantum electrodynamics** (QED). In terms of its accuracy in making predictions, this is one of the most successful physical theories ever created. One could easily spend the whole rest of the course talking about this theory alone, but we want to be able to spend some time on other aspects of the Standard Model as well. So this chapter will provide just an overview of some applications. We will use this theory to construct Feynman diagrams and calculate decay rates or cross sections for just a handful of paradigmatic processes so that you can see how the theory works. In addition to providing insight into the working of QED itself, this will provide a valuable context for studying the rest of the Standard Model.

20.1 Feynman Rules for QED

Let's look more closely at the interaction term $-q\overline{\psi}\gamma^{\mu}\psi A_{\mu}$. Comparing this term to the toy universe term $-\lambda\eta\phi^{2}$ suggests that every vertex will connect two fermion lines with a photon line. But here we have more baggage to consider. The toy universe had a vertex factor of $-i\lambda$; here, we will have a vertex factor of $-iq\gamma^{\mu}$. In QED, the basic fermion fields (the electron, the muon, and the tau) have charge $q = -e$ (where e is the proton charge), so we usually write this factor as $+ie\gamma^{\mu}$.

In our toy universe, the incoming and outgoing lines did not bring in any associated factors. Here, we have polarization vectors ϵ^{μ} associated with photons, factors of u or $\overline{u}$ associated with incoming or outgoing fermions, respectively, and factors of v or $\overline{v}$ associated with incoming or outgoing antifermions, respectively. We also must put these items in the right order so that the various matrix products produce simple numbers. All this is going to make our rules for interpreting Feynman diagrams more complicated than in our toy universe.

In the interest of not getting bogged down in the details, I am going to state the Feynman rules for QED diagrams without formal proof, but the discussion in the previous paragraph will give you perhaps some intuition about where these rules are coming from. Here are the rules for calculating an amplitude $\mathcal{M}$ for a given QED process:

1. **Draw diagrams.** Construct diagrams that connect the incoming and outgoing particle lines with internal particles such that two fermion lines connect with a photon line at each vertex. Charge must also be conserved at each vertex. Draw photon lines with a squiggly line to set them apart from solid fermion lines.

2. **Label lines.** In each diagram, draw arrows next to each particle line to show the direction of momentum flow. Label the arrows for incoming or outgoing particles with four-momenta p_1, p_2, ... and internal arrows with four-momenta q_1, q_2, ...

   In addition, we draw arrows *along the fermion lines* in the direction of motion for fermions but *opposite* the direction of motion for antifermions. This will help us order factors correctly.

   Each *external* line also contributes the factor listed (along with the line appearance) in the table below ($s = +$ or $-$ is a spin index):

Particle	Factor	Drawing
incoming fermion	$u^s(p)$	
outgoing fermion	$\overline{u}^s(p)$	
incoming antifermion	$\overline{v}^s(p)$	
outgoing antifermion	$v^s(p)$	
incoming photon	$\epsilon^{s,\mu}$	
outgoing photon	$(\epsilon^{s,\mu})^*$	

3. **Vertex factors.** Each vertex in the diagram contributes a factor of

$$+ie\gamma^{\mu} \tag{20.2a}$$

4. **Propagators.** Each internal fermion line contributes

$$i\,\frac{\slashed{q}_j + m}{q_j^2 - m^2} \tag{20.2b}$$

(where $\slashed{q}_j \equiv (q_j)_\mu \gamma^\mu$). Each internal photon line contributes

$$\frac{-ig_{\mu\nu}}{q_j^2} \tag{20.2c}$$

5. **Conserve four-momentum at each vertex.** (For simple diagrams, this will determine the four-momenta q_j for internal particles.)

6. **(New!) Follow the arrows.** To put items in the correct order, follow the arrows *on* the fermion lines from start to finish and write the corresponding terms *starting from the right* (backward from English writing).

7. **The result is** $-i\mathcal{M}$. Multiply by i to get $\mathcal{M}$.

8. **Add amplitudes.** To find the total amplitude for a given process up to a certain order, add the amplitudes for all the diagrams up to that order.

9. (New!) Antisymmetrization. Include a minus sign between two diagrams that differ by swapping (between two vertices) two incoming identical fermions, two outgoing identical fermions, an incoming fermion with an outgoing antifermion (corresponding to the same type of fermion), or an incoming antifermion with an outgoing fermion of the same type.

The propagator rules may seem mysterious, but there is actually a method behind the seemingly arbitrary madness, as discussed in Box 20.1. Please note that in Equation 20.2b, I am using the standard convention that when we add a scalar (like m) to a matrix (like $\displaystyle{\not}q_j \equiv q_{j,\mu}\gamma^\mu$), we are to understand that the scalar is multiplied by I: $\displaystyle{\not}q_j + m \equiv {\not}q_j + mI$.

Finally, the new antisymmetrization rule has to do with the fact that initial and final states involving identical fermions must be antisymmetric with respect to interchange of identical fermions (and the fact that an antifermion behaves like a "backward" version of the corresponding fermion).

With these rules in hand, we are now in a position to calculate cross sections in a variety of realistic situations.

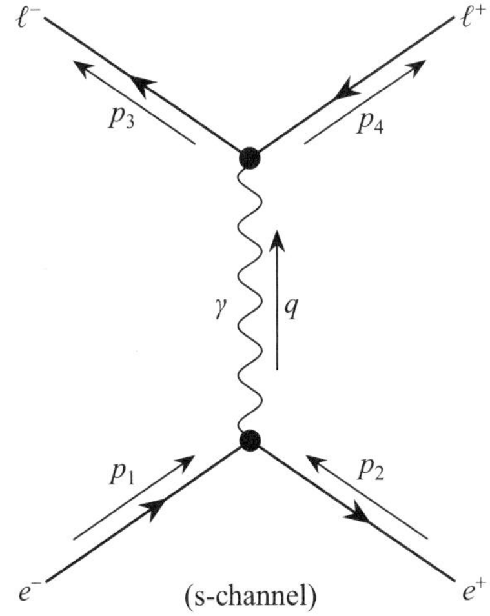

FIGURE 20.1 This figure shows the Feynman diagram for lepton pair production. Note the separate arrows on the fermion lines.

20.2 Lepton Pair Production

A good first example is collisions between an electron-positron pair producing a muon-antimuon or tau-antitau pair, because we have only a single lowest-order Feynman diagram for this process (see Figure 20.1) and we have good experimental data with which to compare.

To construct the formula for the amplitude, we follow the arrows along each fermion line and write the factors associated with each line and vertex right to left ("backward") as we assemble them:

$$-i\mathcal{M} = [\,\overline{v}_2(ie\gamma^\mu)u_1\,]\frac{-ig_{\mu\nu}}{q^2}[\,\overline{u}_3(ie\gamma^\nu)v_4\,], \tag{20.3}$$

where I have used the shorthand notation $u_1 \equiv u^{s_1}(p_1)$ and so on. The square brackets group spinors with a connecting 4×4 gamma matrix: the result in each case is a simple number. Also, note how the $g_{\mu\nu}$ in the photon propagator connects the gamma-matrix indices associated with the two vertices. We cannot have any free indexes or matrices left in the expression for $\mathcal{M}$, as a quantum amplitude must be a simple number.

Conservation of four-momentum at the vertex yields $q = (p_1 + p_2)$. Substituting this in and multiplying by i yields

$$\mathcal{M} = \frac{-e^2}{(p_1 + p_2)^2}[\,\overline{v}_2\gamma^\mu u_1\,][\,\overline{u}_3\gamma_\mu v_4\,]. \tag{20.4}$$

20.3 Casimir's Trick

To go on from here, we would need to know the spin orientations of the incoming and outgoing fermions. If we do know these orientations, then we can substitute in the correct spinors for $u_1, \overline{v}_2, \overline{u}_3,$ and v_4 and calculate the numbers in square brackets, calculate the absolute square of $\mathcal{M}$, and find the cross section.

However, in many experiments, the beams are unpolarized, so the spins are random. We usually do not care about the output spins either: we simply want to know how many particles end up in our detector. In such a case, we need to evaluate $|\mathcal{M}(s_1, s_2, s_3, s_4)|^2$ for all 16 possible spin configurations, *sum* the results, and *average* over the four possible initial spin configurations to get the *average* probability per interaction that particles end up in the detector. For each spin configuration, we must calculate

$$|\mathcal{M}|^2 = \frac{e^4}{(p_1 + p_2)^4}[\,\overline{v}_2\gamma^\mu u_1\,][\,\overline{u}_3\gamma_\mu v_4\,][\,\overline{v}_2\gamma^\nu u_1\,]^*[\,\overline{u}_3\gamma_\nu v_4\,]^*. \tag{20.5}$$

To put it mildly, this will be a lot of work.

However, a nice trick makes this kind of calculation *much* easier. Notice that the spin-dependent parts of Equation 20.5 involve factors of the form $[\,\overline{v}_a\Gamma_1 u_b\,][\,\overline{v}_a\Gamma_2 u_b\,]^*$ or $[\,\overline{u}_a\Gamma_1 v_b\,][\,\overline{u}_a\Gamma_2 v_b\,]^*$, where a and b are indexes that specify the particle and Γ_1 and Γ_2 represent arbitrary 4×4 matrices (usually these are γ^μ matrices, as in Equation 20.5, but the result holds for any matrices). In Box 20.2, you will show that

$$\sum_{\text{all spins}} [\,\overline{v}_a\Gamma_1 u_b\,][\,\overline{v}_a\Gamma_2 u_b\,]^* = \text{Tr}[\Gamma_1(\not{p}_b + m_b)\overline{\Gamma}_2(\not{p}_a - m_a)\,], \qquad (20.6)$$

where $\overline{\Gamma}_2 \equiv \gamma^t\Gamma_2^\dagger\gamma^t$, $\not{p}_a \equiv p_{a\mu}\gamma^\mu$ and $\not{p}_b \equiv p_{b\mu}\gamma^\mu$, and the sum is over all possible spin orientations of particles a and b. (Note that antiparticles get a minus sign in front of the mass term, while particles get a positive sign). This may not *look* much better, but it is actually *huge*: the spinors are gone, and calculating the sum over all spins reduces to multiplying a few gamma matrices and taking the trace.

Also note that when Γ_2 is an ordinary gamma matrix, $\overline{\gamma}^\mu = \gamma^t(\gamma^\mu)^\dagger\gamma^t = \gamma^\mu$ (see Box 20.3) and $\overline{\gamma}_\nu = \gamma^t(g_{\mu\nu}\gamma^\nu)^\dagger\gamma^t = g_{\mu\nu}\gamma^t(\gamma^\nu)^\dagger\gamma^t = g_{\mu\nu}\gamma^\nu = \gamma_\mu$, so we rarely have to worry about the bar.

20.4 Gamma-Matrix and Trace Theorems

Here are some theorems that make applying Casimir's trick easier:

1. $g_{\mu\nu}g^{\nu\mu} = g^{\mu\nu}g_{\nu\mu} = 4$.

This is because $g_{\mu\nu}g^{\nu\alpha} = \delta^\alpha_\mu$, and $\delta^\mu_\mu = 4$ because we are summing the identity matrix down the diagonal. (Since the metric is symmetric, the order of indices does not really matter, so $g_{\mu\nu}g^{\mu\nu} = 4$ also.)

2. $\gamma^\mu\gamma^\nu + \gamma^\nu\gamma^\mu = 2g^{\mu\nu}I$ (see Equation 16.9).

From these two results, we can prove the following (see Box 20.5):

3. $\gamma_\mu\gamma^\mu = 4I$.

4. $\gamma_\mu\gamma^\nu\gamma^\mu = -2\gamma^\nu$.

5. $\gamma_\mu\gamma^\alpha\gamma^\beta\gamma^\mu = 4g^{\alpha\beta}I$.

6. $\gamma_\mu\gamma^\nu\gamma^\alpha\gamma^\beta\gamma^\mu = -2\gamma^\beta\gamma^\alpha\gamma^\nu$.

When A and B are square matrices and c is a (complex) number,

7. $\text{Tr}(A + B) = \text{Tr}(A) + \text{Tr}(B)$.

8. $\text{Tr}(cA) = c\,\text{Tr}(A)$.

9. $\text{Tr}(AB) = \text{Tr}(BA)$.

10. $\text{Tr}(ABC) = \text{Tr}(BCA) = \text{Tr}(CAB) \neq \text{Tr}(ACB) = \text{Tr}(CBA) = \text{Tr}(BAC)$.

Each diagonal element of $A + B$ is the sum of the corresponding diagonal elements of A and B, so adding those elements is the same as adding the traces of A and B: thus Theorem 7. Similarly, multiplying a matrix by c multiplies each of its diagonal elements by c, and we can pull that common factor out of the sum. As for Theorem 9, note that

$$\text{Tr}(AB) = \sum_\mu (AB)_{\mu\mu} = \sum_{\mu\alpha} A_{\mu\alpha}B_{\alpha\mu} = \sum_{\mu\alpha} B_{\alpha\mu}A_{\mu\alpha} = \sum_\alpha (BA)_{\alpha\alpha} \qquad (20.7)$$

and the last is $\text{Tr}(BA)$. Theorem 10 follows from Theorem 9: since $\text{Tr}(ABC) = \text{Tr}(A[BC]) = \text{Tr}([BC]A)$, we can always peel the front matrix off and put it in the back and get the same trace. But because $BC \neq CB$ in general, it will not generally be true that $ABC = ACB$, so $\text{Tr}(ACB) = \text{Tr}(CBA) = \text{Tr}(BAC)$ is an entirely different series of equal traces.

One can now prove the following "trace theorems" (see Box 20.6):

11. The trace of a product of an *odd* number of gamma matrices = 0.

12. $\text{Tr}(I) = 4$.

13. $\text{Tr}(\gamma^\mu \gamma^\nu) = 4g^{\mu\nu}$.

14. $\text{Tr}(\gamma^\mu \gamma^\nu \gamma^\alpha \gamma^\beta) = 4(g^{\mu\nu}g^{\alpha\beta} - g^{\mu\alpha}g^{\nu\beta} + g^{\mu\beta}g^{\nu\alpha})$.

Equation 17.26 introduced the $\gamma^5 \equiv i\gamma^t \gamma^x \gamma^y \gamma^z$ matrix. Since this matrix is a product of an even number of gamma matrices, multiplying γ^5 by an odd number of gamma matrices yields zero.

15. $0 = \text{Tr}(\gamma^5 \gamma^\mu) = \text{Tr}(\gamma^5 \gamma^\mu \gamma^\nu \gamma^\alpha)$, and so on.

When we multiply γ^5 by an *even* number of matrices, we get

16. $\text{Tr}(\gamma^5) = 0$,

17. $\text{Tr}(\gamma^5 \gamma^\mu \gamma^\nu) = 0$,

18. $\text{Tr}(\gamma^5 \gamma^\mu \gamma^\nu \gamma^\alpha \gamma^\beta) = 4i\epsilon^{\mu\nu\alpha\beta}$.

The last symbol represents the completely antisymmetric tensor

$$\epsilon^{\mu\nu\alpha\beta} = \begin{cases} -1 & \text{if } \mu\nu\alpha\beta \text{ is an } even \text{ permutation of } txyz, \\ +1 & \text{if } \mu\nu\alpha\beta \text{ is an } odd \text{ permutation of } txyz, \\ 0 & \text{if any pair of indices is the same.} \end{cases} \quad (20.8)$$

In this situation, an *even* permutation is one that we can arrive at from $txyz$ by interchanging indices an even number of times. For example, $txzy$ is an odd permutation, $xtzy$ is an even permutation and so on.

19. $\{\gamma^5, \gamma^\mu\} = 0 \;\Rightarrow\; \gamma^5 \gamma^\mu = -\gamma^\mu \gamma^5$.

I will show that this very useful theorem is true for γ^y; the proof for the others is similar. Note that $\gamma^\mu \gamma^\nu = -\gamma^\nu \gamma^\mu$ if $\nu \neq \mu$. This means that

$$\gamma^y \gamma^5 = i\gamma^y \gamma^t \gamma^x \gamma^y \gamma^z = -i\gamma^t \gamma^y \gamma^x \gamma^y \gamma^z = +i\gamma^t \gamma^x \gamma^y \gamma^y \gamma^z$$
$$= -i\gamma^t \gamma^x \gamma^y \gamma^z \gamma^y = -\gamma^5 \gamma^y. \quad (20.9)$$

(In other words, moving the γ^y from the front to the back requires swapping a γ^y with three unlike matrices. Each switch results in a minus sign, so the terms in the anticommutator must cancel. This will be true for any γ^μ.)

This is a lot to show and know, but you only need to calculate these theorems once in your life: subsequently, you can simply earmark this page.

20.5 The Averaged Lepton-Production Amplitude

In the case of lepton production, applying Casimir's trick twice yields

$$\sum_{\text{spins}} |\mathcal{M}|^2 = \sum_{\text{spins}} \frac{e^4}{(p_1 + p_2)^4} [\,\bar{v}_2 \gamma^\mu u_1\,][\,\bar{v}_2 \gamma^\nu u_1\,]^* \,[\,\bar{u}_3 \gamma_\mu v_4\,][\,\bar{u}_3 \gamma_\nu v_4\,]^*$$

$$= \frac{e^4}{(p_1 + p_2)^4} \,\text{Tr}[\,\gamma^\mu (\not{p}_1 + m)\gamma^\nu (\not{p}_2 - m)\,]\,\text{Tr}[\,\gamma_\mu (\not{p}_4 - M)\gamma_\nu (\not{p}_3 + M)\,], \quad (20.10)$$

where m is the electron mass and M is the produced lepton mass. To evaluate these traces, we appeal to the trace theorems. First, note that when we multiply out the first trace factor $\text{Tr}[\gamma^\mu (\not{p}_1 - m)\gamma^\nu (\not{p}_2 + m)]$, the first term becomes

$$\mathrm{Tr}(\gamma^{\mu}\not{p}_1\gamma^{\nu}\not{p}_2) = \mathrm{Tr}(\gamma^{\mu}(p_1)_{\alpha}\gamma^{\alpha}\gamma^{\nu}(p_2)_{\beta}\gamma^{\beta})$$

$$= (p_1)_{\alpha}(p_2)_{\beta}\,4[\,g^{\mu\alpha}g^{\nu\beta} - g^{\mu\nu}g^{\alpha\beta} + g^{\mu\beta}g^{\alpha\nu}\,]$$

$$= 4[\,p_1^{\mu}p_2^{\nu} - g^{\mu\nu}(p_1\cdot p_2) + p_2^{\mu}p_1^{\nu}\,] \tag{20.11}$$

by Theorem 14. The cross terms $-\,\mathrm{Tr}(\gamma^{\mu}\not{p}_1\gamma^{\mu}m)$ and $+\,\mathrm{Tr}(\gamma^{\mu}m\gamma^{\mu}\not{p}_2)$ are zero because they involve an odd number of gamma matrices, and the final term is $-\,\mathrm{Tr}(\gamma^{\mu}\gamma^{\nu}m^2) = -m^2 4 g^{\mu\nu}$ by Theorem 13. Therefore,

$$\mathrm{Tr}[\gamma^{\mu}(\not{p}_1 - m)\gamma^{\nu}(\not{p}_2 + m)] = 4[\,p_1^{\mu}p_2^{\nu} - g^{\mu\nu}p_1\cdot p_2 + p_2^{\mu}p_1^{\nu} - g^{\mu\nu}m^2\,]. \tag{20.12}$$

The second trace in Equation 20.10 is the same except M replaces m, 3 replaces 1, 4 replaces 2, and the greek subscripts are lowered. So their product is

$$P \equiv 16[\,p_1^{\mu}p_2^{\nu} - g^{\mu\nu}(p_1\cdot p_2) + p_2^{\mu}p_1^{\nu} - g^{\mu\nu}m^2\,]$$

$$\times [\,(p_4)_{\mu}(p_3)_{\nu} - g_{\mu\nu}(p_4\cdot p_3) + (p_3)_{\mu}(p_4)_{\nu} - g_{\mu\nu}M^2\,]$$

$$= 16[\,(p_1\cdot p_4)(p_2\cdot p_3) - (p_1\cdot p_2)(p_4\cdot p_3) + (p_1\cdot p_3)(p_2\cdot p_4) - (p_1\cdot p_2)M^2$$

$$-\,(p_4\cdot p_4)(p_1\cdot p_2) + 4(p_1\cdot p_2)(p_4\cdot p_3) - (p_1\cdot p_2)(p_3\cdot p_4) + 4(p_1\cdot p_2)M^2$$

$$+\,(p_2\cdot p_4)(p_1\cdot p_3) - (p_2\cdot p_1)(p_4\cdot p_3) + (p_2\cdot p_3)(p_1\cdot p_4) - (p_2\cdot p_1)M^2$$

$$-\,(p_4\cdot p_3)m^2 + 4(p_4\cdot p_3)m^2 - m^2(p_3\cdot p_4) + 4m^2 M^2\,]. \tag{20.13}$$

Whew! However, a number of these terms combine (if we recall that $p\cdot q = q\cdot p$) and all the $(p_1\cdot p_2)(p_4\cdot p_3)$ terms cancel, so the result boils down to

$$P = 16[\,2(p_1\cdot p_4)(p_2\cdot p_3) + 2(p_1\cdot p_3)(p_4\cdot p_2)$$

$$+\,2(p_1\cdot p_2)M^2 + 2(p_4\cdot p_3)m^2 + 4m^2 M^2]. \tag{20.14}$$

So, to get the average squared amplitude, we substitute this for the product of the traces in Equation 20.10 and divide by 4 to take the average over the initial spins (since we have two incoming particles, each with two spin orientations):

$$\langle|\mathcal{M}|^2\rangle = \frac{8e^4}{(p_1+p_2)^4}\left[\,(p_1\cdot p_4)(p_2\cdot p_3) + (p_1\cdot p_3)(p_4\cdot p_2)\right.$$

$$\left. +\,(p_1\cdot p_2)M^2 + (p_4\cdot p_3)m^2 + 2m^2 M^2\,\right]. \tag{20.15}$$

20.6 The Lepton-Production Cross Section

We are finally in a position to calculate an actual cross section. From Equation 15.11, we know that in the center-of-mass frame the differential cross section is

$$d\sigma = \frac{S\,|\mathcal{M}|^2}{(8\pi)^2(E_1+E_2)^2}\,\frac{|\vec{p}_3|}{|\vec{p}_1|}\,d\Omega. \tag{20.16}$$

$S = 1$ here because the outgoing particles are not identical. Because we are ignoring spins, we will also use our spin-averaged $\langle|\mathcal{M}|^2\rangle$ in place of $|\mathcal{M}|^2$.

We will find it much easier to complete the calculation if we also note that the muon-electron mass ratio is almost 207 and the tau-electron ratio is much larger. So, to produce even muons, the electron and positron energies must be so large compared to the electron mass that we can essentially set $m = 0$ compared to the energies involved.

Define axes so that the incoming positron moves along the $+z$ axis and the positive lepton emerges at an angle θ relative to that axis in xz plane (see Figure 20.2). To save writing, define $E \equiv E_1 = E_2$ and $\rho \equiv |\vec{p}_3| = |\vec{p}_4|$. Then

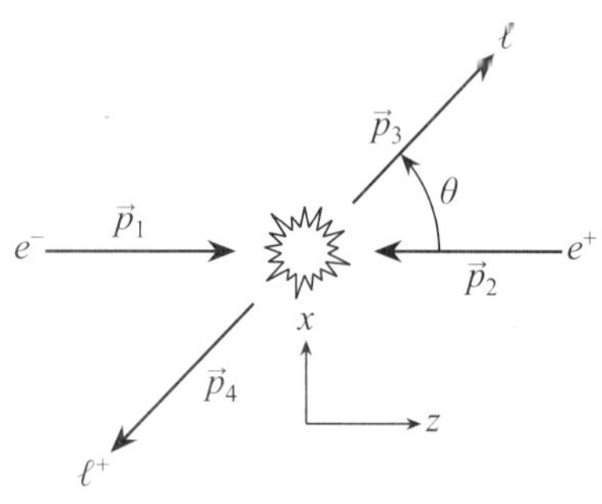

FIGURE 20.2 This diagram specifies momentum assignments for a lepton pair production process.

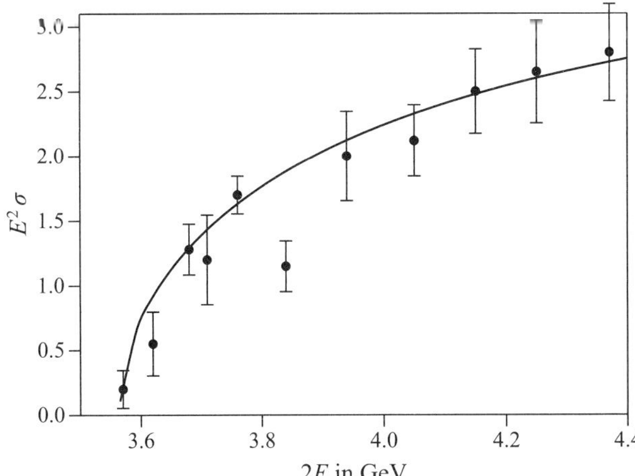

FIGURE 20.3 $E^2\sigma$ in arbitrary units for $e^+e^- \to \tau^+\tau^-$ near the $2E = 2M_\tau$ threshold. The solid curve is Equation 20.19 with $M_\tau = 1.782$ GeV. Adapted from W. Bacino *et al., Phys Rev Lett*, **41** (1978), 13.

$$
p_1 = \begin{bmatrix} E \\ 0 \\ 0 \\ E \end{bmatrix}, \quad
p_2 = \begin{bmatrix} E \\ 0 \\ 0 \\ -E \end{bmatrix}, \quad
p_3 = \begin{bmatrix} E \\ \rho\sin\theta \\ 0 \\ \rho\cos\theta \end{bmatrix}, \quad
p_4 = \begin{bmatrix} E \\ -\rho\sin\theta \\ 0 \\ -\rho\cos\theta \end{bmatrix} \quad (20.17)
$$

(remember that the initial energy must equal the final energy!). In Box 20.4, you will show that the differential cross section becomes

$$
d\sigma = \frac{\alpha^2}{16E^4}\frac{\rho}{E}\left[E^2 + M^2 + \rho^2\cos^2\theta\right]d\Omega, \tag{20.18}
$$

where $\alpha \equiv e^2/4\pi \approx 1/137$ is the well-known **fine-structure constant**. Since the first two terms are independent of θ and ϕ and the last is independent of ϕ, we can easily integrate to find the total cross section for all deflection angles:

$$
\begin{aligned}
\sigma &= \frac{\alpha^2}{16E^4}\frac{\rho}{E}\left[4\pi E^2 + 4\pi M^2 + 2\pi\rho^2\int_0^\pi \cos^2\theta\sin\theta\,d\theta\right] \\
&= \frac{\alpha^2}{16E^4}\frac{\rho}{E}\left[4\pi E^2 + 4\pi M^2 - 2\pi\rho^2\left(\tfrac{1}{3}\cos^3\pi - \tfrac{1}{3}\cos^3 0\right)\right] \\
&= \frac{\pi\alpha^2}{4E^2}\frac{\rho}{E}\left[1 + \frac{M^2}{E^2} + \frac{\rho^2}{3E^2}\right] = \frac{\pi\alpha^2}{4E^2}\sqrt{1 - \frac{M^2}{E^2}}\left[\frac{4}{3} + \frac{2}{3}\frac{M^2}{E^2}\right] \tag{20.19}
\end{aligned}
$$

because (in going from the first to the second step) $\int d\Omega$ over all directions $= 4\pi$, and the integral over all ϕ is 2π. I also used the substitution $u = \cos\theta$, $du = -\sin\theta\,d\theta$. In the next step, I used the fact that $\tfrac{1}{3}\cos^3\pi - \tfrac{1}{3}\cos^3 0 = +\tfrac{2}{3}$, and in the final step, that $\rho = |\vec{p}_3| = (E^2 - M^2)^{1/2}$. The final result agrees very nicely with experimental data (see Figure 20.3).

20.7 Closing Comments

We have done an amazing thing in this chapter: we have actually calculated realistic and physically measurable results from our theory! Of course, we are only calculating the leading-order contribution to this cross section. But note that the amplitudes from higher-order diagrams for the same processes will be suppressed by a factor of $e = \sqrt{4\pi\alpha} \approx 0.30$ for every vertex we add. Most higher-order diagrams for the processes we have considered will have at least four vertices, so (other things being equal) their amplitudes will be suppressed by a factor of $(0.30)^2 \approx 1/10$ relative to two-vertex amplitudes (and the suppression factor is actually larger in many cases). So higher-order terms should be significantly smaller than the leading term we have calculated.

But assuming that the power series actually converges also assumes that the number of diagrams at a given order does not increase faster than the magnitude of each diagram decreases as we go to higher orders. We will see that for certain types of higher-order diagrams, this assumption is demonstrably false, but we can deal with this using a trick called *renormalization*. This technique will (with some small adjustments) allow us to essentially embed some of the most divergent series of higher-order diagrams into the lowest order calculation in a clever way, and so better justify our use of just the lowest-order diagram or diagrams as a decent approximation to the final amplitude. This trick will be our focus in the next chapter.

BOX 20.1 Propagators

We discussed the propagator for the scalar (spin-0) field in some detail in Chapters 10 and 11. But there is a shortcut that arrives at the same end result that we will use to find propagators for other fields. The steps are as follows: (1) Allow the free-field equation of motion to act on the plane-wave solution $e^{-ip\cdot x}$ to get a "momentum-space" version of that equation (making sure p appears positively). (2) The propagator is i divided by the quantity that multiplies the field.

For the scalar field case, the free-field equation is $\partial^{\mu}\partial_{\mu}\phi + m^2\phi = 0$. Evaluating this for the plane wave solution yields $\left[-p^2 + m^2\right]\phi = 0$ or $\left[p^2 - m^2\right]\phi = 0$, so by this rule the propagator should be

$$\text{spin 0 propagator:} \quad \frac{i}{p^2 - m^2}, \tag{20.20}$$

which is exactly what we found in Chapters 10 and 11.

Now, in the case of spin-$\frac{1}{2}$ fermion, the field equation is $i\partial_{\mu}\gamma^{\mu}\psi - m\psi = 0$, whose "momentum space" version is $\left[p_{\mu}\gamma^{\mu} - m\right]\psi = 0$, or more compactly $\left[\not{p} - m\right]\psi = 0$. Therefore, the propagator should be $i/(\not{p} - m)$.

Exercise 20.1.1 But the denominator is the matrix $\not{p} - m \equiv p_{\mu}\gamma^{\mu} - mI$. What does that mean? Show that if we multiply top and bottom by $\not{p} + m$, we get

$$\frac{i}{\not{p} - m} = \frac{i(\not{p} + m)}{p^2 - m^2}, \tag{20.21}$$

which is easier to interpret. (This is the form given in Equation 20.2b.) (*Hints:* Note that $\{\gamma^{\mu}, \gamma^{\nu}\} = 2g^{\mu\nu}$ and that $p_{\mu}p_{\nu} = p_{\nu}p_{\mu} = \frac{1}{2}[p_{\mu}p_{\nu} + p_{\nu}p_{\mu}]$.)

In the photon case, the free-field equation in the Lorenz gauge is $\partial_{\mu}\partial^{\mu}A^{\nu} = 0$. We apply this to the plane-wave solution to get $-p^2 A^{\nu} = 0$ or $p^2 A^{\nu} = 0$. Therefore, we might expect the propagator to be i/p^2. But this neglects the fact that we must sum over the photon polarization possibilities. It turns out that the sum ends up producing a $g_{\mu\nu}$, which has just the right indices to connect with the gamma-matrix indices in the two vertex factors the virtual photon connects. So the photon propagator is $ig_{\mu\nu}/p^2$, as claimed in Equation 20.2c.

BOX 20.2 Casimir's Trick

The goal of this box is to prove Casimir's trick (named for Hendrik Casimir)

$$\sum_{\text{all spins}} [\,\overline{u}_a\Gamma_1 u_b\,][\,\overline{u}_a\Gamma_2 u_b\,]^* = \text{Tr}[\Gamma_1(\not{p}_b + m_b)\overline{\Gamma}_2(\not{p}_a + m_a)\,], \quad (20.22)$$

where Γ_1 and Γ_2 are arbitrary 4×4 matrices and $\overline{\Gamma}_2 \equiv \gamma^t\Gamma_2^\dagger\gamma^t$. We will do this in several steps.

Exercise 20.2.1 First, use the facts that $\gamma^{t\,\dagger} = \gamma^t$, that $(\gamma^t)^2 = I$, and that $(AB)^\dagger = B^\dagger A^\dagger$ when A and B are matrices or vectors to show that

$$[\,\overline{u}_a\Gamma_2 u_b\,]^* = [\,\overline{u}_b\overline{\Gamma}_2 u_a\,]. \quad (20.23)$$

Now we can use the completeness relations 16.39 (written here with $\not{p} \equiv \gamma \cdot p$)

$$\sum_s u^s(p)\overline{u}^s(p) = \not{p} + m \quad \text{and} \quad \sum_s v^s(p)\overline{v}^s(p) = \not{p} - m \quad (20.24)$$

to perform the sum over the spins. We have

$$\sum_{s_b} [\,\overline{u}_a\Gamma_1 u_b\,][\,\overline{u}_a\Gamma_2 u_b\,]^* = \sum_{s_b} \overline{u}_a\Gamma_1 u_b\overline{u}_b\overline{\Gamma}_2 u_a$$

$$= \overline{u}_a\Gamma_1\Big[\sum_{s_b} u^{s_b}(p_b)\overline{u}^{s_b}(p_b)\Big]\overline{\Gamma}_2 u_a = \overline{u}_a\Gamma_1(\not{p}_b + m_b)\overline{\Gamma}_2 u_a. \quad (20.25)$$

Define $B \equiv \Gamma_1(\not{p}_b + m_b)\overline{\Gamma}_2$ to save writing. We also want to sum over particle a's spin index s_a:

$$\sum_{s_a, s_b} [\,\overline{u}_a\Gamma_1 u_b\,][\,\overline{u}_a\Gamma_2 u_b\,]^* = \sum_{s_a} \overline{u}^{s_a}(p_a)Bu^{s_a}(p_a). \quad (20.26)$$

The trick to evaluating this sum involves writing the matrix products in index notation so that we can put the spinors next to each other. Let j and k be matrix-component indices that range from 1 to 4 so that B_{jk} indexes the 16 components of the 4×4 matrix B and $[u^{s_a}(p_a)]_k$ indexes the four components of the spinor $u^{s_a}(p_a)$. If we write the sums implied by the matrix notation on the right side of Equation 20.26 explicitly as sums over j, k, that side becomes

$$\sum_{s_a}\sum_{j,k}[\,\overline{u}^{s_a}(p_a)\,]_j B_{jk}[\,u^{s_a}(p_a)\,]_k = \sum_{j,k} B_{jk}\Big[\sum_{s_a}[\,u^{s_a}(p_a)\,]_k[\,\overline{u}^{s_a}(p_a)\,]_j\Big]$$

$$= \sum_{j,k} B_{jk}A_{kj}, \quad \text{where } A \equiv \not{p}_a + m_a \text{ by Equation 20.24.} \quad (20.27)$$

BOX 20.2 Casimir's Trick (cont.)

In the second step, I have just reordered the factors, as one may freely do when the indexes specify the sums, then applied Equation 20.24.

Exercise 20.2.2 From here, argue that

$$\sum_{s_a, s_b} [\,\overline{u}_a \Gamma_1 u_b\,][\,\overline{u}_a \Gamma_2 u_b\,]^* = \mathrm{Tr}(BA). \tag{20.28}$$

Then substitute in the values of B and A to arrive at Equation 20.22. (*Hint:* Note that if we define $G \equiv BA$, then G's components are $G_{jm} = \sum_k B_{jk} A_{km}$.)

Exercise 20.2.3 How do things change if we are dealing with antiparticles? Specifically, what is the value of $\sum_{s_a, s_b} [\,\overline{v}_a \Gamma_1 v_b\,][\,\overline{v}_a \Gamma_2 v_b\,]^*$ in terms of $\not{p}_a$, m_a, $\not{p}_b$, m_b, Γ_1, and Γ_2?

BOX 20.3 Calculating $\overline{\gamma}^{\mu}$

The goal of this box is to show that

$$\overline{\gamma}^{\mu} \equiv \gamma^{t}(\gamma^{\mu})^{\dagger}\gamma^{t} = \gamma^{\mu}. \tag{20.29}$$

Exercise 20.3.1 Prove this. (*Hints:* Consider the $\mu = t$ and $\mu = x, y, z$ cases separately, write the matrix product out in compact matrix notation, and remember that the Pauli matrices are Hermitian.)

BOX 20.4 The Pair Production Cross Section

Our goal in this box is to use the equations

$$d\sigma = \frac{\langle|\mathcal{M}|^2\rangle}{(8\pi)^2(E_1 + E_2)^2}\frac{|\vec{p}_3|}{|\vec{p}_1|}d\Omega, \quad \text{where} \tag{20.30}$$

$$\langle|M|^2\rangle = \frac{8e^4}{(p_1 + p_2)^4}\Big[(p_1 \cdot p_4)(p_2 \cdot p_3) + (p_1 \cdot p_3)(p_4 \cdot p_2)$$
$$+ (p_1 \cdot p_2)M^2 + (p_4 \cdot p_3)m^2 + 2m^2M^2\Big], \tag{20.31}$$

$$p_1 = \begin{bmatrix} E \\ 0 \\ 0 \\ E \end{bmatrix}, \quad p_2 = \begin{bmatrix} E \\ 0 \\ 0 \\ -E \end{bmatrix}, \quad p_3 = \begin{bmatrix} E \\ \rho\sin\theta \\ 0 \\ \rho\cos\theta \end{bmatrix}, \quad p_4 = \begin{bmatrix} E \\ -\rho\sin\theta \\ 0 \\ -\rho\cos\theta \end{bmatrix} \tag{20.32}$$

to calculate the differential cross section in terms of $E \equiv E_1 = E_2 = E_3 = E_4$, $\rho = |\vec{p}_3| = |\vec{p}_4|$, M, and $\alpha \equiv e^2/4\pi$ in the limit that $E \geq M \gg m$, meaning that we can treat $m \approx 0$.

Exercise 20.4.1 Do this.

BOX 20.5 Gamma-Matrix Theorems

Our goal in this box is to show that

1. $g_{\mu\nu}g^{\,\nu\mu} = g^{\,\mu\nu}g_{\nu\mu} = 4$
2. $\gamma^\mu\gamma^\nu + \gamma^\nu\gamma^\mu = 2g^{\,\mu\nu}I$

imply the following gamma-matrix theorems:

3. $\gamma_\mu\gamma^\mu = 4I$.
4. $\gamma_\mu\gamma^\nu\gamma^\mu = -2\gamma^\nu$.
5. $\gamma_\mu\gamma^\alpha\gamma^\beta\gamma^\mu = 4g^{\alpha\beta}I$.
6. $\gamma_\mu\gamma^\nu\gamma^\alpha\gamma^\beta\gamma^\mu. = -2\gamma^\beta\gamma^\alpha\gamma^\nu$.

Exercise 20.5.1 Do this. (*Hint:* Since $\gamma_\mu = g_{\mu\nu}\gamma^\nu$, we must have $\gamma_\mu\gamma^\mu$ $= g_{\mu\nu}\gamma^\nu\gamma^\mu = g_{\nu\mu}\gamma^\nu\gamma^\mu$ (because the metric is symmetric) $= g_{\mu\nu}\gamma^\mu\gamma^\nu$ (renaming both indices). Use this to prove 3. (Alternatively, just write out the product.) For the others, you can use Theorem 2 to make the γ_μ and γ^μ adjacent.)

BOX 20.6   Trace Theorems

Our goal in this box is to prove the following "trace theorems":

11. The trace of a product of an *odd* number of gamma matrices = 0.

13. $\text{Tr}(\gamma^\mu \gamma^\nu) = 4g^{\mu\nu}$.

14. $\text{Tr}(\gamma^\mu \gamma^\nu \gamma^\alpha \gamma^\beta) = 4(g^{\mu\nu}g^{\alpha\beta} - g^{\mu\alpha}g^{\nu\beta} + g^{\mu\beta}g^{\nu\alpha})$.

16. $\text{Tr}(\gamma^5) = 0$.

(You will prove Theorems 17–18 in a homework problem.) Proving Theorem 11 requires a trick. If we multiply the matrix product by $\gamma^5 \gamma^5 = I$, we see that

$$\text{Tr}(\gamma^\mu \gamma^\nu \cdots \gamma^\sigma) = \text{Tr}(\gamma^\mu \gamma^\nu \cdots \gamma^\sigma \gamma^5 \gamma^5) = \text{Tr}(\gamma^5 \gamma^\mu \gamma^\nu \cdots \gamma^\sigma \gamma^5) \quad (20.33)$$

because we can move one matrix from the back to the front. But since γ^5 anticommutes with every other γ^μ, moving it through the set of matrices back to the final γ^5 will cause a sign flip for each matrix we pass. If the number is odd, then $\text{Tr}(\gamma^\mu \gamma^\nu \cdots \gamma^\sigma \gamma^5 \gamma^5) = \text{Tr}(\gamma^5 \gamma^\mu \gamma^\nu \cdots \gamma^\sigma \gamma^5) = - \text{Tr}(\gamma^\mu \gamma^\nu \cdots \gamma^\sigma \gamma^5 \gamma^5) = 0$, since only zero is equal to negative itself. But since $\gamma^5 \gamma^5 = I$, this also means that $\text{Tr}(\gamma^\mu \gamma^\nu \cdots \gamma^\sigma) = 0$.

Exercise 20.6.1   Prove the others. (*Hints:* For Theorem 13, start by applying Theorem 2, $\text{Tr}(\gamma^\mu \gamma^\nu) = \text{Tr}(2g^{\mu\nu}I - \gamma^\nu \gamma^\mu)$, and go from there. For Theorem 14, apply Theorem 2 to the first pair in the group of four gamma matrices:

$$\text{Tr}(\gamma^\mu \gamma^\nu \gamma^\alpha \gamma^\beta) = \text{Tr}((2g^{\mu\nu}I - \gamma^\nu \gamma^\mu)\gamma^\alpha \gamma^\beta)$$
$$= 2g^{\mu\nu}\,\text{Tr}(\gamma^\alpha \gamma^\beta) - \text{Tr}(\gamma^\nu \gamma^\mu \gamma^\alpha \gamma^\beta). \quad (20.34)$$

Then apply Theorem 2 to the middle pair of matrices in the last term, then to the final pair in the group of four that results. For Theorem 16, note Equation 17.26.)

Homework Problems

P20.1 In this problem, you will prove the remaining trace theorems.

a. Prove Theorem 17: $\mathrm{Tr}(\gamma^5\gamma^\mu\gamma^\nu) = 0$. (*Hint:* Consider the $\mu = \nu$ and $\mu \neq \nu$ cases separately. In the latter case, substitute $\gamma^5 = i\gamma^t\gamma^x\gamma^y\gamma^z$ and appeal to Theorem 13.)

b. Prove Theorem 18: $\mathrm{Tr}(\gamma^5\gamma^\mu\gamma^\nu\gamma^\alpha\gamma^\beta) = 4i\epsilon^{\mu\nu\alpha\beta}$. (*Hints:* First, argue that if any indices are the same, the result is zero. Then argue that exchanging any pair of distinct indices results in a change of sign. Finally, evaluate the case where the indices are t, x, y, z.)

P20.2 For many of the theorems in Section 20.4, there are analogous theorems involving slashed quantities $\not{a} \equiv a_\mu\gamma^\nu$, where a_μ is a covector. For example, for Theorem 2, the corresponding theorem is

$$\gamma^\mu\gamma^\nu + \gamma^\nu\gamma^\mu = 2g^{\mu\nu}I \;\Rightarrow\; \not{a}\not{b} + \not{b}\not{a} = 2(a \cdot b)I. \quad (20.35)$$

This is because if we multiply Theorem 2 by $a_\mu b_\nu$, we get

$$a_\mu b_\nu\gamma^\mu\gamma^\nu + a_\mu b_\nu\gamma^\nu\gamma^\mu = a_\mu b_\nu 2g^{\mu\nu}I$$

$$= \not{a}\not{b} + \not{b}\not{a} = 2(a \cdot b)I. \quad (20.36)$$

(The quantities on the left sides of these equations are clearly matrices; hence the I on the right. But people conventionally omit the I.)

Write down analogous theorems for slashed quantities corresponding to Theorems 4, 5, 6, 13, 14, 17, and 18. (*Hint:* One raises or lowers indices on gamma matrices just as one would indexes on a four-vector.)

P20.3 Use the theorems listed in Section 20.4 to show that if $\not{p} \equiv p_\mu\gamma^\mu$, then $\not{p}\not{p} = p^2$.

P20.4 Numerical results for lepton production cross sections.

a. Show that the all-angle cross section σ for $e^+e^- \to \mu^+\mu^-$ is about $\sigma = (87\text{ nb})\cdot(1\text{ GeV}/2E)^2$, where E is the energy of any of the particles in the center-of-mass frame, in the limit that $E \gg$ the outgoing lepton mass M.

b. Calculate the analogous quantity for $e^+e^- \to \tau^+\tau^-$.

P20.5 Consider the process $\gamma + e^- \to \gamma + e^-$ (this process is called **Compton scattering**.)

a. Draw the lowest-level Feynman diagrams (there are two).

b. Let p_1 and p_2 be the four-momenta of the incoming photon and electron, respectively, and let p_3 and p_4 be the same for the outgoing photon and electron. Compute the amplitude for each diagram and simplify as much you can without making assumptions about spin. (*Hints:* Show that $(p_1 - p_4)^2 - m^2 = -2p_1 \cdot p_4$ and $(p_1 + p_2)^2 - m^2 = 2p_1 \cdot p_2$, where m is the electron mass. You can also use Equation 20.36, the fact that $\epsilon \cdot p = 0$ for a photon, and the fact that $(\not{p}_n - m)u_n = 0$ for an electron to simplify the numerators. Use slash notation.)   (continues →)

c. Let's focus on the diagram with the horizontal virtual electron (the t-channel diagram). In the previous part, you should have found its amplitude to be

$$\mathcal{M}_t = -\frac{e^2}{2p_1 \cdot p_4}\left[\bar{u}_3\not{\epsilon}_2\not{\epsilon}_4\not{p}_2 u 1\right] \quad (20.37)$$

To calculate the spin-averaged amplitude $\langle|\mathcal{M}_t|^2\rangle$ for this diagram, we need to sum over both the electron spins and the photon spins. Casimir's trick handles the electron spins. The analogue to Casimir's trick for photons is

$$\sum_{s_a=+,-} \epsilon^{s_a}_{a,\mu}(\epsilon^{s_a}_{a,\nu})^* = -g_{\mu\nu}, \quad (20.38)$$

where a identifies the photon involved. Apply this rule to sum $|\mathcal{M}|^2$ over the photon spins. (*Hint:* Pull the polarization vector and four-momentum components out in front of the brackets, leaving only gamma matrices inside.)

d. Now use Casimir's trick and the trace theorems to sum over the electron spins and divide by the appropriate factor to get $\langle|\mathcal{M}_t|^2\rangle$. (*Hints:* The Γ_1 and Γ_2 matrices for Casimir's trick in this case will each be products of three gamma matrices. First show that $\overline{\gamma^\mu\gamma^\alpha\gamma^\sigma} = \gamma^\sigma\gamma^\alpha\gamma^\mu$. Then apply Casimir's trick and use trace theorems to simplify the results. You will find Theorems 3–6 helpful. Also prove that $p_3 \cdot p_2 = p_1 \cdot p_4$. You should find the final result to be $\langle|\mathcal{M}_t|^2\rangle = 2e^4(p_1 \cdot p_2)/(p_1 \cdot p_4)$.)

P20.6 Consider the electromagnetic pair production process $\gamma + \gamma \to e^- + e^+$.

a. Draw the lowest-order Feynman diagrams. (There are two because, even though it is possible to distinguish the outgoing particles, we don't know which is connected to which incoming photon.)

b. Let p_3 and p_4 be the four-momenta of the positron and electron, respectively. Use the Feynman rules to show that, to lowest order,

$$\mathcal{M} = e^2\left[\bar{u}_4\not{\epsilon}_2\frac{\not{p}_1 - \not{p}_3 + m}{(p_1 - p_3)^2 - m^2}\not{\epsilon}_1 v_3\right.$$

$$\left. + \bar{u}_4\not{\epsilon}_1\frac{\not{p}_1' - \not{p}_4' + m}{(p_1 - p_4)^2 - m^2}\not{\epsilon}_2 v_3\right]. \quad (20.39)$$

P20.7 Consider the electron-muon scattering process $e^- + \mu^- \to e^- + \mu^-$. (This process is difficult to observe experimentally, but provides a clean application of the Feynman rules for electrodynamics.)

a. Draw the single lowest-order Feynman diagram.

b. Let p_1 and p_2 be the four-momenta of the incoming electron and muon, respectively, and p_3 and p_4 be the same for the outgoing electron and muon. Calculate the scattering amplitude $\mathcal{M}$.   (continues →)

c. Use Casimir's trick to show that $|\mathcal{M}|^2$ summed over all final spins and averaged over the initial spins is

$$\langle\, |\mathcal{M}|^2 \,\rangle = \frac{e^4}{4(p_1 - p_3)^4}\, \mathrm{Tr}[(\gamma_\mu(\not{p}_1 + m)\gamma_\nu(\not{p}_3 + m)]$$
$$\times\, \mathrm{Tr}[(\gamma^\mu(\not{p}_2 + M)\gamma^\nu(\not{p}_4 + M)], \qquad (20.40)$$

where m and M are the electron and muon mass, respectively.

d. Use trace theorems to show that

$$\langle\, |\mathcal{M}|^2 \,\rangle = \frac{8e^4}{(p_1 - p_3)^4} \qquad (20.41)$$
$$\times \Big[(p_1 \cdot p_2)(p_2 \cdot p_4) + (p_1 \cdot p_4)(p_2 \cdot p_3)$$
$$- (p_1 \cdot p_3)M^2 - (p_2 \cdot p_4)m^2 + 2m^2 M^2\Big].$$

e. Analyze this situation in the center-of-mass frame. (This is experimentally crazy, but the cross section is Lorentz invariant, so we can later transform to whatever frame seems experimentally relevant.) The scattering is elastic, so $E_1 = E_3$ and $E_2 = E_4$ in this frame. In the limit that the energies are so large that $m \approx 0$ and $M \approx 0$ in comparison, show that $E_1 = (E^2 - M^2)/2E$ and $E_2 = (E^2 + M^2)/2E$, where $E \equiv E_1 + E_2$ is the total incoming energy in this frame. $E_1 = E_2 = E_3 = E_4$ in this limit. Draw a picture of the scattering event in this frame, and let θ be the angle between direction of the incoming electron (which we can take to be the $+z$ direction) and the outgoing electron. In this limit, evaluate $\langle |\mathcal{M}|^2 \rangle$ in terms of e and θ (the energies should cancel out).

f. Finally, use Equation 20.16 to evaluate the spin-averaged differential cross section $d\sigma$ in terms of e, θ, $d\Omega$, and $E \equiv E_1 + E_2$, the total collision energy in the center-of-mass frame.

What seems to
be the problem today,
Jack?

Doc, it really
hurts when I
do this!

Well then, let's
not do that.

21. RENORMALIZATION

Before going on, we should tie up a loose end that may have been bothering you. Recall that Feynman diagrams express terms in a power series approximation that reflects a perturbative approach to interactions. To make accurate predictions, we need to calculate sufficiently many terms in the perturbation power series so that we can take the remaining terms to be negligible. In general, how many terms do we need to calculate? Are we even certain that the series converges?

So far, we have only calculated the lowest-order terms. We have found, amazingly, that just these *lowest*-order terms have been sufficient to make reasonably accurate predictions. In this chapter, we will see why focusing only on the lowest-order diagrams works well for electromagnetic and weak interactions. This will prepare us for Chapter 30, where it will help us understand why the perturbation approach to the color interaction is much more difficult.

But we will also see that even with QED, we have a serious problem with convergence of certain higher-order terms. We can make the infinities go away using a procedure called *renormalization*. Though this procedure is mathematically difficult (and raises new theoretical puzzles), it does allow us to move forward, and makes new predictions that are supported experimentally.

21.1 Higher-Order Diagrams

We certainly could have constructed higher-order Feynman diagrams earlier in this text: we know the rules. For the simple lepton pair-production process discussed in Chapter 19, for example, we can construct only one two-vertex diagram (see Figure 21.1) and no three-vertex diagrams, but we can construct an entire array of four-vertex diagrams, some of which are shown in Figure 21.2.

One reason that we haven't done this so far is (obviously) that calculating even one or two diagrams is daunting enough. But why didn't we need to do this to get decent results? This is partly because the electromagnetic *coupling constant* is small. Note

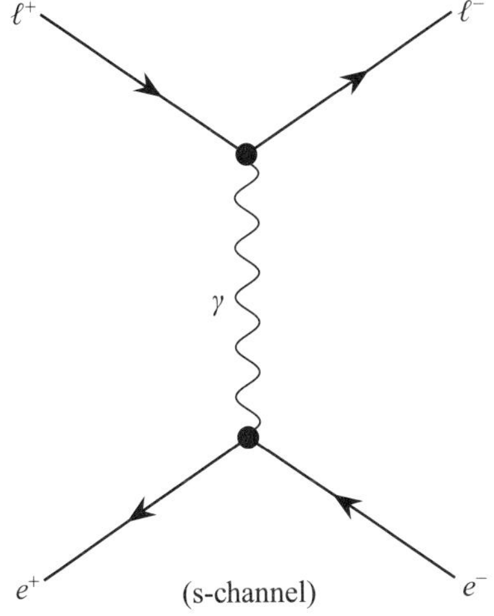

FIGURE 21.1 This figure shows the only possible two-vertex Feynman diagram for lepton pair production.

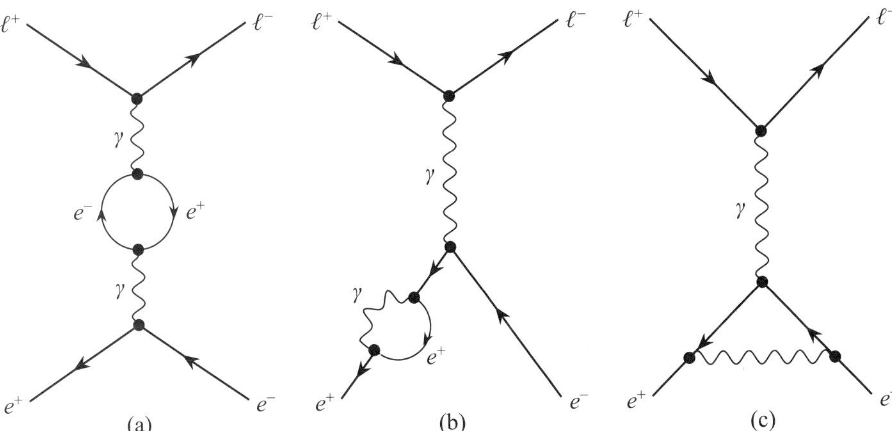

FIGURE 21.2 Some four-vertex Feynman diagrams for lepton pair-production. We can think of the first as modifying the photon propagator, the second as modifying an incoming particle line, and the third as modifying a vertex.

that each vertex in an electromagnetic calculation has an associated factor of the proton charge e, which in natural units is a dimensionless constant ~ 0.3. Therefore (other things being equal), we would expect the amplitude of a four-vertex diagram to be suppressed relative to a two-vertex diagram by a factor of order of magnitude of $(0.3)^2 \sim 1/10$ (and in practice, the suppression is often larger). We will see that the effective weak interaction coupling constant is quite a bit smaller than this at energies below $\sim 10\,\text{GeV}$, so high-order weak interaction diagrams are *very* strongly suppressed.

Another reason we often don't have to worry about these diagrams is that some higher-order diagrams are *automatically handled* as part of the renormalization process. Let's see how this works.

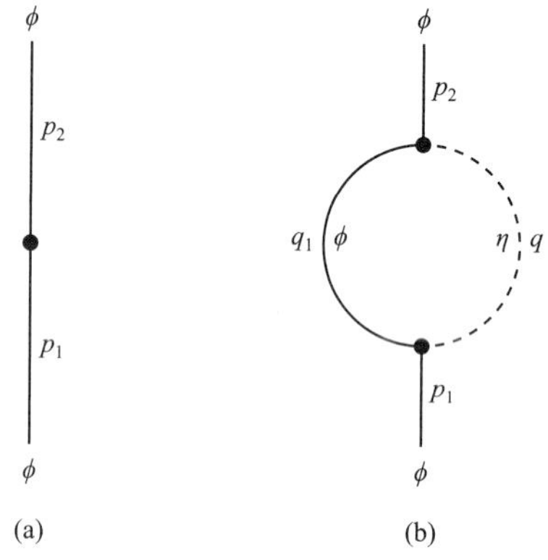

FIGURE 21.3   Panels (a) and (b) show the lowest-order and next-order Feynman diagrams, respectively, for the self interaction of the phion in our toy universe.

21.2 Mass Renormalization

To make things as simple as possible, let's consider again the toy universe we considered in Chapter 13 (or Chapter 10). This universe consisted of two real scalar fields ϕ and η with the following Lagrangian density:

$$\mathcal{L} = \tfrac{1}{2}\partial^\mu\phi\,\partial_\mu\phi - \tfrac{1}{2}m_\phi^2\phi^2 + \tfrac{1}{2}\partial^\mu\eta\,\partial_\mu\eta - \tfrac{1}{2}m_\eta^2\eta^2 - \lambda\eta\phi^2. \tag{21.1}$$

We have been accustomed to thinking of this as two terms describing a free scalar particle ϕ with mass m_ϕ, two terms describing a free scalar particle η with mass m_η, and an interaction term. But we *could* think of a mass term like $\tfrac{1}{2}m_\phi^2\phi^2$ as being a "self-interaction" term. Indeed, we can think of defining a Feynman diagram vertex for this term that looks like that shown in Figure 21.3a. We can think of this vertex as a place where a phion is annihilated and immediately recreated. This one-vertex diagram would have an amplitude of $\mathcal{M}_1 = \tfrac{1}{2}m_\phi^2$ (since $\tfrac{1}{2}m_\phi^2$ plays the same role for this vertex that λ would in a diagram with a single ϕ-ϕ-η vertex, for which $\mathcal{M} = \lambda$, as we saw in Chapter 12).

We can then see diagrams of the type shown in Figure 21.2b as being higher-order modifications of this mass term. The next-order diagram for the phion mass term is shown in Figure 21.3b. Such a virtual loop in an external particle's line would not be directly observable, but would modify the amplitude of the self-interaction and (by extension) the mass of that external particle.

Let's assume that the etaon mass $m_\eta = 0$ so that it is like a photon. Note that conservation of four-momentum at the vertices requires that $q_1 = p_1 - q = p_2 - q$, but does not determine the value of q. To get the total amplitude, we should therefore integrate over $d^4q/(2\pi)^4$ to include the contributions of all the possible values of q (the factor of $(2\pi)^{-4}$ goes with any four-momentum integral to keep things normalized). Applying the Feynman rules for the diagram in Figure 21.3b (see Box 21.1) then yields

$$\mathcal{M}_2 = i\lambda^2 \int \frac{d^4q}{(2\pi)^4}\frac{1}{q^2[\,(q-p_1)^2 - m_\phi^2\,]}. \tag{21.2}$$

Doing this integral can be a bit tricky, but it illustrates methods that will prove useful. (The following tricks come from Goldberg, who got them from Sakurai, who said he got them from Feynman.) First, note that

$$\frac{1}{ab} = \int_0^1 \frac{dx}{[\,ax + b(1-x)\,]^2}, \tag{21.3}$$

as you will show in Box 21.2. (The new variable x is called a **Feynman parameter**.) If we substitute $a = [\,(q-p_1)^2 - m_\phi^2\,]$ and $b = q^2$, then we have

$$
\begin{aligned}
\mathcal{M}_2 &= \frac{i\lambda^2}{16\pi^4} \int_0^1 dx \int \frac{d^4q}{[(q-p_1)^2 x - m_\phi^2 x + q^2(1-x)]^2} \\
&= \frac{i\lambda^2}{16\pi^4} \int_0^1 dx \int \frac{d^4q}{[q^2 x - 2xq\cdot p_1 + p_1^2 x - m_\phi^2 x + q^2 - q^2 x]^2} \\
&= \frac{i\lambda^2}{16\pi^4} \int_0^1 dx \int \frac{d^4q}{[q^2 - 2xq\cdot p_1]^2},
\end{aligned}
\tag{21.4}
$$

where I have used the fact that $p_1^2 = m_\phi^2$. This looks a bit simpler, but at the cost of an extra integral.

Now, the integral over q is over all momentum space, so it will not hurt if we shift the origin a bit: let's change variables to q_s such that $q = q_s + xp_1$. With this recentering, the integral becomes even simpler:

$$
\mathcal{M}_2 = \frac{i\lambda^2}{16\pi^4} \int_0^1 dx \int \frac{d^4q_s}{[q_s^2 - m_\phi^2 x^2]^2}
\tag{21.5}
$$

(see Box 21.3 for the details). This is huge, because it eliminates the directional issues associated with the dot product.

If this were a one-variable integral, it would be simple enough to look up. But we are integrating over all of four-dimensional momentum space. Even if this were over three-dimensional momentum space, it would not be such a problem because the integrand does not depend on direction. But the four-dimensional nature of this integral and the minus signs in the metric make this integral much more complicated than it looks.

However, in such circumstances, one can use a clever trick called a **Wick rotation**, where we set $q_E^t \equiv iq_s^t$ and $q_E^j = q_s^j$. This transformation handles the minus signs in the metric because

$$
\begin{aligned}
-q_s^2 &= -(q_s^t)^2 + (q_s^x)^2 + (q_s^y)^2 + (q_s^z)^2 \\
&= (q_E^t)^2 + (q_E^x)^2 + (q_E^y)^2 + (q_E^z)^2 \equiv q_E^2.
\end{aligned}
\tag{21.6}
$$

We see that after this translation, four-vectors in spacetime become ordinary vectors in a four-dimensional Euclidean space (the E subscript denotes "Euclidean"), which we can then handle much as we would handle an integral over vectors in three-dimensional Euclidean space. This transformation implies that

$$
d^4q_E = dq_E^t \, dq_E^x \, dq_E^y \, dq_E^z = i dq_s^t \, dq_s^x \, dq_s^y \, dq_s^z = i d^4q_s.
\tag{21.7}
$$

Moreover, because the integrand is independent of the direction of q_E in its four-dimensional Euclidean space, we can easily do the "angular" part of the integral. In a Euclidean 3-space, the angular part of the integral delivers a factor of 4π, the area of a sphere divided by r^2. In a Euclidean 4-space, we want the area of a 3D hypersphere. If you type "area of 3D hypersphere" into WolframAlpha, it will tell you that the area is $2\pi^2 r^3$, so doing the angular integral yields a factor of $2\pi^2$. The integral becomes (see Box 21.3)

$$
\mathcal{M}_2 = \frac{1}{2} \left(\frac{\lambda}{2\pi}\right)^2 \int_0^1 dx \int_0^\infty \frac{q_E^3 \, dq_E}{[q_E^2 + m_\phi^2 x^2]^2}.
\tag{21.8}
$$

Now, perhaps you can see what is going to happen. As $q_E \to \infty$, the integrand will approach dq_E/q_E, which will diverge logarithmically. In fact, a careful calculation using a finite upper limit M shows that the leading term in the integral is

$$
\mathcal{M}_2 = \left(\frac{\lambda}{4\pi}\right)^2 \ln \frac{M^2}{m_\phi^2}
\tag{21.9}
$$

in the limit that $M/m_\phi \gg 1$ (see Box 21.3 for the details).

This is a disaster. This is just the next-order correction to the basic mass and already we are getting an infinite result! How are we going to handle this?

The simplest approach to solving the problem is like the joke about the person who goes to the doctor and says, "Doc, it really hurts when I do this," and the doctor says, "Well, don't do that." In this case, it really hurts when we take the upper limit of the integral to infinity. So we simply don't take the upper limit to infinity: instead, we introduce an arbitrary finite upper limit M. We then say that the phion's **bare mass** is m_ϕ, but its **effective mass** $m_{\phi,\text{eff}}$ is such that

$$\tfrac{1}{2}m_{\phi,\text{eff}}^2 = \mathcal{M}_1 + \mathcal{M}_2 + \mathcal{M}_3 + \cdots$$
$$= \tfrac{1}{2}m_\phi^2 + (\lambda/4\pi)\ln(M^2/m_\phi^2) + \cdots, \qquad (21.10)$$

where $\mathcal{M}_n$ is the amplitude of the nth-order diagram. What we observe experimentally is not the bare mass but rather the effective mass.

On the one hand, this is great because it handles all the possible diagrams involving various bubbles along an external particle's worldline. But it also exposes the incomplete nature of our current version of quantum field theory. Presumably, some new physics takes over at extremely high energies that provides an effective cutoff. Maybe, for example, at the scale of the Planck length, space and time are quantized in some manner that limits the energy in the bubbles. For this reason, many people take the cutoff M to be the Planck energy (this is also the only fundamental constant with units of energy floating around). But our choice of cutoff is not really important because it does not appear in any observable quantities. What we *observe* is the effective masses of incoming and outgoing particles, not their "actual" masses.

Physicists call this kind of process **renormalization.**

21.3 Charge Renormalization

Renormalization is not only involved in the masses of particles but also in their interactions. Consider scattering of an electron by a quark (say, the down quark in a proton). Such an interaction would be experimentally relevant in experiments that use high-energy electrons to probe the structure of the proton (a process physicists call DIS, or "deeply inelastic scattering").

One four-vertex diagram for this process is shown in Figure 21.4. The unique *two-vertex* diagram for this process has a single photon line connecting left and right vertices without the bubble in the middle. The amplitude for the basic process without the bubble is such that

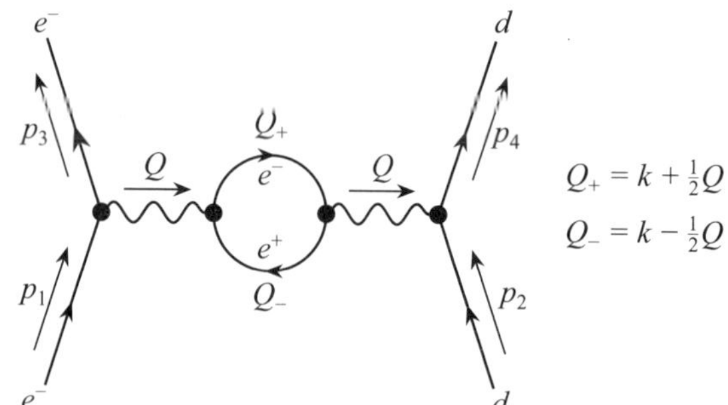

FIGURE 21.4   A Feynman diagram showing one next-order correction to the virtual photon line in electron-quark scattering. The symbol k represents the four-momentum circulating in the bubble.

$$-i\mathcal{M} = [\,\bar{u}_3(ie\gamma^\mu)u_1\,]\frac{-ig_{\mu\nu}}{Q^2}[\,\bar{u}_4(i\tfrac{1}{3}e\gamma^\nu)u_2\,]$$

$$\Rightarrow \ \mathcal{M} = -\tfrac{4}{3}\pi\alpha[\,\bar{u}_3\gamma^\mu u_1\,]\frac{g_{\mu\nu}}{Q^2}[\,\bar{u}_4\gamma^\nu u_2\,]. \tag{21.11}$$

I am using the fine-structure constant $\alpha = e^2/4\pi \approx 1/137$ in place of the electron charge magnitude e along with $Q \equiv p_1 - p_3$ to represent the overall four-momentum carried by the virtual photon. The factor of $\tfrac{1}{3}$ comes from the fact that a d-quark has one-third the charge of an electron.

Applying the Feynman rules to the more complicated diagram in Figure 21.4 involves new rules that explicitly address closed loops in Feynman diagrams.

Feynman Rules for Loops

1. **Undetermined internal momenta.** Include a factor of $d^4q_j/(2\pi)^4$ and integrate over any undetermined four-momenta q_j.

2. **Fermion loops.** For each closed loop involving fermions, include an overall factor of -1, and take the *trace* of the product of the gamma matrices contributed by the loop vertices and propagators.

3. **Boson loops.** For each loop involving spin-1 bosons, include a factor of the metric $g^{\mu\nu}$ and sum over the lower indices contributed by the boson propagators to eliminate all free indices.

We talked about the first rule in the previous section. The other two rules address the problem that the amplitude must be a number, but the vertices and propagators inside the loop contribute gamma matrices in the case of fermions and factors that have two free indices in the case of bosons. A closed fermion implies that we will have a closed loop of gamma matrices, and the implied sum connecting the final matrix to the initial matrix amounts to taking the trace of the matrix product (see Problem P21.1). In the case of a boson loop, connecting the final free index with the first free index amounts to raising one or the other index with $g^{\mu\nu}$ and summing. (The minus sign in the fermion case is because we must do an odd number of operator commutations to get everything to work out, and fermion operators anticommute. But I don't want to get into those weeds here.)

Returning to our particular case, where we had a single photon propagator $-ig_{\mu\nu}/Q^2$ originally, we now have *two* photon propagators, two fermion propagators (each having the form $(\slashed{Q}_i - m)/[Q_i^2 - m^2]$), and two vertex factors of the form $ie\gamma^\mu$. Since the internal loop involves an electron-positron pair, the fermion loop rule requires that we multiply by (-1) and take the trace of the product of the gamma matrices contributed by the loop vertices and propagators.

Conservation of four-momentum at the vertices ensures that $Q = p_1 - p_3 = p_2 - p_4 = Q_+ - Q_-$, but we cannot determine k, the amount of four-momentum circulating in the loop, so the undetermined momentum rule requires that we integrate over k. So the final result for the second-order correction to the amplitude is

$$\mathcal{M}_2 = -\tfrac{4}{3}\pi\alpha[\,\bar{u}_3\gamma^\mu u_1\,]\frac{g_{\mu\beta}}{Q^2}B^{\beta\sigma}\frac{-ig_{\sigma\nu}}{Q^2}[\,\bar{u}_4\gamma^\nu u_2\,],$$

where

$$B^{\beta\sigma} = (-1)\,\mathrm{Tr}\int\frac{d^4k}{(2\pi)^4}(ie\gamma^\beta)\frac{\slashed{Q}_+ - m}{Q_+^2 - m^2}(ie\gamma^\sigma)\frac{\slashed{Q}_- - m}{Q_-^2 - m^2} \tag{21.12}$$

and where we are considering $Q_\pm = k \pm \tfrac{1}{2}Q$ to be functions of k. We can simplify this somewhat by consolidating constants and bringing the trace inside the integral. The result is

$$\mathcal{M}_2 = i\frac{\alpha^2}{3\pi^2}[\,\bar{u}_3\gamma^\mu u_1\,]\frac{C_{\mu\nu}}{Q^4}[\,\bar{u}_4\gamma^\nu u_2\,], \quad \text{where}$$

$$C_{\mu\nu} = \int d^4k\; \frac{g_{\mu\beta}g_{\sigma\nu}\,\mathrm{Tr}[\,\gamma^\beta(\not{Q}_+ - m)\gamma^\sigma(\not{Q}_- - m)\,]}{(Q_+^2 - m^2)(Q_-^2 - m^2)}. \tag{21.13}$$

Now, as you will show in Box 21.4, the numerator in the integral is

$$g_{\mu\beta}g_{\sigma\nu}\,\mathrm{Tr}[\,\gamma^\beta(\not{Q}_+ - m)\gamma^\sigma(\not{Q}_- - m)\,]$$

$$= g_{\mu\nu}(4m^2 - 4k^2 + Q^2) + 8k_\mu k_\nu - 2Q_\mu Q_\nu. \tag{21.14}$$

Problem P21.2 shows that the term $2Q_\mu Q_\nu$ vanishes in the final expression for the amplitude, so we will drop that term. We then follow the same path we did in the phion case (see Box 21.5) to show that

$$C_{\mu\nu} = 4\int_0^1 dx \int d^4k\; \frac{g_{\mu\nu}[\,m^2 - k^2 + Q^2 x(1-x)\,] + 2k_\mu k_\nu}{[\,k^2 - m^2 + Q^2 x(1-x)\,]^2}. \tag{21.15}$$

One can simplify this further by using the integral identity

$$0 = \int d^4k\; \frac{g_{\mu\nu}(\Delta - k^2) + 2k_\mu k_\nu}{(k^2 - \Delta)^2} \tag{21.16}$$

(from Peskin and Schroeder's *An Introduction to Quantum Field Theory*). If we identify $\Delta \equiv m^2 - Q^2 x(1-x)$ so that the denominator has the right form, then the numerator has the form $\Delta - k^2 + 2k_\mu k_\nu + 2Q^2 x(1-x)$, so our expression for $C_{\mu\nu}$ becomes

$$C_{\mu\nu} = 8Q^2 g_{\mu\nu}\int_0^1 dx \int d^4k\; \frac{x(1-x)}{[\,k^2 - m^2 + Q^2 x(1-x)\,]^2}. \tag{21.17}$$

If we now do the Wick rotation (see Box 21.6), we have

$$\frac{C_{\mu\nu}}{Q^4} = -i\,16\pi^2\frac{g_{\mu\nu}}{Q^2}\int_0^1 dx \int \frac{k_E^3\,dk_E\,x(1-x)}{[\,k_E^2 + m^2 - Q^2 x(1-x)\,]^2}. \tag{21.18}$$

Note that this is just a numerical multiple of the same $g_{\mu\nu}/Q^2$ factor that appeared in the original amplitude in Equation 21.11. In fact, we can write the combined amplitudes in the form

$$\mathcal{M} + \mathcal{M}_2 = -\tfrac{4}{3}\pi[\,\bar{u}_3\gamma^\mu u_1\,][\,\bar{u}_4\gamma^\nu u_2\,]\frac{g_{\mu\nu}}{Q^2}\Big[\alpha - \frac{4\alpha^2}{\pi}F(Q^2)\Big],$$

$$\text{where } F(Q^2) \equiv \int_0^1 dx \int_0^\infty \frac{k_E^3\,dk_E\,x(1-x)}{[\,k_E^2 + m^2 - Q^2 x(1-x)\,]^2}, \tag{21.19}$$

as if the whole second-order amplitude provided a correction to the value of the fine structure constant α (or equivalently, to the value of the fundamental *charge*). So, just as a particle has a bare mass but also an effective (experimental) mass that includes corrections due to diagrams that have bubbles along the particle's worldline, a particle also has a bare charge as well as an effective (experimental) charge that includes corrections due to diagrams that have bubbles along the worldlines of virtual interaction photons.

In this case, the effective charge or (equivalently) the effective value of α also depends on the momentum transfer Q, which is connected to the energy involved in the interaction. At zero momentum transfer, we can find the lowest-order correction to α by doing the integrals in Equation 21.19 with Q set to zero. The result (see Box 21.7) is

$$\alpha_R(0) = \alpha\Big[\,1 - \frac{\alpha}{3\pi}\ln\frac{M^2}{m^2}\,\Big], \tag{21.20}$$

where $\alpha_R(0)$ is the renormalized (experimental) fine-structure constant at $Q = 0$, α is the constant's "bare" value, m is the mass of the fermions in the loop, and $M \gg m$ is the arbitrary upper limit on the momentum integral. Because $\alpha/3\pi \approx 1/1292$, this correction is only $\sim 8\%$ even if M is on the Planck scale.

With some work, we can express the Q-dependence in the form

$$\alpha_R(Q^2) = \alpha_R(0)\left[1 + \frac{\alpha_R(0)}{3\pi} f(u) \right] \quad \text{where} \quad u \equiv \frac{-Q^2}{m^2}$$

$$\text{and} \quad f(u) = 6 \int_0^1 x(1-x) \ln[\, 1 + xu(1-x)\,]\, dx. \tag{21.21}$$

(I am not going to make you prove this.) We call $\alpha_R(Q^2)$ a **running coupling constant**, which simply means that its effective value varies with the energy of the interaction. One can show (see Problem P21.5) that the momentum transfer carried by the virtual particle in this interaction is always spacelike, so $-Q^2$ and thus u is positive. Indeed, u is small when the interaction is gentle and becomes large the more momentum is transferred. The positive sign means that $a_R(Q^2)$ (and thus the particle's effective charge) becomes larger during close encounters when a lot of momentum is transferred.

We might understand this intuitively as follows. Remember that we are integrating the vertices in Figure 21.4 over all positions in space and time, so we have nonzero amplitudes for virtual electron-positron bubbles everywhere in space at any given time. We can treat this sea of amplitudes as if ghostly versions of the bubbles were constantly forming and disappearing everywhere around the real electron. During the brief interval when each virtual bubble exists, the virtual positron is attracted to the real electron and the virtual electron is repelled. This effectively polarizes the empty space around the real electron, much as that electron would polarize a dielectric material. This polarization partially cancels the real electron's electric field, making that field appear weaker than it actually is. One can only see the real electron's bare charge in very close encounters (which are necessarily high-energy encounters).

This model is qualitative and not rigorous enough to use by itself to make predictions. But it is useful as a way to visualize and remember the very real result we have calculated above, so long as we don't take the picture too literally.

We have only examined a single higher-order correction diagram so far. We should also consider diagrams with two bubbles, three bubbles, or up to an infinite number of bubbles along the virtual photon line. How can we handle all these corrections?

Fortunately, a two-bubble diagram basically involves squaring the effect of a one bubble diagram and so on, so we can model the sum of all the diagrams' effect on the fine structure constant by the power series $1 + x + x^2 + x^3 + \cdots = 1/(1 - x)$, meaning that the energy dependence due to the entire series is

$$\alpha_R(Q^2) = \frac{\alpha_R(0)}{1 - \alpha_R(0) f(u)/3\pi} . \tag{21.22}$$

This allows us to handle the entire infinite set of diagrams!

21.4 Final Comments about Renormalization

The most important results of this chapter are the *concepts* of mass and charge renormalization and Equations 21.10 and 21.22. Otherwise, this chapter mostly involves clever tricks for doing really nasty integrals. Don't get hung up on the integrals: you won't see them again in this text. The point is for you to believe the results and see where they are coming from.

In this chapter, I have actually only presented a brief and simplified picture of renormalization, which is a vast, complicated, and contentious subject. It points to the fact that quantum field theory is somehow incomplete.

However, we are accustomed in physics to theories (for example, Newtonian physics) that brush over complexities, have built-in inconsistencies (for example, the infinite self-energy of a point particle), and are actually demonstrably wrong in certain cases but are still useful for explaining a broad range of phenomena. We should take quantum field theory and the Standard Model as being analogous. We should be careful in physics not to pretend that a model is the same as the reality it imperfectly describes. The map is not the territory.

Within the framework, it is useful to know when a theory is renormalizable. Again, the subject is complex, but I will state without proof one simple criterion for exposing cases for which renormalization is possible. In a spacetime with dimensions d, a Lagrangian density has units of (energy)^d. A scalar field ϕ or gauge field A^μ has units of energy, and a fermion field ψ has units of (energy)$^{3/2}$. Therefore, the units of a coupling constant λ that connects n_B boson fields and/or n_F fermion fields are

$$\text{units of } \lambda = (\text{energy})^{d-n_B-3n_F/2}. \tag{21.23}$$

It turns out that a quantum field theory is generally renormalizable so long as the exponent $d - n_B - 3n_F/2$ is not negative. For our toy scalar theory, the exponent was 1, and for the electroweak theory, the exponent is 0 (the coupling constant is unitless). Note, however, that spacetime must have at least four dimensions for the Standard Model to be renormalizable. Also, gravity, whose coupling constant would be $G = (E_{Pl})^{-2}$, is not normalizable, which is one of the main reasons we do not yet have a quantum theory of gravitation.

BOX 21.1 Two-Vertex Mass Term Amplitude

The goal in this box is to calculate the amplitude of the Feynman diagram in Figure 21.3b.

Exercise 21.1.1 Evaluate the amplitude and compare to Equation 21.2. (*Hint*: Recall that in the toy theory, we insert a factor $-i\lambda$ and insert a propagator $i/(q_i^2 - m_i^2)$ for each internal line to get $-i\mathcal{M}$. Because the four-momentum q in the loop is undetermined, we also must integrate over $d^4q/(2\pi)^4$.)

BOX 21.2 A Useful Integral

The goal of this box is to show that

$$\frac{1}{ab} = \int_0^1 \frac{dx}{[ax + b(1-x)]^2}.$$

(21.24)

Exercise 21.2.1 Show this. (*Hint:* Substitute $u = ax + b(1-x)$. You will not need all the space provided below.)

BOX 21.3   Simplifying the Toy Universe Integral

The goal of this box is to simplify the momentum-space integral in our two-vertex amplitude for the phion self-interaction. We start with

$$\mathcal{M}_2 = \frac{i\lambda^2}{16\pi^4} \int_0^1 dx \int \frac{d^4q}{[q^2 - 2xq \cdot p_1]^2}. \tag{21.25}$$

Exercise 21.3.1   First, shift the origin by defining $q = q_s + xp_1$, $d^4q = d^4q_s$. Show that we get

$$\mathcal{M}_2 = \frac{i\lambda^2}{16\pi^4} \int_0^1 dx \int \frac{d^4q_s}{[q_s^2 - m^2x^2]^2}. \tag{21.26}$$

Exercise 21.3.2   Now do the Wick rotation, where $q_s^2 = -q_E^2$, $d^4q_s = -id^4q_E = -i(2\pi^2)q_E^3\, dq_E$, where q_E goes from 0 to ∞. Show that the result is

$$\mathcal{M}_2 = \frac{1}{2}\left(\frac{\lambda}{2\pi}\right)^2 \int_0^1 dx \int_0^\infty \frac{q_E^3\, dq_E}{[q_E^2 + m^2x^2]^2}. \tag{21.27}$$

BOX 21.3 Simplifying the Toy Universe Integral (cont.)

Now that this is a one-dimensional integral, we can evaluate it by substitution. Starting with

$$\mathcal{M}_2 = \frac{1}{2}\left(\frac{\lambda}{2\pi}\right)^2 \int_0^1 dx \int_0^\infty \frac{q_E^3\, dq_E}{[q_E^2 + m^2 x^2]^2}, \tag{21.28}$$

we can change variables to $y = q_E^2/m^2 + x^2$.

Exercise 21.3.3 Evaluate the integral over q_E using a finite upper limit of M in the integral over q_E. You should get $\mathcal{M}_2 \approx (\lambda/4\pi)^2 \ln(M^2/m^2)$ in the limit that $M \gg m$. (*Hint:* Don't forget to change the limits of integration when you change variables to y. With the substitution, the only integral you may not know how to do by hand is $\int \ln x\, dx = x \ln x - x$.)

BOX 21.4 Trace Gymnastics

The goal of this box is to use trace theorems to show that

$$g_{\mu\beta}g_{\sigma\nu}\,\mathrm{Tr}[\,\gamma^{\beta}(\not{Q}_{+}-m)\gamma^{\sigma}(\not{Q}_{-}-m)\,]$$

$$= g_{\mu\nu}(4m^{2}-4k^{2}+Q^{2})+8k_{\mu}k_{\nu}-2Q_{\mu}Q_{\nu}, \qquad (21.29)$$

where $Q_{\pm}\equiv k\pm\tfrac{1}{2}Q$.

Exercise 21.4.1 Show this. (*Hint:* Expand the product first, then eliminate terms with an odd number of gamma matrices, then eliminate $Q_{\pm}$.)

BOX 21.5 Simplifying the Interaction Integral

Our goal in this box is to show that applying the identity

$$\frac{1}{ab} = \int_0^1 \frac{dx}{[ax + b(1-x)]^2} \tag{21.30}$$

and the shift $k = k_s + Q(\tfrac{1}{2} - 2x)$ to the expression

$$C_{\mu\nu} = \int d^4k \; \frac{g_{\mu\nu}(4m^2 - 4k^2 + Q^2) + 8k_\mu k_\nu}{(Q_+^2 - m^2)(Q_-^2 - m^2)} \tag{21.31}$$

(where $Q_\pm = k \pm \tfrac{1}{2}Q$) yields (after dropping the subscript on k_s)

$$C_{\mu\nu} = 4 \int_0^1 dx \int d^4k \; \frac{g_{\mu\nu}[m^2 - k^2 + Q^2 x(1-x)] + 2k_\mu k_\nu}{[k^2 - m^2 + Q^2 x(1-x)]^2}. \tag{21.32}$$

Exercise 21.5.1   First, show that if we define $a = Q_+^2 - m^2$ and $b = Q_-^2 - m^2$, then $ax + b(1-x) = 2k \cdot Qx + k^2 - k \cdot Q + \tfrac{1}{4}Q^2 - m^2$.

Exercise 21.5.2   Now let's introduce the shift $k = k_s + Q(\tfrac{1}{2} - x)$. Show that $ax + b(1-x)$ becomes the denominator in Equation 21.32.

BOX 21.5 Simplifying the Interaction Integral (cont.)

Exercise 21.5.3 Now let's introduce the same shift $k = k_s + Q(\frac{1}{2} - x)$ in the numerator. We can drop all odd powers of k_s because the integral will then be odd (and thus yield zero). We can also drop a term involving $Q_{\mu\nu}$ for the reason discussed in Problem P21.2. Show that the result yields the numerator in Equation 21.32 when we drop the subscript on k_s. (You will not need all the space provided.)

BOX 21.6 Wick Rotation of the Interaction Integral

Our goal in this box is to show that doing a Wick rotation means that

$$\int d^4k \, \frac{x(1-x)}{[k^2 - m^2 + Q^2 x(1-x)]^2} = \int_0^\infty \frac{-i2\pi^2 k_E^3 \, dk_E \, x(1-x)}{[k_E^2 + m^2 - Q^2 x(1-x)]^2}. \quad (21.33)$$

Exercise 21.6.1 Prove this. (*Hints:* Remember that a Wick rotation means that $d^4k = -i d^4 k_E$ and $k^2 = -k_E^2$, and that (because the integrand is isotropic) we are able to trivially integrate over the surface of a 3D hypersphere. You will not need all the space provided.)

BOX 21.7   The Corrected Fine-Structure Constant

The goal of this box is to show that

$$F(Q^2) \equiv \int_0^1 dx \int_0^M \frac{k_E^3 \, dk_E \, x(1-x)}{[k_E^2 + m^2 - Q^2 x(1-x)]^2} \approx \frac{1}{12} \ln \frac{M^2}{m^2} \quad (21.34)$$

in the limit where $Q^2 \to 0$ and $M^2 \gg m^2$. This implies that

$$\alpha_R(0) = \alpha \left[1 - \frac{4\alpha}{\pi} F(0) \right] = \alpha \left[1 - \frac{\alpha}{3\pi} \ln \frac{M^2}{m^2} \right], \quad (21.35)$$

as claimed in Equation 21.20.

Exercise 21.7.1   Show this. (*Hint:* It is acceptable to have WolframAlpha do the k_E integral. You will not need all the space below.)

Homework Problems

P21.1 Consider the product ABC of three square matrices A, B, and C. Suppose that we "close the loop" by also summing along the columns of C and down the rows of A (as if we were writing this matrix product on the surface of a cylinder, so that C and A connect to each other on the other sides of their connection to B). Show that the result is the trace of the matrix product ABC. (*Hint:* Using index notation helps.)

P21.2 Consider a term of the form $Q_\mu Q_\nu [\,\bar{u}_3 \gamma^\nu u_1\,][\,\bar{u}_4 \gamma^\mu u_2\,]$ in the context of the amplitude we were considering in Section 21.3. Our goal in this problem is to show that such a term is zero. Note that because $Q = p_1 - p_3 = p_4 - p_2$ in this context, this term is equivalent to

$$[\,\bar{u}_3(\not{p}_1 - \not{p}_3)u_1\,][\,\bar{u}_4(\not{p}_4 - \not{p}_2)u_2\,].\qquad(21.36)$$

Now, consider the Dirac equation

$$0 = (i\gamma^\mu \partial_\mu - m)\psi, \quad \text{where } \psi = e^{-ip\cdot x}u(p).\quad(21.37)$$

a. Show that this implies that $0 = (\not{p} - m)u(p)$.

b. Take the adjoint of both sides of the Dirac equation and show from this that $0 = \bar{u}(p)(\not{p} - m)$. (*Hint:* You need to pay attention to the factor of γ^t implied by the bar. Compute the simple adjoint (dagger) of both sides first, multiply by γ^t on the right, and then use anti-commutation relations to move the γ^t next to the $u^\dagger$, treating the cases of $\mu = t$ and $\mu = x, y, z$ separately.)

c. Combine these expressions to show that the quantity in Equation 21.36 is indeed zero. Note that p_1 and p_3 are both electron momenta, so the m in the Dirac equation for these momenta is the electron mass. Similarly, p_2 and p_4 are both quark momenta.

P21.3 Consider a one-bubble diagram analogous to Figure 21.4 but within the context of phion-phion scattering in our toy universe. Assume that the etaon is massless, and pretend that the outgoing phions are not identical.

a. Use the Feynman rules to show that the amplitude of the one-bubble diagram is

$$\mathcal{M}_2 = \frac{i\lambda^4}{(2\pi)^4 Q^4} \int \frac{d^4 k}{[(k + \tfrac{1}{2}Q)^2 - m_\phi^2][(k - \tfrac{1}{2}Q)^2 - m_\phi^2]},$$

$$(21.38)$$

where $Q = p_1 - p_3$ and k is the four-momentum circulating in the bubble. (Note that $Q^2 < 0$: see Problem P21.5.)

b. Adapt the methods in Boxes 21.5 and 21.6 to show that

$$\mathcal{M}_2 = \frac{\lambda^4}{8\pi^2 Q^4} \int_0^1 dx \int_0^\infty \frac{k_E^3\, dk_E}{[k_E^2 + m_\phi^2 - Q^2 x(1-x)]^2}.$$

$$(21.39)$$

(continues $\rightarrow$)

c. Change the upper limit of the integral over k_E to M, define $b(x) \equiv [\, m_\phi^2 - Q^2 x(1-x)\,]^{1/2}$, and look up the integral to arrive at

$$\mathcal{M}_2 = \frac{\lambda^4}{16\pi^2 Q^4} \int_0^1 dx \left[\frac{b^2}{M^2 + b^2} - 1 + \ln\left(\frac{M^2 + b^2}{b^2}\right) \right].$$

$$(21.40)$$

d. Evaluate $\mathcal{M}_2$ in the limit that $-Q^2 \ll m_\phi^2 \ll M^2$. (*Hint:* Do the approximation first, keeping in mind that $0 \leq x \leq 1$, and integrate the leading term or terms. You will still have an overall factor of $1/Q^4$, which we cannot ignore.)

P21.4 Assuming $\alpha_R(0) = 1/137$, evaluate $\alpha_R(Q^2)$ when $|Q^2| = (10\text{ GeV})^2$ (roughly the value that one might encounter in LHC experiments). Assume that the fermions in the bubble are an electron-positron pair (these will be the easiest to form and have the greatest effect). Are you surprised? (*Hint:* Use the all-bubble Equation 21.22 and note that $f(u)$ is defined in Equation 21.21. First, show that $f(u) \approx \ln u$ when $u \gg 1$. Note that x in Equation 21.21 can be as small as 0, so you can't assume, as you might be tempted to, that $\ln x \ll \ln u$.)

P21.5 If p_1 and p_3 are different four-momenta for the same particle, prove that $Q \equiv p_1 - p_3$ is spacelike (that is, that $Q^2 = Q \cdot Q < 0$.) (*Hint:* Evaluate the invariant dot product $(p_1 - p_3) \cdot (p_1 - p_3)$ in the frame where the particle is at rest when it has four-momentum p_1.)

P21.6 It might seem that a photon could decay to two photons with smaller energy via the following process:

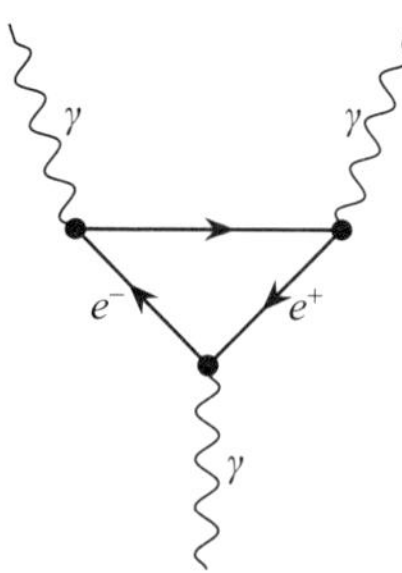

However, a second diagram contributes to this order, one where the electron circulates counterclockwise instead of clockwise. Using the rules for loops in this chapter and the trace theorems in Chapter 20, show that these two diagrams *cancel*, implying that the total amplitude for photon decay is zero. This is a special case of **Furry's theorem**, which states that processes involving closed fermion loops with an odd number of vertices have zero amplitude. (*Hints:* It helps to define momenta in the loop in the direction of the fermion arrows on each diagram, and then compare those momenta between the diagrams. Also you will need the theorem that the trace of a product of gamma matrices is equal to the trace of the same matrices in reverse order.)

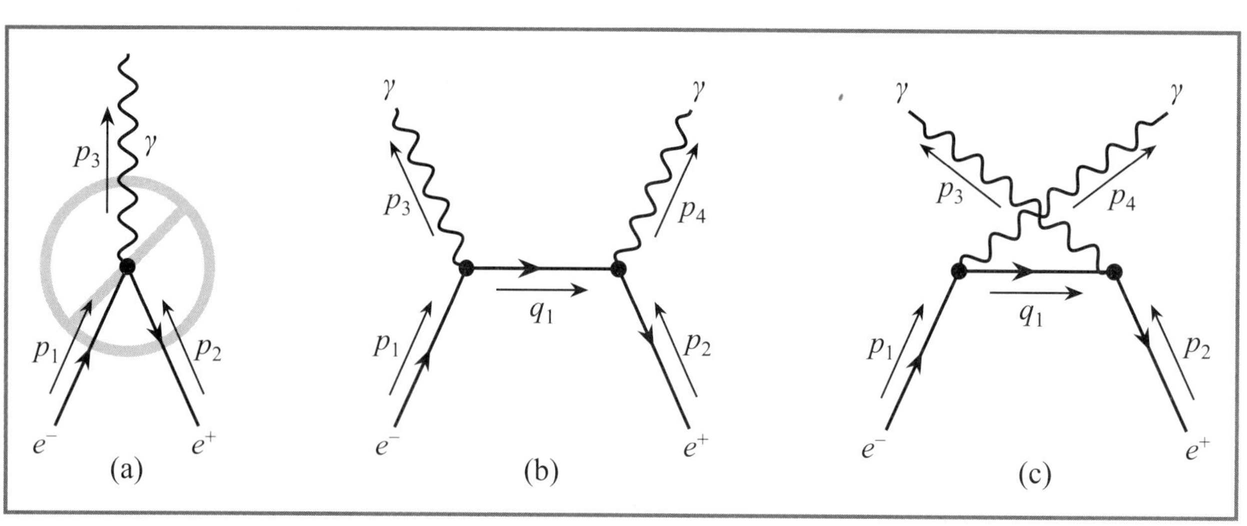

p_3
γ
p_1
p_2
e^-
e^+
(a)
γ
γ
p_3
p_4
p_1
q_1
p_2
e^-
e^+
(b)
γ
γ
p_3
p_4
p_1
q_1
p_2
e^-
e^+
(c)

22. APPLICATIONS OF QED

In this optional (but recommended) chapter, we will apply the Feynman rules we developed for quantum electrodynamics (QED) in Chapter 20 to situations beyond pair production. We will also compare results to experiment. This chapter is not required for any future chapter, but helps to further illustrate how one applies the Feynman rules in new situations, and also how we can make reasonably accurate predictions even when we consider only lowest-order diagrams.

22.1 Bhabha Scattering

In the last chapter, we calculated the cross section for the *lepton pair-production* process $e^- + e^+ \to \ell^- + \ell^+$, where ℓ^- stands for a muon or a tau lepton. Electron-positron collisions are important to understand because several important historical colliders (the SPEAR, PEP-I, and PEP-II colliders at Stanford, the LEP collider at CERN) and a number of currently operating colliders (the VEPP colliders in Russia, the BEPC-II collider in China, the DAFNE collider in Italy, and the SuperKEKB collider in Japan) use colliding e^+e^- beams. Such collisions provide clean tests of a variety of Standard Model features that are easier to analyze than the proton-proton collisions at the Large Hadron Collider.

An important process to consider in such colliders is the **Bhabha scattering process** $e^- + e^+ \to e^- + e^+$ (named after physicist Homi J. Bhabha). Unlike pair production involving muons or taus, analyzing this process to lowest order involves calculating the amplitudes for *two* Feynman diagrams instead of one (see Figure 22.1). Exploring Bhabha scattering will help us better understand how to handle situations involving multiple diagrams of comparable order.

The s-channel diagram is basically the same as for lepton pair production: the incoming electron and positron could annihilate to a virtual photon, then re-form as an electron-positron pair. But when the final particles are an electron-positron pair instead of a muon and antimuon or tau and antitau, we also have the possibility that the incoming particles simply scatter rather than being annihilated and recreated: this is the t-channel diagram shown in Figure 22.1.

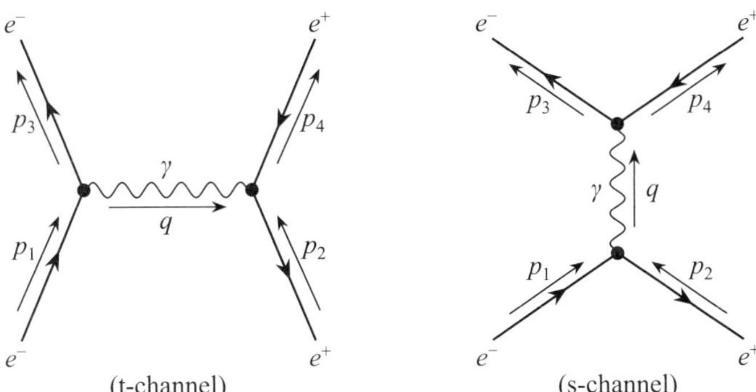

FIGURE 22.1   These are the Feynman diagrams for Bhabha scattering.

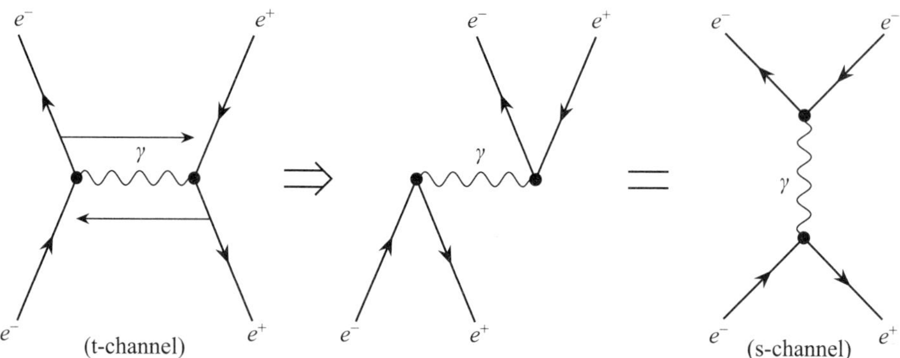

FIGURE 22.2   The "antisymmetrization rule" applies to the two Bhabha scattering diagrams.

The amplitude for the s-channel diagram is exactly the same as for the lepton pair production process calculated in Chapter 20.

$$\mathcal{M}_s = \frac{-e^2}{(p_1 + p_2)^2}[\,\bar{v}_2\gamma^\mu u_1\,][\,\bar{u}_3\gamma_\mu v_4\,] \tag{22.1}$$

(see Equation 20.4). Applying the Feynman rules to the other diagram yields

$$-i\mathcal{M}_t = [\,\bar{u}_3(ie\gamma^\mu)u_1\,]\frac{-ig_{\mu\nu}}{q^2}[\,\bar{v}_2(ie\gamma^\nu)v_4\,]$$

$$\Rightarrow \;\; \mathcal{M}_t = \frac{-e^2}{(p_1 - p_3)^2}[\,\bar{u}_3\gamma^\mu u_1\,][\,\bar{v}_2\gamma_\mu v_4\,]. \tag{22.2}$$

But do we add these amplitudes or subtract them? The "antisymmetrization" rule requires us to "include a minus sign between two diagrams that differ by swapping (between two vertices) two incoming identical fermions, two outgoing identical fermions, an incoming fermion with an outgoing antifermion (corresponding to the same type of fermion), or an incoming antifermion with an outgoing fermion of the same type." In this case, if we take the t-channel diagram and interchange the outgoing electron on one vertex with incoming positron on the other vertex, we get the s-diagram: see Figure 22.2. (It does not matter whether the virtual particle worldline is horizontal or vertical: the calculation of the amplitude from first principles involves integrating over all possible vertex positions anyway, as discussed in either chapter 10 or 12.) So the rule implies that in this case we should *subtract* the amplitudes

$$\mathcal{M} = \mathcal{M}_t - \mathcal{M}_s \quad (\text{or } \mathcal{M}_s - \mathcal{M}_t) \tag{22.3}$$

to lowest order. (Note that the overall sign difference between the two choices disappears when we compute $|\mathcal{M}|^2$.)

If the electron and positron beams are unpolarized and we don't care about the spins of the outgoing particles, then we can use Casimir's trick to calculate the average squared amplitude. But in this case, we have four terms to calculate:

$$\langle\,|\mathcal{M}|^2\,\rangle = \langle\,|\mathcal{M}_t - \mathcal{M}_s|^2\,\rangle = \langle\,(\mathcal{M}_t - \mathcal{M}_s)(\mathcal{M}_t^* - \mathcal{M}_s^*)\,\rangle$$

$$= \langle\,|\mathcal{M})_t|^2\,\rangle - \langle\,\mathcal{M}_t\mathcal{M}_s^*\,\rangle - \langle\,\mathcal{M}_t^*\mathcal{M}_s\,\rangle + \langle\,|\mathcal{M}_s|^2\,\rangle. \tag{22.4}$$

We know from Chapter 20 (Equation 20.15) that

$$\langle\,|\mathcal{M}_s|^2\,\rangle = \frac{8e^4}{(p_1 + p_2)^4}\Big[\,(p_1 \cdot p_4)(p_2 \cdot p_3) + (p_1 \cdot p_3)(p_4 \cdot p_2)$$

$$+ (p_1 \cdot p_2)m^2 + (p_4 \cdot p_3)m^2 + 2m^4\,\Big]. \tag{22.5}$$

(The only part of the formula for the s-channel amplitude that depends on what kind of lepton-antilepton pair is created at the end is the mass M of the outgoing lepton: here $M = m =$ the electron mass.) Now, the relevant limit for currently operating $e^- e^+$ colliders is the extreme relativistic limit where the energies and spatial momenta of the particles are so large that we can treat the electron mass as negligible in comparison, so we will drop all the mass terms in the expression above. Because these colliders collide an electron beam with a positron beam of equal energy, the laboratory frame here is the same as the center of mass frame. In that frame, in the extreme relativistic limit, and using the coordinate system defined in Figure 22.3, we have

$$p_1 = \begin{bmatrix} E \\ 0 \\ 0 \\ E \end{bmatrix}, \quad p_2 = \begin{bmatrix} E \\ 0 \\ 0 \\ -E \end{bmatrix}, \quad p_3 = \begin{bmatrix} E \\ E\sin\theta \\ 0 \\ E\cos\theta \end{bmatrix}, \quad p_4 = \begin{bmatrix} E \\ -E\sin\theta \\ 0 \\ -E\cos\theta \end{bmatrix}, \quad (22.6)$$

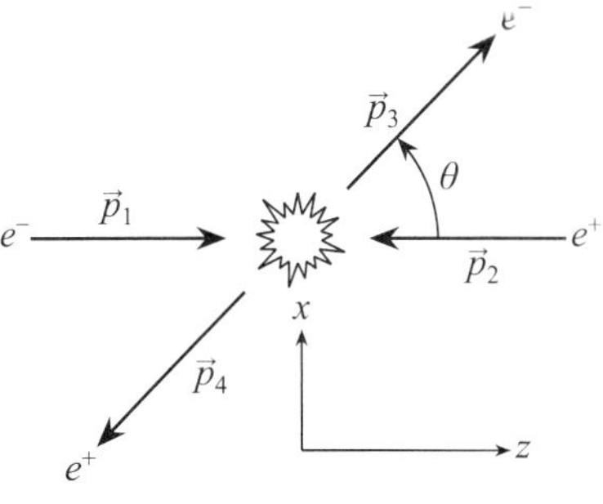

FIGURE 22.3   This diagram displays the coordinate system and momenta definitions for Bhabha scattering.

where E is the energy of the incoming electron. These imply that

$$p_1 \cdot p_2 = p_3 \cdot p_4 = 2E^2, \tag{22.7a}$$

$$p_1 \cdot p_3 = p_2 \cdot p_4 = E^2(1 - \cos\theta), \tag{22.7b}$$

$$p_1 \cdot p_4 = p_2 \cdot p_3 = E^2(1 + \cos\theta). \tag{22.7c}$$

Also, note that in the extreme relativistic limit where $p_1^2 = p_2^2 = m^2 \to 0$,

$$(p_1 + p_2)^2 = p_1^2 + 2p_1 \cdot p_2 + p_2^2 = 0 + 2p_1 \cdot p_2 + 0 = 4E^2. \tag{22.8}$$

So, in this limit, the s-channel scattering amplitude becomes

$$\langle |\mathcal{M}_s|^2 \rangle = \frac{8e^4}{16E^4} \Big[E^4(1 + \cos\theta)^2 + E^4(1 - \cos\theta)^2 \Big]$$

$$= \frac{e^4}{2}(1 + 2\cos\theta + \cos^2\theta + 1 - 2\cos\theta + \cos^2\theta)$$

$$= e^4(1 + \cos^2\theta). \tag{22.9}$$

In Boxes 22.1 and 22.2, you will calculate the other terms in Equation 22.4. The result is that the total scattering amplitude (with $c \equiv \cos\theta$) is

$$\langle |\mathcal{M}|^2 \rangle = \langle |\mathcal{M}_t|^2 \rangle - \langle \mathcal{M}_t \mathcal{M}_s^* \rangle - \langle \mathcal{M}_t^* \mathcal{M}_s \rangle + \langle |\mathcal{M}_s|^2 \rangle$$

$$= 2e^4 \left[\frac{(1+c)^2 + 4}{(1-c)^2} - \frac{(1+c)^2}{1-c} + \frac{1+c^2}{2} \right] = e^4 \left[\frac{3+c^2}{1-c} \right]^2, \tag{22.10}$$

where the last step is discussed in Box 22.3. Now, the general formula for the cross section in the CM frame (Equation 20.16) is

$$d\sigma = \frac{S \langle |\mathcal{M}|^2 \rangle}{(8\pi)^2 (E_1 + E_2)^2} \frac{|\vec{p}_3|}{|\vec{p}_1|} d\Omega. \tag{22.11}$$

Here, $S = 1$ because the outgoing particles are not the same. However, because they have the same (zero) mass, $|\vec{p}_1| = |\vec{p}_2| = E_1 = E_2 = E$, so in our case,

$$d\sigma = \frac{e^4}{(8\pi)^2(2E)^2} \left[\frac{3+c^2}{1-c} \right]^2 d\Omega = \frac{\alpha^2}{16E^2} \left[\frac{3+\cos^2\theta}{1-\cos\theta} \right]^2 d\Omega, \tag{22.12}$$

where $\alpha = e^2/4\pi \approx 1/137$. This prediction matches experimental results well at $E = 17.3$ GeV (see Figure 22.4). When $E = 99.0$ GeV, one needs to include some other diagrams and the energy-dependent value of α (see Figure 22.5).

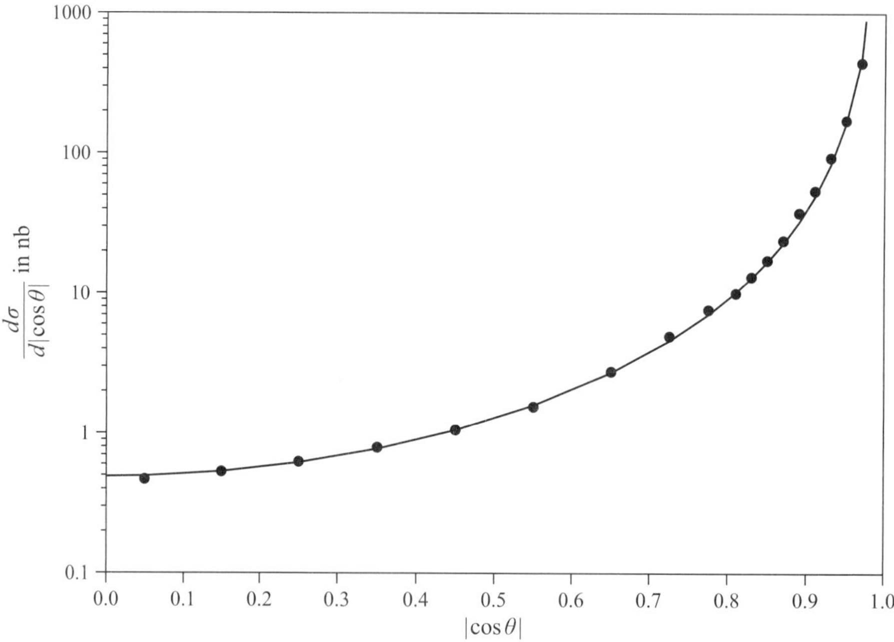

FIGURE 22.4   This figure displays Bhabha cross section measurements obtained at the PETRA collider at collision energy $E = 17.3$ GeV. Dots show experimental measurements, and the solid line shows the prediction of Equation 22.12. The graph was constructed from data in this article by the MARK J Collaboration: *Phys Rept*, 109 (1984) 131–226. The quantity plotted on the vertical axis is $d\sigma/d|\cos\theta| \equiv \left[2\pi\, d\sigma/d\Omega\right]_{\cos\theta>0} + \left[2\pi\, d\sigma/d\Omega\right]_{\cos\theta<0}$.

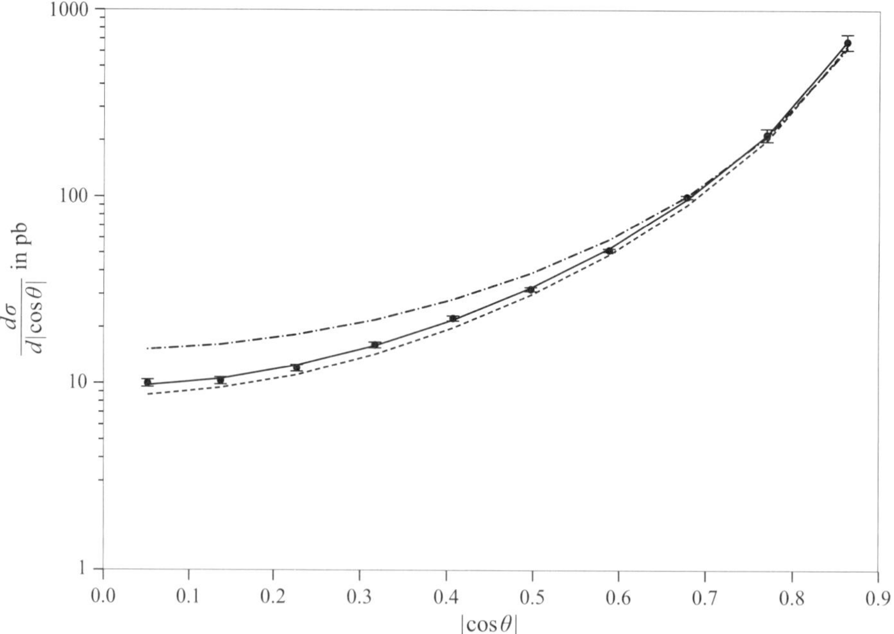

FIGURE 22.5   This figure displays Bhabha cross section measurements obtained at the LEP collider at collision energy $E = 99.0$ GeV. The large-dash line shows the prediction of Equation 22.12, the small-dash line shows a prediction that includes other diagrams, but with α fixed at 1/137.04. The solid line includes the running value of α. The graph was constructed from data in this article by the L3 collaboration: *Phys Lett B*, 623 (2005) 26–36.

22.2 Hadron Production

Another important experimental process in $e^- e^+$ colliders is hadron production $e^- + e^+ \to q^+ + q^-$. The lowest-order Feynman diagram for such a process is shown in Figure 22.6: the electron and positron annihilates to a virtual photon, which produces a quark-antiquark pair. But (except at certain special energies) the quark and antiquark have too much energy to bond and remain at rest; rather, they fly away in opposite directions. Also, since quark or antiquark cannot exist as an isolated particle (for reasons mentioned in Chapter 1 and discussed in depth in Chapter 30), these particles clothe themselves by ripping new quark-antiquark pairs out of the energy (which grows as they depart) in the attractive color field that is trying to hold them together. This creates two opposing jets of mesons and baryons whose momenta are roughly collinear with those of the originally created quark and antiquark.

From the point of view of quantum electrodynamics, the quark-antiquark pair is not different than a lepton-antilepton pair, except that the quark's charge is only the fraction of the charge e of the lepton. We can therefore easily adapt the lepton pair-production result (see Equation 20.19) to this situation

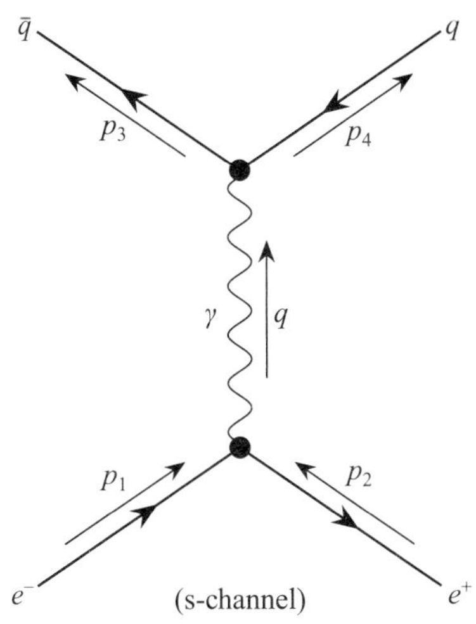

FIGURE 22.6 This diagram shows the lowest-order Feynman diagram for a two-jet hadron production process $(e^- + e^+ \to h^- + h^+)$.

$$\sigma = \frac{\pi Q^2 \alpha^2}{4E^2} \sqrt{1 - \frac{M^2}{E^2}} \left[\frac{4}{3} + \frac{2}{3} \frac{M^2}{E^2} \right], \tag{22.13}$$

where σ is the total cross section for all jet angles, M is the mass of the quark formed, Q is that quark's charge in units of e ($\frac{2}{3}$ for u, c, or t, $\frac{1}{3}$ for d, s, or b), α is the running fine-structure constant, and E is the energy of the incoming electron. (This formula assumes that $E \gg$ the electron mass m, which is usually an excellent approximation in typical experiments.) We have a factor of Q^2 because the upper vertex contributes a factor of $(i Q e \gamma^\mu)$ instead of $(i e \gamma^\mu)$ to the amplitude, the lower vertex still contributes $(i e \gamma^\mu)$, and the whole amplitude gets squared, so we end up with $Q^2 e^4 = (4\pi)^2 Q^2 \alpha^2$, where for lepton pair production we had $e^4 = (4\pi)^2 \alpha^2$.

This formula has a threshold at $E = M$: for $E < M$, the cross section is imaginary, indicating that the process is impossible (there is insufficient energy to create the $q\bar{q}$ pair). When we are significantly above that threshold ($E > M$), note that according to the binomial approximation,

$$\left(1 - \frac{M^2}{E^2}\right)^{1/2} \left(\frac{4}{3} + \frac{2}{3}\frac{M^2}{E^2}\right) \approx \frac{4}{3}\left(1 - \frac{1}{2}\frac{M^2}{E^2}\right)\left(1 + \frac{1}{2}\frac{M^2}{E^2}\right)$$

$$= \frac{4}{3}\left[1 - \frac{1}{2}\frac{M^2}{E^2} + \frac{1}{2}\frac{M^2}{E^2} - \frac{1}{4}\frac{M^4}{E^4}\right]$$

$$\Rightarrow \quad \sigma = \frac{\pi Q^2 \alpha^2}{4E^2} \sqrt{1 - \frac{M^2}{E^2}} \left[\frac{4}{3} + \frac{2}{3}\frac{M^2}{E^2}\right] \approx \frac{\pi Q^2 \alpha^2}{3E^2} \tag{22.14}$$

to an accuracy of order $(M/E)^4$, so E doesn't need to be much above the threshold for the approximation to apply.

As we ramp up the beam energy E, we therefore pass through a sequence of such thresholds: at $E = 106$ MeV, we start to be able to create $\mu^- \mu^+$ pairs, then $u\bar{u}$ and $d\bar{d}$ pairs, then $s\bar{s}$ pairs, then $c\bar{c}$ pairs, then $\tau^- \tau^+$ pairs, and finally $b\bar{b}$ pairs. (Even though the masses in Figure 1.1 for the u, d, and s quarks are each less than the muon mass, the threshold is determined by the masses of the lowest-mass mesons that can be formed from these quark-antiquark pairs, and all such mesons are more massive than the muon.) Since Equation 22.14 applies to each possible pair, passing each threshold therefore causes the cross section to increase in proportion to the *total* number of particles that might be created. An especially practical way to test this is to examine the energy dependence of the ratio

$$R(E) = \frac{\sigma(e^- + e^+ \to \text{hadrons})}{\sigma(e^- + e^+ \to \mu^- + \mu^+)} \tag{22.15}$$

because hadron production is easy to distinguish from muon production. Because the cross section is $\sigma \approx \pi Q^2 \alpha^2/3E^2$ for every quark-antiquark type we can produce but $\sigma \approx \pi\alpha^2/3E^2$ for muons, we would predict that

$$R(E) = \frac{\sigma(e^- + e^+ \to \text{hadrons})}{\sigma(e^- + e^+ \to \mu^- + \mu^+)} \approx 3\sum Q_i^2, \tag{22.16}$$

where the sum is over all types of quark-antiquark pairs we are able to produce at a given energy. The factor of 3 is because each quark-antiquark pair might have any one of three colors, so we are effectively producing *three* new types of particle-antiparticle pairs with each threshold we pass. (Because $\tau^-\tau^+$ pairs decay too rapidly to lay down a track, and decay to hadrons about 60% of the time, we should also add a bit to R when $E > m_\tau = 1.77$ GeV.)

So, for E below the $c\bar{c}$-production threshold at about 1.3 GeV, we should have

$$R(E) \approx 3\left[(\tfrac{2}{3})^2 + (\tfrac{1}{3})^2 + (\tfrac{1}{3})^2\right] = 2. \tag{22.17}$$

Above about 1.8 GeV, we'd expect

$$R(E) \approx 3\left[(\tfrac{2}{3})^2 + (\tfrac{1}{3})^2 + (\tfrac{1}{3})^2 + (\tfrac{2}{3})^2\right] + 0.6 = \tfrac{10}{3} + 0.6 \approx 3.9, \tag{22.18}$$

where the 0.6 is the estimate for the contribution from $\tau^-\tau^+$ production. Above the $b\bar{b}$-production threshold at $E = 4.18$ GeV, we'd expect to add $\tfrac{1}{3}$ to get to 4.2.

Figure 22.7 shows the experimental value of $R(E)$ plotted against the total collision energy $2E$. We do see the clear step from about 2 to just under 4 at $E \approx 2$ GeV, and the smaller step to just above 4 at $E \approx 5$ GeV.

The graph is complicated partly because Equation 22.14 is not accurate when E is close to a threshold, and partly because the produced quark-antiquark pairs are not actually real particles but virtual particles on the way to becoming something else, but especially because at special energies, the produced quark-antiquark can form *real*

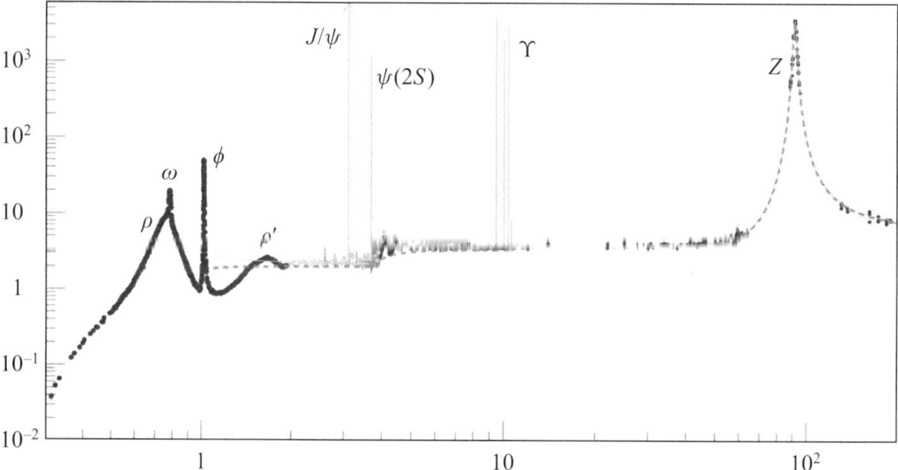

FIGURE 22.7 The hadron production to muon production ratio R plotted as a function of total collision energy $2E$. The dashed line is what we would expect from a naive quark-only model. The solid line reflects a more sophisticated prediction that includes a variety of complications. From the Particle Data Group website `https://pdg.lbl.gov/2022/reviews/contents_sports.html`.

particles at rest The spikes in the cross section at these energies have a width inversely proportional to the corresponding particle's lifetime.

But perhaps the most important feature of this graph is that it provides firm evidence that quarks have three colors. If quarks did not have color, R would be about $\frac{2}{3}$ at $E \approx 1\,\text{GeV}$, a possibility that is clearly excluded.

22.3 Decay of Positronium

In pure quantum electrodynamics, particle decays do not strictly exist: a fermion world-line never terminates at a vertex, and there is no way in QED to convert (say) a muon to an electron. In pure QED, only a bound *system* of elementary fermions can "decay" to something else. The simplest example is positronium "decaying" into two photons. **Positronium** is a quantum system like a hydrogen atom, but where an electron is bound to a positron instead of a proton. People study positronium precisely because it is so simple: it provides an opportunity to test various fundamental aspects of the quantum physics of bound systems as well as details of quantum electrodynamics.

At the fundamental level, the decay is actually the "scattering" process $e^+ + e^- \to \gamma + \gamma$. One might think one could get away with a single vertex, as in Figure 22.8a. We can certainly draw the diagram, but conservation of four-momentum forbids it: in the CM frame, the single outgoing photon would be required to have zero spatial momentum, which is impossible for a photon. The simplest physically possible diagrams have two vertices and two outgoing photons, as shown in Figure 22.8b and 22.8c. Since the photons can be emitted in opposite directions, their total spatial momentum can be zero.

In Box 22.4, you will find the amplitude for the t-channel process in Figure 22.8b to be

$$\mathcal{M}_t = \frac{e^2}{(p_1 - p_3)^2 - m^2}\, \bar{v}_2\, \epsilon_4^*(\not{p}_1 - \not{p}_3 + m)\epsilon_3^*\, u_1. \tag{22.19}$$

The u-channel process in Figure 22.8c is the same except $p_3 \leftrightarrow p_4$:

$$\mathcal{M}_u = \frac{e^2}{(p_1 - p_4)^2 - m^2}\, \bar{v}_2\, \epsilon_3^*(\not{p}_1 - \not{p}_4 + m)\epsilon_4^*\, u_1. \tag{22.20}$$

The total amplitude is the sum of these two amplitudes. A bit of clever simplification (see Box 22.5) shows that in the rest frame of the positronium, the total amplitude reduces to

$$\mathcal{M} = \mathcal{M}_t + \mathcal{M}_u = -\frac{e^2}{2m^2}\bar{v}_2[\,\epsilon_4^*\epsilon_3^*\not{p}_3 + \epsilon_3^*\epsilon_4^*\not{p}_4\,]u_1. \tag{22.21}$$

Now we have to handle the spins. In this case, the positronium is likely to be in its lowest energy state, which happens to be when the electron and positron are in the total spin-0 state ($\sqrt{\frac{1}{2}}[\,|+z, -z\rangle - |-z, +z\rangle\,]$. This means that we will *not* use Casimir's

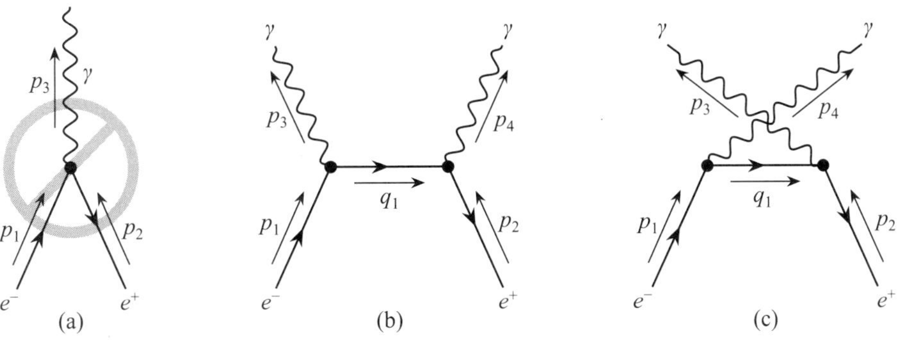

FIGURE 22.8 Feynman diagrams for electron-positron annihilation. (Diagram (a) is forbidden by conservation of four-momentum.)

trick, but (since we know what the spin cases are) instead calculate the cases directly. The scattering amplitude will be

$$\mathcal{M}_0 = \sqrt{\tfrac{1}{2}}[\, \mathcal{M}_{\uparrow\downarrow} - \mathcal{M}_{\downarrow\uparrow}\,], \tag{22.22}$$

where the arrows refer to the spin states of the electron and positron. A bit of calculation (see Box 22.6) yields the following astonishingly simple result:

$$\mathcal{M}_0 = -4e^2. \tag{22.23}$$

From this, we can calculate the differential cross section

$$d\sigma = \frac{|\mathcal{M}_0|^2}{(8\pi)^2(E_1+E_2)^2} \frac{|\,\vec{p}_3\,|}{|\,\vec{p}_1\,|}\, d\Omega = \left(\frac{-4e^2}{16\pi m}\right)^2 \frac{m}{mv}\, d\Omega = \frac{1}{v}\left(\frac{\alpha}{m}\right)^2 d\Omega \tag{22.24}$$

because in the positronium "atom" the electron and positron are basically at rest, so $E_1 = E_2 \approx m$, the photons' common momentum magnitude $|\,\vec{p}_3\,| = |\,\vec{p}_4\,| \approx m$. Moreover, the electron and positron are non-relativistic, so $|\,\vec{p}_1\,| = mv$, where v is the electron's speed when it hits the positron. As $d\sigma$ does not depend on angle, the total cross section for all angles is $\sigma = 4\pi\, d\sigma$.

How do we get a decay rate out of this? If you read Chapter 14, recall that

$$\sigma = \frac{\Gamma V}{v}, \tag{22.25}$$

where Γ is the transitions per unit time for one particle hitting one particle, V is the experimental volume, and v is the relative speed of the interacting particles (see Equation 14.19). We can consider $1/V$ to be the electron's probability density $|\psi(0)|^2$, where $\psi(0)$ is the ordinary electron wavefunction in the positronium atom ground state, which one can calculate using ordinary quantum mechanics. Solving for the decay rate, we have

$$\Gamma = \frac{v}{V}\sigma = 4\pi \left(\frac{\alpha}{m}\right)^2 |\psi(0)|^2. \tag{22.26}$$

(Note how the hard-to-calculate v has dropped out of the calculation.) The positronium average lifetime is $\tau = 1/\Gamma$. When one substitutes in the result for the electron wavefunction, one finds (see Problem P22.4) an average lifetime of about 0.125 ns, which agrees well with experimental results.

22.4 Closing Comments

We have done some amazing things in this chapter: we have actually calculated some realistic and physically measurable results from our theory of quantum electrodynamics! Of course, we are only calculating the leading-order contributions to these cross sections. But in combination with renormalization (which handles many additional diagrams), the facts that each level of complication in a diagram is typically smaller than the previous level by a factor of $e^2 \propto \alpha$ and that $\alpha \approx 1/137$ is pretty small mean that just calculating the lowest-order terms is usually sufficient to make decent predictions, as the experimental evidence presented in this chapter illustrates.

But quantum electrodynamics is only a small (though crucial) part of the Standard Model. The next step is to apply the impressive machinery we have developed to the weak interaction. We will take up this task in the next chapter.

BOX 22.1   Bhabha Scattering: The Value of $\langle |\mathcal{M}_t|^2 \rangle$

Taking the absolute square of Equation 22.2 yields

$$|\mathcal{M}_t|^2 = \frac{e^4}{(p_1 - p_3)^2} [\overline{u}_3 \gamma^\mu u_1][\overline{u}_3 \gamma^\nu u_1]^* \, [\overline{v}_2 \gamma_\mu v_4][\overline{v}_2 \gamma_\nu v_4]^*. \qquad (22.27)$$

(Note how I have renamed the indices to keep the sums in the $\mathcal{M}_t$ and $\mathcal{M}_t^*$ factors in $|\mathcal{M}_t|^2 = \mathcal{M}_t \mathcal{M}_t^*$ distinct.)

Exercise 22.1.1   Verify that applying Casimir's trick to this implies that

$$\langle |\mathcal{M}_t|^2 \rangle = \frac{e^4}{4(p_1 - p_3)^2} \, \mathrm{Tr} \left[\gamma^\mu (\not{p}_1 + m) \gamma^\nu (\not{p}_3 + m) \right]$$

$$\times \mathrm{Tr} \left[\gamma_\mu (\not{p}_4 - m) \gamma_\nu (\not{p}_2 - m) \right]. \qquad (22.28)$$

In particular, where does the factor of 4 come from? Why do we have $+m$ (the electron mass) in some factors, but $-m$ in others?

Since we are taking the limit where all energies and momenta are much larger than the electron mass m, let's save work and drop the mass terms now. The first trace factor therefore becomes

$$\mathrm{Tr}\left[\gamma^\mu(\not{p}_1 + m)\gamma^\nu(\not{p}_3 + m)\right] \approx \mathrm{Tr}\left[\gamma^\mu \not{p}_1 \gamma^\nu \not{p}_3\right]$$
$$= \mathrm{Tr}\left[\gamma^\mu(p_{1\alpha}\gamma^\alpha)\gamma^\nu(p_{3\beta}\gamma^\beta)\right] = p_{1\alpha}p_{3\beta}\,\mathrm{Tr}\left[\gamma^\mu\gamma^\alpha\gamma^\nu\gamma^\beta\right]$$
$$= 4p_{1\alpha}p_{3\beta}\left(g^{\mu\alpha}g^{\nu\beta} - g^{\mu\nu}g^{\alpha\beta} + g^{\mu\beta}g^{\nu\alpha}\right)$$
$$= 4\left(p_1^\mu p_3^\nu - g^{\mu\nu}(p_1 \cdot p_3) + p_1^\nu p_3^\mu\right), \tag{22.30}$$

where I have used the definitions of $\not{p}_1$ and $\not{p}_3$ and Theorem 14.

Exercise 22.1.2   In a similar fashion, show that

$$\mathrm{Tr}\left[\gamma_\mu(\not{p}_4 - m)\gamma_\nu(\not{p}_2 - m)\right] \approx \mathrm{Tr}\left[(g_{\mu\sigma}\gamma^\sigma)\not{p}_4\,(g_{\nu\lambda}\gamma^\lambda)\not{p}_2\right]$$
$$= 4\left(p_{4\mu}p_{2\nu} - g_{\mu\nu}(p_4 \cdot p_2) + p_{4\nu}p_{2\mu}\right). \tag{22.31}$$

Exercise 22.1.3   Multiply Equations 22.30 and 22.31 to get

$$\mathrm{Tr}\left[\gamma^\mu \not{p}_1 \gamma^\nu \not{p}_3\right]\mathrm{Tr}\left[\gamma_\mu \not{p}_4 \gamma_\nu \not{p}_2\right] = 32\left[(p_1 \cdot p_4)(p_3 \cdot p_3) + (p_1 \cdot p_2)(p_3 \cdot p_4)\right]. \tag{22.32}$$

(There is extra space at the top of the next page.)

BOX 22.1   Bhabha Scattering: The Value of $\langle|\mathcal{M}_t|^2\rangle$ (cont.)

So, if we substitute Equation 22.32 into Equation 22.28, we see that in the limit that the electron mass m is negligible compared to the energies and momenta involved, we have

$$\langle|\mathcal{M}_t|^2\rangle = \frac{8e^4}{(p_1 - p_3)^4}\left[(p_1 \cdot p_4)(p_2 \cdot p_3) + (p_1 \cdot p_2)(p_3 \cdot p_4)\right]. \quad (22.33)$$

In the center of mass frame, Equations 22.7 (repeated here) tells us that

$$p_1 \cdot p_2 = p_3 \cdot p_4 = 2E^2, \qquad (22.34a)$$

$$p_1 \cdot p_3 = p_2 \cdot p_4 = E^2(1 - \cos\theta), \qquad (22.34b)$$

$$p_1 \cdot p_4 = p_2 \cdot p_3 = E^2(1 + \cos\theta). \qquad (22.34c)$$

Exercise 22.1.4   Use this and the fact that $(p_1 - p_3)^2 = p_1^2 - 2p_1 \cdot p_3 + p_3^2 \approx 2p_1 \cdot p_3$ in the limit that the electron mass is negligible to show that

$$\langle|\mathcal{M}_t|^2\rangle = 2e^4\frac{(1 + \cos\theta)^2 + 4}{(1 - \cos\theta)^2}. \qquad (22.35)$$

BOX 22.2   Bhabhah Scattering: The Value of $\langle \mathcal{M}_t \mathcal{M}_s^* \rangle$

Multiplying Equations 22.2 and the conjugate of Equation 22.1 yields

$$\mathcal{M}_t \mathcal{M}_s^* = \frac{e^4}{(p_1 - p_3)^2 (p_1 + p_2)^2} [\, \bar{u}_3 \gamma^\mu u_1 \,][\, \bar{v}_2 \gamma_\mu v_4 \,][\, \bar{v}_2 \gamma^\nu u_1 \,]^* [\, \bar{u}_3 \gamma_\nu v_4 \,]^*.$$

$$(22.36)$$

Unfortunately, in this case, none of the factors in brackets pair with any of the conjugated factors to yield something that looks like the pair of terms handled by Casimir's trick as described in Box 20.2, so we must start from scratch. First, note that (as we saw in Box 20.2)

$$[\, \bar{v}_2 \gamma^\nu u_1 \,]^* = \bar{u}_1 \bar{\gamma}^\nu v_2 = \bar{u}_1 \gamma^\nu v_2, \qquad (22.37a)$$

$$[\, \bar{u}_3 \gamma_\nu v_4 \,]^* = \bar{v}_4 \bar{\gamma}_\nu u_3 = \bar{v}_4 \gamma_\nu u_3. \qquad (22.37b)$$

With these substitutions, let's write $P \equiv$ the product of the quantities in square brackets in Equation 22.36 in this order:

$$P = [\, \bar{u}_3 \gamma^\mu u_1 \,][\, \bar{u}_1 \gamma^\nu v_2 \,][\, \bar{v}_2 \gamma_\mu v_4 \,][\, \bar{v}_4 \gamma_\nu u_3 \,] \qquad (22.38)$$

so that each spinor is adjacent to its adjoint. Then we can use the completeness relations (Equations 16.39)

$$\sum_s u^s(p) \bar{u}^s(p) = \not{p} + m \quad \text{and} \quad \sum_s v^s(p) \bar{v}^s(p) = \not{p} - m \qquad (22.39)$$

to show that (with spin indices $s_1, s_2, s_3, s_4 = +, -$)

$$\sum_{s_1, s_2, s_4} P = \bar{u}_3 \gamma^\mu \Big[\sum_{s_1} u^{s_1}(p_1)\, \bar{u}^{s_1}(p_1) \Big] \gamma^\nu \Big[\sum_{s_2} v^{s_2}(p_2)\, \bar{v}^{s_2}(p_2) \Big]$$

$$\times \gamma_\mu \Big[\sum_{s_4} v^{s_4}(p_4)\, \bar{v}^{s_4}(p_4) \Big] \gamma_\nu u_3$$

$$= \bar{u}_3 \big[\gamma^\mu (\not{p}_1 + m) \gamma^\nu (\not{p}_2 - m) \gamma_\mu (\not{p}_4 - m) \gamma_\nu \big] u_3 \equiv \bar{u}_3 B u_3, \qquad (22.40)$$

where B is a 4×4 matrix.

Exercise 22.2.1 Adapt the argument leading to and including Exercise 20.2.2 to show that summing this result over s_3 yields

$$\sum_{\text{all spins}} P = \mathrm{Tr}\big[\gamma^\mu (\not{p}_1 + m) \gamma^\nu (\not{p}_2 - m) \gamma_\mu (\not{p}_4 - m) \gamma_\nu (\not{p}_3 + m) \big]. \quad (22.41)$$

BOX 22.2   Bhabhah Scattering: The Value of $\langle \mathcal{M}_t \mathcal{M}_s^* \rangle$ (cont.)

If we now assume that m is negligible compared to magnitudes of the four-momenta involved, substitute the result we just obtained into Equation 22.36, and divide by 4 to average over the initial spins, we see that we have

$$\langle \mathcal{M}_t \mathcal{M}_s^* \rangle = \frac{e^4}{4(p_1 - p_3)^2 (p_1 + p_2)^2} \, \mathrm{Tr} \left[\gamma^\mu \not{p}_1 \gamma^\nu \not{p}_2 \gamma_\mu \not{p}_4 \gamma_\nu \not{p}_3 \right]. \quad (22.42)$$

The trace part looks pretty awful, with eight gamma matrices:

$$T \equiv \mathrm{Tr} \left[\gamma^\mu \not{p}_1 \gamma^\nu \not{p}_2 \gamma_\mu \not{p}_4 \gamma_\nu \not{p}_3 \right] = p_{1\alpha} p_\beta p_{4\sigma} p_{3\lambda} \, \mathrm{Tr} \left[\gamma^\mu \gamma^\alpha \gamma^\nu \gamma^\beta \gamma_\mu \gamma^\sigma \gamma_\nu \gamma^\lambda \right]. \quad (22.43)$$

But Theorems 6, 5, and 13 in Chapter 20 allow us to make short work of this. They say, respectively, that $\gamma_\mu \gamma^\nu \gamma^\alpha \gamma^\beta \gamma^\mu = -2\gamma^\beta \gamma^\alpha \gamma^\nu$ (note the reversed order), that $\gamma_\mu \gamma^\alpha \gamma^\beta \gamma^\mu = 4g^{\alpha\beta} I$, and that $\mathrm{Tr}(\gamma^\mu \gamma^\nu) = 4g^{\mu\nu}$. Also, note that (for example) $\gamma_\mu B \gamma^\mu = g_{\mu\nu} \gamma^\mu B \gamma^\nu = \gamma^\mu B \gamma_\mu$, where B is any 4×4 matrix.

Exercise 22.2.2   Use these theorems in order to show that

$$T = -32(p_1 \cdot p_4)(p_2 \cdot p_3). \quad (22.44)$$

BOX 22.2   Bhabhah Scattering: The Value of $\langle \mathcal{M}_t \mathcal{M}_s^* \rangle$ (cont.)

Substituting Equation 22.44 into Equation 22.43 yields

$$\langle \mathcal{M}_t \mathcal{M}_s^* \rangle = \frac{-8e^4}{(p_1 - p_3)^2 (p_1 + p_2)^2}(p_1 \cdot p_4)(p_2 \cdot p_3). \qquad (22.45)$$

In the center-of-mass frame, Equations 22.7 (repeated here) say that

$$p_1 \cdot p_2 = p_3 \cdot p_4 = 2E^2, \qquad (22.46a)$$

$$p_1 \cdot p_3 = p_2 \cdot p_4 = E^2(1 - \cos\theta), \qquad (22.46b)$$

$$p_1 \cdot p_4 = p_2 \cdot p_3 = E^2(1 + \cos\theta). \qquad (22.46c)$$

We also have

$$(p_1 - p_3)^2 = p_1^2 - 2p_1 \cdot p_3 + p_3^2 \approx -2p_1 \cdot p_3 \qquad (22.47a)$$

$$(p_1 + p_2)^2 = p_1^2 + 2p_1 \cdot p_2 + p_2^2 \approx 2p_1 \cdot p_2 \qquad (22.47b)$$

when we are taking the limit that $p_1^2 = p_2^2 = p_3^2 = m^2 \to 0$.

Exercise 22.2.3   Use these results to show that

$$\langle \mathcal{M}_t \mathcal{M}_s^* \rangle = e^4 \frac{(1 + \cos\theta)^2}{1 - \cos\theta}. \qquad (22.48)$$

Because this result is real, $\langle \mathcal{M}_s \mathcal{M}_t^* \rangle = \langle \mathcal{M}_t \mathcal{M}_s^* \rangle^* = \langle \mathcal{M}_t \mathcal{M}_s^* \rangle$, so

$$-\langle \mathcal{M}_t \mathcal{M}_s^* \rangle - \langle \mathcal{M}_s \mathcal{M}_t^* \rangle = -2e^4 \frac{(1 + \cos\theta)^2}{(1 - \cos\theta)}, \qquad (22.49)$$

as claimed in Equation 22.10.

BOX 22.3 Bhabha Scattering: Simplifying $\langle |\mathcal{M}|^2 \rangle$.

According to Equation 22.10, we have

$$\langle |\mathcal{M}|^2 \rangle = 2e^4 \left[\frac{(1+c)^2 + 4}{(1-c)^2} - \frac{(1+c)^2}{1-c} + \frac{1+c^2}{2} \right], \qquad (22.50)$$

where $c \equiv \cos \theta$ (saves writing.)

Exercise 22.3.1 Show that we can write this in the simpler form

$$\langle |\mathcal{M}|^2 \rangle = e^4 \left[\frac{3+c^2}{1-c} \right]^2. \qquad (22.51)$$

(*Hint:* Start by putting all terms over a common denominator. It may help to work from both sides.)

BOX 22.4   The Positronium Decay *t*-Channel Amplitude

The goal in this box is to calculate the amplitude for the t-channel process shown in Figure 22.8b, repeated as Figure 22.9.

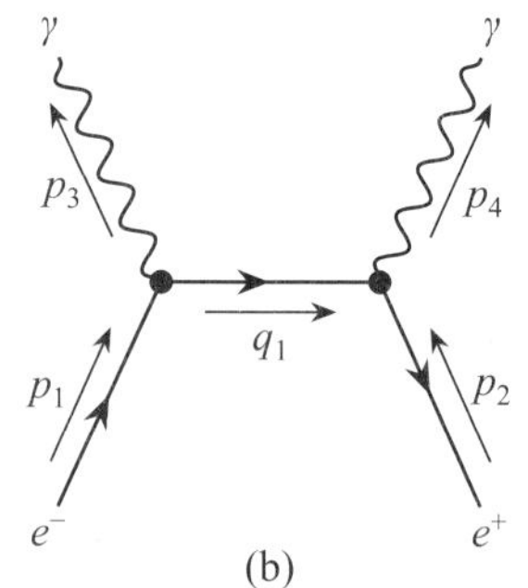

FIGURE 22.9   This figure shows The t-channel process for electron-positron annihlation.

Exercise 22.4.1   Use the Feynman rules to evaluate the amplitude for this process. (*Hint:* At each vertex, you will have a product of a gamma matrix and a polarization vector that must be a simple Lorentz-invariant scalar because the scattering amplitude itself must be a Lorentz-invariant scalar. Therefore, you will need to lower the index on the photon polarization vector so that the product sums correctly to a scalar. The index positions here are not automatically correct as they were in the electron-muon case.)

BOX 22.5 Simplifying the Positronium Amplitude

Our goal in this box is to simplify the positronium decay amplitude $\mathcal{M} = \mathcal{M}_t + \mathcal{M}_u$, where

$$\mathcal{M}_t = \frac{e^2}{(p_1 - p_3)^2 - m^2}\, \bar{v}_2 \not{\epsilon}_4^*(\not{p}_1 - \not{p}_3 + m)\not{\epsilon}_3^* u_1 \qquad (22.52)$$

and $\mathcal{M}_u$ is the same with $3 \leftrightarrow 4$. We will assume that the electron and positron are initially essentially at rest in the positronium bound state, and that the photons come out along the z axis. Then the four-momenta are

$$p_1 = p_2 = \begin{bmatrix} m \\ 0 \\ 0 \\ 0 \end{bmatrix}, \qquad p_3 = \begin{bmatrix} m \\ 0 \\ 0 \\ m \end{bmatrix}, \qquad p_4 = \begin{bmatrix} m \\ 0 \\ 0 \\ -m \end{bmatrix}. \qquad (22.53)$$

Note that the relation $\gamma^\mu \gamma^\nu + \gamma^\nu \gamma^\mu = 2g^{\mu\nu} I$ implies that

$$a_\mu \gamma^\mu b_\nu \gamma^\nu + b_\nu \gamma^\nu a_\mu \gamma^\mu = \not{a}\not{b} + \not{b}\not{a} = 2a_\mu b_\nu g^{\mu\nu} I = 2(a \cdot b) I. \quad (22.54)$$

Let's apply this to $\not{p}_1$ and $\not{\epsilon}_3^*$. If the electron is initially essentially at rest in the positronium bound state, then its four-momentum p_1 has only a time component, while in Coulomb gauge ϵ_3^* has only spatial components. Therefore, $p_1 \cdot \epsilon_3^* = 0$. We also know that $p_3 \cdot \epsilon_3^* = p_4 \cdot \epsilon_4^* = 0$ because an electromagnetic wave must be transverse to its direction of motion. This means that

$$\not{p}_1 \not{\epsilon}_3^* = -\not{\epsilon}_3^* \not{p}_1 \quad \text{and} \quad \not{p}_3 \not{\epsilon}_3^* = -\not{\epsilon}_3^* \not{p}_3 \quad \text{and} \quad \not{p}_4 \not{\epsilon}_4^* = -\not{\epsilon}_4^* \not{p}_4. \quad (22.55)$$

Finally, note that $(\not{p}_1 - m)u_1 = 0$ because $\psi = u_1 e^{-ip \cdot x}$ satisfies the Dirac equation $(i\gamma^\mu \partial_\mu - m)\psi = 0$.

Exercise 22.5.1 Show that under these conditions, we have

$$\mathcal{M} = \mathcal{M}_t + \mathcal{M}_s = -\frac{e^2}{2m^2}\bar{v}_2 \big[\not{\epsilon}_4^* \not{\epsilon}_3^* \not{p}_3 + \not{\epsilon}_3^* \not{\epsilon}_4^* \not{p}_4 \big] u_1. \qquad (22.56)$$

BOX 22.6   Finishing the Positronium Calculation

In this box, we want to calculate $\mathcal{M}_0$ given that

$$\mathcal{M} = \mathcal{M}_t + \mathcal{M}_u = -\frac{e^2}{2m^2}\,\bar{v}_2[\,\epsilon_4^*\epsilon_3^*\not{p}_3 + \epsilon_3^*\epsilon_4^*\not{p}_4\,]u_1, \tag{22.57}$$

assuming a total spin-0 initial state so that

$$\mathcal{M}_0 = \sqrt{\tfrac{1}{2}}[\,\mathcal{M}_{\uparrow\downarrow} - \mathcal{M}_{\downarrow\uparrow}\,], \tag{22.58}$$

where the arrows refer to the spin states of the electron and positron.

If the system's total spin is zero before the decay, then it must be after the decay as well, so we are looking for the amplitude to decay to a photon state of total spin 0. The polarization vectors for photons spinning in the $\pm z$ direction (see Problem 19.5) are $(\epsilon^\pm)^\mu = (0, a, \pm ia, 0)$, where $a = \sqrt{1/2}$ for a normalized state. The polarization state for a pair of photons with zero total spin will therefore be

$$\epsilon_3^\mu \epsilon_4^\nu = \sqrt{\tfrac{1}{2}}\big[\,(\epsilon^+)^\mu(\epsilon^-)^\nu - (\epsilon^-)^\mu(\epsilon^+)^\nu\,\big]. \tag{22.59}$$

Lowering the indices will change the signs of the spatial components, and taking the complex conjugate will change the sign of the imaginary term, so

$$\epsilon_{3,\mu}^* \epsilon_{4,\nu}^* = \sqrt{\tfrac{1}{2}}\big[\,(\epsilon^+)_\mu^*(\epsilon^-)_\nu^* - (\epsilon^-)_\mu^*(\epsilon^+)_\nu^*\,\big], \tag{22.60}$$

where $(\epsilon^\pm)_\mu^* = (0, -a, \pm ia, 0)$. Now, note that $(\not{\epsilon}^\pm)^* = -a\gamma^x \pm ia\gamma^y$, so

$$\epsilon_3^*\epsilon_4^* = \sqrt{\tfrac{1}{2}}a^2\big[(-\gamma^x + i\gamma^y)(-\gamma^x - i\gamma^y) - (-\gamma^x - i\gamma^y)(-\gamma^x + i\gamma^y)\big]. \tag{22.61}$$

Exercise 22.6.1   Multiply this out, simplify using $\gamma^\mu\gamma^\nu + \gamma^\nu\gamma^\mu = 2g^{\mu\nu}I$ and $a^2 = \tfrac{1}{2}$, show that $\epsilon_4^*\epsilon_3^* = -\epsilon_3^*\epsilon_4^*$, and finally show that

$$\epsilon_4^*\epsilon_3^*\not{p}_3 + \epsilon_3^*\epsilon_4^*\not{p}_4 = 2\sqrt{2}mi\gamma^x\gamma^y\gamma^z. \tag{22.62}$$

BOX 22.6 Finishing the Positronium Calculation (cont.)

Let the matrix we have just calculated be B. We need to evaluate

$$\mathcal{M}_0 = \sqrt{\tfrac{1}{2}}[\,\mathcal{M}_{\uparrow\downarrow} - \mathcal{M}_{\downarrow\uparrow}\,] = -\frac{e^2}{2m^2}[\,\sqrt{\tfrac{1}{2}}\,\overline{v}^{\,-}Bu^+ - \sqrt{\tfrac{1}{2}}\,\overline{v}^{\,+}Bu^-\,]. \qquad (22.63)$$

We are assuming that the incoming electron and positron are at rest, so the spin states are

$$u^+ = \sqrt{m}\begin{bmatrix}1\\0\\1\\0\end{bmatrix}, \quad u^- = \sqrt{m}\begin{bmatrix}0\\1\\0\\1\end{bmatrix}, \quad v^+ = \sqrt{m}\begin{bmatrix}0\\1\\0\\-1\end{bmatrix}, \quad v^- = \sqrt{m}\begin{bmatrix}-1\\0\\1\\0\end{bmatrix}.$$

$$(22.64)$$

Remember that $\overline{v} = v^\dagger\gamma^t$, so $\overline{v}Bu = v^\dagger\gamma^tBu$.

Exercise 22.6.2 Evaluate $\mathcal{M}_0$. (*Hint:* It is handy to calculate γ^tB first. Note that $\gamma^5 = i\gamma^t\gamma^x\gamma^y\gamma^z$.)

Homework Problems

P22.1 At high energies, the lowest-order Bhabha scattering diagram is insufficient partly because it becomes possible for the colliding electron and positron to produce a virtual Z (whose rest mass is about 91 GeV) instead of a photon, so we must add an amplitude for that possibility. The propagator for a virtual Z is $-i(g_{\mu\nu} - q_\mu q_\nu/M_Z^2)/(q^2 - M_Z^2)$, in contrast to the photon propagator $-ig_{\mu\nu}/q^2$, where q is the virtual particle's four-momentum. Even so, Figure 22.5 shows that when $\cos\theta$ is large, the lowest-order Bhabha scattering amplitude works pretty well. Explain why neither the Z possibility nor the running value of α is as important at large values of $\cos\theta$ as they are at low values.

P22.2 Why do we not use $\sigma(e^- + e^+ \to e^- + e^+)$ in the denominator for $R(E)$ in Equation 22.15 instead of $\sigma(e^- + e^+ \to \mu^- + \mu^+)$?

P22.3 Suppose that our positronium system was initially in the total spin-1 state $|1, 0\rangle = \sqrt{\frac{1}{2}}[\uparrow\downarrow + \downarrow\uparrow]$. Someone claims that it should be impossible for such a system to decay to two photons, because the outgoing photon spins can only either be aligned or anti-aligned with each other, and thus can have either a total spin of 2 or a total spin of 0, but never 1. Starting with Equation 22.63 in Box 22.6, calculate the amplitude

$$\mathcal{M}_1 = \sqrt{\tfrac{1}{2}}\left[\mathcal{M}_{\uparrow\downarrow} + \mathcal{M}_{\downarrow\uparrow}\right], \qquad (22.65)$$

the amplitude for decaying to two photons from this initial state. Does your result agree with the claim?

P22.4 In an ordinary quantum physics course, one finds that an electron bound to a particle with positive charge $+e$ (as in the case of hydrogen or positronium) has this wavefunction in its lowest-energy state (in our units):

$$\psi(r, \theta, \phi) = 2\sqrt{\frac{1}{4\pi}}\left(\frac{1}{a_0}\right)^{3/2} e^{-r/a_0}, \qquad (22.66)$$

where $a_0 = 1/(\mu\alpha)$ is the Bohr radius, μ is the electron's reduced mass ($= m/2$ in the positronium case), and $\alpha = e^2/4\pi = 1/137.04$ is the fine-structure constant. Use this to calculate the positronium lifetime and compare with the experimentally measured value of 0.1244 ns.

P22.5 (Challenging!) Show that the CM-frame cross section for pair annihilation $e^- + e^+ \to \gamma + \gamma$ in the high-energy limit where $m \approx 0$ is

$$d\sigma = \frac{\alpha^2}{8E^2}\frac{1 + \cos^2\theta}{\sin^2\theta}\, d\Omega, \qquad (22.67)$$

where E is the energy of either incoming particle. (*Hints:* Start with Equations 22.19 and 22.20 but drop the ms. Use Casimir's trick to sum over the electron and positron spins, and the completeness relation

$$\sum_{s_a} \epsilon_{a,\mu}^{s_a}[\epsilon_{a,\nu}^{s_a}]^* = -g_{\mu\nu}, \qquad (22.68)$$

(where a identifies the particle) to sum over the photon spins. Remember that Γ_1 and Γ_2 in Casimir's trick can be any matrix or product of matrices, but also remember that while $\overline{\gamma^\mu} = \gamma^\mu$, $\overline{\gamma^\mu\gamma^\nu\cdots\gamma^\sigma} = \gamma^\sigma\cdots\gamma^\nu\gamma^\mu$ because part of the bar operation involves taking the adjoint. Work on $\mathcal{M}_t$ first, then note that $\mathcal{M}_u$ will be the same with particle 3 $\leftrightarrow$ 4, then work on $\mathcal{M}_t\mathcal{M}_u^*$, and finally use Equations 22.7. You should find the trace part for $\mathcal{M}_t\mathcal{M}_u^*$ to be $\mathrm{Tr}\left[\gamma^\mu(\not{p}_1 - \not{p}_3\gamma^\nu\not{p}_1\gamma_\mu(\not{p}_1 - \not{p}_4)\gamma_\nu\not{p}_2\right]$.)

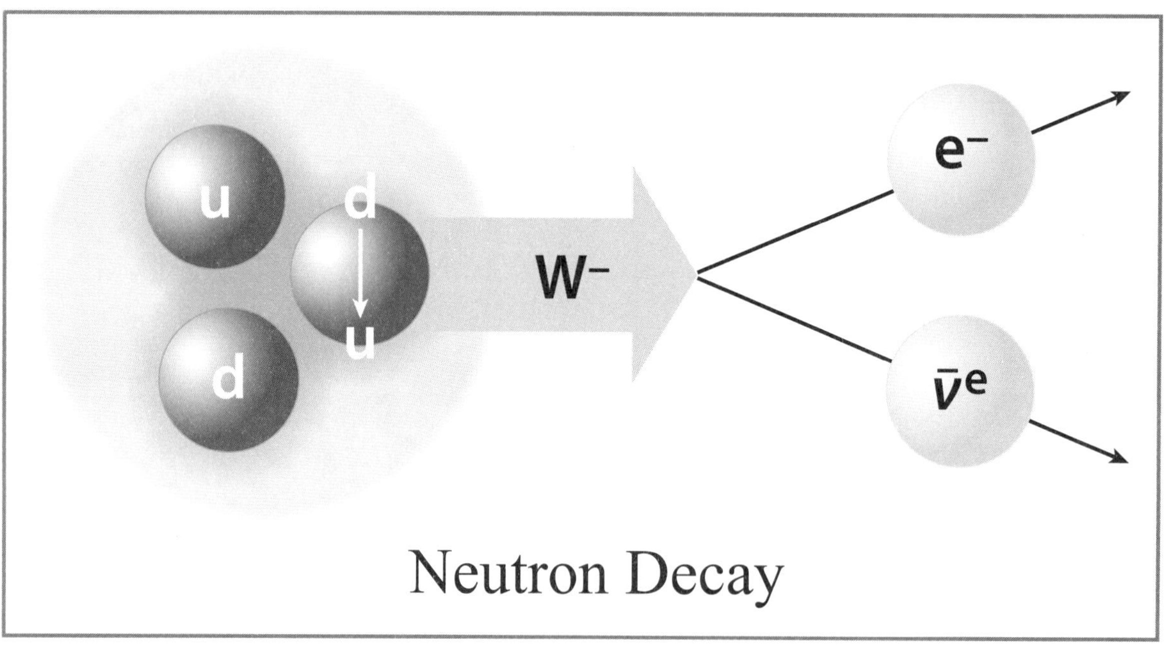

u
d
d
u
W⁻
e⁻
ν̄e
Neutron Decay

The Standard Model is fundamentally based on exploring the consequences of symmetries. In Chapter 19, we saw how the electromagnetic interaction arises from making the U(1) symmetry local. This chapter shows how the weak interaction arises from making the so-called SU(2) symmetry actively local.

The weak interaction only arises in the contexts of particle physics and nuclear physics, and so is likely less familiar to you than the electromagnetic interaction. Even so, the weak interaction has deep consequences for the structure of our universe. If you have studied some nuclear physics, you may know that the **beta decay processes**

$$n \to p + e^- + \overline{v}_e, \qquad p \to n + e^+ + v_e, \quad \text{and} \quad e^- + p \to n + v_e \qquad (23.1)$$

(which are weak-interaction processes) ensure that the numbers of protons and neutrons in small nuclei are nearly equal. These processes therefore determine the masses and thus the behavior of the chemical elements most crucial for life. (Note that conservation of four-momentum forbids the beta decay of an isolated proton, but such decays can happen when the proton is in a nucleus.) Weak interactions also play an important role in radioactive decay and nuclear fusion, which determine how stars evolve. Finally, because the weak interaction alone is asymmetric under the parity operation P, it is implicated in the mystery of why matter dominates over antimatter in the universe. So understanding the weak interaction is important for understanding the fact of our existence.

23.1 Quarks and Leptons

One of the fascinating features of the known elementary fermions is that they come in *pairs*. In weak interactions such as those listed in Equation 23.1, the disappearance of an electron or the appearance of a positron is always accompanied by the appearance of an electron neutrino, as if that neutrino is sort of an alter ego of the electron. Similarly, the disappearance of a positron or the appearance of an electron in a weak interaction is always accompanied by the appearance of an electron antineutrino. Muons and muon neutrinos, taus and tau neutrinos, up and down quarks, strange and charmed quarks, and top and bottom quarks are linked in a similar way in weak interactions.

The complete set of elementary fermions appears in Figure 23.1. The six **leptons** (which do not participate in color interactions) are paired with six **quarks** (which do participate in color interactions). Each group of six consists of three families (or **flavors**) of two particles each. These flavor groups are arranged in the table by successively increasing mass. Of course, each particle has a corresponding and distinct antiparticle.

Though flavors play a central role in the weak interaction (as we will see), the Standard Model does not explain why we have three flavors. (When the muon was discovered in 1936, the physicist Isidor Isaac Rabi famously remarked, "Who ordered that?") As we will also see, the simplest model for these particles also requires that they all be massless. The Higgs interaction eventually tells us *how* these particles acquire mass, but not why they have the masses that they do.

Neutrinos interact *only* through the weak interaction, and so are extraordinarily hard to detect. You would not know it, but neutrinos are extremely common. Every cubic meter in our universe contains roughly 315 million neutrinos from the Big Bang alone, roughly comparable to the number of photons. About 2 million neutrinos per cubic meter near the earth are from the sun, and trillions of solar neutrinos (with energies on the order

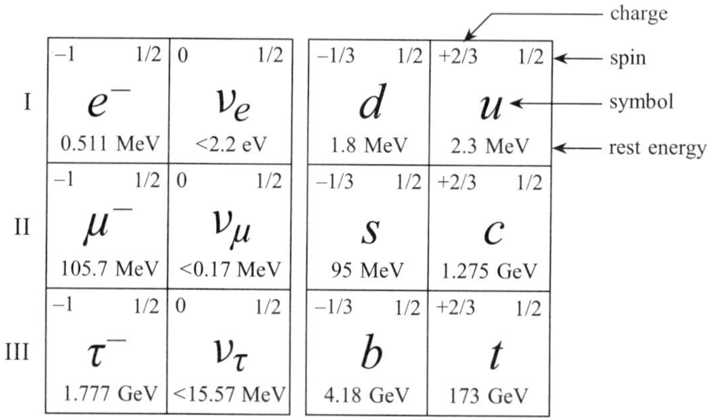

FIGURE 23.1   The fundamental fermions in the Standard Model. The box for the up quark (u) also provides the key for interpreting the numbers.

of MeV!) pass through your body every second, unnoticed because they rarely interact with anything. One of the first experiments to actually measure the solar flux of neutrinos used a 340 m^3 tank of tetrachloroethylene and still observed only a few interactions per day. This clearly shows that, as important as the weak interaction is, it is *really* weak compared to the electromagnetic interaction. We will soon see why.

In the next few sections, we will focus on leptons, because weak interactions involving quarks also involve complicating aspects of the color interaction. We will also focus on only the lightest of the three lepton families: the e^-/ν_e family. This is the family that has the most implications for everyday physics. Handling the other families is mathematically identical: wherever you see e^- and ν_e, you could just as well write μ^- and ν_μ, or d and u, and so on.

23.2 The Global SU(2) Symmetry

Consider a doublet of Dirac spinors

$$\Psi \equiv \begin{bmatrix} \psi_{\nu_e} \\ \psi_e \end{bmatrix} \quad \text{and} \quad \overline{\Psi} = [\, \overline{\psi}_{\nu_e} \quad \overline{\psi}_e \,], \tag{23.2}$$

where each entry in this doublet is itself a four-component Dirac field, one for the electron and one for the electron neutrino. This is actually not the standard notation for this doublet, which (to save writing) is more usually denoted

$$\Psi \equiv \begin{bmatrix} \nu_e \\ e \end{bmatrix} \quad \text{and} \quad \overline{\Psi} = [\, \overline{\nu}_e \quad \overline{e} \,], \tag{23.3}$$

where $e \equiv \psi_e$, $\nu_e \equiv \psi_{\nu_e}$, $\overline{e} \equiv \overline{\psi}_e$, and $\overline{\nu}_e \equiv \overline{\psi}_{\nu_e}$ (*not* the field for an antineutrino, even though in other contexts we often use the symbol $\overline{\nu}_e$ to denote an antineutrino). This notation can sometimes be tricky but is conventional (and does save a lot of writing). The point is, though, that when you *see* e and ν_e you should *think* of the Dirac spinor fields representing those particles.

A weak-interaction Lagrangian density that preserves the magnitude of

$$\overline{\Psi}\Psi = \overline{\nu}_e \nu_e + \overline{e}e \tag{23.4}$$

for "rotations" of the e and ν_e fields would imply that these fields are identical with respect to the weak interaction, making them alter egos. Now, a Lagrangian density for those fields based on what we know so far might be

$$\mathcal{L} = \overline{e}(i\gamma^\mu \partial_\mu - m_e)e + \overline{\nu}_e(i\gamma^\mu \partial_\mu - m_{\nu_e})\nu_e. \tag{23.5}$$

This Lagrangian density does *not* preserve the value of $\overline{\Psi}\Psi$: because the masses of the electron and the electron neutrino are not the same, we cannot "rotate" one field into the other and have the same Lagrangian density. But if the masses of both particles were zero, this Lagrangian density would be

$$\mathcal{L} = i\overline{\Psi}\gamma^{\mu}\partial_{\mu}\Psi, \tag{23.6}$$

which *would* exhibit that symmetry.

The universe does not seem to want to play our little game here, since the electron and neutrino clearly do not have zero mass (though the neutrino's mass is pretty small). But let us put the mass problem on the shelf for a moment and explore *what if* the electron and neutrino had zero mass.

"Rotating" the fields in the doublet given in Equation 23.3 while preserving the value of the product $\overline{\Psi}\Psi$ is mathematically equivalent to rotating an ordinary spin quantum state $|\psi\rangle$ in space while preserving the magnitude $\langle\psi|\psi\rangle$ of that state. In Chapter 9, we found that the operators that rotate a spin state around various coordinate axes are

$$R_x(\theta) = e^{-iJ_x\theta}, \qquad R_y(\theta) = e^{-iJ_y\theta}, \quad \text{and} \quad R_z(\theta) = e^{-iJ_z\theta}, \tag{23.7}$$

where $J_k = \frac{1}{2}\sigma_k$, σ_k are the Pauli matrices, and $k = x, y, z$. The matrices that rotate a spin by θ around a *general* axis $\vec{n}$ (where $\vec{n}$ is a unit vector) are

$$R_{\vec{n}}(\theta) = e^{-i\vec{n}\cdot\vec{J}\theta} = e^{-i\frac{1}{2}\vec{n}\cdot\vec{\sigma}\theta} = e^{-i\frac{1}{2}\vec{\sigma}\cdot\vec{\theta}}, \tag{23.8}$$

where $\vec{\theta} = [\,\theta_x, \theta_y, \theta_z\,] \equiv \vec{n}\theta = [\,n_x\theta, n_y\theta, n_z\theta\,]$. (Note that index positions mean nothing here, as these are not spatial components of four-vectors.) The set of all such matrices form a group called **SU(2)**: "U" because the matrices are unitary ($R^{\dagger}R = e^{+iX}e^{-iX} = I$, where X is the Hermitian matrix $\frac{1}{2}\vec{\sigma}\cdot\vec{\theta}$), "S" (for "special") because det $R = 1$, and "2" because the most basic representation of the group elements (the representation we are using here) are 2×2 matrices.

Now, the "rotations" of the field doublet Ψ we are talking about here are not rotations in three-dimensional physical space. Rather, we simply want to find *the most general* group of matrices M such that $\Psi' = M\Psi$ mix the component fields of Ψ without affecting the value of $\overline{\Psi}\Psi$. This group turns out to be equivalent to the SU(2) group (see Problem P23.1 for more discussion of why this is so).

So, the most general set of transformations of the field doublet Ψ that preserves the value of $\overline{\Psi}\Psi$ can be written

$$\Psi' = e^{-i\frac{1}{2}g_W\vec{\sigma}\cdot\vec{\theta}}\Psi \quad \text{and} \quad \overline{\Psi}' = \overline{\Psi}\,e^{+i\frac{1}{2}g_W\vec{\sigma}\cdot\vec{\theta}}, \tag{23.9}$$

where $\vec{\theta}$ is an arbitrary set of three parameters, g_W is a constant (analogous to charge q in QED) that will (eventually) describe how strongly the flavor doublet is coupled to the weak interaction fields, and $\vec{\sigma}$ are the Pauli matrices, the generators of the SU(2) transformation. Note that pulling out the common factor of g_W amounts to a redefinition of the parameters $\vec{\theta}$ compared to Equation 23.8, but since these parameters are arbitrary anyway (no longer referring to angles in actual space), this doesn't matter.

The partials of Ψ' with respect to the three parameters θ_k at $\vec{\theta} = 0$ are

$$\frac{\partial\Psi'}{\partial\theta_k} = \frac{\partial}{\partial\theta_k}(e^{-i\frac{1}{2}g_W\vec{\sigma}\cdot\vec{\theta}}\,\Psi) = -i\frac{1}{2}g_W\sigma_k\,e^{-i\frac{1}{2}g_W\vec{\sigma}\cdot 0}\Psi = -i\frac{1}{2}g_W\sigma_k\Psi. \tag{23.10}$$

The three conserved currents associated with this transformation are

$$J^{(k)\mu} = \frac{\partial\mathcal{L}}{\partial(\partial_{\mu}\Psi)}\frac{\partial\Psi}{\partial\theta_k} + \frac{\partial\mathcal{L}}{\partial(\partial_{\mu}\overline{\Psi})}\frac{\partial\overline{\Psi}}{\partial\theta_k}$$

$$= i\overline{\Psi}\gamma^{\mu}(-i\tfrac{1}{2}g_W\sigma_k)\Psi + 0 = \tfrac{1}{2}g_W\overline{\Psi}\gamma^{\mu}\sigma_k\Psi. \tag{23.11}$$

We can write the three currents more concretely (see Box 23.1) as

$$J^{(x)\mu} = \tfrac{1}{2}g_W(\overline{\nu}_e\gamma^\mu e + \overline{e}\gamma^\mu \nu_e), \tag{23.12a}$$

$$J^{(y)\mu} = -\tfrac{1}{2}ig_W(\overline{\nu}_e\gamma^\mu e - \overline{e}\gamma^\mu \nu_e), \tag{23.12b}$$

$$J^{(z)\mu} = \tfrac{1}{2}g_W(\overline{\nu}_e\gamma^\mu \nu_e - \overline{e}\gamma^\mu e). \tag{23.12c}$$

Note that I have put the x, y, z superscripts in parentheses to remind us that these are just indices for the three currents, not components of the field (that is what the μ is for). Indeed, most other authors use 1, 2, 3 for these indices. I am using x, y, z for several reasons: because (1) these names correspond with the subscripts we use for the $\sigma_k = (\sigma_x, \sigma_y, \sigma_z)$ generators, and (2) I am going to use symbols for operations involving these currents (and the fields related to these currents) that are analogous to ordinary vector dot products and cross products. Using the same index names makes this analogy a bit easier to translate.

Note that if we didn't have the neutrino term and the minus sign, the $J^{(z)}$ current would look like the current associated with electromagnetism. This has implications that we will explore in Chapter 25.

23.3 Making the SU(2) Symmetry Actively Local

We now use the method we developed in Chapter 19 to make this transformation local. We assume that the parameters $\vec{\theta}(x)$ depend on spacetime position x. This will introduce new derivative terms in the Lagrangian density, but we introduce three new fields $\vec{W}_\mu$ that cancel the new derivative terms and maintain the SU(2) invariance. As with electromagnetism, the quickest way to find the new terms in the Lagrangian density associated with the $\vec{W}_\mu$ fields is to define a "covariant" derivative that (by analogy to the electromagnetic case) is

$$D_\mu = \partial_\mu + i\tfrac{1}{2}g_W\vec{\sigma} \cdot \vec{W}_\mu \tag{23.13}$$

and evaluate the old Lagrangian density with $\partial_\mu \rightarrow D_\mu$:

$$\mathcal{L} = i\overline{\Psi}\gamma^\mu D_\mu\Psi = i\overline{\Psi}\gamma^\mu\partial_\mu\Psi + i\overline{\Psi}\gamma^\mu(i\tfrac{1}{2}g_W\vec{\sigma}\cdot\vec{W}_\mu)\Psi$$

$$= i\overline{\Psi}\gamma^\mu\partial_\mu\Psi - \vec{J}^\mu \cdot \vec{W}_\mu, \tag{23.14}$$

where $\vec{J}^\mu$ are the three currents in Equations 23.12.

Now, to make this work in the case of electromagnetism, we had to transform the A^μ field as well during the transformation: $A'_\mu = A_\mu + \partial_\mu\theta$. By analogy, we would expect that our new fields in this case would transform as $\vec{W}'_\mu = \vec{W}_\mu + \partial_\mu\vec{\theta}$. However, there is a bit of nasty business in this case (having to do with the Pauli matrices failing to commute) that requires an extra term: for an infinitesimal set of transformation parameters $\delta\vec{\theta}$,

$$\vec{W}'_\mu = \vec{W}_\mu + \partial_\mu\delta\vec{\theta} - g_W\vec{W}_\mu \times \delta\vec{\theta}. \tag{23.15}$$

In Box 23.2, you will show that this gauge transformation does make the Lagrangian density locally SU(2)-invariant, but only if we have the new term.

As in the electromagnetic case, we must add a term representing the Lagrangian density for the free field to make the fields "active." We might guess from the electromagnetic analogy that the additional term should be

$$\mathcal{L}_W = -\tfrac{1}{4}\sum_k F^{(k)}_{\mu\nu}F^{(k)\mu\nu} = -\tfrac{1}{4}\vec{F}_{\mu\nu} \cdot \vec{F}^{\mu\nu}, \tag{23.16}$$

and this is correct, but because of the more complicated gauge transformation, the definition of the $\vec{F}_{\mu\nu}$ components is a bit more complicated as well:

$$\vec{F}_{\mu\nu} = \partial_\mu\vec{W}_\nu - \partial_\nu\vec{W}_\mu - g_W\vec{W}_\mu \times \vec{W}_\nu. \tag{23.17}$$

In Box 23.3, you will show that this is appropriately gauge-invariant. (Note that this new tensor is still antisymmetric with respect to the μ and ν indices.) This all looks fine until one realizes that this Lagrangian density has terms involving third and fourth powers of the fields $\vec{W}_\mu$, which means that the Euler-Lagrange equations will be nonlinear (the fields interact with themselves even in the absence of sources). This will make it challenging to even talk about what the "free field" solutions look like. We add this to the list of problems that we are putting on the shelf for the moment.

All this extra complexity arises because SU(2) is a **non-Abelian** group, meaning that the matrices $e^{-\frac{1}{2}ig_W\vec{\sigma}\cdot\vec{\theta}}$ do not commute with each other (in contrast to the **Abelian** group U(1), whose "matrices" $e^{-iq\theta}$ *do* commute, making electromagnetism a simpler gauge theory). Physicists call the theory we have been developing so far a **Yang-Mills** theory, after the two physicists who first explored gauge theories for non-Abelian groups SU(n) in 1954. (Figure 23.2 shows a picture of Yang taken in 1957.)

However, we do finally have a complete, locally SU(2)-invariant Lagrangian density for the e/ν_e lepton flavor (or really any flavor pair):

$$\mathcal{L} = i\overline{\Psi}\gamma^\mu\partial_\mu\Psi - \vec{J}^\mu\cdot\vec{W}_\mu - \tfrac{1}{4}\vec{F}_{\mu\nu}\cdot\vec{F}^{\mu\nu}. \tag{23.18}$$

The three new fields $\vec{W}_\mu$ will (when we quantize the fields) correspond to three different types of photon-like spin-1 particles, which we call **vector bosons**. Since we have no mass terms for these fields (which would look like $M_W^2\,\vec{W}^\mu\cdot\vec{W}_\mu$ and which are forbidden because we cannot make them gauge-invariant), these vector bosons should be massless. This is in direct contradiction to the experimental evidence that these bosons in fact have very large masses. This is yet another problem to shelve temporarily! (Indeed, this problem led physicists to reject the Yang-Mills theory until the concept of symmetry breaking was developed in the 1960s. We will explore that concept in Chapter 26.)

FIGURE 23.2   This is a photograph of Yang Chen-Ning, one of the creators of Yang-Mills theory.

23.4 Rearranging the Currents and Fields

The way we defined the currents in Equation 23.12 was mathematically natural, but we can redefine these currents in a way that is easier to interpret physically. Recall that in ordinary quantum mechanics, we defined angular momentum raising and lowering operators $J_\pm = J_x \pm iJ_y$. Because the SU(2) symmetry obeyed here is analogous to the SU(2) symmetry that describes spin states, the same kind of trick is useful here. Define

$$J^{+\mu} = J^{(x)\mu} + iJ^{(y)\mu} \quad \text{and} \quad J^{-\mu} = J^{(x)\mu} - iJ^{(y)\mu}. \tag{23.19}$$

You can show (see Box 23.4) that these new currents are simply

$$J^{+\mu} = g_W\overline{\nu}_e\gamma^\mu e \quad \text{and} \quad J^{-\mu} = g_W\overline{e}\gamma^\mu \nu_e. \tag{23.20}$$

We also conventionally adjust the definitions of the fields $\vec{W}_\mu$ to connect to these new currents. If we define

$$W_\mu^\pm = \sqrt{\tfrac{1}{2}}(W_\mu^{(x)} \mp iW_\mu^{(y)}), \tag{23.21}$$

then (as you will show in Box 23.5) the interaction term in the Lagrangian density connects W^+ to J^+ and W^- to J^-:

$$\mathcal{L}_{\text{int}} = -\sqrt{\tfrac{1}{2}}(J^{+\mu}W_\mu^+ + J^{-\mu}W_\mu^-) - J^{(z)\,\mu}W_\mu^{(z)}. \tag{23.22}$$

The point of this adjustment is that now each of these interaction terms describes a vertex in a Feynman diagram that we can easily interpret. The $J^{+\mu}W_\mu^+$ term (with $J^{+\mu} = g_W\overline{\nu}_e\gamma^\mu e$) describes a vertex where an electron and the vector boson corresponding to the W_μ^+ field enter and a neutrino leaves. Since the electron has a negative charge and the neutrino has no charge, the boson corresponding to the W_μ^+ field has a positive charge. Similarly, the $J^{-\mu}W_\mu^-$ term (where $J^{-\mu} = g_W\overline{e}\gamma^\mu\nu_e$) describes a vertex where a neutrino meets a negatively charged W^- boson and an electron departs. This is an improvement in simplicity over our previous definitions of the currents and fields, where the interaction terms in the Lagrangian density represented superpositions of these two processes. The remaining term is fine as it stands: because $J^{(z)\mu} = g_W(\bar{\nu}_e\gamma^\mu\nu_e + \bar{e}\gamma^\mu e)$, the interaction term describes a pair of vertices where either a neutrino or electron enters and departs unchanged. The boson associated with the $W^{(z)}$ field therefore has a definite charge of zero.

Note that the $W^\pm_\mu = \sqrt{\frac{1}{2}}\,(W^{(x)}_\mu \mp i\,W^{(y)}_\mu)$ fields are adjoints of each other, so their corresponding bosons are antiparticles of each other. The $W^{(z)}$ field is real, so the corresponding particle is its own antiparticle.

Also, note that the vertex description in the last paragraph could be any permutation of that description with antiparticles entering or leaving substituted for particles leaving or entering, respectively. For example, the $g_W\bar{\nu}_e\gamma^\mu e W^+_\mu$ vertex could alternatively correspond to a W^+ going in and a neutrino and a positron going out (substituting the positron going out for an electron coming in). Figure 23.3 shows the possible permutations for the first vertex.

Physicists conventionally represent the $W^\pm$ and $W^{(z)}$ bosons with wavy lines, like the lines we used for photons in Chapter 20. This might seem confusing, but labels should tell you which is which. (The reason for the convention is that in the unified electroweak theory we will discuss in Chapter 25, the W and photon fields are actually related, so the wavy lines generally represent the electroweak gauge fields.) But to help you distinguish the weak bosons from the photon, I will use wavy lines with a short wavelength for the weak bosons and a significantly longer wavelength for photons.

When vector bosons are virtual, whether we are emitting a W^- or absorbing a W^+ is actually immaterial: the effect is the same. For this reason, many authors draw the

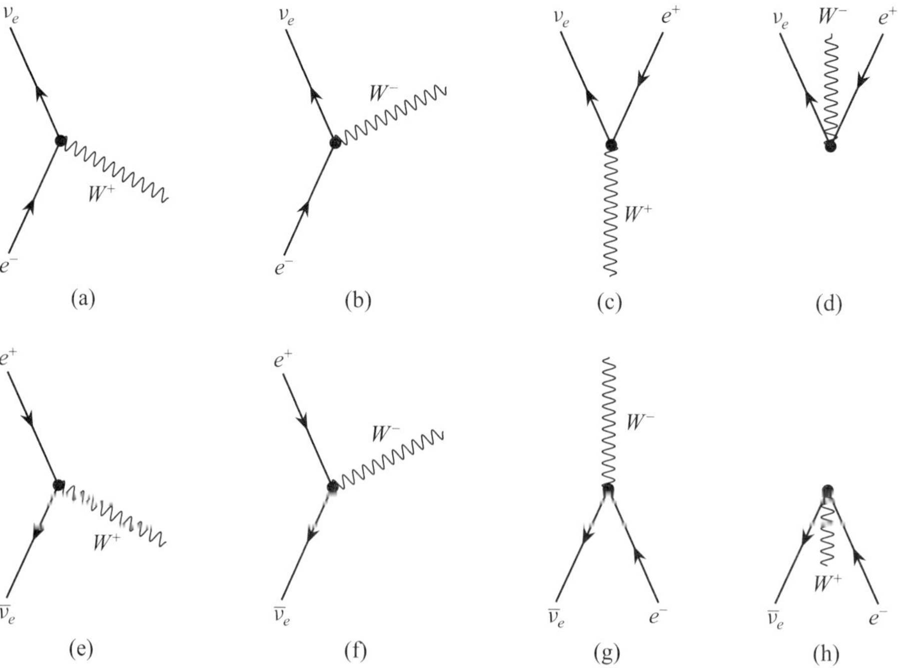

FIGURE 23.3   These panels display permutations of the Feynman-diagram vertex corresponding to the interaction term $J^{+\mu}W^+_\mu = g_W\bar{\nu}_e\gamma^\mu e W^+_\mu$. The disconnected vertices in diagrams (d) and (h) are allowed by that term but cannot conserve four-momentum and thus are not actually allowed in Feynman diagrams. Also, note that $\bar{\nu}_e$ in this context stands for an antineutrino, not the adjoint of the neutrino field. Arrows on the lines are forward for fermions, backward for antifermions.

virtual particle worldline as horizontal and simply label it W, as shown in Figure 23.4. In this diagram, negative charge is flowing to the right. But whether we consider this as a virtual W^- going right or a W^+ going left is arbitrary: the physical result is the same.

Finally (returning to the issue of redefining the fields), note for the record that if we rewrite the free-field Lagrangian density in terms of the new fields, we find (see Box 23.6)

$$\mathcal{L}_W = -\tfrac{1}{4} F^{(z)\mu\nu} F^{(z)}_{\mu\nu} - \tfrac{1}{2} F^{+\mu\nu} F^-_{\mu\nu} \quad \text{where} \quad F^\pm_{\mu\nu} \equiv \sqrt{\tfrac{1}{2}}(F^{(x)}_{\mu\nu} \mp i F^{(y)}_{\mu\nu}). \tag{23.23}$$

However, because of the cross product in the definition of $\vec{F}_{\mu\nu}$, how we should write $F^\pm_{\mu\nu}$ in terms of the $W^\pm_\mu$ fields is tricky (see the box).

23.5 Closing Comments

In this chapter, we have shown how to describe a new interaction that emerges from SU(2) symmetry in the same way that electromagnetism emerges from U(1) symmetry. This interaction looks a lot like the weak interaction we observe experimentally because it allows us to explain processes that involve transmutation of particles within families, something we *do* observe in weak interactions.

However, the theory as it stands has problems. One is that it only makes sense if the interacting fermions and the W bosons are massless, something that we know is not true. We can sort of handle this by constructing the Feynman diagrams that the theory predicts and then substituting the known particle masses into our calculations, even though this is not justified (or even allowed) by the Lagrangian density. This kludge actually works surprisingly well.

But in some cases, the predictions are completely off the mark. The problem is that our theory in its current form is invariant under parity, but the real weak interaction violates parity conservation, in some cases very obviously. Handling this will be our focus in the next chapter.

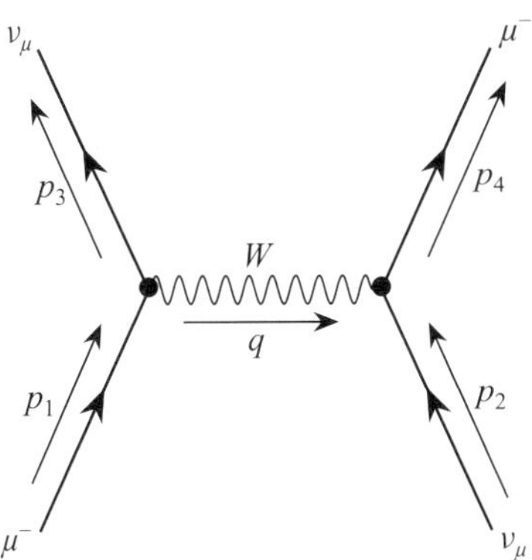

FIGURE 23.4 A Feynman diagram for muon–muon neutrino scattering. Note that the left vertex is the same as that in Figure 23.3a, except I have substituted muon family members for the electron family members. The right vertex represents a $J^{-\mu} W^-_\mu$ interaction term.

BOX 23.1   Explicit Weak Current Equations

In this box, we want to find explicit formulas for the three conserved currents given abstractly by

$$J^{(k)\mu} = \tfrac{1}{2} g_W \overline{\Psi} \gamma^\mu \sigma_k \Psi. \tag{23.24}$$

The notation we are using in this chapter is a bit tricky because the γ^μ and σ_k operators apply in different subspaces. The σ_k operates in the two-dimensional SU(2) subspace of the two-component Ψ and $\overline{\Psi}$ "vectors," each of whose components are themselves four-component Dirac spinor fields e and ν_e. The γ^μ matrix acts in that Dirac spinor subspace. (Since these matrices act in completely different subspaces, these operators commute, so we can treat them essentially as numbers, except in their own subspaces.)

Keeping these different subspaces in mind, consider the $i = x$ case. We have

$$J^{(x)\mu} = \tfrac{1}{2} g_W \overline{\Psi} \gamma^\mu \sigma_x \Psi = \tfrac{1}{2} g_W \begin{bmatrix} \overline{\nu}_e \gamma^\mu & \overline{e}\gamma^\mu \end{bmatrix} \begin{pmatrix} 0 & 1 \\ 1 & 0 \end{pmatrix} \begin{bmatrix} \nu_e \\ e \end{bmatrix}$$

$$= \tfrac{1}{2} g_W \begin{bmatrix} \overline{\nu}_e \gamma^\mu & \overline{e}\gamma^\mu \end{bmatrix} \begin{bmatrix} e \\ \nu_e \end{bmatrix} = \tfrac{1}{2} g_W (\overline{\nu}_e \gamma^\mu e + \overline{e}\gamma^\mu \nu_e). \tag{23.25}$$

Here I have explicitly displayed the SU(2)-subspace aspect of the initial abstract expression. Multiplying the final column vector Ψ by σ_x swaps the components of Ψ, and multiplying the row vector $\overline{\Psi}\gamma^\mu$ by that result yields the final expression. Part of the beauty of that expression is that we have removed the SU(2) superstructure of the Ψ and $\overline{\Psi}$ vectors, so we can interpret the result entirely on the level of Dirac spinors, exactly as we might have in the last chapter.

Exercise 23.1.1   Calculate analogous expressions for the $J^{(y)\mu}$ and $J^{(z)\mu}$ currents and compare to the results quoted in Equations 23.12.

BOX 23.2   Local Gauge Invariance for SU(2)

Our goal in this box is to show that the Lagrangian density $\mathcal{L} = i\overline{\Psi}\gamma^\mu D_\mu \Psi$ with $D_\mu \equiv \partial_\mu + i\frac{1}{2}g_W\vec{\sigma}\cdot\vec{W}_\mu$ is invariant with regard to a local SU(2) transformation $\Psi' = e^{-i\frac{1}{2}g_W\vec{\sigma}\cdot\vec{\theta}}\Psi$, so long as the gauge fields $\vec{W}_\mu$ transform as

$$\vec{W}_\mu' = W_\mu + \partial_\mu\delta\vec{\theta} - g_W\vec{W}_\mu \times \delta\vec{\theta} \tag{23.26}$$

to first order in an infinitesimal set of transformation parameters $\delta\vec{\theta}$.

The Lagrangian will be invariant if $D_\mu\Psi' = e^{-i\frac{1}{2}g_W\vec{\sigma}\cdot\vec{\theta}}D_\mu\Psi$ because then the exponential (which commutes with the γ^μ since they operate on different things) will cancel the exponential coming from the $\overline{\Psi}' = \overline{\Psi}\,e^{+i\frac{1}{2}g_W\vec{\sigma}\cdot\vec{\theta}}$. It is sufficient to demonstrate invariance for an infinitesimal rotation (so long as all *first*-order terms cancel) because we can then construct any finite transformation out of a sequence of infinitesimal transformations. To first order, then, we *want*

$$D_\mu'\Psi' = (I - i\tfrac{1}{2}g_W\vec{\sigma}\cdot\delta\vec{\theta})D_\mu\Psi = (I - i\tfrac{1}{2}g_W\vec{\sigma}\cdot\delta\vec{\theta})(\partial_\mu + i\tfrac{1}{2}g_W\vec{\sigma}\cdot\vec{W}_\mu)\Psi$$

$$= \partial_\mu\Psi + i\tfrac{1}{2}g_W\vec{\sigma}\cdot\vec{W}_\mu\Psi - i\tfrac{1}{2}g_W\vec{\sigma}\cdot\delta\vec{\theta}\,\partial_\mu\Psi + \tfrac{1}{4}g_W^2(\vec{\sigma}\cdot\delta\vec{\theta})(\vec{\sigma}\cdot\vec{W}_\mu)\Psi. \tag{23.27}$$

This must equal (to first order) the actual transformed derivative

$$D_\mu'\Psi' = (\partial_\mu + i\tfrac{1}{2}g_W\vec{\sigma}\cdot\vec{W}_\mu')(I - i\tfrac{1}{2}g_W\vec{\sigma}\cdot\delta\vec{\theta})\Psi$$

$$= (\partial_\mu + i\tfrac{1}{2}g_W\vec{\sigma}\cdot[\vec{W}_\mu + \partial_\mu\delta\vec{\theta} - g_W\vec{W}_\mu \times \delta\vec{\theta}])(I - i\tfrac{1}{2}g_W\vec{\sigma}\cdot\delta\vec{\theta})\Psi. \tag{23.28}$$

Exercise 23.2.1   Multiply out the terms in Equation 23.28, keeping only terms to first order in $\delta\vec{\theta}$, and show that the result is equal to Equation 23.27. (*Hint:* Use the identity $(\vec{a}\cdot\vec{\sigma})(\vec{b}\cdot\vec{\sigma}) = \vec{a}\cdot\vec{b}I + i(\vec{a}\times\vec{b})\cdot\vec{\sigma}$, which was derived in Problem 16.1.)

BOX 23.3   The Field Tensors for the Gauge Fields $\vec{W}_\mu$

Our goal in this box is to show that $\vec{F}^{\mu\nu} \cdot \vec{F}_{\mu\nu}$, where

$$\vec{F}_{\mu\nu} = \partial_\mu \vec{W}_\nu - \partial_\nu \vec{W}_\mu - g_W \vec{W}_\mu \times \vec{W}_\nu \tag{23.29}$$

is gauge invariant for the gauge transformation

$$\vec{W}'_\mu = \vec{W}_\mu + \partial_\mu \delta\vec{\theta} - g_W \vec{W}_\mu \times \delta\vec{\theta} \tag{23.30}$$

to first order in the infinitesimal transformation parameters $\delta\vec{\theta}$.

We begin by finding out what happens to $\vec{F}_{\mu\nu}$ during an infinitesimal gauge transformation. Substituting Equation 23.30 into Equation 23.29 yields

$$\vec{F}'_{\mu\nu} = \partial_\mu \vec{W}'_\nu - \partial_\nu \vec{W}'_\mu - g_W \vec{W}'_\mu \times \vec{W}'_\nu$$

$$= \partial_\mu \vec{W}_\nu + \partial_\mu \partial_\nu \delta\vec{\theta} - g_W \partial_\mu(\vec{W}_\nu \times \delta\vec{\theta}) - \partial_\nu \vec{W}_\mu - \partial_\nu \partial_\mu \delta\vec{\theta} + g_W \partial_\nu(\vec{W}_\mu \times \delta\vec{\theta})$$

$$- g_W(\vec{W}_\mu + \partial_\mu \delta\vec{\theta} - g_W \vec{W}_\mu \times \delta\vec{\theta}) \times (\vec{W}_\nu + \partial_\nu \delta\vec{\theta} - g_W \vec{W}_\nu \times \delta\vec{\theta}). \tag{23.31}$$

Exercise 23.3.1   Multiply out the items in this expression, throw away terms involving $\delta\theta^2$, and cancel some terms to show that

$$\vec{F}'_{\mu\nu} = \vec{F}_{\mu\nu} + g_W \delta\vec{\theta} \times (\partial_\mu \vec{W}_\nu - \partial_\nu \vec{W}_\mu)$$

$$+ g_W^2 \big[\, \vec{W}_\mu \times (\vec{W}_\nu \times \delta\vec{\theta}) + (\vec{W}_\mu \times \delta\vec{\theta}) \times \vec{W}_\nu \,\big]. \tag{23.32}$$

Exercise 23.3.2   Use the identity $\vec{A} \times (\vec{B} \times \vec{C}) + \vec{C} \times (\vec{A} \times \vec{B}) + \vec{B} \times (\vec{C} \times \vec{A}) = 0$ to show that

$$\vec{F}'_{\mu\nu} = \vec{F}_{\mu\nu} + g_W(\delta\vec{\theta} \times \vec{F}_{\mu\nu}). \tag{23.33}$$

BOX 23.3 The Field Tensors for the Gauge Fields $\vec{W}_\mu$ (cont.)

Exercise 23.3.3 Now that we know that

$$\vec{F}'_{\mu\nu} = \vec{F}_{\mu\nu} + g_W(\delta\vec{\theta} \times \vec{F}_{\mu\nu}), \tag{23.34}$$

argue that (to first order in the tiny gauge transformation parameter $\delta\vec{\theta}$)

$$\vec{F}'^{\mu\nu} \cdot \vec{F}'_{\mu\nu} = \vec{F}^{\mu\nu} \cdot \vec{F}_{\mu\nu}. \tag{23.35}$$

(*Hint:* Use the vector identity $\vec{A} \cdot (\vec{B} \times \vec{C}) = \vec{B} \cdot (\vec{C} \times \vec{A}) = \vec{C} \cdot (\vec{A} \times \vec{B})$.)

The $g_W \vec{W}_\mu \times \vec{W}_\nu$ term in Equation 23.29 means that (unlike in the photon case) adding the $\vec{F}^{\mu\nu} \cdot \vec{F}_{\mu\nu}$ term to the Lagrangian density does more than just provide a free-field term for the $\vec{W}$ fields. Note that

$$\vec{F}^{\mu\nu} \cdot \vec{F}_{\mu\nu} = \left(\partial^\mu \vec{W}^\nu - \partial^\nu \vec{W}^\mu - g_W \vec{W}^\mu \times \vec{W}^\nu\right) \cdot \left(\partial_\mu \vec{W}_\nu - \partial_\nu \vec{W}_\mu - g_W \vec{W}_\mu \times \vec{W}_\nu\right)$$

$$= \left(\partial^\mu \vec{W}^\nu - \partial^\nu \vec{W}^\mu\right) \cdot \left(\partial_\mu \vec{W}_\nu - \partial_\nu \vec{W}_\mu\right)$$

$$\quad - g_W\left(\partial^\mu \vec{W}^\nu - \partial^\nu \vec{W}^\mu\right) \cdot \vec{W}_\mu \times \vec{W}_\nu - g_W \vec{W}^\mu \times \vec{W}^\nu \cdot \left(\partial_\mu \vec{W}_\nu - \partial_\nu \vec{W}_\mu\right)$$

$$\quad + g_W^2\left(\vec{W}^\mu \times \vec{W}^\nu\right) \cdot \left(\vec{W}_\mu \times \vec{W}_\nu\right). \tag{23.36}$$

The $\left(\partial^\mu \vec{W}^\nu - \partial^\nu \vec{W}^\mu\right) \cdot \left(\partial_\mu \vec{W}_\nu - \partial_\nu \vec{W}_\mu\right)$ term is the free-field part: the solutions to the Euler-Lagrange equation for this term will be $\vec{\epsilon}^\mu e^{\pm ip\cdot x}$ (where $\vec{\epsilon}^\mu$ is a constant); that is, the standard plane waves associated with free fields. But the remaining terms constitute *self-interaction terms* between the $\vec{W}$ fields. The two terms on the second-to-last line couple three $\vec{W}$ fields to each other and are represented in a Feynman diagram by a vertex connecting three W lines (see Figure 23.5a). The term on the final line couples *four* $\vec{W}$ fields to each other and are represented by a vertex connecting four W lines (see Figure 23.5b). We will see the full Feynman rules for these vertices in Chapter 27.

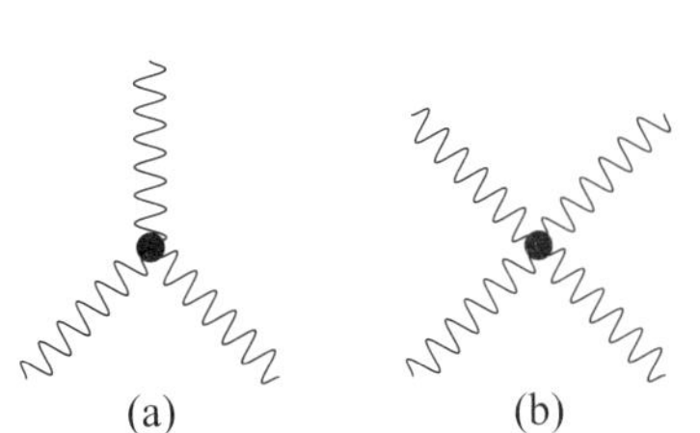

FIGURE 23.5 These diagrams show the self-interaction W vertices introduced by the non-free-field terms in the $\vec{F}^{\mu\nu} \cdot \vec{F}_{\mu\nu}$ Lagrangian density.

BOX 23.4 Charged Currents

Our goal in this box, given the currents

$$J^{(x)\mu} = \tfrac{1}{2} g_W (\overline{v}_e \gamma^\mu e + \overline{e} \gamma^\mu v_e),$$
(23.37a)

$$J^{(y)\mu} = -i \tfrac{1}{2} g_W (\overline{v}_e \gamma^\mu e - \overline{e} \gamma^\mu v_e),$$
(23.37b)

$$J^{(z)\mu} = \tfrac{1}{2} g_W (\overline{v}_e \gamma^\mu v_e - \overline{e} \gamma^\mu e),$$
(23.37c)

is to calculate $J^{\pm\mu} = J^{(x)\mu} \pm i J^{(y)\mu}$.

Exercise 23.4.1 Do this, and compare your results to Equations 23.20. (*Hint:* This should only take a few lines. You will not need all the space provided.)

BOX 23.5 The Charged Vector Bosons

In this box, we will show that

$$\mathcal{L}_{\text{int}} = -\sqrt{\tfrac{1}{2}}(J^{+\mu}W_\mu^+ + J^{-\mu}W_\mu^-) - J^{(z)\mu}W_\mu^{(z)}$$

$$= -J^{(x)\mu}W_\mu^{(x)} - J^{(y)\mu}W_\mu^{(y)} - J^{(z)\mu}W_\mu^{(z)}, \qquad (23.38)$$

given that $J^{\pm\mu} = J^{(x)\mu} \pm i J^{(y)\mu}$ and $W_\mu^{\pm} = \sqrt{\tfrac{1}{2}}(W_\mu^{(x)} \mp i W_\mu^{(y)})$.

Exercise 23.5.1   Show this. (*Hint:* This should only take three or four lines, particularly if one goes in the direction suggested by the equation above.)

BOX 23.6 The Field Lagrangian Density

Our goal in this box is to show that the field Lagrangian density becomes

$$\mathcal{L}_W = -\tfrac{1}{4} F^{(z)\mu\nu} F^{(z)}_{\mu\nu} - \tfrac{1}{2} F^{+\mu\nu} F^-_{\mu\nu} \quad \text{where} \quad F^{\pm}_{\mu\nu} \equiv \sqrt{\tfrac{1}{2}}(F^{(x)}_{\mu\nu} \mp i F^{(y)}_{\mu\nu}), \quad (23.39)$$

and to learn how to write these quantities in terms of the $W^{\pm}_{\mu}$ fields. Now

$$F^{+\mu\nu} F^-_{\mu\nu} = \tfrac{1}{2}(F^{(x)\mu\nu} - i F^{(y)\mu\nu})(F^{(x)}_{\mu\nu} + i F^{(y)}_{\mu\nu})$$

$$= \tfrac{1}{2}(F^{(x)\mu\nu} F^{(x)}_{\mu\nu} + i F^{(x)\mu\nu} F^{(y)}_{\mu\nu} - i F^{(y)\mu\nu} F^{(x)}_{\mu\nu} + F^{(y)\mu\nu} F^{(y)}_{\mu\nu}). \quad (23.40)$$

This will be equal to the $F^{(x)\mu\nu} F^{(x)}_{\mu\nu} + F^{(y)\mu\nu} F^{(y)}_{\mu\nu}$ part of the original Lagrangian density if we multiply by 2 and if the cross terms cancel. But they do indeed cancel, because the fields are not (yet) operators (so the order is not relevant) and we can raise and lower indices to show that $F^{(x)\mu\nu} F^{(y)}_{\mu\nu} = F^{(x)}_{\mu\nu} F^{(y)\mu\nu}$.

Now, how do we find how $F^{\pm}_{\mu\nu}$ and $F^{(z)}_{\mu\nu}$ depend on $W^{\pm}_{\mu}$? The definitions

$$W^{\pm}_{\nu} \equiv \sqrt{\tfrac{1}{2}}(W^{(x)}_{\mu} \mp i W^{(y)}_{\mu}), \qquad F^{\pm}_{\mu\nu} \equiv \sqrt{\tfrac{1}{2}}(F^{(x)}_{\mu\nu} \mp i F^{(y)}_{\mu\nu}), \quad \text{and}$$

$$\vec{F}_{\mu\nu} \equiv \partial_\mu \vec{W}_\nu - \partial_\nu \vec{W}_\mu - g_W \vec{W}_\mu \times \vec{W}_\nu, \quad (23.41)$$

along with the component definition of the cross product, imply that

$$F^+_{\mu\nu} = \sqrt{\tfrac{1}{2}}\big[\partial_\mu W^{(x)}_\nu - \partial_\nu W^{(x)}_\mu - g_W(W^{(y)}_\mu W^{(z)}_\nu - W^{(z)}_\mu W^{(y)}_\nu)\big]$$

$$- i\sqrt{\tfrac{1}{2}}\big[\partial_\mu W^{(y)}_\nu + i\partial_\nu W^{(y)}_\mu + i g_W(W^{(z)}_\mu W^{(x)}_\nu - W^{(x)}_\mu W^{(z)}_\nu)\big]$$

$$= \sqrt{\tfrac{1}{2}}\partial_\mu(W^{(x)}_\nu - i W^{(y)}_\nu) - \sqrt{\tfrac{1}{2}}\partial_\nu(W^{(x)}_\mu - i W^{(y)}_\mu)$$

$$- i g_W \sqrt{\tfrac{1}{2}}(-i W^{(y)}_\mu W^{(z)}_\nu + i W^{(z)}_\mu W^{(y)}_\nu - W^{(z)}_\mu W^{(x)}_\nu + W^{(x)}_\mu W^{(z)}_\nu)$$

$$= \partial_\mu W^+_\nu - \partial_\mu W^+_\nu - i g_W(W^+_\mu W^{(z)}_\nu - W^{(z)}_\mu W^+_\nu). \quad (23.42)$$

Exercise 23.6.1 Similarly, show that

$$F^-_{\mu\nu} = \partial_\mu W^-_\nu - \partial_\nu W^-_\mu - i g_W(W^{(z)}_\mu W^-_\nu - W^-_\mu W^{(z)}_\nu). \quad (23.43)$$

BOX 23.6 The Field Lagrangian Density (cont.)

Exercise 23.6.2 Adding and subtracting $\sqrt{2}W_\mu^\pm = W_\mu^{(x)} \mp i W_\mu^{(y)}$ yields $W_\mu^{(x)} = \sqrt{\frac{1}{2}}\,(W_\mu^+ + W_\mu^-)$ and $W_\mu^{(y)} = i\sqrt{\frac{1}{2}}\,(W_\mu^+ - W_\mu^-)$. Use this to show that

$$F_{\mu\nu}^{(z)} = \partial_\mu W_\nu^{(z)} - \partial_\nu W_\mu^{(z)} - g_W(W_\mu^- W_\nu^+ - W_\mu^+ W_\nu^-). \qquad (23.44)$$

Homework Problems

P23.1 In this problem, we will explore the SU(2) group. Suppose that Ψ is a field doublet and $\overline{\Psi}$ is its adjoint, as illustrated in Equation 23.3. Let $\Psi' = M\Psi$ be a transformation of that field, where M is a 2×2 matrix. The adjoint of that transformation is $\overline{\Psi}' = \overline{\Psi}M^\dagger$. We seek the most general group of transformation matrices M such that $\overline{\Psi}'\Psi' = \overline{\Psi}\Psi$.

a. Prove that the M matrices must be unitary.

b. Show that because the M matrices are unitary, their determinants must be such that $|\det M| = 1$. (*Hints:* First show that $\det M^\dagger = (\det M)^*$. You can do this by explicitly calculating the determinant of a 2×2 matrix with arbitrary components a, b, c, d. Then use the identity $\det(AB) = (\det A)(\det B) = (\det B)(\det A)$.)

c. Suppose that the determinant of a particular matrix M is $e^{i\alpha}$, where α is a real number. Define a new matrix $M_{\text{new}} = (e^{-i\alpha/2}I)M$. The only difference between $M\Psi$ and $M_{\text{new}}\Psi$ is an overall phase, which is physically irrelevant. Show that $\det M_{\text{new}} = 1$. (*Hint:* The identity from part (c) can help. The proof needs only a couple of lines.)

d. So we can choose the physically meaningful subset of unitary M matrices to have $\det M = 1$. Show that this subset is closed under multiplication, so it is also a group.

e. This group of matrices is therefore SU(2) by definition. Now, we will argue that defining an arbitrary member of this group requires three freely chosen real parameters. A completely arbitrary 2×2 matrix is determined by eight real numbers (the real and imaginary parts of its four components or, equivalently, the magnitude and phase of those components). But since the group members are unitary,

$$M^\dagger M = \begin{pmatrix} a^* & c^* \\ b^* & d^* \end{pmatrix}\begin{pmatrix} a & b \\ c & d \end{pmatrix} \qquad (23.45)$$

$$= \begin{pmatrix} |a|^2 + |c^2| & a^*b + c^*d \\ ab^* + cd^* & |b|^2 + |d|^2 \end{pmatrix} = \begin{pmatrix} 1 & 0 \\ 0 & 1 \end{pmatrix},$$

where a, b, c, d are arbitrary. We also require that $\det M = ad - bc = 1$. These equations provide five equations in eight unknowns, so we can (in principle) solve for five of the eight numbers in terms of the other three. Therefore, this group is specified by three free parameters.

But wait! The two equations $a^*b + c^*d = 0$ and $ab^* + cd^* = 0$ are just complex conjugates of each other, so they are not independent. However, show that *either one* provides *two* equations of constraint on our eight real numbers, while the other equations only imply one equation of constraint each. (*Hint:* Remember the result of part (b) when considering the det $M = 1$ equation.)

f. So, how do we know that this group is given by $M \equiv e^{-i\frac{1}{2}g_W\vec{\sigma}\cdot\vec{\theta}}$? Well, this group of 2×2 matrices is described by three free parameters $\vec{\theta}$ and is unitary, so it will be SU(2) so long as it is "special." Show that $\det e^{-i\frac{1}{2}g_W\vec{\sigma}\cdot\vec{\theta}} = 1$. (*Hint:* Use the identity $\det(e^A) = e^{\text{Tr }A}$, which is valid for any complex square matrix A.)

P23.2 Prove that $A^{\mu\nu}B_{\mu\nu} = A_{\mu\nu}B^{\mu\nu}$ (a result that is useful in a number of calculations in this chapter).

P23.3 In this problem, we will find out how to write SU(2) operators as 2×2 matrices. Define $\vec{n} = \vec{\theta}/\theta$, where $\theta \equiv |\vec{\theta}|$. This means that $|\vec{n}| = 1$.

a. Write $\vec{n} \cdot \vec{\sigma}$ as a 2×2 matrix in terms of n^x, n^y, and n^z.

b. Use the above to show that $(\vec{n} \cdot \vec{\sigma})^2 = I$.

c. Let $M = e^{-i\frac{1}{2}g_W\vec{\sigma}\cdot\vec{\theta}}$. Use the above to show that

$$M = \cos(\tfrac{1}{2}g_W\theta)I - i\,\sin(\tfrac{1}{2}g_W\theta)\vec{n} \cdot \vec{\sigma}. \qquad (23.46)$$

d. Write M as a 2×2 matrix in terms of $\frac{1}{2}g_W\theta$, n^x, n^y, and n^z.

P23.4 In this problem, we will explore commutators of covariant derivatives.

a. In the context of U(1) gauge theory (Chapter 19), show that if $D_\mu \equiv \partial_\mu + iqA_\mu$,

$$[D_\mu, D_\nu]\psi = iq(\partial_\mu A_\nu - \partial_\nu A_\mu)\psi = iqF_{\mu\nu}\psi, \qquad (23.47)$$

b. The analogous expression for SU(2) gauge theory would be

$$[D_\mu, D_\nu]\Psi = i\tfrac{1}{2}g_W\vec{\sigma} \cdot \vec{F}_{\mu\nu}\Psi, \qquad (23.48)$$

because $\frac{1}{2}g_W\vec{\sigma}$ performs the same function in the SU(2) gauge transformation $e^{-i\frac{1}{2}g_W\vec{\sigma}\cdot\vec{\theta}}$ as iq does in the U(1) gauge transformation $e^{-iq\theta}$. But is this analogy valid? Show that it is by calculating $[D_\mu, D_\nu]\Psi$ in the SU(2) case and comparing what you get with the definition of $\vec{F}_{\mu\nu}$.

P23.5 Construct the six physically possible vertices (analogous to those shown in Figure 23.3a/b/c/e/f/g) for the $J^{-\mu}W^-_\mu$ interaction term, except use members of the muon family in your diagrams instead of members of the electron family.

P23.6 In our theory, an electron (if it does not remain the same) is converted at a vertex to an electron neutrino or vice versa. We can therefore define a quantity called **lepton number**, where the electron and the electron neutrino have lepton number +1 and the positron and the electron-antineutrino have lepton number −1. Since this number is conserved at any vertex we can construct for our theory (see Figure 23.3), it will be conserved in any weak interaction. It turns out that the complete theory of the weak interaction strictly conserves lepton numbers for each lepton flavor separately. With this in mind, predict the missing particle or particles in each of the following weak interactions. (Also, consider charge conservation and energy conservation.) Note that one really does not need to know the quark composition of the neutron n, proton p, or pion π^- to determine the missing particle.

a. $? + n \to p + e^-$

b. $\mu^- \to e^- + ?$

c. $e^+ + e^- \to \nu_e + ?$

d. $\pi^+ \to \mu^+ + ?$

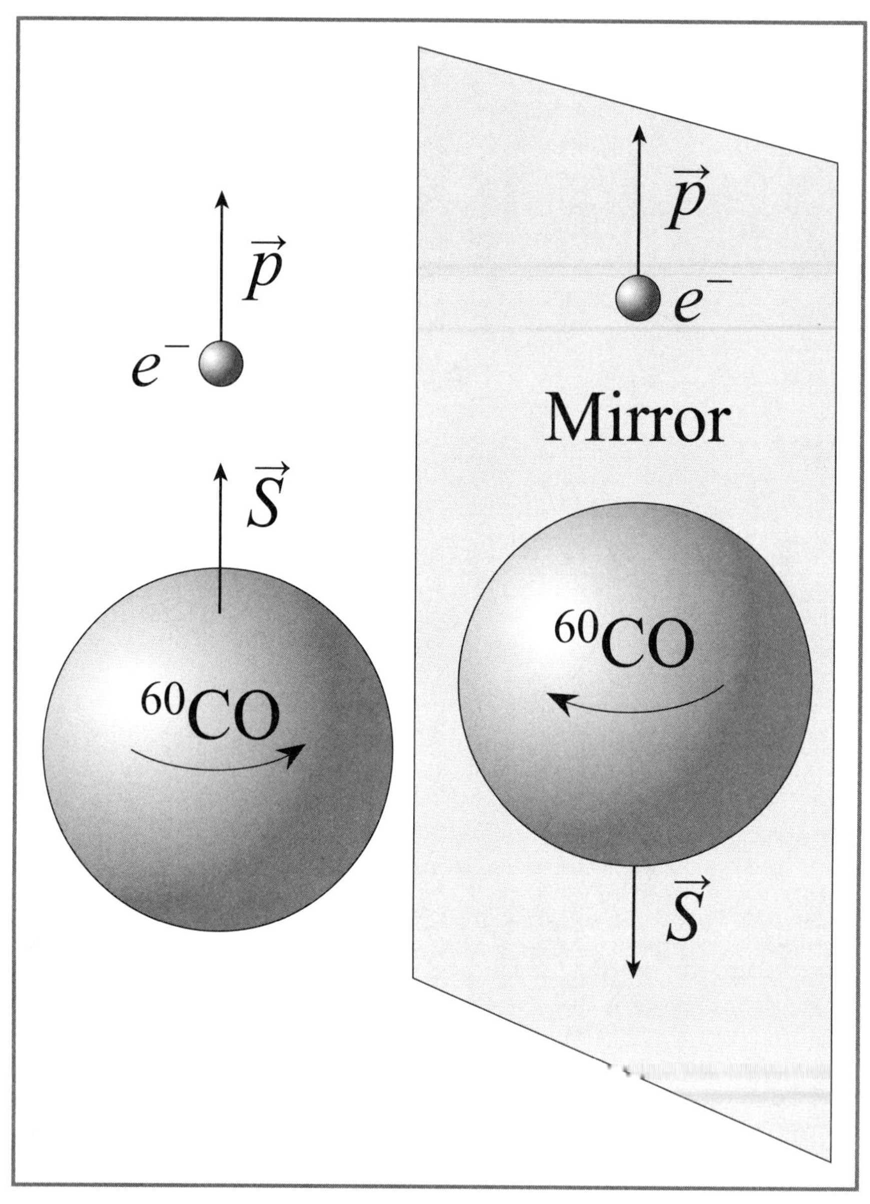

$\vec{p}$
e^-
$\vec{S}$
^{60}CO
$\vec{p}$
e^-
Mirror
^{60}CO
$\vec{S}$

24. THE WEAK INTERACTION'S LEFTIST BIAS

In the last chapter, we developed a Yang-Mills theory of the weak interaction based on making SU(2) a local gauge symmetry. However, this theory is not yet consistent with experimental evidence that the weak interaction is not symmetric with regard to parity. In this chapter, we consider evidence supporting reality's well-known leftist bias (as the joke goes), see how we can modify weak-interaction theory to acknowledge this bias, then do some calculations to see how well the modified theory describes our real universe.

24.1 Feynman Rules for Our Trial Theory

The weak interaction as we have described it so far is much like the electromagnetic interaction, except that we have three photon-like vector bosons W^+, W^-, and $W^{(z)}$, the first two of which are charged. While SU(2) symmetry requires massless particles, we fudge things by adding mass terms to the basic theory's equations that reflect the known particle masses (even though that violates the symmetry). For the moment, we will only consider interactions involving the $W^\pm$ bosons, which have observed masses of $M_W \approx 80.4$ GeV. The propagator for a virtual massive vector boson is given by

$$\text{vector boson propagator:} \quad \frac{-i(g_{\mu\nu} - q_\mu q_\nu/M^2)}{(q^2 - M^2)} \tag{24.1a}$$

(where q is the virtual boson's four-momentum and M is its mass), in place of the photon propagator $-ig_{\mu\nu}/q^2$. When the interaction energies are very small compared to M (as is the case in many particle decays), $q^2 \ll M^2$ and the above reduces to

$$\text{vector boson propagator:} \quad \approx \frac{ig_{\mu\nu}}{M^2} \quad \text{when } q^2 \ll M^2. \tag{24.1b}$$

The only other change is that, because the interaction Lagrangian density is

$$\mathcal{L}_{\text{int}} = -\frac{1}{\sqrt{2}}(J^{+\mu}W^+_\mu + J^{-\mu}W^-_\mu) - J^{(z)\mu}W^{(z)}_\mu \tag{24.2}$$

instead of the QED $-J^\mu A_\mu$, the weak vertex factor would be

$$\text{weak vertex factor } (W^\pm): \quad \frac{-ig_W\gamma^\mu}{\sqrt{2}} \tag{24.3}$$

for vertices involving the $W^\pm$ bosons instead of the corresponding QED factor $-iq_e\gamma^\mu = ie\gamma^\mu$ for vertices involving photons.

Our trial theory (like QED) is ambidextrous (parity-invariant). The goal in the next few sections is to use this theory to make predictions that we can check experimentally and compare to reality.

24.2 The Trial Theory Muon Decay Amplitude

One of the cleanest cases we can study with our theory is the weak decay of the muon: $\mu^- \to \nu_\mu + e^- + \overline{\nu}_e$. Unlike the positronium case considered in Chapter 22, this is a true decay, not a case of scattering in disguise. In our theory, we describe the decay using the

"

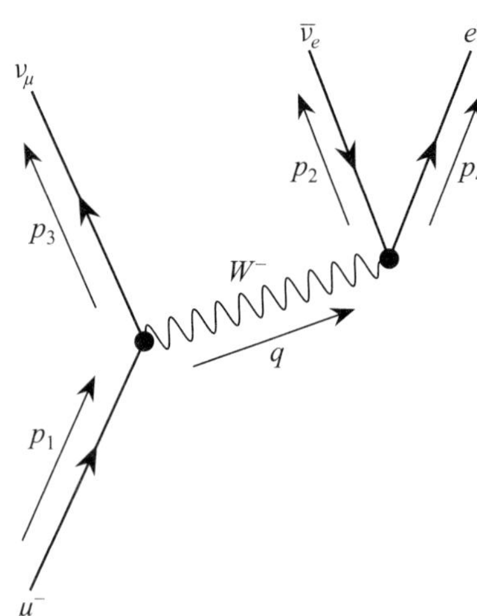

FIGURE 24.1 This figure shows the Feynman diagram for muon decay.

Feynman diagram shown in Figure 24.1. Because the electron is the only charged particle with lower rest energy than the muon, this is the only possible decay in our theory. This is also essentially the only way muons actually decay, which supports the theory.

Applying the Feynman rules to Figure 24.1, we find that

$$-i\mathcal{M} = [\,\bar{u}_3(-ig_W\sqrt{\tfrac{1}{2}}\gamma^\mu)u_1\,]\left(\frac{ig_{\mu\nu}}{M_W^2}\right)[\,\bar{u}_4(-ig_W\sqrt{\tfrac{1}{2}}\gamma^\nu)v_2\,],\tag{24.4}$$

where I am using the low-energy approximation for the propagator. Pulling out numerical factors and multiplying by i yields

$$\mathcal{M} = \frac{g_W^2}{2M_W^2}[\,\bar{u}_3\gamma^\mu u_1\,]\,g_{\mu\nu}\,[\,\bar{u}_4\gamma^\nu v_2\,].\tag{24.5}$$

Assuming the neutrinos are massless (they aren't quite, but their masses are more than 10^8 times smaller than the energies involved here), Casimir's trick tells us that

$$\langle\,|\mathcal{M}|^2\,\rangle = \frac{1}{8}\left(\frac{g_W}{M_W}\right)^4 \mathrm{Tr}[\,\gamma^\mu(\not{p}_1 + m_\mu)\gamma^\alpha\not{p}_3\,]\,g_{\mu\nu}g_{\alpha\beta}$$

$$\times\,\mathrm{Tr}[\,\gamma^\nu\not{p}_2\gamma^\beta(\not{p}_4 + m_e)\,],\tag{24.6}$$

where the extra factor of 2 comes from averaging over the two spin states of the initial muon. A spirited application of trace theorems (see Box 24.1) yields

$$\langle\,|\mathcal{M}|^2\,\rangle = 4\left(\frac{g_W}{M_W}\right)^4\big[\,(p_1\cdot p_2)(p_3\cdot p_4) + (p_1\cdot p_4)(p_2\cdot p_3)\,\big].\tag{24.7}$$

In the rest frame of the muon, this reduces to

$$\langle\,|\mathcal{M}|^2\,\rangle = 2\left(\frac{g_W}{M_W}\right)^4 m_\mu^2\,[\,E_2(m_\mu - 2E_2) + E_4(m_\mu - 2E_4)\,]\tag{24.8}$$

(see Box 24.2), where I have simplified the math by taking $m_e \to 0$ (which is actually also an excellent approximation here).

If this were a decay to two particles, the energies would be determined and we could calculate the cross section easily. But because the muon decays to three particles, we will have a range of possible energies and momenta, so we need to go back to Fermi's Golden Rule (see Chapter 10 or 14) to calculate the decay rate Γ from scratch. In this case, the golden rule tells us that

$$\Gamma = \frac{S}{2E_1}\int\left(\frac{d^3p_2}{(2\pi)^3 2E_2}\right)\left(\frac{d^3p_3}{(2\pi)^3 2E_3}\right)\left(\frac{d^3p_4}{(2\pi)^3 2E_4}\right)$$

$$\times\,(2\pi)^4\delta^4(p_1 - p_2 - p_3 - p_4)\langle\,|\mathcal{M}|^2\,\rangle.\tag{24.9}$$

In the muon rest frame, $E_1 = m_\mu$, and we have no identical outgoing particles, so $S = 1$. In order to calculate the cross section, we must do three momentum-space integrals. After some serious effort (see Box 24.3), we find that

$$d\Gamma = \frac{3m_\mu^2}{(4\pi)^3}\left(\frac{g_W}{M_W}\right)^4 E^2\left(1 - \frac{16E}{9m_\mu}\right)dE,\tag{24.10}$$

where $E \equiv E_4$ is the electron energy.

One might have expected the decay rate to go to zero when $E = \tfrac{1}{2}m_\mu$ since that is the maximum energy that the electron can have. At this limit, the two neutrinos both move opposite to the electron and carry energy E and momentum E in the opposite direction. If we had only one emitted neutrino, its momentum state would be uniquely determined at that limit and the probability *would* go to zero compared to other energies

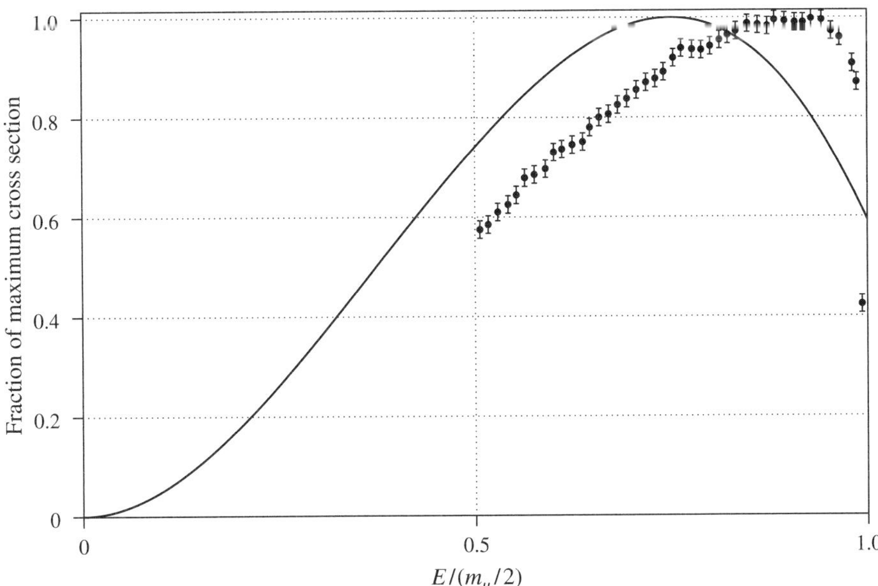

FIGURE 24.2   This graph shows the electron energy spectrum for muon decay. The curve is the spectrum predicted by our trial theory (Equation 24.10), though the height is arbitrary. The data are from Bardon et al., *Phys. Rev. Lett.* **14**, 449 (1965).

(since there are enormously more possible momentum states at other energies). But with two outgoing neutrinos, we still have many ways to *split* the total energy E between the neutrinos, meaning that $d\Gamma \neq 0$ even at the limit.

We can easily integrate this to find the total decay rate

$$\Gamma = \frac{3m_\mu^2}{(4\pi)^3} \left(\frac{g_W}{M_W}\right)^4 \left[\frac{E^3}{3} - \frac{4E^4}{9m_\mu}\right]_0^{\frac{1}{2}m_\mu} = \frac{3m_\mu^2}{(4\pi)^3} \left(\frac{g_W}{M_W}\right)^4 \left[\frac{m_\mu^3}{24} - \frac{m_\mu^3}{36}\right]$$

$$= \frac{3m_\mu^5}{72(4\pi)^3} \left(\frac{g_W}{M_W}\right)^4 = \frac{3m_\mu^5}{3(8\pi)^3} \left(\frac{g_W}{M_W}\right)^4 . \tag{24.11}$$

Unfortunately, we cannot compare an experimental decay rate to this formula unless we know g_W/M_W, which we can only determine by experiment. But we *can* look at the electron energy *spectrum* predicted by Equation 24.10.

The failure is spectacular. The data (shown in Figure 24.2) do not match the prediction at all. Something is seriously wrong with our trial theory.

24.3 Making the Weak Interaction Left-Handed

In 1956, Robert Marshak gave his graduate student E. C. G. Sudarshan the task of studying the weak interaction. At the time, physicists were considering all sorts of possible expressions for the weak interaction. Sudarshan carefully studied the experimental data. By 1957, Sudarshan had figured out an expression for the weak interaction that was consistent with all the data, but was not allowed to present it at the Rochester Conference on High-Energy Nuclear Physics that spring because he was only a graduate student. His advisor Marshak arranged for him to present his findings (which included a comprehensive analysis of the experimental data) to physicists at the RAND Corporation that summer, including Murray Gell-Mann. (The talk had to be held at a restaurant because Sudarshan, an Indian national, could not enter the RAND building.)

Marshak asked Sudarshan to write it up immediately, which Sudarshan did, but Marshak decided to wait to present the results at a conference in September. But just before

Marshak's talk, Feynman and Gell-Mann submitted a paper to *Physical Review* that asserted the expression Sudarshan had proposed (crediting Sudarshan only with "important discussions"). Feynman later publicly admitted in 1963 that the "theory was discovered by Sudarshan and Marshak [and] publicized by Feynman and Gell-Mann . . . ," but the fact that Feynman and Gell-Mann published first may have cost Sudarshan (and/or Marshak, given the subordination of graduate students at the time) a Nobel Prize. Sudarshan was also passed over for a Nobel Prize in 2005 that was awarded for work based on his contributions to theoretical quantum optics. (This history is primarily from an anonymous document posted on `http://quest.ph.utexas.edu/Reviews/VA/VA.pdf`.)

Sudarshan's insight can be expressed in the following way: *the weak interaction only operates between particles with left-handed chirality*. We saw in Chapter 17 (see Equation 17.27) that the operator

$$P_L = \tfrac{1}{2}(1 - \gamma^5) = \begin{pmatrix} I & 0 \\ 0 & 0 \end{pmatrix} \tag{24.12}$$

operates on any Dirac spinor to isolate the part with left-handed chirality. So we modify the interaction terms in the Lagrangian density to contain this operator:

$$\tfrac{1}{\sqrt{2}} J^{+\mu} W^+_\mu = \tfrac{1}{2\sqrt{2}} g_W \big[\, \bar{\nu}_e \gamma^\mu (1 - \gamma^5) e \, W^+_\mu \, \big], \tag{24.13a}$$

$$\tfrac{1}{\sqrt{2}} J^{-\mu} W^-_\mu = \tfrac{1}{2\sqrt{2}} g_W \big[\, \bar{e} \gamma^\mu (1 - \gamma^5) \nu_e \, W^-_\mu \, \big], \tag{24.13b}$$

$$J^{(z)\mu} W^{(z)}_\mu = \tfrac{1}{2} g_W \big[\, \bar{\nu}_e \gamma^\mu (1 - \gamma^5) \nu_e + \bar{e} \gamma^\mu (1 - \gamma^5) e \, \big] W^{(z)}_\mu. \tag{24.13c}$$

This reflects a local gauge symmetry that is not local SU(2), but rather SU(2)$_L$, a symmetry that does not apply to chirally right-handed particles (meaning that such particles cannot participate in weak interactions).

How does this change our calculations? The Feynman rule for the weak vertex factor now becomes

$$\text{weak vertex factor } (W^\pm): \quad \frac{-i g_W \gamma^\mu (1 - \gamma^5)}{2\sqrt{2}} \tag{24.14}$$

for interactions involving $W^\pm$, and the muon decay amplitude becomes

$$\mathcal{M} = \frac{g_W^2}{8 M_W^2} [\, \bar{u}_3 \gamma^\mu (1 - \gamma^5) u_1 \,] \, g_{\mu\nu} \, [\, \bar{u}_4 \gamma^\nu (1 - \gamma^5) v_2 \,]. \tag{24.15}$$

Though $\overline{\gamma^5} \equiv \gamma^t \gamma^{5\dagger} \gamma^t \neq \gamma^5$, it turns out that $\overline{\gamma^\mu (1 - \gamma^5)} = \gamma^\mu (1 - \gamma^5)$ (see Problem P24.7). With this in mind, we can apply Casimir's trick (and divide by 2 to average over initial spins) to get

$$\langle \, |\mathcal{M}|^2 \, \rangle = \frac{1}{128} \left(\frac{g_W}{M_W} \right)^4 \text{Tr}[\, \gamma^\mu (1 - \gamma^5)(\not{p}_1 + m_\mu) \gamma^\alpha (1 - \gamma^5) \not{p}_3 \,] \, g_{\mu\nu} g_{\alpha\beta}$$

$$\times \text{Tr}[\, \gamma^\nu (1 - \gamma^5) \not{p}_2 \gamma^\beta (1 - \gamma^5)(\not{p}_4 + m_e) \,]. \tag{24.16}$$

This looks horrible, but it actually leads to a simpler expression for the average squared amplitude (see Box 24.4):

$$\langle \, |\mathcal{M}|^2 \, \rangle = 2 \left(\frac{g_W}{M_W} \right)^4 (p_1 \cdot p_2)(p_3 \cdot p_4). \tag{24.17}$$

We can integrate this much as before (see Problem P24.1) to get

$$d\Gamma = \frac{m_\mu^2}{2(4\pi)^3} \left(\frac{g_W}{M_W} \right)^4 E^2 \left(1 - \frac{4E}{3m_\mu} \right) dE. \tag{24.18}$$

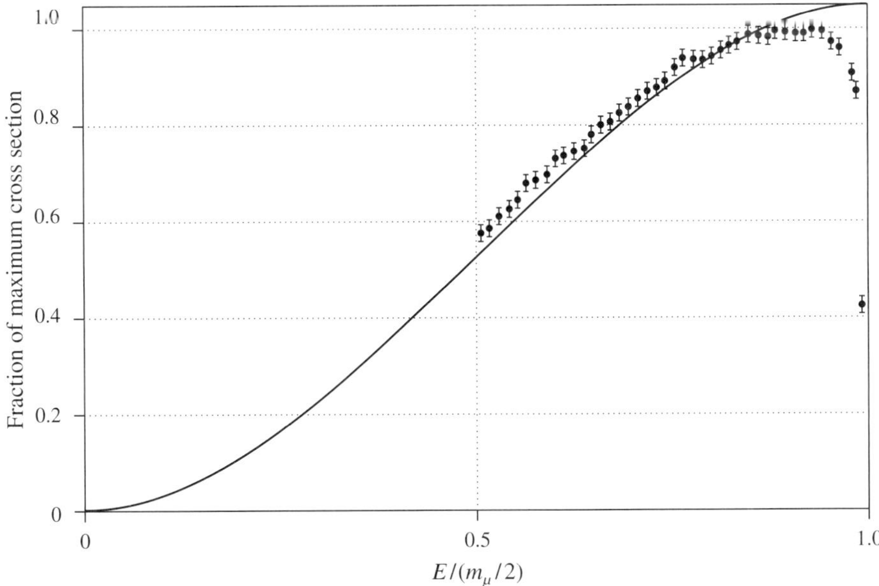

FIGURE 24.3   The experimental electron energy spectrum for muon decay (from Bardon et al., *Phys. Rev. Lett.* **14**, 449, 1965), where the solid line shows the spectrum predicted by Equation 24.18 with the magnitude adjusted to give a best fit.

This curve has a maximum (!) at the maximum possible electron energy, which is $\frac{1}{2}m_\mu = 52.83\,\text{MeV}$ (again, because there are many possible ways to split energy between the two outgoing neutrinos). The total decay rate is

$$\Gamma = \frac{m_\mu^2}{2(4\pi)^3}\left(\frac{g_W}{M_W}\right)^4 \left[\frac{E^3}{3} - \frac{E^4}{3m_\mu}\right]_0^{\frac{1}{2}m_\mu} = \frac{m_\mu^2}{2(4\pi)^3}\left(\frac{g_W}{M_W}\right)^4\left[\frac{m_\mu^3}{24} - \frac{m_\mu^3}{48}\right]$$

$$= \frac{m_\mu^5}{96(4\pi)^3}\left(\frac{g_W}{M_W}\right)^4 = \frac{m_\mu^5}{12(8\pi)^3}\left(\frac{g_W}{M_W}\right)^4. \tag{24.19}$$

As Figure 24.3 shows, the agreement between this spectrum and the experimental data is *much* better. The remaining discrepancy is mostly due to the fact that the emitted electrons sometimes also emit photons that carry away some of the electrons' energy before that energy is measured. This has the effect of lowering the probability that we will observe high-energy electrons and enhances the probability of observing lower-energy electrons compared to the idealized spectrum. (For a diagram whose theoretical curve incorporates these effects, see Bardon et al., *Phys. Rev. Lett.* **14**, 449 (1965).) More recent experiments (although lacking pretty graphs) even more strongly support Sudarshan's thesis. The weak interaction really is left-handed.

24.4 The Puzzle of Pion Decay

When I first began teaching particle physics, I recall being perplexed by the decay of the π^- meson. This meson consists of a down quark paired with an anti-up quark, and can decay via a weak interaction to either $\mu^- + \overline{\nu}_\mu$ or $e^- + \overline{\nu}_e$. Figure 24.4 shows a Feynman diagram for this decay process. (Like the decay of positronium, this decay is really a scattering process in disguise, where the colliding particles happen to be in a bound state before the interaction.)

We have seen in chapter 13 that, other things being equal, the decay rate depends on the momentum space available to its targets, meaning that decays in which a lot of energy is released should be much more probable. If we assume, quite reasonably, that

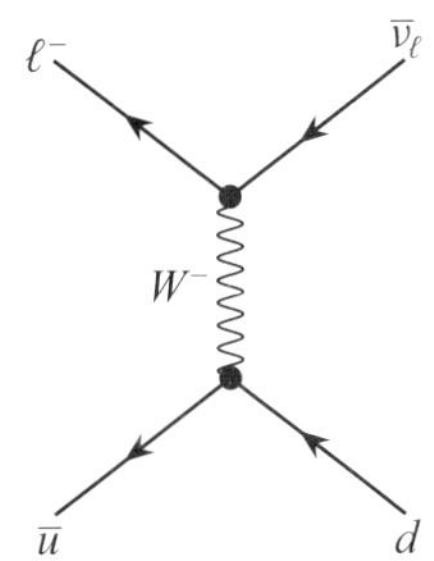

FIGURE 24.4   A Feynman diagram for pion decay as a weak scattering event. The ℓ^- stands for a charged lepton (either e^- or μ^-) and $\overline{\nu}_\ell$ for the lepton's corresponding antineutrino. In a weak interaction, the d and u quarks play the same roles as an electron and an electron neutrino, respectively.

the amplitude should formally be the same for both decay processes, this principle would lead us to expect that the pion should preferentially decay to $e^- + \bar{\nu}_e$. Indeed, because the ratio of the momentum magnitude $|\vec{p}_F|$ of the outgoing leptons is about 2.34 to 1 in favor of the decay to $e^- + \bar{\nu}_e$ (see Box 24.5), the decay to electrons should be decisively favored.

However, this is *not* what we observe experimentally: the pion decays to $\mu^- + \bar{\nu}_\mu$ virtually all the time. Indeed, the decay-rate ratio is experimentally

$$\frac{\Gamma(\pi^- \rightarrow e^- + \bar{\nu}_e)}{\Gamma(\pi^- \rightarrow \mu^- + \bar{\nu}_\mu)} = 1.23 \times 10^{-4}. \tag{24.20}$$

This drove me crazy when I first began to teach particle physics, as it seemed to contradict an important basic principle that I was trying to teach!

However, another principle is at play here. Pions have spin 0, which means the spins of the outgoing leptons must be opposed along any axis we choose. Choose the axis to be the axis along which the departing leptons move. Since these leptons must move in opposite directions and have opposite spins along that axis, they must have the *same helicity*: the decay must produce leptons whose helicities are either both left-handed or both right-handed.

Now, *massless* particles have *definite* helicity connected to their chirality. Again, assume neutrinos are massless. As discussed in Chapter 17, massless antiparticles with left-handed chirality have right-handed helicity, so if only chirally left-handed particles participate in weak interactions, the antineutrino emitted in this decay must have right-handed helicity, implying that the lepton must also have right-handed helicity (see Figure 24.5). But we have just said that the weak interaction does not couple to particles with right-handed chirality, so such a decay would be *impossible* if chirality were the same as helicity.

But helicity isn't quite the same thing as chirality for *massive* particles (as we discussed in Chapter 17), so these processes *can* go forward. However, the energy released during the decay to an electron ($m_\pi - m_e \approx 139$ MeV) is so large compared to the electron's mass ($m_e = 0.511$ MeV) that the electron is going to behave much more like a massless particle in this context than the muon. This means that our hypothetical left-handed weak process is much less likely to produce a right-handed electron than a right-handed muon. This explains why the decay to an electron is so strongly suppressed. (See Problem P24.4 for an actual calculation.)

However, unlike the muon decay considered earlier, this particular decay does not prove that the weak interaction is left-handed: our trial theory actually yields the same ratio. The suppression here in both cases is really due to trying to connect the (nearly) massless neutrino to a charged lepton with the *same* helicity. This would be impossible if the charged lepton was also massless, regardless of whether the neutrino and that lepton were both right-handed or both left-handed, so this interaction is strongly suppressed when the charged lepton's mass is small in either case. However, it does emphasize the important role that chirality plays in fermion interactions.

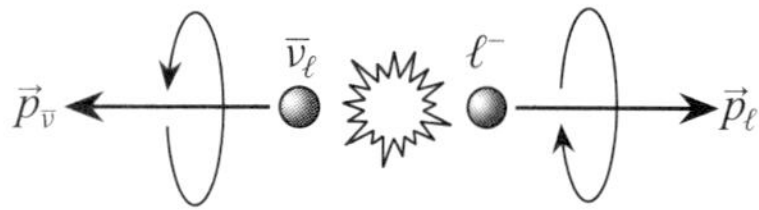

FIGURE 24.5   Since the original pion has spin 0, the spins of the outgoing particles must be opposite. But since their momenta are also opposite, their helicities must be the *same* (the diagram shows the case where both have right-handed helicities, as you can verify with that hand).

24.5 Charge Conjugation and Parity

How does the weak interaction's leftist bias affect behavior under the charge conjugation and parity operators? We saw in Chapter 17 that the charge conjugation and parity operators when operating on a Dirac spinor are

$$C\psi = -i\gamma^y \psi^* \quad \text{and} \quad P\psi = \gamma^t \psi. \tag{24.21}$$

The charge conjugation operator acting on $W_\mu^\pm$ is just the complex conjugate (no spinor to mess with), so

$$CW_\mu^\pm = \sqrt{\tfrac{1}{2}}(W_\mu^{(x)} \mp i W_\mu^{(y)})^* = \sqrt{\tfrac{1}{2}}(W_\mu^{(x)} \pm i W_\mu^{(y)}) = W_\mu^\mp, \tag{24.22}$$

as one might expect. The parity operator has the effect of flipping the signs of the spatial components of four-vectors

$$PW_\mu^\pm = (W_t^\pm, -W_x^\pm, -W_y^\pm, -W_z^\pm). \tag{24.23}$$

In Box 24.6, we will see that terms in the interaction Lagrangian density with the form $\overline{\psi}_2\gamma^\mu\psi_1$ and $\overline{\psi}_2\gamma^\mu\gamma^5\psi_1$ transform under charge conjugation as follows:

$$C(\overline{\psi}_2\gamma^\mu\psi_1) = \overline{\psi}_1\gamma^\mu\psi_2 \quad \text{and} \quad C(\overline{\psi}_2\gamma^\mu\gamma^5\psi_1) = -\overline{\psi}_1\gamma^\mu\gamma^5\psi_2, \tag{24.24}$$

implying that our interaction Lagrangian density will no longer be invariant under charge conjugation, because it contains a sum of terms of this type. Similarly, under parity, terms of this form will transform as

$$P(\overline{\psi}_2[\,\gamma^t, \gamma^x, \gamma^y, \gamma^z\,]\psi_1) = \overline{\psi}_2[\,\gamma^t, -\gamma^x, -\gamma^y, -\gamma^z\,]\psi_1, \tag{24.25a}$$

$$P(\overline{\psi}_2[\,\gamma^t, \gamma^x, \gamma^y, \gamma^z\,]\gamma^5\psi_1) = \overline{\psi}_2[\,-\gamma^t, \gamma^x, \gamma^y, \gamma^z\,]\gamma^5\psi_1, \tag{24.25b}$$

meaning that $\overline{\psi}_2\gamma^\mu\psi_1$ transforms like an ordinary four-vector, while $\overline{\psi}_2\gamma^\mu\gamma^5\psi_1$ transforms like an axial four-vector (its spatial components do not change under a parity transformation). This means that an interaction term like $\overline{\psi}_2\gamma^\mu\psi_1 W_\mu^\pm$ would be parity-invariant (the sign changes in the spatial terms of $\overline{\psi}_2\gamma^\mu\psi_1$ cancel those in the $W_\mu^\pm$), but adding terms involving γ^5 will spoil parity-invariance.

In 1957, Chien-Shiung Wu (see Figure 24.6) and her collaborators showed experimentally that the decay of cobalt 60 (which involves the weak interaction) is *not* invariant under a parity transformation (mirror reflection). Specifically, Wu found that electrons emitted by the cobalt 60 decay were preferentially emitted in the direction *opposite* to the nuclear spin, behavior that is not consistent with the process' mirror image, as shown in Figure 24.7.

However, you will see in Box 24.6 that the theory of the weak interaction presented in this chapter is invariant under CP.

24.6 Parting Comments

Our $SU(2)_L$ theory now fits experimental reality, at least for interactions involving leptons exchanging virtual $W^\pm$ bosons. But we are not yet done. The relevant symmetry for the Standard Model is actually $SU(2)_L \times U(1)$, which involves combining the electromagnetic and weak interactions into a unified theory describing an "electroweak interaction." We need to understand this more fully to correctly describe weak interactions that involve quarks and/or do not involve exchanging $W^\pm$ bosons. We begin that task in the next chapter.

FIGURE 24.6 Chien-Shiung Wu in her laboratory. (From Smithsonian Institution Archives.)

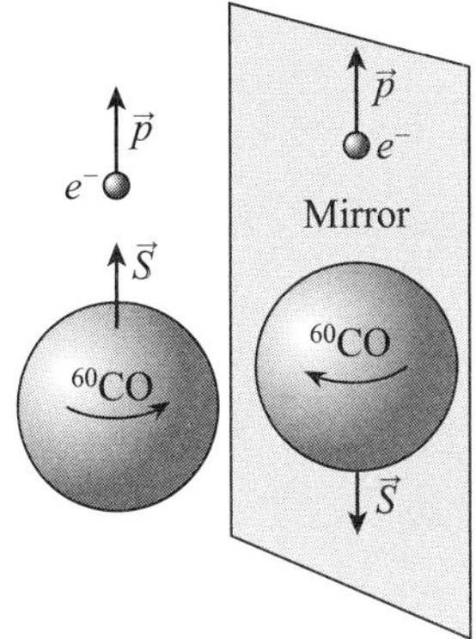

FIGURE 24.7 A process that emits an electron preferentially in the direction of nuclear spin does not look the same in a mirror. (Note: A full parity transformation actually involves a mirror reflection *and* a rotation of 180° around the mirror's normal. But because Lagrangian densities are generally rotationally invariant, the crucial part is the mirror reflection, and it is simpler to draw just that part.)

BOX 24.1   The Trial-Theory Average Squared Amplitude

Our goal in this box is to use the trace theorems from Chapter 20 to calculate the average squared amplitude given by

$$\langle \, |\mathcal{M}|^2 \, \rangle = \frac{1}{8} \left(\frac{g_W}{M_W} \right)^4 \text{Tr}[\, \gamma^\mu (\not{p}_1 + m_\mu) \gamma^\alpha \not{p}_3 \,] \, g_{\mu\nu} g_{\alpha\beta}$$
$$\times \, \text{Tr}[\, \gamma^\nu \not{p}_2 \gamma^\beta (\not{p}_4 + m_e) \,] \qquad (24.26)$$

in terms of p_1, p_2, p_3, and p_4. (Note that the subscript on the m_μ indicates that this is the mass of the muon; it is not an index we are summing over.)

Let's look at the first trace term. We can write this as

$$\text{Tr}[\, \gamma^\mu (\not{p}_1 + m_\mu) \gamma^\alpha \not{p}_3 \,] = \text{Tr}[\, \gamma^\mu p_{1\sigma} \gamma^\sigma \gamma^\alpha p_{3\lambda} \gamma^\lambda \,] + \text{Tr}[\, \gamma^\mu m_\mu \gamma^\alpha p_{3\lambda} \gamma^\lambda \,].$$
$$(24.27)$$

The second trace is zero because it involves an odd number of gamma matrices. If we pull the numerical values $p_{1\sigma} p_{3\lambda}$ out in front of the first and apply trace theorem 14, we get

$$\text{Tr}[\, \gamma^\mu (\not{p}_1 + m_\mu) \gamma^\alpha \not{p}_3 \,] = p_{1\sigma} p_{3\lambda} \, \text{Tr}[\, \gamma^\mu \gamma^\sigma \gamma^\alpha \gamma^\lambda \,]$$
$$= 4 p_{1\sigma} p_{3\lambda} [\, g^{\mu\sigma} g^{\alpha\lambda} - g^{\mu\alpha} g^{\sigma\lambda} + g^{\mu\lambda} g^{\sigma\alpha} \,]$$
$$= 4 [\, p_1^\mu p_3^\alpha - g^{\mu\alpha} (p_1 \cdot p_3) + p_1^\alpha p_3^\mu \,]. \qquad (24.28)$$

Exercise 24.1.1   Calculate the other trace similarly, multiply to get $\langle \, |\mathcal{M}|^2 \, \rangle$, and compare with Equation 24.7. (*Hints:* For ease in comparison, use η and ρ for the other two indices you will need. Also, remember that $g^{\mu\alpha} g_{\mu\alpha} = 4$.)

BOX 24.2   The Trial-Theory Average Squared Amplitude II

Our goal in this box is to calculate

$$\langle |\mathcal{M}|^2 \rangle = 4 \left(\frac{g_W}{M_W} \right)^4 \left[(p_1 \cdot p_2)(p_3 \cdot p_4) + (p_1 \cdot p_4)(p_2 \cdot p_3) \right] \quad (24.29)$$

in the rest frame of the muon, where $p_1^\mu = (m_\mu, 0, 0, 0)$. First, note that as p_1 has only a time component, $p_1 \cdot p_2 = m_\mu E_2$. Since $p_1 = p_2 + p_3 + p_4$ by conservation of four-momentum, we also have $(p_3 + p_4)^2 = (p_1 - p_2)^2$. So

$$(p_3 + p_4)^2 = p_3^2 + p_4^2 + 2 p_3 \cdot p_4 = 0 + m_e^2 + 2 p_3 \cdot p_4$$

$$= (p_1 - p_2)^2 = p_1^2 + p_2^2 - 2 p_1 \cdot p_2 = m_\mu^2 - 2 p_1 \cdot p_2. \quad (24.30)$$

Solving for $p_3 \cdot p_4$ yields

$$p_3 \cdot p_4 = \frac{m_\mu^2 - m_e^2}{2} - p_1 \cdot p_2 = \frac{m_\mu^2 - m_e^2}{2} - m_\mu E_2. \quad (24.31)$$

We can simplify this expression by taking $m_e \to 0$. This is an excellent approximation because m_μ^2 is almost 50,000 times larger than m_e^2. In this limit,

$$p_3 \cdot p_4 = \tfrac{1}{2} m_\mu (m_\mu - 2 E_2). \quad (24.32)$$

Exercise 24.2.1   Do the same kind of calculation with $p_1 \cdot p_4$ and $(p_2 + p_3)^2$ to calculate the other term in the expression for $\langle |\mathcal{M}|^2 \rangle$. Compare your result with Equation 24.8.

BOX 24.3   The Trial-Theory Muon Decay Rate

The goal in this box is to apply Fermi's Golden Rule, which says that

$$\Gamma = \frac{1}{2E_1} \int \left(\frac{d^3 p_2}{(2\pi)^3 2E_2} \right) \left(\frac{d^3 p_3}{(2\pi)^3 2E_3} \right) \left(\frac{d^3 p_4}{(2\pi)^3 2E_4} \right)$$

$$\times (2\pi)^4 \delta^4(p_1 - p_2 - p_3 - p_4) \langle |\mathcal{M}|^2 \rangle \tag{24.33}$$

to calculate the muon decay rate.

First, note that $E_1 = m_\mu$ for a muon at rest. Next, we write

$$\delta^4(p_1 - p_2 - p_3 - p_4) = \delta(m_\mu - E_2 - E_3 - E_4)\delta^3(\vec{p}_2 + \vec{p}_3 + \vec{p}_4) \tag{24.34}$$

and perform the integral over $\vec{p}_3$. The result is

$$\Gamma = \int \frac{\langle |\mathcal{M}|^2 \rangle}{16(2\pi)^5 m_\mu} \frac{d^3 p_2 \, d^3 p_4}{E_2 E_3 E_4} \delta(m_\mu - E_2 - E_3 - E_4), \tag{24.35}$$

where integrating over the momentum part of the Dirac delta has had the effect of setting $|\vec{p}_3| = |\vec{p}_2 + \vec{p}_4|$. Since particle 3 is a massless neutrino, $E_3 = |\vec{p}_3|$. To do the integral over $\vec{p}_2$, first note that

$$E_3^2 = |\vec{p}_2 + \vec{p}_4|^2 = |\vec{p}_2|^2 + |\vec{p}_4|^2 + 2\vec{p}_2 \cdot \vec{p}_4$$

$$= E_2^2 + E_4^2 + 2E_2 E_4 \cos\theta, \tag{24.36}$$

where θ is the angle between $\vec{p}_2$ and $\vec{p}_4$ (remember that we are treating both particle 2—a neutrino—and particle 4—the electron—as having negligible mass, meaning that $E_2 = |\vec{p}_2|$ and $E_4 \approx |\vec{p}_4|$). Since we take the direction of $\vec{p}_4$ as fixed when we are integrating over $\vec{p}_2$, we can write

$$d^3 p_2 = E_2^2 \, dE_2 \, \sin\theta \, d\theta \, d\phi. \tag{24.37}$$

The integral over ϕ yields 2π because nothing depends on ϕ. To do the integral over θ, it helps to change variables to E_3, as the Dirac delta depends on E_3.

Exercise 24.3.1   Take the derivative of both sides of 24.36 with respect to θ to show that $dE_3 = -E_2 E_4 \sin\theta \, d\theta / E_3$, and from there, show that

$$\int_0^\pi \frac{\sin\theta \, d\theta}{E_3} \delta(m_\mu - E_2 - E_3 - E_4) = \frac{1}{E_2 E_4} \quad \text{or } 0, \tag{24.38}$$

where the first result applies if the Dirac delta spikes between $E_3(\pi)$ and $E_3(0)$.

BOX 24.3 The Trial-Theory Muon Decay Rate (cont.)

Note that the integration limits are

$$E_3(0), E_3(\pi) = \sqrt{E_2^2 + E_4^2 \pm 2E_2E_4} = \mid E_2 \pm E_4 \mid. \qquad (24.39)$$

So the region in which the integration is nonzero is given by

$$\mid E_2 - E_4 \mid < m_\mu - E_2 - E_4 < E_2 + E_4. \qquad (24.40)$$

If we add $E_2 + E_4$ to both sides and divide by 2, we get

$$\tfrac{1}{2}\big[\mid E_2 - E_4 \mid + E_2 + E_4 \big] < \tfrac{1}{2}m_\mu < E_2 + E_4. \qquad (24.41)$$

Now, note that if $E_2 < E_4$, then the quantity in brackets is $-E_2 + E_4 + E_2 + E_4 = 2E_4$. If $E_2 > E_4$, then it is $E_2 - E_4 + E_2 + E_4 = 2E_2$. So the quantity on the left is simply the larger of E_2 and E_4. This means that E_2 and E_4 must *each* be less than $\tfrac{1}{2}m_\mu$, while their sum must be *larger* than $\tfrac{1}{2}m_\mu$.

This specifies the limits on the remaining integrals. We will save the E_4 integral for last, so the range of integration for E_2 goes from $\tfrac{1}{2}m_\mu - E_4$ up to $\tfrac{1}{2}m_\mu$, while E_4 goes from 0 to $\tfrac{1}{2}m_\mu$. Since

$$\langle \mid \mathcal{M} \mid^2 \rangle = 2 \left(\frac{g_W}{M_W} \right)^4 m_\mu^2 \, [\, E_2(m_\mu - 2E_2) + E_4(m_\mu - 2E_4) \,] \qquad (24.42)$$

does not depend on angles, the angular integrals will be trivial.

Exercise 24.3.2 Do the integral over E_2 and the angular part of $\vec{p}_4$ and compare to Equation 24.10. (*Hint:* Let $x \equiv E_2$, $a \equiv \tfrac{1}{2}m_\mu$, and $b \equiv E_4$.)

BOX 24.4   The Correct Muon Decay Amplitude

In this box, we will use trace theorems to evaluate the correct expression for the average squared muon decay amplitude:

$$\langle\, |\mathcal{M}|^2\,\rangle = \frac{1}{128}\left(\frac{g_W}{M_W}\right)^4 \mathrm{Tr}[\,\gamma^\mu(1-\gamma^5)(\not{p}_1+m_\mu)\gamma^\alpha(1-\gamma^5)\not{p}_3\,]\,g_{\mu\nu}g_{\alpha\beta}$$

$$\times \mathrm{Tr}[\,\gamma^\nu(1-\gamma^5)\not{p}_2\gamma^\beta(1-\gamma^5)(\not{p}_4+m_e)\,]. \quad (24.43)$$

As we did before, we will evaluate each trace term separately.

Consider $\mathrm{Tr}[\,\gamma^\mu(1-\gamma^5)(\not{p}_1+m_\mu)\gamma^\alpha(1-\gamma^5)\not{p}_3\,]$. As $\gamma^5 = i\gamma^t\gamma^x\gamma^y\gamma^z$ is a shorthand for a product of an even number of gamma matrices, $(1-\gamma^5)$ also counts as an even number of gamma matrices. Therefore, any product involving m_μ instead of $\not{p}_1$ will be a product of an odd number of gamma matrices, meaning that its trace will be zero. So this term reduces to

$$\mathrm{Tr}_1 \equiv \mathrm{Tr}[\,\gamma^\mu(1-\gamma^5)\not{p}_1\gamma^\alpha(1-\gamma^5)\not{p}_3\,]$$

$$= p_{1\sigma}p_{3\lambda}\Big(\mathrm{Tr}[\gamma^\mu\gamma^\sigma\gamma^\alpha\gamma^\lambda] - \mathrm{Tr}[\gamma^\mu\gamma^5\gamma^\sigma\gamma^\alpha\gamma^\lambda]$$

$$- \mathrm{Tr}[\gamma^\mu\gamma^\sigma\gamma^\alpha\gamma^5\gamma^\lambda] + \mathrm{Tr}[\gamma^\mu\gamma^5\gamma^\sigma\gamma^\alpha\gamma^5\gamma^\lambda]\Big). \quad (24.44)$$

Now, γ^5 anticommutes with all other γ^μ, so in the middle terms, we can bring the γ^5 to the front at the cost of changing the sign in each case. In the last term, we can bring the second γ^5 forward two places where it cancels with the first (since $[\gamma^5]^2 = I$). Therefore, this trace reduces to

$$\mathrm{Tr}_1 = 2p_{1\sigma}p_{3\lambda}\Big(\mathrm{Tr}[\gamma^\mu\gamma^\sigma\gamma^\alpha\gamma^\lambda] + \mathrm{Tr}[\gamma^5\gamma^\mu\gamma^\sigma\gamma^\alpha\gamma^\lambda]\Big)$$

$$= 8p_{1\sigma}p_{3\lambda}(g^{\mu\sigma}g^{\alpha\lambda} - g^{\mu\alpha}g^{\sigma\lambda} + g^{\mu\lambda}g^{\alpha\sigma} + i\epsilon^{\mu\sigma\alpha\lambda}) \quad (24.45)$$

$$= 8(p_1^\mu p_3^\alpha - g^{\mu\alpha}(p_1\cdot p_3) + p_1^\alpha p_3^\mu + ip_{1\sigma}p_{3\lambda}\epsilon^{\mu\sigma\alpha\lambda}), \quad (24.46)$$

where I have used trace theorems 14 and 18. Only the last term is new.

Exercise 24.4.1   Similarly, calculate the second trace term in Equation 24.43.

BOX 24.4 The Correct Muon Decay Amplitude (cont.)

The total amplitude depends on the product of these traces:

$$\text{Tr}_1 g_{\mu\nu} g_{\alpha\beta} \text{Tr}_2 = 8(p_1^\mu p_3^\alpha - g^{\mu\alpha}(p_1 \cdot p_3) + p_1^\alpha p_3^\mu + ip_{1\sigma} p_{3\lambda}\epsilon^{\mu\sigma\alpha\lambda})g_{\mu\nu}g_{\alpha\beta}$$

$$\times\ 8(p_2^\nu p_4^\beta - g^{\nu\beta}(p_2 \cdot p_4) + p_2^\beta p_4^\nu + ip_{2\eta} p_{4\rho}\epsilon^{\nu\eta\beta\rho})$$

$$= 64(p_1^\mu p_3^\alpha - g^{\mu\alpha}(p_1 \cdot p_3) + p_1^\alpha p_3^\mu + ip_{1\sigma} p_{3\lambda}\epsilon^{\mu\sigma\alpha\lambda})$$

$$\times\ (p_{2\mu} p_{4\alpha} - g_{\mu\alpha}(p_2 \cdot p_4) + p_{2\alpha} p_{4\mu} + ip_2^\eta p_4^\rho \epsilon_{\mu\eta\alpha\rho}). \qquad (24.47)$$

Now, if we compare this product with the trace product in Box 24.1, everything is the same except that the product is larger by an overall factor of 4, and we have a new term in each trace involving the ϵ-tensor. But notice that the tensor quantity $S_{13}^{\mu\alpha} \equiv p_1^\mu p_3^\alpha - g^{\mu\alpha}(p_1 \cdot p_3) + p_1^\alpha p_3^\mu$ is a *symmetric* tensor ($S_{13}^{\mu\alpha} = S_{13}^{\alpha\mu}$). Similarly, $S_{\mu\alpha,24} \equiv p_{2\mu} p_{4\alpha} - g_{\mu\alpha}(p_2 \cdot p_4) + p_{2\alpha} p_{4\mu}$ is symmetric.

Exercise 24.4.2 Prove that $S_{13}^{\mu\alpha}\epsilon_{\mu\eta\alpha\rho} = 0$. (*Hint:* Remember that $\epsilon_{\mu\eta\alpha\rho}$ is antisymmetric under interchange of *any* pair of indices. So swap the μ and α indices in both tensors, then rename indices.)

Similarly, $S_{\mu\alpha,24}\epsilon^{\mu\sigma\alpha\lambda} = 0$, so both cross terms involving the new ϵ-tensor term are zero. The only new term that survives is $-p_{1\sigma} p_{3\lambda} p_2^\eta p_4^\rho \epsilon^{\mu\sigma\alpha\lambda}\epsilon_{\mu\eta\alpha\rho}$, which the identity $\epsilon^{\mu\nu\sigma\lambda}\epsilon_{\mu\nu\eta\rho} = -2(\delta_\eta^\sigma \delta_\rho^\lambda - \delta_\rho^\sigma \delta_\eta^\lambda)$ tells us is equal to

$$+2p_{1\sigma} p_{3\lambda} p_2^\eta p_4^\rho (\delta_\eta^\sigma \delta_\rho^\lambda - \delta_\rho^\sigma \delta_\eta^\lambda) = 2p_{1\eta} p_{3\rho} p_2^\eta p_4^\rho - 2p_{1\rho} p_{3\sigma} p_2^\eta p_4^\rho$$

$$= 2(p_1 \cdot p_2)(p_3 \cdot p_4) - 2(p_1 \cdot p_4)(p_3 \cdot p_2). \ (24.48)$$

Exercise 24.4.3 With this in mind, calculate $\text{Tr}_1 g_{\mu\nu} g_{\alpha\beta} \text{Tr}_2$ and show that $\langle |\mathcal{M}|^2 \rangle = 2(g_W/M_W)^4 (p_1 \cdot p_2)(p_3 \cdot p_4)$.

BOX 24.5 Pion Decay Kinematics

In this box, we will use conservation of four-momentum to determine the energy E_ℓ of the emitted lepton and the magnitude of the spatial momentum $|\vec{p}_F|$ carried by the final particles in the decay of a π^- to either $\mu^- + \overline{\nu}_\mu$ or $e^- + \overline{\nu}_e$. Assume that the pion is initially at rest, that the charged lepton moves in the $+z$ direction after the decay, and that the neutrinos have zero mass. Note that $m_\pi = 139.57\,\mathrm{MeV}$, $m_\mu = 105.66\,\mathrm{MeV}$, and $m_e = 0.511\,\mathrm{MeV}$.

Exercise 24.5.1   Show that the common final spatial momentum is

$$|\vec{p}_F| = \frac{m_\pi^2 - m_\ell^2}{2m_\pi}, \tag{24.49}$$

where m_ℓ is the charged lepton's mass. Find a similar formula for the charged lepton's energy E_ℓ. Evaluate both of these quantities in MeV for both decay to an electron and decay to a muon. (*Hints:* To save writing, define $q = |\vec{p}_F|$, and note that $E_\ell = \sqrt{m_\ell^2 + q^2}$. Use the latter to eliminate E_ℓ in favor of q.)

BOX 24.6   The C and P Operations

In this box, we examine the effects of C and P operations on terms in the weak-interaction Lagrangian density term

$$\mathcal{L}_{\text{int}} = -\frac{1}{\sqrt{2}}(J^{+\mu}W_\mu^+ + J^{-\mu}W_\mu^-) - J^{(z)\mu}W_\mu^{(z)}$$

$$= -g_W\frac{1}{2\sqrt{2}}\left[\bar{v}_e\gamma^\mu(1-\gamma^5)e\,W_\mu^+ + \bar{e}\gamma^\mu(1-\gamma^5)v_e\,W_\mu^-\right]$$

$$- g_W\frac{1}{2}\left[\bar{v}_e\gamma^\mu(1-\gamma^5)v_e + \bar{e}\gamma^\mu(1-\gamma^5)e\right]W_\mu^{(z)}. \qquad (24.50)$$

This involves terms of the form $\bar{\psi}_2\gamma^\mu\psi_1$ and $\bar{\psi}_2\gamma^\mu\gamma^5\psi_1$.

Note that C acting on a Dirac spinor field ψ yields $C\psi = -i\gamma^y\psi^*$. The adjoint of this is $\overline{(C\psi)} = +i\psi^T(\gamma^y)^\dagger\gamma^t = -i\psi^T\gamma^y\gamma^t$ because $(\gamma^y)^\dagger = -\gamma^y$. Therefore, the action of C on a term such as $\bar{\psi}_2\gamma^\mu\psi_1$ will be

$$C(\bar{\psi}_2\gamma^\mu\psi_1) = (-i\psi_2^T\gamma^y\gamma^t)\gamma^\mu(-i\gamma^y\psi_1^*) = -\psi_2^T[\gamma^y\gamma^t\gamma^\mu\gamma^y]\psi_1^*. \qquad (24.51)$$

We can put Equation 24.51 in more recognizable notation by noting that for vectors a and b and matrix M, we have $a^T M b = b^T M^T a$ because the transpose of a scalar is equal to the scalar. Therefore, we can rewrite Equation 24.51 as

$$C(\bar{\psi}_2\gamma^\mu\psi_1) = -\psi_1^\dagger[\gamma^y\gamma^t\gamma^\mu\gamma^y]^T\psi_2 = -\psi_1^\dagger(\gamma^y)^T(\gamma^\mu)^T(\gamma^t)^T(\gamma^y)^T\psi_2. \qquad (24.52)$$

To simplify this, note that $(\gamma^t)^T = +\gamma^t$, $(\gamma^x)^T = -\gamma^x$, $(\gamma^y)^T = +\gamma^y$, and $(\gamma^z)^T = -\gamma^z$. Also, note that $\gamma^\mu\gamma^\nu = -\gamma^\nu\gamma^\mu$ if $\mu \neq \nu$ and $(\gamma^y)^2 = -I$. So

$$C(\bar{\psi}_2\gamma^\mu\psi_1) = -\psi_1^\dagger\gamma^y\begin{bmatrix}\gamma^t\\-\gamma^x\\\gamma^y\\-\gamma^z\end{bmatrix}\gamma^t\gamma^y\psi_2 = \psi_1^\dagger\gamma^y\begin{bmatrix}\gamma^t\\-\gamma^x\\\gamma^y\\-\gamma^z\end{bmatrix}\gamma^y\gamma^t\psi_2$$

$$= \psi_1^\dagger(-I)\begin{bmatrix}-\gamma^t\\+\gamma^x\\+\gamma^y\\+\gamma^z\end{bmatrix}\gamma^t\psi_2 = -\psi_1^\dagger\gamma^t\begin{bmatrix}-\gamma^t\\-\gamma^x\\-\gamma^y\\-\gamma^z\end{bmatrix}\psi_2 = +\bar{\psi}_1\gamma^\mu\psi_2, \qquad (24.53)$$

where the column vector keeps track of the four cases for γ^μ individually. Note that the effect of the charge conjugation is to reverse the order of the spinors.

Exercise 24.6.1 In the same manner, calculate $C(\bar{\psi}_2\gamma^\mu\gamma^5\psi_1)$. (*Hint:* Note that $(\gamma^5)^T = \gamma^5$ and $\gamma^\mu\gamma^5 = -\gamma^5\gamma^\mu$ for all μ. Just tweak what we have above.)

BOX 24.6 The C and P Operations (cont.)

Now, consider the parity operator. $P\psi = \gamma^t\psi$, so $P(\overline{\psi}) = \psi^\dagger(\gamma^t)^\dagger\gamma^t = \psi^\dagger$ since $(\gamma^t)^\dagger = \gamma^t$ and $(\gamma^t)^2 = I$. Therefore,

$$P(\overline{\psi}_2\gamma^\mu\psi_1) = \psi_2^\dagger\gamma^\mu\gamma^t\psi_1. \tag{24.54}$$

There is no transpose here, but pulling the γ^t to its proper place in front flips the signs of the spatial γ^μ matrices. If we have a γ^5 after the γ^μ, then pulling the γ^t to the left past the γ^5 also results in an initial overall sign flip. So

$$P(\overline{\psi}_2\gamma^\mu\psi_1) = \overline{\psi}_2 \begin{bmatrix} \gamma^t \\ -\gamma^x \\ -\gamma^y \\ -\gamma^z \end{bmatrix} \psi_1 \quad \text{and} \quad P(\overline{\psi}_2\gamma^\mu\gamma^5\psi_1) = \overline{\psi}_2 \begin{bmatrix} -\gamma^t \\ +\gamma^x \\ +\gamma^y \\ +\gamma^z \end{bmatrix} \gamma^5\psi_1. \tag{24.55}$$

Finally, let us consider a combined CP operation on a term of the form $\overline{\psi}_2\gamma^\mu(1-\gamma^5)\psi_1$. In a minute, you will show that

$$CP(\overline{\psi}_2\gamma^\mu(1-\gamma^5)\psi_1) = \overline{\psi}_1 \begin{bmatrix} \gamma^t \\ -\gamma^x \\ -\gamma^y \\ -\gamma^z \end{bmatrix} (1-\gamma^5)\psi_2. \tag{24.56}$$

In conjunction with what CP does to the $W_\mu^\pm$ ($C(W^\pm) = W^\mp$ and $P(W_\mu^\pm) = +W_t^\pm, -W_j^\pm$), this means that the first term in Equation 24.50 becomes the second term and vice versa. Since charge conjugation does nothing to the electrically neutral $W^{(z)}$, the third term remains the same. Therefore, our current weak-interaction Lagrangian density is invariant under CP.

To see where Equation 24.56 comes from, note that $CP\psi = -i\gamma^y\gamma^t\psi^*$, which implies that $CP\overline{\psi} = (-i\gamma^y\gamma^t\psi^*)^\dagger\gamma^t = +i\psi^T\gamma^t(-\gamma^y)\gamma^t$. Therefore,

$$CP(\overline{\psi}_2\gamma^\mu(1-\gamma^5)\psi_1) = -\psi_2^T\gamma^t\gamma^y\gamma^t\gamma^\mu(1-\gamma^5)\gamma^y\gamma^t\psi_1^*$$

$$= -\psi_1^\dagger\,[\,\gamma^t\gamma^y\gamma^t\gamma^\mu(1-\gamma^5)\gamma^y\gamma^t\,]^T\,\psi_2, \tag{24.57}$$

where in the last step I used the $a^T M b = b^T M^T a$ trick.

Exercise 24.6.2 Starting from here, use the methods we have developed in this box (including the column vector trick) to verify Equation 24.56.

Homework Problems

P24.1 Carefully go through the calculation in Box 24.3 to see what changes if the average squared muon amplitude is as given in Equation 24.17 instead of 24.7, and show that Equation 24.18 is the correct result for $d\Gamma$.

P24.2 The observed average muon lifetime is 2.197×10^{-6} s, and it has an observed mass of 0.10566 GeV.
a. From this information, calculate $(g_W m_\mu / M_W)^2$. (The amplitude of a higher order muon-decay diagram turns out to be smaller than that for the next lower order by about this factor.)
b. By analogy, estimate the lifetime of the τ^-, whose mass is $m_\tau = 1.7769$ GeV, assuming that it can only decay to $\nu_\tau e^- \bar{\nu}_e$ or $\nu_\tau \mu^- \bar{\nu}_\mu$, and compare to its actual lifetime of 2.9×10^{-13} s. (The discrepancy is because, about 65% of the time, the τ^- decays to other things.)

P24.3 What is the *average* electron energy in muon decay?

P24.4 The pion decay rate is not a simple weak-interaction calculation, because determining the bound state of the quarks involved in Figure 24.4 involves tricky issues having to do with the color interaction between the quarks. But we can gather all the unknowns in this problem into a simple package as follows. Redraw the Feynman diagram with a blob representing the complex physics involved when the pion's quarks interact to produce a W^- boson (see Figure 24.8). The Feynman rules then imply that

$$\mathcal{M} = F^\mu \frac{g_W^2 g_{\mu\nu}}{8 M_W^2} [\bar{u}_\ell \gamma^\nu (1 - \gamma^5) v_\nu], \qquad (24.58)$$

where the F^μ contains everything we don't know about the pion vertex.

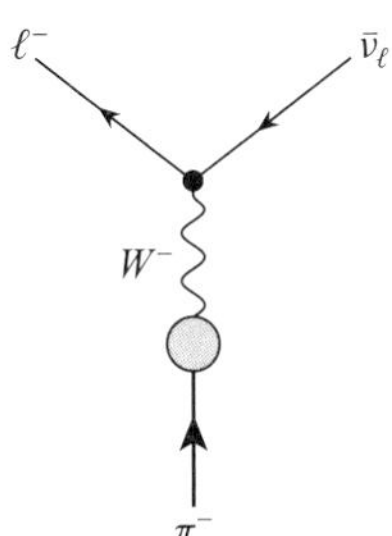

FIGURE 24.8   This drawing shows a Feynman diagram of pion decay with a blob representing the complicated quark interactions with the W^-. The ℓ^- stands for a negative lepton (either e^- or μ^-), and $\bar{\nu}_\ell$ stands for the lepton's corresponding antineutrino.

We do know that this factor must be a four-vector to connect with the $g_{\mu\nu}\gamma^\nu$ in the rest of the expression. Now, because the pion's spin is zero, the only four-vector associated with the pion is its four-momentum p^μ, so we must have $F^\mu = f_\pi p^\mu$, where f_π is some scalar. The only scalar quantity that we can construct out of p^μ is $p^2 = m_\pi^2$, which is a constant. Therefore, the quantity f_π cannot depend on anything having to do

with the pion's motion or energy: It must be the same constant for all pion decays. Physicists call it the **pion decay constant**.
a. Show that the decay rate is

$$\Gamma = \frac{f_\pi^2}{\pi m_\pi^3} \left(\frac{g_W}{4 M_W}\right)^4 m_\ell^2 (m_\pi^2 - m_\ell^2)^2, \quad (24.59)$$

where m_ℓ is the mass of the ℓ^- lepton. (*Hint:* Use the results from Box 24.4 to help determine $\langle |\mathcal{M}|^2 \rangle$.)
b. Calculate the branching ratio

$$\frac{\Gamma(\pi^- \to e^- + \bar{\nu}_e)}{\Gamma(\pi^- \to \mu^- + \bar{\nu}_\mu)}. \qquad (24.60)$$

c. The pion lifetime is 2.6×10^{-8} s. Using $M_W = 80.4$ GeV and $f_\pi \approx 0.1304$ GeV, estimate the value of g_W.

P24.5 The IceCube Neutrino Observatory experiment seeks to observe high-energy cosmic neutrinos. A muon-antineutrino hitting one of the u quarks in a proton in an ice molecule can convert that u to a d (converting the proton to a neutron), and convert the antineutrino to an antimuon: in short, $\bar{\nu}_\mu + p \to n + \mu^+$. In the aftermath of a sufficiently high-energy collision, the μ^+ moves faster than the speed of light in ice, creating **Cherenkov radiation**. A huge array of photomultiplier tubes buried in the superclear ice about a mile below the surface of the South Pole detects this light, and from that light one can calculate the energy and the direction of the outgoing μ^+.
a. Draw the Feynman diagram for this interaction.
b. Let the $\bar{\nu}_\mu$'s direction of motion define the $+z$ direction. Let E be that antineutrino's energy, θ be the angle between the antimuon's momentum and the z axis in what we will define to be the yz plane, and assume (for simplicity) that the proton and neutron have the same mass m, and that the masses of the antimuon and antineutrino are negligible. Use conservation of four-momentum to calculate the antimuon's energy E_1 in terms of E, m, and θ. (*Hint:* You should find it is something divided by $m + E[1 - \cos\theta]$.)
c. Between 2010 and 2012, the IceCube Collaboration detected three antimuons with energies exceeding 1 PeV of energy; one of these (dubbed "Big Bird") had an energy of $E_1 \approx 2.0$ PeV. (1 PeV $= 10^{15}$ eV ≈ 160 μJ. This is an enormous energy for a single particle!) Now, the energy E of the incoming neutrino must be bigger than E_1 and less than infinity. Show that the range of antimuon angles θ relative to the initial antineutrino's direction must be quite small, and put an approximate upper limit on θ in degrees. (*Hint:* For small angles, $\cos\theta \approx 1 - \frac{1}{2}\theta^2$ if θ is in radians.)

This result means that by tracking the antimuon, we can determine the direction of the incoming antineutrino fairly precisely. Measurements like this have enabled astrophysicists to claim with some degree of certainty that Big Bird came from the supermassive black hole at the center of a galaxy (M. Kadler et al., *Nature Phys* **12**, 807, 2016).

P24.6 Calculate the total cross section for the "inverse muon decay" process $\nu_\mu + e^- \to \mu^- + \nu_e$ in the center-of-mass frame, treating the electron mass as negligible. (*Hints:* You can adapt the muon decay amplitude for this situation. When averaging over initial spin states, remember that because a neutrino must be left-handed it has only one spin state instead of two. You should find the differential cross section to be isotropic.)

P24.7 Prove that $\overline{\gamma^\mu(1 - \gamma^5)} = \gamma^\mu(1 - \gamma^5)$. (*Hints:* Remember that in the context of Casimir's trick, $\overline{\Gamma} \equiv \gamma^t \Gamma^\dagger \gamma^t$ for an arbitrary matrix Γ. The definition of γ^5 as a matrix is given by Equation 17.26. Also, recall that we showed in Box 20.3 that $\overline{\gamma^\mu} = \gamma^\mu$.)

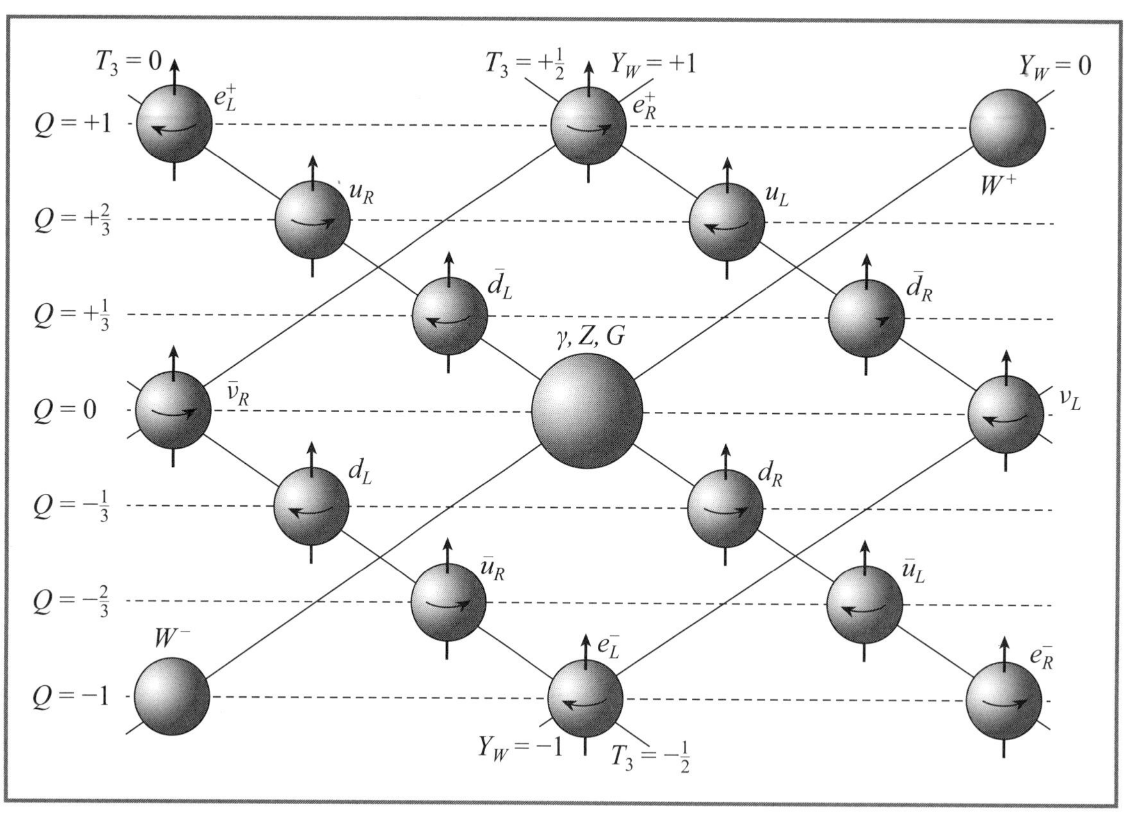

T_3 = 0
e_L^+
T_3 = +1/2
Y_W = +1
e_R^+
Y_W = 0
Q = +1
W^+
u_R
u_L
Q = +2/3
d̄_L
d̄_R
Q = +1/3
γ, Z, G
ν̄_R
ν_L
Q = 0
d_L
d_R
Q = −1/3
ū_R
ū_L
Q = −2/3
W^−
e_L^−
e_R^−
Q = −1
Y_W = −1
T_3 = −1/2

25. THE UNIFIED ELECTROWEAK THEORY

The $SU(2)_L$ theory of the weak interaction we have presented so far represents a major step forward in our ability to make realistic predictions, but it also has major flaws. We have not yet talked about interactions involving the Z vector boson. Even more importantly, we have not justified why the masses of the charged leptons and the vector bosons $W^\pm$, Z are nonzero when the gauge symmetry specifically forbids them from being nonzero.

We will deal with these issues in this chapter and the next. The unified **electroweak theory** developed by Glashow, Weinberg, and Salam in the 1960s (which earned them the 1979 Nobel Prize in physics) provided a full and experimentally accurate picture of the weak interaction by more fully understanding its deep connection to the electromagnetic interaction. In this chapter, we will explore how to unify the weak and electromagnetic interactions into one complete and coherent picture. In the next chapter, we will explore how the Higgs mechanism and symmetry breaking explain how particle masses enter the picture.

In many ways, electroweak theory is the beating heart of the Standard Model. It provides a complete and plausible theoretical framework for these interactions and is also very successful experimentally. By unifying these interactions into a coherent whole, it also set the agenda for subsequent research, building hopes that one might find a similar **Grand Unified Theory** (GUT) embracing the color interaction. The success of electroweak theory has been very influential in the field.

25.1 Combining $SU(2)_L$ and $U(1)$

In Chapter 23, we began by noting that $\overline{\Psi}\Psi$, where Ψ is a flavor doublet of fermion fields such as

$$\Psi = \begin{bmatrix} \nu_e \\ e \end{bmatrix}, \quad \begin{bmatrix} \nu_\mu \\ \mu \end{bmatrix}, \quad \begin{bmatrix} \nu_\tau \\ \tau \end{bmatrix}, \quad \begin{bmatrix} u \\ d \end{bmatrix}, \quad \begin{bmatrix} c \\ s \end{bmatrix}, \quad \text{or} \quad \begin{bmatrix} t \\ b \end{bmatrix}, \quad (25.1)$$

was symmetric under a global $SU(2)$ transformation. When we made this symmetry actively local, the weak interaction (in its original trial form) emerged.

But $\overline{\Psi}\Psi$ is also symmetric under a global $U(1)$ transformation. One might suppose that making this symmetry local would yield a completely independent electromagnetic interaction, and this would be the case if the weak interaction was ambidextrous, as it was in our failed trial version. But the actual weak interaction arises from a local $SU(2)_L$ symmetry, not $SU(2)$. This entangles the interactions, as we will see.

The first step to exploring the $SU(2)_L \otimes U(1)$ symmetry is to write down the complete transformation. For chirally left-handed fields, the transformation is

$$\Psi'_L = e^{-i\frac{1}{2}g'Y_L\theta^0} e^{-\frac{1}{2}g_W\vec{\sigma}\cdot\vec{\theta}}\Psi_L, \quad (25.2a)$$

and for chirally right-handed fields, only the $U(1)$ symmetry applies, so

$$\Psi'_R = e^{-i\frac{1}{2}g'Y_R\theta^0}\Psi_R. \quad (25.2b)$$

(Remember that we can extract the chirally right- and left-handed fields from the full Dirac field using the $\frac{1}{2}(1\pm\gamma^5)$ operators.) The g_W is the weak interaction coupling constant we have seen before, but these equations introduce new "charges" $g'Y_L$ and $g'Y_R$ for the $U(1)$ transformation: Y_L is the **weak hypercharge** of the left-handed doublet (the same Y_L must apply to both fields in the doublet to preserve the SU(2)$_L$ symmetry), while the Y_R is the hypercharge of the right-handed fields (in principle, each separate right-handed field might have its own Y_R). The g' extracts any common magnitude and will allow us to assign convenient integer values to the weak hypercharges. The θ^0 is the variable parameter for the U(1) symmetry.

The next step, as usual, is to write down the currents corresponding to these symmetries. Because the U(1) symmetry applies to the right-handed current, by analogy to what we found in Chapter 19, it is simply

$$J_\mu^{(0,R)} = \tfrac{1}{2}g'Y_R\overline{\Psi}_R\gamma_\mu\Psi_R = \tfrac{1}{2}g'Y_R[\ \overline{\nu}_R\quad \overline{e}_R\ ]\gamma_\mu \begin{bmatrix} \nu_R \\ e_R \end{bmatrix}$$

$$= \tfrac{1}{2}g'Y_R(\overline{\nu}_R\gamma_\mu\nu_R + \overline{e}_R\gamma_\mu e_R), \tag{25.3}$$

(where for concreteness I am focusing on the ν_e, e doublet, and I have dropped the e on ν_e to save writing). The left-handed U(1) current is similar:

$$J_\mu^{(0,L)} = \tfrac{1}{2}g'Y_L(\overline{\nu}_L\gamma_\mu\nu_L + \overline{e}_L\gamma_\mu e_L), \tag{25.4a}$$

while the left-handed SU(2) currents (copying from Equation 23.12) are

$$J_\mu^{(x,L)} = \tfrac{1}{2}g_W(\overline{\nu}_L\ \gamma_\mu e_L + \overline{e}_L\gamma_\mu\nu_L), \tag{25.4b}$$

$$J_\mu^{(y,L)} = -i\tfrac{1}{2}g_W(\overline{\nu}_L\gamma_\mu e_L - \overline{e}_L\gamma_\mu\nu_L), \tag{25.4c}$$

$$J_\mu^{(z,L)} = \tfrac{1}{2}g_W(\overline{\nu}_L\gamma_\mu\nu_L - \overline{e}_L\gamma_\mu e_L). \tag{25.4d}$$

The interaction Lagrangian density then becomes (as before)

$$\mathcal{L}_{\text{int}} = -[\ J^{(0,R)} + J^{(0,L)}\ ]\cdot B - J^{(x,L)}\cdot W^{(x)} - J^{(y,L)}\cdot W^{(y)} - J^{(z,L)}\cdot W^{(z)}, \tag{25.5}$$

where B^μ is the four-vector gauge field associated with the U(1) symmetry. (Why is this not the electromagnetic field A^μ? We are about to see why it is not.) Again, as we did in Chapter 23, we transform from $W^{(x)}$ and $W^{(y)}$ to $W^\pm$, making the interaction term

$$\mathcal{L}_{\text{int}} = -[\ J^{(0,R)} + J^{(0,L)}\ ]\cdot B - J^{(z,L)}\cdot W^{(z)}$$

$$- \tfrac{1}{\sqrt{2}}g_W\big[\ W_\mu^+\,\overline{\nu}_L\gamma^\mu e_L + W_\mu^-\,\overline{e}_L\gamma^\mu\nu_L\ \big]. \tag{25.6}$$

For the moment, focus on the first line of this, which involves interactions where the charge does not change ($e \rightarrow e$ or $\nu_e \rightarrow \nu_e$), meaning that the carrier of the interaction is electrically neutral. If we also focus only on the part having to do with left-handed particles, and gather the factors in front of the $\overline{e}_L\gamma_\mu e_L$ and $\overline{\nu}_L\gamma_\mu\nu_L$ terms in Equations 25.4, we have

$$\mathcal{L}_{\text{neutral},L} = -\tfrac{1}{2}\big[\ g'Y_L B_\mu + g_W W_\mu^{(z)}\ \big]\overline{\nu}_L\gamma^\mu\nu_L$$

$$- \tfrac{1}{2}\big[\ g'Y_L B_\mu - g_W W_\mu^{(z)}\ \big]\overline{e}_L\gamma^\mu e_L. \tag{25.7}$$

25.2 The Weinberg Angle

The interesting thing is that neutrinos interact not only with the SU(2)$_L$ $W^{(z)}$ field but also with the U(1) B field, and electrons also interact with both. But what we call the electromagnetic interaction should *not* interact with neutrinos, because they are electrically neutral.

Fortunately, we can solve this problem with what amounts to a "rotation" of how we define the two neutral fields. Consider the following transformation:

$$\begin{bmatrix} B \\ W^{(z)} \end{bmatrix} = \begin{pmatrix} \cos\theta_W & \sin\theta_W \\ -\sin\theta_W & \cos\theta_W \end{pmatrix}\begin{bmatrix} A \\ Z \end{bmatrix}. \tag{25.8}$$

The A field will become the electromagnetic field (the part of the electroweak interaction that ignores neutrinos), while the Z field will be the "purely weak" part of the neutral electroweak interaction. Note that this transformation in no way changes the physics of our system: we are just defining new fields that are linear superpositions of the old fields. This is analogous to orienting our coordinate system in mechanics to separate what is going on in the x direction from what is going on in the y direction. Physicists call the angle θ_W the **Weinberg angle**.

In Box 25.1, we will see that in terms of the new fields, the interaction Lagrangian density becomes

$$\mathcal{L}_{\text{neutral}, L} = -c_1 \bar{\nu}_L \gamma^\mu \nu_L A_\mu - c_2 \bar{e}_L \gamma^\mu e_L A_\mu$$
$$- c_3 \bar{\nu}_L \gamma^\mu \nu_L Z_\mu - c_4 \bar{e}_L \gamma^\mu e_L Z_\mu, \tag{25.9a}$$

$$\text{where} \quad c_1 = \tfrac{1}{2} g' Y_L \cos \theta_W - \tfrac{1}{2} g_W \sin \theta_W, \tag{25.9b}$$

$$c_2 = \tfrac{1}{2} g' Y_L \cos \theta_W + \tfrac{1}{2} g_W \sin \theta_W, \tag{25.9c}$$

$$c_3 = \tfrac{1}{2} g' Y_L \sin \theta_W + \tfrac{1}{2} g_W \cos \theta_W, \text{ and} \tag{25.9d}$$

$$c_4 = \tfrac{1}{2} g' Y_L \sin \theta_W - \tfrac{1}{2} g_W \cos \theta_W. \tag{25.9e}$$

In order to match up with our expectations, we want $c_1 = 0$ and $c_2 = |q_e|$ so that the electromagnetic field ignores neutrinos, and the coupling between the electron and the electric field (that is, the electron's electric charge) has magnitude $|q_e|$. We conventionally define $Y_L = -1$ for any lepton doublet. A bit of work (see Box 25.2) then shows that

$$g' = -g_W \tan \theta_W \quad \text{and} \quad |q_e| = g_W \sin \theta_W. \tag{25.10}$$

Various cross-section ratios allow one to determine $\sin^2 \theta_W$ experimentally:

$$\sin^2 \theta_W = 0.2234(1) \quad \Rightarrow \quad \sin \theta_W = 0.4727. \tag{25.11}$$

At low energies, $e = 0.3029$, so we can infer that $g_W = 0.641$. (A technical note: At the level we are considering here, the definition of these quantities seems straightforward, but renormalization and the energy dependence of various quantities make things more complicated. In particular, there are several ways to define θ_W: I am using the so-called "on-shell" definition here.)

25.3 The Charges of Leptons

Once we have defined the Weinberg angle and the value $Y_L = -1$ for leptons, we can (see Box 25.3) rewrite Equation 25.9a as

$$\mathcal{L}_{\text{neutral}, L} = -|q_e| A_\mu [\bar{e}_L \gamma^\mu e_L]$$
$$- g_W \frac{Z_\mu}{2 \cos \theta_W} \left([\bar{\nu}_L \gamma^\mu \nu_L] - \cos(2\theta_W)[\bar{e}_L \gamma^\mu e_L] \right). \tag{25.12}$$

The weak term shows that neutrinos and electrons respond differently to the Z field. We can express this difference in terms of a quantity T_3 we call the **weak isospin**: in analogy with spinors in quantum mechanics, we assign the upper field in a flavor doublet (the neutrino in the case of leptons) the isospin value $T_3 = \tfrac{1}{2}$ (isospin "up"), and the lower field the isospin value $T_3 = -\tfrac{1}{2}$ (isospin "down"). With these assignments of value for Y_L and T_3, the charge of any left-handed lepton (in units of the proton charge $|q_e|$) is given by

$$Q = \tfrac{1}{2} Y_L + T_3 \tag{25.13}$$

since $T_3 = +\frac{1}{2}$ and $Y_L = -1$ yields $Q = 0$ for neutrinos, and $T_3 = -\frac{1}{2}$ and $Y_L = -1$ yields $Q = -1$ for e^-, μ^-, and τ^-. As you will see in Box 25.4, we can also write the neutral weak interaction term for either $\overline{\nu}_L \gamma^\mu \nu_L$ or $\overline{e}_L \gamma^\mu e_L$ in the generalized form

$$\mathcal{L}_{\text{int},Z} = -g_W \frac{Z_\mu}{\cos \theta_W} [\, T_3 - Q \sin^2 \theta_W \,]\overline{\psi}_{L,R} \gamma^\mu \psi_{L,R}. \tag{25.14}$$

It turns out that this equation also works for right-handed fields (the subscripts on the particle fields should be either both L or both R). Remember that right-handed particles do not participate in the $SU(2)_L$ symmetry. Right-handed neutrinos and right-handed massive leptons e^-, μ^-, and τ^- are not part of doublets, but are **singlets**; that is, fields that stand alone, not part of doublets. We therefore assign such fields a value of $T_3 = 0$ (they are not part of an isospin doublet), then assign a value of Y_R that yields the same charge as their left-handed sibling in the generalized equation

$$Q = \tfrac{1}{2} Y_W + T_3, \tag{25.15}$$

where the **weak hypercharge** Y_W now represents either Y_L or Y_R. For example, we see that right-handed electrons (since $T_3 = 0$) must have $Y_R = -2$ instead of $Y_L = -1$ to have the correct charge. Right-handed neutrinos have $Y_R = 0$. These designations also work in Equation 25.14. In particular, we see that right-handed neutrinos (with $Q = 0$ and $T_3 = 0$) don't participate in the weak interaction at all, though right-handed electrons do. In Box 25.4, you will verify that Equation 25.14 also gives the correct expressions for right-handed electron interactions.

Note that since chirally right-handed neutrinos can interact with other particles neither weakly nor electromagnetically, it is not clear that they exist at all. If neutrinos were strictly massless, it would be impossible to create chirally right-handed neutrinos, or to detect them if they do exist (unless they are connected in some manner with the gravitational interaction and were created primordially somehow). Neutrinos do have a small mass (which we will talk about later), but exactly what this means and where this mass comes from is not clear. We currently do not have any solid experimental evidence that chirally right-handed neutrinos exist.

A different way of expressing Equation 25.14 without explicitly specifying whether we are talking about right- or left-handed fields is

$$\mathcal{L}_{\text{int},Z} = -g_W \frac{Z_\mu}{2 \cos \theta_W} \overline{\psi} \gamma^\mu [\, c_V - c_A \gamma^5 \,]\psi,$$

$$\text{with} \quad c_V \equiv T_{3L} - 2Q \sin^2 \theta_W \quad \text{and} \quad c_A \equiv T_{3L}, \tag{25.16}$$

where T_{3L} is the value of T_3 for the left-handed version of the particle. You will show that this is correct in Problem P25.1. (The conventional V and A coefficient subscripts reflect that, as we saw in the last chapter, $\overline{\psi} \gamma^\mu \psi$ and $\overline{\psi} \gamma^\mu \gamma^5 \psi$ transform like a *vector* and an *axial* vector, respectively, under parity.)

25.4 Quarks and the Weak Interaction

The way that quarks are involved with the weak interaction is a bit more complicated than it is for leptons. From the context of the *neutral* weak interaction (interactions involving the Z_μ field), the main distinction is that quarks have a hypercharge of $Y_L = \frac{1}{3}$ instead of $Y_L = -1$ for the left-handed doublet. Table 25.1 shows the isospin and hypercharge assignments for all particles in the lowest-mass lepton and quark doublets.

There are mysteries here that the Standard Model does not explain. Why are quark charges simple fractions of the electron charge ($\frac{2}{3}$ and $-\frac{1}{3}$) but not equal to the electron charge? Is the fact that quark triplets are stable (and thus have integer charge) related?

Symbol	Chirality	Isospin (T_3)	Hypercharge (Y_W)	Charge (Q)
ν_L	left	1/2	-1	0
ν_R	right	0	0	0
e_L	left	$-1/2$	-1	-1
e_R	right	0	-2	-1
u_L	left	1/2	1/3	2/3
u_R	right	0	4/3	2/3
d_L	left	$-1/2$	1/3	$-1/3$
d_R	right	0	$-2/3$	$-1/3$

TABLE 25.1   This table lists the properties of particles in the lowest-mass electroweak doublets. (Note that right-handed neutrinos may not actually exist.)

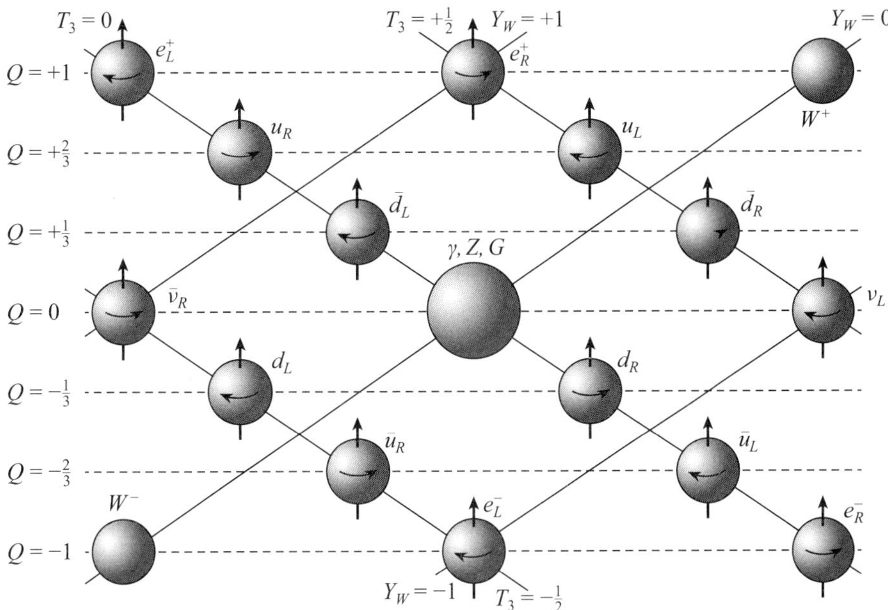

FIGURE 25.1   This chart shows one way to arrange all the particles in the lowest-mass electroweak doublets. The arrows depict how a particle's momentum would be related to its spin if it were massless (that is, they display *chirality*, not helicity). Overbars represent antiparticles, e^+ is a positron, γ is a photon, and G is a gluon. (Adapted from Goldberg, *The Standard Model in a Nutshell*, Princeton, 2017.)

The assignment of hypercharges in Table 25.1 does yield the correct observed charges, and yields a pretty pattern (see Figure 25.1), but the Standard Model does not explain the pattern.

Equation 25.16 applies not only to leptons but also to quarks with the values of T_3 and Q given in Table 25.1. This means that in a Feynman diagram, the vertex factor for a neutral weak interaction involving leptons or quarks is

$$\text{Vertex factor for } Z: \quad \frac{-ig_W}{2\cos\theta_W}\gamma^\mu\left[\,c_V - c_A\gamma^5\,\right],$$

$$\text{with} \quad c_V \equiv T_{3L} - 2Q\sin^2\theta_W \quad \text{and} \quad c_A \equiv T_{3L}. \tag{25.17}$$

Fermion	c_V	c_A
ν_e, ν_μ, ν_τ	$+\frac{1}{2}$	$\frac{1}{2}$
e^-, μ^-, τ^-	$-\frac{1}{2} + 2\sin^2\theta_W$	$-\frac{1}{2}$
u, c, t	$+\frac{1}{2} - \frac{4}{3}\sin^2\theta_W$	$\frac{1}{2}$
d, s, b	$-\frac{1}{2} + \frac{2}{3}\sin^2\theta_W$	$-\frac{1}{2}$

TABLE 25.2   This table provides a reference for the values of c_V and c_A for various types of particles involved in a weak interaction with a Z. Note that the experimental value of (the "on-shell" definition of) $\sin^2\theta_W = 0.2234$.

Table 25.2 gives a handy list of the coefficients for various kinds of fermions.

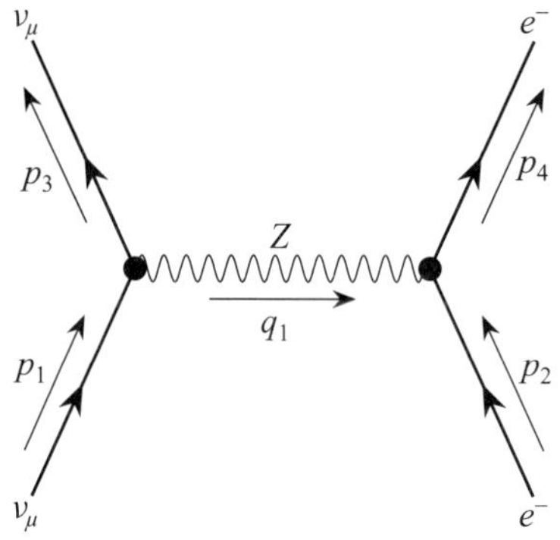

FIGURE 25.2   This is the Feynman diagram for electron neutrino scattering.

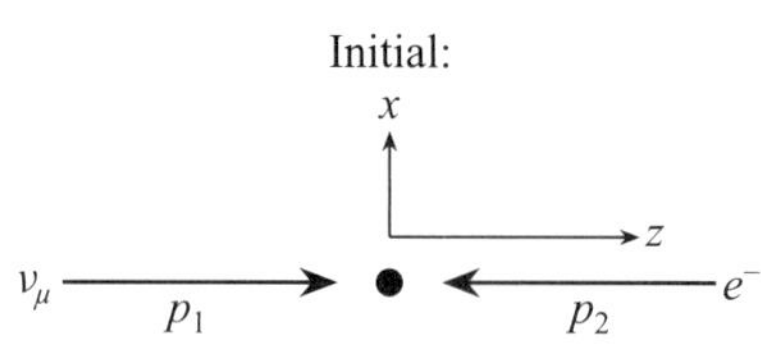

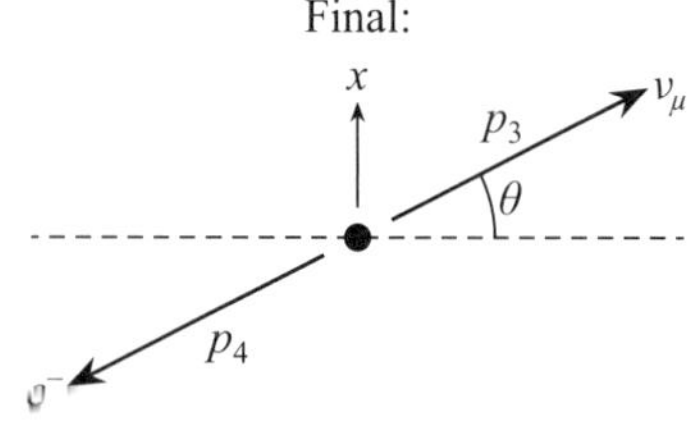

FIGURE 25.3   This drawing defines coordinate axes and the angle θ for neutrino-electron scattering in the CM frame.

25.5 Elastic Neutrino-Electron Scattering

As an example of the success of the electroweak theory, consider the case of elastic neutrino-electron scattering. The Feynman diagram for this process is shown in Figure 25.2. I hope that practice has made you pretty good at reading scattering amplitudes from Feynman diagrams, so I won't go over all the steps. The amplitude for this diagram (assuming the energy is $\ll M_Z$) is

$$\mathcal{M} = \frac{1}{8}\left(\frac{g_W}{M_Z\cos\theta_W}\right)^2 \left[\bar{u}_3\gamma^\mu(1-\gamma^5)u_1\right] g_{\mu\nu}\left[\bar{u}_4\gamma^\nu(c_V - c_A\gamma^5)u_2\right], \quad (25.18)$$

where M_Z is the mass of the Z. Applying Casimir's trick yields

$$\langle|\mathcal{M}|^2\rangle = \frac{1}{128}\left(\frac{g_W}{M_Z\cos\theta_W}\right)^4 \mathrm{Tr}[\,\gamma^\mu(1-\gamma^5)\slashed{p}_1\gamma^\alpha(1-\gamma^5)\slashed{p}_3\,]g_{\mu\nu}g_{\alpha\beta}$$

$$\times \mathrm{Tr}[\,\gamma^\nu(c_V - c_A\gamma^5)(\slashed{p}_2 + m)\gamma^\beta(c_V - c_A\gamma^5)(\slashed{p}_4 + m)\,], \quad (25.19)$$

where m is the electron mass and c_V and c_A have the values listed for electrons. Note that I have divided by a factor of 2 to average over the incoming electron spins (the neutrino has only one spin state). A calculation similar to ones we have done before (see Box 25.5) yields the following:

$$\langle|\mathcal{M}|^2\rangle = \frac{1}{2}\left(\frac{g_W}{M_Z\cos\theta_W}\right)^4 \Big[\,(c_V + c_A)^2(p_1\cdot p_2)(p_3\cdot p_4)$$

$$+ (c_V - c_A)^2(p_1\cdot p_4)(p_2\cdot p_3) - m^2(c_V^2 - c_A^2)(p_1\cdot p_3)\,\Big]. \quad (25.20)$$

In the center-of-mass frame, suppose that the neutrino and electron approach each other along the z axis and depart in such a way that each particle has been deflected through an angle θ as shown in Figure 25.3. The calculation simplifies a bit more if we assume that the collision energy is large compared to the electron mass, so that $E_n^2 \gg m^2$ and $E_n^2 \approx |\vec{p}_n|^2$ (where $n = 1,2,3,4$). This means that all particles involved in the collision will have approximately the same energy, which we will call E. Therefore,

$$p_1\cdot p_2 = E_1 E_2 - p_{1z}p_{2z} = E^2 - E(-E) = 2E^2, \quad (25.21a)$$

$$p_3\cdot p_4 = E_3 E_4 - p_{3z}p_{4z} - p_{3x}p_{4x}$$

$$= E^2 - E\cos\theta(-E\cos\theta) - E\sin\theta(-E\sin\theta) = 2E^2, \quad (25.21b)$$

$$p_1\cdot p_4 = E_1 E_4 - p_{1z}p_{4z} = E^2 - E(-E\cos\theta)$$

$$= E^2(1 + \cos\theta) = 2E^2\cos^2\tfrac{1}{2}\theta, \quad (25.21c)$$

$$p_2\cdot p_3 = E_2 E_3 - p_{2z}p_{3z} = E^2 - (-E)(E\cos\theta) = 2E^2\cos^2\tfrac{1}{2}\theta, \quad (25.21d)$$

$$p_1\cdot p_3 = E_1 E_3 - p_{1z}p_{3z} = E^2 - E^2\cos\theta = 2E^2\sin\tfrac{1}{2}\theta, \quad (25.21e)$$

where I have used the identities $\cos^2 x = \frac{1}{2}(1 + \cos 2x)$ and $\sin^2 x = \frac{1}{2}(1 - \cos 2x)$. Note that the first two terms in Equation 25.20 are proportional to E^4, while the last is proportional to $m^2 E^2 \ll E^4$. Dropping that term yields

$$\langle |\mathcal{M}|^2 \rangle = 2 \left(\frac{g_W}{M_W \cos \theta_W} \right)^4 E^4 \left[(c_V + c_A)^2 + (c_V - c_A)^2 \cos^4 \tfrac{1}{2}\theta \right]. \quad (25.22)$$

Using Equation 15.11, we find that the differential cross section is

$$d\sigma = \frac{S \langle |\mathcal{M}|^2 \rangle}{(8\pi)^2 (E_1 + E_2)^2} \frac{|\vec{p}_3|}{|\vec{p}_1|} d\Omega = \frac{S \langle |\mathcal{M}|^2 \rangle}{(8\pi)^2 (2E)^2} \frac{\not{E}}{\not{E}} d\Omega$$

$$= \frac{1}{128\pi^2} \left(\frac{g_W}{M_Z \cos \theta_W} \right)^4 E^2 \left[(c_V + c_A)^2 + (c_V - c_A)^2 \cos^4 \tfrac{1}{2}\theta \right] d\Omega \quad (25.23)$$

since $S = 1$ (no identical outgoing particles) and the E^2 in the denominator cancels two powers of E in the numerator. WolframAlpha says that

$$\int_0^\pi \cos^4(\tfrac{1}{2}\theta) \sin \theta \, d\theta = \frac{2}{3}, \quad (25.24)$$

so the total cross section for all angles is

$$\sigma = \frac{1}{128\pi^2} \left(\frac{g_W}{M_Z \cos \theta_W} \right)^4 E^2 \left[4\pi (c_V + c_A)^2 + \frac{4\pi}{3} (c_V - c_A)^2 \right]. \quad (25.25)$$

A bit of algebra (see Box 25.6) leads to

$$\sigma = \frac{1}{32\pi} \left(\frac{g_W}{M_Z \cos \theta_W} \right)^4 E^2 \left[1 - 4 \sin^2 \theta_W + \tfrac{16}{3} \sin^4 \theta_W \right]. \quad (25.26)$$

Now, no one is actually going to perform this experiment in the center-of-mass frame: one is much more likely to do this by directing a beam of muon neutrinos at electrons at rest in a target. Fortunately, as we have discussed in Chapter 14, the cross section is itself Lorentz-invariant. We have expressed it in terms of the particle energies E in the center-of-mass frame, but we can easily use Equation 25.21a to express it in an obviously Lorentz-invariant form:

$$\sigma = \frac{1}{64\pi} \left(\frac{g_W}{M_Z \cos \theta_W} \right)^4 (p_1 \cdot p_2) \left[1 - 4 \sin^2 \theta_W + \tfrac{16}{3} \sin^4 \theta_W \right]. \quad (25.27)$$

In the frame where the electron is at rest, $p_1 \cdot p_2 = p_1^t p_2^t - p_1^x p_2^x - p_1^y p_2^y - p_1^z p_2^z = E_\nu m_e - 0 - 0 - 0$, where E_ν is the energy of the incoming neutrino in the lab frame, m_e is the mass of the electron, and the spatial components of the electron's four-momentum p_2 are zero. A bit of number crunching (see Problem P25.3) leads to a predicted cross section of $\sigma/E_\nu = 1.58 \times 10^{-46}$ m^2/GeV. The measured result (from Ahrens et al., *Phys Rev D*, **41**, 3297 (1990)) is $\sigma/E_\nu = (1.80 \pm 0.36) \times 10^{-46}$ m^2/GeV, which is comfortably consistent with this result.

One can also calculate ratios with other processes. In the next chapter, we will find that $M_Z \cos \theta_W = M_W$. This makes it easy to compare this cross section with the cross section for the "inverse muon decay" process $\nu_\mu + e^- \rightarrow \nu_e + \mu^-$ (see also Box 25.6), at least in the limit that the muon mass is also negligible:

$$\frac{\sigma(\nu_\mu + e^- \rightarrow \nu_\mu + e^-)}{\sigma(\nu_\mu + e^- \rightarrow \nu_e + \mu^-)} = \tfrac{1}{4} - \sin^2 \theta_W + \tfrac{4}{3} \sin^4 \theta_W = 0.093. \quad (25.28)$$

The experimental value for this ratio is 0.11 ± 0.02 (see Vilain, P. et al., *Phys Lett B*, **364**, 121 (1995)), so this result is also within experimental uncertainties.

25.6 Closing Comments

The unified electroweak theory is a huge step forward. It resolves almost all of the puzzles associated with the weak interaction and has (to date) been quite consistent with experiment. But we have one major problem left to resolve. The electroweak theory still only makes sense theoretically if all particles have zero mass, when they obviously do not. How can we fix this glaring problem? That is the subject of the next chapter.

BOX 25.1   Implications of the Weinberg Angle

The goal of this box is to show that, given

$$\mathcal{L}_{\text{neutral}, L} = -\tfrac{1}{2}\big[\, g'Y_L B_\mu + g_W W^{(z)}_\mu \,\big]\overline{\nu}_L \gamma^\mu \nu_L$$

$$-\tfrac{1}{2}\big[\, g'Y_L B_\mu - g_W W^{(z)}_\mu \,\big]\overline{e}_L \gamma^\mu e_L \qquad (25.29)$$

and the rotation associated with the Weinberg angle

$$\begin{bmatrix} B \\ W^{(z)} \end{bmatrix} = \begin{pmatrix} \cos\theta_W & \sin\theta_W \\ -\sin\theta_W & \cos\theta_W \end{pmatrix} \begin{bmatrix} A \\ Z \end{bmatrix}, \qquad (25.30)$$

the neutral, left-handed weak-interaction term becomes

$$\mathcal{L}_{\text{neutral}, L} = -c_1 \overline{\nu}_L \gamma^\mu \nu_L A_\mu - c_2 \overline{e}_L \gamma^\mu e_L A_\mu \qquad (25.31a)$$

$$- c_3 \overline{\nu}_L \gamma^\mu \nu_L Z_\mu - c_4 \overline{e}_L \gamma^\mu e_L Z_\mu, \qquad (25.31b)$$

$$\text{where} \quad c_1 = \tfrac{1}{2}g'Y_L \cos\theta_W - \tfrac{1}{2}g_W \sin\theta_W, \qquad (25.31c)$$

$$c_2 = \tfrac{1}{2}g'Y_L \cos\theta_W + \tfrac{1}{2}g_W \sin\theta_W, \qquad (25.31d)$$

$$c_3 = \tfrac{1}{2}g'Y_L \sin\theta_W + \tfrac{1}{2}g_W \cos\theta_W, \qquad (25.31e)$$

$$\text{and} \quad c_4 = \tfrac{1}{2}g'Y_L \sin\theta_W - \tfrac{1}{2}g_W \cos\theta_W. \qquad (25.31f)$$

Exercise 25.1.1   Show this.

BOX 25.2 Making A^μ the Electromagnetic Field

Our goal in this box is to show that, given that we want

$$c_1 = \tfrac{1}{2} g' Y_L \cos\theta_W - \tfrac{1}{2} g_W \sin\theta_W = 0 \quad \text{and} \tag{25.32a}$$

$$c_2 = \tfrac{1}{2} g' Y_L \cos\theta_W + \tfrac{1}{2} g_W \sin\theta_W = |\,q_e\,|, \tag{25.32b}$$

and given that we are choosing $Y_L = -1$ for lepton doublets, then we must have

$$g' = -g_W \tan\theta_W \quad \text{and} \quad |\,q_e\,| = g_W \sin\theta_W. \tag{25.33}$$

Exercise 25.2.1 Show this.

BOX 25.3 Eliminating g'

In this box, we will see that we can use $g' = -g_W \tan\theta_W$ and the fact that $Y_L = -1$ for lepton doublets to rewrite

$$\mathcal{L}_{\text{neutral},\,L} = -c_1 \bar{\nu}_L \gamma^\mu \nu_L A_\mu - c_2 \bar{e}_L \gamma^\mu e_L A_\mu$$

$$- c_3 \bar{\nu}_L \gamma^\mu \nu_L Z_\mu - c_4 \bar{e}_L \gamma^\mu e_L Z_\mu, \tag{25.34a}$$

$$\text{where} \quad c_1 = \tfrac{1}{2} g' Y_L \cos\theta_W - \tfrac{1}{2} g_W \sin\theta_W, \tag{25.34b}$$

$$c_2 = \tfrac{1}{2} g' Y_L \cos\theta_W + \tfrac{1}{2} g_W \sin\theta_W, \tag{25.34c}$$

$$c_3 = \tfrac{1}{2} g' Y_L \sin\theta_W + \tfrac{1}{2} g_W \cos\theta_W, \tag{25.34d}$$

$$\text{and} \quad c_4 = \tfrac{1}{2} g' Y_L \sin\theta_W - \tfrac{1}{2} g_W \cos\theta_W, \tag{25.34e}$$

in the form of

$$\mathcal{L}_{\text{neutral},\,L} = -|q_e|\, A_\mu [\bar{e}_L \gamma^\mu e_L]$$

$$- g_W \frac{Z_\mu}{2\cos\theta_W} \left([\,\bar{\nu}_L \gamma^\mu \nu_L\,] - \cos(2\theta_W)[\,\bar{e}_L \gamma^\mu e_L\,] \right). \tag{25.35}$$

Exercise 25.3.1 Show this. (*Hint:* $\cos^2 x - \sin^2 x = \cos 2x$.)

BOX 25.4   The Z Interaction in Terms of Q and T_3

Our goal in this box is to show that the expression

$$\mathcal{L}_{\text{int},Z} = -g_W \frac{Z_\mu}{\cos\theta_W}[\,T_3 - Q\sin^2\theta_W\,]\overline{\psi}\gamma^\mu\psi \qquad (25.36)$$

correctly describes how particle fields connect to the neutral weak mediator Z for both right-handed and left-handed fields. We begin with the left-handed fields, where Equation 25.12 told us that

$$\mathcal{L}_{LZ} = -g_W \frac{Z_\mu}{2\cos\theta_W}\left(\left[\,\overline{\nu}_L\gamma^\mu\nu_L\,\right] - \cos(2\theta_W)\left[\,\overline{e}_L\gamma^\mu e_L\,\right]\right), \qquad (25.37)$$

and where $T_3 = \frac{1}{2}$ and $Q = 0$ for neutrinos, and $T_3 = -\frac{1}{2}$ and $Q = -1$ for the massive leptons (e^-, μ^-, τ^-).

Exercise 25.4.1   Show that Equation 25.36 correctly describes each term in Equation 25.37. (*Hint:* $\cos^2 x - \sin^2 x = \cos 2x$.)

BOX 25.4 The Z Interaction in Terms of Q and T_3 (cont.)

Now let's look at the interaction involving right-handed particles. Right-handed neutrinos shouldn't interact at all, and since $T_3 = 0$ and $Q = 0$ for right-handed neutrinos, Equation 25.36 does just fine. Now, for right-handed electrons, note that Equation 25.3 tells us that

$$J_\mu^{(0,R)} = \tfrac{1}{2}g'Y_R\left(\overline{\nu}_R\gamma_\mu\nu_R + \overline{e}_R\gamma_\mu e_R\right), \tag{25.38}$$

and Equation 25.6 tells us that $\mathcal{L}_{\mathrm{int},B} = -\left[\, J^{(0,R)} + J^{(0,L)} \,\right] \cdot B$, so the interaction term involving only right-handed electrons is

$$-J_e^{(0,R)} \cdot B = \tfrac{1}{2}g'Y_R\overline{e}_R\gamma^\mu e_R B_\mu. \tag{25.39}$$

Equation 25.8 also tells us that

$$B_\mu = \cos\theta_W A_\mu + \sin\theta_W Z_\mu. \tag{25.40}$$

Exercise 25.4.2 In Equation 25.39, first eliminate the $g'Y_R$ in favor of g_W (remember that $Y_R = -2$ for electrons) and combine with Equation 25.40 to get two terms: one involving A_μ and one Z_μ. Show that the A_μ term has the same form as for left-handed electrons. Also, show that the Z_μ term is the same as what we get if we substitute $T_3 = 0$ and $Q = -1$ (the correct numbers for right-handed electrons) into Equation 25.36.

BOX 25.5 Elastic Electron Neutrino Scattering

The goal of this box is to use trace theorems to compute the amplitude given in Equation 25.19 (repeated here for easy reference):

$$\langle |\mathcal{M}|^2 \rangle = \frac{1}{128} \left(\frac{g_W}{\cos \theta_W M_Z} \right)^4 \mathrm{Tr}[\, \gamma^\mu (1 - \gamma^5) \not{p}_1 \gamma^\alpha (1 - \gamma^5) \not{p}_3 \,] g_{\mu\nu} g_{\alpha\beta}$$

$$\times \mathrm{Tr}[\, \gamma^\nu (c_V - c_A \gamma^5)(\not{p}_2 + m) \gamma^\beta (c_V - c_A \gamma^5)(\not{p}_4 + m) \,]. \quad (25.41)$$

For ease in writing, call the first trace term Tr_1 and the second Tr_2.

We already calculated the first trace factor in Box 23.4. The result is

$$\mathrm{Tr}_1 = 8\big(p_1^\mu p_3^\alpha - g^{\mu\alpha}(p_1 \cdot p_3) + p_1^\alpha p_3^\mu + i p_{1\sigma} p_{3\lambda} \epsilon^{\mu\sigma\alpha\lambda} \big). \quad (25.42)$$

Exercise 25.5.1 Evaluate the other term. (*Hint:* Study the calculation in Box 24.4 carefully and note differences. Besides the factors of c_V and c_A, note that terms involving m^2 will contribute.)

BOX 25.5 Elastic Electron Neutrino Scattering (cont.)

The product of the traces is

$$
\mathrm{Tr}_1 \, g_{\mu\nu} g_{\alpha\beta} \mathrm{Tr}_2 = 8\left[\, p_1^\mu p_3^\alpha - g^{\mu\alpha}(p_1 \cdot p_3) + p_1^\alpha p_3^\mu + i p_{1\sigma} p_{3\lambda} \epsilon^{\mu\sigma\alpha\lambda} \, \right]
$$
$$
\times 4\left[\, (c_V^2 + c_A^2)(p_{2\mu} p_{4\alpha} - g_{\mu\alpha}(p_2 \cdot p_4) + p_{2\alpha} p_{4\mu}) \right.
$$
$$
\left. + 2 i c_V c_A p_2^\eta p_4^\rho \epsilon_{\mu\eta\alpha\rho} + m^2(c_V^2 - c_A^2) g_{\mu\alpha} \, \right]
$$
$$
= 32\left[\, S^{\mu\alpha} + i p_{1\sigma} p_{3\lambda} \epsilon^{\mu\sigma\alpha\lambda} \, \right]\left[R_{\mu\alpha} + 2 i c_V c_A p_2^\eta p_4^\rho \epsilon_{\mu\eta\alpha\rho} \, \right], \quad (25.43a)
$$

where

$$
S^{\mu\alpha} \equiv p_1^\mu p_3^\alpha - g^{\mu\alpha}(p_1 \cdot p_3) + p_1^\alpha p_3^\mu, \tag{25.43b}
$$
$$
R_{\mu\alpha} \equiv (c_V^2 + c_A^2)(p_{2\mu} p_{4\alpha} - g_{\mu\alpha}(p_2 \cdot p_4) + p_{2\alpha} p_{4\mu}) + m^2(c_V^2 - c_A^2) g_{\mu\alpha}. \tag{25.43c}
$$

Because $S^{\mu\alpha}$ and $R_{\mu\alpha}$ are symmetric, we have $S^{\mu\alpha} \epsilon_{\mu\eta\alpha\rho} = 0 = R_{\mu\alpha} \epsilon^{\mu\sigma\alpha\lambda}$ for the reasons discussed in Box 24.4, so the cross terms vanish. The final term is

$$
-2 c_V c_A p_{1\sigma} p_{3\lambda} \epsilon^{\mu\sigma\alpha\lambda} p_2^\eta p_4^\rho \epsilon_{\mu\eta\alpha\rho} = +4 c_V c_A p_{1\sigma} p_{3\lambda} p_2^\eta p_4^\rho (\delta_\eta^\sigma \delta_\rho^\lambda - \delta_\rho^\sigma \delta_\eta^\lambda)
$$
$$
= 4 c_V c_A (p_1 \cdot p_2)(p_3 \cdot p_4) - 4 c_V c_A (p_1 \cdot p_4)(p_2 \cdot p_3), \quad (25.44)
$$

where I have used the identity mentioned in Box 24.4.

Exercise 25.5.2 Now calculate $R_{\mu\alpha} S^{\mu\alpha}$, then $\mathrm{Tr}_1 \, g_{\mu\nu} g_{\alpha\beta} \mathrm{Tr}_2$, and compare your result with Equation 25.20.

BOX 25.6   Neutrino-Electron Scattering Cross Section

The goal of this box is to simplify the expression for the neutrino-electron scattering cross section given in Equation 25.25 (repeated here for easy reference):

$$\sigma = \frac{1}{128\pi^2} \left(\frac{g_W}{M_Z \cos\theta_W} \right)^4 E^2 \left[4\pi (c_V + c_A)^2 + \frac{4\pi}{3} (c_V - c_A)^2 \right]. \quad (25.45)$$

Exercise 25.6.1   Substitute the values of c_V and c_A for the electron from Table 25.2 into this expression and simplify. Also, compare this result to the total cross section for inverse muon decay, which is

$$\sigma_{\text{imd}} = \frac{1}{8\pi} \left(\frac{g_W}{M_W} \right)^4 E^2 \quad (25.46)$$

in the limit that $E^2 \gg m_\mu^2$ (see Problem 24.6).

Homework Problems

P25.1 Show that Equation 25.16 is equivalent to Equation 25.14 for both left-handed and right-handed terms. (*Hints:* (1) Take Equation 25.14 and write the version for left-handed fields. Label T_3 by T_{3L} to indicate that we are applying this equation to left-handed fields. (2) In that equation, replace $\overline{\psi}_L \gamma^\mu \psi_L$ by $\frac{1}{2}\overline{\psi}\gamma^\mu(1-\gamma^5)\psi$, which is the same thing. (3) Multiply out the binomials $(T_{3L} - Q\sin^2\theta_W)(1-\gamma^5)$, keeping everything between the γ^μ on the left and the ψ on the right. (4) Repeat the previous three steps for right-handed fields, noting that $\overline{\psi}_R \gamma^\mu \psi_R = \frac{1}{2}\overline{\psi}\gamma^\mu(1+\gamma^5)\psi$ and $T_{3R}=0$. (5) Add the equations to get something that correctly handles both types of fields. (6) Between the γ^μ and ψ, group terms that don't multiply the γ^5 and those that do. The first group of terms is c_V, the second is c_A.)

P25.2 Consider the doublet of Dirac spinors

$$\Psi = \begin{bmatrix} \psi_u \\ \psi_d \end{bmatrix} \tag{25.47}$$

before any gauge transformation. Assume that the left-handed components and right-handed components transform under a global gauge transformation as

$$\Psi'_L = e^{-i\frac{1}{2}g'Y_L\theta^0} e^{-i\frac{1}{2}g_W\vec{\sigma}\cdot\vec{\theta}} \Psi_L, \tag{25.48a}$$

$$\Psi'_R = e^{-i\frac{1}{2}g'Y_R\theta^0} \Psi_R, \tag{25.48b}$$

and that the left-handed doublet has a common hypercharge of Y_L, while the independent right-handed fields have separate hypercharges Y_{uR} and Y_{dR}. Assume that $\theta^x = \theta^y = 0$.
a. Show that the left-handed doublet transforms as

$$\Psi'_L = \begin{bmatrix} \psi'_{uL} \\ \psi'_{dL} \end{bmatrix} = \begin{bmatrix} e^{-i\frac{1}{2}(g'Y_L\theta^0 + g_W\theta^z)}\psi_{uL} \\ e^{-i\frac{1}{2}(g'Y_L\theta^0 - g_W\theta^z)}\psi_{dL} \end{bmatrix}. \tag{25.49}$$

b. Argue that

$$\overline{\psi}\psi = \overline{\psi}_L\psi_R + \overline{\psi}_R\psi_L. \tag{25.50}$$

(*Hint:* Work from right to left. Expand each ψ as a four-component spinor, and remember that ψ_L has zeros for the lower two components and ψ_R has zeros for the upper two components.)

c. Use the result of the previous part to write

$$\overline{\Psi}'\Psi' = [\,\overline{\psi}'_u \quad \overline{\psi}'_d\,]\begin{bmatrix} \psi'_u \\ \psi'_d \end{bmatrix} \tag{25.51}$$

in terms of ψ_{uL}, ψ_{uR}, ψ_{dL}, and ψ_{dR}. (*Hint:* Invent some shorthand for the exponentials to save writing.)
d. What constraint on how Y_L must be related to Y_{uR} and Y_{dR} arises if we insist that if the transformation phase for $\overline{\psi}_{uL}\psi_{uR}$ in the expansion in the previous part has the form $e^{i\alpha}$ for some α, then the transformation phase for the $\overline{\psi}_{dL}\psi_{dR}$ has the form $e^{-i\alpha}$? Is this the same constraint that would arise if we required the same relationship between the $\overline{\psi}_{uR}\psi_{uL}$ and $\overline{\psi}_{dR}\psi_{dL}$ terms?

P25.3 Calculate the numerical value of the cross section σ for the elastic $v_\mu + e^- \to v_\mu + e^-$ scattering process given in Equation 25.27, and verify that the result is what is quoted below that equation. (The cross section for this process is difficult to measure because it is so small: cross sections for neutrinos interacting with nucleons are tens of thousands of times larger.)

P25.4 Check that the numerical value given in Equation 25.28 is correct, given the experimental value of $\sin\theta_W$ provided in the text.

P25.5 When we considered elastic scattering of a muon neutrino by an electron, why did we not consider elastic scattering of an *electron*-neutrino by an electron: $v_e + e^- \to v_e + e^-$? The answer is simplicity. There is a second Feynman diagram for the latter process that does not appear in the first case. What is that diagram, and why doesn't it appear in the muon-scattering case?

P25.6 One can also consider elastic scattering of a muon-*anti*neutrino by an electron: $\overline{v}_\mu + e^- \to \overline{v}_\mu + e^-$.
a. Explain why the cross section for this reaction will not be the same as for $v_\mu + e^- \to v_\mu + e^-$.
b. Calculate the ratio $\sigma(v_\mu e^-)/\sigma(\overline{v}_\mu e^-)$.
c. Compare to the experimental ratio $1.37(+0.65,-0.44)$ (Bergsma et al., *Phys Lett* **117B**, 272, (1982)).

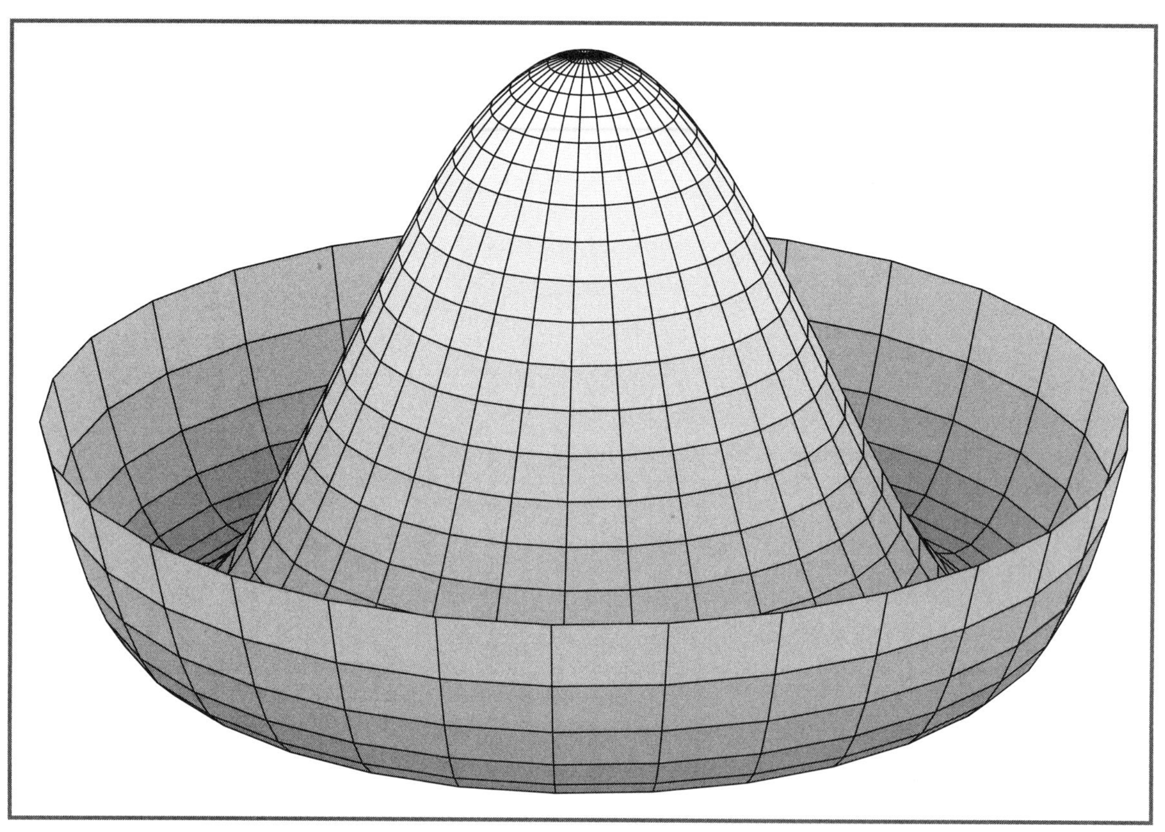

26. THE HIGGS MECHANISM

In spite of the success of electroweak theory, the entire logic of the theory requires both the theory's fundamental fermions and the interaction-mediating bosons to be massless. So far, we have been simply inserting the observed masses by hand, as if this made any kind of sense. But it does *not* make sense, and represents an internal contradiction that cannot be allowed to stand. Historically, this led physicists to reject Yang-Mills-type theories based on gauge transformations as unrealistic.

However, this changed in 1964 when Brout and Englert, then Higgs, and then Guralnik, Hagen, and Kibble published papers outlining a solution to the problem, a solution that has come to be known as the **Higgs mechanism**. (Englert and Higgs won the 2013 Nobel Prize for this work.) In the broadest strokes, the mechanism shows how fundamentally massless particles gain apparent mass through a process known as "spontaneous symmetry breaking" and interaction with a new field called the "Higgs field." The purpose of this chapter is to describe how the Higgs mechanism works.

26.1 Mass Terms in a Lagrangian Density

Before we begin, it behooves us to think very carefully about how we can recognize mass terms in a Lagrangian density. The Lagrangian densities we have considered so far have had very explicit mass terms. For example, in the complex Klein-Gordon Lagrangian density

$$\mathcal{L}_{KG} = \partial^\mu \phi^* \, \partial_\mu \phi - m^2 \phi^* \phi, \tag{26.1}$$

the second term is the mass term. We know that this is a mass term because plane-wave solutions $e^{-ip\cdot x}$ to this field's Euler-Lagrange equation

$$\partial^\mu \partial_\mu \phi + m^2 \phi = 0 \tag{26.2}$$

have four-momenta such that $-p^2 + m^2 = 0 \Rightarrow p^2 = m^2$. We see that to enter the Euler-Lagrange equation in the correct way, a mass term in $\mathcal{L}$ must (1) involve the absolute square of the field, and (2) be negative. Any such term in a Lagrangian density will cause the field to behave as if it has a mass.

Sometimes mass terms can be hidden. To see how this might happen, consider the (completely artificial) Lagrangian density

$$\mathcal{L} = \partial^\mu \phi^* \partial_\mu \phi + V_0^4 e^{-|b\phi|^2}, \tag{26.3}$$

where V_0 is a real constant with units of energy and b is a real constant with units of inverse energy. At first glance, this does not seem to have a mass term at all: it looks like we have a massless field with an interaction term $V_0^4 e^{-|b\phi|^2}$ that involves the field coupling to itself in various ways. But if you expand the exponential term, you will expose a mass term in hiding:

$$\mathcal{L} = \partial^\mu \phi^* \, \partial_\mu \phi + V_0^4 - V_0^4 b^2 |\phi|^2 + \tfrac{1}{2} V_0^4 b^4 |\phi|^4 - \cdots . \tag{26.4}$$

The constant V_0^4 term is physically irrelevant because it will disappear when we calculate the Euler-Lagrange equations for the field. But the $-V_0^4 b^2 |\phi|^2$ term is a mass term. So we can interpret the first three terms as the Lagrangian density of a free scalar field with mass m such that

$$m^2|\phi|^2 = V_0^4 b^2|\phi|^2 \quad \Rightarrow \quad m = V_0^2 b, \tag{26.5}$$

and the *remaining* terms (in a quantum field theory) would describe perturbing interaction terms involving the field coupling to itself.

Another issue can arise that is at the heart of the Higgs mechanism. Suppose we add to our array of particles a new spin-0 boson described by the following Higgs Lagrangian density:

$$\begin{aligned}
\mathcal{L} &= \partial^\mu\phi^* \, \partial_\mu\phi - \tfrac{1}{4}\lambda^2(|\phi|^2 - v^2)^2 \\
&= \partial^\mu\phi^* \, \partial_\mu\phi - \tfrac{1}{4}\lambda^2\,|\phi|^4 + \tfrac{1}{2}\lambda^2 v^2\,|\phi|^2 - \tfrac{1}{4}\lambda^2 v^4,
\end{aligned} \tag{26.6}$$

where λ is a real and unitless constant and v is a real constant with units of energy. The $\tfrac{1}{2}\lambda^2 v^2\,|\phi|^2$ term looks at first glance like a mass term, but it has the wrong sign. So this Lagrangian density apparently describes a massless spin-0 boson with some self-interaction terms.

But this is *not* the correct interpretation due to a more subtle issue. The entire development of quantum field theory rests on the assumption that the field represents a disturbance away from the system's lowest-energy state, which we define to be the *vacuum*. (When we move to quantize the field, the field becomes an operator that creates and annihilates particles, and the vacuum by definition is where further annihilation is impossible.) In all of the Lagrangian densities we have considered so far, the ground state (minimum-energy state) has been where $\phi = 0$. But this is not so for the Lagrangian density above. To see what the vacuum state really is, write the Lagrangian density in the form $\mathcal{L} = \mathcal{K} - \mathcal{U}$, where (by analogy to classical physics) $\mathcal{K}$ is the "kinetic energy" term ($\partial^\mu\phi^* \, \partial_\mu\phi$ in this case) and $\mathcal{U}$ is a "potential energy" term ($\tfrac{1}{4}\lambda^2[\,|\phi|^2 - v^2\,]^2$ in this case). Figure 26.1 shows a plot of $\mathcal{U}(\phi)$. You can see from either the graph or from the mathematical form of $\mathcal{U} = \tfrac{1}{4}\lambda^2(|\phi|^2 - v^2)^2$ that $\mathcal{U}$ has a minimum not at $\phi = 0$ but rather at the circular set of points in the complex plane where $|\phi| = v$ (conveniently, $\mathcal{U} = 0$ at these points). We call v the **vacuum expectation value** (VEV) of the theory.

So, to fulfill the assumptions of the quantum field theory program, our field *should* reflect a disturbance from the system's *lowest-energy state*, which represents the true vacuum. In this case, we actually have an entire continuous set of vacuum states to choose from (anywhere on the circular minimum of the potential energy). We can arbitrarily choose our vacuum to be $\phi_g = v$ (meaning that we are choosing our vacuum state such that ϕ_g is purely real). Then define a pair of real fields $H(x)$ and $G(x)$ such that

$$\phi(x) = \left(\frac{H(x)}{\sqrt{2}} + v\right) e^{iG(x)/\sqrt{2}v} = \frac{1}{\sqrt{2}}(H + a)e^{iG/a}, \tag{26.7}$$

where I have defined $a \equiv \sqrt{2}v$ to save writing. The two real fields H and G replace the two degrees of freedom $\phi = \phi_1 + i\phi_2$ in our original complex field and describe displacements from the vacuum in the radial direction and around the circular trough, respectively. (We will see the point of the $\sqrt{2}$ and $\sqrt{2}v$ factors in the denominators shortly.) If we substitute this into the Lagrangian density (see Box 26.1), we get

$$\mathcal{L} = \left[\tfrac{1}{2}\partial^\mu H \, \partial_\mu H - \tfrac{1}{2}\lambda^2 v^2 H^2\right] + \left[\tfrac{1}{2}\partial^\mu G \, \partial_\mu G\right]$$
$$- \left[\frac{1}{16}\lambda^2(H^4 + 4\sqrt{2}vH^3)\right] + \left[\frac{1}{4v^2}\partial^\mu G \, \partial_\mu G(H^2 + 2\sqrt{2}vH)\right]. \tag{26.8}$$

The first term in brackets is the Klein-Gordon Lagrangian density for a real scalar field called a **Higgs field** whose associated spin-0 particle has mass $m_H = \lambda v$, and the second term gives a Klein-Gordon Lagrangian density for a massless scalar field whose

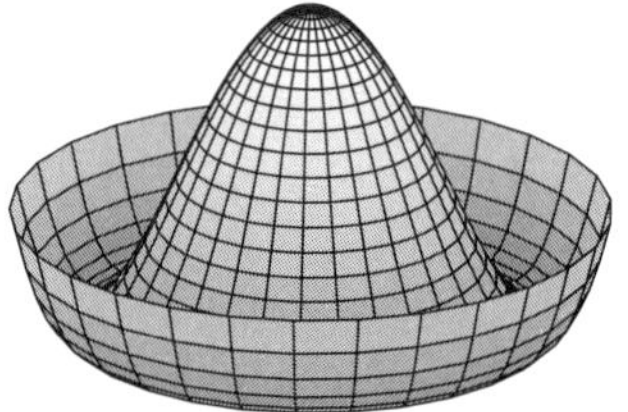

FIGURE 26.1   This is a graph of the Higgs potential energy function $\mathcal{U}(\phi)$. The horizontal axes on this graph are the real and imaginary parts of the complex value of ϕ.

associated spin-0 particle is called a **Goldstone boson**. (Now you see the point of the factors of $\sqrt{2}$ and $\sqrt{2}v$ in Equation 26.7: they scale the fields to give the usual factors of $\frac{1}{2}$ that we expect in the Lagrangian densities for real fields, though a field's scale is actually physically irrelevant for the physics.) The third term in brackets describes a set of self-interactions involving the Higgs field, and the odd-looking fourth term can be interpreted as describing interactions between the two fields H and G. The main point is that this adjustment to the correct vacuum state has led to a massive H field emerging from the apparently massless ϕ field.

Please note that the Lagrangian densities in Equations 26.6 and 26.8 represent the *same physics* written in different notation. That being said, only the second formulation is appropriate for a quantum field theory, which *assumes* that the system's ground state is the vacuum.

26.2 Spontaneous Symmetry Breaking

Equation 26.7 illustrates the concept of **spontaneous symmetry breaking**. The original Lagrangian density had a U(1) symmetry because it is invariant under the transformation $\phi' = \phi e^{-iq\theta}$. But our new Lagrangian density does *not* have an analogous symmetry for the H and G fields, because we had to choose which of the ground states $|\phi| = v$ we consider to be our vacuum state, and making a choice breaks the original symmetry. Physicists call this "spontaneous" symmetry breaking because nothing in the system *itself* forces us to choose one vacuum state or the other. A ball at the top of a symmetrical hill would randomly roll down the hill's right or left side, and so spontaneously choose which of its "ground states" to settle down in, but this introduced asymmetry has nothing to do with the symmetric hill. The system's underlying symmetry is *hidden* by our choosing an arbitrary ground state.

Examples of spontaneous symmetry breaking are common even in everyday physics. One is the phenomenon of **ferromagnetism**. The atomic spins in a ferromagnet have no *intrinsic* reason to point in any particular direction, and so long as the magnet's temperature is higher than its **Curie point** the atomic spins remain truly random. But in a ferromagnetic material, the atomic spins interact with each other in such a way that having neighboring spins aligned is energetically preferred. When the magnet's temperature falls below the Curie point, regions (called **domains**) in the ferromagnet spontaneously arise where the spins have aligned themselves in a particular common direction, a direction that emerges simply by random chance. Within such a domain, the system's initial intrinsic symmetry has been broken: the aligned spins now indicate a certain special direction in space that hides the system's original isotropy.

This symmetry breaking necessarily comes with the massless Goldstone boson field (as Goldstone proved), and at first glance, that fact looked fatal to the theory. If $m_H = \lambda v$ is large enough, people in 1964 might not have been able to see the particle associated with the Higgs field H, but should have already observed any massless particles the field predicts. But as has often been the case in the history of quantum field theory, suspending disbelief for the moment pays off down the road.

26.3 The Higgs Doublet

The Higgs Lagrangian density given in Equation 26.6 illustrates the basic idea behind symmetry breaking, but we need a slightly more complicated version to provide what we need to solve the mass problems in electroweak theory. Our electroweak theory so far has *assumed* a set of distinct massless fermion doublets as given, and derived a set of interaction fields (representing the implications of local gauge symmetries applied to the fermion Lagrangian density) represented by spin-1 bosons. We complete this model by adding one more fundamental doublet to the theory: a *scalar* doublet

$$\Phi = \begin{bmatrix} \phi_1 \\ \phi_2 \end{bmatrix}, \tag{26.9}$$

where ϕ_1 and ϕ_2 are *complex* scalar fields. We simply insert this doublet and its Higgs potential energy along with the fermion doublets into a Lagrangian density that looks like this:

$$\mathcal{L}_u = \partial^\mu \Phi^\dagger \partial_\mu \Phi - \frac{1}{4}\lambda^2(\Phi^\dagger\Phi - v^2)^2 + \sum_{j=1}^{6} i\overline{\Psi}^{(j)}\gamma^\mu\partial_\mu\Psi^{(j)} + \mathcal{L}_{\Phi,\Psi}. \tag{26.10}$$

The sum is over the six (three lepton and three quark) fermion doublets, and the $\mathcal{L}_{\Phi,\Psi}$ term is a possible set of interactions between the scalar doublet and the fermion fields (which we will talk about later).

We have constructed this Lagrangian density to be symmetric under an $SU(2)_L \otimes U(1)$ transformation. Since spin-0 scalar fields have no chirally right- or left-handed parts, we claim as part of our working hypothesis that the Φ doublet has a weak hypercharge $Y_H = 1$ and transforms under $SU(2)_L \otimes U(1)$ as

$$\Phi' = e^{-i\frac{1}{2}g'\theta^0}e^{-i\frac{1}{2}g_W\vec{\sigma}\cdot\vec{\theta}}\Phi. \tag{26.11}$$

Note that the potential energy term $\mathcal{U} \equiv -\frac{1}{4}\lambda^2(\Phi^\dagger\Phi - v^2)^2$ is symmetric under this transformation. But I should note that the theory otherwise provides no explanation for the particular form of this potential energy term: it is simply the least-complicated one with both the required symmetry and true vacuum states some distance from $\Phi = 0$, as required for symmetry breaking.

We semi-arbitrarily choose the vacuum state for the scalar doublet Φ to be

$$\Phi_v = \begin{bmatrix} 0 \\ v \end{bmatrix}, \tag{26.12}$$

so that Φ in terms of the fields H and G relative to the true vacuum becomes

$$\Phi = \begin{bmatrix} \phi_1 \\ \frac{H+a}{\sqrt{2}}e^{iG/a} \end{bmatrix}. \tag{26.13}$$

I say semi-arbitrarily because according to our usual expression for determining the charge Q associated with the bottom component of Φ is

$$Q = \frac{Y_W}{2} + T_3 = +\frac{1}{2} - \frac{1}{2} = 0 \tag{26.14}$$

because we have defined the weak hypercharge Y_W of this particular doublet to be $Y_H = 1$ and because the lower member of doublet has isospin $T_3 = -\frac{1}{2}$. Therefore, the Higgs field that emerges from symmetry breaking will be disconnected from the electromagnetic field.

26.4 Applying the Gauge Symmetry

We now make the $SU(2)_L \otimes U(1)$ symmetry local and active by converting all derivatives in the Lagrangian density to covariant derivatives

$$\partial_\mu \to D_\mu = \partial_\mu + i\frac{1}{2}g'B_\mu + i\frac{1}{2}g_W\vec{\sigma}\cdot\vec{W}_\mu \tag{26.15}$$

and adding free-field terms for the new fields generated. For the fermion part of the universal Lagrangian (after rotating the fields by the Weinberg angle), this produces exactly the electroweak results we saw in the last chapter. The new part consists of terms associated with the new scalar doublet:

$$\mathcal{L}_s = D^\mu\Phi^\dagger D_\mu\Phi - \frac{\lambda^2}{4}(\Phi^\dagger\Phi - v^2)^2. \tag{26.16}$$

In Box 26.2, you will show that expanding the first term yields the following:

$$D^\mu \Phi^\dagger D_\mu \Phi = \partial^\mu \Phi^\dagger \partial_\mu \Phi + i\tfrac{1}{2}g' B_\mu \left[(\partial^\mu \Phi^\dagger)\Phi - \Phi^\dagger(\partial^\mu \Phi) \right]$$
$$+ i\tfrac{1}{2}g_W \vec{W}_\mu \cdot \left[(\partial^\mu \Phi^\dagger)\vec{\sigma}\Phi - \Phi^\dagger \vec{\sigma}(\partial^\mu \Phi) \right]$$
$$+ \tfrac{1}{4}\Phi^\dagger \left(g'^2 B^\mu B_\mu + 2g' g_W B^\mu \vec{\sigma} \cdot \vec{W}_\mu + g_W^2 \vec{\sigma} \cdot \vec{W}^\mu \, \vec{\sigma} \cdot \vec{W}_\mu \right)\Phi. \quad (26.17)$$

The two terms in square brackets are conserved currents for the U(1) and SU(2) symmetries, and the final term describes relationships between the B_μ and $\vec{W}_\mu$ gauge fields. The first term expands to

$$\partial^\mu \Phi^\dagger \partial_\mu \Phi = \begin{bmatrix} \partial^\mu \phi_1^* & \partial^\mu \phi_2^* \end{bmatrix} \begin{bmatrix} \partial_\mu \phi_1 \\ \partial_\mu \phi_2 \end{bmatrix} = \partial^\mu \phi_1^* \partial_\mu \phi_1 + \partial^\mu \phi_2^* \partial_\mu \phi_2. \quad (26.18)$$

These are "kinetic energy" terms for the fields ϕ_1 and ϕ_2. We do not see any mass terms for these fields in the remaining terms in Equation 26.17, so the fields ϕ_1 and ϕ_2 seem to describe two massless spin-0 bosons.

Now comes the first amazing trick. The SU(2) symmetry allows us to rotate the doublet fields. We can choose a particular SU(2) gauge (that is, a particular rotation of the doublet) to eliminate the scalar doublet's top component ϕ_1 entirely. The U(1) symmetry allows us to rotate the phase of the remaining bottom component: we can therefore choose a particular U(1) gauge where the remaining bottom component is real. These gauge choices require us to transform the $\vec{W}$ and B_μ fields accordingly, but the *form* of these terms in the Lagrangian density does not change (because the Lagrangian density has been deliberately constructed to be *invariant* under these gauge transformations!). Remember that choosing a particular gauge does not change any of the actual physics in the situation; it is just choosing what amounts to a convenient coordinate system in which to analyze that physics.

You may be concerned, however, that we are apparently eliminating degrees of freedom from our Lagrangian density. The original doublet had four degrees of freedom (the real and imaginary parts of ϕ_1 and ϕ_2), while the new version has only one. We will see shortly, however, that when we make this gauge choice, three degrees of freedom appear elsewhere that we did not have before. So, we are not actually eliminating these degrees of freedom, we are just expressing them in a different way. Patience will be rewarded.

Now we break the symmetry and reexpress the now-real ϕ_2 field in terms of the Higgs field H:

$$\phi_2(x) = \frac{1}{\sqrt{2}}(H + a), \quad (26.19)$$

where $a = \sqrt{2}v$ as before. Because we have removed the field's imaginary part by choosing a U(1) gauge, the $e^{iG/a}$ term is gone and we have therefore eliminated the entire Goldstone boson field G we had worried about before. After the gauge choice and symmetry breaking, the $\partial^\mu \Phi^\dagger \partial_\mu \Phi - \tfrac{1}{4}\lambda^2(\Phi^\dagger\Phi - v^2)^2$ terms in the original Lagrangian density become

$$\tfrac{1}{2}\partial^\mu H \, \partial_\mu H - \tfrac{1}{2}\lambda^2 v^2 H^2 - \left[\tfrac{1}{16}\lambda^2(H^4 + 4\sqrt{2}v H^3) \right] \quad (26.20)$$

(just eliminate the G terms from Equation 26.8). This describes a massive Higgs field with some self-interaction terms. After the gauge choice, you can show (see Box 26.3) that the two conserved current terms in the expansion of $D^\mu \Phi^\dagger D_\mu \Phi$ in Equation 26.17 become zero, and the final term becomes

$$\tfrac{1}{8}(H + a)^2 \left[(g'B - g_W W^{(z)})^2 + g_W^2(W^{(x)})^2 + g_W^2(W^{(y)})^2 \right], \quad (26.21)$$

where each square is the dot product of the enclosed four-vector with itself. This simplifies further when we incorporate the Weinberg rotation of the fields and the fact that we need $g' = -g_W \tan\theta_W$ to decouple the photon from the weak interaction. You will show in Box 26.4 that this implies that

$$g' B_\mu - g_W W_\mu^{(z)} = -g_W \frac{Z_\mu}{\cos\theta_W}. \tag{26.22}$$

This means that the A_μ field (representing the photon) becomes completely decoupled from the Higgs field (that is, the Higgs field is electrically neutral, as we previously anticipated). Gathering the remaining terms, we see that the entire scalar term in the Lagrangian density has become

$$\mathcal{L}_s = \tfrac{1}{2}\partial^\mu H\,\partial_\mu H - \tfrac{1}{2}\lambda^2 v^2 H^2 - \left[\frac{1}{16}\lambda^2(H^4 + 4\sqrt{2}v H^3)\right]$$
$$+ \frac{(a^2 + 2aH + H^2)}{8} g_W^2 \left[\frac{(Z)^2}{\cos^2\theta_W} + (W^{(x)})^2 + (W^{(y)})^2\right]. \tag{26.23}$$

The full universal Lagrangian density also contains the free-field Lagrangian densities for the originally massless A_μ, Z_μ. and $W_\mu^{(x,y)}$ four-vector fields. For the A_μ field, this term looks like

$$\mathcal{L}_{A,\text{massless}} = -\tfrac{1}{4}F^{\mu\nu}F_{\mu\nu} \quad \text{where} \quad F_{\mu\nu} = \partial_\mu A_\nu - \partial_\nu A_\mu, \tag{26.24}$$

and the expressions for the other fields are similar (though the definition of $F_{\mu\nu}$ is a bit more complicated because of the cross products). Now, the Lagrangian density for a *massive* four-vector field A_μ looks like

$$\mathcal{L}_{A,\text{mass}} = -\tfrac{1}{4}F^{\mu\nu}F_{\mu\nu} + \tfrac{1}{2}m^2 A^2 \tag{26.25}$$

(see Box 26.5). So we recognize the $\tfrac{1}{8}a^2 g_W^2(Z)^2/\cos^2\theta_W$, $\tfrac{1}{8}a^2 g_W^2(W^{(x)})^2$, and $\tfrac{1}{8}a^2 g_W^2(W^{(y)})^2$ terms in Equation 26.23 as *mass* terms for their respective fields, with masses

$$M_W = \frac{a g_W}{2} = \frac{v g_W}{\sqrt{2}}, \tag{26.26a}$$

$$M_Z = \frac{M_W}{\cos\theta_W}. \tag{26.26b}$$

These are also connected to the Higgs mass:

$$M_H = v\lambda = \frac{\sqrt{2}\lambda}{g_W} M_W. \tag{26.26c}$$

(The other terms involving aH and H^2 combined with the Z and W fields represent interaction terms between the Higgs boson and the weak bosons.) Since an analogous mass term does not appear in Equation 26.23 for the electromagnetic A^μ field, the photon remains massless after symmetry breaking.

(Note that we prefer to deal with the $W^\pm = \sqrt{\tfrac{1}{2}}(W^{(x)} \mp i W^{(y)})$ bosons, but since each is in a sense just a rotation of the original fields, these new $\pm$ fields have the same mass M_W as the old x, y fields.)

This is huge. Adding the scalar doublet and the effect of symmetry breaking has given the weak vector bosons masses! Experimentally, we find that

$$M_W = 80.38 \pm 0.01 \text{ GeV}, \tag{26.27a}$$

$$M_Z = 91.188 \pm 0.002 \text{ GeV}, \tag{26.27b}$$

$$\text{and} \quad M_H = 125.10 \pm 0.14 \text{ GeV}. \tag{26.27c}$$

Note that $(M_W/M_Z)^2 = \cos^2 \theta_W = 1 - \sin^2 \theta_W$, and in fact this is how the "on-shell" value of $\sin^2 \theta_W$ is defined. From these values and the value of $g_W = 0.641$ found in the last chapter, one can calculate v and λ.

Note that this explains why the weak interaction is so much weaker than the electromagnetic interaction in low-energy particle decays. Though its coupling constant $g_W = 0.641$ is actually larger than the electromagnetic coupling constant $|q_e| = 0.3029$, the terms in low-energy, weak-interaction scattering amplitudes involve factors like $(g_W/M_W)^2$ compared to electromagnetic factors like $(q_e/q)^2$, where q is the magnitude of the off-shell virtual photon's four-momentum. Since q is of the order of magnitude of the energies involved in the interaction, which might be on the order of MeV in a typical particle decay, $q \ll M_W$, meaning that the weak interaction during particle decays is strongly suppressed by the large mass of the vector bosons. The electromagnetic and weak interactions only become comparable at energies $\gtrsim M_W$.

We also now know where the degrees of freedom have gone. Note that massless spin-1 (four-vector) bosons like the photon have only two spin states, while massive spin-1 bosons have three states. We lost three degrees of freedom in applying our choice of gauge, but the same process gave the vector bosons (the Z, the $W^{(x)}$, and the $W^{(y)}$) masses, which means that each gains an additional degree of freedom they did not possess initially. Our gauge choice therefore converts Goldstone-field degrees of freedom into an extra spin degree of freedom for each of the three weak fields, supporting the idea that this is just a "change of coordinates," nothing more. The metaphor is that the vector bosons $W^{\pm}$ and Z "eat" the Goldstone bosons and become massive from the meal.

26.5 Interactions Between the Higgs and Vector Bosons

The new Higgs contribution to the Lagrangian density does involve a variety of interaction terms between the Higgs boson and the vector bosons (see the last line of Equation 26.23). If we eliminate the factors of a in these terms in favor of the experimentally observable values of M_W, M_Z, and g_W, we find (see Box 26.6) that these interaction terms become

$$\mathcal{L}_{\text{int}} = g_W M_W H W^+ \cdot W^- + g_W \frac{M_Z}{2 \cos \theta_W} H(Z)^2$$

$$+ \frac{g_W^2}{4} H^2 W^+ \cdot W^- + \frac{g_W^2}{8 \cos^2 \theta_W} H^2(Z)^2. \tag{26.28}$$

These interaction terms yield new possible vertices in Feynman diagrams, as illustrated in Figure 26.2. (There are also vertices corresponding to the Higgs self-interaction terms on the first line of Equation 26.23, but energies required to see the effects of these terms are hard to access experimentally.)

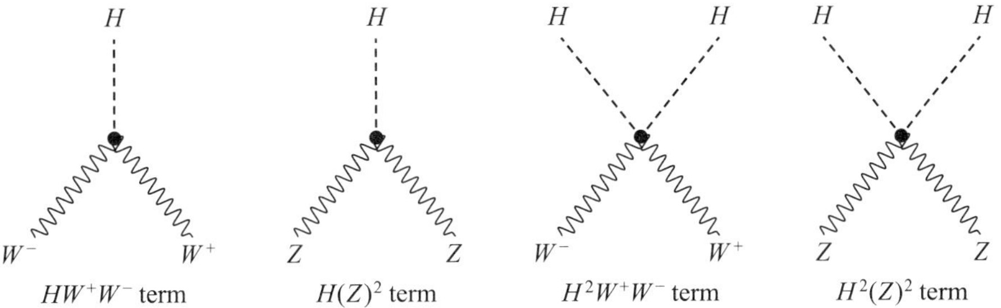

FIGURE 26.2   These pictures show the Feynman-diagram vertices for the interaction terms between the vector bosons and the Higgs bosons that appear in Equation 26.28.

The vertex factors for each vertex shown in Figure 26.2 are

$$g_W M_W H W^+ \cdot W^- \text{ term:} \quad i g_W M_W g_{\mu\nu}, \tag{26.29a}$$

$$g_W \frac{M_Z}{2\cos\theta_W} H(Z)^2 \text{ term:} \quad i g_W \frac{M_Z}{\cos\theta_W} g_{\mu\nu}, \tag{26.29b}$$

$$\frac{g_W^2}{4} H^2 W^+ \cdot W^- \quad \text{ term:} \quad i \frac{g_W^2}{2} g_{\mu\nu}, \tag{26.29c}$$

$$\frac{g_W^2}{8\cos^2\theta_W} H^2(Z)^2 \quad \text{ term:} \quad i \frac{g_W^2}{2\cos^2\theta_W} g_{\mu\nu}. \tag{26.29d}$$

The $g_{\mu\nu}$ comes from the fact that each vertex has to connect two vector bosons, one of which will have a μ index and the other a ν index. It also looks like the last three vertex factors are off by factors of two, but we get a factor of two for each pair of identical bosons involved in the vertex.

26.6 Parting Comments

We have come an enormous distance in this chapter. We now understand how the Higgs field and symmetry breaking can give the initially massless $SU(2)_L$ gauge fields masses, and how those masses should be related to each other and to the Higgs mass. Though verification of the $M_Z = M_W / \cos\theta_W$ prediction was already compelling evidence, the discovery of the predicted Higgs boson in 2012 by physicists working at the Large Hadron Collider represented nature's crowning endorsement of the theory (which the Nobel committee acknowledged by awarding Englert and Higgs the 2013 Nobel Prize in physics).

Our work with the weak interaction, however, is not quite complete. We have given the weak vector bosons masses, but all the fermions are still massless, as required by the $SU(2)_L$ symmetry. Also, there is the curious experimental fact that a weak interaction can connect quarks in different doublets, something that is impossible with lepton doublets. We will deal with both of these issues in the next chapter.

BOX 26.1   Expanding ϕ Around the True Vacuum

The purpose of this box is to substitute

$$\phi(x) = \frac{1}{\sqrt{2}}(H + \sqrt{2}v)e^{iG/\sqrt{2}v} = \frac{1}{\sqrt{2}}(H + a)e^{iG/a} \qquad (26.30)$$

(where $a \equiv \sqrt{2}v$) into the Higgs Lagrangian density given in Equation 26.6

$$\mathcal{L} = \partial^{\mu}\phi^* \, \partial_{\mu}\phi - \tfrac{1}{4}\lambda^2(|\phi|^2 - v^2)^2 \qquad (26.31)$$

to show that we end up with Equation 26.8.

Exercise 26.1.1   Perform this substitution and compare with Equation 26.8.

BOX 26.2 Expanding the Covariant Derivative

Our goal in this box is to show that $D^\mu \Phi^\dagger D_\mu \Phi$, where Φ is a complex scalar doublet and

$$D_\mu = \partial_\mu + i\tfrac{1}{2}g' B_\mu + i\tfrac{1}{2}g_W \vec{\sigma} \cdot \vec{W}_\mu \qquad (26.32)$$

expands to the result in Equation 26.17 (repeated here):

$$D^\mu \Phi^\dagger D_\mu \Phi = \partial^\mu \Phi^\dagger \partial_\mu \Phi + i\tfrac{1}{2}g' B_\mu \big[\, (\partial^\mu \Phi^\dagger)\Phi - \Phi^\dagger(\partial^\mu \Phi) \,\big]$$
$$+ i\tfrac{1}{2}g_W \vec{W}_\mu \cdot \big[\, (\partial^\mu \Phi^\dagger)\vec{\sigma}\,\Phi - \Phi^\dagger \vec{\sigma}(\partial^\mu \Phi) \,\big]$$
$$+ \tfrac{1}{4}\Phi^\dagger\big(\, g'^2 B^\mu B_\mu + 2g' g_W B^\mu \,\vec{\sigma}\cdot\vec{W}_\mu + g_W^2 \vec{\sigma}\cdot\vec{W}^\mu\,\vec{\sigma}\cdot\vec{W}_\mu \,\big)\Phi. \qquad (26.33)$$

The first step is to substitute Equation 26.32 into $D^\mu \Phi^\dagger D_\mu \Phi$ (and to note that $D^\mu \Phi^\dagger$ is really a shorthand for $[\, D^\mu \Phi\,]^\dagger$):

$$D^\mu \Phi^\dagger D_\mu \Phi = (\partial^\mu \Phi^\dagger - i\tfrac{1}{2}g' \Phi^\dagger B^\mu - i\tfrac{1}{2}g_W \Phi^\dagger \vec{\sigma}\cdot\vec{W}^\mu)$$
$$\times (\partial_\mu \Phi + i\tfrac{1}{2}g' B_\mu \Phi + i\tfrac{1}{2}g_W \vec{\sigma}\cdot\vec{W}_\mu \Phi). \qquad (26.34)$$

From here, it is really a matter of expanding out the nine terms and then combining a few of them. Remember that because $\vec{\sigma}$ acts in the doublet space, it must always appear between the $\Phi^\dagger$ and the Φ, which must appear in that order. The other factors are simple numbers, so order does not matter.

Exercise 26.2.1 Take it from here to Equation 26.33.

BOX 26.3   Simplifying the Locally Invariant Lagrangian Density

Our goal in this box is to show that if we choose our gauge so that Φ has only a real lower component $\phi_2 = (H + a)/\sqrt{2}$, then in the expansion

$$D^\mu \Phi^\dagger D_\mu \Phi = \partial^\mu \Phi^\dagger \partial_\mu \Phi + i\tfrac{1}{2}g' B_\mu \big[(\partial^\mu \Phi^\dagger)\Phi - \Phi^\dagger(\partial^\mu \Phi) \big]$$

$$+ i\tfrac{1}{2}g_W \vec{W}_\mu \cdot \big[(\partial^\mu \Phi^\dagger)\vec{\sigma}\Phi - \Phi^\dagger \vec{\sigma}(\partial^\mu \Phi) \big]$$

$$+ \tfrac{1}{4}\Phi^\dagger \big(g'^2 B^\mu B_\mu + 2g' g_W B^\mu \vec{\sigma} \cdot \vec{W}_\mu + g_W^2 \vec{\sigma} \cdot \vec{W}^\mu \vec{\sigma} \cdot \vec{W}_\mu \big)\Phi \tag{26.35}$$

the two conserved current terms become zero and the final term becomes

$$\tfrac{1}{8}(H + a)^2 \big[(g'B - g_W W^{(z)})^2 + g_W^2 (W^{(x)})^2 + g_W^2 (W^{(y)})^2 \big], \tag{26.36}$$

where each square is the dot product of the enclosed four-vector with itself.

Exercise 26.3.1   First, argue that the mere fact that ϕ_2 is real is sufficient to imply that the two current terms in square brackets are zero.

Now, let M be a 2×2 matrix constructed out of σ matrices. In our gauge,

$$\begin{bmatrix} 0 & \phi_2 \end{bmatrix} \begin{pmatrix} M_{11} & M_{12} \\ M_{21} & M_{22} \end{pmatrix} \begin{bmatrix} 0 \\ \phi_2 \end{bmatrix} = \phi_2^2 M_{22}. \tag{26.37}$$

Note that only σ_z has a σ_{22} component. In working with the $\vec{\sigma} \cdot \vec{W}^\mu \, \vec{\sigma} \cdot \vec{W}_\mu$ term, it helps to remember that $\sigma_i^2 = I$ and $\sigma_i \sigma_j + \sigma_j \sigma_i = 0$ when $i \neq j$.

Exercise 26.3.2   Show that with our choice of gauge, the final line in Equation 26.35 becomes Equation 26.36. (*Hint:* Note that the quantity in parentheses in that last line is really a 2×2 matrix that involves a linear combination of I and the Pauli matrices. That is why the previous paragraph is relevant.)

BOX 26.4   Applying the Weinberg Rotation

We saw in the last chapter that a clever rotation of fields given by the Weinberg formula

$$\begin{bmatrix} B \\ W^{(z)} \end{bmatrix} = \begin{pmatrix} \cos\theta_W & \sin\theta_W \\ -\sin\theta_W & \cos\theta_W \end{pmatrix} \begin{bmatrix} A \\ Z \end{bmatrix}, \tag{26.38}$$

with the Weinberg angle chosen that so that $g' = -g_W \tan\theta_W$, decoupled the photon field A_μ from the weak interaction. Our goal in this box is to show that this implies that

$$g' B_\mu - g_W W^{(z)}_\mu = -g_W \frac{Z_\mu}{\cos\theta_W}. \tag{26.39}$$

Exercise 26.4.1   Do this. (*Hint:* Simply substitute in the values of B_μ and $W^{(z)}_\mu$ from the Weinberg formula into the left side of this equation, and use $g' = -g_W \tan\theta_W$ to eliminate g'.)

BOX 26.5 The Lagrangian Density for a Massive Vector Field

I claimed in the body of the chapter that the Lagrangian density for a *massive* four-vector field A_μ looks like

$$\mathcal{L}_{A,\,\text{mass}} = -\tfrac{1}{4} F^{\mu\nu} F_{\mu\nu} + \tfrac{1}{2} m^2 A^2, \tag{26.40}$$

where $F_{\mu\nu} = \partial_\mu A_\nu - \partial_\nu A_\mu$ and $A^2 = g^{\mu\nu} A_\mu A_\nu$. We can show that this is correct by finding the Euler-Lagrange equation for the field, substituting in a plane-wave trial solution $A^\mu = \epsilon^\mu e^{-ip\cdot x}$, and showing that the solution works if we have $p^2 = m^2$ (that is, the solution describes a field with mass m).

Exercise 26.5.1 Do this. (*Hint:* Recall from Box 19.4 that

$$\frac{\partial(F^{\alpha\beta} F_{\alpha\beta})}{\partial(\partial_\mu A_\nu)} = 4 F^{\mu\nu}, \tag{26.41}$$

so there is no need to redo this calculation. Be careful to avoid index collisions. Finally, remember that we normally use the Lorenz gauge, where $\partial_\mu A^\mu = 0$.)

BOX 26.6   Vector-Higgs Interaction Terms

The goal in this box is to show that the Vector-Higgs interaction terms in the last line of Equation 26.23, which are

$$\mathcal{L}_{\text{int}} = \frac{(2aH + H^2)}{8} g_W^2 \left[\frac{(Z)^2}{\cos^2 \theta_W} + (W^{(x)})^2 + (W^{(y)})^2 \right], \qquad (26.42)$$

become the terms given in Equation 26.28 (repeated here)

$$\mathcal{L}_{\text{int}} = g_W M_W H W^+ \cdot W^- + g_W \frac{M_Z}{2 \cos \theta_W} H(Z)^2$$

$$+ \frac{g_W^2}{4} H^2 W^+ \cdot W^- + \frac{g_W^2}{8 \cos^2 \theta_W} H^2 (Z)^2 \qquad (26.43)$$

when we eliminate a, $W^{(x)}$, and $W^{(y)}$ in favor of the experimentally observable values $M_W = a g_W/2$, $M_Z = M_W / \cos \theta_W$, and $W^{\pm} = \sqrt{\frac{1}{2}}(W^{(x)} \mp i W^{(y)})$.

Exercise 26.6.1   Do this. (*Hint:* This is mostly a matter of substitution, but you should begin by showing that $W^+ \cdot W^- = \frac{1}{2}[\,(W^{(x)})^2 + (W^{(y)})^2]$.)

Homework Problems

P26.1 Consider the following Lagrangian density for a real scalar field ϕ:

$$\mathcal{L} = \tfrac{1}{2}(\partial_\mu \phi)(\partial^\mu \phi) + V_0^4 \cos b\phi, \qquad (26.44)$$

where V_0 and b are positive constants. Does the ϕ field have mass? If so, what is the mass in terms of V_0 and b?

P26.2 Determine the values of v, λ, and $\sin^2 \theta_W$ from the measured values of M_W, M_Z, M_H, and $g_W = 0.641$.

P26.3 To see how the mass of the Z has the effect of suppressing the weak interaction relative to the electromagnetic interaction, consider the decay of a muon-antimuon pair to an electron-positron pair. This can go either electromagnetically (via a virtual photon) or weakly (via a virtual Z).

a. Draw the lowest-order Feynman diagrams for these decay processes.

b. Write down the amplitudes for each process.

c. One can use Casimir's trick and trace theorems to show that the average squared amplitude for the weak process (treating the electron mass as negligible compared to the muon mass m) is

$$\langle|\mathcal{M}_Z|^2\rangle = 8 \left(\frac{g_W}{2 \cos \theta_W M_Z} \right)^4$$

$$\times \Big[(c_V^2 + c_A^2)^2 [(p_1 \cdot p_3)(p_2 \cdot p_4) + (p_1 \cdot p_4)(p_2 \cdot p_3)$$

$$+ m^2(p_3 \cdot p_4)]$$

$$+ 4(c_V c_A)^2 [(p_1 \cdot p_3)(p_2 \cdot p_4) - (p_1 \cdot p_4)(p_2 \cdot p_3)] \Big].$$

$$(26.45)$$

You do not have to show this. Starting from here, though, calculate $\langle|\mathcal{M}_\gamma|^2\rangle$ for the electromagnetic process by comparing the expressions for the amplitudes and making appropriate adjustments. Then calculate the ratio of the weak decay rate to the electromagnetic decay rate for a muon and antimuon that are initially essentially at rest in the center-of-mass frame (and where, again, we treat the electron's mass as negligible). For handy reference, $g_W = 0.641$, $q = 0.303$, $\sin^2 \theta_W = 0.2234$, $M_Z = 91.2 \text{ GeV}$, and the muon mass is $m = 0.1057 \text{ GeV}$.

P26.4 Before the Higgs boson was actually discovered, people considered it possible that it might have a sufficiently high mass to decay to Ws. Estimate the transition rate Γ for the Higgs particle at rest decaying to a W^+ and W^- if the Higgs were to have a mass of 200 GeV. How would the transition rate to two Zs compare? (*Hint:* The Feynman diagram for each decay process will consist of a single vertex; see Figure 26.2.)

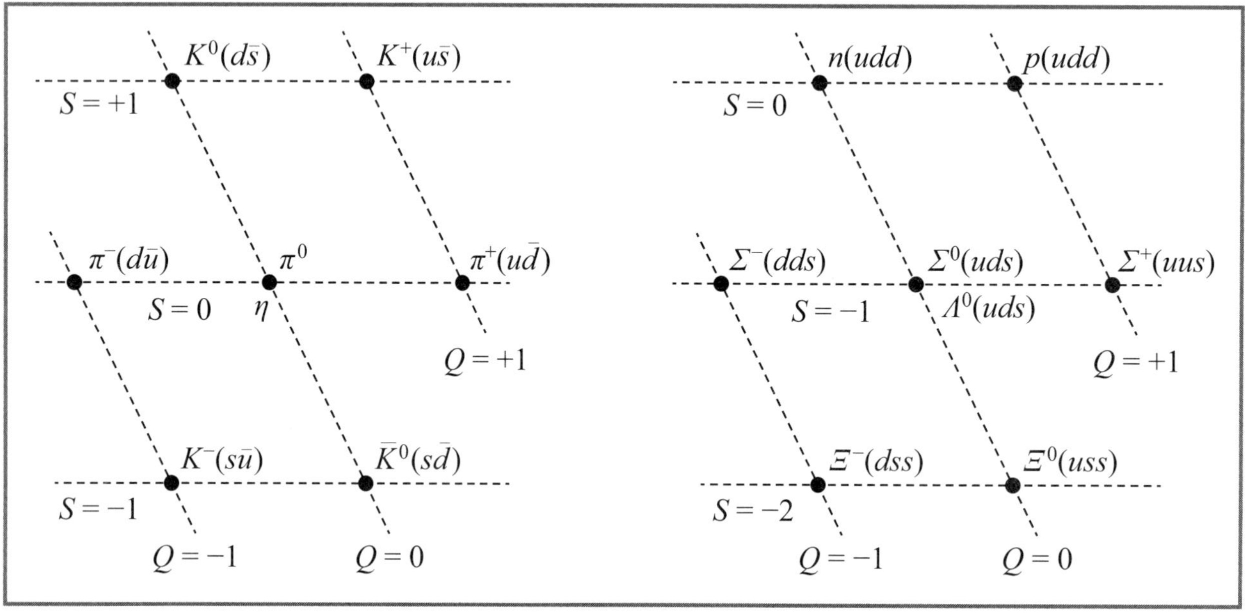

27. FERMION MASSES AND MIXING

In this chapter, we will put some finishing touches on the Standard Model's picture of the weak interaction. We will first see how the Higgs mechanism enables fermions in the Standard Model to have nonzero masses. Then we will see how the Standard Model understands and handles how the weak interaction connects quark families but not lepton families.

27.1 Fermion Masses

We have seen that fermion masses were required to be zero to maintain SU(2) symmetry. However, nothing prevents us from adding terms to the Lagrangian density that represent interactions between fields, so long as the terms themselves maintain the $SU(2)_L \otimes U(1)$ symmetry of the theory before symmetry breaking. Adding the Higgs field to our universal Lagrangian density provides an opportunity to add terms describing interactions between the Higgs field and fermion fields. In this section, we will see that it is possible to introduce Higgs-fermion terms that lead to fermion masses after symmetry breaking.

To begin, it helps to better understand what we are looking for. The Dirac Lagrangian density for a fermion field with mass is

$$\mathcal{L}_f = \overline{\psi}(i\gamma^\mu \partial_\mu - m)\psi. \tag{27.1}$$

We will see in Box 27.1 that, when expressed in terms of the chirally left- and right-handed parts of the field, this becomes

$$\mathcal{L}_f = i\overline{\psi}_L \gamma^\mu \partial_\mu \psi_L + i\overline{\psi}_R \gamma^\mu \partial_\mu \psi_R - m(\overline{\psi}_L \psi_R + \overline{\psi}_R \psi_L). \tag{27.2}$$

(Remember that ψ_L has zeros in the lower two slots of its four-component Dirac spinor and ψ_R has zeros in the upper two slots.) So what we are looking for are terms containing $(\overline{\psi}_L \psi_R + \overline{\psi}_R \psi_L)$ multiplied by a negative constant.

Expressing this in terms of the chirally left- and right-handed parts of the fields is important because the way these parts transform under the $SU(2)_L \otimes U(1)$ symmetry is different, and any new terms that we add must be invariant with regard to such transformations.

Now, consider a term of the form

$$\mathcal{L}_{\text{int}} = -b[\,\overline{\Psi}_L \Phi \psi_{2R} + \overline{\psi}_{2R} \Phi^\dagger \Psi_L\,], \tag{27.3}$$

where b is a positive constant, Φ is the Higgs doublet, Ψ is a fermion doublet, and ψ_2 is just the lower component of that doublet (for example, the electron as opposed to the electron neutrino, or the down quark as opposed to the up quark). In Box 27.2, we will see that an interaction term of this form is indeed invariant under a $SU(2)_L \otimes U(1)$ transformation. When we break the symmetry and make our usual choice of gauge, the Higgs doublet's top component ϕ_1 is zero and the bottom component ϕ_2 is real. This means that the doublet product $\overline{\Psi}_L \Phi$ becomes simply $\overline{\psi}_{2L} \phi_2$. Substituting this and $\phi_2 = (H + a)/\sqrt{2}$ into the previous equation yields

$$\mathcal{L}_{\text{int}} = -\frac{b(H + a)}{\sqrt{2}}[\,\overline{\psi}_{2L} \psi_{2R} + \overline{\psi}_{2R} \psi_{2L}\,]$$

$$= -\frac{ba}{\sqrt{2}}[\,\overline{\psi}_{2L} \psi_{2R} + \overline{\psi}_{2R} \psi_{2L}\,] - \frac{b}{\sqrt{2}} H[\,\overline{\psi}_{2L} \psi_{2R} + \overline{\psi}_{2R} \psi_{2L}\,]. \tag{27.4}$$

The first term in the last expression is precisely the mass term we are looking for ($m = ba/\sqrt{2}$), while the other term just gives us a new interaction between the lower fermion in the doublet and the Higgs field.

This seems to give us a mass for only the lower component of the doublet. This might seem acceptable for the lepton doublets (since the upper-component neutrinos have negligible mass), but it is not acceptable for quarks, since both quarks have significant mass. It is not even really acceptable for leptons, because we now know that neutrinos do not have strictly zero mass (even though we often make that approximation). However, one can show (see Box 27.3) that a term of the form

$$\mathcal{L}_{\text{int}} = -b \left[\, \overline{\Psi}_L \Phi_c \psi_{1R} + \overline{\psi}_{1R} \Phi_c^\dagger \Psi_L \, \right]$$

$$\text{where} \quad \Phi_c \equiv i\sigma_y \Phi^* = \begin{pmatrix} 0 & 1 \\ -1 & 0 \end{pmatrix} \begin{bmatrix} \phi_1^* \\ \phi_2^* \end{bmatrix} = \begin{bmatrix} \phi_2^* \\ -\phi_1^* \end{bmatrix} \qquad (27.5)$$

is *also* invariant under $\text{SU}(2)_L \otimes \text{U}(1)$. When we elect the gauge where ϕ_1 is zero and ϕ_2 is real and apply symmetry breaking, this ends up generating an equation like Equation 27.4 that involves ψ_{1R} and ψ_{1L} instead of ψ_{2R} and ψ_{2L}.

So, if we add to our universal Lagrangian density six interaction terms of the form given in Equation 27.3 and six terms of the form given in Equation 27.5 (with different values of the parameter b for each), we do not violate the theory's basic symmetry, but symmetry breaking then gives all the theory's fundamental quarks and leptons any masses that we like!

There are several unsatisfying parts to this. One is that all of these coupling constants b seem totally arbitrary: no one knows why they have the values they do. Another is that the coupling constants b for neutrinos are very small compared to the others since neutrino masses are smaller than 1 eV (maybe much smaller), while the mass of even the electron is 0.5 MeV. Moreover, there is no comparable discrepancy between the fields in a quark doublet (indeed, in the c, s and t, b doublets, it is the top component that has the greater mass!). These observational facts beg for an explanation that the Standard Model cannot supply.

27.2 Quark Families

We have seen that the weak interaction is based on the $\text{SU}(2)_L$ symmetry of doublets, or families, of fermions, for example, the electron with the electron neutrino, the muon with the muon neutrino, and so on. Experimentally, the *lepton* families are isolated: a weak interaction only connects the leptons within each family. Therefore, for example, the weak interaction can never convert a muon to an electron: when the muon decays, the weak interaction converts it to a muon neutrino and creates an $e^- \overline{v}_e$ pair out of the released energy. But the analogous statement does not apply to quarks. This was discovered in the following way.

In the 1950s and 1960s, physicists were able to use particle accelerators to create the whole series of exotic mesons and baryons shown in Table 1.1. At the time, the lowest-mass mesons—the pions—were thought to be the carriers of the strong nuclear force acting between protons and neutrons, but nobody saw a place for the other particles, which were therefore "strange."

In particular, there was the K^0 meson, which was much like a more massive version of the π^0 meson. But they behaved very differently: the π^0 decayed very rapidly to two photons, while the K^0 lasted (on average) more than 8 orders of magnitude longer, and decayed most often into a pair of pions. In 1953, Gell-Mann and Nishijima independently proposed that the anomaly could be explained if the electromagnetic and strong interactions obeyed a law of conservation of "strangeness," which the weak interaction could violate. By observing which decays were allowed and which were forbidden by

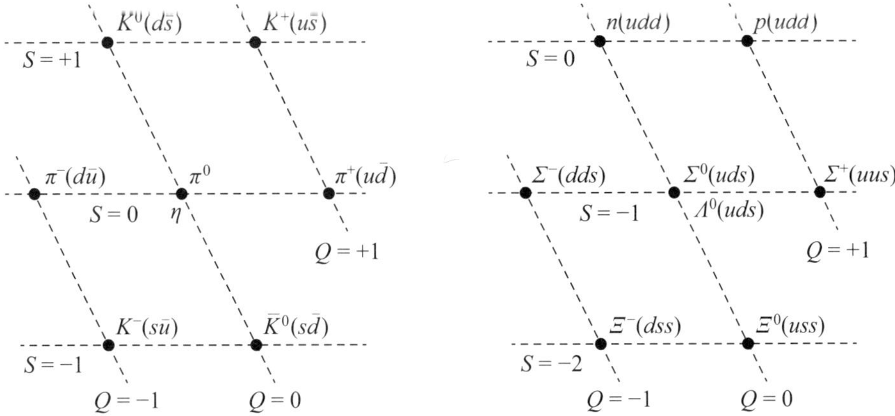

FIGURE 27.1   The "Eightfold Way" arrangement of the lowest-mass mesons and baryons. Quark configurations are provided in parentheses, except for the $\pi^0 = (u\bar{u} - d\bar{d})/\sqrt{2}$ and the η, which is a complicated superposition of $u\bar{u}$, $d\bar{d}$, and $s\bar{s}$.

the strong and electromagnetic interactions, physicists were able to assign integer values for strangeness S to every particle then known.

In 1962, Gell-Mann and Ne'eman independently showed that the lowest-mass baryons and the lowest-mass mesons formed symmetric sets of eight that mirrored a mathematical structure associated with the SU(3) group. (Gell-Mann called this the "Eightfold Way" after a religious concept in Buddhism concerning the path to enlightenment.) In 1964, Gell-Mann and Zweig independently noted that these structures (see Figure 27.1) could be most easily explained by assuming that baryons and mesons were constructed of three more basic particles, which Gell-Mann dubbed "quarks." This hypothesis was confirmed in 1968, when experiments that used 20-GeV electrons to probe the proton's inner structure revealed it to be a composite structure of more fundamental particles. We now understand the particles of Gell-Mann's "Eightfold Way" as being particular combinations (see Figure 27.1) of the up, down, and strange (u, d, s) quarks, and that "conservation of strangeness" simply means that strong and electromagnetic interaction vertices don't allow conversion of a strange quark to something else.

However, the fact that the weak interaction *does* allow this means that quark families are not completely independent, as lepton families seem to be. So how do we handle this mathematically?

27.3 The Cabibbo Angle

An especially simple weak interaction involving quarks is the decay of the pion ($\pi^- \rightarrow \mu^- + \bar{\nu}_\mu$). Figure 27.2 shows the fundamental Feynman diagram for this process considered as scattering of the quark constituents of π^-. Calculation of the actual cross section is complicated by the color interaction between the quarks, but the decay process is conceptually straightforward.

Now consider the analogous decay of the negative kaon ($K^- \rightarrow \mu^- + \bar{\nu}_\mu$). The negative kaon has a quark composition of $\bar{u}s$, so in the Feynman diagram in Figure 27.2, we would replace the d by an s.

Now, if the weak interaction respected quark family boundaries (or equivalently, if it conserved strangeness), this decay would be completely impossible, but this decay mode actually represents the majority (about 64%) of negative kaon decays, and the decay rate is quite comparable to that of the pion decay.

How can we understand this? In quantum mechanics, we encounter many situations where the particle states that are eigenstates of a certain operator (and thus have well-defined values of the operator's corresponding observable) are not the same as the

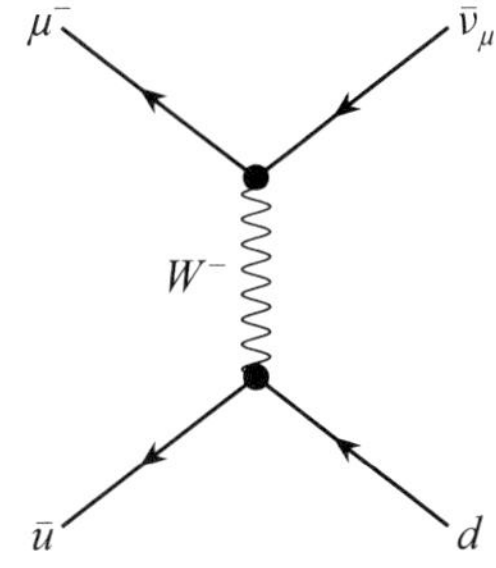

FIGURE 27.2   This is the Feynman diagram for a pion decay understood as a weak scattering event.

eigenstates of a different operator. We have also encountered situations in quantum field theory where a superposed combination of fields makes more sense to use than the original fields. For example, the vector boson fields $W_\mu^\pm = \sqrt{\tfrac{1}{2}}(W_\mu^{(x)} \mp W_\mu^{(y)})$ have a much more straightforward physical interpretation than the original fields $W_\mu^{(x)}$ and $W_\mu^{(y)}$ that emerged naturally from localizing the SU(2) symmetry.

What if the fields that couple to the $W_\mu^\pm$ and Z weak vector bosons are not the same but are rather superpositions of the fields that connect to the Higgs boson and thus have definite mass? Let's call the quark fields that have definite masses u, d, c, s, t, b, and let's call the fields that connect to the weak interaction u', d', c', s', t', b'. To preserve the SU(2) symmetry, top-component fields in the various family doublets will have to rotate into each other, and bottom component fields will have to rotate into each other separately. So (ignoring for the moment the t, b family) we might have something like

$$
\begin{bmatrix} u' \\ c' \end{bmatrix} = \begin{pmatrix} \cos\theta_+ & \sin\theta_+ \\ -\sin\theta_+ & \cos\theta_+ \end{pmatrix} \begin{bmatrix} u \\ c \end{bmatrix},
$$

$$
\begin{bmatrix} d' \\ s' \end{bmatrix} = \begin{pmatrix} \cos\theta_- & \sin\theta_- \\ -\sin\theta_- & \cos\theta_- \end{pmatrix} \begin{bmatrix} d \\ s \end{bmatrix}. \tag{27.6}
$$

The left equation, for example, allows us to write the weak-interaction fields u', c' as linear combinations of the definite-mass fields u, c in a way that preserves their overall magnitude. (Note that the column vectors here are not the usual quark doublets: the left equation mixes the *top* components of those doublets and the right equation mixes their *bottom* components.) The weak interaction terms in the universal Lagrangian density then look like

$$
\mathcal{L}_W = -\frac{g_W}{2\sqrt{2}} W_\mu^+ \left[\bar{u}' \gamma^\mu (1 - \gamma^5) d' + \bar{c}' \gamma^\mu (1 - \gamma^5) s' + \cdots \right] + \cdots, \tag{27.7}
$$

where the dots stand for similar terms involving the other quark and lepton doublets and other weak fields.

It turns out (see Box 27.4) that we only need to rotate *either* the u, c, t fields *or* the d, s, b fields. By convention, we do the latter so that

$$
u' = u, \qquad c' = c, \qquad \begin{bmatrix} d' \\ s' \end{bmatrix} = \begin{pmatrix} \cos\theta_C & \sin\theta_C \\ -\sin\theta_C & \cos\theta_C \end{pmatrix} \begin{bmatrix} d \\ s \end{bmatrix}. \tag{27.8}
$$

We call θ_C the **Cabibbo angle** (after its inventor Nicola Cabibbo).

The negative kaon (a particle with definite mass) has composition $\bar{u}s$. If the definite-mass quark fields u, s were the *same* as the weak interaction fields u', s', the kaon could not decay, because $\mathcal{L}_W$ does not connect fields in different weak interaction doublets. But Equation 27.8 says that the definite-mass s-quark is a mixture of the d' and s' weak-interaction quark fields, and $\mathcal{L}_W$ *does* connect d' to $u' = u$, so kaon decay is allowed. Indeed, the ratio of the vertex factors for the kaon decay and the pion decay (where a d is decaying to a u) must be

$$
\frac{\text{vertex factor}(s \to u)}{\text{vertex factor}(d \to u)} = \frac{\sin\theta_C}{\cos\theta_C} = \tan\theta_C. \tag{27.9}
$$

If we also factor in the kinematic aspects (see Problem P27.1), we find that

$$
\frac{\Gamma(K^- \to \mu^- + \bar{\nu}_\mu)}{\Gamma(\pi^- \to \mu^- + \bar{\nu}_\mu)} = \tan^2\theta_C \left(\frac{m_\pi}{m_K}\right)^3 \left(\frac{m_K^2 - m_\mu^2}{m_\pi^2 - m_\mu^2}\right)^2. \tag{27.10}
$$

By measuring this ratio, one can determine θ_C. Experimental evidence from this and other ratios implies that $\theta_C \approx 13°$.

27.4 The CKM Matrix

However, we actually have three quark families, not two. In 1973, Makoto Kobayashi and Toshihide Maskawa extended Cabibbo's approach to all three quark families. The **Cabibbo-Kobayashi-Maskawa Matrix** (or **CKM matrix**) describes how to "rotate" the fields with definite masses to those that connect via the weak interaction:

$$
\begin{bmatrix} d' \\ s' \\ b' \end{bmatrix} = \begin{pmatrix} V_{11} & V_{12} & V_{13} \\ V_{21} & V_{22} & V_{23} \\ V_{31} & V_{32} & V_{33} \end{pmatrix} \begin{bmatrix} d \\ s \\ b \end{bmatrix}.
\tag{27.11}
$$

In Box 27.5, we will see that this matrix has *four* degrees of freedom, which can be expressed in terms of three real parameters and one complex phase. One standard way of expressing the CKM matrix is as three successive rotations:

$$
\begin{aligned}
V &= \begin{pmatrix} 1 & 0 & 0 \\ 0 & c_{23} & s_{23} \\ 0 & -s_{23} & c_{23} \end{pmatrix} \begin{pmatrix} c_{13} & 0 & s_{13}e^{-i\delta} \\ 0 & 1 & 0 \\ -s_{13}e^{i\delta} & 0 & c_{13} \end{pmatrix} \begin{pmatrix} c_{12} & s_{12} & 0 \\ -s_{12} & c_{12} & 0 \\ 0 & 0 & 1 \end{pmatrix} \\[2mm]
&= \begin{pmatrix} c_{12}c_{13} & s_{12}c_{13} & s_{13}e^{-i\delta} \\ -s_{12}c_{23} - c_{12}s_{23}s_{13}e^{i\delta} & c_{12}c_{23} - s_{12}s_{23}s_{13}e^{i\delta} & s_{23}c_{13} \\ s_{12}s_{23} - c_{12}c_{23}s_{13}e^{i\delta} & -c_{12}s_{23} - s_{12}c_{23}s_{13}e^{i\delta} & c_{23}c_{13} \end{pmatrix},
\end{aligned}
\tag{27.12}
$$

where the real parameters are θ_{12}, θ_{13}, and θ_{23}, and c_{jk} and s_{jk} are shorthands for $\cos\theta_{jk}$ and $\sin\theta_{jk}$, respectively. If we assume the Standard Model's constraints (V_{jk} is unitary and only three quark families exist), then experimentally

$$
s_{12} = 0.2265(7), \quad s_{13} = 0.0036(1), \quad s_{23} = 0.0405(9), \quad \delta = 1.14(3) \approx 65°, \tag{27.13}
$$

where the uncertainty in parentheses in each case is in the last digit. To get a big-picture sense of what is going on, it also helps to see the magnitudes of the CKM matrix components:

$$
\begin{pmatrix} |V_{11}| & |V_{12}| & |V_{13}| \\ |V_{21}| & |V_{22}| & |V_{23}| \\ |V_{31}| & |V_{32}| & |V_{33}| \end{pmatrix} = \begin{pmatrix} 0.9740 & 0.2265 & 0.0036 \\ 0.2264 & 0.9732 & 0.0405 \\ 0.0085 & 0.0398 & 0.99917 \end{pmatrix}
\tag{27.14}
$$

(again, the uncertainty is in the last digit). We see that the weak-interaction states are *almost* the same as the those with well-defined masses, but not quite. Note also that because s_{13} and s_{23} are so small, any product of these in Equation 27.12 is so small as to be negligible. Therefore, only the V_{13}, V_{31}, and V_{32} components have non-negligible complex phases.

Kobayashi and Maskawa shared the 2008 Nobel Prize for the advances they made in describing quark mixing. (Interestingly, they did their work long before there was any experimental evidence for a third family of quarks or leptons.)

27.5 Vertex Factors and CP Violation

The Lagrangian density for the weak interaction term that converts a d' to a $u' = u$ therefore looks something like this:

$$
\begin{aligned}
\mathcal{L}_{ud'} &= -\frac{g_W}{2\sqrt{2}} \left[\bar{u}\gamma^\mu (1 - \gamma^5) d' W^+_\mu \right] \\[2mm]
&= -\frac{g_W}{2\sqrt{2}} \left[\bar{u}\gamma^\mu (1 - \gamma^5)(V_{11}\, d + V_{12}\, s + V_{13}\, b) W^+_\mu \right],
\end{aligned}
\tag{27.15}
$$

where (as per convention) I have dropped the prime on the u, since $u' = u$. To have a more general notation, let's give lower-component quarks d, s, b the abstract names d_1, d_2, d_3 and give the upper-component quarks u, c, t the names u_1, u_2, u_3. Then a generalized version of Equation 27.15 would look like

$$\mathcal{L}_{u_j d'_j} = -\frac{g_W}{2\sqrt{2}}\left[\overline{u}_j \gamma^\mu (1-\gamma^5) d'_j W^+_\mu\right]$$

$$= -\frac{g_W}{2\sqrt{2}}\left[\overline{u}_j \gamma^\mu (1-\gamma^5)(V_{j1}\,d_1 + V_{j2}\,d_2 + V_{j3}\,d_3) W^+_\mu\right]. \quad (27.16)$$

Thus, where we had one interaction term before, we now have three. This means that the vertex factor in an amplitude calculation that corresponds to a vertex in a Feynman diagram that connects a d_k to a u_j should be

$$(\text{vertex factor for } d_k \to u_j) = -i\frac{g_W}{2\sqrt{2}}\gamma^\mu (1-\gamma^5) V_{jk}. \quad (27.17)$$

Similarly, the Lagrangian density for the weak-interaction term that converts a u_j to a d_k would look like

$$\mathcal{L}_{d'_j u_j} = -\frac{g_W}{2\sqrt{2}}\left[\overline{d}'_j \gamma^\mu (1-\gamma^5) u_j W^-_\mu\right]$$

$$= -\frac{g_W}{2\sqrt{2}}\left[(\overline{d}_1 V^*_{j1} + \overline{d}_2 V^*_{j2} + \overline{d}_3 V^*_{j3})\gamma^\mu (1-\gamma^5) u_j W^-_\mu\right] \quad (27.18)$$

because the $\overline{d}_j$ involves the adjoint of the field, and taking the adjoint of

$$\begin{bmatrix} d'_1 \\ d'_2 \\ d'_3 \end{bmatrix} = \begin{pmatrix} V_{11} & V_{12} & V_{13} \\ V_{21} & V_{22} & V_{23} \\ V_{31} & V_{32} & V_{33} \end{pmatrix} \begin{bmatrix} d_1 \\ d_2 \\ d_3 \end{bmatrix} \quad \text{yields}$$

$$\begin{bmatrix} d'^\dagger_1 & d'^\dagger_2 & d'^\dagger_3 \end{bmatrix} = \begin{bmatrix} d^\dagger_1 & d^\dagger_2 & d^\dagger_3 \end{bmatrix} \begin{pmatrix} V^*_{11} & V^*_{21} & V^*_{31} \\ V^*_{12} & V^*_{22} & V^*_{32} \\ V^*_{13} & V^*_{23} & V^*_{33} \end{pmatrix}. \quad (27.19)$$

(Note that I have chosen the j index to always correspond to the u, c, or t quark.) So the vertex factor for any one of these interactions must be

$$(\text{vertex factor for } u_j \to d_k) = -i\frac{g_W}{2\sqrt{2}}\gamma^\mu (1-\gamma^5) V^*_{jk}. \quad (27.20)$$

In our model, quark mixing does *not* apply to weak interactions with the Z: that vertex factor is the same as for leptons, as seen in Equation 21.17.

The Lagrangian density for all the possible quark interactions with $W^\pm$ vector bosons will then be

$$\mathcal{L}_{q,W} = -\sum_{jk} \frac{g_W}{2\sqrt{2}}\left[\overline{u}_j \gamma^\mu (1-\gamma^5) d_k W^+_\mu V_{jk} + \overline{d}_k \gamma^\mu (1-\gamma^5) u_j W^-_\mu V^*_{jk}\right]. \quad (27.21)$$

Now, remember from Box 24.6 that under a combined CP operation,

$$\overline{\psi}_1 \gamma^\mu (1-\gamma^5)\psi_2 W^\pm_\mu \quad \longleftrightarrow \quad \overline{\psi}_2 \gamma^\mu (1-\gamma^5)\psi_1 W^\mp_\mu. \quad (27.22)$$

So, under a combined CP operation, our Lagrangian density becomes

$$\mathcal{L}^{CP}_{q,W} = -\sum_{jk} \frac{g_W}{2\sqrt{2}}\left[\overline{d}_k \gamma^\mu (1-\gamma^5) u_j W^-_\mu V_{jk} + \overline{u}_j \gamma^\mu (1-\gamma^5) d_k W^+_\mu V^*_{jk}\right]. \quad (27.23)$$

(The V_{jk} matrix elements don't change in a CP operation because they are simple numbers.) If we compare this to Equation 27.21, we see that the Lagrangian density will be CP-invariant if and only if V_{jk} is real. This is actually a crucially important statement. You will show in Box 27.5 that, though a real mixing matrix is possible for two quark families, it is not possible for three quark families. Therefore, we must have quark mixing and at least three quark families to have the weak interaction violate CP. Since CP violation is one of the necessary conditions for having a matter-antimatter asymmetry, quark mixing and three quark families are necessary for us to exist at all!

27.6 Feynman Rules for the Electroweak Theory

We are finally in a position to state a complete set of Feynman rules for the electroweak interaction in the Standard Model.

1. **Draw diagrams.** Using the possible vertices listed in item 3 below, construct diagrams that connect the ingoing and outgoing particle lines with internal particles.

2. **Label lines.** Draw arrows next to each particle line to show the direction of momentum flow. Label the arrows for ingoing or outgoing particles with four-momenta $p_1, p_2, \ldots$ and internal particles with four-momenta $q_1, q_2, \ldots$. Also, draw arrows *along the fermion lines* in the direction of motion for fermions but *opposite* the direction of motion for antifermions. Each *external* line contributes the factor that is listed (along with the line appearance) in the table below.

Particle	Factor	Drawing
incoming fermion	$u^s(p)$	
outgoing fermion	$\bar{u}^s(p)$	
incoming antifermion	$\bar{v}^s(p)$	
outgoing antifermion	$v^s(p)$	
incoming photon, $W^\pm$, or Z	$\epsilon^{\mu,s}$	
outgoing photon, $W^\pm$, or Z	$(\epsilon^{\mu,s})^*$	
incoming or outgoing Higgs	(none)	

3. **Vertex factors.** Each vertex in the diagram contributes the appropriate factor listed below.

Vertex	Factor	Drawing
fermion with photon	$-iq\gamma^\mu$ (q is fermion's charge)	
lepton with W	$-i\dfrac{g_W}{2\sqrt{2}}\gamma^\mu(1-\gamma^5)$	
quark with W	$-i\dfrac{g_W}{2\sqrt{2}}\gamma^\mu(1-\gamma^5)V_{jk}$	
quark with W	$-i\dfrac{g_W}{2\sqrt{2}}\gamma^\mu(1-\gamma^5)V^*_{jk}$	
fermion with Z	$\dfrac{-ig_W}{2\cos\theta_W}\gamma^\mu(c_V - c_A\gamma^5)$	
fermion with Higgs	$-\dfrac{i}{2}\dfrac{m_f}{M_W}g_W$	

W with Higgs	$ig_W M_W g_{\mu\nu}$	
Z with Higgs	$ig_W \dfrac{M_Z}{\cos\theta_W} g_{\mu\nu}$	
W with 2 Higgs	$i\dfrac{g_W^2}{2} g_{\mu\nu}$	
Z with 2 Higgs	$i\dfrac{g_W^2}{2\cos^2\theta_W} g_{\mu\nu}$	
W with photon	$ie[g_{\nu\sigma}(q_1-q_2)_\mu$ $+g_{\sigma\mu}(q_2-q_3)_\nu$ $+g_{\mu\nu}(q_3-q_1)_\sigma]$	
W with 2 photons	$-ie^2[2g_{\mu\nu}g_{\lambda\sigma}$ $-g_{\mu\lambda}g_{\nu\sigma}$ $-g_{\mu\sigma}g_{\nu\lambda}]$	
W with photon and Z	$ieg_W\cos\theta_W[2g_{\mu\nu}g_{\lambda\sigma}$ $-g_{\mu\lambda}g_{\nu\sigma}$ $-g_{\mu\sigma}g_{\nu\lambda}]$	

For a table of c_V and c_A values, see Table 23.2. The final three interaction vertices come from the extra terms we had to add to $-\frac{1}{4}\vec{F}^{\mu\nu}\cdot\vec{F}_{\mu\nu}$ to ensure local gauge invariance beyond the $(\partial^\mu\vec{W}^\nu - \partial^\nu\vec{W}^\mu)\cdot(\partial_\mu\vec{W}_\nu - \partial_\nu\vec{W}_\mu)$ terms that actually describe the free $\vec{W}$ fields (see Box 23.3).

4. **Propagators.** Each internal line contributes

Particle	Propagator	Drawing
fermion	$i\dfrac{\slashed{q}_j + m}{q_j^2 - m^2}$, where $\slashed{q}_j \equiv (q_j)_\mu \gamma^\mu$	
photon	$-i\dfrac{g_{\mu\nu}}{q_j^2}$	
$W^\pm$	$-i\dfrac{g_{\mu\nu} - (q_{j,\mu}q_{j,\nu})/M_W^2}{q_j^2 - M_W^2}$	
Z	$-i\dfrac{g_{\mu\nu} - (q_{j,\mu}q_{j,\nu})/M_Z^2}{q_j^2 - M_Z^2}$	
Higgs	$\dfrac{i}{q_j^2 - M_H^2}$	

Note that we can neglect the second term in the numerator and the first term in the denominator of the propagators for the $W^\pm$ and Z if $q_j^2 \ll M^2$.

5. **Conserve four-momentum and charge at each vertex.** (For simple diagrams, this determines the four-momenta q_j for internal particles.)

6. **Follow the arrows.** To put items in the correct order, follow the arrows *on the* fermion lines from start to finish, and write the corresponding terms *starting from the right* (backward from English writing).

7. **Undetermined internal momenta.** Include a factor of $d^4 q_j/(2\pi)^4$, and integrate over any undetermined four-momenta q_j (use an upper cutoff if necessary).

8. **Fermion loops.** For each closed loop involving fermions, include an overall factor of -1, and take the trace of the product of the gamma matrices contributed by loop vertices and propagators.

9. **Boson loops.** For each loop involving spin-1 bosons, include a factor of $g^{\mu\nu}$, and sum over the lower indices contributed by the boson propagators to eliminate all free indices.

10. **The result is** $-i\mathcal{M}$. Multiply by i to get $\mathcal{M}$.

11. **Add amplitudes.** To find the total amplitude for a given process up to a certain order, add the amplitudes for all the diagrams up to that order.

12. **Antisymmetrization.** Include a minus sign between two diagrams that differ by swapping two incoming identical fermions, two outgoing identical fermions, an incoming fermion with an outgoing antifermion (corresponding to the same type of fermion), or an incoming antifermion with an outgoing fermion of the same type.

With these rules, we can calculate the amplitude $\mathcal{M}$ for almost all electroweak Feynman diagrams.

BOX 27.1   The Mass Term for a Fermion Field

The goal in this box is to show that

$$\mathcal{L}_f = \overline{\psi}(i\gamma^\mu \partial_\mu - m)\psi$$

$$= i\overline{\psi}_L \gamma^\mu \partial_\mu \psi_L + i\overline{\psi}_R \gamma^\mu \partial_\mu \psi_R - m(\overline{\psi}_L \psi_R + \overline{\psi}_R \psi_L). \qquad (27.24)$$

It is easiest to work backward. Consider the first of the four final terms. Note that $\psi_L = \frac{1}{2}(1 - \gamma^5)\psi$ and $\psi_R = \frac{1}{2}(1 + \gamma^5)\psi$. Also, remember that γ^5 anticommutes with every other γ^μ, that $(\gamma^5)^2 = I$, and that $(\gamma^5)^\dagger = \gamma^5$. So

$$i\overline{\psi}_L(\gamma^\mu \partial_\mu)\psi_L = \tfrac{1}{4}i\psi^\dagger (1 - \gamma^5)^\dagger \gamma^t \gamma^\mu \partial_\mu (1 - \gamma^5)\psi$$

$$= \tfrac{1}{4}i\psi^\dagger \gamma^t \gamma^\mu (1 - \gamma^5)(1 - \gamma^5)\partial_\mu \psi$$

$$= \tfrac{1}{4}i\overline{\psi}\gamma^\mu(1 - 2\gamma^5 + 1)\partial_\mu \psi = \tfrac{1}{2}i\overline{\psi}\gamma^\mu(1 - \gamma^5)\partial_\mu \psi, \quad (27.25)$$

where, in going to the second line, I moved the left $(1 - \gamma^5)$ past the γ^t and the γ^μ, flipping twice the sign in front of the γ^5 that I was moving. Similarly,

$$i\overline{\psi}_R \gamma^\mu \partial_\mu \psi_R = \tfrac{1}{2}i\overline{\psi}\gamma^\mu(1 + \gamma^5)\partial_\mu \psi, \qquad (27.26)$$

so these two terms will add to $i\overline{\psi}\gamma^\mu \partial_\mu \psi$.

You might wonder why our expansion does not include cross terms such as $i\overline{\psi}_L \gamma^\mu \partial_\mu \psi_R$ or $i\overline{\psi}_R \gamma^\mu \partial_\mu \psi_L$. Note that

$$i\overline{\psi}_L(\gamma^\mu \partial_\mu)\psi_R = \tfrac{1}{4}i\psi^\dagger (1 - \gamma^5)^\dagger \gamma^t \gamma^\mu \partial_\mu (1 + \gamma^5)\psi$$

$$= \tfrac{1}{4}i\psi^\dagger \gamma^t \gamma^\mu (1 - \gamma^5)(1 + \gamma^5)\partial_\mu \psi$$

$$= \tfrac{1}{4}i\overline{\psi}\gamma^\mu(1 - [\gamma^5]^2)\,\partial_\mu \psi = 0. \qquad (27.27)$$

So including such a term would add nothing.

Exercise 27.1.1   Similarly, show that $m(\overline{\psi}_L \psi_R + \overline{\psi}_R \psi_L) = m\overline{\psi}\psi$ and that including terms like $m\overline{\psi}_L \psi_L$ would add nothing.

BOX 27.2 Interaction Term Invariance I

Our goal in this box is to show that a term of the form

$$\mathcal{L}_{\text{int}} = -b[\,\overline{\Psi}_L \Phi \psi_{2R} + \overline{\psi}_{2R} \Phi^\dagger \Psi_L\,] \tag{27.28}$$

is invariant under a $\text{SU}(2)_L \otimes \text{U}(1)$ transformation. Remember that b is a positive constant, Ψ is a fermion doublet, and ψ_2 is the doublet's lower component.

Because $Y_H = 1$ for the Higgs doublet, the Φ doublet transforms as

$$\Phi' = e^{-i\frac{1}{2}g'Y_H\theta^0} e^{-i\frac{1}{2}gw\vec{\sigma}\cdot\vec{\theta}} \Phi = e^{-i\frac{1}{2}g'\theta^0} e^{-i\frac{1}{2}gw\vec{\sigma}\cdot\vec{\theta}} \Phi. \tag{27.29}$$

Assume for the sake of argument that the fermion doublet is a lepton. For a (left-handed) lepton doublet, $Y_L = -1$ (see Table 25.1), so this doublet transforms as

$$\Psi' = e^{-i\frac{1}{2}g'Y_L\theta^0} e^{-i\frac{1}{2}gw\vec{\sigma}\cdot\vec{\theta}} \Psi = e^{+i\frac{1}{2}g'\theta^0} e^{-i\frac{1}{2}gw\vec{\sigma}\cdot\vec{\theta}} \Psi. \tag{27.30}$$

Finally, since the right-handed lower component has isospin $T_3 = 0$ and its charge $Q = \frac{1}{2}Y_R + T_3$ must be -1, it must have $Y_R = -2$. Right-handed fields do not participate in the $\text{SU}(2)_L$ symmetry, so this field transforms as

$$\psi'_{2R} = e^{+ig'\theta^0} \psi_{2R} \tag{27.31}$$

under $\text{U}(1)$ alone. So the package $\overline{\Psi}_L \Phi \psi_{2R}$ transforms as

$$\begin{aligned}
\overline{\Psi}'_L \Phi' \psi'_{2R} &= (\Psi'_L)^\dagger \gamma^t \Phi' \psi'_{2R} \\
&= (\Psi_L^\dagger e^{-i\frac{1}{2}g'\theta^0} e^{+i\frac{1}{2}gw\vec{\sigma}\cdot\vec{\theta}}) \gamma^t (e^{-i\frac{1}{2}g'\theta^0} e^{-i\frac{1}{2}gw\vec{\sigma}\cdot\vec{\theta}} \Phi)(e^{+ig'\theta^0} \psi_{2R}) \\
&= \overline{\Psi}_L e^{-i\frac{1}{2}g'\theta^0} e^{-i\frac{1}{2}g'\theta^0} e^{+ig'\theta^0} e^{+i\frac{1}{2}gw\vec{\sigma}\cdot\vec{\theta}} e^{-i\frac{1}{2}gw\vec{\sigma}\cdot\vec{\theta}} \Phi \psi_{2R} \\
&= \overline{\Psi}_L I \Phi \psi_{2R}. \tag{27.32}
\end{aligned}$$

In the next-to-last step, I used the facts that the $\text{U}(1)$ exponentials are just numbers and so commute with everything, and that the γ^t acts in the spinor space and so commutes with the exponentials $e^{\pm i\frac{1}{2}gw\vec{\sigma}\cdot\vec{\theta}}$, which are 2×2 matrices in the doublet space. Since $\overline{\psi}_{2R} \Phi^\dagger \Psi_L = (\overline{\Psi}_L \Phi \psi_{2R})^\dagger$, the second term in Equation 27.28 will be invariant if the first is.

Exercise 27.2.1 Show that $\overline{\Psi}_L \Phi \psi_{2R}$ is invariant for a quark doublet.

BOX 27.3 Interaction Term Invariance II

Our goal in this box is to show that a term of the form

$$\mathcal{L}_{\text{int}} = -b\,[\,\overline{\Psi}_L \Phi_c \psi_{1R} + \overline{\psi}_{1R}\Phi_c^\dagger \Psi_L\,], \quad \text{where } \Phi_c \equiv i\sigma_y \Phi^*, \quad (27.34)$$

is invariant under $SU(2)_L \otimes U(1)$. The tricky part here is determining the transformation properties of the Φ_c doublet. From Equation 27.29,

$$\Phi' = e^{-i\frac{1}{2}g'\theta^0} e^{-i\frac{1}{2}gw\vec{\sigma}\cdot\vec{\theta}} \Phi \quad \Rightarrow \quad \Phi'^* = e^{+i\frac{1}{2}g'\theta^0} e^{+i\frac{1}{2}gw\vec{\sigma}^*\cdot\vec{\theta}} \Phi^*. \quad (27.35)$$

Note that taking the complex conjugate of $e^{-i\frac{1}{2}gw\vec{\sigma}\cdot\vec{\theta}}$ does not merely reverse the sign of i; σ_y also has imaginary components. In fact, $\sigma_y^* = -\sigma_y$, while σ_x and σ_z are unaffected. However, also note that σ_y anticommutes with σ_x and σ_z but commutes with itself. This means that

$$\Phi'_c = (i\sigma_y \Phi^*)' = i\sigma_y\, e^{+i\frac{1}{2}g'\theta^0} e^{+i\frac{1}{2}gw\vec{\sigma}^*\cdot\vec{\theta}} \Phi^*$$

$$= e^{+i\frac{1}{2}g'\theta^0} e^{-i\frac{1}{2}gw\vec{\sigma}\cdot\vec{\theta}} (i\sigma_y \Phi^*) = e^{+i\frac{1}{2}g'\theta^0} e^{-i\frac{1}{2}gw\vec{\sigma}\cdot\vec{\theta}} \Phi_c \quad (27.36)$$

because pulling σ_y to the right past the σ_x and σ_z terms in the exponent changes the signs of those terms. Doing this does not change the sign of the σ_y in the exponent, but that was already negated by taking the complex conjugate. The result is that Φ_c has the same $SU(2)$ transformation as Φ but the inverse $U(1)$ transformation.

Exercise 27.3.1 With this information, show that $\overline{\Psi}_L \Phi_c \psi_{1R}$ is invariant under $SU(2)_L \otimes U(1)$ for both lepton and quark doublets.

BOX 27.4 Simplifying Quark Mixing

Consider quark weak-interaction terms such as

$$\mathcal{L}_W = -\frac{g_W}{2\sqrt{2}} W_\mu^+ \big[\, \overline{u}\,'\gamma^\mu(1-\gamma^5)d' + \overline{c}\,'\gamma^\mu(1-\gamma^5)s' \,\big]. \qquad (27.37)$$

The concept of quark mixing asserts that the quark field doublets (u', d') and (c', s') that are the independent doublets as seen by the weak interaction are in fact not the doublets (u, d) and (c, s) that have well-defined masses. This concept would generally seem to require that we rotate *both* the upper-component quark fields u, c and the lower-component fields d, s:

$$\begin{bmatrix} u' \\ c' \end{bmatrix} = \begin{pmatrix} \cos\phi & \sin\phi \\ -\sin\phi & \cos\phi \end{pmatrix} \begin{bmatrix} u \\ c \end{bmatrix}, \quad \begin{bmatrix} d' \\ s' \end{bmatrix} = \begin{pmatrix} \cos\theta & \sin\theta \\ -\sin\theta & \cos\theta \end{pmatrix} \begin{bmatrix} d \\ s \end{bmatrix}.$$

$$(27.38)$$

The purpose of this box is to prove that in fact we need to rotate only one or the other (we conventionally choose to rotate the lower one only).

Exercise 27.4.1 Expand the two weak interaction Lagrangian-density terms above into four terms involving u, d, s, and c, and argue that the coefficients in front of these terms are either $\cos(\theta - \phi)$ or $\pm\sin(\theta - \phi)$. Therefore, only the *difference* in rotations matters physically, so we might as well define $\theta_C = \theta - \phi$ and assume that the rotation only affects the lower-component quarks.

BOX 27.5   Free Parameters for the CKM Matrix

Our goal in this box is to determine the number of parameters required to describe the CKM matrix V. This 3×3 matrix has nine complex components, and therefore 18 degrees of freedom. However, the matrix must be unitary to preserve the fields' magnitudes. The unitarity condition $V^\dagger V = I$ imposes nine conditions on these components (one equation for each diagonal component of I and two for each component above the diagonal), leaving nine degrees of freedom.

Now, the point of V_{jk} is to specify the value of the Lagrangian density term

$$\mathcal{L}_{jk} = -\frac{g_W}{2\sqrt{2}} W_\mu^+ \bar{u}_j \gamma^\mu (1 - \gamma^5)\, d_k\, V_{jk}. \tag{27.39}$$

Now, the quark fields' phases are physically meaningless. If we adjust the $u_1 = u$ field's phase, we must adjust the phase of the components V_{1k} to compensate. Similarly, if we adjust the $d_3 = t$ field's phase, we must adjust the phases of the components V_{j3} to compensate. Since we can set any of the six quark-field phases to be anything we like, it would seem that we could use this freedom to set six components of V_{jk} to have any phases we like. But changing the phases of *all* the fields by the *same* amount doesn't actually do anything to V_{jk}: it is really the *phase differences* that matter. So, if we define the overall phase to be the phase choice we give to the first quark, only the degree to which we choose the other five quark-field phases to be *different* than this actually changes the values of V_{jk}. Therefore, our ability to choose quark phases allows us to impose five more conditions on the components V_{jk} (not six), leaving four degrees of freedom.

So a complete description of the CKM matrix requires four free parameters. Is it possible to choose the components of the CKM matrix to be real? If the components V_{jk} were to be real, then unitarity by itself would require that

$$V^\dagger V = \begin{pmatrix} a & d & g \\ b & e & h \\ c & f & q \end{pmatrix} \begin{pmatrix} a & b & c \\ d & e & f \\ g & h & q \end{pmatrix} = \begin{pmatrix} 1 & 0 & 0 \\ 0 & 1 & 0 \\ 0 & 0 & 1 \end{pmatrix}, \tag{27.40}$$

where a, b, c, d, e, f, g, h, and q are real numbers.

Exercise 27.5.1   Argue that this equation imposes six *independent* constraints on the nine components a, b, c, d, e, f, g, h, and q, so we have only three degrees of freedom left over. (*Hint:* One of the constraints is that $a^2 + d^2 + g^2 = 1$.)

BOX 27.5 Free Parameters for the CKM Matrix (cont.)

Since the most general real 3×3 unitary matrix has only three degrees of freedom but the most general CKM matrix must have four, it follows that we *cannot* choose the phases of our quark fields to make the CKM matrix entirely real. We can describe the fourth parameter as a complex phase $e^{i\delta}$ that we attach to at least one matrix component. (In the conventional format for the matrix, we append this phase factor or its complex conjugate to four off-diagonal components; see Equation 27.12.) As the text notes, it is these complex elements that create violations of CP symmetry.

What if we only had two quark families? A 2×2 complex matrix has four components, and so eight degrees of freedom. Being unitary imposes four conditions on those eight degrees of freedom.

Exercise 27.5.2 We have four quarks in two quark families. Our freedom to choose quark phases allows us to put additional constraints on the components of the 2×2 matrix V. How many? How many degrees of freedom do we have left afterward? Show that a real matrix V *does have* this many degrees of freedom, meaning that we *could* choose V to be completely real in this case.

So, if we had just two quark families, we could choose quark field phases to make the CKM matrix entirely real without changing any physical predictions of the theory. Since complex components of the CKM matrix are required to create violations of CP symmetry, we would not in such a case observe violations of CP symmetry in nature. But if we have three quark families, we *cannot* make the CKM matrix purely real simply by manipulating the phases of the quark fields. Therefore, we *can* have violations of CP symmetry, as we indeed observe in nature. To put it another way, having the *possibility* of CP symmetry violations (which are required for our existence) requires at least three quark families.

Homework Problems

P27.1 Consider the $\pi^- \to \mu^- + \bar{\nu}_\mu$ and $K^- \to \mu^- + \bar{\nu}_\mu$ decay processes discussed in Section 27.2. Given the ratio of the amplitudes given in Equation 27.9, verify that the ratio of the decay rates is indeed given by Equation 27.10. (*Hints:* Use Equation 24.59. Remember that the f_π factor contains all of the information about the quark vertex, so it includes the appropriate factor involving the Cabibbo angle. Assume that the analogous factor f_K for the kaon should be the same except for differences in the factors that multiply the quark vertex in each case. As a result, argue that $f_K/f_\pi = \tan\theta_C$.)

P27.2 The Λ^0 baryon has a mass of 1.116 GeV.

a. How massive would the Λ^0 have to be to decay if the weak interaction respected quark families (that is, if the families $[\,u, d\,]$, $[\,c, s\,]$, and $[\,t, b\,]$ were completely independent in the weak interaction like the lepton families $[\,\nu_e, e^-\,]$, $[\,\nu_\mu, \mu^-\,]$, and $[\,\nu_\tau, \tau\,]$ are)? Note that because this mass is larger than 1.116 GeV, if quark mixing did not exist, the Λ^0 would be *stable*, so it would be a particle like the neutron that could appear in atomic nuclei, making chemistry much more complicated. (The Ξ^0 would also be stable.)

b. Because of quark mixing, the Λ^0 does decay, commonly to $p + \pi^-$. Draw the lowest-order Feynman diagram for this decay. (*Hint:* Two quark worldlines will simply coast through the diagram, because they are not involved in the weak decay. Draw these lines on the left.)

P27.3 Before anyone knew about the c quark, the simple Cabibbo angle transformation given in Equation 27.8 worked very well to predict a number of useful decay rates. But it had a stark failure: it predicted that K^0 could decay to $\mu^+ + \mu^-$, a decay mode not seen in nature at anywhere near the rate predicted.

a. Draw the lowest-order Feynman diagram for this decay process. (*Hint:* The diagram will have four vertices, and its central part will look like a box. Be sure to conserve charge at the vertices.)

b. In 1970, Glashow, Iliopoulos, and Maiani (GIM) proposed a solution to this problem. Suppose (they suggested) that in addition to the u, d, and s quarks, a fourth quark (now called the c) existed. If this were true, a second diagram (this time with a virtual c instead of a virtual u) would also exist. Draw this diagram.

c. Argue that the amplitudes from these diagrams should nearly cancel, explaining the almost vanishing decay rate observed. (*Hint:* Equations 27.17 and 27.20 state that each quark vertex will be weighted by an appropriate component of V_{jk} or V_{jk}^*. If we ignore the third quark family, V is just the matrix appearing in Equation 27.8. Use this to write out the Cabibbo angle factor associated with each vertex. If the mass of the u was the same as that of the c, all other factors in the amplitude would be the same and the cancel-

lation would be exact. Even though the c is significantly more massive than the u, these virtual particles are so far off their mass shells that the masses don't matter much, so the amplitudes very nearly cancel.) Inventing a whole new quark just to resolve an esoteric theoretical issue was a bold prediction at the time, but within a few years, people had found more direct experimental evidence for the c quark.

P27.4 Since the top quark has a mass of 173 GeV, it can decay to a bottom quark (4.18 GeV) and a *real* W^+ boson (80.4 GeV); all other quarks must decay to a *virtual* W^+, which not only will be far off-shell (because of its large mass compared to the other particles involved) but also must decay to something else, requiring at least a two-vertex diagram instead of the single-vertex diagram needed for the top decay. This means that the top quark will decay quite rapidly compared to other quarks. Estimate the lifetime of the top quark, assuming (for simplicity) that other decay modes are negligible and that the bottom quark is massless compared to the top and W^+ (both good approximations). Your answer should be on the order of 10^{-25} s.

P27.5 As argued in Box 27.5, the CKM matrix should be unitary.

a. Prove that if we consider the columns of a 3×3 unitary matrix to be column vectors, then different columns must be orthogonal to each other. (*Hint:* Do the same kind of calculation as in Exercise 27.5.1, except assume that the matrix elements might be complex.)

b. Use Equations 27.12 and 27.13 to write each of the components of the CKM matrix in the form of $a + ib$, where a and b are real numbers accurate to five decimal places.

c. The unitary nature of the CKM matrix can be tested experimentally. For example, orthogonality of the first and third columns requires that $0 = V_{11}^* V_{13} + V_{21}^* V_{23} + V_{31}^* V_{33}$, or $0 = 1 + z_1 + z_2$, where

$$z_1 = \frac{V_{21}^* V_{23}}{V_{11}^* V_{13}} \quad \text{and} \quad z_2 = \frac{V_{31}^* V_{33}}{V_{11}^* V_{13}}. \qquad (27.41)$$

If we were to draw the terms 1, z_1, and z_2 on the complex plane, we should get a closed triangle, which physicists call a **unitarity triangle**. Given the values of the CKM components you calculated in part (b), check whether $1 + z_1 + z_2$ is really close to zero experimentally. Be sure to consider uncertainties.

P27.6 We can determine each component of the CKM matrix by examining hadron decay or creation processes that involve changing quark flavors. In this problem, we will estimate the Cabibbo angle θ_C by examining decays of the D meson. The D^+ meson consists of a charm quark and an anti-down quark. There are many ways that this meson can decay, but let's focus on the following decay modes:

$$D^+ \to \pi^0 + e^+ + \nu_e, \quad \text{and} \quad D^+ \to \overline{K}^0 + e^+ + \nu_e. \quad (27.42)$$

a. Draw the Feynman diagrams for these decays. (*Hint:* The $\bar{d}$ quark is essentially a "spectator" in this decay. You can indicate this by drawing it as a straight line from the bottom of the diagram to the top without connecting it to anything.)

b. Identify the difference in the amplitudes associated with these diagrams that depends on the Cabibbo angle.

c. The mass of the D^+ is about 1.870 GeV. Use the observed ratio

$$\frac{\Gamma(\overline{K}^0 + e^+ + \nu_e)}{\Gamma(D^+ \rightarrow \pi^0 + e^+ + \nu_e)} = (20 \pm 6) \quad (27.43)$$

to estimate $\sin\theta_C$ and its uncertainty.

P27.7 Even though the Higgs particle does not directly interact with the massless photon, it can decay to two photons. The virtual particles will be top quarks so that they can be as close to being on-shell as possible. Draw the lowest-order Feynman diagram for this decay and go as far as you can toward calculating the amplitude without worrying about summing photon spins or integrating over the internal loop. (*Hints:* The Feynman diagram has three vertices. Note that conservation of four-momentum at the vertices will constrain all but one of the three virtual-particle four-momenta. Work toward expressing the amplitude in terms of dot products of four-momenta and/or photon polarization four-vectors integrated over the undetermined virtual-particle four-momentum. Remember that the dot product of a photon's four-momentum with its own polarization four-vector is zero.)

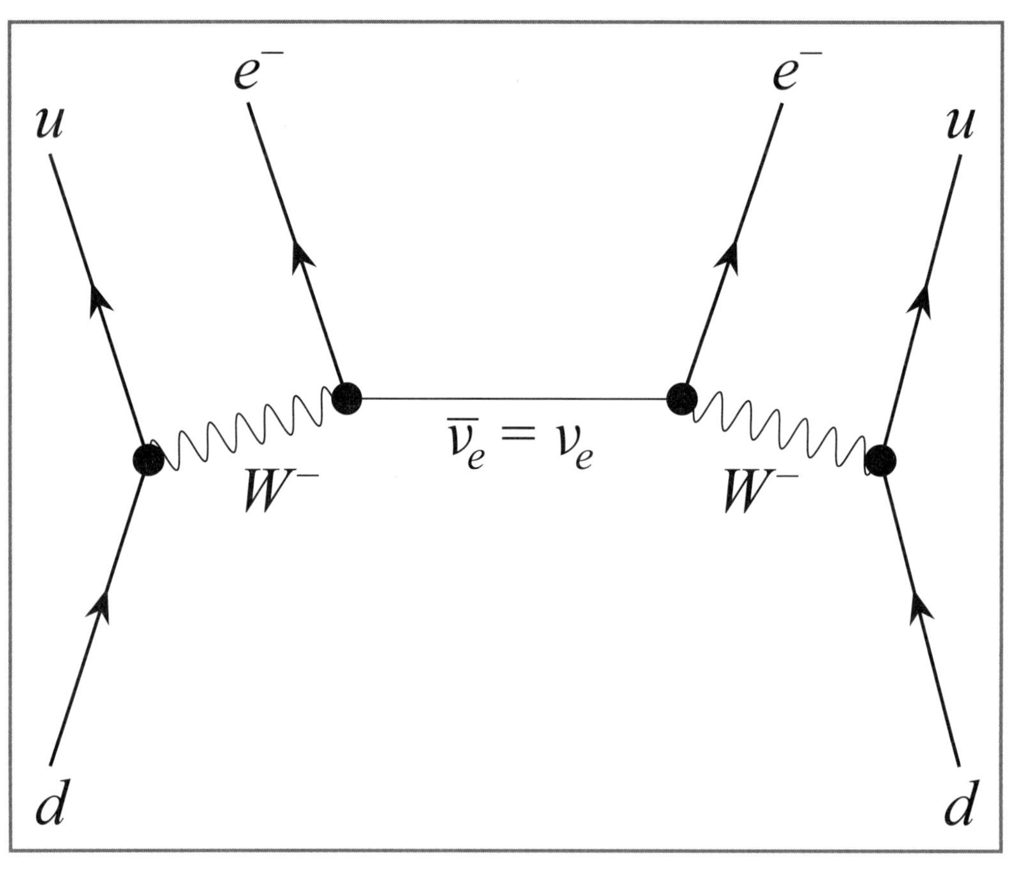

u
e^-
e^-
u
W^-
$\overline{\nu}_e = \nu_e$
W^-
d
d

28. NEUTRINO MASSES AND MIXING

In this chapter, we will complete our discussion of the weak interaction in the Standard Model by exploring the question of lepton masses and lepton mixing. The same Higgs mechanism that provides masses to quarks can also provide masses to all the leptons, and certainly does in the case of the electron, muon, and tau at least. But the neutrino masses are so much smaller, one wonders if another mechanism is at work. We will examine the experimental data regarding neutrino masses and look at several theoretical possibilities.

28.1 Lepton Mixing Issues

Recall that quarks and the massive leptons get their masses from Lagrangian density terms that look like

$$\mathcal{L}_{\text{int}} = -b[\,\overline{\Psi}_L \Phi \psi_R + \overline{\psi}_R \Phi^\dagger \Psi_L\,]\,, \tag{28.1}$$

where b is a coupling constant that ends up defining the mass, Ψ_L is the fermion doublet (with only the chirally left-handed parts selected), ψ_R is the right-handed part of the specific fermion in question, and Φ is the Higgs doublet (to be replaced by its charge-conjugate Φ_c if ψ_R is the upper component field in its fermion doublet).

When applying this to neutrinos, the first question to answer is whether chirally right-handed neutrinos exist. If they do exist, they cannot participate in any Standard Model interactions at all. So how would they be created, and how might we detect them? Many theoretical physicists therefore declare that chirally right-handed neutrinos (called **sterile neutrinos** because of their inability to interact) are simply not part of the Standard Model. (This is not an issue with the other particles, because right-handed electrons, muons, taus, and quarks can be created by electromagnetic and/or color interactions.)

If right-handed neutrinos do not exist, then Lagrangian density terms like Equation 28.1 are impossible, so the Higgs mechanism that gives masses to other particles cannot be applied to neutrinos. By this reasoning, neutrino masses in the Standard Model should be exactly zero.

If the neutrino masses were exactly zero, then mixing analogous to quark mixing would also be impossible. Remember from the last chapter that we only need to mix either the top components of our SU(2) doublets or the bottom components. So let's mix the neutrinos. However, since they all have zero mass, we can't tell those with definite mass from those who participate in interactions with $W^\pm$ bosons. Indeed, the *only* distinction between massless electron, muon, and tau neutrinos is that they are the neutrinos connected by weak interactions to electrons, muons, and taus, respectively. In such a case, the whole concept of weak mixing is irrelevant: the lepton doublets are independent by definition.

28.2 Neutrino Oscillations

The earliest evidence for neutrino mass (though it was not understood as that originally) came from the search for neutrinos from the sun. The sun should produce electron neutrinos prodigiously because in converting four protons to helium, two of the protons have to be converted to neutrons, a process that involves converting a u-quark to a d-quark, which will produce a positron and an electron neutrino (production of a muon and muon neutrino is not energetically possible). A calculation based on calculating

"

the energy released by each fusion reaction and correlating that with the power radiated by the sun implies that the neutrino flux at the earth's surface should be about 6.3×10^{14} m^2/s.

This may seem large, but neutrinos are fiendishly difficult to detect because of their tiny cross section. In the 1960s, Raymond Davis Jr. and John N. Bahcall built a neutrino detector consisting of a 380-m^3 tank containing perchloroethylene (dry-cleaning fluid) deep underground in an abandoned mine (the detector was put deep underground so that the overlying rock would shield the detector from solar or cosmic particles of any other type). An electron neutrino striking one of the chlorine atoms in the perchloroethylene could change a d-quark in one of its neutrons to a u-quark and an electron, changing the chlorine atom to an argon atom. Since argon is very different chemically from chlorine, it was possible to chemically separate the handful of argon atoms that might be generated in the tank per day. (I personally was aware of this experiment much earlier than most people because Davis's daughter was a classmate of mine at Carleton College in the early 1970s. Davis came to Carleton to deliver a talk about this experiment that I vividly remember.)

Davis found that the neutrino count was only about a third of the rate expected, which caused some consternation in the discipline. Over the following decades, the result was verified experimentally and no reasonable adjustment to the predicted rate could account for the discrepancy. (Davis won a Nobel Prize for this work in 2002.)

Eventually, the evidence became overwhelming that the solution had to be a phenomenon known as **neutrino oscillation**, an idea proposed in 1957 by Bruno Pontecorvo, but which languished because it required neutrinos to have mass, something not easily reconciled with the Standard Model (as we have seen). If neutrinos do have mass, then particle mixing would apply to leptons as well as quarks. The **Pontecorvo-Maki-Nakagawa-Sakata matrix** (or **PMNS matrix**) is the lepton analogue of the CKM matrix for quarks and has the same form mathematically, except that by convention we rotate the upper-component doublet fields (the neutrinos) instead of the lower-component fields:

$$\begin{bmatrix} \nu_e \\ \nu_\mu \\ \nu_\tau \end{bmatrix} = \begin{pmatrix} U_{e1} & U_{e2} & U_{e3} \\ U_{\mu 1} & U_{\mu 2} & U_{\mu 3} \\ U_{\tau 1} & U_{\tau 2} & U_{\tau 3} \end{pmatrix} \begin{bmatrix} \nu_1 \\ \nu_2 \\ \nu_3 \end{bmatrix}, \quad \text{where}$$

$$U = \begin{pmatrix} c_{12}c_{13} & s_{12}c_{13} & s_{13}e^{-i\delta} \\ -s_{12}c_{23} - c_{12}s_{23}s_{13}e^{i\delta} & c_{12}c_{23} - s_{12}s_{23}s_{13}e^{i\delta} & s_{23}c_{13} \\ s_{12}s_{23} - c_{12}c_{23}s_{13}e^{i\delta} & -c_{12}s_{23} - s_{12}c_{23}s_{13}e^{i\delta} & c_{23}c_{13} \end{pmatrix}, \quad (28.2)$$

where ν_1, ν_2, and ν_3 are neutrino fields, and c_{mn} and s_{mn} are shorthands for sines and cosines of the three angular parameters θ_{12}, θ_{13}, and θ_{23}.

But how do neutrino masses lead to "oscillations," and how do they explain the Davis result? To simplify things, let's ignore the tau neutrino for the moment, and assume that the electron neutrino quantum wavefunction is a superposition of definite-mass wavefunctions ν_1 and ν_2:

$$\begin{bmatrix} |\nu_e\rangle \\ |\nu_\mu\rangle \end{bmatrix} = \begin{pmatrix} \cos\theta & \sin\theta \\ -\sin\theta & \cos\theta \end{pmatrix} \begin{bmatrix} |\nu_1\rangle \\ |\nu_2\rangle \end{bmatrix} \qquad (28.3a)$$

$$\Rightarrow \begin{bmatrix} |\nu_1\rangle \\ |\nu_2\rangle \end{bmatrix} = \begin{pmatrix} \cos\theta & -\sin\theta \\ \sin\theta & \cos\theta \end{pmatrix} \begin{bmatrix} |\nu_e\rangle \\ |\nu_\mu\rangle \end{bmatrix}. \qquad (28.3b)$$

(The inverse of a real unitary matrix is simply its transpose.) Suppose that in a laboratory we use a weak-interaction process at the origin event $x_0^\alpha = 0$ in our frame to create a beam of pure electron neutrinos. From Equation 28.3a, this means that

$$|\nu(0)\rangle = |\nu_e\rangle = \cos\theta|\nu_1\rangle + \sin\theta|\nu_2\rangle. \qquad (28.4)$$

Now free-particle states with definite mass also are states of definite four-momentum. If we approximate the definite-mass neutrino states as plane-wave solutions to their respective Dirac equations, then the neutrino beam's state at a detector some spacetime position x_D^α away is

$$|\nu(t)\rangle = \cos\theta\, e^{-ip_1 \cdot x_D}|\nu_1\rangle + \sin\theta\, e^{-ip_2 \cdot x_D}|\nu_2\rangle. \tag{28.5}$$

From here, one can show (see Box 28.1) that the probability of measuring a neutrino that was originally an electron neutrino to be a muon neutrino a distance L from its creation point will be

$$\left|\langle\nu_\mu|\nu(t)\rangle\right|^2 \approx \left[\ \sin 2\theta \,\sin\left(\frac{m_1^2 - m_2^2}{4E}L\right)\right]^2, \tag{28.6}$$

assuming that the neutrino energy $E \approx E_1 \approx E_2$ is much larger than either m_1 or m_2. We can see that the probability oscillates from being 0 at certain positions downstream to being as large as $\sin^2 2\theta$ at other positions.

Therefore, the low result of the Davis experiment could be explained if the sun is at the right distance for most of the original electron neutrinos to have morphed themselves into muon or tau neutrinos, which would not interact with the chlorine atoms in the Davis detector and would therefore not be observed.

This transmutation is enhanced by an effect known as the **MSW effect** (after physicists Stanislav Mikheyev, Alexei Smirnov, and (independently) Lincoln Wolfenstein, who first described it). Electron neutrinos propagating through matter can interact with electrons in such a manner as to enhance the oscillation rate for electron neutrinos (but not muon neutrinos because normal matter contains no muons). The MSW effect does affect solar neutrinos as they travel through the sun and as they go through the rock on the way to the detector. It turns out that the MWS effect puts additional constraints on the oscillation rate and (because electron and muon neutrinos are affected differently) can actually be used to determine whether $m_2 > m_1$ (the vacuum oscillation only allows one to determine the difference in the squared mass, not which is larger). See Christian Y. Cardall and Daniel J. H. Chung *Phys. Rev. D* **60**, 073012 (1999) for a deeper discussion of the MSW effect than I can provide here.

The Sudbury Neutrino Observatory (in another mine in Sudbury, Ontario) provided the conclusive evidence that neutrino oscillation was real in 2001, while also resolving the solar neutrino problem. This detector observed neutrinos using three different processes involving neutrinos interacting with heavy water (D_2O), one of which had the same probability for all neutrino flavors (unlike other detectors that were mostly sensitive to electron neutrinos). Data from this detector showed conclusively that the total flux of neutrinos from the sun (considering all types of neutrinos) was indeed consistent with theoretical predictions. The experiment's director Arthur B. McDonald won a 2015 Nobel Prize for this work.

There are other sources of neutrinos that can be used to put constraints on neutrino mixing. High-energy cosmic rays hitting the earth's atmosphere can produce high-energy pions. A π^-, for example, can then decay to either $e^- + \overline{\nu}_e$ or $\mu^- + \overline{\nu}_\mu$. Though not quite as conclusive as the results from Sudbury, observation of atmospheric muons by the Super-Kamiokande detector offered the *first* evidence of neutrino oscillations by measuring the difference between the flux of downward atmospheric neutrinos and upward atmospheric neutrinos (which have traveled through the earth). Takaaki Kajita shared the 2015 Nobel Prize with McDonald for this research.

Also, a number of currently running experiments pair an artificial source of neutrinos (a particle accelerator or nuclear reactor) with detectors located at different distances from the source. Such experiments can observe neutrino oscillations in a more direct and controlled way than other experiments.

$\theta_{12} = 34°$	$\theta_{13} = 48°$	$\theta_{23} = 8.5°$
$\delta = 220°$	$\Delta m_{21}^2 = 7.6 \times 10^{-5}\,\mathrm{eV}^2$	$\Delta m_{32}^2 = 2.42 \times 10^{-3}\,\mathrm{eV}^2$

TABLE 28.1   Experimental values of neutrino parameters (from the 2020 Particle Data Group review article on neutrino mixing). Uncertainties are in the last nonzero digit. The values quoted assume $m_3 > m_2$ (though they change by at most a few percent if one assumes $m_3 < m_1$).

The current data from all of these sources assure us that $m_2 > m_1$, but do not tell us whether m_3 is bigger than both or smaller than both. Assuming that m_3 is bigger, the experimental constraints imply the results listed in Table 28.1. These results do imply clear mixing between neutrino types: indeed, θ_{12} and θ_{13} are so large that ν_e is not even approximately the same as ν_1.

The experimental parameters in the table specify the *differences* between the squared masses ($\Delta m_{21}^2 \equiv m_2^2 - m_1^2$, $\Delta m_{32}^2 \equiv m_3^2 - m_2^2$), because this is what experimentally measuring Equation 28.6 tells us. But experimentally determining the absolute mass of m_1 (or any of the others) is very difficult. In principle, one could measure m_1 by carefully measuring the recoil of particles in neutron decay ($n \to p + e^- + \overline{\nu}_e$), or (more practically) the radioactive decay of tritium ($^3_1\mathrm{H} \to {}^3_2\mathrm{He} + e^- + \overline{\nu}_e$). But the difficulty of such measurements means that we know only that $m_e^{\mathrm{eff}} \lesssim 1.1\,\mathrm{eV}$ from the KATRIN experiment (remember that the mass of the ν_e is not well-defined, so m_e^{eff} is a sort of superposition of m_1 and m_2). We can better estimate the effective mass of the ν_μ from examining $\pi^- \to \mu^- + \overline{\nu}_\mu$: measurements constrain $m_\mu^{\mathrm{eff}} \lesssim 0.19\,\mathrm{eV}$. So it seems likely that m_1 and m_2 are in the range of hundredths to maybe tenths of an eV. Improving these measurements continues to be an active area of research.

28.3 The Seesaw Mechanism

Even if we don't exactly know what the masses are, the experimental data make it abundantly clear that neutrinos have mass. But, as we have seen, it is difficult to see how this could come from the Higgs mechanism, and even if it did, why the neutrino masses are so much smaller than those of other fermions. This suggests that the mechanism for generating neutrino masses might be something different than the Higgs mechanism. This section explores one such possibility, known as the **seesaw mechanism**.

The mass term in a Dirac equation typically has the form $-m\overline{\psi}\psi$. Since the charge conjugation operator C converts a particle field into the field of its antiparticle, and since the basic structure of quantum field theory requires a particle and its antiparticle to have the same mass, we might write this term in the form

$$\mathcal{L}_{m_D} = -\tfrac{1}{2}m_D[\,\overline{\psi}\psi + \overline{\psi}^{\,c}\psi^c\,],\tag{28.7}$$

where $\psi^c \equiv C\psi = -i\gamma^y\psi$ (see Chapter 17). However, this is not the most general relativistically invariant expression imaginable. One can also consider terms that mix the particle and antiparticle fields such as

$$\mathcal{L}_{m_M} = -\tfrac{1}{2}m_M[\,\overline{\psi}^{\,c}\psi + \overline{\psi}\psi^c\,].\tag{28.8}$$

These terms are called **Majorana terms** after Ettore Majorana, the physicist who first discussed them. So the most general possible mass term would include both of these cases. We could write this most general term as a matrix equation:

$$\mathcal{L}_m = -\tfrac{1}{2}\,[\,\overline{\psi}\ \ \overline{\psi}^{\,c}\,]\begin{pmatrix} m_D & m_M \\ m_M & m_D \end{pmatrix}\begin{bmatrix} \psi \\ \psi^c \end{bmatrix}.\tag{28.9}$$

We can check the viability of such an expression while also setting ourselves up to consider consistency with $\mathrm{SU}(2)_\mathrm{L}$ by expressing this in terms of chirally left- and right-handed fields:

$$\psi_L \equiv \tfrac{1}{2}(1-\gamma^5)\psi = \begin{bmatrix} w_L \\ 0 \end{bmatrix} \quad \text{and} \quad \psi_R \equiv \tfrac{1}{2}(1+\gamma^5)\psi = \begin{bmatrix} 0 \\ w_R \end{bmatrix}, \quad (28.10)$$

where w_L and w_R are the two-component Weyl spinors representing the chirally left- and right-handed components of the entire spinor. (Note: I have denoted the Weyl spinors ψ_L and ψ_R previously, but here I want to make a distinction: ψ_L and ψ_R are four-component Dirac spinors with the other-handed part erased.) Now, in Box 28.2 you will show that

$$(\psi^c)_L = \begin{bmatrix} -i\sigma_y w_R \\ 0 \end{bmatrix} = (\psi_R)^c \quad \text{and} \quad (\psi^c)_R = \begin{bmatrix} 0 \\ i\sigma_y w_L \end{bmatrix} = (\psi_L)^c, \quad (28.11)$$

which (in words) tells us that the chirally left-handed part of the charge-conjugated field is constructed out of the field's right-handed part (it is in fact the *conjugate* of that right-handed part) and vice versa.

This is important because we saw in Chapter 27 that

$$\overline{\psi}\psi = \overline{\psi}_L\psi_R + \overline{\psi}_R\psi_L \quad \text{and} \quad \overline{\psi}_L\psi_L = \overline{\psi}_R\psi_R = 0. \quad (28.12)$$

This is because the γ^t in $\overline{\psi}\psi = \psi^\dagger\gamma^t\psi$ connects the left-handed part of $\psi^\dagger$ with the right-handed part of ψ and vice versa (and remember that we needed $\overline{\psi}\psi$ instead of $\psi^\dagger\psi$ to preserve Lorentz invariance). The point is that only terms that connect chirally left-handed parts to right-handed parts will be nonzero. So we can expand Equation 28.9 into a version that explicitly displays the chiral parts as follows:

$$\mathcal{L}_m = -\tfrac{1}{2}\,[\,\overline{\psi}_R \quad (\overline{\psi}^c)_R \quad \overline{\psi}_L \quad (\overline{\psi}^c)_L\,] \begin{pmatrix} 0 & 0 & m_D & m_R \\ 0 & 0 & m_L & m_D \\ m_D & m_L & 0 & 0 \\ m_R & m_D & 0 & 0 \end{pmatrix} \begin{bmatrix} \psi_R \\ (\psi^c)_R \\ \psi_L \\ (\psi^c)_L \end{bmatrix}. $$

$$(28.13)$$

The zeros in the diagonal blocks clarify that chirally right-handed parts don't couple to right-handed parts, and likewise for left-handed parts. Note that m_R couples the terms ψ_R and $(\psi^c)_L = (\psi_R)^c$ that are based on the chirally right-handed field, and similarly m_L connects ψ_L with $(\psi^c)_R = (\psi_L)^c$.

Such a term will be Lorentz-invariant, but we also must be careful to ensure that the term is consistent with the $\mathrm{SU}(2)_L \otimes \mathrm{U}(1)$ symmetry of the Standard Model. That symmetry requires that the left-handed neutrino (and right-handed antineutrino)—the one linked to the charged lepton in an SU(2) doublet—be massless before symmetry breaking. But a *right*-handed neutrino is, as we've seen, disconnected from the symmetry, and so could exist and have a mass independent of anything else. So a mass term that would be consistent with relativity and the $\mathrm{SU}(2)_L \otimes \mathrm{U}(1)$ symmetry could be

$$\mathcal{L}_m = -\tfrac{1}{2}\,[\,\overline{\psi}_R \quad (\overline{\psi}^c)_R \quad \overline{\psi}_L \quad (\overline{\psi}^c)_L\,] \begin{pmatrix} 0 & 0 & m_D & m_M \\ 0 & 0 & 0 & m_D \\ m_D & 0 & 0 & 0 \\ m_M & m_D & 0 & 0 \end{pmatrix} \begin{bmatrix} \psi_R \\ (\psi^c)_R \\ \psi_L \\ (\psi^c)_L \end{bmatrix}, $$

$$(28.14)$$

where I have set $m_L = 0$ and renamed $m_R = m_M$, the right-handed neutrino's **Majorana mass.**

In this formulation, the left-handed neutrino (the one involved in the weak interaction) and the right-handed neutrino are not the neutrinos with definite masses. We find the masses by finding the eigenvalues of the matrix. In Box 28.3, we will see that the positive eigenvalues are

$$m_\pm = \tfrac{1}{2}\left(\sqrt{4m_D^2 + m_M^2} \pm m_M\right), \quad (28.15)$$

and the corresponding (unnormalized) eigenvectors are

$$
\text{for } m_+ : \quad \mathrm{ev}_+ = \begin{bmatrix} 1 \\ m_-/m_D \\ m_D/m_+ \\ 1 \end{bmatrix}, \quad \text{and for } m_- : \quad \mathrm{ev}_- = \begin{bmatrix} -1 \\ m_+/m_D \\ -m_D/m_- \\ 1 \end{bmatrix}. \tag{28.16}
$$

The approximation thought to be experimentally relevant here is when $m_M \gg m_D$. In such a case (again, see Box 28.3), we have

$$
m_+ \approx m_M, \quad \mathrm{ev}_+ \approx \begin{bmatrix} 1 \\ 0 \\ 0 \\ 1 \end{bmatrix}, \quad m_- \approx m_D \frac{m_D}{m_M}, \quad \mathrm{ev}_- \approx \begin{bmatrix} 0 \\ 1 \\ -1 \\ 0 \end{bmatrix}. \tag{28.17}
$$

We see that if the left- and right-handed neutrino system gets a Dirac mass of m_D about equal to the masses of other particles, then one of the neutrino solutions is exploded to be approximately m_M instead, and the other is suppressed by a factor of m_D/m_M. (This is the "seesaw" part.) Imagine, for example, that the Dirac mass is $m_D \sim 100\,\text{GeV}$ (roughly the mass of the top quark), while the Majorana mass is $m_M \sim 10^{15}\,\text{GeV}$ (about the Planck mass)—both credible mass scales that might be involved. Then

$$
m_- \sim \frac{(100\,\text{GeV})^2}{10^{15}\,\text{GeV}} = 10^{-11}\,\text{GeV} = 0.01\,\text{eV}, \tag{28.18}
$$

which is of the order of magnitude we want!

28.4 Is the Neutrino Really a Majorana Particle?

The fermion with mass m_- is a strange version of the neutrino, though: its eigenvector implies that its field is something like

$$
\psi = (\psi^c)_R + \psi_L = (\psi_L)^c + \psi_L. \tag{28.19}
$$

Note that the conjugate version of this neutrino is itself:

$$
\psi^c = ((\psi_L)^c + \psi_L)^c = \psi_L + (\psi_L)^c = \psi. \tag{28.20}
$$

This is a reconceptualization of the neutrino where it serves as its own antiparticle, just as the photon is its own antiparticle. You can see that this is constructed out of our original chirally left-handed neutrino, so this is the one that participates in weak interactions. But the original neutrino field had nothing in the right-handed slot, whereas this one has the conjugate of its left-handed field providing the Weyl spinor for its right-handed slot. This makes the field self-conjugate. We call a fermion that is its own antiparticle a **Majorana fermion**.

Note that any particle that is its own antiparticle must have zero charge. Of all the fermions we know about, the neutrino is the *only one* that could possibly be a Majorana fermion.

The neutrino whose mass is m_M is a Majorana version of the original right-handed neutrino field, which is not connected to the weak interaction. This neutrino is therefore sterile. Even if it wasn't, its mass is so large as to be inaccessible experimentally, so we are not going to prove this model by looking for this particle.

But a low-mass Majorana particle is physically different than our originally separate left-handed neutrino and right-handed antineutrino. In particular, certain nuclei can undergo what is called a "double beta" decay process. For example,

$$
{}^{76}\text{Ge} \rightarrow {}^{76}\text{Se} + 2e^- + 2\bar{\nu}_e. \tag{28.21}
$$

The Feynman diagram for this process is illustrated in Figure 28.1. In this process, the down quark in each of two neutrons decays to an up quark, converting each neutron to a proton. Normally, this process would emit two electron antineutrinos, and because there are four light particles involved in the decay, the electrons will have a spread of energies ranging from zero to half the energy released in the decay. (The final nucleus is so massive it essentially has no recoil kinetic energy.)

But if the electron neutrino is really a Majorana neutrino, it is its own antiparticle, and therefore the Feynman diagram shown in Figure 28.2 is possible, where no neutrinos are emitted. In this case, the decay involves only two light electrons, which should get energies exactly equal to half the decay energy released. Therefore, we should see a spike in the probability that electrons are emitted with half the decay energy in addition to the curve of energies from the normal decay with neutrinos. A number of experiments are currently running to look for this distinctive decay process (and earn the Nobel Prize that will inevitably follow its discovery), but as of this writing, no such decay has ever been observed. On the other hand, the process will be very rare, so the model is not yet ruled out by any means.

I have also completely glossed over the theoretical difficulty of explaining where such a mass term would come from. If one allows for the existence of the sterile right-handed neutrino, one can generate normal Dirac neutrino masses using the Higgs mechanism. But one cannot generate Majorana mass terms this way at all, and other ways to generate the term are complicated and inelegant. While the seesaw mechanism seems to yield the right order of magnitude of masses (thus avoiding the question of why neutrino masses are so different than other fermion masses), any explanation of where the mass term comes from must be from beyond the Standard Model.

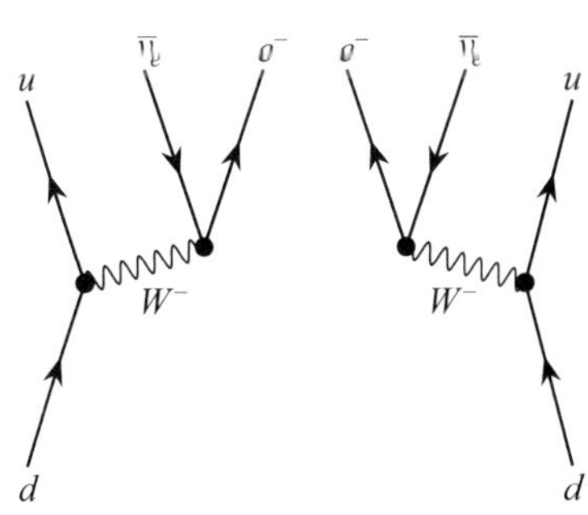

FIGURE 28.1   This is the Feynman diagram for a normal double beta decay. This process is allowed by the Standard Model.

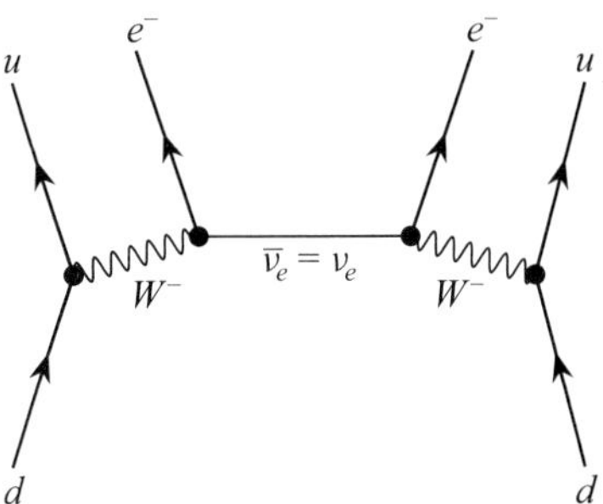

FIGURE 28.2   This is the lowest-order Feynman diagram for a neutrino-less double beta decay. This process is only allowed if the neutrino is a Majorana fermion.

BOX 28.1   Neutrino Oscillation

I argued in Section 28.2 that if we create a neutrino beam at the spactime origin $x_0^\alpha = 0$, the neutrino beam's state at a detector some spacetime position x_D^α away is

$$|v(t)\rangle = \cos\theta \, e^{-ip_1\cdot x_D}|v_1\rangle + \sin\theta \, e^{-ip_2\cdot x_D}|v_2\rangle. \qquad (28.22)$$

Our goal in this box is to calculate the probability $\left|\langle v_\mu|v(t)\rangle\right|^2$ that the neutrino will be found to be a muon neutrino in the detector.

The calculation is a bit tricky because v_1 and v_2 neutrinos have different masses and therefore different spatial momenta for the same energy, or different energies for the same spatial momenta. In most experiments, the neutrino energies ($\sim$ MeV) are so large compared to their masses (fractions of eV) that $E \approx |\vec{p}|$. Assume that definite-mass states $|v_1\rangle$ and $|v_2\rangle$ were formed by a process that gave them the same energy E. They will therefore have slightly different spatial momenta $|\vec{p}_1 \neq |\vec{p}_2|$ by virtue of their tiny but distinct masses. At the detector a distance L away, let's measure this neutrino state at the time T corresponding to L/v_{avg}, where v_{avg} is *average* of the speeds of the $|v_1\rangle$ and $|v_2\rangle$ neutrinos (treating them as classical particles): this makes the calculation simplest. We have

$$v_{\text{avg}} = \frac{1}{2}\left(\frac{|\vec{p}_1|}{E} + \frac{|\vec{p}_2|}{E} \right) \;\Rightarrow\; L = v_{\text{avg}}T = \frac{|\vec{p}_1| + |\vec{p}_2|}{2E}T. \qquad (28.23)$$

The four-displacement x_D^α of the detector measurement relative to the neutrino creation event (assuming we define the x axis to go through both) is therefore

$$x_D^\alpha = \begin{bmatrix} T \\ L \\ 0 \\ 0 \end{bmatrix} = \frac{T}{2E}\begin{bmatrix} E + E \\ |\vec{p}_1| + |\vec{p}_2| \\ 0 \\ 0 \end{bmatrix} = \frac{T(p_1 + p_2)^\alpha}{2E}. \qquad (28.24)$$

If we pull an overall phase out of Equation 28.22, we have

$$|v(T)\rangle = e^{-ip_1\cdot x_D}\Big[\cos\theta \, |v_1\rangle + \sin\theta \, e^{i(p_1-p_2)\cdot x_D}|v_2\rangle \Big]$$

$$= e^{-ip_1\cdot x_D}\Big[\cos\theta \, |v_1\rangle + \sin\theta \, e^{iT(m_1^2-m_2^2)/2E}|v_2\rangle \Big]. \qquad (28.25)$$

Exercise 28.1.1   Verify the last step.

Now that we have taken care to expose the implications of the tiny *difference* between the neutrinos' four-momenta, we can take advantage of the fact that in any likely experimental situation, the neutrino energies are so large compared to their masses that the neutrinos move at essentially the speed of light. The errors we introduce by saying that $L \approx T$ are therefore negligible at this point. So

$$|\nu(T)\rangle \approx e^{-ip_1 \cdot x_D} \left[\cos\theta \, |\nu_1\rangle + \sin\theta \, e^{iL(m_1^2 - m_2^2)/2E} |\nu_2\rangle \right]. \qquad (28.26)$$

From here it takes pretty much an ordinary quantum mechanics calculation to calculate the probability that the neutrino will be observed to be a muon neutrino. Remember from Equation 28.3a that

$$\begin{bmatrix} |\nu_e\rangle \\ |\nu_\mu\rangle \end{bmatrix} = \begin{pmatrix} \cos\theta & \sin\theta \\ -\sin\theta & \cos\theta \end{pmatrix} \begin{bmatrix} |\nu_1\rangle \\ |\nu_2\rangle \end{bmatrix}. \qquad (28.27)$$

Exercise 28.1.2   Calculate the probability $\left| \langle \nu_\mu | \nu(T) \rangle \right|^2$ and compare your result to Equation 28.6. (*Hints:* To save writing, define $b \equiv L(m_1^2 - m_2^2)/2E$. Have a table of trigonometric identities handy.)

BOX 28.2 Handedness of the Conjugate

The goal of this box is to prove Equation 28.11 (repeated here):

$$(\psi^c)_L = \begin{bmatrix} -i\sigma_y w_R \\ 0 \end{bmatrix} = (\psi_R)^c \quad \text{and} \quad (\psi^c)_R = \begin{bmatrix} 0 \\ i\sigma_y w_L \end{bmatrix} = (\psi_L)^c. \quad (28.28)$$

Exercise 28.2.1 Prove this. (*Hints:* Remember that $\psi_{L,R} = \frac{1}{2}(1 \mp \gamma^5)\psi$,

$$\tfrac{1}{2}(1 - \gamma^5) = \begin{pmatrix} 1 & 0 \\ 0 & 0 \end{pmatrix}, \qquad \tfrac{1}{2}(1 + \gamma^5) = \begin{pmatrix} 0 & 0 \\ 0 & 1 \end{pmatrix}, \qquad \gamma^y = \begin{pmatrix} 0 & \sigma_y \\ -\sigma_y & 0 \end{pmatrix},$$

$$(28.29)$$

and $\psi^c = -i\gamma^y\psi$. Work toward the middle from each end.)

BOX 28.3 Mass Matrix Eigenvalues and Eigenvectors

The goal in this box is show that the Majorana mass matrix

$$M = \begin{pmatrix} 0 & 0 & m_D & m_M \\ 0 & 0 & 0 & m_D \\ m_D & 0 & 0 & 0 \\ m_M & m_D & 0 & 0 \end{pmatrix} \tag{28.30}$$

has positive eigenvalues and corresponding eigenvectors

$$m_\pm = \tfrac{1}{2}\left(\sqrt{4m_D^2 + m_M^2} \pm m_M\right), \qquad \mathrm{ev}_\pm = \begin{bmatrix} \pm 1 \\ m_\mp/m_D \\ \pm m_D/m_\pm \\ 1 \end{bmatrix}. \tag{28.31}$$

(The masses $m_\pm$ must be positive, so we only want the positive eigenvalues.)

Exercise 28.3.1 Enter eigenvalues {{0,0,d,m}, {0,0,0,d}, {d,0,0,0}, {m, d,0,0}} into WolframAlpha and simplify the results you see to verify Equation 28.31. (Note that I have used $m \equiv m_M$ and $d \equiv m_D$ for simplicity.)

Exercise 28.3.2 Evaluate these quantities in the limit that $m_M \gg m_D$. (*Hint:* Use the binomial approximation for the square root so that you can say something more informative than "$m_- = 0$".)

Homework Problems

P28.1 Consider a nuclear reactor that produces 2 GW of power, about 2% of which comes out in the form of 4-MeV electron antineutrinos. Assume that the cross section for the neutrino interaction $\bar{\nu}_e + p \to n + e^+$ is about 10^{-47} m^{-2}. Assume also the two-neutrino model described in Section 28.2 (for small distances and low energies, this is a decent model because oscillation to a tau neutrino is not very likely).

The goal is to design an experiment to investigate neutrino oscillation. You will place a detector close to the reactor to measure the neutrino flux before oscillation, and another detector at a distance where you expect the number of muon neutrinos to be maximized.

a. How many neutrinos per second does the reactor produce?

b. How far away should you place the far detector? (Ignore any possible MSW effect.)

c. Suppose you want the far detector's mass to be large enough so that it would register about 1000 events per year if there were no neutrino oscillation. What should the detector's mass be in kilograms? (Assume the neutrinos are emitted equally in all directions.)

P28.2 Suppose we determine, somehow, that the the neutrino with mass m_1 has the smallest mass and that its mass is about 0.005 eV. According to the data in Table 28.1, what are the masses of the other two neutrinos?

P28.3 Assume that all three neutrinos (the electron, muon, and tau neutrinos) are Majorana neutrinos with m_D for each equal to the mass of the corresponding up-type quark, with $m_M \approx 10^{15}$ GeV. Calculate the mass of each neutrino, Δm_{12}^2 and Δm_{23}^2.

P28.4 Consider the possibility that dark matter might be neutrinos. Measurements imply that the density of dark matter in the universe is about 2.1×10^{-27} kg/m^3, and physical conditions during the Big Bang imply a present neutrino number density of roughly 1.4×10^8 /m^3. If dark matter is entirely massive neutrinos, what is the average mass of a neutrino (averaging over all three types)? Explain why this scenario is inconsistent with available experimental information about neutrino masses.

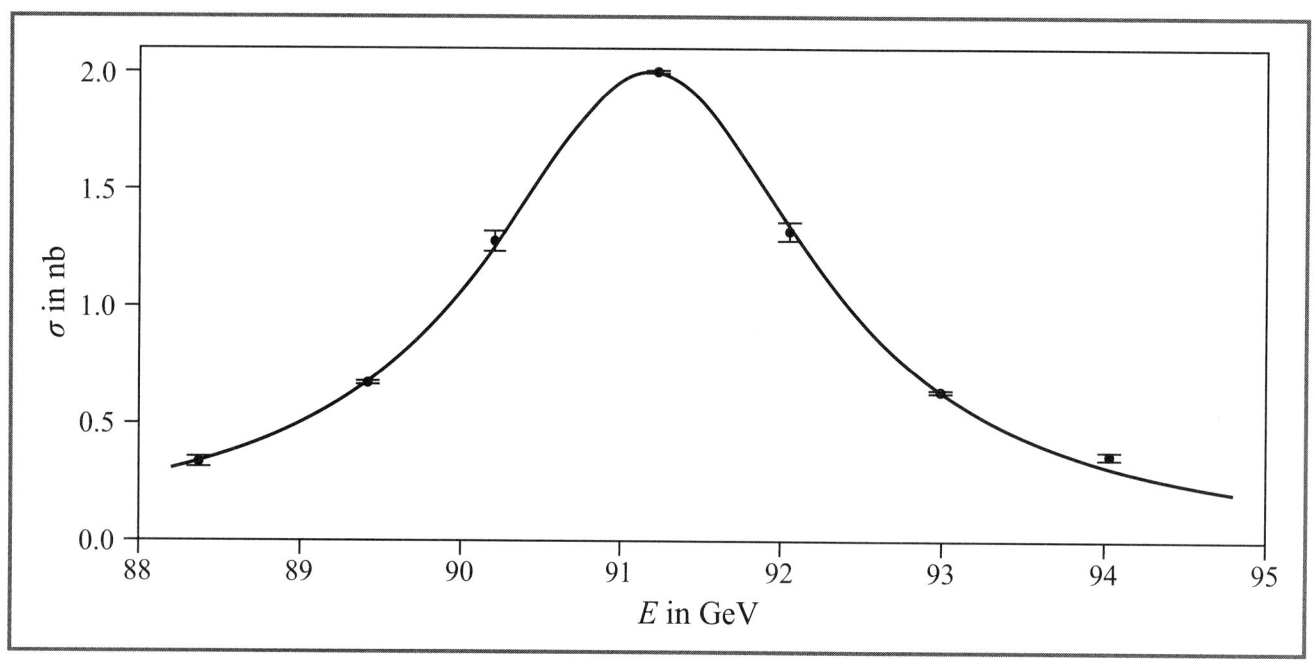

σ in nb
2.0
1.5
1.0
0.5
0.0
E in GeV
88
89
90
91
92
93
94
95

29. APPLICATIONS OF ELECTROWEAK THEORY

The goal of this optional chapter is to discuss applications of electroweak theory that allow us to test aspects of the theory and measure quantities such as the masses and lifetimes of the Z and W and the number of neutrino types. We will aso explore the concept of a particle *resonance*.

29.1 Z Production

One of the most important tests we can perform regarding the weak interaction is to actually produce the weak-interaction bosons $W^\pm$ and Z and measure their characteristics. If you have read Chapter 22, you have seen the value of electron-positron colliders for investigating the electromagnetic interaction. But the main purpose of the Large Electron-Positron Collider (LEP) at CERN was to be a "factory" for producing $W^\pm$ and Z bosons. Between 1989 and 2000, this collider and its associated detectors ALEPH, DELPHI, OPAL, and L3 provided crucial information about these bosons.

Let's begin by looking at the production of Z bosons. The Feynman diagram for the $e^- + e^+ \to Z$ consists of a single vertex (see Figure 29.1). Using the rules, we see that the amplitude for this process is

$$-i\mathcal{M} = [\bar{v}_2 \left(\frac{-ig_W}{2\cos\theta_W} \gamma^\mu [c_V - c_A \gamma^5] \right) u_1](\epsilon_{3,\mu})^*,$$

$$\Rightarrow \quad \mathcal{M} = \frac{g_W}{2\cos\theta_W}[\bar{v}_2 \gamma^\mu (c_V - c_A \gamma^5) u_1]\,\epsilon^*_{3,\mu}. \tag{29.1}$$

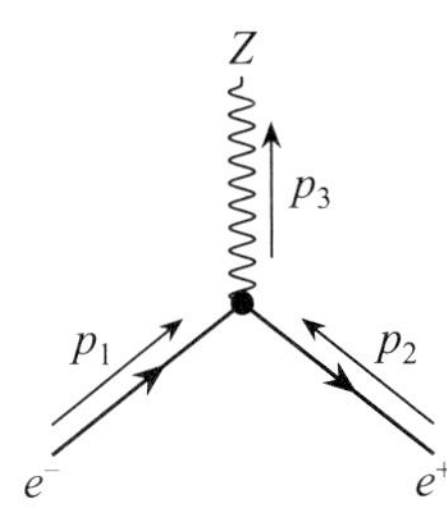

FIGURE 29.1 This is the Feynman diagram for $e^- + e^+ \to Z$.

From Table 25.2, the values of c_V and c_A for electrons are $c_V = -\frac{1}{2} + 2\sin\theta_W$ and $c_A = -\frac{1}{2}$. If we assume that the electron and positron beams are unpolarized, then we can sum over the spins of the electron and positron using Casimir's trick, sum over the Z spin using the massive-boson completeness relation

$$\sum_{s_3}(\epsilon^{s_3}_{3,\mu})^* \epsilon^{s_3}_{3,\nu} = -g_{\mu\nu} + \frac{p_{3,\mu}p_{3,\nu}}{M_Z^2}, \tag{29.2}$$

divide by 4 to average over the electron and positron spins, and use Fermi's Golden Rule to calculate the cross section. In the limit that the electron mass is negligible (an excellent approximation in this case), the result in the center-of-mass frame (see Box 29.1) is

$$\sigma = \frac{\pi g_W^2}{8\cos^2\theta_W M_Z}\left(\tfrac{1}{2} - 2\sin\theta_W + 4\sin^2\theta_W\right)\delta(E - M_Z), \tag{29.3}$$

where $E = E_1 + E_2$ is the total collision energy in that frame. The Dirac delta here is needed to enforce conservation of energy, but since it goes to infinity, it spoils our chances of calculating the cross section's actual magnitude.

But the calculation assumes that the Z is a completely *real* particle. In truth, however, the Z decays quite quickly to some kind of fermion antifermion pair—so quickly that we can never observe it directly. So the Z here is really more like a virtual particle in the diagram shown in Figure 29.2. To really understand how to handle this kind of situation, we need to better understand how to handle an internal particle that is close to being real but also is unstable (so far, we have implicitly been assuming that virtual particles in our diagrams have been the virtual versions of *stable* real particles). Addressing this question involves exploring the concept of *resonance*.

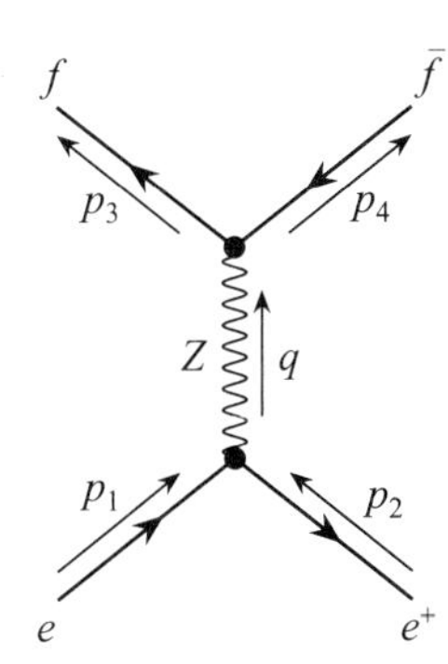

FIGURE 29.2 This shows a more realistic Feynman diagram for production of a Z by colliding an e^- and e^+. The f and $\bar{f}$ could be any fermion-antifermion pair.

485

29.2 The Propagator for a Decaying Particle

Consider a plane-wave solution to either the Klein-Gordon or Dirac equation for a particle at rest. The exponential part of the wave is

$$e^{-ip\cdot x} = e^{-iEt}. \tag{29.4}$$

This is the quantum amplitude for a particle at rest to be found at an arbitrary position at time t. The *probability* that this particle would be found within a block of positions with volume d^3x at time t is proportional to $|e^{-iEt}|^2 d^3x$.

Now, if the particle were to decay with an average lifetime of τ, we would want this probability to decay as $e^{-t/\tau}$. We could engineer this to happen by changing the amplitude of the particle at rest to be

$$e^{-ip\cdot x} = e^{-i(E-i\Gamma/2)t}. \tag{29.5}$$

The absolute square of this is

$$e^{+i(E-i\Gamma/2)^*t}e^{-i(E-i\Gamma/2)t} = e^{(iE-\Gamma/2-iE-\Gamma/2)t} = e^{-\Gamma t}, \tag{29.6}$$

which is what we want if we identify $\tau = 1/\Gamma$.

In Box 20.1, we learned how to find the denominator of a particle's propagator. The first step is to apply the free field's Euler-Lagrange equation. In this case, the free-field equation in the Lorenz gauge is

$$\partial_\mu \partial^\mu Z^\nu + M_Z^2 Z^\nu = 0. \tag{29.7}$$

If we substitute $Z^\nu = \epsilon^\nu e^{-i(E-i\Gamma_Z/2)t}$ into this, we find that (since $\partial^t = \partial_t$)

$$\begin{aligned}
0 &= \partial_\mu \partial^\mu \epsilon^\nu e^{-i(E-i\Gamma_Z/2)t} + M_Z^2 \epsilon^\nu e^{-i(E-i\Gamma_Z/2)t} \\
&= \partial_t \partial_t \epsilon^\nu e^{-i(E-i\Gamma_Z/2)t} + M_Z^2 \epsilon^\nu e^{-i(E-i\Gamma_Z/2)t} \\
&= [-i(E-i\Gamma_Z/2)]^2 \epsilon^\nu e^{-i(E-i\Gamma_Z/2)t} + M_Z^2 \epsilon^\nu e^{-i(E-i\Gamma_Z/2)t} \\
&= -(E^2 - iE\Gamma_Z - \tfrac{1}{4}\Gamma_Z^2 - M_Z^2)Z^\nu.
\end{aligned} \tag{29.8}$$

The denominator of the propagator is the negative of the quantity that multiplies the field, so in this case,

$$\text{propagator} \propto \frac{1}{E^2 - M_Z^2 - iE\Gamma_Z - \tfrac{1}{4}\Gamma_Z^2}. \tag{29.9}$$

(This only applies in the frame where the virtual Z is at rest.)

To calculate the value of $\langle |\mathcal{M}|^2 \rangle$ for unpolarized beams of electrons and positrons, we need to use the correct expression for the propagator's numerator, multiply by the vertex factors, take the absolute square, sum over spins, and average over initial spins. This is a complicated calculation, and one must repeat the process for each possible outgoing fermion-antifermion pair, but we can get a quick estimate as follows. The denominator of the expression for $\langle |\mathcal{M}|^2 \rangle$ will be the absolute square of the expression in Equation 29.9. But for scattering of two particles to two particles, $\mathcal{M}$ must be unitless, so the numerator must have units of (energy)^4. Consider the situation where the virtual Z is approximately real ($E \approx M_Z$), and let's assume $\Gamma_Z \ll M_Z$. Also, note that the masses of the incoming electron-positron pair as well as all possible outgoing particles are essentially zero in comparison to M_Z. In such a case, the only relevant energy is M_Z, so we would expect that for a given pair of outgoing particles $f\bar{f}$,

$$\langle |\mathcal{M}_f|^2 \rangle \approx \frac{K_f(\theta)M_Z^4}{|E^2 - M_Z^2 - iE\Gamma_Z - \tfrac{1}{4}\Gamma_Z^2|^2} \approx \frac{K_f(\theta)M_Z^4}{(E^2 - M_Z^2)^2 + M_Z^2 \Gamma_Z^2}, \tag{29.10}$$

where K_f is some unitless function of the scattering angle θ (Box 29.2 justifies the final step). The all-angle cross section for a given outcome is therefore

$$\sigma = \int \frac{S\langle|\mathcal{M}_f|^2\rangle}{(8\pi)^2(E_1 + E_2)^2} \frac{|\vec{p}_3|}{|\vec{p}_1|}\, d\Omega \approx \sigma_{0f} \frac{M_Z^2\Gamma_Z^2}{(E^2 - M_Z^2)^2 + M_Z^2\Gamma_Z^2}, \quad (29.11)$$

where the integral is over all solid angles (see Box 29.3 for the details). The value of the constant $\sigma_{0f} \equiv [\int K_f(\theta)d\Omega]/64\pi^2\Gamma_Z^2$ (which is the total cross section when the collision energy $E \equiv E_1 + E_2$ is equal to M_Z) will depend on the output particle f, but we see that *all* the cross sections have the E-dependence illustrated in Figure 29.3. This bump in the cross section is called a **resonance.** Indeed, the same function of E applies to *any* process that creates a short-lived particle of mass M at rest, so long as $E \approx M$ and $\Gamma \ll M$. The width of the curve at half its peak value (see Figure 29.3 and Box 29.4) is Γ.

The point is that we can measure both the mass of the Z and its lifetime by measuring the energy of the resonance curve's peak and its width at half maximum, respectively. In practice, we actually find both by determining the curve parameters that best fit the whole set of available data. Figure 29.4 shows that the fit to measured cross sections from the LEP collider is excellent for $M_Z = 91.2$ GeV and $\Gamma_Z = 2.5$ GeV. This value of Γ_Z corresponds to an average lifetime of about 2.6×10^{-25} s—far too short for the Z to leave a measurable track even if it was fully real.

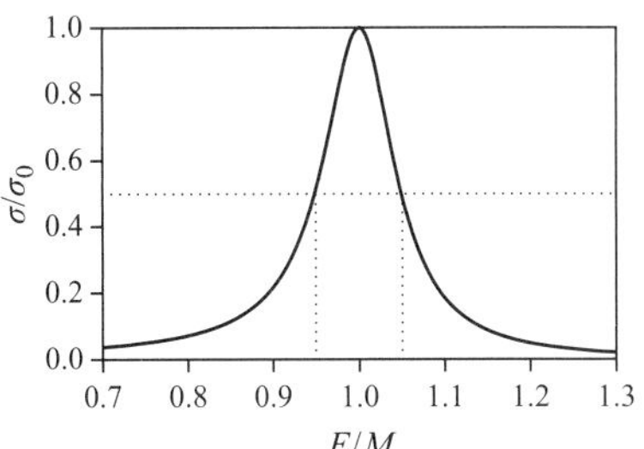

FIGURE 29.3 A graph of the resonance curve given in Equation 29.11 for $\Gamma = M/10$, where M is the unstable particle's mass. Note that the curve's width at half maximum is Γ.

29.3 The Mass of the $W^\pm$

Unfortunately, the same approach does not work for the $W^\pm$ particle. The process $e^- + e^+ \to W^\pm$ is forbidden by charge conservation, so we cannot observe a resonance arising from creation of a single particle at rest. But we can experimentally determine M_W another way. At sufficiently high energies, the process $e^- + e^+ \to W^- + W^+$ becomes possible. The cross section for this process in the center-of-mass frame has the form

$$\sigma = \int \frac{S\langle|\mathcal{M}|^2\rangle}{(8\pi)^2(E_1 + E_2)^2} \frac{|\vec{p}_3|}{|\vec{p}_1|}\, d\Omega = \frac{1}{(8\pi)^2 E^2} \frac{|\vec{p}_3|}{|\vec{p}_1|} \int \langle|\mathcal{M}|^2\rangle\, d\Omega. \quad (29.12)$$

The integral of the spin-averaged squared scattering amplitude integrated over all solid angles will just be a unitless number K that might depend on the collision energy E, but for a sufficiently narrow range of energies, it will be approximately constant. Also, note that the momentum magnitude of the incoming electron is $|\vec{p}_1| \approx E_1 = \frac{1}{2}E$ because the electron's mass is negligible compared to the energies involved. The value of the

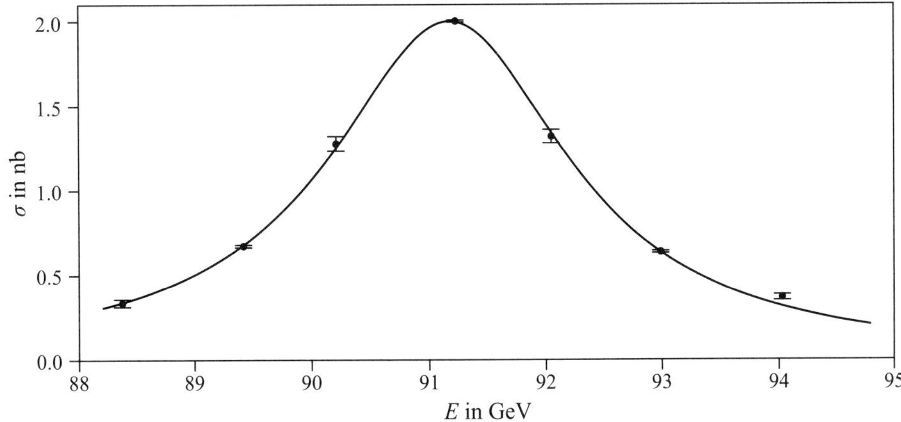

FIGURE 29.4 This graph shows the Z resonance curve for $M_Z = 91.2$ GeV and $\Gamma_Z = 2.5$ GeV compared to data for $e^- + e^+ \to \mu^- + \mu^+$ cross sections reported by the ALEPH collaboration (R. Barate, et al., *Phys Rev B*, **399**, 329, 1997).

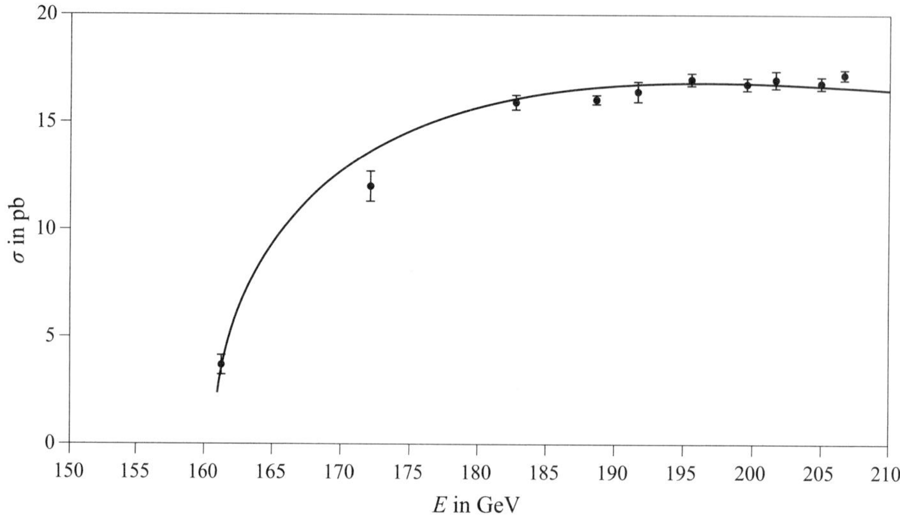

FIGURE 29.5 This graph shows the curve given by Equation 29.13 with $M_W = 80.38$ GeV (solid line) compared to combined cross section data from the ALEPH, DELPHI, L3, and OPAL detectors at LEP (the data are from S. Schael *et al.*, *Phys Rept* **532**, 119, (2013)).

outgoing W^- momentum is $|\vec{p}_3| = (E_W^2 - M_W^2)^{1/2} = (\frac{1}{4}E^2 - M_W^2)^{1/2}$ since each W gets half the collision energy in the center-of-mass frame. This means that the cross section in a narrow energy range where $E \approx 2M_W$ will have the form

$$\sigma \approx \frac{K}{(16\pi)^2} \frac{\sqrt{E^2 - 4M_W^2}}{E^3},$$

(29.13)

independent of the details of the scattering amplitude. Note that the cross section for $E < 2M_W$ is forbidden by energy conservation but has a characteristic shape for $E \gtrsim 2M_W$. We can determine M_W by finding the value of M_W that yields the best match to cross section measurements near that critical energy. Figure 29.5 shows that combined cross section data from the four LEP detectors match Equation 29.13 well when $M_W = 80.38$ GeV.

29.4 The Lifetime of the W

We can predict the lifetime of the W directly from the Feynman rules. Figure 29.6 shows the Feynman diagram for the decay $W^- \rightarrow e^- + \bar{\nu}_e$. Applying the Feynman rules to this diagram leads to the following results:

$$-i\mathcal{M} = \epsilon_{1\mu}\bar{u}_2 \left[-i\frac{g_W}{2\sqrt{2}}\gamma^\mu(1 - \gamma^5) \right] v_3 \implies$$

$$|\mathcal{M}|^2 = \frac{g_W^2}{8}\epsilon_{1\mu}\epsilon_{1\nu}^*[\bar{u}_2\gamma^\mu(1 - \gamma^5)v_3][\bar{u}_2\gamma^\nu(1 - \gamma^5)v_3]^*.$$

(29.14)

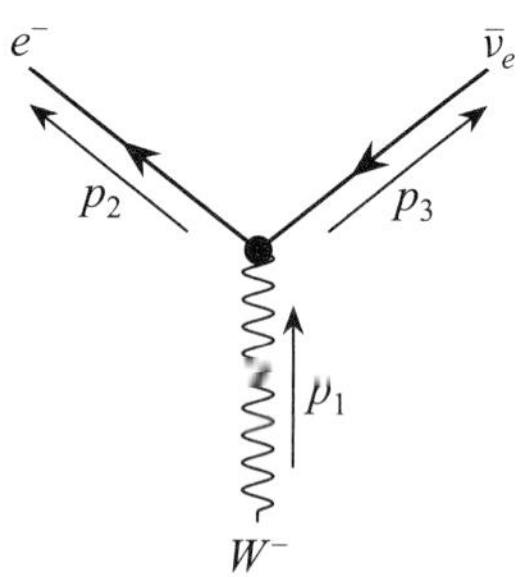

FIGURE 29.6 This is the Feynman diagram for $W^- \rightarrow e^- + \bar{\nu}_e$.

Now, we could go through the normal process of summing over all spins and averaging over initial spins (see Problem P29.3), but we can estimate the result much more quickly. The squared amplitude for a decay must have units of (energy)^2, and because the electron and antineutrino have negligible masses, the only energy available in the rest frame of the W^- is its own mass M_W. So we might credibly guess that $|\mathcal{M}|^2 \approx g_W^2 M_W^2$ for each spin case. But the weak interaction only connects left-handed particles, so there is really only one possible spin orientation for the final particles, and therefore (by conservation of angular momentum) only one spin orientation for the W^- that can lead to any results. So, when we sum over all the initial and final spin possibilities, only one term will be nonzero. If we then average over the three possible W^- spin states in an unpolarized beam, we might guess that

$$\langle|\mathcal{M}|^2\rangle = \tfrac{1}{3}g_W^2 M_W^2. \tag{29.15}$$

This guess turns out to be exactly right! Of course, this is pure luck, since we could have multiplied our original guess that $|\mathcal{M}|^2 = g_W^2 M_W^2$ for each spin term by any unitless number and still have satisfied the dimensional argument.

Substituting this result into Equation 15.5 to get the decay rate yields

$$\Gamma_W = \frac{S|\vec{p}_F|}{8\pi M_W^2}\langle|\mathcal{M}|^2\rangle = \frac{1 \cdot \tfrac{1}{2}M_W}{8\pi M_W^2}\tfrac{1}{3}g_W^2 M_W^2 = \frac{g_W^2}{48\pi}M_W \tag{29.16}$$

because the output particles are not identical, and because for a massless output particle $|\vec{p}_F| = $ its energy $= \tfrac{1}{2}M_W$. So, once we have determined that $M_W = 80.38$ GeV and $g_W = 0.641$, we can predict $\Gamma_W = 0.219$ GeV for $W^- \to e^- + \bar{v}_e$.

But Equation 29.14 tells us that so long as we can treat the outgoing particles as massless compared to the W^-, the decay rate to *any* of the possible results $\mu^-\bar{v}_\mu$, $\tau^-\bar{v}_\tau$, $d\bar{u}$, and $s\bar{c}$ should be *identical* (decay to $b\bar{t}$ is forbidden by energy conservation). Since the quark-antiquark pairs could have any color/anticolor combination, we should count each of the quark-antiquark decay rates three times. Therefore, we predict that the W's total decay rate should be

$$\Gamma_W = (1 + 1 + 1 + 3 + 3)(0.219\,\text{GeV}) = 1.97\,\text{GeV}, \tag{29.17}$$

and that it should decay to hadrons twice as often as to leptons.

The latter prediction is quite easy to test: according to the Particle Data Group (`https://pdg.lbl.gov/2022/tables/contents_tables.html`), the W decays to hadrons $67.41 \pm 0.27\%$ of the time. Though very close, the experimental result is just a bit higher than our quick prediction because the outgoing particle masses are not *exactly* zero. But this result clearly demonstrates experimentally that quarks indeed have three colors!

Measuring Γ_W is more difficult, but one can extract it from the total cross section for the process $e^- + e^+ \to$ hadrons (see S. Schael *et al., Phys Rept* **532**, 119 (2013). The analysis is complicated, but the current Particle Data Group result is $\Gamma_W = 2.085 \pm 0.042$ GeV. This is a bit higher than our prediction because our calculation was too simplistic (partly because it does not include masses, but for other reasons as well).

29.5 The Number of Lepton Families

One weak interaction that might occur in an e^-e^+ collider is $e^- + e^+ \to v + \bar{v}$. This *must* be a purely weak process: since the neutrinos have no charge, we can completely ignore the electromagnetic process $e^- + e^+ \to \gamma \to f + \bar{f}$ that would be significant at most energies when the final fermions are charged (that is, when they are anything *but* neutrinos). This makes the analysis of neutrino production especially simple, and also makes such processes good for testing aspects of the weak interaction in particular. (By the way, the reason that we could ignore the electromagnetic process when considering Z production above is that energies near the Z resonance are one of the few collision energy ranges where the cross section involving a virtual Z completely dominates that involving a virtual photon.)

On the other hand, *detecting* a $e^- + e^+ \to v + \bar{v}$ process is tricky, because the neutrinos cannot be observed, making it difficult to tell whether a collision has occurred at all. Directly measuring the cross section for this process is therefore impractical. But we *can* determine the total cross section in two different ways.

One is to note that on the way to producing the neutrinos, one or other of the electrons might also radiate a photon. Since direct annihilation of e^- and e^+ must produce *two* photons, production of a single photon and *nothing else* must have come from a neutrino production event, particularly if the photon comes off at an angle, as only neutrinos could invisibly carry away the opposing momentum. One can both predict and measure this cross section. The measurement is particularly easy for energies near the Z resonance peak since the cross section will be especially large there.

One of the more interesting things that one can do with such a measurement is to measure the number of lepton families that exist. Suppose that more lepton families exist than the three families we know about, but we have not found them yet because the corresponding charged leptons are too massive to have been created in current particle accelerators. The calculation for the $e^- + e^+ \to \nu + \overline{\nu}$ cross section is the same for all neutrino types, and since all neutrinos are virtually massless, the total cross section for producing neutrinos should be a simple multiple of *all* the types that exist, including those from undetected families. By measuring the cross section for single photon events and comparing it with the theoretical calculation, we should be able to determine the number of families. Using this approach, physicists using the L3 detector at LEP were able to show in 1992 that the number of neutrino types is 3.14 ± 0.27 (O. Adriani *et al.*, *Phys Lett B*, **292**, 463, 1992).

A clever alternative process that does not require knowing the magnitude of either the total cross section for Z production or the total decay rate Γ_Z goes as follows. Γ_Z is the sum of the decay rates to various outcomes:

$$\Gamma_Z = \Gamma_{ee} + \Gamma_{\mu\mu} + \Gamma_{\tau\tau} + \Gamma_{\text{had}} + \Gamma_{\nu\nu}, \tag{29.18}$$

where Γ_{ee} is the rate for a Z-mediated process producing an $e^- e^+$ pair; $\Gamma_{\mu\mu}$, $\Gamma_{\tau\tau}$, and $\Gamma_{\nu\nu}$ are the corresponding rates for $\mu^- \mu^+$, $\tau^- \tau^+$, and $\nu\overline{\nu}$ production, respectively; and Γ_{had} is the rate for producing any quark-antiquark pair (which will ultimately lead to production of various hadrons as the quark and antiquark separate). The Standard Model predicts that the rates to any lepton-antilepton pair should be the same (in the limit that the masses are negligible), so we can simplify this equation to read

$$\Gamma_Z = 3\Gamma_{\ell\ell} + \Gamma_{\text{had}} + \Gamma_{\nu\nu}, \tag{29.19}$$

where $\Gamma_{\ell\ell} = \Gamma_{ee} = \Gamma_{\mu\mu} = \Gamma_{\tau\tau}$ is the common decay rate to any one lepton type.

We can't measure the partial decay rates directly, because these rates represent only the final vertex in the two-vertex diagram in Figure 29.2. But we can easily measure ratios of cross sections to various outcomes, and since the initial vertex ($e^- + e^+ \to Z$) is the same for all, the ratio of the cross sections will be the same as the ratios of the decay rates:

$$R_{\text{had}} \equiv \frac{\sigma_{\text{had}}}{\sigma_{\ell\ell}} = \frac{\Gamma_{\text{had}}}{\Gamma_{\ell\ell}}, \qquad R_{\nu\nu} = \frac{\sigma_{\nu\nu}}{\sigma_{\ell\ell}} = \frac{\Gamma_{\nu\nu}}{\Gamma_{\ell\ell}}. \tag{29.20}$$

Since the decay rate to any specific neutrino type should be the same as $\Gamma_{\ell\ell}$, we should have $\Gamma_{\nu\nu} = N_\nu \Gamma_{\ell\ell}$, where N_ν is the number of (assumedly low-mass) neutrino types. Therefore, we can reexpress Equation 29.19 in terms of ratios:

$$\frac{\Gamma_Z}{\Gamma_{\ell\ell}} = R_{\text{had}} + 3 + R_{\nu\nu} = R_{\text{had}} + 3 + N_\nu. \tag{29.21}$$

The trouble with this expression is that while we can measure the ratio R_{had} by measuring $\sigma_{\text{had}}/\sigma_{\ell\ell}$, we still have three unknowns in this equation: Γ_Z, $\Gamma_{\ell\ell}$ and N_ν. In particular, we'd like to express the left side of the equation in terms of measurable quantities. Box 29.5 shows a very clever trick for doing this, assuming that we can accurately measure σ_{had} at the peak of the Z resonance. Under those conditions,

$$\frac{\Gamma_Z}{\Gamma_{\ell\ell}^0} = \left(\frac{12\pi R_{\text{had}}^0}{\sigma_{\text{had}}^0 M_Z^2} \right)^{1/2}, \tag{29.22}$$

where the 0 superscript indicates that we are evaluating these quantities at a collision energy corresponding to the resonance maximum. Inserting this into Equation 29.21 and solving for N_ν yields

$$N_\nu = \left(\frac{12\pi R^0_{\text{had}}}{\sigma^0_{\text{had}} M^2_Z}\right)^{1/2} - R^0_{\text{had}} - 3. \tag{29.23}$$

Everything on the right side of this expression is measurable, again, by fitting the data with a curve based on these parameters. The original result using this method (with a small correction for the nonzero mass of the τ) was published in S. Schael *et al., Phys Rept* **427**, 257 (2006), and was

$$N_\nu = 2.9840 \pm 0.0082, \tag{29.24}$$

which was (disturbingly) two standard deviations away from the expected result of 3. This remained a puzzle for a long time, but in 2020, a more accurate prediction of the Bhabha cross section required revising this number to

$$N_\nu = 2.9963 \pm 0.0074 \tag{29.25}$$

(see P. Janot and S. Jadach, *Phys Lett B*, **803**, 135319, 2020). This is convincing evidence that only three lepton families exist (that have light neutrinos at least).

BOX 29.1   The Z Production Cross Section

Our goal in this box is to use Equations 29.1 and 29.2 (repeated here)

$$\mathcal{M} = \frac{g_W}{2\cos\theta_W}[\bar{v}_2\gamma^\mu(c_V - c_A\gamma^5)u_1]\epsilon^*_{3,\mu} \quad \text{and} \qquad (29.26a)$$

$$\sum_{s_3} (\epsilon^{s_3}_{3,\mu})^*\epsilon^{s_3}_{3,\nu} = -g_{\mu\nu} + \frac{p_{3,\mu}p_{3,\nu}}{M_Z^2} \qquad (29.26b)$$

to arrive at Equation 29.3 (repeated here)

$$\sigma = \frac{\pi g_W^2}{8\cos^2\theta_W M_Z}\left(\tfrac{1}{2} - 2\sin\theta_W + 4\sin^2\theta_W\right)\delta(2E - M_Z). \quad (29.26c)$$

The absolute square of Equation 29.26a is

$$|\mathcal{M}|^2 = \frac{g_W^2}{4\cos^2\theta_W}[\bar{v}_2\gamma^\mu(c_V - c_A\gamma^5)u_1]\epsilon^*_{3,\mu}[\bar{v}_2\gamma^\nu(c_V - c_A\gamma^5)u_1]^*\epsilon_{3,\nu}. \quad (29.27)$$

Applying Casimir's trick—noting that $\overline{\gamma^\mu(c_V - c_A\gamma^5)} = \gamma^\mu(c_V - c_A\gamma^5)$, and that we are treating the electron and positron masses as negligible—to sum over electron and positron spins and Equation 29.26b to sum over the Z spins yields

$$\sum_{s_1,s_2,s_3} |\mathcal{M}|^2 = \frac{g_W^2}{4\cos^2\theta_W}\,\text{Tr}[\gamma^\mu(c_V - c_A\gamma^5)\not{p}_1\gamma^\nu(c_V - c_A\gamma^5)\not{p}_2]$$

$$\times\left(-g_{\mu\nu} + \frac{p_{3,\mu}p_{3,\nu}}{M_Z^2}\right) \equiv \frac{g_W^2}{4\cos^2\theta_W}B^{\mu\nu}S_{\mu\nu}, \qquad (29.28)$$

where $B^{\mu\nu}$ is the trace part and $S_{\mu\nu}$ is the Z spin part (the part in parentheses). Let's work on the trace part first.

Exercise 29.1.1   Let $\not{p}_1 \equiv p_{1,\alpha}\gamma^\alpha$ and $\not{p}_2 \equiv p_{2,\beta}\gamma^\beta$. Use the trace theorems in Chapter 20 and the fact that $(\gamma^5)^2 = I$ to show that

$$B^{\mu\nu} = 4(c_V^2 + c_A^2)(p_1^\mu p_2^\nu - g^{\mu\nu}(p_1\cdot p_2) + p_2^\mu p_1^\nu) + 8ic_V c_A p_{1,\alpha}p_{2,\beta}\epsilon^{\mu\alpha\nu\beta}. \quad (29.29)$$

(There is extra space on the next page if you need it.)

Exercise 29.1.2 Show that

$$B^{\mu\nu}S_{\mu\nu} = 4(c_V^2 + c_A^2)\left[(p_1 \cdot p_2) + \frac{2(p_1 \cdot p_3)(p_2 \cdot p_3)}{M_Z^2}\right]. \qquad (29.30)$$

(*Hint:* Recall that $S_{\mu\nu}\epsilon^{\mu\alpha\nu\beta} = 0$ for any symmetric tensor $S_{\mu\nu}$.)

BOX 29.1 The Z Production Cross Section (cont.)

Fermi's Golden Rule (see Chapter 10 or 14) for two input particles and one output particle is

$$\sigma = \frac{S}{4\sqrt{(p_1 \cdot p_2)^2 - (m_1 m_2)^2}} \int \frac{d^3 p_3}{(2\pi)^3 2E_3} (2\pi)^4 \delta^4(p_1 + p_2 - p_3) |\mathcal{M}|^2. \quad (29.31)$$

If we substitute Equation 29.30 for $B^{\mu\nu} S_{\mu\nu}$ into Equation 29.28 for $\sum |\mathcal{M}|^2$ and divide by 4 to average over all electron and positron spin possibilities, we get

$$\langle |\mathcal{M}|^2 \rangle = \frac{g_W^2}{4 \cos^2 \theta_W} (c_V^2 + c_A^2) \left[(p_1 \cdot p_2) + \frac{2(p_1 \cdot p_3)(p_2 \cdot p_3)}{M_Z^2} \right]. \quad (29.32)$$

Exercise 29.1.3 Combine these expressions in the limit that the electron and positron masses m_1 and m_2 are negligible, and do the integral to arrive at

$$\sigma = \frac{\pi g_W^2}{16 \cos^2 \theta_W E_3} (c_V^2 + c_A^2) \left[1 + \frac{2(p_1 \cdot p_3)(p_2 \cdot p_3)}{(p_1 \cdot p_2) M_Z^2} \right] \delta(E_1 + E_2 - E_3), \quad (29.33)$$

subject to the requirement that $\vec{p}_1 + \vec{p}_2 = \vec{p}_3$.

Exercise 29.1.4   Finally, evaluate the dot products in the center-of-mass frame, and express the cross section in that frame in terms of the collision energy $E = E_1 + E_2$ and the appropriate values of c_V and c_A from Table 25.2. Compare with Equation 29.3.

BOX 29.2   Calculating $|E^2 - M_Z^2 - iE\Gamma_Z - \frac{1}{4}\Gamma_Z^2|^2$

Our goal in this box is to show that

$$|E^2 - M_Z^2 - iE\Gamma_Z - \tfrac{1}{4}\Gamma_Z^2|^2 \approx (E^2 - M_Z^2)^2 + M_Z^2\Gamma_Z^2 \qquad (29.34)$$

in the limit that $\Gamma_Z \ll M_Z$ and $E \approx M_Z$, as claimed in Equation 29.10.

Exercise 29.2.1   Show this. (*Hints:* Define $A \equiv E^2 - M_Z^2 - \frac{1}{4}\Gamma_Z^2$. Keep only the leading term in Γ_Z, and note that $E - M_Z$ is the same order as Γ_Z because Γ_Z is the width of the curve at half maximum, and we are interested in the curve's shape within a few widths of the peak. Finally, note that $E^2 - M_Z^2 = (E + M_Z)(E - M_Z)$.)

BOX 29.3   The Z Cross Section at the Resonance Peak

The goal of this box is to show that substituting Equation 29.10 (repeated here)

$$\langle |\mathcal{M}_f|^2 \rangle \approx \frac{K_f(\theta) M_Z^4}{(E^2 - M_Z^2)^2 + M_Z^2 \Gamma_Z^2} \tag{29.35}$$

into the left side of Equation 29.11 for the cross section in the center-of-mass frame (repeated here)

$$\sigma = \int \frac{S\langle |\mathcal{M}_f|^2 \rangle}{(8\pi)^2 (E_1 + E_2)^2} \frac{|\vec{p}_3|}{|\vec{p}_1|} \, d\Omega \tag{29.36}$$

yields the right side of that equation (repeated here)

$$\sigma \approx \sigma_{0f} \frac{M_Z^2 \Gamma_Z^2}{(E^2 - M_Z^2)^2 + M_Z^2 \Gamma_Z^2}, \quad \text{where } \sigma_{0f} \equiv \frac{1}{64\pi^2 \Gamma_Z^2} \int K_f(\theta) \, d\Omega \tag{29.37}$$

in the limit that $E \approx M_Z$ (that is, near the peak of the resonance) and assuming that the masses of the initial particles and final particles are all negligible compared to M_Z.

Exercise 29.3.1   Show this.

BOX 29.4   The Resonance Width

The purpose of this box is to show that a resonance curve of the form

$$f(E) = \frac{M^2\Gamma^2}{(E^2 - M^2)^2 + M^2\Gamma^2} \tag{29.38}$$

has half its peak value of 1 when $E = M \pm \frac{1}{2}\Gamma$ (in the limit that $\Gamma \ll M$).

Exercise 29.4.1   Show this. (*Hint:* Focus on $(E^2 - M^2)^2$.)

BOX 29.5 A Clever Trick for Calculating $\Gamma_Z/\Gamma_{\ell\ell}$

The goal of this box is to justify Equation 29.22 (repeated here)

$$\frac{\Gamma_Z}{\Gamma_{\ell\ell}} = \left(\frac{12\pi R^0_{\text{had}}}{\sigma^0_{\text{had}} M_Z^2} \right)^{1/2}, \tag{29.39}$$

where $R^0_{\text{had}} = \sigma^0_{\text{had}}/\sigma^0_{\ell\ell} = \Gamma_{\text{had}}/\Gamma_{\ell\ell}$ is a measurable cross section ratio. (As before, the 0 superscript indicates that we are evaluating Z resonance peak.)

We begin with the fact that one can express the peak cross section for the process $e^- + e^+ \to Z \to f + \overline{f}$ in the form

$$\sigma^0_{ff} = 12\pi \frac{\Gamma_{ee}\Gamma_{ff}}{M_Z^2 \Gamma_Z^2}, \tag{29.40}$$

where Γ_{ee} and Γ_{ff} are the decay rates for $Z \to e^- + e^+$ and $Z \to f + \overline{f}$, respectively. This takes some effort to prove (see Problem P29.4 if you really want to get into it), but the basic idea is that Γ^0_{ee} and Γ_{ff} are each proportional to the absolute squares of their respective vertex factors in the squared amplitude for the complete process at the resonance peak, the $M_Z^2\Gamma_Z^2$ is what is left of the propagator denominator when we evaluate it at that peak, and the 12π handles the difference between the product of two decay rates and a single cross section. (The tricky part is handling all of the spin summations and proving that one can in fact separate the contributions from the vertices, which are entangled in the squared amplitude for the complete process.)

Let's accept this as true, and note that the decay rate $\Gamma^0_{ee} = \Gamma^0_{\ell\ell}$ because all leptons are equivalent when interacting with the Z.

Exercise 29.5.1 Substitute σ^0_{had} (expressed in the form given in Equation 29.40) and $R^0_{\text{had}} = \Gamma_{\text{had}}/\Gamma_{\ell\ell}$ into the right side of Equation 29.39 and show that one does indeed get the left side.

Homework Problems

P29.1 Consider the decay of the Z.
a. Verify that the lifetime of the Z is 2.6×10^{-25} s, as claimed in the text.
b. In $e^- e^+$ colliders, the Z should be created at rest, so it would not leave a track that we could observe even if it lasted a long time. But suppose that we create a highly relativistic 100-TeV Z in some other process. How far would it travel before decaying?

P29.2 One can show that the spin-averaged differential cross section for the process $e^- + e^+ \to Z \to f + \bar{f}$ in the center of mass frame and when $E = E_1 + E_2 = M_Z$ is

$$d\sigma = \frac{g_Z^4}{1024\pi^2 \Gamma_Z^2}\Big[(c_V^2 + c_A^2)_e (c_V^2 + c_A^2)_f (1 + \cos^2\theta)$$

$$+ 8(c_V c_A)_e (c_V c_A)_f \cos\theta\Big] d\Omega, \quad (29.41)$$

where $g_Z \equiv g_W / \cos\theta_W$, θ is the angle between the momenta of the incoming electron and outgoing fermion, respectively, and the $(\ )_e$ and $(\ )_f$ notations mean that one should evaluate the enclosed for electrons and final fermions, respectively. (You do not need to show this.) Show that the all-angle cross section evaluated at this energy is

$$\sigma_0 = \frac{g_Z^4}{192\pi \Gamma_Z^2}\Big[(c_V^2 + c_A^2)_e (c_V^2 + c_A^2)_f\Big], \quad (29.42)$$

and evaluate this in barns if the final products are leptons.

P29.3 Compute $\langle |\mathcal{M}|^2 \rangle$ for the squared amplitude given in Equation 29.14 and verify that Equation 29.15 is correct. (*Hint:* Use Equation 29.2 to sum over the three possible W states.)

P29.4 Verify that Equation 29.40 is correct. (*Hints:* Start by adapting the results from Box 29.1 to calculate the decay rates Γ_{ee} and Γ_{ff} for a Z at rest decaying to an electron-positron pair or other fermion-antifermion pair, respectively. Remember that one must average over the three spin possibilities for the initial Z instead of the four possible initial spins considered in the box. Then work backward from equation 29.40 to equation 29.42.)

P29.5 Compare the cross section for Higgs production compared to the cross section for Z production in an $e^- e^+$ collider, and speculate why the Higgs was discovered using the LHC $p\bar{p}$ collider and not LEP. (*Hint:* So that you can compare these cross sections more precisely, express the energy-conserving Dirac delta in each case as $\delta(x - 1)$, where $x = E/M$, E is the total energy of the colliding particles, and M is the mass of the particle that its produced.)

P29.6 Calculate the theoretical decay rate for $W^+ \to u + \bar{d}$ in terms of g_W, M_W, and an appropriate component of the CKM matrix.

P29.7 Consider the process $e^- + e^+ \to Z \to f + \bar{f}$.
a. Use Equation 29.41 to calculate what is called the **forward-backward asymmetry** $\mathcal{A}_{FB} \equiv (\sigma_F - \sigma_B)/(\sigma_F + \sigma_B)$, where σ_F and σ_B are the cross sections for the negative final fermion being scattered in the forward or backward direction ($\cos\theta > 0$ or $\cos\theta < 0$, respectively) relative to the incoming electron. You should find that

$$\mathcal{A}_{FB} = \frac{3}{4}\mathcal{A}_e \mathcal{A}_f \text{ where } \mathcal{A}_f = \left(\frac{2c_V/c_A}{1 + (c_V/c_A)^2}\right)_f, \quad (29.43)$$

where $(\ )_f$ instructs one to evaluate the enclosed for the fermion f.
b. Evaluate $\mathcal{A}_{FB}$ for leptons, u quarks, and d quarks in terms of $x \equiv \sin^2\theta_W$. Submit a computer-generated graph of $\mathcal{A}_{FB}$ for all three cases as a function of x for $0 \le x \le 1$.
c. In which case (leptons, u quarks, or d quarks) will measuring $\mathcal{A}_{FB}$ provide the most sensitive determination of $x \equiv \sin^2\theta_W$ and why?

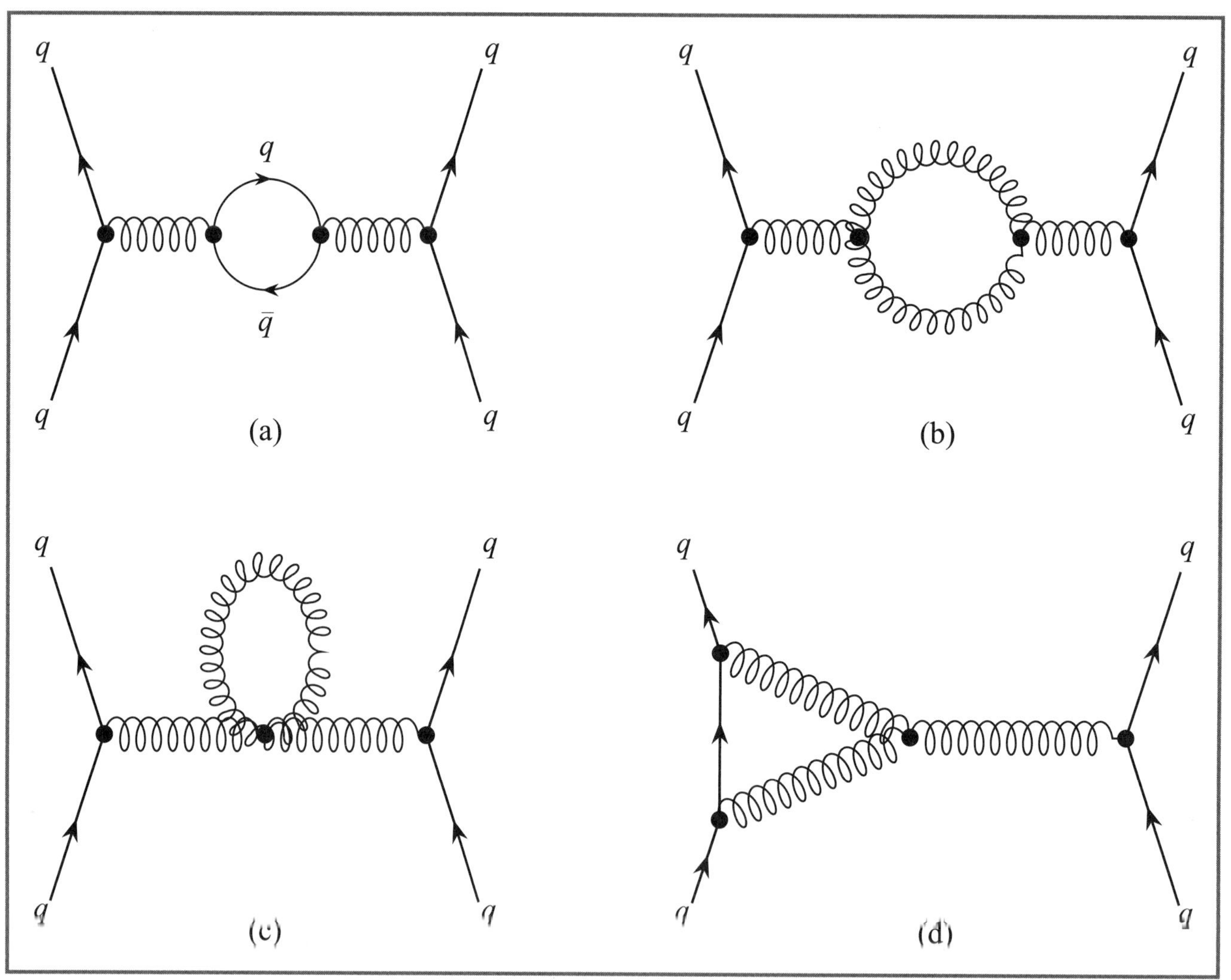

30. THE COLOR INTERACTION

The last piece of the Standard Model is **quantum chromodynamics** (QCD), the theory describing the **color interaction**. The color interaction binds quarks into color-neutral mesons (such as pions) and baryons (such as protons and neutrons). The imperfect cancellation of these forces (when protons and neutrons get sufficiently close) binds those color-neutral baryons into nuclei, much as the imperfect cancellation of electromagnetic forces causes electrically neutral atoms to bind together into molecules. (People often call this residual force that binds nucleons the "strong nuclear interaction.")

In many regards, the color interaction is one of the simplest in the Standard Model. The color interaction naturally emerges from making an SU(3) symmetry local, the symmetry is unbroken (meaning that the gluons carrying the interaction remain massless), and (unlike the weak interaction) it is ambidextrous. The color interaction also treats all quarks identically and utterly ignores all leptons. However, we will see that this interaction is difficult to apply to the real world, for two important reasons: (1) gluons interact with themselves, making their field theory nonlinear in a complicated way (think about how complicated electrodynamics would be if photons were charged), and (2) the color coupling constant is large and varies greatly with energy, making our whole perturbation approach problematic. We will see in this chapter how these complications arise from the simple SU(3) symmetry.

30.1 The SU(3) Group

As the name implies, the set of matrices that comprise the SU(3) group are unitary 3×3 matrices that have determinants equal to 1 (instead of being an arbitrary complex phase $e^{i\phi}$, as would be the case for a general unitary matrix). These matrices act on three-component column vectors in some abstract, complex three-dimensional space in such a way as to "rotate" them in that space without affecting either their magnitude or overall phase.

As you will see in Box 30.1, the set of SU(3) matrices has eight free parameters, and therefore can be written in the general form

$$U(\vec{\theta}) = e^{-i\frac{1}{2}\vec{\lambda}\cdot\vec{\theta}}, \quad \text{where } \lambda_j = 2i \left[\frac{dU}{d\theta_j}\right]_0. \tag{30.1}$$

Note that in this context, the "vector" $\vec{\theta}$ has *eight* components (one for each free parameter), the $\frac{1}{2}$ is conventional, and $\vec{\lambda}$ stands for the set of eight 3×3 generator matrices, each of which is the derivative of $e^{-i\frac{1}{2}\vec{\lambda}\cdot\vec{\theta}}$ with respect to one of the parameters, evaluated where that parameter is zero. These generator matrices are to the SU(3) group what the three Pauli matrices are to the SU(2) group. These eight matrices (called the **Gell-Mann matrices**) are

$$\lambda_1 = \begin{pmatrix} 0 & 1 & 0 \\ 1 & 0 & 0 \\ 0 & 0 & 0 \end{pmatrix}, \quad \lambda_2 = \begin{pmatrix} 0 & -i & 0 \\ i & 0 & 0 \\ 0 & 0 & 0 \end{pmatrix}, \quad \lambda_3 = \begin{pmatrix} 1 & 0 & 0 \\ 0 & -1 & 0 \\ 0 & 0 & 0 \end{pmatrix},$$

$$\lambda_4 = \begin{pmatrix} 0 & 0 & 1 \\ 0 & 0 & 0 \\ 1 & 0 & 0 \end{pmatrix}, \quad \lambda_5 = \begin{pmatrix} 0 & 0 & -i \\ 0 & 0 & 0 \\ i & 0 & 0 \end{pmatrix},$$

$$\lambda_6 = \begin{pmatrix} 0 & 0 & 0 \\ 0 & 0 & 1 \\ 0 & 1 & 0 \end{pmatrix}, \quad \lambda_7 = \begin{pmatrix} 0 & 0 & 0 \\ 0 & 0 & -i \\ 0 & i & 0 \end{pmatrix}, \quad \lambda_8 = \frac{1}{\sqrt{3}} \begin{pmatrix} 1 & 0 & 0 \\ 0 & 1 & 0 \\ 0 & 0 & -2 \end{pmatrix}. \tag{30.2}$$

As you will see in Box 30.2, the matrix $\vec{\lambda} \cdot \vec{\theta}$ must be Hermitian and traceless. You will show in that box that *any* 3×3 Hermitian and traceless matrix can be written as a linear combination of the Gell-Mann matrices.

The Pauli matrices obey the commutation relation $[\sigma_j, \sigma_k] = 2i \sum_m \epsilon_{jkm} \sigma_m$, where ϵ_{jkm} is the **Levi-Civita symbol** (see Equation 16.3). The Gell-Mann matrices have an analogous commutation relation:

$$[\lambda_j, \lambda_k] = 2i \sum_m f_{jkm} \lambda_m, \tag{30.3}$$

where the indices j, k, m in this case range over the values 1 through 8. The $8 \times 8 \times 8$ values f_{jkm} are called the **structure constants** for the SU(3) symmetry. For SU(2), the equivalent set of values ϵ_{jkm} was completely antisymmetric and had values 1, -1, or 0. The SU(3) structure constants f_{jkm} are also antisymmetric in that f_{jkm} changes sign if we switch any two indices, but the nonzero values are not ± 1. You can obtain the value of any nonzero constant by applying antisymmetry to the following list:

$$f_{147} = f_{246} = f_{257} = f_{345} = f_{516} = f_{637} = \tfrac{1}{2}$$

$$f_{123} = 1, \qquad f_{458} = f_{678} = \tfrac{\sqrt{3}}{2}. \tag{30.4}$$

So, for example, the constant f_{135}, though it is not obviously zero, cannot be found by permuting the indices on any of the quantities listed above, so it must be zero. On the other hand, $f_{367} = -f_{637} = -\tfrac{1}{2}$. You will calculate some of these constants in Box 30.3.

30.2 The Color Property

The Standard Model assumes that quarks have a property called **color**. "Color" is analogous to electric charge: it specifies how each quark couples to the color interaction. Unlike electric charge, though, there are *three* types of color charge, conventionally called *red, green,* and *blue*. These names are metaphors describing how these charges combine: just as equal amounts of red, green, and blue light combine to make colorless white light, so equal amounts of red, green, and blue color charge (combined in the correct way to yield what is called a **color singlet**) create something that is color neutral (insensitive to the color interaction), analogous to something with zero net charge. However, unlike with electric charge (where any simple sum of charges with net charge zero is electrically neutral), we must combine colors in a special way to yield something that is color neutral: simply combining equal amounts of red, green, and blue (or combining red with anti-red) will not do. We will talk about this more in a bit.

30.3 Global SU(3) Symmetry

The theory of quantum chromodynamics describes a particular quark flavor (say, an up quark) by a triplet of three fields, one for each color:

$$\Psi_u = \begin{bmatrix} \psi_r \\ \psi_g \\ \psi_b \end{bmatrix}. \tag{30.5}$$

You can think of this as being analogous to the SU(2) doublet that describes the electron and the neutrino. In electroweak theory, the SU(2) symmetry is broken, so the properties of the electron and neutrino are quite distinct. In quantum chromodynamics, the SU(3) symmetry remains unbroken: red, green, and blue up quarks are identical in all aspects except for their colors. The free-field Lagrangian density for the six different quark flavors is

$$\mathcal{L} = \sum_f \left[i\overline{\Psi}_f \gamma^\mu \partial_\mu \Psi_f - m_f \overline{\Psi}_f \Psi_f \right], \tag{30.6}$$

where f is an index that spans the six quark flavors (u, d, c, s, t, b). This Lagrangian density is unchanged by an SU(3) transformation

$$\Psi' = e^{-i\frac{1}{2}g_s \vec{\theta}\cdot\vec{\lambda}}\Psi \qquad \overline{\Psi}' = \overline{\Psi}e^{+i\frac{1}{2}g_s\vec{\theta}\cdot\vec{\lambda}}, \tag{30.7}$$

where we have added the factor g_s to the basic SU(3) transformation to provide an overall color-interaction coupling constant.

As you will show in Box 30.4, the conserved current for this symmetry is

$$J_k^\mu = \sum_f \tfrac{1}{2}g_S \overline{\Psi}_f \gamma^\mu \lambda_k \Psi_f, \tag{30.8}$$

where k ranges from 1 to 8—one current for each parameter.

30.4 Making SU(3) a Local Symmetry

The color interaction emerges from making the SU(3) symmetry a *local* symmetry. The process for generating a local gauge theory should, I hope, be familiar by now: we replace any partial derivatives ∂_μ in the free-field Lagrangian density with the covariant derivative

$$D_\mu = \partial_\mu + i\tfrac{1}{2}g_S\vec{\lambda}\cdot\vec{G}_\mu, \tag{30.9}$$

where $\vec{G}_\mu$ stands for the eight gauge fields (one for each parameter) that we must introduce to make the symmetry local. In quantum field theory, these fields describe eight spin-1 **gluons** that in Feynman diagrams carry the color interaction between quarks. Because the SU(3) symmetry is unbroken in this case, the gluons are required to be massless (any gluon mass term in the Lagrangian density would not be invariant under the local SU(3) transformation).

If we take our free-field Lagrangian density (Equation 30.6) and replace ∂_μ with D_μ, the first term becomes

$$\begin{aligned}
\mathcal{L}_{\text{first}} &= \sum i\overline{\Psi}_f \gamma^\mu D_\mu \Psi_f = \sum \left[i\overline{\Psi}_f\gamma^\mu\partial_\mu\Psi_f + i\overline{\Psi}_f\gamma^\mu(i\tfrac{1}{2}g_S\vec{\lambda}\cdot\vec{G}_\mu)\Psi_f \right] \\
&= \sum i\overline{\Psi}_f\gamma^\mu\partial_\mu\Psi_f - \vec{J}^\mu\cdot\vec{G}_\mu \quad \text{(see Equation 30.8),} \tag{30.10}
\end{aligned}$$

where the sum is over the quark flavors f. The new term adds an interaction between two quark-field terms and a gluon field, which we would represent in a Feynman diagram by a vertex that connects two quark lines with a gluon line, as shown in Figure 30.1.

In addition, the gauge fields have their own gauge transformation law, which by analogy to the SU(2) case (see Equation 23.15) is

$$\vec{G}'_\mu = \vec{G}_\mu + \partial_\mu\delta\vec{\theta} - g_S\vec{G}_\mu \times \delta\vec{\theta}, \tag{30.11}$$

where $\delta\vec{\theta}$ represents an infinitesimal change in the transformation parameters, and the

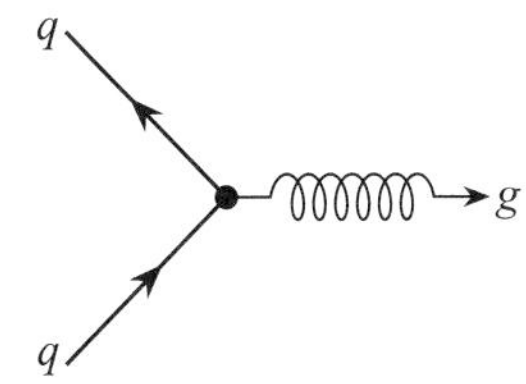

FIGURE 30.1 This diagram shows a typical quark-gluon vertex. (Physicists conventionally represent gluon worldlines by "springs.")

eight-dimensional "cross product" is defined by

$$(\vec{A} \times \vec{B})_m \equiv \sum_{ij} f_{jkm} A_j B_k \tag{30.12}$$

because the structure constants f_{jkm} play the same role for SU(3) that the completely antisymmetric tensor ϵ_{jkm} that defines the ordinary cross product plays for SU(2). We also must define a Lagrangian density for the $\vec{G}_\mu$ gluon fields, which (again by analogy to the SU(2) case) must have the form

$$\mathcal{L}_{\text{gluon}} = -\tfrac{1}{4}\vec{F}^{\mu\nu}\cdot\vec{F}_{\mu\nu} \quad \text{where } \vec{F}_{\mu\nu} = \partial_\mu \vec{G}_\nu - \partial_\nu \vec{G}_\mu - g_S \vec{G}_\mu \times \vec{G}_\nu \tag{30.13}$$

for the term to be gauge-invariant.

As discussed in Box 23.3, only *part* of $\mathcal{L}_{\text{gluon}}$ actually describes a free field (whose solution is a plane wave). If we multiply out the terms in $\mathcal{L}_{\text{gluon}}$, we see that

$$
\begin{aligned}
\mathcal{L}_{\text{gluon}} = -\tfrac{1}{4}\big[\, & (\partial_\mu \vec{G}_\nu - \partial_\nu \vec{G}_\mu)\cdot(\partial^\mu \vec{G}^\nu - \partial^\nu \vec{G}^\mu) \\
& - g_S(\partial_\mu \vec{G}_\nu - \partial_\nu \vec{G}_\mu)\cdot(\vec{G}^\mu \times \vec{G}^\nu) - g_S(\partial^\mu \vec{G}^\nu - \partial^\nu \vec{G}^\mu)\cdot(\vec{G}_\mu \times \vec{G}_\nu) \\
& + g_S^2(\vec{G}^\mu \times \vec{G}^\nu)\cdot(\vec{G}_\mu \times \vec{G}_\nu)\,\big].
\end{aligned}
\tag{30.14}
$$

Only the first of these four terms (the one in the top line) leads to an Euler-Lagrange equation whose solution is a free field. The other three terms should be considered *self-interaction* terms for the gluon field. In a Feynman diagram, we would represent the middle two terms (which involve a product of three $\vec{G}_\mu$ fields) by a vertex where three gluon worldlines meet, and the final term (which involves a product of four $\vec{G}_\mu$ fields) by a vertex where four gluon lines meet (see Figure 30.2).

Similar terms (and similar Feynman diagram vertices) exist in electroweak theory for the $W^\pm$ and Z bosons (see Box 23.3), but these terms become experimentally relevant only at energies comparable to the masses of these bosons (at lower energies, they can't be formed as real particles, and are so far off-shell as virtual particles that such interactions are strongly suppressed). But gluons are massless, so these diagrams are experimentally relevant even at the lowest energies. Indeed, they are the cause of the nonlinear field equations that makes calculating the *behavior* of the color interaction so complicated (even though the theory itself is simpler than the electroweak theory).

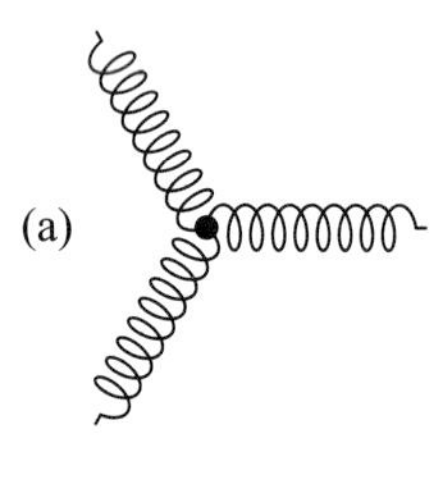
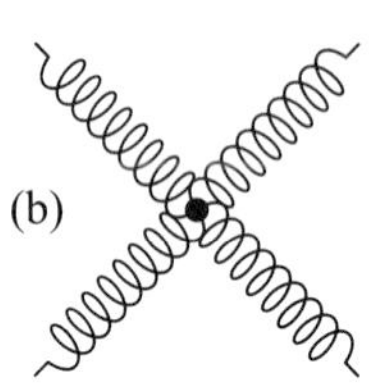

FIGURE 30.2 These are the vertices in a Feynman diagram that represent the gluon self-interaction terms in the gauge-invariant $\mathcal{L}_{\text{free}}$.

30.5 Quark Confinement and Asymptotic Freedom

These extra interaction terms ultimately mean that we will never observe quarks as free particles. Let's see how this works.

Figure 30.3 shows the lowest-order Feynman diagram for a color interaction between two quarks. Consider the process of adding bubbles to the gluon line. Adding a quark bubble (as shown in Figure 30.4a) is directly analogous to adding an electron bubble in the case of electrodynamics, and has a similar effect: it causes **screening** of the color charge, effectively making the color charge larger at smaller distances (or higher energies).

However, other diagrams exist in chromodynamics because gluons can interact with themselves. Some of these diagrams are illustrated in Figure 30.4b–d. It turns out that these diagrams exert an **anti-screening** effect, where the apparent color charge gets *stronger* with larger distances (smaller energies) and *weaker* at smaller distances (larger energies). The total formula for the renormalized running color coupling constant analogous to Equation 20.22 is

$$\alpha_S(Q^2) = \frac{\alpha_S(\mu^2)}{1 + \frac{1}{12\pi}\alpha_S(\mu^2)(11n - 2f)\ln(|Q^2|/\mu^2)} \quad \text{for } |Q^2| \gg \mu^2, \tag{30.15}$$

where μ is a parameter I will describe shortly, α_S is the color interaction equivalent to the electromagnetic fine-structure constant, n is the number of colors (three in the Standard

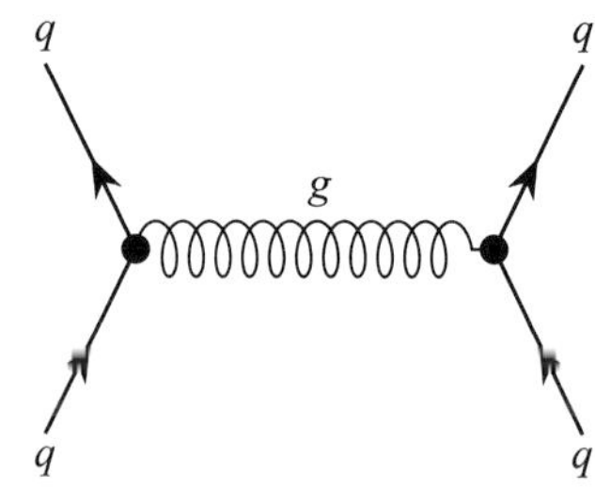

FIGURE 30.3 This is the lowest-order Feynman diagram for a color interaction between two quarks.

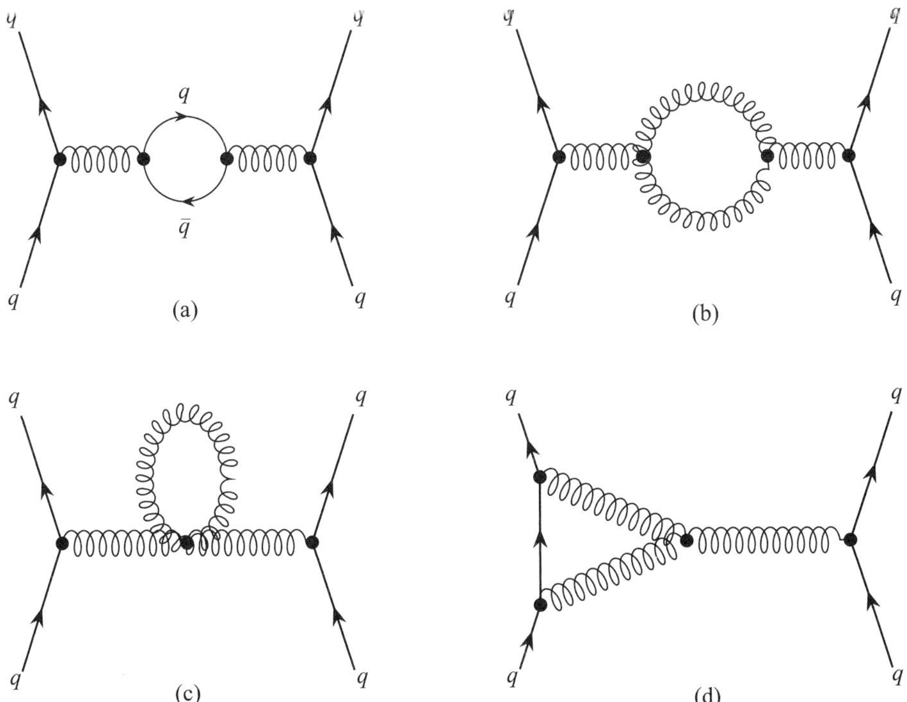

FIGURE 30.4   Some of the many higher-order diagrams that contribute to the color interaction between two quarks. Diagrams (b)–(d) are possible only because gluons self-interact.

Model), and f is the number of quark flavors (six in the Standard Model). So, in any theory where $11n > 2f$ (such as the Standard Model, where $33 > 12$), anti-screening wins over screening, so a quark's effective color becomes *smaller* at increasing energy (smaller distances) rather than larger. (Note that this is entirely due to the gluon self-interaction diagrams: without these diagrams, we would not have the part proportional to $11n$.)

Physicists call this phenomenon **asymptotic freedom** because, at very high energies, quarks behave as if they were ordinary free particles, and we can usefully apply the perturbation-based calculations of quantum field theory. The same calculations break down at low energies and require difficult and computationally intensive numerical calculations to make any headway at all.

I promised to describe the parameter μ^2 that appears in this formula. In quantum electrodynamics, it makes sense to measure "the" charge of a particle (and thus the fine-structure constant $\alpha_R(0) = e^2/4\pi$) to be its fully screened value at large distances (low energies). This is what we would do experimentally in classical physics. It therefore also makes sense to use $\alpha_R(0)$ as the reference value to which we compare $\alpha_R(Q^2)$ at other energies. But this will not work for chromodynamics, because zero energy is precisely where the value blows up (and Equation 30.15, which is based on perturbation calculations, does not really apply). But we can use the value of α_S at any Q^2 as a reference; it does not change the physics of the formula. So, in practice, we choose a reference four-momentum magnitude μ so that $\alpha(\mu^2) \ll 1$ and the perturbation approach is valid. So long as this condition is satisfied, it does not matter which value one uses (see Box 30.5). To use the formula, however, one does need to know the experimental value of α_S at *some* value of μ. It has become conventional to evaluate α_S at an energy equal to the mass M_Z of the Z vector boson. The accepted value (according to the 2020 Particle Data Group report) is

$$\alpha_S(M_Z^2) = 0.1179 \pm 0.0010. \tag{30.16}$$

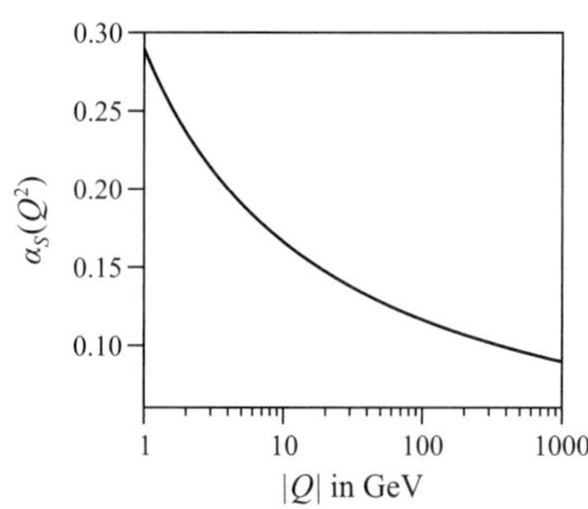

FIGURE 30.5   This graph shows how α_S depends on energy.

The way that $\alpha_S(Q^2)$ varies with Q^2 (see Figure 30.5) is extremely important physically. First of all, $\alpha_S(Q^2)$ varies much more sharply with Q^2 than does the electromagnetic fine-structure constant (see Box 30.6). But the fact that the interaction strength blows up at low energies (large distances) is even more important at the everyday level. It means qualitatively that if we were to try to separate quarks at low energy, the interaction becomes stronger and stronger until the interaction potential energy becomes large enough to create quark-antiquark pairs that combine with the existing quarks to create color-neutral particles. This is why quarks can only exist in color-neutral combinations, a phenomenon called **quark confinement.**

But using quantum chromodynamics to actually calculate the details of quark confinement, low-energy interactions, or even such a seemingly simple thing as the mass of the proton, is extremely difficult. The most promising technique at present is **lattice gauge theory**, which involves solving the equations of quantum chromodynamics numerically on a discrete grid of spacetime points. The main problem with this approach is that the number of grid points required to get good results is prohibitively large. Still, calculations with on the order of 10^6 grid points (only about 32 points per coordinate direction!) yield tolerably good predictions of many meson and baryon masses.

30.6 Color Singlets

A color-neutral combination is called a **color singlet**. Why this terminology? The analogy is to what happens when we combine two spins.

We saw in Chapter 9 that the four spin states that one would expect when combining two spin-$\frac{1}{2}$ particle states divide into three superpositions that have total angular momentum 1 and a single (singlet) superposition state with total angular momentum 0:

Spin 1 : $|1, 1\rangle = |++\rangle,$ $|1, 0\rangle = \sqrt{\tfrac{1}{2}}|+-\rangle + \sqrt{\tfrac{1}{2}}|-+\rangle,$ $|1, -1\rangle = |--\rangle,$

Spin 0 : $|0, 0\rangle = \sqrt{\tfrac{1}{2}}|+-\rangle - \sqrt{\tfrac{1}{2}}|-+\rangle,$ (30.17)

where I am using the notation $|\pm\pm\rangle$ as a shorthand for the spin-combination basis states $|\tfrac{1}{2}, \pm\tfrac{1}{2}\rangle \otimes |\tfrac{1}{2}, \pm\tfrac{1}{2}\rangle$. SU(2) rotations rotate the spin-1 kets into linear combinations of each other, and the $|0, 0\rangle$ state to itself. So, though both the $|1, 0\rangle$ and $|0, 0\rangle$ states have zero projection on the z axis, an appropriate SU(2) rotation can convert $|1, 0\rangle$ to a $|1, 1\rangle$ state that has nonzero projection on that axis, showing that this state still has nonzero spin. But no SU(2) rotation can convert the $|0, 0\rangle$ state to any state with nonzero spin projection on any axis: the singlet $|0, 0\rangle$ state is truly "spinless."

Similarly, if we form a meson by combining a quark and an antiquark, each with three possible colors, it seems that we should have nine possible combinations ($|r\bar{r}\rangle$, $|r\bar{g}\rangle$, etc.). But one can find eight mutually orthogonal superpositions of these color states that an SU(3) rotation converts to linear combinations of the others:

$$|1\rangle \equiv \sqrt{\tfrac{1}{2}}|r\bar{g}\rangle + \sqrt{\tfrac{1}{2}}|g\bar{r}\rangle, \qquad |2\rangle \equiv \sqrt{\tfrac{1}{2}}|r\bar{g}\rangle - \sqrt{\tfrac{1}{2}}|g\bar{r}\rangle,$$

$$|3\rangle \equiv \sqrt{\tfrac{1}{2}}|r\bar{r}\rangle - \sqrt{\tfrac{1}{2}}|g\bar{g}\rangle, \qquad |4\rangle \equiv \sqrt{\tfrac{1}{2}}|r\bar{b}\rangle + \sqrt{\tfrac{1}{2}}|b\bar{r}\rangle,$$

$$|5\rangle \equiv \sqrt{\tfrac{1}{2}}|r\bar{b}\rangle - \sqrt{\tfrac{1}{2}}|b\bar{r}\rangle, \qquad |6\rangle \equiv \sqrt{\tfrac{1}{2}}|g\bar{b}\rangle + \sqrt{\tfrac{1}{2}}|b\bar{g}\rangle,$$

$$|7\rangle \equiv \sqrt{\tfrac{1}{2}}|g\bar{b}\rangle - \sqrt{\tfrac{1}{2}}|b\bar{g}\rangle, \qquad |8\rangle \equiv \sqrt{\tfrac{1}{6}}\left(|r\bar{r}\rangle + |g\bar{g}\rangle - 2|b\bar{b}\rangle\right). \qquad (30.18)$$

(These conventional forms, which are somewhat difficult to calculate, mirror the structure of the Gell-Mann matrices in Equation 30.2.) One might think that the third and final superpositions would be color-neutral because they involve combinations of a color with its anticolor, but an SU(3) rotation can rotate them to another color in the list (see Problem P30.1.) But we can construct a ninth state that is orthogonal to all of these:

$$\text{meson color singlet} = \sqrt{\tfrac{1}{3}}\left(|r\bar{r}\rangle + |g\bar{g}\rangle + |b\bar{b}\rangle\right). \tag{30.19}$$

This is the color singlet. We might intuit that this state, which weights each color-anticolor pair equally, would be especially symmetric, and that intuition is justified: an arbitrary SU(3) rotation does indeed rotate this state to itself. Any meson must therefore have this as its color state.

When we combine three quarks into a baryon, we have 27 possible color states. It turns out that out of the basis color states $|rrr\rangle$, $|rrg\rangle$, $|rgg\rangle$, and so on, one can construct a group of 10 superpositions that rotate into one another under an SU(3) transformation, two distinct groups of 8 states that are similarly closed under SU(3) rotations, and one color singlet state that always rotates to itself:

$$\begin{aligned}\text{baryon color singlet state} = \sqrt{\tfrac{1}{6}}\Big(&|rgb\rangle + |gbr\rangle + |brg\rangle \\ &- |rbg\rangle - |grb\rangle - |bgr\rangle\Big).\end{aligned} \tag{30.20}$$

Any baryon must have this as its color state.

30.7 Feynman Rules for Chromodynamics

Each particle entering a color-interaction vertex has a color state that we must somehow describe. For quarks, we attach to the Dirac spinor $u^s(p)$ a three component column vector c whose components specify the quark's color mix:

$$\Psi = \begin{bmatrix} \psi_r \\ \psi_g \\ \psi_b \end{bmatrix} = \begin{bmatrix} c_r \\ c_g \\ c_b \end{bmatrix} u^s(p), \tag{30.21}$$

where c_r, c_g, c_b specify the amount of red, green, and blue, respectively, while the $u^s(p)$ Dirac spinor describes the quark's momentum and spin state (which is the same for all three color fields).

As the current term $\vec{J}^\mu \cdot \vec{G}_\mu = \sum_{k,f} \tfrac{1}{2} g_s \overline{\Psi}_f \gamma^\mu \lambda_{(k)} \Psi_f G_{(k)\mu}$ in Equation 30.10 indicates, there is one gluon field for each λ matrix, and therefore eight different types of basic gluons. But a given gluon entering or leaving a vertex might be a superposition of these basic states. So, for gluons, in addition to the photon-like polarization four-vector ϵ^μ, we need to specify which of the orthogonal color combinations it represents by an eight-component vector $a_{(k)}$. (I'll put these eight-item indices in parentheses from now on to help distinguish them from other indices.)

With these things in mind, the rules for basic Feynman diagrams involving gluons are as follows:

1. **Draw diagrams.** Using the possible vertices listed in item 3 below, construct diagrams that connect incoming and outgoing particle lines with internal particles.

2. **Label lines.** Draw arrows next to each particle line to show the direction of momentum flow. Label the arrows for incoming or outgoing particles with four-momenta p_1, p_2, and so on and internal particles with four-momenta q_1, q_2, and so on. Also, draw arrows *along* the fermion lines in the direction of motion for fermions but opposite to the direction of motion for antifermions. Each *external* line contributes the factor that is listed (along with a diagram of the line appearance) in the table below:

Particle	Factor	Drawing
incoming quark	$u^s(p)c$	
outgoing quark	$\overline{u}^{\,s}(p)c^\dagger$	
incoming antiquark	$\overline{v}^{\,s}(p)c^\dagger$	
outgoing antiquark	$v^s(p)c$	
incoming gluon	$\epsilon_{\mu,s}a_{(k)}$	
outgoing gluon	$\epsilon^*_{\mu,s}a^*_{(k)}$	

3. **Vertex factors.** Each vertex in the diagram below contributes the corresponding factor shown.

Particle	Factor	Drawing
quark-gluon	$\sum_k -i\frac{1}{2}g_S\gamma^\mu\lambda_{(k)}$	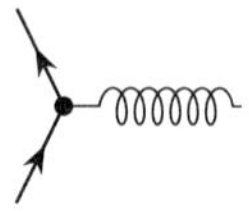
3-gluon	$-g_S f_{(ijk)}\big[\, g_{\mu\nu}(q_{1\sigma}-q_{2\sigma})$ $+g_{\nu\sigma}(q_{2\mu}-q_{3\mu})+g_{\sigma\mu}(q_{3\nu}-q_{1\nu})\,\big]$	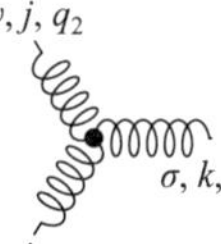 ν, j, q_2; σ, k, q_3; μ, i, q_1
4-gluon	$-ig_S^2\sum_{(k)}\big[\, f_{(ijk)}f_{(mnk)}(g_{\mu\sigma}g_{\nu\rho}-g_{\mu\rho}g_{\nu\sigma})$ $+f_{(ink)}f_{(jmk)}(g_{\mu\nu}g_{\sigma\rho}-g_{\mu\sigma}g_{\nu\rho})$ $+f_{(imk)}f_{(njk)}(g_{\mu\rho}g_{\nu\sigma}-g_{\mu\nu}g_{\rho\sigma})\,\big]$	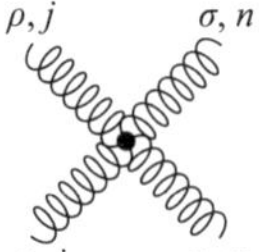 ρ, j; σ, n; μ, i; ν, m

(In the 3-gluon diagram, the four-momenta q_1, q_2, q_3 come from the fact that the interaction terms in the second line of Equation 30.14 involve derivatives of the gluon fields. Taking the derivative of a plane-wave field solution $e^{-iq_\mu x^\mu}$ multiplies the field by $-iq_\mu$. The vertex factor in that diagram assumes that the four-momenta q_1, q_2, q_3 all point *toward* the vertex, so flip the sign of the four-momentum if it describes an outgoing particle.)

4. **Propagators.** Each internal quark line contributes

$$i\,\frac{\not{q}_j + m}{q_i^2 - m^2},\tag{30.22a}$$

(where $\not{q}_j \equiv (q_j)_\mu\gamma^\mu$ and the index j indicates which internal line we are talking about). Each internal gluon line contributes

$$-i\,\frac{g_{\mu\nu}\delta_{(kn)}}{q_j^2}.\tag{30.22b}$$

5. **Conserve four-momentum at each vertex.** (In two-vertex diagrams, this determines the four-momentum q of the internal particle.)

6. **Follow the arrows.** To put the items in the correct order, follow the arrows on the fermion lines from start to finish, and write the corresponding terms *starting from the right* (backward from English writing).

7. **Undetermined internal momenta.** Include a factor of $d^4 q_j/(2\pi)^4$ and integrate over any undetermined four-momenta q_j (use an upper cutoff if necessary).

8. **Fermion loops.** For each closed loop involving fermions, include an overall factor of -1, and take the trace of the product of the gamma matrices contributed by loop vertices and propagators.

9. **Boson loops.** For each closed loop involving spin-1 bosons, include a factor of $g_{\mu\nu}$ and sum over the upper indices contributed by the boson propagators to eliminate all free indices. (There are additional rules for gluon loops that require concepts beyond our scope here.)

10. **The result is** $-i\mathcal{M}$. Multiply by i to get $\mathcal{M}$.

11. **Add the amplitudes.** To find the total amplitude $\mathcal{M}$ for a given process up to a certain order, add the amplitudes for all diagrams up to that order.

12. **Antisymmetrization.** Include a minus sign between two diagrams that differ by swapping (between two vertices) two incoming identical fermions, two outgoing identical fermions, an incoming fermion with an outgoing antifermion (corresponding to the same type of fermion), or an incoming antifermion with an outgoing fermion of the same type.

With these rules, we can calculate the amplitude $\mathcal{M}$ for most **tree-level** Feynman diagrams (that is, diagrams that do not involve loops). The main difference between chromodynamic amplitudes and electroweak amplitudes is that chromodynamic amplitudes include additional **color factors** of the form $c_j^\dagger \lambda_{(k)} c_n$, where the $c_j^\dagger$ and c_n come from the quarks and/or antiquarks (with momenta p_j and p_n) involved in a vertex and the $\lambda_{(k)}$ from the gluon connected to that vertex. Otherwise, chromodynamic amplitude calculations are pretty similar to the ones we have done for electroweak interactions.

But, as a fair warning, these simple rules are not sufficient to capture the complexity of chromodynamic diagrams that involve loops. Summing over gluon spin states also involves new rules that don't apply when we sum over photon spin states. Addressing these issues, sadly, would take more time than we can spare here, but we can at least do some basic calculations with the rules as stated.

BOX 30.1 How Many Parameters Does SU(3) Need?

Our goal in this box is to determine the number of parameters required to completely specify an SU(3) matrix. Remember that such a matrix must be a 3×3 matrix that is unitary and has determinant 1.

The matrices in this case are 3×3 matrices with 9 total complex components, equivalent to 18 real numbers. Requiring the matrices to be unitary will put 9 constraints on these 18 numbers:

$$
M^{\dagger}M = \begin{pmatrix} a^* & d^* & g^* \\ b^* & e^* & h^* \\ c^* & f^* & k^* \end{pmatrix} \begin{pmatrix} a & b & c \\ d & e & f \\ g & h & k \end{pmatrix} = \begin{pmatrix} 1 & 0 & 0 \\ 0 & 1 & 0 \\ 0 & 0 & 1 \end{pmatrix}
$$

$$
= \begin{pmatrix} |a|^2 + |d|^2 + |g|^2 & a^*b + d^*e + g^*h & a^*c + d^*f + g^*k \\ b^*a + e^*d + h^*g & |b|^2 + |e|^2 + |h|^2 & b^*c + e^*f + h^*k \\ c^*a + f^*d + k^*g & c^*b + f^*e + k^*h & |c|^2 + |f|^2 + |k|^2 \end{pmatrix}.
$$

Think of each complex number in the matrix as having the form of an amplitude times a phase: $c = Ae^{i\delta}$. We see that requiring each diagonal element of $M^{\dagger}M$ to be 1 puts an equation of constraint on the amplitudes but not the phases of the elements in M involved in each expression. Therefore, the equations along the diagonal in the final expression above put 3 constraints on the 18 numbers required to specify M.

Exercise 30.1.1 The top middle off-diagonal equation has the form $a^*b + d^*e + g^*h = 0$, and the other off-diagonal equations have a similar form. Note that every equation below the diagonal is the complex conjugate of an equation above the diagonal, so these equations are not really independent: if one is true, then so is the other. Nonetheless, the off-diagonal equations impose a total of six constraints on the components of M. Explain why. (*Hint:* Here it is helpful to think of the sum on the left side as producing some complex number with a real and imaginary part.)

Therefore, requiring M to be unitary puts 9 constraints on the 18 numbers required to describe the matrix. Now one can prove that for *any* unitary matrix M, $|\det M| = 1 \Rightarrow \det M = e^{i\delta}$, where δ is arbitrary. Therefore, constraining $\det M = 1$ implies that we are requiring that $\delta = 0$: this is one more equation of constraint. Therefore, we have a total of 10 constraints on the 18 numbers needed to define the elements of M. This means that we need $18 - 10 = 8$ free parameters to define the elements of an SU(3) matrix M.

BOX 30.2 Why the Lambdas Must Be Traceless

We have already seen that if a matrix $U \equiv e^{-iH\theta}$ is unitary, then its generator H must be Hermitian. One can also prove quite generally that $\det(e^A) = e^{\mathrm{Tr}\,A}$ for any square matrix A. If we require our unitary SU(3) matrix $M \equiv e^{-i\frac{1}{2}\vec{\lambda}\cdot\vec{\theta}}$ to have $\det M = 1$, it follows that the matrix $\vec{\lambda}\cdot\vec{\theta}$ must have zero trace for any arbitrary eight-component real "vector" $\vec{\theta}$.

The Gell-Mann matrices are

$$\lambda_1 = \begin{pmatrix} 0 & 1 & 0 \\ 1 & 0 & 0 \\ 0 & 0 & 0 \end{pmatrix}, \quad \lambda_2 = \begin{pmatrix} 0 & -i & 0 \\ i & 0 & 0 \\ 0 & 0 & 0 \end{pmatrix}, \quad \lambda_3 = \begin{pmatrix} 1 & 0 & 0 \\ 0 & -1 & 0 \\ 0 & 0 & 0 \end{pmatrix},$$

$$\lambda_4 = \begin{pmatrix} 0 & 0 & 1 \\ 0 & 0 & 0 \\ 1 & 0 & 0 \end{pmatrix}, \quad \lambda_5 = \begin{pmatrix} 0 & 0 & -i \\ 0 & 0 & 0 \\ i & 0 & 0 \end{pmatrix}, \tag{30.23}$$

$$\lambda_6 = \begin{pmatrix} 0 & 0 & 0 \\ 0 & 0 & 1 \\ 0 & 1 & 0 \end{pmatrix}, \quad \lambda_7 = \begin{pmatrix} 0 & 0 & 0 \\ 0 & 0 & -i \\ 0 & i & 0 \end{pmatrix}, \quad \lambda_8 = \frac{1}{\sqrt{3}}\begin{pmatrix} 1 & 0 & 0 \\ 0 & 1 & 0 \\ 0 & 0 & -2 \end{pmatrix}.$$

Exercise 30.2.1 Show that one can write *any* 3×3 Hermitian and traceless matrix H in the form $\vec{\lambda}\cdot\vec{\theta}$ with real coefficients θ_1 through θ_8 by calculating those coefficients in terms of the components of H.

BOX 30.3   SU(3) Structure Constants

In this box, you will check the values of some of the structure constants listed in Equation 30.4. Given that f_{jkn} is completely antisymmetric, the nonzero structure constant values must have distinct indices. So we have 8 possible values for the first index, 7 for the second (since it can't be the same as the first), and 6 for the third, for a total of 336 possibly nonzero values. Of course, there are $3 \times 2 = 6$ ways to arrange the indices, so there are only 56 *independent* values. But most are zero, as the following illustrates.

Exercise 30.3.1   Evaluate $[\lambda_1, \lambda_2]$, show that $f_{12n} = 0$ for $n \neq 3$, and verify that $f_{123} = 1$. (*Hint:* The λ_k matrices are listed in Equation 30.2, repeated below.)

$$\lambda_1 = \begin{pmatrix} 0 & 1 & 0 \\ 1 & 0 & 0 \\ 0 & 0 & 0 \end{pmatrix}, \quad \lambda_2 = \begin{pmatrix} 0 & -i & 0 \\ i & 0 & 0 \\ 0 & 0 & 0 \end{pmatrix}, \quad \lambda_3 = \begin{pmatrix} 1 & 0 & 0 \\ 0 & -1 & 0 \\ 0 & 0 & 0 \end{pmatrix},$$

$$\lambda_4 = \begin{pmatrix} 0 & 0 & 1 \\ 0 & 0 & 0 \\ 1 & 0 & 0 \end{pmatrix}, \quad \lambda_5 = \begin{pmatrix} 0 & 0 & -i \\ 0 & 0 & 0 \\ i & 0 & 0 \end{pmatrix}, \tag{30.28}$$

$$\lambda_6 = \begin{pmatrix} 0 & 0 & 0 \\ 0 & 0 & 1 \\ 0 & 1 & 0 \end{pmatrix}, \quad \lambda_7 = \begin{pmatrix} 0 & 0 & 0 \\ 0 & 0 & -i \\ 0 & i & 0 \end{pmatrix}, \quad \lambda_8 = \frac{1}{\sqrt{3}} \begin{pmatrix} 1 & 0 & 0 \\ 0 & 1 & 0 \\ 0 & 0 & -2 \end{pmatrix}.$$

BOX 30.3 SU(3) Structure Constants (cont.)

Exercise 30.3.2 In a similar way, calculate $[\lambda_1, \lambda_8]$ and $[\lambda_3, \lambda_4]$ and determine the associated structure constants. (*Hint:* The λ matrices are all linearly independent, so no linear combination of them can add to the zero matrix.)

BOX 30.4 The SU(3) Conserved Current

Our goal in this box is to evaluate the conserved current for the SU(3) symmetry. From Chapter 5, the formula for calculating the conserved current is

$$J^{\mu} = \sum_{n} \frac{\partial \mathcal{L}}{\partial(\partial_{\mu}\psi_{n})}\left[\frac{d\psi_{n}}{d\theta}\right]_{0},\qquad(30.29)$$

where the sum is over the different fields in the Lagrangian density that are affected by the transformation, and where we evaluate the derivative at $\theta = 0$. In the case in question, the terms in our Lagrangian density that involve field derivatives are $\mathcal{L} = \sum_{f} i\overline{\Psi}_{f}\gamma^{\mu}\partial_{\mu}\Psi_{f}$. We see that partial derivatives of only Ψ_{f} (not $\overline{\Psi}_{f}$) appear in the Lagrangian density.

Exercise 30.4.1 Use the expression above to evaluate the conserved current for the SU(3) symmetry. (*Hint:* This is pretty straightforward—you will not need all the space below.)

BOX 30.5 The Value of μ Does Not Matter

Our goal in this box is to show that our choice of the value of μ in Equation 30.15 (repeated below)

$$\alpha_S(Q^2) = \frac{\alpha_S(\mu^2)}{1 + \frac{1}{12\pi}\alpha_S(\mu^2)(11n - 2f)\ln(|Q^2|/\mu^2)} \quad \text{for } \left|Q^2\right| \gg \mu^2 \quad (30.30)$$

does not matter, in the sense that if the formula is valid for choice μ_1, one can prove that it is true for choice μ_2 (so long as $\alpha_S(\mu^2) \ll 1$ for both choices).

Exercise 30.5.1 Prove this. (*Hint:* It is easiest to work with the inverse

$$\frac{1}{\alpha_S(Q^2)} = \frac{1}{\alpha_S(\mu_1^2)} + \frac{11n - 2f}{12\pi}\ln\left(\frac{|Q^2|}{\mu_1^2}\right). \quad (30.31)$$

Assuming that this is true for μ_1, use it to calculate $\alpha_S(\mu_2^2)$ (by setting $Q^2 = \mu_2^2$), and solve the resulting equation for $\alpha_S(\mu_1^2)$ in terms of $\alpha_S(\mu_2^2)$. Then substitute the result into Equation 30.31, and show that you get the same equation as before but with $\mu_1 \to \mu_2$.)

BOX 30.6   How the Color Fine-Structure Constant Varies

Our goal in this box is to examine how the color-interaction fine-structure constant $\alpha_S(Q^2)$ varies and compare that variation to that of the electromagnetic fine-structure constant $\alpha_e(Q^2)$. We have $11n - 2f = 33 - 12 = 21$ (where n is the number of colors and f the number of quarks), and taking $\alpha_S(\mu^2)$ to be $\alpha_S(M_Z^2) = 0.1179$, Equation 30.15 implies that

$$\alpha_S([10\,\text{GeV}]^2) = \frac{\alpha_S(M_Z^2)}{1 + \frac{21}{12\pi}\alpha_S(M_Z^2)2\ln(10\,\text{GeV}/91.2\,\text{GeV})}$$

$$= \frac{0.1179}{1 + \frac{21}{12\pi}0.1179 \cdot 2\ln(0.110)} = 0.1661. \qquad (30.32)$$

Exercise 30.6.1   Similarly, calculate α_S at $|Q| = 100$ GeV. Also, calculate the EM fine-structure constant, which is

$$\alpha_e(Q^2) = \frac{\alpha_e(0)}{1 - \frac{1}{3\pi}\alpha_e(0)f(u)}, \qquad (30.33)$$

(where $m = 0.00051$ GeV is the electron mass) at $|Q| = 10$ GeV and 100 GeV, and compare how much α_S and α vary over that range. (Note: At the energies involved here, the $f(u) \approx \ln(|Q^2|/m^2)$. Remember that $\ln x^2 = 2\ln x$. For simplicity, assume that $\alpha_e(0) = 1/137$.)

Homework Problems

P30.1 An SU(3) transformation matrix M transforms a three-component color column vector c to a new color vector: $c' = Mc$. For example, the matrix

$$M = \begin{pmatrix} 0 & 0 & 1 \\ 1 & 0 & 0 \\ 0 & 1 & 0 \end{pmatrix} \qquad (30.34)$$

yields
$$\begin{bmatrix} c'_r \\ c'_g \\ c'_b \end{bmatrix} = \begin{pmatrix} 0 & 0 & 1 \\ 1 & 0 & 0 \\ 0 & 1 & 0 \end{pmatrix} \begin{bmatrix} c_r \\ c_g \\ c_b \end{bmatrix} = \begin{bmatrix} c_b \\ c_r \\ c_g \end{bmatrix},$$

meaning that it converts whatever amount c_r of red that was in c to the same amount of green in c', and similarly converts green to blue and blue to red. This amounts to one possible "rotation" in color space.

a. Show that M is unitary and has determinant 1.

b. As this M basically converts $r \to g$, $g \to b$, and $b \to r$, it converts the meson color state $|1\rangle = \sqrt{\tfrac{1}{2}}|r\overline{g}\rangle + \sqrt{\tfrac{1}{2}}|g\overline{r}\rangle$ in Equation 30.18 to

$$|1'\rangle = \sqrt{\tfrac{1}{2}}|g\overline{b}\rangle + \sqrt{\tfrac{1}{2}}|b\overline{g}\rangle = |6\rangle.$$

Show that it rotates $|3\rangle$ to the linear combination $|3'\rangle = \alpha|3\rangle + \beta|8\rangle$, and find α and β.

c. What is $|8'\rangle$?

d. Show that this operator rotates the meson singlet state to itself.

P30.2 Pretend that quarks are free particles and consider the quark-antiquark scattering diagram in Figure 30.6a. Assume the quarks are of different families so that annihilation is not a possibility.

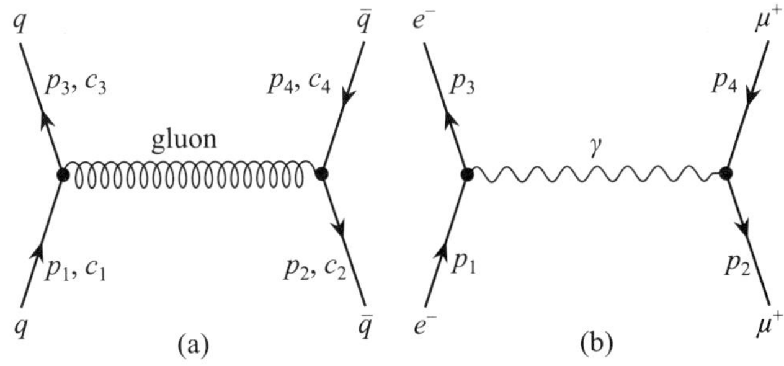

FIGURE 30.6   These are the Feynman diagrams for quark-antiquark scattering (a) and electron-antimuon scattering (b).

a. Use the Feynman rules to evaluate the amplitude of this process.

b. Compare the amplitude to that for the electromagnetic electron-antimuon scattering process in Figure 30.6b. If we express the ratio as $f\alpha_S/\alpha_e$, what is f?

c. Now, we know that at low energies the potential energy for the electromagnetic interaction is $V = -\alpha_e/r$, so we might suspect that at low energies the potential energy for the color interaction would be $V = -f\alpha_S/r$ (if quarks were free particles like electrons and α_S did not blow up at low energies). Assume that the initial quark colors are red and anti-green, and calculate the color factor. Argue that in this case, the quark interaction is *repulsive*. (If we were to repeat the calculation assuming that the quarks were initially and finally in the color singlet state, which I won't ask you to do, we would find the interaction to be *attractive*. This is another reason why the meson bound state of a quark and antiquark must be a color singlet.)

P30.3 We have eight gluons, one for each of the eight color-anticolor states that rotate into one another. If we had a ninth gluon corresponding to the color singlet state, it would be a color-neutral massless particle. All the gluons are electrically neutral, so this hypothetical color-neutral gluon would be a lot like a photon. Wait, could it *be* the photon? Explain why it cannot. (*Hint:* The color-interaction Lagrangian density discussed in this chapter implies that gluons connect with all quarks—and thus with all baryons—the same way. Why is that a problem for a hypothetical photon, considering what we observe in nature?)

P30.4 When a quark interacts with an antiquark, as in Figure 30.6a, the complete color factor is:

$$f = \frac{1}{4}\sum_k (c_3^\dagger \lambda_{(k)} c_1)(c_2^\dagger \lambda_{(k)} c_4).$$

Suppose that I have a red quark interacting with an anti-green antiquark. The net color must be unchanged in this interaction, the antiquark cannot have a normal color, and the quark cannot have an anticolor, so the quark must remain red and the antiquark must remain anti-green:

$$c_1 = c_3 = \begin{bmatrix} 1 \\ 0 \\ 0 \end{bmatrix}, \qquad c_2 = c_4 = \begin{bmatrix} 0 \\ 1 \\ 0 \end{bmatrix}. \qquad (30.35)$$

a. Show that the color factor in this case evaluates to $-\tfrac{1}{6}$.

b. What would be the color factor in the case where the system has the state $|\psi\rangle_c = \sqrt{\tfrac{1}{2}}|r\overline{r}\rangle - \sqrt{\tfrac{1}{2}}|g\overline{g}\rangle$? (*Hint:* You can think of calculating the color factor for this superposition state as evaluating

$$\langle\psi_c|f|\psi_c\rangle = \left[\sqrt{\tfrac{1}{2}}\langle r\overline{r}| - \sqrt{\tfrac{1}{2}}\langle g\overline{g}|\right]f\left[\sqrt{\tfrac{1}{2}}|r\overline{r}\rangle - \sqrt{\tfrac{1}{2}}|g\overline{g}\rangle\right]$$

$$= \tfrac{1}{2}\langle r\overline{r}|f_a|r\overline{r}\rangle - \tfrac{1}{2}\langle r\overline{r}|f_b|g\overline{g}\rangle$$

$$- \tfrac{1}{2}\langle g\overline{g}|f_c|r\overline{r}\rangle + \tfrac{1}{2}\langle g\overline{g}|f_d|g\overline{g}\rangle, \qquad (30.36)$$

where f_a, f_b, f_c, and f_d are the color factors for the initial state given by the ket and the final state given by the bra in each of the four terms. The total color factor f is therefore $f = \tfrac{1}{2}[f_a - f_b - f_c + f_d]$. You should find the answer to be $-\tfrac{1}{6}$.)

P30.5 Consider a quark and antiquark with four-momenta p_1 and p_2, respectively, that meet and annihilate to a virtual gluon that subsequently splits into a lower-mass quark-antiquark pair with momenta p_3 and p_4 (an s-channel process). Suppose that the quark and antiquark are in the meson color-singlet state.

a. Find the amplitude $\mathcal{M}$ for this process.

b. You should have found that the color factor is

$$f = \tfrac{1}{4} \sum_k (c_2^\dagger \lambda_{(k)} c_1)(c_3^\dagger \lambda_{(k)} c_4). \qquad (30.37)$$

Show that this factor is zero, indicating that this process cannot occur. (This is basically because any gluon color state is orthogonal to the color singlet state.) (*Hint:* See the hint for Problem 30.4b for how to handle superposition color states.)

P30.6 A convenient expression for the running color-coupling constant is

$$\alpha_S(Q^2) = \frac{12\pi}{(11n - 2f)\ \ln(|Q^2|/\Lambda^2)}, \qquad (30.38)$$

where Λ is a parameter defined by the relation

$$\ln \Lambda^2 = \ln \mu^2 - \frac{12\pi}{(11n - 2f)\,\alpha_S(\mu^2)}. \qquad (30.39)$$

a. Show that Equation 30.38 follows from Equations 30.15 and 30.39.

b. Use the experimental value of $\alpha_S(M_Z^2)$ to determine the value of Λ.

c. Submit a computer-generated graph of $\alpha_S(Q^2)$ versus $|Q|$ from 1 GeV to 1000 GeV (use a logarithmic scale for $|Q|$).

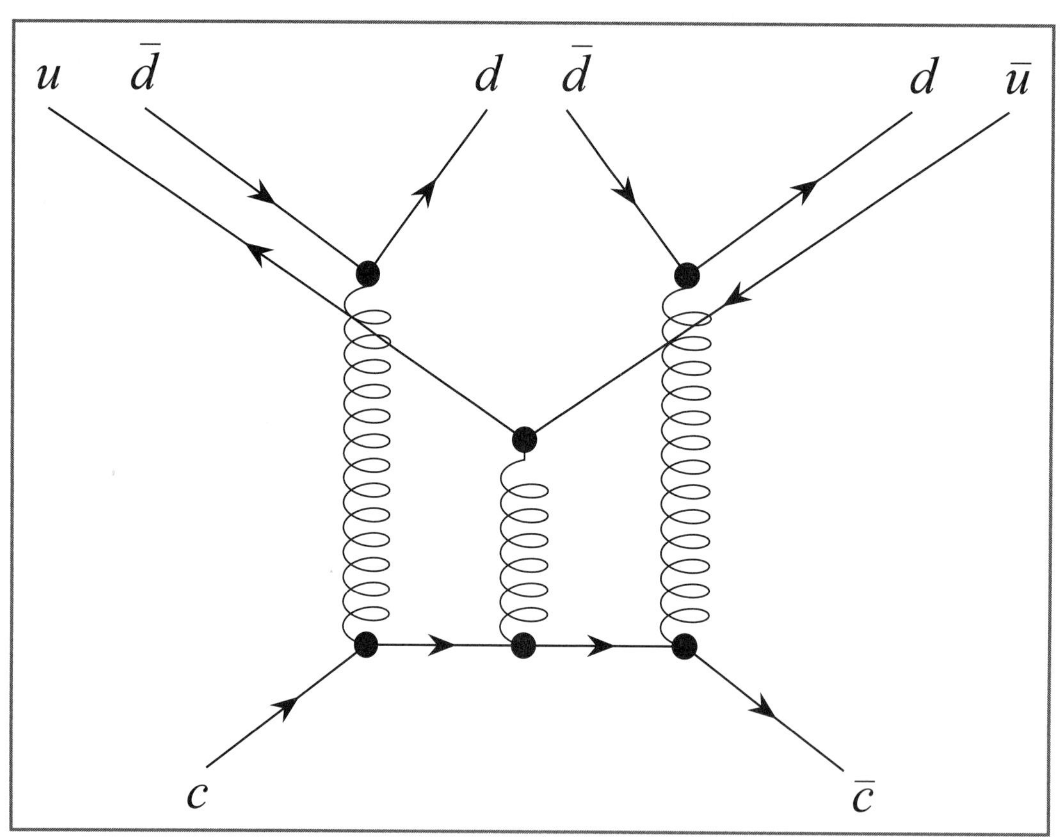

u
$\bar{d}$
d
$\bar{d}$
d
$\bar{u}$
c
$\bar{c}$

31. APPLICATIONS OF CHROMODYNAMICS

In this optional chapter, we will explore several applications of quantum chromodynamics, both to illustrate how the methods can be used and to examine certain historically important experimental results.

31.1 A Short History of the Quark Model

As mentioned in Chapter 27, Gell-Mann and Zweig independently proposed in 1964 that the structure of low-mass hadrons displayed in Figure 27.1 could be elegantly explained if they were constructed out of three quarks, which we now call u, d, and s. By 1968, experiments had also established that the proton had an internal structure instead of being a point-like particle. Even so, most physicists of the time were dubious about the quark model. One of the challenges was the discovery of the Δ^{++} baryon, which (according to the quark model) must have quark composition uuu. Because it has a spin of $\frac{3}{2}$, all its quarks seemingly must be in the same quantum state, violating the Pauli exclusion principle. When O. Greenberg proposed in 1964 that a way out would be to assume that the quarks had three colors (each u would then have a unique color state and therefore not violate the principle), some saw this crazy idea as evidence that the quark model was too contrived to be real.

However, this all changed in 1974. In November of that year, two groups, one from Brookhaven and one from the Stanford Linear Accelerator Center (SLAC), announced the existence of a new particle. One group called it the J and the other group called it the ψ, so the particle is now known as the J/ψ. It had the peculiar property of lasting more than three orders of magnitude longer than other mesons in a similar mass range. As David J. Griffiths memorably put it, this was as surprising as discovering a village in the jungle where people lived an average of 70,000 years. The J/ψ therefore pointed to some fundamentally new physics beyond the models that were then accepted. Physicists called this the **November Revolution** (a wry reference to the October Revolution in 1917 that established the communist Bolsheviks as the ruling party in Russia).

But the physics needed for explaining the anomalous lifetime of the J/ψ was actually already in the literature. James Bjorken and Sheldon Glashow had speculated in 1964 about the existence of a fourth quark (which would make the set of four quarks parallel to the set of four leptons then known), and in 1970, Glashow, John Iliopoulos, and Luciano Maiani pointed to indirect evidence of its existence (see Problem P31.1). After discovery of the J/ψ, people quickly saw that interpreting the J/ψ as a $c\bar{c}$ meson in the context of quantum chromodynamics explained its anomalous lifetime. Moreover, if the charm quark existed, then other charmed mesons and baryons should exist, and these were soon discovered: evidence supporting the existence of charmed baryons was announced in 1975, the $D^0 = c\bar{u}$ and $D^+ = c\bar{d}$ mesons were discovered in 1976, and the $D_s^+ = c\bar{s}$ meson in 1977. The success of the quark model in explaining the J/ψ's anomalous lifetime and in predicting the existence of these new particles effectively resuscitated the then-moribund quark model.

However, the τ lepton was also discovered in 1975, ruining Glashow's lovely symmetry between quarks and leptons. But the success of the quark model inspired people to look for two more quarks to restore the symmetry. The $\Upsilon = b\bar{b}$ meson was discovered in 1977. But because this meson had zero net "bottomness," firm evidence of a new conserved quark type required discovering particles with "bare bottom" that (unlike the Υ)

could not decay via a strong or electromagnetic interaction. The $B^- = b\bar{u}$ and $B^0 = b\bar{d}$ mesons fitting the bill were discovered in 1983. The top quark, however, is so massive that it was tough to find, and it also turned out to be so short-lived that it does not have time to form bound mesons or baryons at all. It took until 1995 to more indirectly but definitively establish the top quark's existence.

31.2 The Decay of the J/ψ

We see that the discovery of the $J/\psi = c\bar{c}$ was historically crucial in establishing the reality of the quark model. But why was its lifetime so long? Quantum chromodynamics explains this nicely.

All the previously known hadrons in this mass range were excited states of the basic hadrons constructed of u, d, and s quarks shown in Figure 27.1. But because these hadrons were bound by color interactions, their decays involved color interactions that rearranged those quarks into lower-energy configurations, with the excess energy carried away by newly created quark-antiquark pairs. At low energies, the vertex factor α_S is very large, so the decay rates were large and lifetimes were short (typically $\sim 10^{-23}$ s).

So how can the J/ψ decay? Feynman diagrams for the lowest-order color interactions that one might *think* would allow the J/ψ to decay are shown in Figure 31.1. But none of these are actually possible. Why?

The diagram in Figure 31.1a seems straightforward but has several problems. The most important is that the J/ψ must be a color singlet, but no gluon is a color singlet. Therefore, the J/ψ cannot annihilate to a single gluon. The diagram in Figure 31.1b is forbidden by conservation of energy: the masses of the outgoing $D^0 = c\bar{u}$ and $\overline{D}^0 = u\bar{c}$ together are larger than the mass of the J/ψ, and these are the lightest mesons that contain c quarks. The diagram shown in Figure 31.1c gets around the color singlet problem (because *two* gluons can collectively carry the colors associated with a color singlet state) and the energy conservation problem (because many mesons exist that have lower masses than half of a J/ψ), but this diagram would violate conservation of angular momentum: the J/ψ has spin 1, but two massless gluons with spin 1 can only have a total spin of 2 or 0, not 1.

The lowest-order diagram that *is* actually possible is shown in Figure 31.2. Three gluons can collectively carry the colors associated with a color singlet state and also have total spin 1, and many mesons exist whose masses are small enough to conserve energy. The problem with this diagram is that the decay rate that it predicts is proportional to α_S^6 (six vertices), while the lowest-order diagram describing the decay of an excited meson or baryon constructed of u, d, and s quarks would (like 31.1b) involve two vertices and so is proportional to α_S^2. Moreover, in a diagram like 31.1b, the virtual gluon would

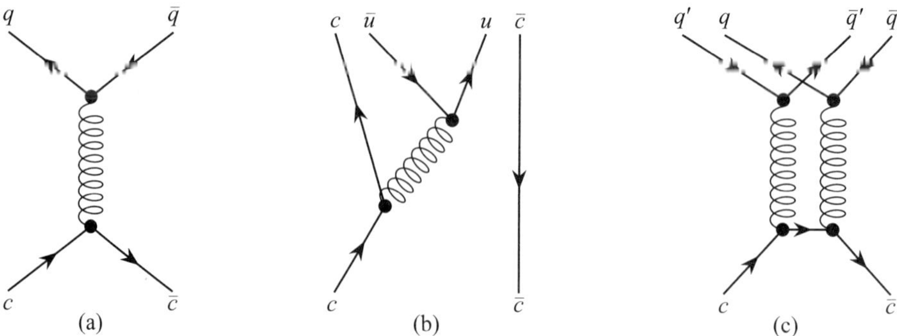

FIGURE 31.1   These Feynman diagrams for conceivable J/ψ decay processes are all actually forbidden.

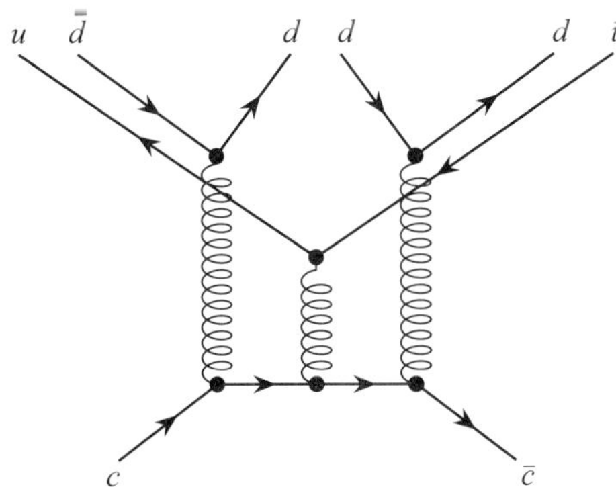

FIGURE 31.2   This is the lowest-order Feynman diagram for an actually possible J/ψ decay ($J/\psi \rightarrow \pi^+ + \pi^0 + \pi^-$). Many other low-mass meson products are possible: the crucial feature of any lowest order diagram is the three virtual gluons.

have a relatively small value of $|Q^2|$, so the running value of α_S would be large (≈ 1). However, each quark in Figure 31.2 must carry (on average) about a third of the J/ψ's mass of 3.097 GeV, where the running value of $\alpha_S \approx 0.29$. So we would expect the decay rate of the J/ψ to be on the order of $(0.29)^6 \approx 6 \times 10^{-4}$ times smaller than the decays of excited mesons of a similar mass, implying that its lifetime would be on the order of $(6 \times 10^{-4})^{-1} = 1.7 \times 10^3$ times longer. This explains the J/ψ's surprisingly long lifetime.

31.3 States of Charmonium

The J/ψ is just one example of a $c\bar{c}$ bound state. Such states are collectively called states of **charmonium,** a nod to positronium, the name coined earlier to describe a bound electron and a positron. One of the reasons that charmonium and its more massive cousin bottomonium $b\bar{b}$ are interesting is that their quarks are massive enough that they are non-relativistic in bound states, and the distance between them is small enough in such states that the anti-screening effect is small and the quarks behave like relatively free particles. This means that the system behaves physically like positronium, whose states in turn are the same as the well-known states of the hydrogen atom (after adjusting the reduced mass of the electron). By examining the difference between the states of positronium, charmonium, and bottomonium, we can learn about how the color-interaction and electromagnetic potential energy formulas differ.

The lowest-energy charmonium state is the η_c (2.984 GeV), which has total spin 0 instead of 1 like the J/ψ, though both are in the $n = 1$, $\ell = 0$ hydrogen-like orbital ground state. The fact that these states differ in energy by about 0.113 GeV means that the part of the color interaction that depends on the quarks' relative spin orientation (analogous to the magnetic interaction between the electron and positron spins in positronium) is very strong (in positronium, the energy difference is only 0.84 meV). The η_c' (3.637 GeV) and ψ' (3.686 GeV) particles correspond to charmonium states with $n = 2$, $\ell = 0$, and spin 0 and 1, respectively (the energy difference is smaller because the quarks are farther apart in the $n = 2$ state than in the $n = 1$ state). The χ_0, χ_1, and χ_2 particles (3.415, 3.511, and 3.556 GeV, respectively) correspond to the $n = 2$, $\ell = 1$ states, where the two quark spins are aligned to yield a total spin of 1, which then combines with the orbital angular momentum to yield total angular momenta of 0, 1, and 2, respectively. The h_c (3.525 GeV) is also an $n = 2$, $\ell = 1$ state with total angular momentum 1, but in this case, the two quark spins are anti-aligned so that their total spin is zero. All states with $n = 3$ or higher are more massive than the $D^0 + \overline{D}^0$ and therefore decay so rapidly they are only quasi-bound.

These energy levels are consistent with the potential energy formula

$$V(r) = -\frac{4}{3}\frac{\alpha_S}{r} + F_0 r, \tag{31.1}$$

with $a_S \approx 0.2$ and $F_0 \approx 0.9\,\text{GeV/fm}$ (though having only two accessible energy levels makes the fit a bit loose). The $F_0 r$ term expresses the anti-screening effect that makes the color interaction energy increase as the quark separation increases. This means that at large r the force between the quarks approaches a constant value of $F_0 = 16$ metric tons in more normal units (making it clear why separating quarks is energetically impossible!). While one cannot find analytic solutions for the potential energy function in Equation 31.1, as we can for the hydrogen atom, one can find the energy levels numerically for given values of α_S and F_0 and so find the values that give the best fit (see Problem P31.2 to see how to do this yourself).

31.4 The Decay Rate of the η_c

Figure 31.2 is so complicated that calculating the decay rate from quantum chromodynamics is a daunting task. Calculating the decay rate of the η_c is simpler and has parallels to the decay of positronium (see Chapter 22). The η_c has spin 0, and so (unlike the J/ψ) can decay to two gluons as in Figure 31.1b. Though calculating the full decay rate would involve calculating the amplitude for the entire diagram, let's focus on finding the amplitude for just the decay to two gluons (this will emphasize the parallels to the decay of positronium to two photons).

Figure 31.3 shows the t-channel and u-channel Feynman diagrams for this decay (a two-vertex s-channel diagram is also possible but is forbidden in this case; see Problem P31.3). According to the Feynman rules listed in the last chapter, the amplitude for the t-channel diagram is such that

$$-i\mathcal{M}_t = \overline{v}_2 c_2^\dagger \sum_k \left[-i\tfrac{1}{2}g_S\gamma^\mu\lambda_{(k)}\right]\left(\epsilon_{4,\mu}^* a_{4(k)}^*\right)\frac{i(\slashed{q}-m)}{q^2-m^2}$$

$$\sum_n \left[-i\tfrac{1}{2}g_S\gamma^\nu\lambda_{(n)}\right]\left(\epsilon_{3,\nu}^* a_{3(n)}^*\right)u_1 c_1, \tag{31.2}$$

where m is the mass of the c quark (the factors in square brackets are the vertex factors, the factors in parentheses describe the outgoing gluons, and the final factor on the first line

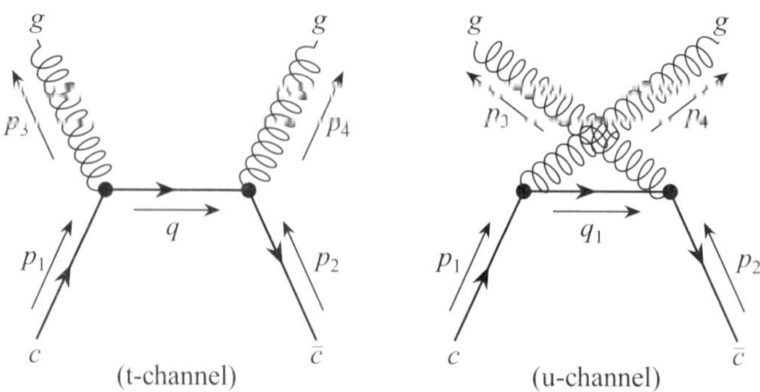

FIGURE 31.3 These are the possible Feynman diagrams for $\eta_c \to g + g$.

describes the virtual quark). Conservation of four-momentum at the left vertex requires that $q = p_1 - p_3$, so we can simplify

$$q^2 - m^2 = p_1^2 - 2p_1 \cdot p_3 - p_3^2 - m^2 = -2p_1 \cdot p_2 \tag{31.3}$$

because $p_1^2 = m^2$ and $p_3^2 = 0$ for a massless gluon. Gathering associated factors together and simplifying yields

$$\mathcal{M}_t = \frac{-g_S^2}{8p_1 \cdot p_3}\left[\bar{v}_2 \not{\epsilon}_4^*(\not{p}_1 - \not{p}_3 + m)\not{\epsilon}_3^* u_1\right] \sum_{k,n} a_{4(k)}^* a_{3(n)}^* [\, c_2^\dagger \lambda_{(k)}\lambda_{(n)} c_1]. \tag{31.4}$$

Note that the first quantity in brackets groups the factors residing in the four-dimensional spinor space, while the second quantity in brackets groups the factors that involve the three-dimensional color space. Each quantity is a simple number. The sums indexed by (k) and (n) are over the eight 3×3 SU(3) generator matrices $\lambda_{(k)}$.

The result for the second diagram is the same except with $3 \leftrightarrow 4$:

$$\mathcal{M}_u = \frac{-g_S^2}{8p_1 \cdot p_4}\left[\bar{v}_2 \not{\epsilon}_3^*(\not{p}_1 - \not{p}_4 + m)\not{\epsilon}_4^* u_1\right] \sum_{k,n} a_{3(n)}^* a_{4(k)}^* [\, c_2^\dagger \lambda_{(n)}\lambda_{(k)} c_1]. \tag{31.5}$$

Here I have kept the k index with gluon 3 and the n index with gluon 4 to make it clearer that the factor that we are summing over k and n is the same in both amplitudes, except that the lambda matrices are reversed.

The total amplitude is the sum of these two amplitudes. A bit of clever simplification (see Box 31.1) shows that in the rest frame of the η_c, the total amplitude $\mathcal{M} = \mathcal{M}_t + \mathcal{M}_u$ reduces to

$$\mathcal{M} = \frac{-g_S^2}{8m^2} \sum_{k,n} a_{4(k)}^* a_{3(n)}^* \big[\, \bar{v}_2 \not{\epsilon}_4^* \not{\epsilon}_3^* \not{p}_3 u_1 (\, c_2^\dagger \lambda_{(k)}\lambda_{(n)} c_1)$$

$$+ \bar{v}_2 \not{\epsilon}_3^* \not{\epsilon}_4^* \not{p}_4 u_1 (c_2^\dagger \lambda_{(n)}\lambda_{(k)} c_1)\,\big]. \tag{31.6}$$

Now we have to handle the spins. In this case, the quark spins are in a total spin-0 state $\sqrt{\frac{1}{2}}[\,|+z, -z\rangle - |-z, +z\rangle\,]$. This means that we will *not* use Casimir's trick, but (since we know what the spin cases are) will instead calculate the cases directly. The scattering amplitude will be

$$\mathcal{M}_0 = \sqrt{\tfrac{1}{2}}[\,\mathcal{M}_{\uparrow\downarrow} - \mathcal{M}_{\downarrow\uparrow}\,], \tag{31.7}$$

where the arrows refer to the spin states of the c and $\bar{c}$, respectively. A bit of calculation (see Box 31.2) shows that

$$\bar{v}_2^- \not{\epsilon}_4^* \not{\epsilon}_3^* \not{p}_3 u_1^+ = \bar{v}_2^- \not{\epsilon}_3^* \not{\epsilon}_4^* \not{p}_4 u_1^+ = +2\sqrt{2}m^2, \tag{31.8a}$$

$$\bar{v}_2^+ \not{\epsilon}_4^* \not{\epsilon}_3^* \not{p}_3 u_1^- = \bar{v}_2^+ \not{\epsilon}_3^* \not{\epsilon}_4^* \not{p}_4 u_1^- = -2\sqrt{2}m^2. \tag{31.8b}$$

This means that

$$\mathcal{M}_0 = \sqrt{\tfrac{1}{2}}\frac{-g_S^2}{8m^2}\sum_{k,n} a^*_{4(k)}a^*_{3(n)}\Big[\,\bar{v}_2^{\,-}\not{\epsilon}_4^*\not{\epsilon}_3^*\not{p}_3 u_1^+(\,c_2^\dagger\lambda_{(k)}\lambda_{(n)}c_1)$$

$$+\,\bar{v}_2^{\,-}\not{\epsilon}_3^*\not{\epsilon}_4^*\not{p}_4 u_1^+(c_2^\dagger\lambda_{(n)}\lambda_{(k)}c_1)\,\Big]$$

$$-\sqrt{\tfrac{1}{2}}\frac{-g_S^2}{8m^2}\sum_{k,n} a^*_{4(k)}a^*_{3(n)}\Big[\,\bar{v}_2^{\,+}\not{\epsilon}_4^*\not{\epsilon}_3^*\not{p}_3 u_1^-(\,c_2^\dagger\lambda_{(k)}\lambda_{(n)}c_1)$$

$$+\,\bar{v}_2^{\,+}\not{\epsilon}_3^*\not{\epsilon}_4^*\not{p}_4 u_1^-(c_2^\dagger\lambda_{(n)}\lambda_{(k)}c_1)\,\Big]$$

$$=\sqrt{\tfrac{1}{2}}\frac{-g_S^2}{8m^2}\sum_{k,n} a^*_{4(k)}a^*_{3(n)}\Big[\,2\sqrt{2}m^2(\,c_2^\dagger\lambda_{(k)}\lambda_{(n)}c_1)$$

$$+\,2\sqrt{2}m^2(c_2^\dagger\lambda_{(n)}\lambda_{(k)}c_1)\,\Big]$$

$$+\sqrt{\tfrac{1}{2}}\frac{-g_S^2}{8m^2}\sum_{k,n} a^*_{4(k)}a^*_{3(n)}\Big[\,2\sqrt{2}m^2(\,c_2^\dagger\lambda_{(k)}\lambda_{(n)}c_1)$$

$$+\,2\sqrt{2}m^2(c_2^\dagger\lambda_{(n)}\lambda_{(k)}c_1)\,\Big]$$

$$=\frac{-g_S^2}{2}\sum_{k,n} a^*_{4(k)}a^*_{3(n)}(c_2^\dagger\{\lambda_{(k)},\,\lambda_{(n)}\}c_1). \tag{31.9}$$

So the remaining task is to evaluate the sum. To save writing, let $\Lambda_{(kn)}\equiv\{\lambda_{(k)},\,\lambda_{(n)}\}$. As the quarks are in the color singlet state $(r\bar{r}+bg\bar{g}+b\bar{b})/\sqrt{3}$,

$$c_2^\dagger\Lambda_{(kn)}c_1 = \frac{1}{\sqrt{3}}\left([\,1\ \ 0\ \ 0\,]\,\Lambda_{(kn)}\begin{bmatrix}1\\0\\0\end{bmatrix}+[\,0\ \ 1\ \ 0\,]\,\Lambda_{(kn)}\begin{bmatrix}0\\1\\0\end{bmatrix}\right.$$

$$\left.+[\,0\ \ 0\ \ 1\,]\,\Lambda_{(kn)}\begin{bmatrix}0\\0\\1\end{bmatrix}\right)=\frac{1}{\sqrt{3}}\,\mathrm{Tr}[\Lambda_{(kn)}]. \tag{31.10}$$

But (as you will show in Box 31.3)

$$\mathrm{Tr}[\Lambda_{(kn)}] = \mathrm{Tr}\{\lambda_{(k)},\,\lambda_{(n)}\} = 4\delta_{(kn)}. \tag{31.11}$$

This leaves us with

$$\mathcal{M}_0 = \frac{-2g_S^2}{\sqrt{3}}\sum_{k} a^*_{4(k)}a^*_{3(k)}. \tag{31.12}$$

The color singlet state for a combination of two gluons turns out to be

$$\text{two-gluon color singlet} = \frac{1}{\sqrt{8}}\Big(|1\rangle|1\rangle + |2\rangle|2\rangle + |3\rangle|3\rangle + |4\rangle|4\rangle + |5\rangle|5\rangle$$

$$+\,|6\rangle|6\rangle + |7\rangle|7\rangle + |8\rangle|8\rangle\Big). \tag{31.13}$$

The proof is somewhat complicated (see Problem P31.6 if you are interested), but it is certainly plausible that an SU(3) rotation (which rotates each gluon state $|n\rangle$ into a normalized superposition of other gluon states) might indeed rotate this equally weighted superposition into itself. If we accept that this is the singlet state, then the sum over the a's that specify that state will be $\sum a^*_{4(k)}a^*_{3(k)} = (1/\sqrt{8})(1\cdot 1 + 1\cdot 1 + \cdots + 1\cdot 1) = 8/\sqrt{8}$. Therefore,

$$\mathcal{M}_0 = -\frac{2}{\sqrt{3}}g_S^2\sqrt{8} = -\sqrt{\frac{32}{3}}\,g_S^2. \tag{31.14}$$

From this, we can use the standard two-to-two scattering formula to calculate the differential cross section

$$d\sigma = \frac{|\mathcal{M}_0|^2}{(8\pi)^2(E_1+E_2)^2}\frac{|\vec{p}_3|}{|\vec{p}_1|}d\Omega = \frac{32g_S^4/3}{(8\pi)^2(2m)^2}\frac{m}{mv}d\Omega$$

$$= \frac{2}{3}\left(\frac{g_S^2}{4\pi m}\right)^2\frac{1}{v}d\Omega = \frac{2}{3}\frac{1}{v}\left(\frac{\alpha_S}{m}\right)^2 d\Omega, \qquad (31.15)$$

because in the charmonium "atom" the two quarks are non-relativistic, so $E_1 = E_2 \approx m$, the gluons' common momentum magnitude $|\vec{p}_3| = |\vec{p}_4| \approx m$, and $|\vec{p}_1| = mv$, where v is the quarks' relative speed when they collide and annihilate. As $d\sigma$ does not depend on angle, the total cross section for all gluon departure angles is $\sigma = 4\pi\, d\sigma$.

How do we get a decay rate out of this? If you read Chapter 14, recall that

$$\sigma = \frac{\Gamma V}{v}, \qquad (31.16)$$

where Γ is the transitions per unit time for one particle hitting one particle, V is the experimental volume, and v is the relative speed of the interacting particles (see Equation 14.19). We can consider $1/V$ to be the c-quark's probability density $|\psi(0)|^2$, where $\psi(0)$ is the c's wavefunction in the charmonium "atom's" ground state, which one can calculate using ordinary quantum mechanics. Solving for the decay rate, we have

$$\Gamma = \frac{v}{V}\sigma = \frac{8\pi}{3}\left(\frac{\alpha_S}{m}\right)^2|\psi(0)|^2. \qquad (31.17)$$

(Note how the hard-to-calculate v has dropped out of the calculation.)

This is not the end of the story, because we can only numerically calculate $|\psi(0)|^2$, and (since we cannot have bare gluons flying about) we should also include in our calculation the part of the Feynman diagrams that involves the creation of outgoing hadrons (see Figure 31.1c). The complete calculation is very difficult because there are many possible outgoing hadrons, and the result could involve additional exchanges of low-energy virtual gluons between the departing quarks. This makes the calculation of a total decay rate impractical.

But our calculation is not completely without experimental consequence. For one thing, we have verified that a decay to two gluons is possible. Moreover, these gluons must fly away from the annihilation point in opposite directions, so we would expect to see two opposed jets of hadrons emerging from the annihilation event that carry away the gluons' opposed momenta.

Moreover, even though we don't know $|\psi(0)|^2$ for the η_c "atom," the decay of the η_c to two photons must involve the same factor, so we can predict the branching ratio with some confidence. One can tweak the previous calculation to produce the decay rate to two photons: the result (see Box 31.4) is

$$\Gamma_{\gamma\gamma} = \frac{64\pi}{81}\left(\frac{\alpha}{m}\right)^2|\psi(0)|^2. \qquad (31.18)$$

We would therefore predict the branching ratio to be

$$\frac{\Gamma_{\gamma\gamma}}{\Gamma_{gg}} = \frac{64\pi/81}{8\pi/3}\left(\frac{\alpha}{\alpha_S}\right)^2 = \frac{8}{27}\left(\frac{\alpha}{\alpha_S}\right)^2. \qquad (31.19)$$

Each gluon gets the energy of half the mass of the η_c, or about 1.5 GeV. The running values of α_S and α at this energy are about 0.26 and $\frac{1}{135.3}$, respectively. Therefore, we would predict that the ratio would be

$$\frac{\Gamma_{\gamma\gamma}}{\Gamma_{gg}} = \frac{8}{27}\left(\frac{1/135.3}{0.26}\right)^2 = 2.4\times 10^{-4}. \qquad (31.20)$$

The measured result for $\Gamma_{\gamma\gamma}/\Gamma$ according to the 2022 Particle Data Group table for the η_c is $1.61 \pm 0.12 \times 10^{-4}$. The order of magnitude is correct, which is pretty good considering we are not actually calculating the complete diagram for the $\eta_c \to$ hadrons case.

31.5 The Discovery of the Higgs Boson

For a number of years, the Higgs boson was the most important predicted but unseen particle in the Standard Model. Its discovery in 2012 was therefore historic in establishing the validity of the Higgs mechanism, which is crucial for making the Standard Model work. The importance of this discovery was underlined by the awarding of the 2013 Nobel Prize to Peter Higgs and François Englert for their theoretical work on this mechanism.

In principle, it would seem easy to create a Higgs boson. Fermion-antifermion collisions can create Higgs bosons, as shown in Figure 31.4 (the fermion and antifermion must be in a spin singlet and color-singlet state because the Higgs has zero spin and no color). The amplitude for this diagram is

$$\mathcal{M}_{ffH} = \frac{g_W}{2} \frac{m_f}{M_W} [\bar{v}_2 u_1], \qquad (31.21)$$

where m_f is the mass of the fermion, M_W is the mass of the $W^\pm$ boson, and g_W is the weak coupling constant. For comparison, we see that the amplitude for producing a Z is

$$\mathcal{M}_{ffZ} = \frac{g_W}{2\cos\theta_W} [\bar{v}_2 \epsilon_Z^*(c_V - c_A \gamma^5) u_1]. \qquad (31.22)$$

The quantities in square brackets in both equations are just numbers, which will have units of mass because of the way the spinors are normalized. Since these numbers (in the high-energy limit required to create either particle) will be on the order of the collision energy $E = E_1 + E_2$ and will be comparable in the two cases, the ratio of the production cross sections will be of the order of

$$\frac{\sigma_H}{\sigma_Z} \sim \left| \frac{\mathcal{M}_{ffH}}{\mathcal{M}_{ffZ}} \right|^2 \sim \left(\frac{m_f}{M_W} \right)^2. \qquad (31.23)$$

Since the electron mass is tiny compared to that of the $W^\pm$, and it was hard enough to produce Z's in $e^- e^+$ colliders, producing Higgs bosons in such a collider is going to be virtually impossible. Even a pp collider like the LHC would have trouble producing Higgs bosons this way. Not only would we have to produce an antiquark somehow and have it meet a quark in a color and spin singlet state, but the masses of the u and d quarks that we would typically find in a proton are on the order of MeV and therefore also tiny compared to M_W.

However, Figure 31.5 shows another process that can much more efficiently create a Higgs boson in a pp collider. Because many gluons will be produced in a high-energy $p + p$ collision that has sufficiently high energies to produce nearly on shell top quarks, and the top quark's mass is in fact larger than M_W, this process is *much* more likely to create a Higgs boson than colliding fermions.

Calculating the amplitude of this diagram from the Feynman rules is very difficult, but we can make some estimates. The result must have the form of the product of the vertex factors times three for the three possible quark colors times some function $f(E)$ of the collision energy in the center-of-mass frame:

$$\mathcal{M} \propto \frac{g_S^2}{4} \frac{g_W}{2} \frac{m_t}{M_W} 3 f(E) = \tfrac{3}{2}\pi\alpha_S g_W \frac{m_t}{M_W} f(E), \qquad (31.24)$$

where m_t is the mass of the top quark and $\alpha_S \equiv g_S^2/4\pi$. The function $f(E)$ contains all of the complexities associated with the virtual quark loop as well as spin factors and color

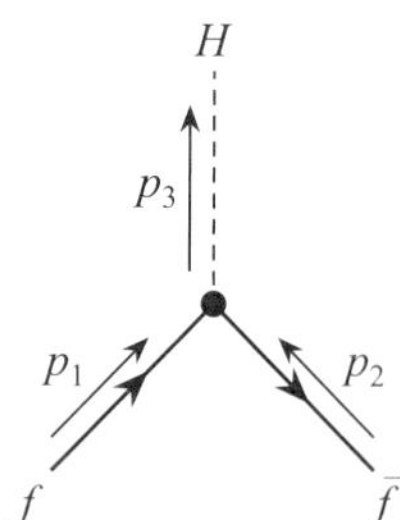

FIGURE 31.4   This is the lowest-order Feynman diagram for production of a Higgs boson from the collision of a fermion-antifermion pair.

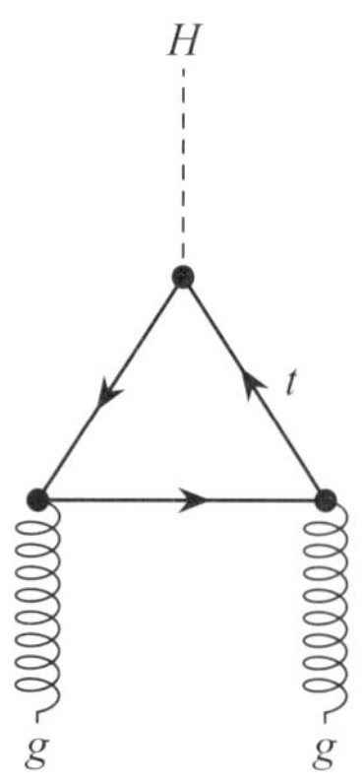

FIGURE 31.5   This Feynman diagram shows a practical process for producing a Higgs boson in a pp collider. The circulating particle is a top quark.

factors (beyond the factor of 3 already included). Now, $f(E)$ must have units of energy because Fermi's Golden Rule for two particles going to one particle requires $\mathcal{M}$ to have such units. In the center-of-mass frame, the collision energy $E = m_H$. We can estimate the unitless numerical factor associated with the loop using a method known as **naive dimensional analysis** (Manohar and Georgi, *Nucl Phys* B **234**, 189, 1984), which states that every closed loop in a Feynman diagram contributes a unitless factor of $(1/4\pi)^2$ (see, for example, Equation 21.9). So, if we estimate that the other unitless factors in $f(E)$ are of order 1, then $f(E) \sim E/(4\pi)^2 = m_H/(4\pi)^2$ and

$$\mathcal{M} \sim \frac{3\alpha_s g_W}{32\pi} \frac{m_t}{M_W} m_H^2.$$

(31.25)

From this, one can show (see Box 31.5) that the total cross section should have the form $\sigma = \sigma_0 \delta(E - m_H)$, where

$$\sigma_0 \sim \frac{9\alpha_s^2 g_W^2}{2048\pi m_H} \left(\frac{m_t}{M_W}\right)^2 = \frac{3.7 \times 10^{-7}}{\text{GeV}}.$$

(31.26)

The delta function means that there will be a spike in the cross section at $E = m_H$. In practice, this will be spread out in a bell-shaped curve $R(E - m_H)$ called a *resonance* (see Section 29.2 for a full discussion). This function (technically a distribution) is (like the Dirac delta itself) essentially a probability per unit energy range and should integrate to 1 if we integrate over the energy range where it is nonzero. So $R(E - m_H)\Delta E \approx 1$, where ΔE is the width of the curve. The total cross section should therefore be

$$\sigma \sim \sigma_0 \frac{1}{\Delta E}\left[R(E - m_H)\Delta E\right] = \sigma_0 \frac{1}{\Delta E}(1) = \left(\frac{3.7 \times 10^{-7}}{\text{GeV}}\right)\frac{1}{\Delta E}.$$

(31.27)

The observed width is $\sim 5\,\text{GeV}$, so we would predict a cross section of

$$\sigma \sim \left(\frac{3.7 \times 10^{-8}}{\text{GeV}}\right)\left(\frac{1}{5\,\text{GeV}}\right)\left(\frac{0.1973\,\text{GeV}}{1/\,\text{fm}}\right)^2\left(\frac{1\,\text{b}}{100\,\text{fm}^2}\right) = 30\,\text{pb}.$$

(31.28)

A significantly more complicated calculation that includes some higher-order diagrams (see `twiki.cern.ch/twiki/bin/view/LHCPhysics/LHCHWGGGF`) yields $\sigma \approx 15$ pb, so our very crude estimate is not too bad!

The discovery of the Higgs boson was announced at a time when the integrated beam luminosity $\int \mathcal{L}\,dt$ = (number of possible colliding particle pairs per unit area) was about 11 fb^{-1} = 11000 pb^{-1}, so we might have expected roughly 300,000 events by our estimate. However, our estimate does not account for the fact that one must multiply by the probability of having gluons of the appropriate energy in a given collision, and that only a certain subset of decay outcomes can be easily distinguished from the background. The latter in particular will greatly reduce the number of events we have to work with.

The discoverers focused on two specific decay process (called the "golden channels"): $H \rightarrow ZZ \rightarrow e^- + e^+ + e^- + e^+$ or $\mu^- + \mu^+ + \mu^- + \mu^+$ and $H \rightarrow \gamma + \gamma$. Because the Z is massive, it couples to the Higgs fairly well, and decays to easily identified $e^- e^+$ or $\mu^- \mu^+$ pairs about 7% of the time. The advantage of a decay to two photons (via a diagram that is essentially the reverse of Figure 31.5, with photons replacing the gluons) is that the background of $p + p \rightarrow \gamma + \gamma$ is completely smooth, with no other possible resonances. But since many other Higgs decay processes exist, these golden channels will represent only a small fraction of events.

On July 4, 2012, research teams at CERN announced the discovery of the Higgs boson by observing a resonance in each of these two channels (see Figures 31.6 and 31.7). One can see that the resonances amounted to only a relative handful of events, a tiny fraction of the likely Higgs production events that took place during that time.

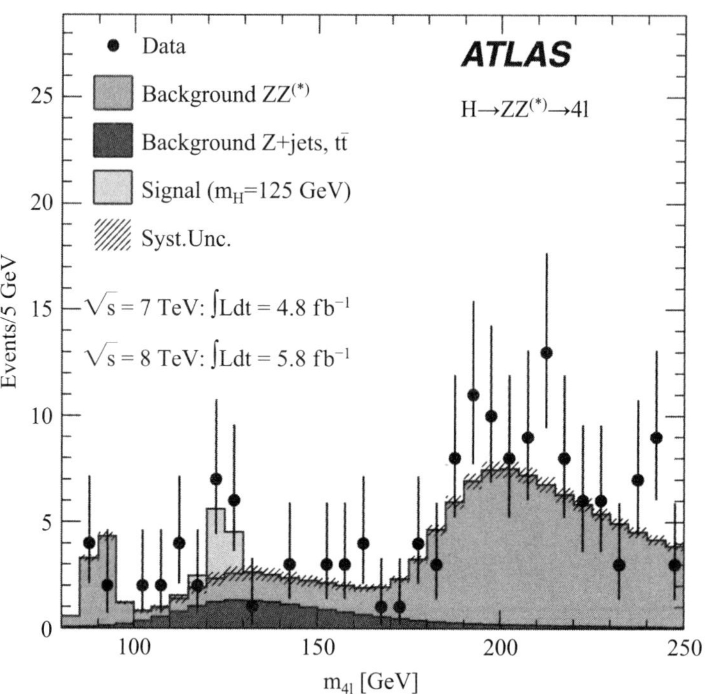

FIGURE 31.6 The Higgs resonance, as it appears in the measured cross section of $p + p \rightarrow 4$ leptons. Credit: Atlas Collaboration, *Phys Lett* B, **716**, 1 (2012). (Reprinted with permission.)

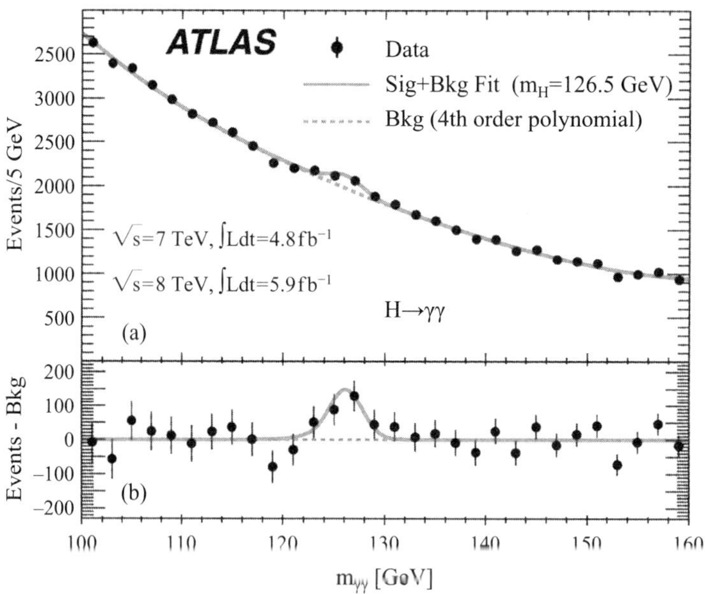

FIGURE 31.7 The Higgs resonance, as it appears in the measured cross section of $p + p \rightarrow \gamma + \gamma$. Credit: Atlas Collaboration, *Phys Lett* B, **716**, 1 (2012). (Reprinted with permission.)

BOX 31.1 Simplifying the η_c Decay Amplitude

We seek to simplify the η_c decay amplitude $\mathcal{M} = \mathcal{M}_t + \mathcal{M}_u$, where

$$\mathcal{M}_t = \frac{-g_S^2}{8p_1 \cdot p_3}\left[\bar{v}_2\,\epsilon\!\!\!/_4^*(p\!\!\!/_1 - p\!\!\!/_3 + m)\epsilon\!\!\!/_3^*\,u_1\right]\sum_{k,n} a_{3(n)}^* a_{4(k)}^* [\,c_2^\dagger \lambda_{(k)}\lambda_{(n)}c_1], \quad (31.29a)$$

$$\mathcal{M}_u = \frac{-g_S^2}{8p_1 \cdot p_4}\left[\bar{v}_2\,\epsilon\!\!\!/_3^*(p\!\!\!/_1 - p\!\!\!/_4 + m)\epsilon\!\!\!/_4^*\,u_1\right]\sum_{k,n} a_{3(n)}^* a_{4(k)}^* [\,c_2^\dagger \lambda_{(n)}\lambda_{(k)}c_1]. \quad (31.29b)$$

We will assume that the quarks are initially essentially at rest in the charmonium ground state, and that the gluons come out along the z axis. Then

$$p_1 = p_2 = \begin{bmatrix} m \\ 0 \\ 0 \\ 0 \end{bmatrix}, \qquad p_3 = \begin{bmatrix} m \\ 0 \\ 0 \\ m \end{bmatrix}, \qquad p_4 = \begin{bmatrix} m \\ 0 \\ 0 \\ -m \end{bmatrix}. \quad (31.30)$$

Note that the relation $\gamma^\mu\gamma^\nu + \gamma^\nu\gamma^\mu = 2g^{\mu\nu}I$ implies that

$$a_\mu\gamma^\mu b_\nu\gamma^\nu + b_\nu\gamma^\nu a_\mu\gamma^\mu = a\!\!\!/\,b\!\!\!/ + b\!\!\!/\,a\!\!\!/ = 2a_\mu b_\nu g^{\mu\nu}I = 2(a \cdot b)I. \quad (31.31)$$

Let's apply this in this case to $p\!\!\!/_1$ and $\epsilon\!\!\!/_3^*$. If the c quark is initially essentially at rest in the η_c, then its four-momentum p_1 has only a time component, while in Coulomb gauge ϵ_3^* has only spatial components. Therefore, $p_1 \cdot \epsilon_3^* = 0$. We also know that $p_3 \cdot \epsilon_3^* = p_4 \cdot \epsilon_4^* = 0$ because the gluon's polarization, like the photon's, must be transverse to its direction of motion. This means that

$$p\!\!\!/_1\epsilon\!\!\!/_3^* = -\epsilon\!\!\!/_3^* p\!\!\!/_1 \quad \text{and} \quad p\!\!\!/_3\epsilon\!\!\!/_3^* = -\epsilon\!\!\!/_3^* p\!\!\!/_3 \quad \text{and} \quad p\!\!\!/_4\epsilon\!\!\!/_4^* = -\epsilon\!\!\!/_4^* p\!\!\!/_4. \quad (31.32)$$

We also know that $(p\!\!\!/_1 - m)u_1 = 0$ because $\psi = u_1 e^{-ip\cdot x}$ satisfies the Dirac equation $(i\gamma^\mu\partial_\mu - m)\psi = 0$.

Exercise 31.1.1 Show that under these conditions, $\mathcal{M} = \mathcal{M}_t + \mathcal{M}_u$ is

$$\mathcal{M} = \frac{-g_S^2}{8m^2}\sum_{k,n} a_{4(k)}^* a_{3(n)}^* \bar{v}_2\big[\,\epsilon\!\!\!/_4^*\epsilon\!\!\!/_3^* p\!\!\!/_3(\,c_2^\dagger\lambda_{(k)}\lambda_{(n)}c_1)$$

$$+ \epsilon\!\!\!/_3^*\epsilon\!\!\!/_4^* p\!\!\!/_4(c_2^\dagger\lambda_{(n)}\lambda_{(k)}c_1)\,\big]u_1. \quad (31.33)$$

BOX 31.2 The Spin Part of the η_c Decay Amplitude

In this box, we want to calculate $\mathcal{M}_0$ given that

$$\mathcal{M} = \frac{-g_S^2}{8m^2} \sum_{k,n} a_{4(k)}^* a_{3(n)}^* \bar{v}_2 \Big[\epsilon_4^* \epsilon_3^* \!\!\not{p}_3 (c_2^\dagger \lambda_{(k)} \lambda_{(n)} c_1)$$

$$+ \epsilon_3^* \epsilon_4^* \!\!\not{p}_4 (c_2^\dagger \lambda_{(n)} \lambda_{(k)} c_1) \Big] u_1, \qquad (31.34)$$

assuming a total spin-0 initial state, so that

$$\mathcal{M}_0 = \sqrt{\tfrac{1}{2}} [\, \mathcal{M}_{\uparrow\downarrow} - \mathcal{M}_{\downarrow\uparrow} \,], \qquad (31.35)$$

where the arrows refer to the spin states of the c and $\bar{c}$ quarks, respectively.

If the system's total spin is zero before the decay, then it must be after the decay as well, so we are looking for the amplitude to decay to a photon state of total spin 0. The polarization vectors for spin in the $\pm z$ direction turn out to be $(\epsilon^\pm)^\mu = (0, a, \pm ia, 0)$, where $a = \sqrt{\tfrac{1}{2}}$ for a normalized state. The polarization state for a photon with zero total spin will therefore be

$$\epsilon_3^\mu \epsilon_4^\nu = \sqrt{\tfrac{1}{2}} \Big[(\epsilon^+)^\mu (\epsilon^-)^\nu - (\epsilon^-)^\mu (\epsilon^+)^\nu \Big]. \qquad (31.36)$$

Lowering the indices will change the signs of the spatial components, and taking the complex conjugate will change the sign of the imaginary term, so

$$\epsilon_{3,\mu}^* \epsilon_{4,\nu}^* = \sqrt{\tfrac{1}{2}} \Big[(\epsilon^+)_\mu^* (\epsilon^-)_\nu^* - (\epsilon^-)_\mu^* (\epsilon^+)_\nu^* \Big], \qquad (31.37)$$

where $(\epsilon^\pm)_\mu^* = (0, -a, \pm ia, 0)$. Now note that $(\not{\epsilon}^\pm)^* = -a\gamma^x \pm ia\gamma^y$, so

$$\not{\epsilon}_3^* \not{\epsilon}_4^* = \sqrt{\tfrac{1}{2}} a^2 \Big[(-\gamma^x + i\gamma^y)(-\gamma^x - i\gamma^y) - (-\gamma^x - i\gamma^y)(-\gamma^x + i\gamma^y) \Big]. \ (31.38)$$

Exercise 31.2.1 Multiply this out, simplify using $\gamma^\mu \gamma^\nu + \gamma^\nu \gamma^\mu = 2g^{\mu\nu} I$ and $a^2 = \tfrac{1}{2}$, show that $\not{\epsilon}_4^* \not{\epsilon}_3^* = -\not{\epsilon}_3^* \not{\epsilon}_4^*$, and finally show that

$$\not{\epsilon}_4^* \not{\epsilon}_3^* \!\!\not{p}_3 = i\sqrt{2}\, m\gamma^x \gamma^y (\gamma^t + \gamma^z), \qquad \not{\epsilon}_3^* \not{\epsilon}_4^* \!\!\not{p}_4 = i\sqrt{2}\, m\gamma^x \gamma^y (-\gamma^t + \gamma^z). \quad (31.39)$$

BOX 31.2   The Spin Part of the η_c Decay Amplitude (cont.)

Note that $\bar{v}_2\gamma^x\gamma^y(\pm\gamma^t + \gamma^z)u_1 = v_2^\dagger\gamma^t\gamma^x\gamma^y(\pm\gamma^t + \gamma^z)u_1$. Now,

$$\gamma^t\gamma^x\gamma^y\gamma^t = -\gamma^t\gamma^x\gamma^t\gamma^y = +\gamma^t\gamma^t\gamma^x\gamma^y = \gamma^x\gamma^y \tag{31.40}$$

because γ^t anticommutes with spatial gammas and $\gamma^t\gamma^t = I$. Also, note that $i\gamma^t\gamma^x\gamma^y\gamma^z \equiv \gamma^5$. This means that

$$i\sqrt{2}\,mv_2^\dagger\gamma^t\gamma^x\gamma^y(\pm\gamma^t + \gamma^z)u_1 = \pm i\sqrt{2}\,mv_2^\dagger\gamma^x\gamma^y u_1 + \sqrt{2}\,mv_2^\dagger\gamma^5 u_1. \tag{31.41}$$

We want to evaluate this for cases where the c and $\bar{c}$ have opposite spins. Because we are assuming that quarks are at rest, their spin states are

$$u^+ = \sqrt{m}\begin{bmatrix}1\\0\\1\\0\end{bmatrix}, \quad u^- = \sqrt{m}\begin{bmatrix}0\\1\\0\\1\end{bmatrix}, \quad v^+ = \sqrt{m}\begin{bmatrix}0\\1\\0\\-1\end{bmatrix}, \quad v^- = \sqrt{m}\begin{bmatrix}-1\\0\\1\\0\end{bmatrix}. \tag{31.42}$$

Exercise 31.2.2 Evaluate the four quantities $iv^{+\dagger}\gamma^x\gamma^y u_-$, $iv^{-\dagger}\gamma^x\gamma^y u_+$, $v^{+\dagger}\gamma^5 u_-$, and $v^{-\dagger}\gamma^5 u_+$. (*Hint:* Evaluate the 4×4 matrix for $i\gamma^x\gamma^y$ first.)

BOX 31.3 Evaluating $\mathrm{Tr}\{\lambda_{(k)}, \lambda_{(n)}\}$

First, note that for *any* pair of matrices A and B, $\mathrm{Tr}[AB] = \mathrm{Tr}[BA]$ by Theorem 9 in Chapter 20. This, in combination with Theorem 7, implies that

$$\mathrm{Tr}\{A, B\} = \mathrm{Tr}[AB + BA] = 2\,\mathrm{Tr}[AB]. \tag{31.44}$$

Now, Gell-Mann matrices are

$$\lambda_1 = \begin{pmatrix} 0 & 1 & 0 \\ 1 & 0 & 0 \\ 0 & 0 & 0 \end{pmatrix}, \qquad \lambda_2 = \begin{pmatrix} 0 & -i & 0 \\ i & 0 & 0 \\ 0 & 0 & 0 \end{pmatrix}, \qquad \lambda_3 = \begin{pmatrix} 1 & 0 & 0 \\ 0 & -1 & 0 \\ 0 & 0 & 0 \end{pmatrix},$$

$$\lambda_4 = \begin{pmatrix} 0 & 0 & 1 \\ 0 & 0 & 0 \\ 1 & 0 & 0 \end{pmatrix}, \qquad \lambda_5 = \begin{pmatrix} 0 & 0 & -i \\ 0 & 0 & 0 \\ i & 0 & 0 \end{pmatrix}, \tag{31.45}$$

$$\lambda_6 = \begin{pmatrix} 0 & 0 & 0 \\ 0 & 0 & 1 \\ 0 & 1 & 0 \end{pmatrix}, \qquad \lambda_7 = \begin{pmatrix} 0 & 0 & 0 \\ 0 & 0 & -i \\ 0 & i & 0 \end{pmatrix}, \qquad \lambda_8 = \frac{1}{\sqrt{3}}\begin{pmatrix} 1 & 0 & 0 \\ 0 & 1 & 0 \\ 0 & 0 & -2 \end{pmatrix}.$$

Gell-Mann actually *defined* these matrices so that $\mathrm{Tr}[\lambda_{(k)}\lambda_{(n)}] = 2\delta_{(kn)}$, as it makes a number of expressions simpler. But one must compute 35 matrix products to verify that he did it right (see Problem P31.4), so let's just check (to pick an arbitrary pair) that this works for λ_4 and λ_7.

Exercise 31.3.1   Verify that $\mathrm{Tr}[\lambda_4\lambda_4] = \mathrm{Tr}[\lambda_7\lambda_7] = 2$ and that $\mathrm{Tr}[\lambda_4\lambda_7] = 0$. Also, test another arbitrary pair of your choice.

BOX 31.4 The Decay Rate for $\eta_c \to \gamma + \gamma$

Applying the Feynman rules for electrodynamics to the diagrams in Figure 31.3 (but with the gluon lines changed to photon lines) yields

$$-i\mathcal{M}_t = \bar{v}_2 \big[-i\tfrac{2}{3}e\gamma^\mu \big] \epsilon^*_{4,\mu} \frac{i(\slashed{q} + m)}{q^2 - m^2} \big[-i\tfrac{2}{3}e\gamma^\nu \big] \epsilon^*_{3,\mu} u_1$$

$$\Rightarrow \quad \mathcal{M}_t = \frac{-2e^2}{9 p_1 \cdot p_3} \big[\bar{v}_2 \slashed{\epsilon}_4 (\slashed{p}_1 - \slashed{p}_3 + m) \slashed{\epsilon}_3 u_1 \big], \tag{31.46}$$

where I have used $q^2 - m^2 = (p_1 - p_3)^2 - m^2 = -2 p_1 \cdot p_3$, and in the vertex factors, I have used the fact that charge of the c quark is $-\tfrac{2}{3}$ that of an electron. The expression for $\mathcal{M}_u$ is the same except $3 \leftrightarrow 4$.

If we compare this with Equation 31.4, repeated here,

$$\mathcal{M}_t = \frac{-g_S^2}{8 p_1 \cdot p_3} \big[\bar{v}_2 \slashed{\epsilon}_4^* (\slashed{p}_1 - \slashed{p}_3 + m) \slashed{\epsilon}_3^* u_1 \big] \sum_{k,n} a^*_{4(k)} a^*_{3(n)} \big[c_2^\dagger \lambda_{(k)} \lambda_{(n)} c_1 \big], \tag{31.47}$$

we see that the expressions would be the same if we were to replace g_S in the latter with $4e/3$ and simply set all the color stuff to 1.

Exercise 31.4.1 With this in mind, adjust the remainder of the derivation leading to Equation 31.17 to find the decay rate of the η_c to two photons. (*Hint:* Be especially careful around Equation 31.9.)

BOX 31.5 The Cross Section for Higgs Production

Fermi's Golden Rule (see Chapter 10 or 14) for two input particles and one output particle is

$$\sigma = \frac{S}{4\sqrt{(p_1 \cdot p_2)^2 - (m_1 m_2)^2}} \int \frac{d^3 p_3}{(2\pi)^3 2E_3} (2\pi)^4 \delta^4(p_1 + p_2 - p_3)|\mathcal{M}|^2. \quad (31.48)$$

In this case, particles 1 and 2 are gluons, and particle 3 is the Higgs. We have estimated the amplitude in the center-of-mass frame to be roughly

$$\mathcal{M} \sim \frac{\alpha_S g_W}{32\pi} \frac{m_t}{M_W} m_H. \quad (31.49)$$

Exercise 31.5.1 The final Higgs cross section will have the form $\sigma = \sigma_0 \delta(E - m_H)$, where $E = E_1 + E_2$ is the collision energy (the delta enforces energy conservation). Calculate σ_0 both symbolically and numerically, using $\alpha_S \approx 0.13$ and $g_W \approx 0.65$ at the energy in question. (Note: You will find that σ_0 has units of GeV^{-1}, which might be surprising since the cross section should have units of area or GeV^{-2}. This is because the Dirac delta carries units of GeV^{-1} [as $\int \delta(E - m_H)dE = 1$], so σ will indeed have units of GeV^{-2}.)

Homework Problems

P31.1 (This problem is the same as Problem 27.3.) Before anyone knew about the c quark, the simple Cabibbo angle transformation given in Equation 27.8 worked very well to predict a number of useful decay rates. But it had a stark failure: it predicted that K^0 could decay to $\mu^+ + \mu^-$, a decay mode not seen in nature at anywhere near the rate predicted.

a. Draw the lowest-order Feynman diagram for this decay process. (*Hint:* The diagram will have four vertices, and its central part will look like a box. Be sure to conserve charge at the vertices.)

b. In 1970, Glashow, Iliopoulos, and Maiani (GIM) proposed a solution to this problem. Suppose (they suggested) that in addition to the u, d, and s quarks, a fourth quark (now called the c) existed. If this were true, a second diagram (this time with a virtual c instead of a virtual u) would also exist. Draw this diagram.

c. Argue that the amplitudes from these diagrams should nearly cancel, explaining the almost vanishing decay rate observed. (*Hint:* Equations 27.27 and 27.20 state that each quark vertex will be weighted by an appropriate component of V_{jk} or V_{jk}^*. If we ignore the third quark family, V is just the matrix appearing in Equation 27.8. Use this to write out the Cabibbo-angle factor associated with each vertex. If the mass of the u was the same as that of the c, all other factors in the amplitude would be the same and the cancellation would be exact. Even though the c is significantly more massive than the u, these virtual particles are so far off their mass shells that the masses don't matter much, so the amplitudes nearly cancel.)

Inventing a whole new quark just to resolve an esoteric theoretical issue was a bold prediction at the time, but within a few years people had found more direct experimental evidence for the c quark.

P31.2 In this problem, we will investigate the potential energy formula given in Equation 31.1.

a. The charmonium states corresponding to the particles η_c (2.984 GeV) and J/ψ (3.098 GeV) both involve systems of $c\bar{c}$ quarks in the same orbital energy level $n = 1$, but the energy of the system is lower when the quark spins are anti-aligned rather than aligned (as would be the case with ordinary magnets). Classically, this spin-spin interaction energy would be proportional to $\vec{S}_1 \cdot \vec{S}_2$, and in quantum mechanics, the value of that dot product is $+\frac{1}{4}$ when the spins are aligned and $-\frac{3}{4}$ when they are anti-aligned. Using this information, find the value of the charmonium $n = 1$ orbital energy level *excluding* the spin-spin interaction.

b. Do the same for the $c\bar{c}$ systems in the $n = 2$ energy level, given that the η_c' has a mass of 3.638 GeV and the ψ' has a mass of 3.686 GeV. (Note that the difference in mass is smaller, because in the $n = 2$ level, the quarks are farther apart.)

c. The charmonium states corresponding to systems in the $n = 3$ orbital energy level are only quasi-bound, but the state with total spin 1 seems to have an energy of 4.040 GeV. Given this, and that the spin energy will be smaller yet, make a guess about the energy of the $n = 3$ orbital state for charmonium excluding the spin effect.

d. The web app SchroSolver numerically solves the non-relativistic Schrödinger equation for selected potential energy functions. Run it by entering pages.pomona.edu/~tmoore/QuarkSchroSolver/ in your browser, choose the "charmonium" option from the "Configuration" menu to choose a potential energy of the form given in Equation 31.1, and experiment with the values of m, α_S, and F_0 to find the best fit to the $n = 1, 2, 3$ energy levels. (You may find that an "effective mass" that is a bit higher than the c quark's nominal mass of 1.275 GeV works a bit better in this crude model.)

e. Also, try this with the bottomonium system $b\bar{b}$, where the spin-0 and spin-1 states for the $n = 1$ orbital energy level have masses of 9.399 GeV and 9.460 GeV, respectively, the corresponding values for the $n = 2$ states are 9.999 GeV and 10.023 GeV, and for $n = 3$ the spin-1 state has a mass of 10.355 GeV. The nominal mass of the bottom quark is 4.18 GeV. Again, you may find that a higher effective mass works better.

P31.3 The fact that three gluons can meet in a vertex means that an s-channel diagram exists for the decay of the η_c to two gluons.

a. Draw this diagram.

b. Why is this diagram *not* possible for the decay of the η_c in particular? (List two reasons.)

P31.4 Consider exhaustively verifying that Gell-Mann correctly chose the λ matrices so that $\text{Tr}[\lambda_{(k)}\lambda_{(n)}] = 2\delta_{(kn)}$.

a. How do we know that we must verify 36 matrix products to determine that this is correct? (Why not 64?)

b. Use Mathematica or some other mathematical tool or programming language to calculate all 36 (or 64) products, and so verify the theorem above. Submit your code and a printout of your results.

P31.5 Rework the proof for the decay amplitude $\eta_c \rightarrow g + g$ to calculate the amplitude for the decay of particle with initial spin 1 (like the J/ψ): $\mathcal{M}_1 = \sqrt{\frac{1}{2}}\,[\mathcal{M}_{\uparrow\downarrow} + \mathcal{M}_{\downarrow\uparrow}]$. Show that in fact $\mathcal{M}_1 = 0$, verifying that the J/ψ cannot decay to two gluons.

P31.6 Our goal in this problem is to determine the color singlet state for two gluons, that is, the linear combination of paired gluon states $\sum A_{jk}|j\rangle|k\rangle$ (where $j, k = 1, 2, \ldots, 8$) that is invariant under an SU(3) rotation.

Gluons carry a combination of color and anticolor. To see how such a combination transforms, define

$$c \equiv \begin{bmatrix} r \\ g \\ b \end{bmatrix}, \quad \bar{c} \equiv \begin{bmatrix} \bar{r} \\ \bar{g} \\ \bar{b} \end{bmatrix}, \quad M = c\bar{c}^{\dagger} = \begin{pmatrix} r\bar{r} & r\bar{g} & r\bar{b} \\ g\bar{r} & g\bar{g} & g\bar{b} \\ b\bar{r} & b\bar{g} & b\bar{b} \end{pmatrix}.$$
$$(31.50)$$

The value of writing these color-anticolor combinations in the form of a matrix is that because we know how c and $\bar{c}$ transform under an SU(3) rotation, we can figure out how M must transform. Let $c' = Uc$, where U is an SU(3) rotation matrix; then $\bar{c}' = U\bar{c}$ also (since colors and anticolors rotate the same way). This means that

$$M' = c'(\bar{c}')^{\dagger} = (Uc)(U\bar{c})^{\dagger} = Uc\bar{c}^{\dagger}U^{\dagger} = UMU^{\dagger}. \ (31.51)$$

The only problem is that this matrix represents nine color-anticolor combinations but there are only eight gluons. The extra color-anticolor combination is the color-singlet combination $r\bar{r} + g\bar{g} + b\bar{b}$, which does not correspond to a gluon. We can get rid of this combination by subtracting one-third of the matrix's trace from its diagonal elements:

$$G \equiv M - \tfrac{1}{3}\,\mathrm{Tr}[M]\,I. \qquad (31.52)$$

Note that the transformation of G is given by

$$G' = M' - \tfrac{1}{3}\,\mathrm{Tr}[M']I = UMU^{\dagger} - \tfrac{1}{3}\,\mathrm{Tr}[UMU^{\dagger}]I$$
$$= UGU^{\dagger}. \qquad (31.53)$$

a. Fill in the missing steps leading to the last equality. (*Hints:* Note that $\mathrm{Tr}[ABC] = \mathrm{Tr}[CAB]$ and that $U^{\dagger}U = UU^{\dagger} = I$ because the SU(3) transformation matrix U is unitary.)

The traceless matrix G therefore represents the set of color-anticolor combinations that might appear in a single gluon, written in a matrix form that we know how to transform. The question now is to see how we might find a sum of these combinations for a *pair* of gluons that is unchanged by a color rotation (the definition of a color singlet). The quantity $s = \mathrm{Tr}[G_1 G_2]$ turns out to fit the bill because

$$s' = \mathrm{Tr}[G_1'G_2'] = \mathrm{Tr}[UG_1U^{\dagger}UG_2U^{\dagger}] = \mathrm{Tr}[UG_1G_2U^{\dagger}]$$
$$= \mathrm{Tr}[U^{\dagger}UG_1G_2] = \mathrm{Tr}[G_1G_2] = s. \qquad (31.54)$$

b. Show that $s = \mathrm{Tr}[M_1M_2] - \tfrac{1}{3}\,\mathrm{Tr}[M_1]\,\mathrm{Tr}[M_2]$.

c. Show that, as a sum of color-anticolor combinations,

$$s = \tfrac{2}{3}\Big[(r\bar{r})_1(r\bar{r})_2 + (g\bar{g})_1(g\bar{g})_2 + (b\bar{b})_1(b\bar{b})_2\Big]$$

$$- \tfrac{1}{3}\Big[(r\bar{r})_1(g\bar{g})_2 + (r\bar{r})_1(b\bar{b})_2 + (g\bar{g})_1(r\bar{r})_2$$

$$+ (g\bar{g})_1(b\bar{b})_2 + (b\bar{b})_1(r\bar{r})_2 + (b\bar{b})_1(g\bar{g})_2\Big]$$

$$+ (r\bar{g})_1(g\bar{r})_2 + (r\bar{b})_1(b\bar{r})_2 + (g\bar{r})_1(r\bar{g})_2$$

$$+ (g\bar{b})_1(b\bar{g})_2 + (b\bar{r})_1(r\bar{b})_2 + (b\bar{g})_1(g\bar{b})_2. \ (31.55)$$

d. The gluon color states are the same as the eight color-anticolor combinations listed in Equation 30.18, repeated here:

$$|1\rangle \equiv \sqrt{\tfrac{1}{2}}|r\bar{g}\rangle + \sqrt{\tfrac{1}{2}}|g\bar{r}\rangle, \quad |2\rangle \equiv \sqrt{\tfrac{1}{2}}|r\bar{g}\rangle - \sqrt{\tfrac{1}{2}}|g\bar{r}\rangle,$$

$$|3\rangle \equiv \sqrt{\tfrac{1}{2}}|r\bar{r}\rangle - \sqrt{\tfrac{1}{2}}|g\bar{g}\rangle, \quad |4\rangle \equiv \sqrt{\tfrac{1}{2}}|r\bar{b}\rangle + \sqrt{\tfrac{1}{2}}|b\bar{r}\rangle,$$

$$|5\rangle \equiv \sqrt{\tfrac{1}{2}}|r\bar{b}\rangle - \sqrt{\tfrac{1}{2}}|b\bar{r}\rangle, \quad |6\rangle \equiv \sqrt{\tfrac{1}{2}}|g\bar{b}\rangle + \sqrt{\tfrac{1}{2}}|b\bar{g}\rangle,$$

$$|7\rangle \equiv \sqrt{\tfrac{1}{2}}|g\bar{b}\rangle - \sqrt{\tfrac{1}{2}}|b\bar{g}\rangle,$$

$$|8\rangle \equiv \sqrt{\tfrac{1}{6}}\big(|r\bar{r}\rangle + |g\bar{g}\rangle - 2|b\bar{b}\rangle\big). \qquad (31.56)$$

So, for example, the two-gluon color state $|1\rangle|1\rangle$ is

$$|1\rangle|1\rangle = \tfrac{1}{2}\Big[|r\bar{g}\rangle + |g\bar{r}\rangle\Big]_1\Big[|r\bar{g}\rangle + |g\bar{r}\rangle\Big]_2$$

$$= \tfrac{1}{2}\Big[|r\bar{g}\rangle_1|r\bar{g}\rangle_2 + |r\bar{g}\rangle_1|g\bar{r}\rangle_2$$

$$+ |g\bar{r}\rangle_1|r\bar{g}\rangle_2 + |g\bar{r}\rangle_1|g\bar{r}\rangle_2\Big]. \qquad (31.57)$$

By expanding the other gluon kets similarly, show that

$$\sqrt{8}|s\rangle \equiv |1\rangle|1\rangle + |2\rangle|2\rangle + |3\rangle|3\rangle + |4\rangle|4\rangle$$
$$+ |5\rangle|5\rangle + |6\rangle|6\rangle + |7\rangle|7\rangle + |8\rangle|8\rangle \quad (31.58)$$

is the same set of color-anticolor combinations as in Equation 31.55 (with $|\ \rangle_{1,2} \to (\ )_{1,2}$). So, if s in Equation 31.55 is color-invariant, then $|s\rangle$ must be as well. (Dividing Equation 31.58 by $\sqrt{8}$ ensures that the $|s\rangle$ state is normalized.)

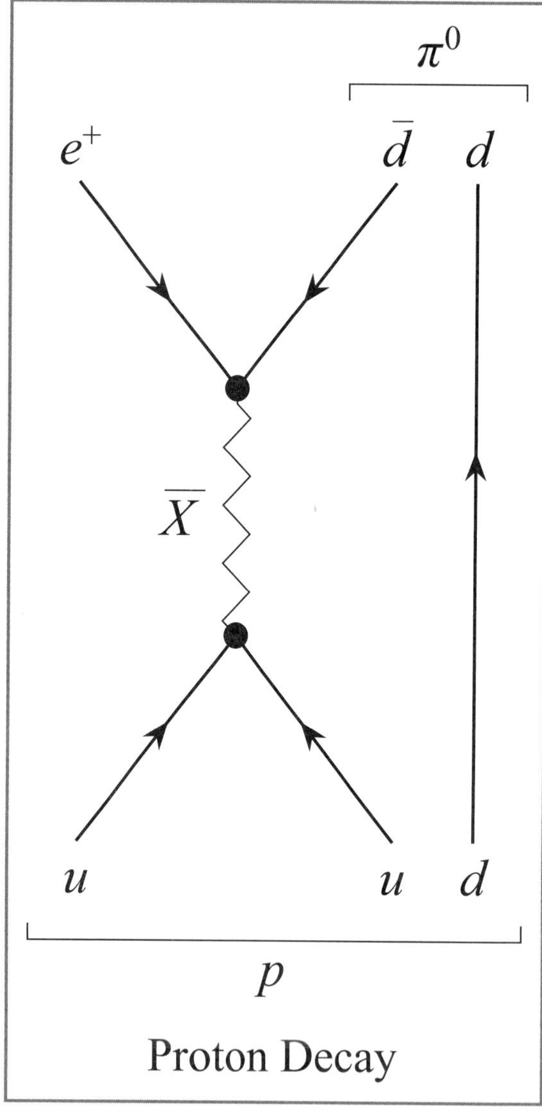

π⁰
e⁺
d̄
d
X̄
u
u
d
p
Proton Decay

32. BEYOND THE STANDARD MODEL

As successful as the Standard Model has been, we know that as a physical theory it is incomplete and presents some glaring problems that need resolution. This final chapter will discuss some of those problems and explore some of the various proposals for theories that might enable us to move beyond the Standard Model to something more complete.

The Standard Model exhibits a number of puzzles that mostly exercise theorists (who like their theories to be tidy, unambiguous, self-consistent, and complete), but there are a few experimental puzzles that may point to physics beyond the Standard Model. This section will explore some of the problems of both types.

This chapter has three major parts. Section 32.1 reviews some of the theoretical problems with the Standard Model. Section 32.2 explores some of the experimental results that seem to contradict the Standard Model. Unlike previous chapters in this book, these sections will have subsections that discuss each of the various issues in each section in turn. I think that this organization will help make the structure of these sections clearer. Sections 32.3 through 32.6 will discuss four different approaches to moving beyond the Standard Model: the **SU(5) Grand Unified Theory**, **Supersymmetry**, **Axions**, and **String Theory**. Each section will discuss not only the benefits that each theory offers but also the reasons that that theory has not yet replaced the Standard Model.

The chapter closes with some personal reflections on the beauties of particle physics, looking forward to new discoveries.

32.1 Theoretical Issues with the Standard Model

The major theoretical issues with the Standard Model include the following:

- Normal ordering and renormalization
- Too many free parameters
- Neutrino masses
- The Higgs mass problem
- The asymmetry between matter and antimatter
- Dark matter and dark energy
- Gravity

In this section, we will discuss each issue in turn.

32.1.1 Normal Ordering and Renormalization

Chapter 12 discusses the "normal ordering" problem in some depth. To briefly summarize, the issue is that in quantum field theory a particle is an excitation of an *operator* field. We can calculate a field's energy and charge density the same way we did with classical scalar fields, but to get physically meaningful results, we must put the field operators in a different order than a natural calculation seems to deliver. But since the operators don't commute, putting them in the right order seems to deliver infinite results, which we ignore.

As discussed in Chapter 12, most physicists do not consider this to be a real problem but rather an artifact of moving from a theory where the field values commute to one where they don't. My purpose in bringing it up here is to say that not all theorists agree that brushing aside this issue makes sense. They would say that it is a bit too convenient to ignore commutation issues when they don't deliver physically meaningful results and yet depend on them in other contexts to predict other (experimentally verified!) results. My point is that this is an open issue.

Chapter 21 discusses some of the issues regarding renormalization and offers a seemingly successful resolution. Imposing a simple finite upper limit on integrals over the momentum in loops is simple conceptually, has a decent physical interpretation (we have an upper limit because the theory breaks down at high energies), and nicely displays the basic logic of renormalization (this is why I presented it in Chapter 21). My purpose here is to admit that this method can't be right, because it is consistent with neither Lorentz symmetry or gauge symmetry. People have proposed various alternative renormalization schemes that preserve both, but most schemes lack a clear physical motivation other than providing a way to get the right answer. Each scheme also has its strengths and weaknesses. The method that seems to handle the issue in a way that preserves Lorentz and gauge invariance and works for all interactions is **lattice gauge theory**, which imagines spacetime to be divided up into discrete points. However, this approach does not lend itself to analytic solutions but rather requires numerical solutions that are extremely computationally intensive. If this is really the fundamental solution to the renormalization problem, then we could consider quantum field theory to be a continuum limit of this approach. But why is spacetime divided up into discrete points? How many points per unit volume and why? My point is that the full story behind how and why renormalization works still needs to be told.

32.1.2 Too Many Free Parameters

The Standard Model involves a number of quantities whose values are not determined by the theory but can be determined only by experiment. Exactly how many depends on what you think is part of the Standard Model and what is not, but a reasonable list includes these 19 parameters:

- nine quark and lepton masses
- one color interaction coupling constant (g_S)
- two electroweak parameters (e, θ_W)
- two Higgs mechanism parameters (m_H, v)
- four free parameters in the CKM matrix (see Chapter 27)
- one color CP violation angle θ_{QCD} (more about this later).

(Note that $9 + 1 + 2 + 2 + 4 + 1 = 19$.) In addition, we have seven parameters associated with neutrino masses (one can debate whether these should be counted as part of the Standard Model, but they certainly are related):

- three neutrino masses
- four free parameters in the neutrino PMNS matrix (see Chapter 28).

Finally, there are four numbers that are essential as background assumptions in the Standard Model:

- the number of spacetime dimensions
- the number of quark and lepton families
- the number of fundamental symmetries
- the quark-electron charge ratio.

This is a total of 30 parameters.

Ideally, one might argue that a theory ought to determine all or at least some of these numbers, particularly simple quantities like the number of quark and lepton families (and why these are the same), or the quark-electron charge ratio. The fact that neutrino masses are *almost* zero (compared to other particle masses) but not *exactly* zero also begs for some kind of explanation that the Standard Model cannot provide.

Some of these numbers may be constrained or even predicted using what is known as the **anthropic principle**: if the numbers were significantly different, then the universe could not support life, so we would not be here to observe it. I am going to talk about this more later, but for now let me simply say that this is precisely why having all these loose parameters bothers people: we want to understand *why* our universe happens to be one that supports life.

However, one could take a more positive perspective on these numbers. Physical theories to date have *always* involved quantities that we must measure, and most involve many more experimentally determined numbers (for example, tables of atomic masses). *In principle* (according to the Standard Model) we can understand *everything* about the quantum behavior of all ordinary matter using only 30 parameters! How great is that!

32.1.3 The Neutrino Mass Problem

As discussed in Chapter 28, we know from experiment that neutrinos have tiny but definitely nonzero masses. Nonzero neutrino masses in the Standard Model would require that chirally right-handed neutrinos exist. But *how* can they exist? Neutrinos are uncharged and have no color, so they cannot be created by color or electromagnetic interactions. But a chirally right-handed neutrino also cannot be created by the weak interaction, since that interaction only connects chirally left-handed particles. So how could we create *any* chirally right-handed neutrinos out of the energy of the Big Bang?

People have thought of various mechanisms for creating neutrino mass terms (one of which was presented in Chapter 28). But such schemes are technically outside the Standard Model. According to the Standard Model, chirally right-handed neutrinos should not exist, and neutrinos should be massless. The fact that they are not massless means that the Standard Model is not complete.

This puzzle is actually part of a larger problem that physicists call the **Hierarchy Problem**. A number of unitless ratios in the theory (for example, $m_e/m_t \sim 10^{-6}$) seem very different than zero. In most physical theories, unitless numbers that appear in the various equations are usually of order 1, and that is true for *most* unitless numbers in the Standard Model, but not all. Why are some unitless numbers so very different?

32.1.4 Higgs Mass Issues

In Chapter 21 on renormalization, we saw that bubble diagrams along particle lines mean that a particle's experimental mass is not equal to its true or "bare" mass. For all particles other than the Higgs boson, such bubbles produce corrections that are proportional to the logarithm of the cutoff energy used as the high-energy limit in the integrals over particle loops. This means that even if the cutoff energy is very high (on the order of the Planck energy), the difference between a particle's bare and experimental mass is not that large.

However, when computing the same corrections to the Higgs mass, it turns out that calculations like those we did in Chapter 21 show that boson loops produce corrections to the Higgs mass that are of the same order as the cutoff energy. If we take the cutoff to be the Planck mass (1.2×10^{19} GeV), then the correction to the Higgs' bare mass is 10^{17} times larger than the Higgs' experimental mass (126 GeV). This means that the corrections must almost entirely cancel the Higgs' bare mass to 17 decimal places, but not quite. Even worse, the M_W mass is linked to the Higg's bare mass, as we have seen, so M_W should be huge, and it is not. The only way to avoid this in the Standard Model is to "fine-tune" the divergent terms by hand so that they cancel at each order of the perturbation. But this seems quite unsatisfactory.

Along with the Higgs mass, there is the problem with the Higgs potential energy. In Chapter 26, we just picked the simplest potential energy formula that had a nonzero vacuum expectation value. It is quite challenging to verify experimentally that we have chosen the correct potential energy formula. But even if we did, where does this potential energy formula come from? Why does the Higgs field have this peculiar potential energy form while other fields have simpler forms?

32.1.5 Matter-Antimatter Asymmetry

The universe is full of matter but hardly any antimatter. In 1967, Andrei Sakharov argued that three conditions must be met at some point soon after the Big Bang for *any* theory to produce an excess of baryons over anti-baryons:

- Quark number conservation must be violated.
- C symmetry and CP symmetry must be violated.
- The interactions must take place out of thermal equilibrium.

The argument for these conditions is basically as follows. Your theory must allow you to create more quarks than antiquarks and thus violate conservation of quark number. Charge conjugation (C) symmetry must also be broken, because otherwise even processes that create or destroy quarks would create or destroy equally many antiquarks. We must also violate T (time-order) symmetry, because the future must have more quarks in it than the past, but if T symmetry is maintained, then the future cannot be fundamentally different than the past. But CPT symmetry must be preserved by all known quantum field theories (including those in the Standard Model). So, in addition to breaking the C symmetry, our theory must also break the CP symmetry so that the T symmetry can be broken. Finally, thermal equilibrium is essentially defined as the state where nothing changes, so if we want something fundamental like the number of quarks in the universe to change, the universe must not be in thermal equilibrium.

The Standard Model has no mechanism for quark number violation at the Feynman diagram level, but in 1976, Gerard 't Hooft proposed a mechanism within the Standard Model. This original proposal proved unworkable, but further developments of the idea have proved more credible. The problem is that the strength of the effect cannot be derived analytically; one must use lattice gauge theory to make estimates. (The mechanism only operates at extremely high temperatures, so it also cannot be tested experimentally.) While the present estimates are rough, and further work is necessary, this mechanism *might* work, but definitive proof remains elusive.

The Standard Model does exhibit C and CP violations arising from the nonzero phase in the CKM matrix (see Chapter 27). However, the current consensus in the community is that the amount of violation is not sufficient (by many orders of magnitude) to create the observed matter-antimatter asymmetry during the Big Bang by itself.

The Standard Model does predict that at certain times during the first few moments of expansion (for example, at the time of electroweak symmetry breaking, which occurs at about $t \sim 10^{-12}$ s), the universal fluid is out of thermal equilibrium. So, if sufficient quark creation with sufficient C and CP violation could occur during one of these transitions, the matter-antimatter asymmetry might be explained. The insufficient amount of C and CP violation in the Standard Model currently appears to be the issue that requires physics beyond the Standard Model to resolve.

Another possible method (see Shuve and Yavin, *Phys Rev D* **89**, 075014, 2014) requiring only minimal changes to the Standard Model is to generate quarks through their connections to leptons. A right-handed neutrino and neutrino oscillations both make this possible, but some very fine tuning is necessary.

32.1.6 Dark Matter and Dark Energy

Cosmological observations have firmly established the existence of what physicists call *dark matter* and *dark energy*. **Dark matter** is an unknown kind of matter that interacts with ordinary matter gravitationally but only very weakly (and perhaps not at all) otherwise. The existence of dark matter has been established from a variety of observations (beginning in 1932) regarding the orbital speeds of stars in galaxies, the orbital speeds of galaxies in clusters, dark matter's imprint on the cosmic microwave background, as well as gravitational lensing studies. In all cases, the evidence indicates that galaxies contain more matter (on the order of 5 to 10 times as much matter) than can possibly be consistent with the amount of matter in observed stars and gas clouds.

One of the most vivid illustrations of dark matter and its nature is the composite image shown in Figure 32.1. This image shows two clusters of galaxies that have moved though each other, basically horizontally. The stars in the galaxies passed through the collision unscathed, but the gas in the clusters interacted significantly, and because of friction slowed down and remained roughly at the collision point. This gas radiated some of the energy from its collision as X-rays that were observed by the Chandra X-Ray Observatory. In the image, these gas regions appear as two glowing lobes (colored pink in the original) near the image center. By observing the gravitational lensing caused by the gravitational fields of the clusters that passed through each other, one could infer the location and the amount of gravitating mass associated with the clusters, which was indicated in blue on the original image and indicated in Figure 32.1 by the glowing regions circled in white, roughly centered on the remains of the clusters. The image makes it clear that the dark matter also passed through the collision (meaning that it is not ordinary gas), and the amount of lensing indicates that it is much greater than the mass in stars, particularly now that the gas has been stripped away. This indicates that the dark matter, though dispersed like a gas surrounding the cluster, is not ordinary gas but at most weakly interacts both with ordinary matter and with itself. It also illustrates that the dark matter is "cold" in the sense that its particles are not sufficiently energetic to escape the gravitational field of a galaxy or galactic cluster, which in turn means that its particles must be non-relativistic.

FIGURE 32.1   This is a composite image of the Bullet Cluster. The regions occupied by dark matter are circled in white. (You can view this in color at apod.nasa.gov/apod/image/0608/ bulletcluster_comp_f2048.jpg.) (Credit: X-ray: NASA/CXC/ Markevitch et al., Optical: NASA/STScI; Magellan/U.Arizona/Clowe et al., Lensing Map: NASA/STScI; ESO WFI; Magellan/U.Arizona/Clowe et al.)

The cosmic microwave data imply that, in round numbers, about 27% of the energy in the universe is dark matter and only a bit less than 5% is the kind of ordinary matter that is the subject of the Standard Model. Dark matter cannot be ordinary Standard Model matter in some hidden form: cosmological measurements put strong upper limits on the total density of Standard Model matter in the universe (including neutrinos and/or matter trapped in black holes). The Standard Model simply does not have a place for dark matter.

Similarly, cosmological measurements imply that almost 69% of the universe's energy density is in the form of what is called **dark energy**. This form of energy is very strange in that the pressure it exerts is *negative* and apparently equal in magnitude to its energy density, unlike any ordinary matter or energy that we know of (including a photon gas, whose pressure is positive and equal to only one-third the gas's energy density in natural units). In the gravitational field equations, this large negative pressure has the effect of overwhelming the attractive gravitational field associated with the energy density and creating a gravitational field that has a net repulsive effect. This manifests itself in an increase in the universe's expansion rate over time.

General relativity does allow for a "vacuum energy" term with this character to appear as an intrinsic part of the gravitational field equations, and the concept of "vacuum energy" could have a quantum explanation associated with the zero-point energy of a quantum field. But when one tries to crunch the numbers (see Problem P32.1), one gets a number that is 120 orders of magnitude too large to be consistent with observations.

The fact that the Standard Model has no place for either dark matter or dark energy means that some completely different and still unknown physics must describe more than 95% of the energy in the universe.

32.1.7 Gravity

This brings us to the subject of gravity. General relativity has its own issues (mostly having to do with black holes) that point toward the need for a quantum theory of gravity. However, the Standard Model has no place for gravity. I mentioned in Chapter 21 that a straightforward quantum field theory for gravity cannot be normalized, so gravity is not likely to be included in any future theory that is based on our current understanding of quantum field theory.

The development of a successful quantum theory of gravity is one of the most important unresolved problems to be addressed by physics beyond the Standard Model. People have worked hard for decades on possible theories, but no one has yet made any experimentally verified predictions contradicting general relativity. Observations of gravitational waves from colliding black holes have also failed to exhibit any deviations from the predictions of general relativity, even though these observations are the first to test the theory in the strong-field limit, so we have no guidance from nature either.

32.2 Experimental Problems

Given the long list of theoretical problems I have just described (which, in combination, indicate a critical need for physics beyond the Standard Model), it is ironic that one of the difficulties moving forward from the Standard Model is precisely its success. People had hoped that the LHC would provide some solid experimental evidence for physics beyond the Standard Model, but it has, at best, only produced tantalizing hints of measurements that are maybe a tiny bit different from what the Standard Model maybe predicts. This is complicated by the fact that what the Standard Model exactly predicts is tough to know: the calculations are difficult, and some predictions involve using numerical results from lattice gauge theory that are computationally difficult and intrinsically uncertain (because the lattices cannot be made very fine), and yet it is difficult to determine how uncertain these calculations are.

That being said, three recent bits of evidence announced in 2021 and 2022 provide *possible* indications of physics beyond the Standard Model.

32.2.1 Muon $g - 2$ Measurements

Physicists have been measuring the muon's gyromagnetic ratio g for decades, hoping to either vindicate or contradict the Standard Model. The gyromagnetic ratio gives the constant of proportionality that indicates the strength of a spinning particle's magnetic moment $\vec{\mu}$ (as indicated by its response to a known magnetic field), and the particle's spin angular momentum $\vec{S}$:

$$\vec{\mu} = g\frac{q}{2m}\vec{S}, \tag{32.1}$$

where m is the muon's mass and $q = -e$ is its charge. The Dirac equation predicts that $g = 2$, but higher-order diagrams in the Standard Model predict a slight increase in this value. This difference is usually expressed in terms of what physicists call the **gyromagnetic anomaly** a_μ:

$$a_\mu \equiv \tfrac{1}{2}(g - 2). \tag{32.2}$$

In April of 2022, researchers at Fermilab announced a new measurement of a_μ that was 4.2 standard deviations away from the theoretical prediction for that number. According to their measurements (see B. Abi et al., *Phys Rev Lett* **126**, 141801, 2021):

- experimental result: $a_\mu = 0.001\ 165\ 920\ 61(41)$,
- theoretical prediction: $a_\mu = 0.001\ 165\ 918\ 10(43)$,

where the numbers in parentheses are the uncertainties in the last two digits. The difference is not quite at the 5σ level conventionally required for a "discovery" but is almost there.

This announcement received an enormous amount of attention in the media, which enthusiastically presented it as clear experimental evidence of physics beyond the Standard Model. But some physicists remain skeptical. A number of past near-discoveries have evaporated under greater scrutiny (one of the most notorious recent cases was the claim that neutrinos were observed moving faster than the speed of light, a result that turned out to be due to a cable being disconnected). In this particular case, most media stories didn't quite mention that there are at least two groups working on the theoretical estimate from the Standard Model (T. Aoyama, et al., *Phys Rept* **887**, 1, 2020, and S. Borsanyi, et al., *Nature* **593**, 51–55, 2021) that disagree on methods. The Fermilab paper used the first result, but the second group's result, which is much closer to the experimental value, was not addressed. It therefore appears that there remains some work to do to clarify what "the Standard Model prediction" actually is.

Others have noted that even if the result is true, this does not entirely upend the Standard Model: consistency with experiment may require merely the existence of one more charged particle. So the import of this result remains to be seen as of this writing.

32.2.2 B-Meson Decay

Also in 2021, physicists using the LHCb detector announced an apparent discrepancy in the ratio R_K of the decay rate of $B^0 \to K^0 + \mu^- + \mu^+$ to that of $B^0 \to K^0 + e^- + e^+$. Other than small corrections at low energy due to the difference between the electron and muon mass, the Standard Model predicts $R_K = 1$ (all leptons should behave the same). The LHCb collaboration reported $R_K = 0.846^{+0.044}_{-0.041}$, which is about 3σ away from the expected result (LHCb Collaboration, *Nature Physics,* **18**, 277–282, 2022).

This result, unlike the muon g–2 experiment, is a clean test of a basic principle that does not involve any complicated calculations. However, it is noteworthy that results from the Belle detector at the KEKB accelerator in Tsukuba, Japan, for the same reactions in the same energy range are completely consistent with the Standard Model: $R_K = 1.03^{+0.28}_{-0.24}$ (Belle Collaboration, *JHEP* **2021**, 105, 2021). The wide uncertainty range on the latter result means that the Belle and LHCb results are not inconsistent for that energy range, but the Belle results including data from all energy ranges examined is $R_K = 1.10^{+0.16}_{-0.15}$, which *is* inconsistent with the LHCb result (see Figure 32.2).

One of the reasons that the particle physics community repeats experiments using different detectors is to rule out possible detector-dependent systematic errors. The discrepant results between these detectors needs to be resolved and the error ranges reduced before we can be certain that a violation of the Standard Model actually exists. Experiments are currently in progress as of this writing that may help resolve these questions.

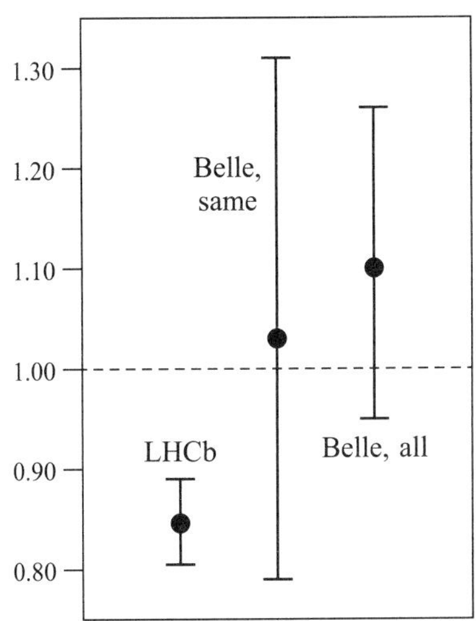

FIGURE 32.2   Values of R_K measured by different detectors. The "Belle, same" uncertainty bar is for the same energy range as measured by LHCb; the "Belle, all" is for all energy ranges measured by the Belle detector.

32.2.3 The Mass of the W

In April of 2022, physicists at Fermilab announced results from the CDF II detector at the Fermilab Tevatron accelerator regarding the mass of the W that were inconsistent with the Standard Model at the 7σ level. Their result (CDF Collaboration, *Science*, **376**, 6589, pp. 170–176, 2022) stated a value of 80.4335(94) GeV compared to a predicted Standard Model result of 80.357(6).

However, as Figure 32.3 illustrates, results from other detectors differ, and some are inconsistent with the new result (particularly the data from L3, ATLAS, and Fermilab's own D0 II detector). Again, the community needs to understand these discrepancies better.

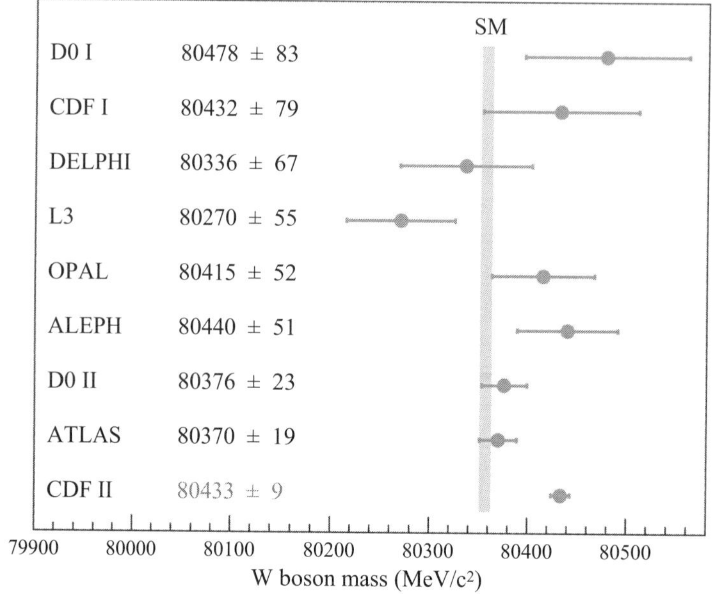

FIGURE 32.3   This diagram compares W mass results from various detectors. (Credit: CDF Collaboration, *Science*, **376**, 6589, pp. 170–176 (2022). Reprinted according to the terms of the CC BY 4.0 license.)

32.3 SU(5) GUT

We now turn to considering possible theories beyond the Standard Model that might help resolve some of the theoretical and experimental difficulties we have noted.

Perhaps the most straightforward extension of the Standard Model is the **SU(5) Grand Unified Theory**. Taking its inspiration from the successful electroweak unification discussed in Chapter 25, this theory seeks to unify the electroweak and color interactions into a single theory.

It turns out that the SU(5) group is the smallest group that contains SU(2) and SU(3) as subgroups and so preserves those symmetries. But before we dig into how the SU(5) group operates, let's review the number of fields we seek to explain. The basic structure will be simplest before symmetry breaking, so we will consider all the fields to be massless, which means that chirally left-handed fields are completely distinct from chirally right-handed fields. Also, we will only look at the lowest-mass quark and lepton families, as presumably all the families will be equivalent.

With these things in mind, we have eight left-handed fields to consider: ν_e, e_L^-, u_{rL}, u_{gL}, u_{bL}, d_{rL}, d_{gL}, d_{bL}. The lepton fields formed an $SU(2)_L$ doublet, as did the u_L and d_L fields. In addition, we have seven right-handed fields that do not participate in the $SU(2)_L$ symmetry: e_R^+, u_{rR}, u_{gR}, u_{bR}, d_{rR}, d_{gR}, d_{bR} (note that we do not have a right-handed neutrino). But the charge conjugates (antiparticles) of these right-handed fields are left-handed fields, so we can describe the whole set of fields in terms of the following 15 left-handed fields (now dropping the R, L subscript):

$$\text{basic fields} = \nu_e, e^-, u_r, u_g, u_b, d_r, d_g, d_b, e^+, \overline{u}_r, \overline{u}_g, \overline{u}_b, \overline{d}_r, \overline{d}_g, \overline{d}_b. \quad (32.3)$$

The reason I am focusing on left-handed fields is that, since the SU(2) symmetry is actually an $SU(2)_L$ symmetry, our SU(5) symmetry will actually have to be an $SU(5)_L$ symmetry to have $SU(2)_L$ as a subgroup.

Now, the basic representation of the $SU(5)_L$ group will be a set of 5×5 matrices that rotate some five-component column vector we will call Ψ. Which subset of the fields above can serve as the elements of this vector? The $SU(5)_L$ contains $SU(2)_L$ and $SU(3)_L$ as subgroups, so an $SU(2)_L$ rotation must mix two of the fields but leave the others alone. An $SU(3)_L$ rotation must mix the three remaining fields but leave the first two alone. Moreover, the total charge of the five fields must be zero for group theoretic reasons having to do with the fact that the generators of the $SU(5)_L$ must be traceless. The only five fields that fit all these restrictions are $v_e, e^-, \overline{d}_r, \overline{d}_g, \overline{d}_b$. The basic $SU(5)_L$ transformation therefore looks like

$$
U\Psi =
\left(
\begin{array}{cc|ccc}
SU(2) & SU(2) & ? & ? & ? \\
SU(2) & SU(2) & ? & ? & ? \\
\hline
? & ? & SU(3) & SU(3) & SU(3) \\
? & ? & SU(3) & SU(3) & SU(3) \\
? & ? & SU(3) & SU(3) & SU(3)
\end{array}
\right)
\begin{bmatrix}
v_e \\ e^- \\ \overline{d}_r \\ \overline{d}_g \\ \overline{d}_b
\end{bmatrix}.
\tag{32.4}
$$

The question marks indicate components in the transformation that are giving the physics beyond the $SU(2)_L$ and $SU(3)_L$ symmetries. The full $SU(5)$ transformation must have $5^2 - 1 = 24$ generators, and when we make the symmetry actively local, this will mean adding 24 gauge fields. We already know 12 of these fields: the photon, the Z, the W^+, the W^-, and the 8 gluon fields. The remaining 12 fields are new fields that mix leptons with quarks. Since they convert a $\overline{d}$ quark to an electron or a neutrino, they must carry either a charge of $+\frac{1}{3}$ (for a $\overline{d}$ going to a neutrino) or $+\frac{4}{3}$ (for a $\overline{d}$ going to an electron), as well as whatever the color of the $\overline{d}$ was to conserve color. We call the three gauge bosons with charge $+\frac{1}{3}$ and various colors "Y bosons," and those with charge $+\frac{4}{3}$ and various colors "X bosons." These, along with their antiparticles, comprise the 12 remaining gauge fields.

But what about the other 10 particle fields? How do they fit in? It turns out that we can arrange these fields into an antisymmetric matrix M that transforms under an $SU(5)_L$ rotation as

$$
M' = U^\dagger M U, \quad \text{where } M =
\begin{pmatrix}
0 & e^+ & d_r & d_g & d_b \\
-e^+ & 0 & u_r & u_g & u_b \\
-d_r & -u_r & 0 & \overline{u}_b & \overline{u}_g \\
-d_g & -u_g & -\overline{u}_b & 0 & \overline{u}_r \\
-d_b & -u_b & -\overline{u}_g & -\overline{u}_r & 0
\end{pmatrix}.
\tag{32.5}
$$

Why don't we see quarks morphing into electrons or vice versa? The $SU(5)_L$ symmetry is clearly a broken symmetry, since the particles are not massless. We would expect this symmmetry-breaking to occur in the same way as the $SU(2)_L$ symmetry-breaking, via some kind of Higgs mechanism. If the energy associated with this symmetry-breaking is large enough, then the masses of the X and Y bosons will be so large that the interactions they mediate would be strongly suppressed in the same way that weak interactions are suppressed at energies $E \ll M_W$. For us to see *no* interactions at experimentally accessible energies, these masses must be quite high.

The theory actually predicts an approximate energy for this symmetry-breaking, but that energy depends somewhat on the exact formulation of the Higgs-type mechanism used to break the symmetry. But we can get a basic sense of what the energy scale might be by looking at the energy-dependencies of the color α_S coupling constant and the electroweak coupling constant α. Remember that α gets larger as the interaction energy increases, whereas α_S gets smaller. We might expect that the symmetry between the color and electroweak interactions would be unbroken at roughly the energy scale where these coupling constants become equal. That turns out to be at about 10^{15} GeV, which is indeed roughly what we get from more complicated calculations. This is so high that processes that convert electrons to quarks or vice versa will indeed be highly suppressed.

What does this theory give us? For one thing, perhaps you noticed that it provides a natural explanation as to why the quark charges must be $-\frac{1}{3}$ and $+\frac{2}{3}$! We have seen that to cancel the negative charge of the electron, the charges of the three quarks in the basic

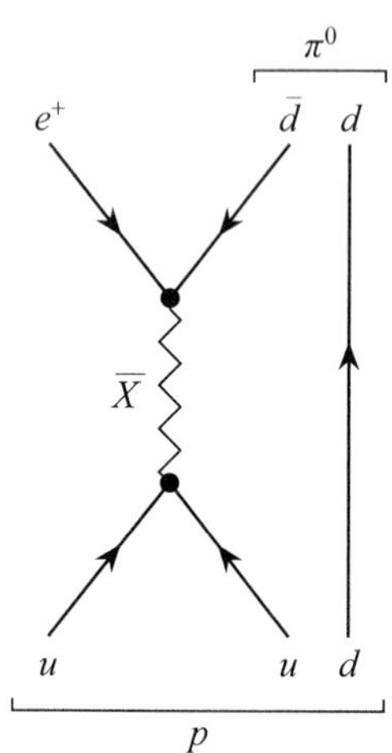

FIGURE 32.4 This is the Feynman diagram describing proton decay in the SU(5) GUT.

vector Ψ must add up to $+1$. Since we have three colors, and the charges of a given type of quark with different colors must be the same, the charge of each quark we are calling a $\bar{d}$ must be $+\frac{1}{3}$. Therefore, the d must have a charge of $-\frac{1}{3}$. Since the u is the upper component of the SU(2) doublet with the d, and since the upper component's charge is one elementary charge unit higher than that of the lower, the charge of the u must be $1 + (-\frac{1}{3}) = +\frac{2}{3}$.

Another nice thing about SU(5) theory is that there is only one coupling constant, which we can call g_5. This turns out to actually enable a *prediction* of the value of $\sin^2 \theta_W$. A simplistic calculation leads to $\sin^2 \theta_W = \frac{3}{8}$, which was in agreement with experimental results at the time it was first done in the mid-1970s. But as more data came in the experimental results moved downward, eventually settling on a value of about 0.23. On the theory side, people soon realized that one needed to include how the couplings depended on energy, and that revised the theoretical result downward as well. The current prediction from the most basic version of the SU(5) theory is that $\sin^2 \theta_W = 0.21$, which is a bit too low (well outside the uncertainty range of the measured value, unfortunately).

The most significant problem with the simplest version of the SU(5) GUT is its predictions regarding proton decay. In this theory, the proton can decay to $e^+ + \pi^0$ using the mechanism shown in Figure 32.4. The amplitude of this process would be proportional to g_5^2/M_X^2. The decay rate is proportional to $|\mathcal{M}|^2$ and thus to g_5^4/M_X^4 but must have units of energy. Since the electron and pion mass are small compared to the proton, the most relevant energy is the mass of the proton m_p, so the decay rate must be

$$\Gamma \sim \frac{g_5^4 m_p^5}{M_X^4}. \tag{32.6}$$

Note that this calculation is very sensitive to the value of M_X, but if we assume that $g_5 \sim 1$ and $M_X \sim 10^{15}\,\text{GeV}$, then the proton lifetime is about $10^{28}\,\text{y}$ (see Problem P32.2). A more accurate calculation that accounts for the probability of the up quarks being close enough to interact yields something more like $10^{31}\,\text{y}$.

This might seem impossible to measure, but one merely needs to gather enough protons in one place and watch them very carefully. A cube of water 10 m on a side contains about 3×10^{32} protons (see Problem P32.3). So, if you could watch that carefully for a year, you should see maybe 30 proton decays.

Many experiments have been done and others are currently in progress to detect proton decay. The proton lifetime for the particular SU(5) decay pictured is now known to be longer than $1.6 \times 10^{33}\,\text{y}$, which is widely considered to be inconsistent with the simplest versions of the SU(5) GUT.

Even if the proton decay prediction were to work out, the SU(5) GUT has other problems. One is that (like the Standard Model) it does not allow for a right-handed neutrino, which is necessary for neutrinos to have mass. It also does not resolve the Higgs mass problem, or explain the mass mixing matrices, and so on.

The various inadequacies of the SU(5) GUT has prompted people to look for alternatives in the same vein. One possibility that is still viable is an **SO(10) GUT** that arises when we make the SO(10) group a local symmetry. The basic representation of the SO(10) group is the set of matrices that rotate vectors in a 10-dimensional Euclidean space. It might seem that such a group would be ill-suited to handling rotations of complex-valued fields, but many groups have different representations that are mathematically equivalent. As a simple example, the U(1) group is mathematically equivalent to the SO(2) group that describes rotations in a two-dimensional Euclidean space.

One representation of this group rotates 16-component complex vectors into each other. A 16-component complex vector can represent the 15 fields we considered earlier with the addition of a right-handed neutrino, which allows neutrinos to have mass (and indeed, the 16th field in this theory has all the right properties to be that neutrino). This theory also has 45 total gauge bosons (23 new bosons in addition to the 12 we already know), so it has more flexibility than the SU(5) to fit experimental results. This theory also predicts proton decay but with a larger lifetime that is still consistent with measured limits.

Other possible GUT theories in the same spirit as the SU(5) include the so-called "flipped" SU(5) theory, SU(8) theory, and others. No theory of this type has yet won acceptance as a theory to replace the Standard Model.

32.4 Supersymmetry (SUSY)

A curious asymmetry exists in the Standard Model in that all the "basic" fields (leptons and quarks) are fermions, while all the bosons in the theory (except for the Higgs) arise from the fields that emerge from making symmetries actively local. Supersymmetric theories are based on the idea that every fermion should have a corresponding spin-0 boson with the same color and charge, and every boson should have a corresponding spin-$\frac{1}{2}$ fermion with the same color and charge. We call the supersymmetric partner to each known field its **superpartner**, and we indicate it by putting a tilde over the symbol for the corresponding normal field. The name for the superpartner corresponding to a given fermion is the same as that of the corresponding fermion with an "s" in front (to indicate a scalar boson). Therefore, the superpartner to the electron is the $\widetilde{e}$ "selectron" boson, and the name for the superpartner to the u quark is the $\widetilde{u}$ squark. The name for the fermion superpartner to a boson is the same as that of the boson plus a final "-ino". So the photon's superpartner is the $\widetilde{\gamma}$ "photino" and that for the W^+'s superpartner is the $\widetilde{W}^+$ "wino" (pronounced "weeno").

If this symmetry between a field and its superpartner were not broken somehow, then we would readily observe the consequences. In particular, chemistry would not exist, since any electron in an atom could decay to the atom's orbital ground state by emitting a photino and becoming a selectron. Since selectrons are bosons, *all* the atom's electrons would eventually decay this way and we would end up with atoms that are chemically the same but with different masses. There would be no molecules, no life, and no one to mourn the lack of interesting chemistry.

But we already know how symmetry breaking can give different masses to symmetric partners. To fit with experiment, symmetry breaking must give the known fields' supersymmetric partners masses high enough that they have not been observed so far. It also turns out that there is an interesting distinction between the particles we have already observed and those we have not. We have seen that all the particles we have observed (except for the Higgs) must have zero mass to satisfy the constraints of the U(1), SU(2)$_L$, and SU(3) symmetries. These particles therefore must gain the masses that we observe through the Higgs mechanism. But when one writes the Lagrangian density for a supersymmmetric theory, the superpartners of these known fields can have the usual kinds of mass terms in their Lagrangian densities without violating these symmetries. This provides a natural distinction between these two kinds of fields.

The Feynman rules in a supersymmetric version of the Standard Model would be the same as for the Standard Model, except that we add to each vertex new vertices and we replace the partners appearing in the original vertex by their superpartners in pairs, using the same coupling constants. See Figure 32.5 for vertex diagrams that would appear in addition to the basic weak vertex involving a W, an electron, and a neutrino. Note that the superpartners must come in pairs because otherwise angular momentum cannot be conserved (in Figure 32.5b, for example, the outgoing wino has spin $\frac{1}{2}$ and carries away the spin of the neutrino, so the other outgoing particle must be a selectron, not an electron).

These rules have three important consequences:

- Superpartners produced from normal particles will come in pairs.
- A superpartner will decay to a normal particle and another superpartner.
- The lowest-mass superpartner (LSP) must be stable.

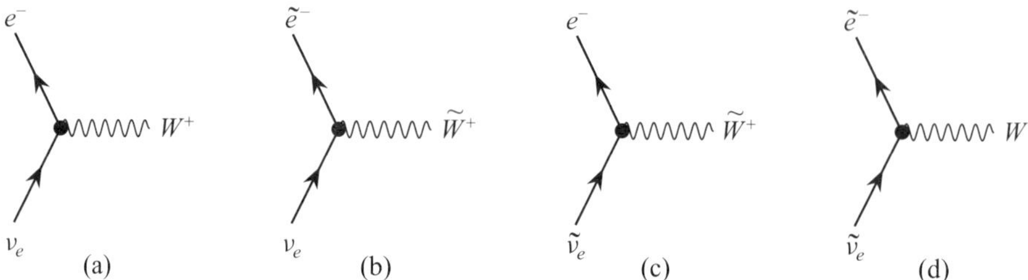

FIGURE 32.5   In a supersymmetric theory, the Standard Model vertex shown in diagram (a) would be supplemented by the vertices shown in diagrams (b)–(d).

The LSP could be any of the superpartners, though it is likely to correspond to one of the gauge bosons (or some linear combination of those fields). Say, for simplicity, that the LSP is the photino. It will interact with normal particles exactly as a photon would, except that its large mass would mean that the amplitude would be suppressed at low energies for the same reason that the large mass of the W suppresses the amplitude of weak interactions at low energies. Therefore, we would expect the photino to interact only weakly with ordinary matter, somewhat like a neutrino. (See Problem P32.4.)

Why would we be interested in SUSY? (1) Probably the most important reason is that it eliminates the Higgs mass problem. Because fermions and bosons come in pairs, the bubbles contributed by each almost exactly cancel, meaning that the Higgs bare and experimental masses will be close. (2) SUSY also provides a more natural explanation for the Higgs potential, and eliminates some subtle problems with the Higgs potential formula used in the Standard Model. (3) The LSP could explain dark matter, since it will be a massive particle that interacts only weakly with ordinary matter, and it should have been produced in abundance in the Big Bang. (4) Since quark number is not conserved at high energies, it might solve the matter-antimatter asymmetry problem. (5) It also provides a large mass scale that might support the seesaw mechanism for neutrino masses (see Chapter 28). These are terrific strengths!

An observation of even one of the superpartners would be very strong evidence for this theory. Moreover, the masses of the superpartners cannot be too much larger than the Higgs, or SUSY's solution to the Higgs mass problem falls apart. At high energies, the cross sections for producing superpartners should be reasonably large. One of the main strikes against the SUSY model is that none of the superpartners has yet been observed.

The SUSY model also introduces a large number of new parameters, including (but not limited to) the masses of the superpartners. This large parameter space means that all possible SUSY models are hard to explore. But current experimental lower limits on the possible superpartner masses mean that a number of the simplest SUSY models have already been eliminated.

However, we can apply the supersymmetry principle to other theories beyond the Standard Model. In fact, supersymmetry nicely complements some of theories we have already discussed. So we can construct a SUSY SU(5) theory, or a SUSY SO(10) theory, and so on.

Any SUSY model will also allow proton decay, but some versions predict a sufficiently long lifetime to be consistent with current observations. Currently, the most viable models with regard to the proton decay problem in particular are SO(10) (without SUSY), a SUSY version of SO(10), and a SUSY version of flipped SU(5).

(Anyone interested in learning more about SUSY as a beginner, though at a somewhat higher level than in this book, might look at Stephen P. Martin's "A Supersymmetry Primer," `arxiv.org/pdf/hep-ph/9709356.pdf`. One might also consult "Superspace or one thousand and one lessons in supersymmetry" by renowned MIT theorist S. James Gates Jr. [see Figure 32.6] et al., `arxiv.org/pdf/hep-th/0108200.pdf`.)

FIGURE 32.6   Theoretical physicist S. James Gates Jr. (Credit: John Consoli, Creative Director/Photography Director, University of Maryland, College Park, College Park, MD. Used with permission of Dr. Gates.)

32.5 The Strong CP Problem

It turns out that in the ordinary Standard Model, one is free to add a term to the color interaction Lagrangian density that is consistent with SU(3) symmetry and Lorentz invariance. Such a term would depend on an arbitrarily chosen parameter θ_{QCD} (see the list of parameters back in Section 31.1.2). This term looks like

$$\mathcal{L}_{CP} = \theta_{QCD} \frac{\alpha_S}{8\pi} \sum_j F^{(j)\mu\nu} \widetilde{F}^{(j)}_{\mu\nu}, \quad \text{where } \widetilde{F}^{(j)}_{\mu\nu} \equiv \tfrac{1}{2}\epsilon_{\mu\nu\alpha\beta} F^{(j)\alpha\beta} \qquad (32.7)$$

(the sum is over the eight gluon fields). Since nature seems to make mandatory anything that is mathematically possible, maybe we should include this term.

This term, however, makes the color interaction violate CP. To see how this works, note that the analogous electromagnetic term would be

$$\mathcal{L}_{CP,QED} = \theta_{QED} \frac{\alpha}{8\pi} \vec{E} \cdot \vec{B}. \qquad (32.8)$$

This term would violate U(1), so we can't add this to the Standard Model, but it will help us understand the behavior of the analogous color term. Under a parity transformation, $\vec{E} \to -\vec{E}$ and $\vec{B} \to \vec{B}$, so this term violates parity. However, if we change all positive charges to negative and vice versa, both $\vec{E}$ and $\vec{B}$ change sign, so the term is invariant under C. Therefore, this term will violate CP. The color term in Equation 32.7 will behave similarly.

Could this be the answer to the matter-antimatter asymmetry problem? Unfortunately, no. Such a term would cause the neutron to have an electric dipole moment, which is not observed. This puts a strong constraint on the magnitude of θ_{QCD}: $|\theta_{QCD}| < 10^{-10}$. But if this term is allowed, why is it so small? This is called the **Strong CP problem**.

In 1977, Helen Quinn (Figure 32.7) and Roberto Peccei proposed a solution. Suppose we add a new symmetry to our list: $\psi' = e^{-i\beta\gamma^5}\psi$. This would add a new term to the mass part of the Lagrangian density for any particle:

$$m\overline{\psi}\psi \to m\overline{\psi}\psi - 2i\beta\overline{\psi}\gamma^5\psi. \qquad (32.9)$$

This means that the Lagrangian density is not invariant under the transformation. But, as usual, we can cancel this change by introducing a new field called an **axion** field that has the effect of canceling the new mass term. They showed that it also canceled the θ_{QCD} term, explaining why it is so small.

This might seem to be an unnecessarily complicated way to deal with a term we did not need to include in the first place, but the axion particle turns out to have properties that would make it an excellent candidate to explain dark matter. Therefore, people have searched diligently for this particle for years.

Unfortunately, these searches have turned up nothing, and the limits that they have imposed have greatly reduced the parameter space of possible axion cross sections and masses, making this theory seem less and less likely.

FIGURE 32.7　Theoretical physicist Helen R. Quinn. Credit: AIP.

32.6 Superstring Theory

String theory represents a completely different approach to the particle physics problem. This theory (which people first began to explore in the late 1960s) imagines all fundamental particles to be either tiny strings and/or membranes vibrating in a 10- or 11-dimensional spacetime. All of the dimensions except for the four we observe are "compactified" so that they appear to us large-scale beings to be infinitesimal. The original versions of string theory only handled bosonic particles, but in the 1980s people discovered that combining string theory with supersymmetry eliminated a number of problems. The combination became known as **superstring theory**. But people also came to realize that there seemed to be five seemingly distinct versions of the theory. In 1995,

Edward Witten showed that the five versions were special limiting cases of a general 11-dimensional theory called **M-theory**. That particular superstring theory has been the focus of most of the research since.

A common way to visualize the compactification of dimensions is to imagine a two-dimensional space in the form of a tube that is very long compared to its diameter. At a large enough scale, this two-dimensional surface would look like a one-dimensional line. A string wrapped around this tube would like a point particle. A *vibrating* string of this type would have energy and therefore look (at a sufficiently large scale) like a particle with mass.

The main advantage of superstring theory is that it promises (unlike the other theories we have considered) to unify gravity with the other interactions and to provide a consistent approach to quantum gravity. It also seems to provide a coherent understanding of why the integrals we encountered in Chapter 21 are finite. The theory also is simpler and more elegant conceptually (though not mathematically) than many of the other theories.

However, superstring theory has many problems as a practical replacement for the Standard Model. The energy scale (on the order of the Planck mass) where superstring theory might be directly tested is completely inaccessible, both for current particle accelerators and for any detector that one might imagine constructing. This means practical applications of superstring theory to accessible physics involve taking a low-energy limit. The usual method for dealing with this kind of issue is to perturb the theory around some background that would apply in the zero-energy limit, and expand in a power series.

The problem is that the background (which would correspond to the "vacuum" state in quantum field theory) is not uniquely defined. Rather, one must solve equations of motion and select a solution that best approximates the vacuum with which we are familiar. The space of possible solutions is poorly understood at present, and no one knows how to find the right solution. A brute-force search is not possible (some have estimated the possible number of solutions as 10^{500}), but no technique has emerged just yet to better zero in on the useful solutions. (See `ncatlab.org/nlab/show/string+theory+FAQ` for a more in-depth discussion.)

Superstring theory also apparently has issues working with dark energy, which rules out a large number of potential versions of superstring theory. Viable solutions to this problem remain to be found.

Earlier in this chapter, I commented that the 30 or so free parameters associated with the Standard Model seemed a bit much. Superstring theory doesn't have parameters as such (and so promises to be a "final" theory in that sense), but unless some method of finding the right vacuum state for doing perturbation theory emerges, randomly sifting through 10^{500} vacuum solutions is impossible. Maybe superstring theory does offer a "final" theory of physics, but maybe we will never find out.

After more than 60 years of development, superstring theory is still not able to duplicate the successes of the Standard Model, or make a single new prediction that can be experimentally verified. The core distinction between a scientific theory and mere metaphysical speculation is whether the theory is testable and falsifiable. At present, one can argue that superstring theory is not testable or falsifiable, because although one can falsify a given version, there are always more versions to explore. Metaphysical concepts are a perfectly reasonable place to begin in formulating a new idea, but the goal is to produce a scientific theory. At what point does one give up pursuing a metaphysical idea that one cannot actually use to do physics?

32.7 An Unscientific Postscript

Our universe is wonderfully rich and complex. One of the fascinating features of the Standard Model is that our actual existence depends on many of its specific features, and also on the specific values of many of its 30 or so parameters. That some of these features or parameters are highly constrained by our existence is called the **anthropic principle**.

As an example, I mentioned in Chapter 21 that spacetime must have four dimensions or quantum electrodynamics is not renormalizable. (Four spacetime dimensions is also the minimum number of dimensions in which a gravitational field can exist.) If the weak interaction did not exist, there would be no way to convert protons to neutrons or vice versa, implying that a chemical element might have any number of neutrons, which would make chemistry much different and probably incompatible with life. If the color interaction parameter α_S were a bit smaller, fusion would be impossible and only hydrogen would exist. If α_S were somewhat larger (physicists don't agree on exactly how much of an increase is required, but less than 50%), no hydrogen would exist at all, because it would have all fused during the Big Bang.

A classic example that has been investigated at some depth has to do with the formation of carbon in stars. In the cores of massive stars, carbon is formed in a two-step process from helium: $^4\text{He} + {}^4\text{He} \to {}^8\text{Be}$ and $^8\text{Be} + {}^4\text{He} \to {}^{12}\text{C}$. However, this process is problematic because ^8Be is unstable, and so does not last very long. Astronomer Sir Fred Hoyle noted in the 1950s that not much carbon would be produced at all unless there were some kind of resonance in its nuclear energy levels that would enhance its production. On that basis, he *predicted* that an excited nuclear energy level for carbon would be found at an energy of 7.65 MeV above the ground state, at just the combined energy that the same nucleons in the $^8\text{Be} + {}^4\text{He}$ state have. This would provide an enhanced cross section that would allow the carbon nucleus to form and then settle into its ground state before flying apart again. This nuclear state was subsequently found. Moreover, there must *not* be a similar resonance for ^{16}O, or all the carbon thus formed would be converted into oxygen. It turns out that oxygen does have an energy level at 7.12 MeV, but this is far enough below the energy of the same set of nucleons in $^{12}\text{C} + {}^4\text{He}$ (7.19 MeV) that the formation of oxygen from carbon is *not* enhanced in the same way that carbon's formation from beryllium is. This ensures that such stars form roughly equal amounts of carbon and oxygen, a condition necessary for carbon-based life.

These energy levels are determined by the color and electromagnetic coupling constants α_S and α. If the values of these constants were even a few percent different, we would not exist. (My main source for this discussion is from McGrath, *A Fine-Tuned Universe,* WKJ Press, 2009, pp. 132–134.)

How one *interprets* this fine tuning is metaphysics, not physics. One might take this to be evidence of multiverses, God, some kind of evolution of universes, or even that we are living in a made-up simulation. Or perhaps we are just incredibly lucky.

But (and here is the unscientific part) the truth is that we are living in an amazingly lovely and variegated universe that did not have to be this way, and part of what lies behind its intricate beauty are the turning wheels of the Standard Model. I hope that your journey through this book has given you some perspective and appreciation of just how awesome the universe is, and enhance your joy at being alive on a beautiful planet in an exuberantly (and perhaps unnecessarily) rich and complex universe. May we all be grateful!

Homework Problems

P32.1 Find a combination of the constants G, $\hbar$, and c that in SI units has units of mass density. This naively should be the order of magnitude of a vacuum energy due to some kind of quantum zero-point energy. Calculate its magnitude and compare with the observed dark-energy density of roughly 7×10^{-27} kg/m^3.

P32.2 Calculate from Equation 32.6 the average life of the proton in years.

P32.3 Calculate the number of protons in a cube of water 10 m on a side. (*Hints:* The atomic mass of water is 18 g/mol, and a gram of water occupies about 1 cm^3 of volume.)

P32.4 Consider a photino scattering from an electron.
a. Draw the two Feynman diagrams for this process (there will be a t-channel and an s-channel diagram). Assign momenta so that the incoming photino has four-momentum p_1 and the outgoing electron has four-momentum p_3.
b. Draw the same diagrams for a photon scattering from an electron (Compton scattering).
c. Let's focus on the t-channel diagram. Applying the Feynman Rules to the photon scattering case yields

$$\mathcal{M}_\gamma = i\epsilon_{1\mu}\bar{u}_3(-ie\gamma^\mu)\frac{i(\not{q}+m)}{q^2-m^2}(-ie\gamma^\nu)u_2\epsilon^*_{4\nu}, \quad (32.10)$$

where m is the mass of an electron. Rather than do a complicated calculation, let's do an estimate. Let's ignore the gamma matrices and assume that the collision energy is high enough that $|q| \gg m$. Since the electron spinors are normalized to $\sqrt{m}$ and the photon spinors to 1, this amplitude will be of order

$$\mathcal{M}_\gamma \sim -e^2\sqrt{m}\frac{1}{|q|}\sqrt{m} = -e^2\frac{m}{|q|} \quad (32.11)$$

within some unitless factors involving spin averages. (The minus sign comes from the fact that $q^2 < 0$ for a t-channel diagram.) Note that this is unitless, as a $2 \to 2$ amplitude must be. Now let's do the same thing for the photino scattering case. Since the photino is a fermion, it will have a spinor whose magnitude will be of order $\sqrt{M}$, where M is the photino mass. Let's also assume that the selectron has a comparable mass ($M \sim$ hundreds of GeV), and note that the selectron is a scalar (spin-0) particle. The coupling constant in the vertex will still be e by the supersymmetry principle. Assuming that $M \gg |q|$, explain why the photino scattering amplitude is of order

$$\mathcal{M}_{\tilde{\gamma}} \sim -e^2\frac{m}{M}. \quad (32.12)$$

d. Compute the ratio in cross sections given these conditions and assuming that $|q| \approx 30$ MeV and $M \approx 300$ GeV.

Appendix A. POTENTIALS FROM INTERACTIONS

The purpose of this appendix is to illustrate how an interaction term in a Lagrangian density for a field can lead (in a classical context) to a potential energy of interaction between particles. We will do this in the context of the simple "toy universe" model discussed in Chapters 10–15, but one can easily generalize the approach to other kinds of fields.

A.1 Euler-Lagrange Equations for Our Theory

Our toy universe consists of phions, which are the excitations of the scalar field $\phi(x)$, and etaons, which are the excitations of the scalar field $\eta(x)$. Suppose that phions have mass m_ϕ and etaons have m_η, and that they interact in a way that is analogous to the way electrons interact with photons. The Lagrangian density for these fields is

$$\mathcal{L} = \tfrac{1}{2}\partial_\mu\phi\,\partial^\mu\phi - \tfrac{1}{2}m_\phi^2\phi^2 + \tfrac{1}{2}\partial_\mu\eta\,\partial^\mu\eta - \tfrac{1}{2}m_\eta^2\eta^2 - \lambda\phi^2\eta. \tag{A.1}$$

The interaction term is the final term, with λ specifying the strength of the interaction. Since scalar fields have units of energy, λ also has units of energy. The Euler-Lagrange equations for these fields are the coupled equations

$$\partial_\mu\partial^\mu\phi + m_\phi^2\phi = -2\lambda\eta\phi \quad\text{and}\quad \partial_\mu\partial^\mu\eta + m_\eta^2\eta = -\lambda\phi^2. \tag{A.2}$$

A.2 Lumps in the Phion Field

To consider how these equations might affect particles that have a clear location in space, we will define a "lump" in the ϕ-field such that

$$\phi^2 \propto \delta^3(\vec{r} - \vec{r}_0), \quad\text{meaning that } \phi \propto \sqrt{\delta^3(\vec{r} - \vec{r}_0)}. \tag{A.3}$$

This will play the role of a particle at position $\vec{r}_0$. We must be careful with units here, though. The units of $\delta^3(\vec{r} - \vec{r}_0)$ must be (energy)^3 because the integral of this delta function over all space must be 1 (which is unitless), and the volume element d^3x for this integral has units of (distance)$^3 = $ (energy)$^{-3}$. Therefore, let's define the constant of proportionality so that

$$\phi^2 = \frac{1}{m_\phi}\delta^3(\vec{r} - \vec{r}_0). \tag{A.4}$$

This will give our lump the right units.

In fact, let's consider two lumps: one at the origin and the other at $\vec{r}_2$, described by the expressions

$$\phi_1^2 = \frac{1}{m_\phi}\delta^3(\vec{r}) \quad\text{and}\quad \phi_2^2 = \frac{1}{m_\phi}\delta^3(\vec{r} - \vec{r}_2). \tag{A.5}$$

We will consider the phion at the origin to create an etaon field, and the other phion will respond to that field.

A.3 The Etaon Field

Consider the effect of the lump at the origin on the η field. According to that field's Euler-Lagrange equation, we have

$$\partial_\mu \partial^\mu \eta + m_\eta^2 \eta = -\frac{\lambda}{m_\phi} \delta^3(\vec{r}). \tag{A.6}$$

If the situation is essentially stationary, we can ignore the time derivatives, meaning that the Euler-Lagrange equation becomes

$$\left[-\vec{\nabla}^2 + m_\eta^2\right]\eta = -\frac{\lambda}{m_\phi}\delta^3(\vec{r}) \quad or \quad \left[\vec{\nabla}^2 - m_\eta^2\right]\eta = \frac{\lambda}{m_\phi}\delta^3(\vec{r}). \tag{A.7}$$

This becomes a static field equation for the η field. We can solve this equation by expressing $\eta(\vec{r})$ in momentum space (which amounts to doing a Fourier transformation):

$$\eta(\vec{r}) = \int \frac{d^3p}{(2\pi)^3} e^{i\vec{p}\cdot\vec{r}} \eta(\vec{p}). \tag{A.8}$$

Then note that

$$\left[\vec{\nabla}^2 - m_\eta^2\right]\eta = \int \frac{d^3p}{(2\pi)^3}\left[\vec{\nabla}^2 - m_\eta^2\right]e^{i\vec{p}\cdot\vec{r}}\eta(\vec{p})$$

$$= \int \frac{d^3p}{(2\pi)^3}\left[-|\vec{p}|^2 - m_\eta^2\right]e^{i\vec{p}\cdot\vec{r}}\eta(\vec{p}) \tag{A.9}$$

since ∂_x^2 will pull down two factors of ip_x from the exponent and the partials with respect to y and z will behave analogously. But one of the Dirac delta identities is

$$\int \frac{d^3p}{(2\pi)^3} e^{i\vec{p}\cdot\vec{r}} = \delta^3(\vec{r}). \tag{A.10}$$

So, if we substitute Equations A.9 and A.10 back into equation A.7, we find that

$$\int \frac{d^3p}{(2\pi)^3}\left[-|\vec{p}|^2 - m_\eta^2\right]e^{i\vec{p}\cdot\vec{r}}\eta(\vec{p}) = \frac{\lambda}{m_\phi}\int \frac{d^3p}{(2\pi)^3}e^{i\vec{p}\cdot\vec{r}}$$

$$0 = \int \frac{d^3p}{(2\pi)^3}\left[\left(-|\vec{p}|^2 - m_\eta^2\right)\eta(\vec{p}) - \frac{\lambda}{m_\phi}\right]e^{i\vec{p}\cdot\vec{r}}. \tag{A.11}$$

The only way that this can be true for all values of $\vec{r}$ is for the quantity in square brackets to vanish:

$$0 = \left(-|\vec{p}|^2 - m_\eta^2\right)\eta(\vec{p}) - \frac{\lambda}{m_\phi} \quad \Rightarrow \quad \eta(\vec{p}) = -\frac{\lambda}{m_\phi}\frac{1}{|\vec{p}|^2 + m_\eta^2}. \tag{A.12}$$

Now we can substitute this back into Equation A.8 to get

$$\eta(\vec{r}) = -\frac{\lambda}{m_\phi}\int \frac{d^3p}{(2\pi)^3}\frac{e^{i\vec{p}\cdot\vec{r}}}{|\vec{p}|^2 + m_\eta^2}. \tag{A.13}$$

We can write $d^3p = |\vec{p}|^2 \sin\theta\, d|\vec{p}|\, d\theta\, d\phi'$, where the θ specifies the angle between $\vec{p}$ and $\vec{r}$, and ϕ' specifies the angle of $\vec{p}$ around that axis. Since the integrand does not depend on ϕ', we can quickly integrate over that. If we define $\rho \equiv |\vec{p}|$ to save writing, the integrals quickly become

$$\eta(\vec{r}) = -\frac{\lambda}{m_\phi(2\pi)^3}2\pi \int \frac{\rho^2\, d\rho}{\rho^2 + m_\eta^2}\int_0^\pi \sin\theta\, d\theta\, e^{i\rho r \cos\theta}. \tag{A.14}$$

We can calculate the final integral easily by substitution of variables: since $u = \cos\theta$ and $du = -\sin\theta\, d\theta$, we have

$$\int_0^\pi \sin\theta \, d\theta \, e^{i\rho r \cos\theta} = + \int_{-1}^{+1} e^{i\rho r u} = \frac{1}{i\rho r}\left[e^{i\rho r} - e^{-i\rho r}\right]$$

$$= \frac{1}{i\rho r} 2i \sin \rho r = \frac{2}{\rho r}\sin \rho r. \tag{A.15}$$

Substituting this back into Equation A.14 yields

$$\eta(\vec{r}) = -\frac{\lambda}{m_\phi(2\pi)^2}\frac{2}{r}\int \frac{\rho \, d\rho}{\rho^2 + m_\eta^2}\sin \rho r. \tag{A.16}$$

This is a nastier integral, but WolframAlpha can handle it:

$$\eta(\vec{r}) = -\frac{\lambda}{m_\phi(2\pi)^2}\frac{2}{r}\left(\frac{\pi}{2}e^{-m_\eta r}\right) = -\frac{\lambda}{m_\phi}\frac{e^{-m_\eta r}}{4\pi r}. \tag{A.17}$$

So we see that our bump ϕ_1 creates an exponentially damped inverse-r η-field in a static situation.

A.4 The Interaction Potential Energy

We can find the energy associated with this field's interaction with the ϕ_2 bump by integrating over all space (excluding the origin) the interaction term in the Hamiltonian, which is $+\lambda\phi^2\eta$:

$$E_{\text{int}} = \int d^3x \, (+\lambda\phi^2\eta) = -\frac{\lambda^2}{m_\phi^2}\int d^3x \, \delta(\vec{r} - \vec{r}_2)\frac{e^{-m_\eta r}}{4\pi r}$$

$$= -\frac{\lambda^2}{m_\phi^2}\frac{e^{-m_\eta r_2}}{4\pi r_2}. \tag{A.18}$$

This is essentially the interaction potential energy for the interaction of two static phions through the etaon field. We can see that this is an attractive interaction, but the nonzero mass of the etaon means that the potential energy decreases exponentially with distance: if $m_\eta = 0$, the field would die off as $1/r$, just like the electrostatic potential energy.

Homework/Activity Problems

PA.1 Use WolframAlpha to check the step from Equation A.16 to Equation A.17.

PA.2 Using the methods of Chapter 6, check that the interaction portion of the energy density for the Lagrangian density in Equation A.1 is indeed $+\lambda\phi^2\eta$.

PA.3 Does changing the sign of the interaction term in the Lagrangian density make it a repulsive interaction?

PA.4 A simplified version of the electrodynamics interaction term (but still ignoring spin) would be $+e\psi^2 A$, where e is the fundamental charge, ψ is the electron field, and A is the photon field. Assume that ψ and A are still scalar fields but that ψ (like the real electron field) has units of $(\text{energy})^{3/2}$. Assume that A has units of energy and that its particles have zero mass. How would these modifications change the derivation?

Appendix B. SHORT ANSWERS TO SELECTED PROBLEMS

(Note that many other problems have answers provided in the problem text. Some of the answers below are only partial answers.)

Chapter 1.

1.2 Here is another example:

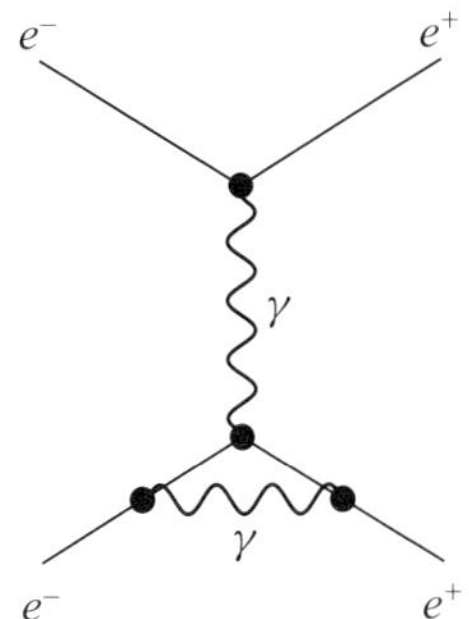

1.4 b. 0.388 GeV for the first case.

1.6 b. Conservation of energy would be violated.

1.8 a. $\nu_e + n \to p + e^-$.

Chapter 2.

2.1 d. $4.32 \times 10^{-18}\,\text{GeV}^4$.

2.2 f. $40/41$ in the $-x$ direction.

2.3 b. In the lab frame, the dot product is $2Em + m^2$.

2.4 e. 0.2714 in the $-x$ direction.

2.6 c. 0.757.

2.8 δ is of the order of magnitude of 10^{-8}.

2.10 c. In this case, x is indicating a coordinate axis (which is *kind* of like a vector component, I guess).

Chapter 3.

3.2 c. $2at + 2bx$.

3.5 One of the six equations is $E'_x = E_x$.

Chapter 4.

4.4 b. σ has units of energy.

4.5 d. Note that $k_\mu x^\mu = \pm \omega t - \omega z$ if $k_z \equiv \omega$.

4.6 e. $B = \frac{1}{32}A$. **f.** $D = 0$.

Chapter 5.

5.3 c. $R = \sqrt{5}, \theta = -26°$.

Chapter 6.

6.2 c. Repulsive.

6.6 c. $J^\mu = i\frac{1}{2}\partial^\mu \Phi^T X \Phi - i\frac{1}{2}\Phi^T X \partial^\mu \Phi$.

Chapter 7.

7.4 $\theta = 45°$ (as one might intuit, but be sure to argue that other possible extrema are smaller).

7.10 b. $p_x = k\hbar$.

Chapter 8.

8.2 $e^{-iS_x\theta} = \begin{pmatrix} \cos\frac{1}{2}\theta & -i\sin\frac{1}{2}\theta \\ -i\sin\frac{1}{2}\theta & \cos\frac{1}{2}\theta \end{pmatrix}$.

8.7 c. $\omega = eB_z/m$.

Chapter 9.

9.6 The answer to the final question should be yes.

9.7 One of the eigenvectors is $\sqrt{\frac{1}{2}} \begin{bmatrix} 1 \\ 0 \\ -1 \end{bmatrix}$.

Chapter 10.

10.1 b. 4.56×10^{-23} s.

10.3 b. 2×10^{10}/s.

10.4 $\theta = \frac{1}{2}\tan^{-1}\left(\frac{2\lambda}{m_\phi^2 - m_\eta^2}\right)$.

10.5 d. Decay to two etaons is *not* possible.

10.6 c. Three-vertex diagrams are *not* possible.

Chapter 11.

11.2 $\langle 0|a_k a_k a_k^\dagger a_k^\dagger|0\rangle = 2$.

Chapter 12.

12.3 e. $\epsilon \cdot p = 0$.

Chapter 13.

13.1 $\theta = \frac{1}{2}\tan^{-1}\left(\frac{2\lambda}{m_\phi^2 - m_\eta^2}\right)$.

Chapter 14.

14.1 b. Unitless.

14.2 $t_{1/2} = 4.56 \times 10^{-23}$ s.

14.5 e. $\sigma = \pi R^2$.

14.7 b. 2×10^{10} /s

Chapter 15.

15.1 $\tau = 1.10 \times 10^{-17}$ s.

15.6 e. $\sigma = 9.7 \times 10^{-14}$ barn $= 97$ fb.

Chapter 16.

16.4 d. $\overline{\psi}^{+} = \begin{bmatrix} 0 \\ \sqrt{m/3} \\ 0 \\ \sqrt{3m} \end{bmatrix} e^{ip \cdot x}, \overline{\psi}^{-} = \begin{bmatrix} \sqrt{3m} \\ 0 \\ -\sqrt{m/3} \\ 0 \end{bmatrix} e^{ip \cdot x}.$

Chapter 17.

17.4 c. $\overline{\psi} \gamma^{\mu} \gamma^{5} \psi$ is a pseudo-vector.

Chapter 18.

18.4 b. $J^{\mu} = \overline{\psi} \gamma^{\mu} \gamma^{5} \psi.$

Chapter 19.

19.4 b. $E_{z} = i M^{2} e^{-ip \cdot x}.$

19.6 b. $\vec{E} \times \vec{B} = 2\omega^{2} A^{2} \cos^{2}(\omega t - \omega z)$ in the $+z$ direction.

Chapter 20.

20.2 The analogous theorem to Theorem 14 is

$$\text{Tr}(\not{a}\not{b}\not{c}\not{d}) = 4(a \cdot b)(c \cdot d) - 4(a \cdot c)(b \cdot d) + 4(a \cdot d)(b \cdot c).$$

20.5 b. The result is the same as for part a.

20.7 f. $d\sigma = \dfrac{e^{4}}{32\pi^{2}E^{2}} \left[\dfrac{4 + (1 + \cos^{2}\theta)^{2}}{(1 - \cos\theta)^{2}} \right].$

Chapter 21.

21.4 $\alpha_{R}([10 \text{ GeV}]^{2}) = 1/135.1.$

Chapter 23.

23.6 b. (*Two* particles will replace the question mark.)

Chapter 24.

24.2 b. lifetime for this set of decay products $= 1.633 \times 10^{-12}$ s.

24.3 $\frac{7}{20} m_{\mu}.$

24.4 b. $1.283 \times 10^{-4}.$

24.5 c. $\theta < 97$ mrad.

24.6 $\sigma = \dfrac{E_{1}^{2}}{8\pi} \left(\dfrac{g_{W}}{M_{W}} \right)^{4} \left[1 - \left(\dfrac{m_{\mu}}{2E_{1}} \right)^{2} \right]$ in the CM frame, where E_{1} is the neutrino energy.

Chapter 25.

25.6 b. Ratio is 1.235.

Chapter 26.

26.2 b. $\sin^2 \theta_W = 0.2230$, a bit different than the "on-shell" value of 0.2239 because of complications having to do with exactly how one should define this value when considering renormalization.

26.3 c. $\Gamma_Z / \Gamma_\gamma \approx 4.2 \times 10^{-12}$.

26.4 Lifetime for decay to Ws would be about 6.4×10^{-25} s in this scenario.

Chapter 27.

27.6 c. $\sin \theta_C = 0.22 \pm 0.04$.

Chapter 28.

28.1 c. About 10,000 metric tons.

28.4 $m_3 \approx 0.0299$ eV.

Chapter 29.

29.1 b. ~ 35 proton-widths.

29.5 $\sigma_H / \sigma_Z \sim 1.7 \times 10^{-11}$. This becomes a lot larger when the collision involves quarks instead of electrons.

Chapter 30.

30.2 b. $f = -\frac{1}{6}$.

30.3 b. $\Lambda = 0.045$ GeV.

Chapter 31.

31.2 b. $E_2 = 3.674$ GeV.

Chapter 32.

32.2 b. $3.0 \times 10^{28} y$.

32.4 d. $\sim 10^{-8}$.

Appendix A.

A.3 No.

CREDITS

Frontispiece Chapter 3. Albert Einstein writing on blackboard in Pasadena (1931).
Associated Press

Frontispiece/Figure 6.1. Amalie "Emmy" Noether (1882–1935), c. 1910.
Image courtesy of Professor Emiliana P. Noether, Agnes Scott College, Decatur, Georgia

Figure 23.2. Yang Chen-Ning, one of the creators of Yang-Mills theory.
The Nobel Foundation.

Figure 24.6. Chien-Shiung Wu in her laboratory.
Image Courtesy of Smithsonian Institution Archives. Accession 90-105, Science Service
Records, Image No. SIA2010-1509.

Figures 31.6 and 31.7. The Higgs resonance.
These figures were published in Physics Letters B, 716; G. Aad, T. Abajyan, B. Abbott, J.
Abdallah, S. Abdel Khalek, A.A. Abdelalim, O. Abdinov, R. Aben, B. Abi, M. Abolins,
O.S. AbouZeid, H. Abramowicz, H. Abreu, B.S. Acharya, L. Adamczyk, D.L. Adams,
T.N. Addy, J. Adelman, S. Adomeit, P. Adragna, T. Adye, S. Aefsky et al.; "Observation
of a New Particle in the Search for the Standard Model Higgs Boson with the ATLAS
Detector at the LHC"; pp. 1–29, Copyright Elsevier.

Figure 32.1. A composite image of the Bullet Cluster.
Credit: X-ray: NASA/CXC/Markevitch et al.; Optical: NASA/STScI, Magellan/
University of Arizona/Clowe et al.; Lensing Map: NASA/STScI; ESO WFI; Magel-
lan/University of Arizona/Clowe et al.

Figure 32.6. Theoretical physicist S. Jim Gates, Jr.
Credit: John Consoli
Creative Director/Photography Director
University of Maryland, College Park
College Park, MD

Figure 32.7. Theoretical physicist Helen R. Quinn.
Image courtesy of the American Institute of Physics.

INDEX

Page numbers with an "f" indicate figures; with a "p" indicate problems; with a "t" indicate tables.

MAXWELL'S EQUATIONS

The Faraday Tensor: $F^{\mu\nu} \equiv \partial^\mu A^\nu - \partial^\nu A^\mu = \begin{pmatrix} 0 & -E^x & -E^y & -E^z \\ E^x & 0 & -B^z & B^y \\ E^y & B^z & 0 & -B^x \\ E^z & -B^y & B^x & 0 \end{pmatrix}$

The definition of $F^{\mu\nu} \;\Rightarrow\; 0 = \partial^\mu F^{\alpha\beta} + \partial^\alpha F^{\beta\mu} + \partial^\beta F^{\mu\alpha} \;\Rightarrow\; \vec{\nabla}\cdot\vec{B} = 0, \quad \vec{\nabla}\times\vec{E} + \dfrac{\partial\vec{B}}{\partial t} = 0.$

The remaining Maxwell equations are $\partial_\mu F^{\mu\nu} = J^\mu \;\Rightarrow\; \vec{\nabla}\cdot\vec{E} = \rho, \quad \vec{\nabla}\times\vec{B} - \dfrac{\partial\vec{E}}{\partial t} = \vec{J}.$

We also typically use **Lorenz Gauge:** $\partial_\mu A^\mu = 0$. (The units we are using are such that $\varepsilon_0 = (4\pi k)^{-1} = 1$.)

DIRAC DELTA

Definition: $\quad \delta(x' - x) = 0$ except at $x = x'$, $\quad \int dx\, \delta(x'-x) f(x) = f(x')$

3D version: $\quad \delta^3(\vec{x}' - \vec{x}) \equiv \delta(x'-x)\delta(y'-y)\delta(z'-z)$

4D version: $\quad \delta^4(x'-x) \equiv \delta(t'-t)\delta(x'-x)\delta(y'-y)\delta(z'-z)$

Identities: $\quad (2\pi)^3 \delta^3(\vec{p}'-\vec{p}) = \int d^3x\, e^{\pm i(\vec{p}'-\vec{p})\cdot\vec{x}}, \quad \delta^3(\vec{r}'-\vec{r}) = \int \dfrac{d^3p}{(2\pi)^3}\, e^{\pm i(\vec{r}'-\vec{r})\cdot\vec{p}}$

$\qquad\qquad (2\pi)^4 \delta^4(p'-p) = \int d^4x\, e^{\pm i(p'-p)\cdot x}, \quad \delta^4(x'-x) = \int \dfrac{d^4p}{(2\pi)^4}\, e^{\pm i(x'-x\cdot p)}$

Properties: $\quad \delta(\alpha(x'-x)) = \frac{1}{|\alpha|}\delta(x'-x), \quad \delta(g(x)) = \sum_k \frac{\delta(x_k-x)}{(dg/dx)_{x_k}}, \;$ for x_k such that $g(x_k) = 0$

RULES OF QUANTUM MECHANICS

See pages 112–113, 128, 130.

ADJOINT RULES

- Number: $c^\dagger = c^*$, Ket: $|\psi\rangle^\dagger = (|\psi\rangle^T)^* = \langle\psi|$, Operator: $A^\dagger = (A^T)^*$. (T represents the transpose.)
- If $|\psi_2\rangle = A|\psi_1\rangle$, then $\langle\psi_2| = \langle\psi_1|A^\dagger; \quad (AB)^\dagger = B^\dagger A^\dagger$.
- A Hermitian operator A is **self-adjoint:** $A^\dagger = A$, has real eigenvalues, and has orthogonal eigenvectors.
- A **unitary** operator U's adjoint is its inverse: $U^\dagger U = I$. Its eigenvalues have unit magnitude: $|u_n| = 1$.
- If observable $\mathcal{A}$ has eigenvalues/vectors a_n and $|a_n\rangle$, the corresponding operator is $A = \sum_n a_n|a_n\rangle\langle a_n|$.

COMMUTATOR/ANTICOMMUTATOR RULES

$[A, B] \equiv AB - BA \qquad\qquad \{A, B\} \equiv AB + BA$

$[A, B] = -[B, A] \qquad\qquad \{A, B\} = +\{B, A\}$

$[AB, C] = A[B, C] + [A, C]B \qquad \{AB, C\} = A[B, C] + \{A, C\}B$

ANGULAR MOMENTUM RULES

- AM operators are J_x, J_y, J_z such that $[J_x, J_y] = i\hbar J_z$, $[J_y, J_z] = i\hbar J_x$, $[J_z, J_x] = i\hbar J_y$.
- AM eigenstates are $|j, m\rangle$, where $j = 0, \frac{1}{2}, 1, \frac{3}{2}$, etc. and m goes from $-j$ to j in integer steps.
- $J^2|j, m\rangle = (J_x^2 + J_y^2 + J_z^2)|j, m\rangle = j(j+1)\hbar^2|j, m\rangle$, $J_z|j, m\rangle = m\hbar|j, m\rangle$.
- Total AM of two particles: $j_{\text{tot}} = j_1 + j_2$ down to $|j_1 - j_2|$ in integer steps.

LAGRANGIAN DENSITIES $\mathcal{L}$

- $\mathcal{L}$ must be a Lorentz scalar having units of E^4.
- Neither a constant term nor a term linear in the field has physical consequences.
- A negative term that depends on the square of the field (for example. $-\frac{1}{2}m^2\phi^2$) gives the field a mass.
- A term that couples different fields (for example, $\lambda\phi^2\eta$, where ϕ and η are fields) is an interaction term, which is represented in a Feynman diagram by a vertex where lines representing the coupled fields meet.

EULER-LAGRANGE EQUATIONS

(one equation for each field ϕ_n):

$$0 = \partial_\mu \left(\frac{\partial \mathcal{L}}{\partial(\partial_\mu \phi_n)} \right) - \frac{\partial \mathcal{L}}{\partial \phi_n}$$

CONSERVED CURRENT

for a transformation $\phi \to \phi'$ that preserves the form of $\mathcal{L}$:

$$J^\mu = \sum_n \frac{\partial \mathcal{L}}{\partial(\partial_\mu \phi_n)} \left[\frac{df_n}{d\epsilon} \right]_{\epsilon=0} \qquad \text{when } \phi'_n = f_n(\epsilon, \phi_n) \text{ and } \phi'_n = f_n(0, \phi_n) = \phi_n.$$

$$J^\mu = \sum_n \frac{\partial \mathcal{L}}{\partial(\partial_\mu \phi_n)} \left[\frac{d\phi_n}{d\epsilon} \right]_{\epsilon=0} - \mathcal{L} \left[\frac{dx'^\mu}{d\epsilon} \right]_{\epsilon=0} \qquad \text{when } \phi'_n = \phi_n(x'^\mu(\epsilon)), \ x'^\mu(0) = x^\mu, \text{ and } \partial_\mu x'^\mu = 0.$$

STRESS-ENERGY TENSOR

the conserved current when $x'^\mu(\epsilon^\nu) = x^\mu + \epsilon^\mu$:

$$T^{\mu\nu} = \sum_n \frac{\partial \mathcal{L}}{\partial(\partial_\mu \phi_n)} \partial^\nu \phi_n - g^{\mu\nu}\mathcal{L}, \qquad \text{where} \quad \begin{array}{l} T^{tt} = \rho = \text{energy density, } T^{tj} = \text{density of } j\text{-momentum,} \\ T^{jk} = \text{flux of } j\text{-momentum in the } k \text{ direction.} \end{array}$$

GENERATORS

If $M(\theta) = e^{-i\theta X}$ is a transformation, then X is the generator.

Example: $R_z(\theta) = e^{-i\theta J_z}$ is a rotation of θ around the z axis; J_z is the generator.

BASIC FREE-FIELD LAGRANGIAN DENSITIES

Name	Lagrangian Density	E-L Equation	Spin	Field Units	Note
Klein-Gordon	$\mathcal{L} = \frac{1}{2}\partial_\mu\phi^*\partial_\mu\phi - \frac{1}{2}m^2\phi^*\phi$	$0 = \partial_\mu\partial^\mu\phi + m^2\phi$	0	$[\phi] = E$	
Dirac	$\mathcal{L} = i\bar{\psi}\gamma^\mu\partial_\mu\psi - m\bar{\psi}\psi$	$0 = i\gamma^\mu\partial_\mu\psi - m\psi$	1/2	$[\psi] = E^{3/2}$	$\bar{\psi} \equiv \psi^\dagger\gamma^t$
Maxwell	$\mathcal{L} = -\frac{1}{4}F_{\mu\nu}F^{\mu\nu}$	$0 = \partial_\mu F^{\mu\nu}$	1	$[A^\mu] = E$	

(where $F_{\mu\nu} \equiv \partial_\mu A_\nu - \partial_\nu A_\mu$)

QUANTIZING THE FIELDS

(where $a_{\vec{p}}^\dagger$ and $a_{\vec{p}}$ are creation and annihilation operators for a particle with momentum $\vec{p}$, $b_{\vec{p}}^\dagger$, and $b_{\vec{p}}$ are the same for the corresponding antiparticle; r, s are spin indices.)

	Operator Expansion of Field	Operator Relations
Scalar	$\phi = \displaystyle\int \frac{d^3p}{(2\pi)^3} \frac{1}{\sqrt{2E_{\vec{p}}}} \left[a_{\vec{p}}\, e^{-ip\cdot x} + b_{\vec{p}}^\dagger\, e^{+ip\cdot x} \right]$	$[\, a_{\vec{p}}^\dagger, a_{\vec{p}}\,] = (2\pi)^3\delta(\vec{p} - \vec{q})$
Dirac	$\psi = \displaystyle\int \frac{d^3p}{(2\pi)^3} \frac{1}{\sqrt{2E_{\vec{p}}}} \sum_s \left[u^s(p)a_{s\vec{p}}\, e^{-ip\cdot x} + v^s(p)b_{s\vec{p}}^\dagger\, e^{+ip\cdot x} \right]$	$\{ a_{r\vec{p}}^\dagger, a_{s\vec{p}} \} = \delta_{rs}(2\pi)^3\delta(\vec{p} - \vec{q})$
Photon	$A^\mu = \displaystyle\int \frac{d^3p}{(2\pi)^3} \frac{1}{\sqrt{2E_{\vec{p}}}} \sum_s \epsilon_{s\vec{p}}^\mu \left[a_{s\vec{p}}\, e^{-ip\cdot x} a_{s\vec{p}}^\dagger\, e^{+ip\cdot x} \right]$	$[\, a_{r\vec{p}}^\dagger, a_{s\vec{p}} \,] = \delta_{rs}(2\pi)^3\delta(\vec{p} - \vec{q})$

GENERAL PROCESS for making a GLOBAL SYMMETRY LOCAL

If $\mathcal{L}$ is globally symmetric under $\Psi' = e^{-i\frac{1}{2}g \sum \theta_n X_n}\Psi$ (where g is a coupling constant, θ_n represents N parameters, X_n represents N generators, and we drop the $\frac{1}{2}$ if $N = 1$), then

1. Convert $\partial_\mu \to D_\mu = \partial_\mu + i\frac{1}{2}g \sum_n X_n A_n^\mu$, where A_n^μ represents N new gauge boson fields.
2. This will generate a new conserved current term $\sum_n -J_n^\mu A_{\mu,n}$.
3. This term will specify interaction vertices connecting gauge bosons with particles in the original $\mathcal{L}$.
4. Define a tensor $F_n^{\mu\nu} \equiv \partial^\mu A_n^\nu - \partial^\nu A_n^\mu - q \sum_n f_{jkn} A_j^\mu A_k^\nu$, where f_{jkn} is such that $[X_j, X_k] = 2i \sum_n f_{jkn}X_n$.
5. Add a new term $-\frac{1}{4}\sum_n F_n^{\mu\nu}F_{\mu\nu,n}$. The $-\frac{1}{4}\sum_n(\partial^\mu A_n^\nu - \partial^\nu A_n^\mu)(\partial_\mu A_{\nu,n} - \partial_\nu A_{\mu,n})$ part specifies how the free gauge boson fields evolve in time, and other parts describe self-interactions between those fields.

TABLE OF c_V, c_A VALUES FOR Z VERTICES

Fermion	c_V	c_A	
ν_e, ν_μ, ν_τ	$+\frac{1}{2}$	$\frac{1}{2}$	
e^-, μ^-, τ^-	$-\frac{1}{2} + 2\sin^2\theta_W$	$-\frac{1}{2}$	and $\sin^2\theta_W \approx 0.23,\ \ g_W \approx 0.64$
u, c, t	$+\frac{1}{2} - \frac{4}{3}\sin^2\theta_W$	$\frac{1}{2}$	
d, s, b	$-\frac{1}{2} + \frac{2}{3}\sin^2\theta_W$	$-\frac{1}{2}$	

THE CKM MATRIX

$$V_{jk} = \begin{pmatrix} c_{12}c_{13} & s_{12}c_{13} & s_{13}e^{-i\delta} \\ -s_{12}c_{23} - c_{12}s_{23}s_{13}e^{i\delta} & c_{12}c_{23} - s_{12}s_{23}s_{13}e^{i\delta} & s_{23}c_{13} \\ s_{12}s_{23} - c_{12}c_{23}s_{13}e^{i\delta} & -c_{12}s_{23} - s_{12}c_{23}s_{13}e^{i\delta} & c_{23}c_{13} \end{pmatrix},$$

where the real parameters are θ_{12}, θ_{13}, and θ_{23}, and c_{jk} and s_{jk} are shorthands for $\cos\theta_{jk}$ and $\sin\theta_{jk}$ respectively. If we assume V_{jk} is unitary and only three quark families exist, then experimentally

$$s_{12} = 0.2265(7), \qquad s_{13} = 0.0036(1), \qquad s_{23} = 0.0405(9), \qquad \delta = 1.14(3) \approx 65^\circ,$$

where the uncertainty in parentheses in each case is in the last digit.

$$|V_{jk}| = \begin{pmatrix} 0.9740 & 0.2265 & 0.0036 \\ 0.2264 & 0.9732 & 0.0405 \\ 0.0085 & 0.0398 & 0.99917 \end{pmatrix}$$

GELL-MANN MATRICES for SU(3)
(analogous to Pauli matrices for SU(2)).

$$\lambda_1 = \begin{pmatrix} 0 & 1 & 0 \\ 1 & 0 & 0 \\ 0 & 0 & 0 \end{pmatrix}, \qquad \lambda_2 = \begin{pmatrix} 0 & -i & 0 \\ i & 0 & 0 \\ 0 & 0 & 0 \end{pmatrix}, \qquad \lambda_3 = \begin{pmatrix} 1 & 0 & 0 \\ 0 & -1 & 0 \\ 0 & 0 & 0 \end{pmatrix},$$

$$\lambda_4 = \begin{pmatrix} 0 & 0 & 1 \\ 0 & 0 & 0 \\ 1 & 0 & 0 \end{pmatrix}, \qquad \lambda_5 = \begin{pmatrix} 0 & 0 & -i \\ 0 & 0 & 0 \\ i & 0 & 0 \end{pmatrix},$$

$$\lambda_6 = \begin{pmatrix} 0 & 0 & 0 \\ 0 & 0 & 1 \\ 0 & 1 & 0 \end{pmatrix}, \qquad \lambda_7 = \begin{pmatrix} 0 & 0 & 0 \\ 0 & 0 & -i \\ 0 & i & 0 \end{pmatrix}, \qquad \lambda_8 = \frac{1}{\sqrt{3}}\begin{pmatrix} 1 & 0 & 0 \\ 0 & 1 & 0 \\ 0 & 0 & -2 \end{pmatrix}.$$

$$[\lambda_j, \lambda_k] = 2i \sum_{m=1 \to 8} f_{jkm}\lambda_n,$$

The structure constants f_{jkm} change sign if we switch any two indices, and one can obtain the value of any nonzero constant by applying antisymmetry to the following list:

$$f_{147} = f_{246} = f_{257} = f_{345} = f_{516} = f_{637} = \tfrac{1}{2}f_{123} = 1, \qquad f_{458} = f_{678} = \tfrac{\sqrt{3}}{2}.$$

RUNNING COUPLING CONSTANTS

EM: $\quad \alpha_R(Q^2) = \dfrac{\alpha_R(0)}{1 - \alpha_R(0)f(u)/3\pi}, \quad$ where $u \equiv \dfrac{-Q^2}{m_e^2} \quad$ and $\quad f(u) = 6\displaystyle\int_0^1 x(1-x)\ln[\,1 + xu(1-x)\,]\,dx.$

Color: $\quad \alpha_S(Q^2) = \dfrac{\alpha_S(\mu^2)}{1 + \frac{1}{12\pi}\alpha_S(\mu^2)(11n - 2f)\ln(|Q^2|/\mu^2)} \quad$ for $\quad |Q^2| \gg \mu^2,$

where $n = $ number of colors $= 3$, $f = $ number of quarks $= 6$, and $\alpha_S(M_Z^2) = 0.118(1)$.

PAULI MATRICES

$$\sigma_x = \begin{pmatrix} 0 & 1 \\ 1 & 0 \end{pmatrix}, \qquad \sigma_y = \begin{pmatrix} 0 & -i \\ i & 0 \end{pmatrix}, \qquad \sigma_z = \begin{pmatrix} 1 & 0 \\ 0 & -1 \end{pmatrix}, \qquad \sigma^\mu \equiv (I, \sigma_x, \sigma_y, \sigma_z), \qquad \bar\sigma^\mu \equiv (I, -\sigma_x, -\sigma_y, -\sigma_z).$$

PAULI MATRIX IDENTITIES

$$[\sigma_j, \sigma_k] = 2i \sum_m \epsilon_{jkm}\sigma_m, \quad \{\sigma_j, \sigma_k\} = 2\delta_{jk}, \quad (\vec{a}\cdot\vec{\sigma})(\vec{b}\cdot\vec{\sigma}) = \vec{a}\cdot\vec{b}\,I + i(\vec{a}\times\vec{b})\cdot\vec{\sigma},$$

$$p\cdot\sigma \equiv p^\mu g_{\mu\nu}\sigma^\mu, \quad p\cdot\overline{\sigma} \equiv p^\mu g_{\mu\nu}\overline{\sigma}^\mu, \quad (p\cdot\sigma)(p\cdot\overline{\sigma}) = (p\cdot\overline{\sigma})(p\cdot\sigma) = m^2 I, \quad p\cdot\sigma + p\cdot\overline{\sigma} = 2E_{\vec{p}}.$$

For particles moving in the $\pm z$ direction:

$$\sqrt{p\cdot\sigma} = \begin{pmatrix} \sqrt{E_{\vec{p}} - |\vec{p}|} & 0 \\ 0 & \sqrt{E_{\vec{p}} + |\vec{p}|} \end{pmatrix}, \qquad \sqrt{p\cdot\overline{\sigma}} = \begin{pmatrix} \sqrt{E_{\vec{p}} + |\vec{p}|} & 0 \\ 0 & \sqrt{E_{\vec{p}} - |\vec{p}|} \end{pmatrix}.$$

GAMMA MATRICES

$$\gamma^t = \begin{pmatrix} 0 & 0 & 1 & 0 \\ 0 & 0 & 0 & 1 \\ 1 & 0 & 0 & 0 \\ 0 & 1 & 0 & 0 \end{pmatrix}, \qquad \gamma^x = \begin{pmatrix} 0 & 0 & 0 & 1 \\ 0 & 0 & 1 & 0 \\ 0 & -1 & 0 & 0 \\ -1 & 0 & 0 & 0 \end{pmatrix}, \qquad \gamma^y = \begin{pmatrix} 0 & 0 & 0 & -i \\ 0 & 0 & i & 0 \\ 0 & i & 0 & 0 \\ -i & 0 & 0 & 0 \end{pmatrix}, \qquad \gamma^z = \begin{pmatrix} 0 & 0 & 1 & 0 \\ 0 & 0 & 0 & -1 \\ -1 & 0 & 0 & 0 \\ 0 & 1 & 0 & 0 \end{pmatrix}$$

$$\gamma^5 = \begin{pmatrix} -1 & 0 & 0 & 0 \\ 0 & -1 & 0 & 0 \\ 0 & 0 & 1 & 0 \\ 0 & 0 & 0 & 1 \end{pmatrix} = i\gamma^t\gamma^x\gamma^y\gamma^z. \qquad \textbf{Compact notation: } \gamma^\mu = \begin{pmatrix} 0 & \sigma^\mu \\ \overline{\sigma}^\mu & 0 \end{pmatrix} \quad \gamma^5 = \begin{pmatrix} -I & 0 \\ 0 & I \end{pmatrix}$$

DIRAC SPINORS

for spin-$\frac{1}{2}$ particles (r and s are spin indices with values $+$, $-$):

For	Plane Wave	Spinor	Spin Up	Spin Down	Normalization
Particle:	$\psi = u(p)e^{-ip\cdot x}$	$u(p) = \begin{bmatrix} \sqrt{p\cdot\sigma}\,\xi \\ \sqrt{p\cdot\overline{\sigma}}\,\xi \end{bmatrix}$	$\xi^+ = \begin{bmatrix} 1 \\ 0 \end{bmatrix}$	$\xi^- = \begin{bmatrix} 0 \\ 1 \end{bmatrix}$	$u^{r\dagger}u^s = +2E_{\vec{p}}\,\delta^{rs}$ $\overline{u}^r u^s = +2m\delta^{rs}$
Antiparticle:	$\psi = v(p)e^{+ip\cdot x}$	$v(p) = \begin{bmatrix} \sqrt{p\cdot\sigma}\,\eta \\ -\sqrt{p\cdot\overline{\sigma}}\,\eta \end{bmatrix}$	$\eta^+ = \begin{bmatrix} 0 \\ 1 \end{bmatrix}$	$\eta^- = -\begin{bmatrix} 1 \\ 0 \end{bmatrix}$	$v^{r\dagger}v^s = +2E_{\vec{p}}\,\delta^{rs}$ $\overline{v}^r v^s = -2m\delta^{rs}$

Spinor product identities and completeness relations:

$$\overline{u}^r(p)\,v^s(p) = \overline{u}^r(p)\,v^s(p) = 0, \qquad\qquad u^{r\dagger}(\vec{p})v^s(-\vec{p}) = v^{r\dagger}(\vec{p})u^s(-\vec{p}) = 0,$$

$$\sum_s u^s(p)\overline{u}^s(p) = p_\mu\gamma^\mu + mI \equiv \not{p} + mI, \qquad \sum_s v^s(p)\overline{v}^s(p) = p_\mu\gamma^\mu - mI \equiv \not{p} - mI$$

OPERATORS ON SPINORS

Spin AM	Boost Generator	Helicity	Chirality	Charge Conj.	Parity
$\vec{\Sigma} = \dfrac{1}{2}\begin{pmatrix} \vec{\sigma} & 0 \\ 0 & \vec{\sigma} \end{pmatrix}$	$-\dfrac{i}{2}\begin{pmatrix} \vec{\sigma} & 0 \\ 0 & -\vec{\sigma} \end{pmatrix}$	$\dfrac{\vec{p}\cdot\vec{\Sigma}}{|\vec{p}|}$	$P_L = \frac{1}{2}(1-\gamma^5)$ $P_R = \frac{1}{2}(1+\gamma^5)$	$C\psi = -i\gamma^y\psi^\dagger$	$P\psi = \gamma^t\psi$

NOTES ABOUT CALCULATING $\mathcal{M}$

- $\mathcal{M}$ must be a (complex) Lorentz scalar (that is, a *number*).
- This means that all greek indices must connect in such a way that there are no free indexes left.
- Also, all matrix products must end up producing simple numbers.
- Arrows *along* fermion lines in a Feynman diagram point in the direction of motion for particles, but opposite to the direction of motion for antiparticles.
- To construct a factor in $\mathcal{M}$, follow these arrows as you write down the terms *right to left* (backwards).
- Arrows *next to* lines on a Feynman diagram indicate the direction of the momenta that label the lines.
- One can arbitrarily choose the direction of motion for a horizontal virtual particle ($\longrightarrow$ is conventional).
- The *units* of $\mathcal{M}$ depend on the number of ingoing or outgoing particles.

FEYNMAN RULES

- For the scalar toy theory: p. 169–170, 218–219
- For electroweak theory: pp. 459–461
- For quantum chromodynamics: pp. 509–511